Spectroscopy and Dynamics of Collective Excitations in Solids

NATO ASI Series

Advanced Science Institutes Series

A series presenting the results of activities sponsored by the NATO Science Committee, which aims at the dissemination of advanced scientific and technological knowledge, with a view to strengthening links between scientific communities.

The series is published by an international board of publishers in conjunction with the NATO Scientific Affairs Division

A	Life Sciences	Plenum Publishing Corporation
B	Physics	New York and London
C	Mathematical and Physical Sciences	Kluwer Academic Publishers
		Dordrecht, Boston, and London
D	Behavioral and Social Sciences	
E	Applied Sciences	
F	Computer and Systems Sciences	Springer-Verlag
G	Ecological Sciences	Berlin, Heidelberg, New York, London,
H	Cell Biology	Paris, Tokyo, Hong Kong, and Barcelona
I	Global Environmental Change	

PARTNERSHIP SUB-SERIES

1. Disarmament Technologies	Kluwer Academic Publishers
2. Environment	Springer-Verlag
3. High Technology	Kluwer Academic Publishers
4. Science and Technology Policy	Kluwer Academic Publishers
5. Computer Networking	Kluwer Academic Publishers

The Partnership Sub-Series incorporates activities undertaken in collaboration with NATO's Cooperation Partners, the countries of the CIS and Central and Eastern Europe, in Priority Areas of concern to those countries.

Recent Volumes in this Series:

Volume 353 — Hadron Spectroscopy and the Confinement Problem
edited by D. V. Bugg

Volume 354 — Physics and Chemistry of Low-Dimensional Inorganic Conductors
edited by Claire Schlenker, Jean Dumas, Martha Greenblatt, and Sander van Smaalen

Volume 355 — Stability of Materials
edited by A. Gonis, P. E. A. Turchi, and Josef Kudrnovský

Volume 356 — Spectroscopy and Dynamics of Collective Excitations in Solids
edited by Baldassare Di Bartolo

Series B: Physics

Spectroscopy and Dynamics of Collective Excitations in Solids

Edited by

Baldassare Di Bartolo

Boston College
Chestnut Hill, Massachusetts

Assistant Editor

Stamatios Kyrkos

Boston College
Chestnut Hill, Massachusetts

Springer Science+Business Media, LLC

Proceedings of a NATO Advanced Study Institute
and International School of Atomic and Molecular Spectroscopy Workshop on
Spectroscopy and Dynamics of Collective Excitation in Solids,
held June 17 – July 1, 1995,
in Erice, Italy

NATO-PCO-DATA BASE

The electronic index to the NATO ASI Series provides full bibliographical references (with keywords and/or abstracts) to about 50,000 contributions from international scientists published in all sections of the NATO ASI Series. Access to the NATO-PCO-DATA BASE is possible in two ways:

—via online FILE 128 (NATO-PCO-DATA BASE) hosted by ESRIN, Via Galileo Galilei, I-00044 Frascati, Italy

—via CD-ROM "NATO Science and Technology Disk" with user-friendly retrieval software in English, French, and German (©WTV GmbH and DATAWARE Technologies, Inc. 1989). The CD-ROM contains the AGARD Aerospace Database.

The CD-ROM can be ordered through any member of the Board of Publishers or through NATO-PCO, Overijse, Belgium.

Additional material to this book can be downloaded from http://extra.springer.com.

Library of Congress Cataloging-in-Publication Data

```
Spectroscopy and dynamics of collective excitations in solids / edited
  by Baldassare Di Bartolo, assistant editor Stamatios Kyrkos.
      p.   cm. -- (NATO ASI series. B, Physics ; v. 356)
    "Proceedings of a NATO Advanced Study Institute and International
  School of Atomic and Molecular Spectroscopy Workshop on Spectroscopy
  and Dynamics of Collective Excitation in Solids, held June 17-July
  1, 1995, in Erice, Italy"--CIP verso t.p.
    "Published in cooperation with NATO Scientific Affairs Division."
    Includes bibliographical references and index.
    ISBN 978-1-4613-7675-0    ISBN 978-1-4615-5835-4 (eBook)
    DOI 10.1007/978-1-4615-5835-4
    1. Exciton theory--Congresses.  2. Collective excitations-
  -Congresses.  3. Solids--Spectra--Congresses.   I. Di Bartolo,
  Baldassare.  II. North Atlantic Treaty Organization.  Scientific
  Affairs Division.  III. NATO Advanced Research Study Institute and
  International School of Atomic and Molecular Spectroscopy Workshop
  on Spectroscopy and Dynamics of Collective Excitation in Solids
  (1995 : Erice, Italy)  IV. Series.
  QC176.8.E9S673  1997
  530.4'16--dc21                                    96-47008
                                                       CIP
```

ISBN 978-1-4613-7675-0

© 1997 Springer Science+Business Media New York
Originally published by Plenum Press, New York in 1997
Softcover reprint of the hardcover 1st edition 1997
http://www.plenum.com

10 9 8 7 6 5 4 3 2 1

Science is not to be regarded as a storehouse of facts to be used for material purposes, but as one of the great human endeavours to be ranked with arts and religion as the guide and expression of man's fearless quest for truth.

Sir Richard Arman Gregory

PREFACE

This book presents the proceedings of the course "Spectroscopy and Dynamics of Collective Excitations in Solids" held in Erice, Italy from June 17 to July 1, 1995. This meeting was organized by the International School of Atomic and Molecular Spectroscopy of the "Ettore Majorana" Centre for Scientific Culture.

The purpose of this course was to present and discuss physical models, mathematical formalisms, experimental techniques and applications relevant to the subject of collective excitations in solids. By bringing together specialists in the field of solid state spectroscopy, this course provided a much needed forum for the critical assessment and evaluation of recent and past developments in the physics of solids.

A total of 83 participants came from 57 laboratories and 20 different countries (Austria, Belgium, Brazil, Denmark, Finland, France, Germany, Greece, Israel, Italy, Japan, The Netherlands, Norway, Portugal, Russia, Spain, Switzerland, Turkey, the United Kingdom, and the United States).

The secretaries of the course were Stamatios Kyrkos and Daniel Di Bartolo.

45 lectures divided in 13 series were given. In addition 8 (one or two-hour) "long seminars," 1 "special lecture," 2 interdisciplinary lectures, 29 "short seminars," and 16 posters were presented. The sequence of lectures was in accordance with the logical development of the subject of the meeting. Each lecturer started at a rather fundamental level and ultimately reached the frontier of knowledge in the field.

Two round-table discussions were held. The first round-table discussion took place after three and a half days of lectures in order to evaluate the work done in the first days of the course and consider suggestions and proposals regarding the organization, format, and presentation of the lectures. The second one was held at the conclusion of the course, so that the participants could comment on the work done during the entire meeting and discuss various proposals for the next course of the International School of Atomic and Molecular Spectroscopy.

I wish to express my sincere gratitude to Ms. Zaini and Dr. Gabriele, to Ms. Savalli and Mr. Pilarski, and to all the personnel of the "Ettore Majorana" Centre, who contributed to create a congenial atmosphere for our meeting. I also wish to acknowledge the sponsorship of the meeting by the NATO Scientific Affairs Division, the ENEA Organization, Boston College, the European Physical Society, the Italian Ministry of Education, the Italian Ministry of University and Scientific Research, and the Sicilian Regional Government.

I would like to thank the members of the Organizing Committee (Doctors Auzel, Baldacchini, and Macfarlane, Professor Klingshirn, and Mr. Glezer), the secretaries of the course (Stamatios Kyrkos and Daniel Di Bartolo), Angela Siraco, and John Di Bartolo for their valuable help. I would also like to thank Nino La Francesca, a past secretary of seven courses of the School of Spectroscopy, who was this year part of the staff and helped a lot with the running of the meeting.

It would be difficult to describe the spirit of true collaboration and the interest with which everybody participated in this Institute and the atmosphere of cordiality and friendship that pervaded this meeting. I feel privileged for having been able to encounter in my native land of Trapani so many fine people who participated in this course, to share the Erice experience with them and to have had the opportunity to see again several friends and to establish new friendships.

I am already looking forward to the next 1997 meeting of the International School of Atomic and Molecular Spectroscopy, where I am sure I will see again many of you, my friends. Arrivederci a presto!

Baldassare (Rino) Di Bartolo
Director of the International School of Atomic
and Molecular Spectroscopy of the
"Ettore Majorana" Centre

CONTENTS

LIGHT-MATTER INTERACTION - EXPERIMENTAL ASPECTS
C. Klingshirn

THEORETICAL DESCRIPTION OF COLLECTIVE EXCITATIONS:
BLOCH EQUATIONS AND RELAXATION MECHANISMS
R. Zimmermann

LINEAR AND NONLINEAR OPTICAL SPECTROSCOPY:
SPECTRAL, TEMPORAL AND SPATIAL RESOLUTION
J.M. Hvam

THE STUDY OF COLLECTIVE EXCITATIONS IN SOLIDS
BY INELASTIC NEUTRON SCATTERING
T. Riste

EXCITATION DYNAMICS IN ORGANIC MOLECULES, SOLIDS, FULLERENES AND POLYMERS
P. Prasad

THE IR VIBRATIONAL PROPERTIES OF COMPOSITE SOLIDS AND PARTICLES:
THE LYDDANE-SACHS-TELLER RELATION REVISITED
A.J. Sievers

INTRINSIC LOCALIZED MODES IN ANHARMONIC LATTICES
A.J. Sievers, S.R. Bickham, and S.A. Kiselev

COLLECTIVE EXCITATIONS IN MAGNETIC MATERIALS
G.A. Gehring

PLASMONS AND SURFACE PLASMONS IN BULK METALS, METALLIC CLUSTERS,
AND METALLIC HETEROSTRUCTURES
R.v. Baltz

ENLIGHTENMENT ON LUMINESCENT MATERIALS
C.R. Ronda

TECHNIQUES OF ULTRAFAST SPECTROSCOPY
E.N. Glezer

INTERACTION OF ULTRASHORT LASER PULSES WITH SOLIDS
E. Mazur

ENERGY TRANSFER AND MIGRATION OF EXCITATION IN SOLIDS
AND CONFINED STRUCTURES
F. Auzel

BROAD BAND CENTERS APPLIED FOR LASER MATERIALS:
EXAMPLE OF TETRAHEDRALLY COORDINATED CENTERS
G. Boulon

INTERDISCIPLINARY LECTURES

SHORT SEMINARS

POSTERS

Spectroscopy and Dynamics of Collective Excitations in Solids

NONLINEAR OPTICS AND COLLECTIVE EXCITATIONS

N. Bloembergen

Pierce Hall
Harvard University
Cambridge, Massachusetts 02138

ABSTRACT

A broad overview of light scattering and nonlinear optical experiments is presented which gives information about the structure and dynamics of collective excitations in condensed matter. The general ideas are illustrated with examples taken from various types of polaritons, from excitons in quantum-well semiconductors, and from excitons in polymers.

I. INTRODUCTION

Acoustic waves are obviously a collective excitation of kinetic and elastic energy of all atoms in a crystalline lattice or a fluid. Brillouin[1] first discussed theoretically how a driven acoustic wave in a transparent medium will act as a diffractive grating for light. He had this idea just before he was drafted for military service in World War I, but it was published eight years later as part of his Ph.D. dissertation in 1922. The physical idea is clear for longitudinal acoustic waves which produce periodic density and concomitantly optical density or refraction index variations. A traveling acoustic wave acts as a moving grating. The corresponding fundamental quantum process may be described as the inelastic scattering of one photon with frequency ω_1 and wave vector $\mathbf{k}_1$, into another with frequency ω_2 and wave vector $\mathbf{k}_2$, while simultaneously an acoustic phonon is created or destroyed. Energy and momentum conservation require

$$\omega_1 - \omega_2 = \pm\omega_{ac} \quad \text{and} \quad \mathbf{k}_1 - \mathbf{k}_2 = \pm\mathbf{k}_{ac} \quad . \tag{1}$$

Since the light wave length is large compared to the interatomic distance, light scattering experiments can only probe conditions near the center of the Brillouin zone in k-space. For isotropic media, the maximum acoustic phonon wave number occurs for backward scattering and is about $|2\mathbf{k}_{light}|$. In this direction the maximum frequency shift in the scattered light occurs. Its magnitude is $\Delta\omega = 2(v_{ac}n/c)\omega_1$, where $v_{ac} = \omega_{ac}/k_{ac}$ is the acoustic phase velocity and c/n is light phase velocity in a medium with refractive index n.

This shift in scattered light as a function of the scattering angle was fist observed by E.F. Gross[2] in 1930. The Brillouin scattering was induced by thermally excited

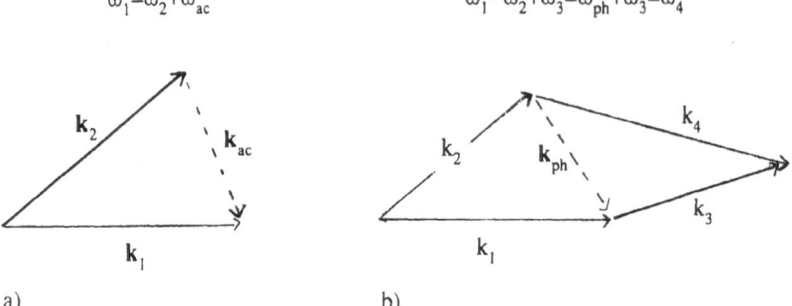

Figure 1. (a) Momentum diagram of Brillouin scattering, **(b)** Momentum diagram of four-wave light mixing (CARS)

acoustic phonons. Light diffraction by a coherent ultrasonic acoustic wave was demonstrated by Debye and Sears,[3] and by Lucas and Biguard[4] more than a decade after Brillouin's theoretical work. Conversely, one can excite a coherent acoustic wave by mixing of two light waves with slightly different frequencies and the correct angle for momentum matching. Such stimulated Brillouin scattering was first demonstrated with one incident laser beam.[5] A new light wave at $\omega_2 = \omega_1 - \omega_{ac}$ is created in the backward direction, while, simultaneously, a forward acoustic wave is generated. This process builds up exponentially from the spontaneous generation of one backward photon with a thermally excited (or spontaneously emitted) forward phonon, provided the incident light wave exceeds a certain critical threshold. Clearly stimulated Brillouin scattering and any process involving two intense light beams with well-defined frequencies and wave vectors are part of nonlinear optics.

Inelastic light scattering is a manifestation of the same basic physical process, represented by the diagram of Figure 1. It is understood that a photon or a phonon may be emitted by a spontaneous or a thermally activated process, or by an externally applied coherent wave. Thus such light scattering processes are subsumed to be part of nonlinear optics, although historically they preceded this field. One now has the capability of choosing two, three or more incident light signals with well-defined tunable frequencies, directions of propagation and polarization. Thus nonlinear optics is a versatile tool for studying collective excitations, although it is necessarily restricted to the vicinity of the origin in momentum space. Neutron scattering does not have this limitation.

The next closely related example is Raman scattering, which historically preceded the experimental detection of Brillouin scattering. Raman active vibrations, or optical phonon modes, scatter the light. In this case, because the dispersion curve for optical phonons is horizontal near the origin in k-space, the frequency of the scattered light, $\omega = \omega_1 \pm \omega_{\text{vib}}$, does not depend on the scattering angle, at least in centrosymmetric crystals. Antistokes scattering can occur if the Raman-active mode is thermally excited. These modes cannot be excited by external mechanical means, as was possible for acoustic modes. Coherent excitation of the Raman modes is possible by stimulated Raman scattering—a process that was experimentally demonstrated a few years before Brillouin scattering.[6]

Closely related to Raman scattering is the two-photon absorption process.[7,8] In this case, two photons are annihilated while a material excitation with energy $\hbar(\omega_1 + \omega_2)$ and momentum $\hbar(\mathbf{k}_1 + \mathbf{k}_2)$ is created. With the availability of broadly

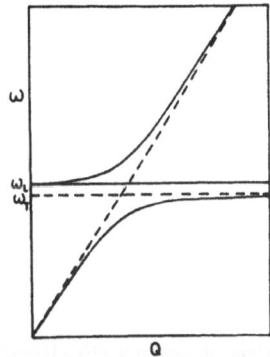

Figure 2. Dispersion relations for uncoupled (dashed lines) and coupled (continuous lines) electomagnetic wave and crystal excitation (polariton) (after reference 15).

tunable dye lasers, semiconductor lasers, and optical parametric oscillators, ω_1 and ω_2 can be varied from the ultraviolet to the infrared. At fixed ω_1 and ω_2, the wave vector $\mathbf{k}_1 + \mathbf{k}_2$ can be varied by changing the angle between the light beams.

A very versatile nonlinear optical experimental method is the four-wave mixing configuration in which three light beams are incident with frequencies ω_1, ω_2 and ω_3; wave vectors $\mathbf{k}_1$, $\mathbf{k}_2$ and $\mathbf{k}_3$; and polarizations $\hat{e}_1$, $\hat{e}_2$ and $\hat{e}_3$. A new wave at $\omega_4 = \omega_1 + \omega_2 - \omega_3$, wave vector $\mathbf{k}_4 = \mathbf{k}_1 + \mathbf{k}_2 - \mathbf{k}_3$ and polarization direction $\hat{e}_4$ is created. This process may be enhanced by intermediate two-photon or Raman-type resonances. Furthermore, time-resolved techniques with pulsed input beams can give additional information. For example, two coincident light pulses at ω_1 and ω_3 may create a material excitation with energy $\hbar(\omega_1 - \omega_3)$ at $t = 0$. At a later time τ_d, a pulse at ω_2 may interrogate this excitation. Thus the damping and propagation characteristics of this excitation may be probed directly. Examples of other collective excitations that may be probed by these nonlinear optical scattering techniques include plasma waves or plasmons, spin waves or magnons and bound electron-hole pairs or excitons. These excitons may be confined to two dimensions at surfaces or interfaces, or even to one dimension along conjugated chains.

In the following section, the case of polaritons will be described in detail. It received a great deal of attention in the sixties and seventies. In the eighties and up to the present time, excitons in semiconductor quantum-well structures and in polymeric chains have been the focus of much activity. A brief introduction to these topics is presented in the last two sections of this overview.

II. NONLINEAR SPECTROSCOPY OF POLARITONS

A polariton is a mixed mode of a collective material excitation and an electromagnetic or photonic excitation. They occur for all crystal excitations which have transverse electric-dipole coupling to the electromagnetic field. The material excitation may be modeled as a harmonic oscillator with resonant frequency ω_T. For phonon vibrations, ω_T corresponds to an infrared frequency, while for excitons, ω_T lies just below the bottom of the conduction band of a semiconducting or insulating crystal. A dispersion relation may be derived by the standard procedure of the linear response to an externally applied electric field of frequency ω, which will produce a one-photon absorption process in the vicinity of ω_T. This response is well described in a classical textbook by Born and Huang.[10] The uncoupled electromagnetic wave and the material excitation are described by dispersion curves indicated by the dashed

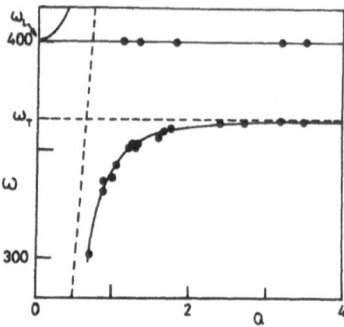

Figure 3. Theoretical phonon-polariton dispersion relation for GaP with experimental points from Reference 17. Units are $2\pi cm^{-1}$ for w, and $10^6 m^{-1}$ for Q (after Reference 15).

lines in Figure 2. The dispersion relation for the coupled modes is indicated by the continuous lines. The lines should not be used for the linear optical process from which they have been obtained. They are, however, very useful in the discussion of light scattering and nonlinear processes mentioned in the introduction.

In the case of non-centrosymmetric crystals, the lack of inversion symmetry implies that both one-photon and two-photon processes have electric dipole coupling to the same mode. The two-photon process may also show a resonance at the longitudinal mode frequency ω_L. When lowest order nonlinear terms of polarization are added to the Maxwell wave equation and also to the driving force of the material oscillator, detailed calculations confirm that the processes of Raman scattering, two-photon absorption, second harmonic generation and four-wave light mixing will exhibit resonances along the undamped polariton dispersion curves represented by continuous lines in Figure 2. The special situation for Raman scattering in non-centrosymmetric crystals was first recognized theoretically by Huang,[10] and experimentally by Pouillet.[11] Mills and Burstein[12] as well as Barker and Loudon[13] have written excellent review articles on the polariton problem. The nonlinear optical properties of polaritons have been reviewed in a comprehensive manner by Loudon[14,15] and by De Martini.[16] Some highlights of the experimental investigations, which were predominantly carried out in the period 1965–1980, will be recapitulated here.

Henry and Hopfield[17] determined experimental points on the phonon polariton lower branch in a GaP single crystal by Raman scattering of a HeNe laser beam, as shown in Figure 3. This was soon followed by an experiment of Faust and Henry,[18] in which the polariton was externally excited by infrared laser beams at several different frequencies. Thus, sum and difference frequencies corresponding to Stokes and anti-Stokes components on the HeNe beam were created.

The lower polariton branch in GaP was also probed in four-wave mixing experiments by Coffinet and DeMartini.[19] This is the first experiment in which the polariton resonance was observed by a continuous variation in k-spaces. Laser beams at two fixed frequencies were derived by stimulated Raman oscillators in pairs of liquids pumped by a ruby laser. The angle between the two beams impinging on the GaP crystal could be varied. The observed signal at $2\omega_1 - \omega_2$ showed a resonance when $|\mathbf{k}_1 - \mathbf{k}_2|$ crossed the lower branch polariton curve at $\omega_1 - \omega_2$. By interchanging the fluids in the Raman oscillators, points for several values of $\omega_1 - \omega_2$ were obtained. DeMartini subsequently expanded this technique by introducing a time delay between the first pair of pulses at ω_1 and ω_2, and subsequent probing pulse at ω_1. Thus the damping and propagation characteristics of the polariton may be investigated.

The same experimental method was used to study surface polaritons.[20]. In this case, $\omega_1 - \omega_2$ is chosen to lie in the forbidden gap between ω_L and ω_T. Four-wave

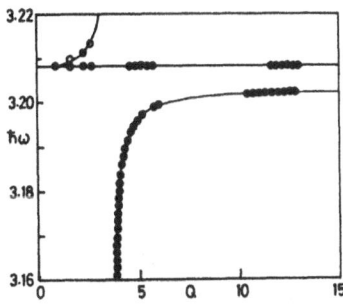

Figure 4. Exciton-polariton dispersion relation (after Reference 15) in CuCl.

mixing is enhanced when the component of $k_1 - k_2$ parallel to the surface is chosen correctly to lie on the dispersion curve for the surface polariton.

The cubic symmetry of GaP and other III-V or II-VI direct-gap semiconductor compounds admits only one phonon-polariton isotropic in k-space. In crystals with lower symmetry, the situation is considerably more complex, as discussed by Barker and Loudon[13]. Experimental results on four-wave mixing in a crystal of $LiNbO_3$ exhibit resonances with several polariton branches.

Turning now to nonlinear optical studies of exciton polaritons, the discussion will be limited to CuCl, which has the same crystal symmetry as GaP. Cuprous chloride exhibits a very sharp exciton resonance at low temperatures in the near UV, with $\hbar\omega_T = 3.208$ eV. The upper branch of the transverse exciton polariton and the longitudinal polariton have been studied by two-photon absorption. Frohlich et al.[9] used one laser beam from an Nd-glass laser and measured the absorption line of the exciton resonance with a high resolution spectrometer as the angle between the two beams was varied. They obtained the experimental points shown as open circles in Figure 4. The intensity of the two-photon resonance is a sensitive function of the sum frequency and demonstrates the destructive interference between the electronic nonresonant nonlinearity and the excitonic resonant nonlinearity. The two-photon signal vanishes when $\hbar(\omega_1 + \omega_2)$ equals 3.218 eV, as shown in Figure 5. Loudon discusses this point in detail.[14] Second harmonic generation experiments by Haueisen and Mahr[21] also demonstrate the exciton resonance on the upper branch.

The lower exciton polariton branch is not accessible by difference frequency mixing, as the probing frequencies would have to be chosen in the strongly absorbing ultraviolet region. This branch has been successfully probed by hyper-Raman scattering. In this process, two incident quanta at ω_1 are absorbed and two scattered excitations result. If one excitation is scattered in the backward direction, it has almost pure photon character, while the other forward excitation with a momentum nearly equal to $|3k_1|$ has virtually no photon character and remains unobserved. For near forward scattering, excitations with mixed character are emitted; both yield detectable photon signals. The scattering geometries are shown in Figure 6. In contrast to the luminescence signals, the hyper-Raman scattering depends on the exciting frequency ω_1, when $2\omega_1 = 2\hbar\omega_T - \Delta$. In this case, real bi-excitons are created with a binding energy $\Delta = 2q$ meV. The elegant experiments of Honerlage et al.[22] have resulted in an extremely precise determination of the lower exciton polariton dispersions, as shown by the solid dots in Figure 4.

The two-photon creation of bi-excitons has a giant cross section as predicted by Hanamura,[23] because of the strong dipole coupling to the intermediate virtual creation of one exciton. Very high concentrations of bi-excitons, up to 10^{19} cm^3, can be realized. This could lead to Bose-Einstein condensation in k-space, as suggested and

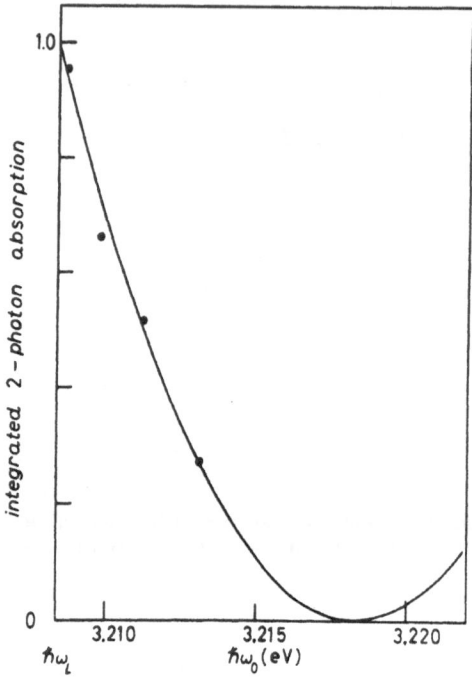

Figure 5. Variation of integrated two-photon absorption coefficient with polariton energy in CuCl (after Reference 14).

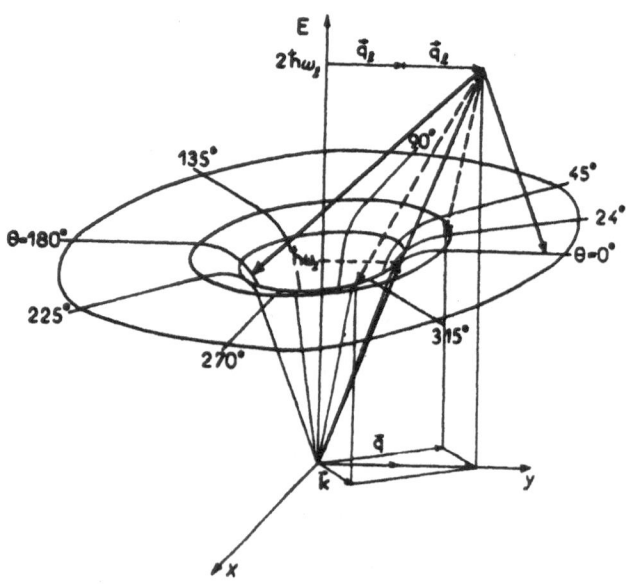

Figure 6. Resonant hyper-Raman scattering with dispersion of the phonon-polariton branch in CuCl (after Reference 22).

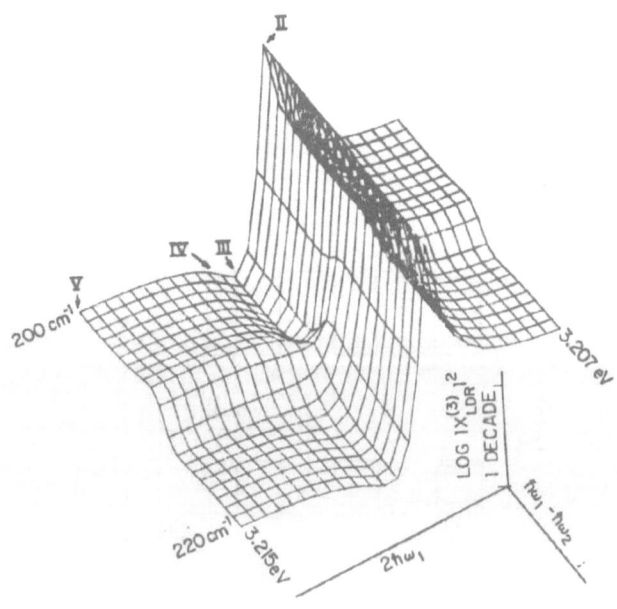

Figure 7. Disperson of the nonlinear susceptibility for four-wave mixing in CuCl with $2w$ in the vicinity of the exciton-polariton resonance, and $w_1 - w_2$ in the vicinity of the phonon-polariton resonance (after Reference 26).

reviewed by Mysyrovitcz.[24] The experimental evidence remains ambiguous, however. Klingshorn and Haug[25] have given a comprehensive review of high density excitation in a variety of direct-gap semiconductors. Their survey includes extensive references to the literature.

Finally, a four-wave mixing experiment is mentioned in which simultaneously the resonances of the exciton polariton and the phonon polariton in CuCl are probed[26]. In Figure 7, the dispersion characteristics of the four-wave mixing signal are presented in a two-dimensional frequency space, when $2\omega_1$ is chosen in the vicinity of the phonon of the longitudinal polariton resonance, and simultaneously $\omega_1 - \omega_2$ is chosen in the vicinity of the phonon polariton at cm^{-1}. These experiments could, in principle, be extended by varying the angle between the light beams to investigate the dependence on wave vectors.

In conclusion, light scattering and nonlinear optical experiments are ideally suited to probe the polariton dispersion of modes with mixed electromagnetic and excitonic or phonon-like character.

III. NONLINEAR OPTICAL PROPERTIES OF SEMICONDUCTOR QUANTUM WELLS

Semiconductor quantum-well structures, as shown in Figure 8, have been the subject of intense investigation during the past fifteen years[27,28,29]. An important characteristic is exhibited in the difference in the linear absorption spectrum near the band gap between the bulk material and the quantum well of the same material, GaAs, shown in Figure 9. In the quantum well, resonances of two types of excitons, consisting of electron-light hole pairs and electron heavy hole pairs respectively, are clearly resolved at room temperature. The excitons in the confinement of the quantum

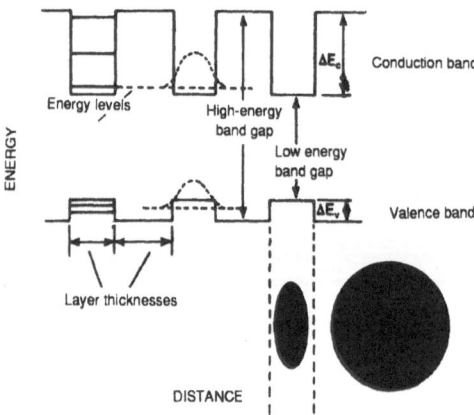

Figure 8. Quantum-well structure and corresponding real space energy band structure and exciton structure (after Reference 29).

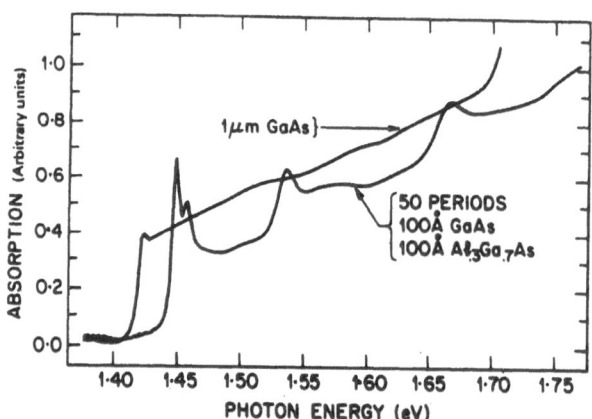

Figure 9. Comparison of room-temperature absorption spectra of bulk GaAs and a quantum-well structure (after Reference 27).

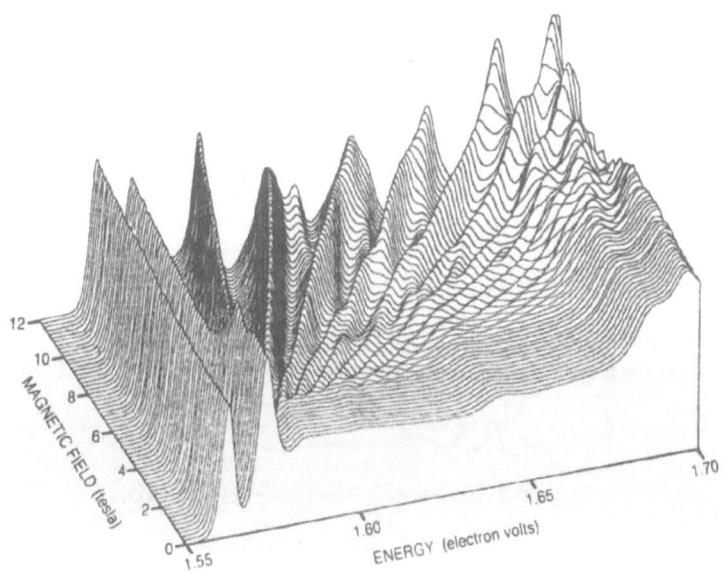

Figure 10. Progressive confinement from two-dimensional quantum-well structure to zero-dimensional regime in a strong perpendicular magnetic field (after Reference 28).

well have a larger Coulomb binding energy. The charge distribution of excitons in the quantum well and in the bulk material is schematically sketched in Figure 8. The absorption edge is shifted towards higher frequencies due to confinement. There are stepwise increases in the density states because of the quantization of the kinetic energy associated with the motion normal to the walls of the quantum well. An electric field applied normal to the wall will tend to push electrons toward one wall and the holes toward the other. There are pronounced Stark shifts in the absorption spectrum. If a magnetic field is applied normal the walls, the kinetic energy associated with the two-dimensional motion parallel to the walls is quantized in Landau levels. Thus, a progressive confinement from a quasi-two-dimensional regime in zero magnetic field to a quasi-zero-dimensional regime of isolated energy levels occurs as the magnetic field is increased. This behavior is evident from linear absorption spectra, shown in Figure 10.

The sensitivity of the position of the absorption at the band edge to an electric field may be used to achieve a very fast electro-optic modulator, or a self-electric-optic device (SEED). In either case, the quantum wells are incorporated in p-i-n junction. They are sandwiched between a p-type and n-type material. An electric field across the quantum wells may be maintained by a reverse bias across the junction. Light at a wavelength just below the absorption edge is transmitted through the device. A sudden shift in field may cause the exciton absorption peak to coincide with the light frequency. The process is fast and requires very low switching energy. In the SEED device, the switching is initiated by a change in intensity of the light beam. An external bias voltage causes a voltage drop across the diode structure so that a low level light beam undergoes some absorption. As the light intensity is increased, the resistance of the p-i-n structure decreases as the density of photo-excited carriers increases. Thus the voltage across the well structure drops. The exciton absorption peak shifts, causing more light to be absorbed. This increases the photocurrent, causing still more light to be absorbed. Thus the device can switch to a new stable

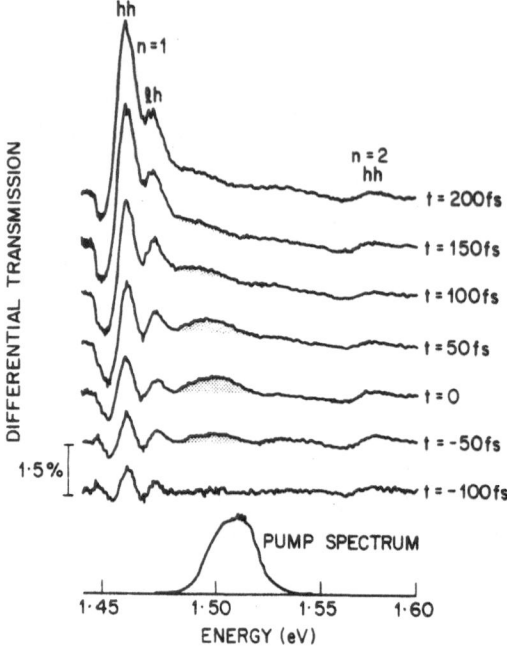

Figure 11. Differential transmission of multiple quantum-well structure as a function of pump-probe delay. The pump pulse has a duration or 6 *ps* and is tuned 25 meV below the heavy hole exciton resonance. The probe pulses also have 6 *ps* duration and are tuned 1 meV below the heavy resonance in (a), and above this resonance in (b) (after Reference 27).

position. It will switch back the to initial position if the light is decreased. Thus a very fast bistable optical element of small size with low power consumption may be constructed.

The large oscillator strength associated with the exciton resonance in these small devices also leads to striking nonlinear optical response for light frequencies in the vicinity of the absorption edge. Several excellent review articles concerning both the linear and nonlinear optical characteristics of quantum-well devices have been published.[27,28] In this section, only a brief extract of the salient nonlinear characteristics will be presented.

Three regimes should be distinguished. If the frequency of the incident light is chosen well below the absorption peaks of the exciton resonances, the excitons will be virtually excited. The discussion of nonlinearities in this case is very similar to that of an atomic or molecular vapor where some excited state configuration is admixed to the ground state. The admixture is inversely proportional to the detuning from resonance and proportional to the intensity of the incident light. The response time is very fast, on the order of the inverse of the detuning. A strong pulsed pump field will produce a nearly instantaneous dynamic Stark shift of the exciton resonance. A weak probe pulse in the vicinity of the exciton resonance will either be less absorbed or more absorbed depending on the sign of its detuning from resonance. The change in transmission of a probe pulse due to the Stark shift induced by a 6 *ps* pump pulse is shown in Figure 11. The long time changes in transmission are caused by real exciton population changes, caused by the pump in the tail of the absorption edge. For very short, very intense femtosecond pulses, the Rabi frequency may become comparable

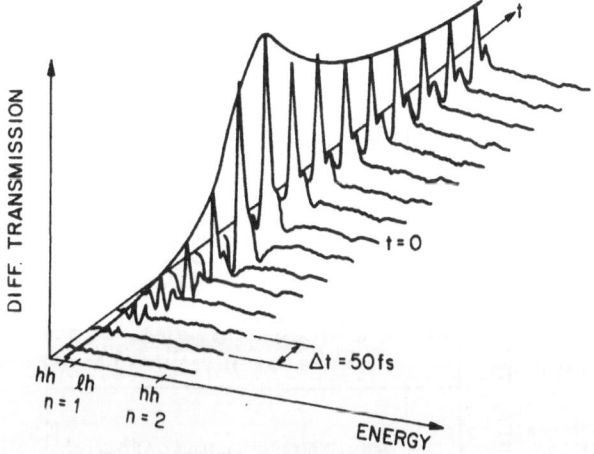

Figure 12. Room-temperature differential transmission spectra in a multiple quantum-well structure for various time delays between the broad-band 50 *fs* probe pulse and the 100 *fs* pump pulse applied above resonance (after Reference 27).

to the detuning and the width of the exciton resonance. Then more complex changes in the probe absorption or emission may be expected in analogy to the situation in atomic spectroscopy.

If the pump pulse frequency is chosen well above the band gap and above the exciton resonances, the effect of free carrier electron-hole pairs becomes evident. The changes in transmission of a weak broad-band probing pulse following a 100 *fs* narrow-band pump pulse are shown in Figure 12. The spectral hole burning at the pump frequency due to saturation relaxation in about 100 *fs* are caused by phase space filling. Due to fast relaxation processes in the dense carrier plasma, the occupation of states in *k*-space adjusts rapidly to a thermal distribution. The states near the origin in *k*-space from which excitons are formed are occupied. This is called phase space filling. Thus absorption at the exciton resonances is suppressed, as evidenced by the increased differential transmission of the probe at the exciton resonant frequencies. The data show that relaxation in the occupation of states in dense carrier plasma occurs on a time scale of 100 *fs*. The effect of Coulomb screening and the corresponding band gap renormalization and renormalization of exciton states is of minor importance in the two-dimensional geometry of quantum wells. In three-dimensional bulk material of GaAs, by contrast, these renormalization effects dominate the behavior of the changes in the complex dielectric function of a probe.

If a pump field is applied at an exciton resonant frequency itself, the transmission increases corresponding to direct saturation or phase space filling of the resonance. The data in Figure 13 show that the exciton population relaxes partially as they are ionized by collisions with thermally excited LO-phonons at room temperature. This process has a time constant of about 300 *fs* and determines the width of the exciton resonances.

These examples demonstrate that nonlinear effects in samples of very small dimensions are readily detectable. This may lead to fast optical devices with low switching power and compatible with solid state electronics. The various relaxation processes occurring in a dense electron-hole plasma and at high concentrations of excitons are accessible to experimental investigation.

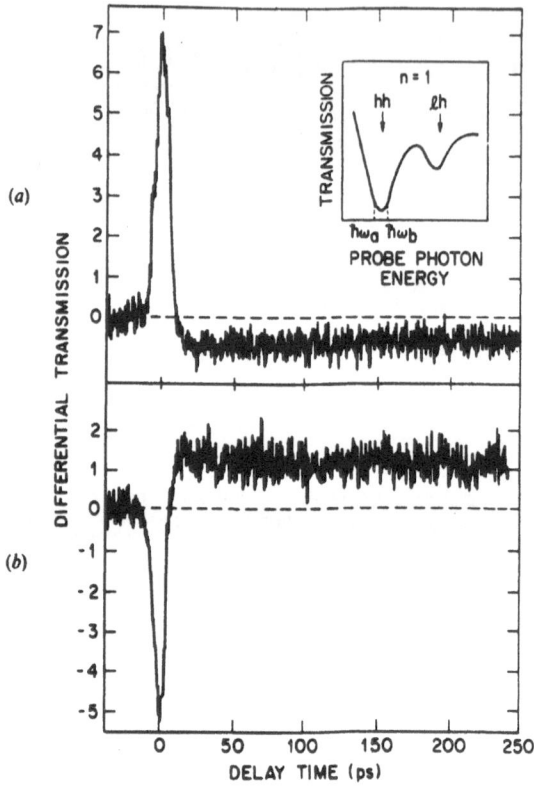

Figure 13. Room-temperature differential transmission spectra in a multiple quantum-well structure, when the pump pulse is applied at the exciton resonance (after Reference 27).

IV. NONLINEAR OPTICAL PROPERTIES OF POLYMERIC EXCITATIONS

The nonlinear optics of large organic molecules and polymeric materials has been the subject of intense investigation during the past two decades. The results of this activity have been reviewed in several books and review articles.[30-33] In the present context, attention will be focused on the fact that in polydiacetylenes (PDA) a dominant excited electron configuration may be described as a self-trapped one-dimensional electron-hole pair. The nature of the electronic transition has been described by Kobayashi,[34] who carried out time-resolved four-wave mixing experiments. The pertinent energy diagram is schematically represented in Figure 14. The coupling between the electronic excitation and vibrational degrees is quite strong in PDA's, and this leads to a conformational change, trapping the exciton. Experiments on induced changes in the optical absorption spectrum following a strong pump pulse at the free exciton transition exhibit the behavior of saturation (or bleaching) due to phase-space-filling.[35] The size of the collective excitation along the one-dimensional chain is about 4 nm. Pumping below the resonance leads to an optical Stark shift shown in Figure 15. Another feature indicative of large electron-phonon coupling is the resonant effect when the sum of the pump and phonon frequencies lies in the vicinity of the exciton resonance.

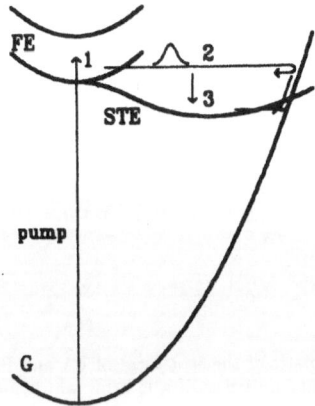

Figure 14. Qualitative energy level diagram for the exciton resonance in polydiacetylenes (PDA). FE is the free exciton potential curve, STE is the self-trapped exciton potential curve. There is no barrier in going from 1 to 2 and a phonon is emitted in going from 2 to 3 (after Reference 34).

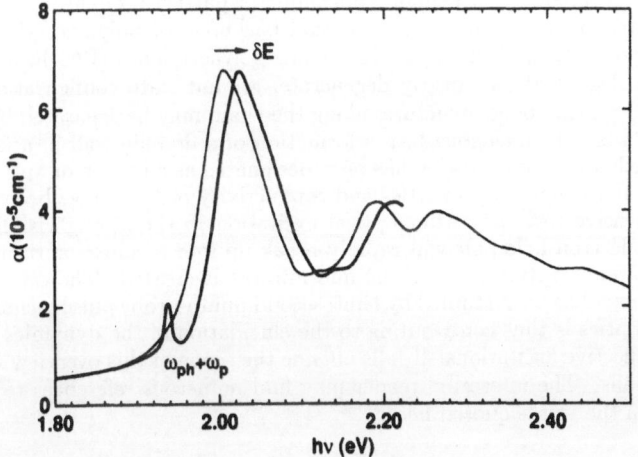

Figure 15. Schematic description of the optical Stark shift of the one-dimensional exciton in PDA. A phonon-mediated resonant structure is indicated (after Reference 35).

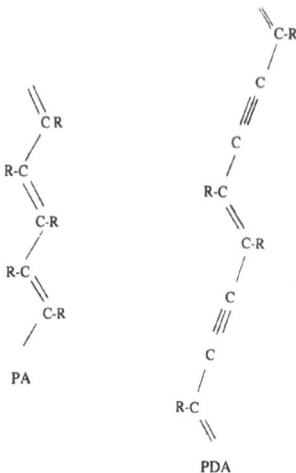

Figure 16. Schematic of the structures of PA and PDA conjugated bond chains. R represents a variety of side arms which may be substituted for hydrogen atoms.

A theoretical discussion of these effects has been given by Greene *et al.*[36] The approach used for semiconductor well structures must be augmented by the explicit addition of a vibrational excitation. There is nevertheless a remarkable parallel when the PDA characteristics are compared with the quantum well optical properties. Mukamel *et al.*[37,38] have related the magnitude of the nonlinear susceptibility to the electronic structure of conjugated polyenes. They also discuss the (coherence) size of the excitation and the analogy with small semiconductor particles (quantum dots).

The structures of the one-dimensional backbone of polydiacetylene and polyacetylene are compared in Figure 16. Trans-polyacetylenes (PA) have, in contrast to PDA, a degenerate, or nearly degenerate, ground state configuration. The two possible alternating bond structures along the chain may be separated by a region of transition, which is analogous to the formation of a domain wall in a ferromagnetic crystal. Such a transition region has been designated as a soliton or anti-soliton. The interesting linear optical properties and conductivity of PA's have been discussed in a comprehensive review.[39] After optical excitation in the strong visible absorption band, the electron-hole pair will rapidly break up into a soliton-antisoliton pair. A new mid-gap absorption band in the mid-infrared is created. The time evolution of these processes has been studied by femtosecond pump-probe pulse techniques.[40,41,42] Nonlinear optics is thus contributing to the elucidation of the dynamics of these fascinating collective excitations. It falls outside the scope of this overview to reproduce further details. The interested reader may find numerous references to the original literature in the works quoted here.[30−42]

V. CONCLUSION

Nonlinear optical investigations have contributed significantly to our understanding of collective excitations in insulating crystal, semiconductors and organic materials. Conversely, their nonlinear optical properties may be used in devices such as optical parametric up-and-down converters, in all-optical-logic switches for information processing, and in three-dimensional holograms for information storage and retrieval. Much further development work is needed before large-scale technological applications could become practical.

REFERENCES

1. L. Brillouin, *Ann. de Physique.* 17:103 (1921).
2. E.F. Gross, *Nature* 126:201,400–603 (1930).
   Zeitschr. für Physik. 63:685 (1930).
3. P. Debye and F.W. Sears, *Proc. Nat. Acad. Sci.,* Washington. 18:409 (1932).
4. R. Lucas and P. Biquard, *J. Phys Radium.* 3:464 (1932).
5. R. Chiao and B.P. Stoicheff, *Phys. Rev. Lett.* 12:290 (1964).
6. E.J. Woodbury and W.K. Ng, *Proc IRE.* 50:2367 (1962).
   G. Eckhardt, R.W. Hellworth, F.J. McClung, S.E. Schwarz, D. Weiner, and E.J. Woodbury, *Phys. Rev. Lett.* 9:455 (1962).
7. M. Goeppert-Mayer, *Ann. der Physik.* 9:273 (1931).
8. W. Kaiser, and G.C.B. Garrett, *Phys. Rev. Lett.* 7:229 (1961).
9. D. Frohlich, E. Mohler, and P. Wiesner, *Phys. Rev. Lett.* 26:554 (1971).
10. M. Born and H. Kun, "Dynamical Theory of Crystal Lattices," Clarendon, Oxford (1964).
11. H. Poulet, *Ann. der Physik.* (Paris), 10:908 (1955).
12. D.L. Mills and E. Burstein, *Rep. Prog. Phys.* 37:817 (1974).
13. A.S. Barker, Jr. and R. Loudon, *Rev. Mod. Phys.* 44:18 (1972).
14. R. Loudon, "Nonlinear Spectroscopy," N. Bloembergen, ed., North Holland Publ., Amsterdam (1977), p. 296.
15. R. Loudon, "Collective Excitations in Solids," B. Di Bartolo, ed., Plenum Press, New York (1981), p. 479.
16. F. De Martini, "Nonlinear Spectroscopy," N. Bloembergen, ed., North Holland Publ., Amsterdam (1977), p. 319.
17. C.H. Henry and J.J. Hopfield, *Phys. Rev. Lett.* 15:964 (1965).
18. W.L. Faust and C.H. Henry, *Phys. Rev. Lett.* 17:1265 (1966).
19. J.P. Coffinet and F. De Martini, *Phys. Rev. Lett.* 22:60 (1969).
20. F. De Martini, P. Mataloni, E. Palange, and Y.R. Shen, *Phys. Rev. Lett.* 37:440 (1976).
21. D.C. Haueisen and H. Mahr, *Phys. Rev. Lett.* 26:838 (1971).
22. B. Hoenerlage, A. Bivas, and Vu Duy Phach, *Phys. Rev. Lett.* 41:49 (1978).
23. E. Hanamura, *Solid State Comm.* 12:951 (1973).
24. A. Mysyrowicz, D. Hulin, and L.L. Chase, "Collective Excitations in Solids," B. DiBartolo, ed., Plenum Press, New York (1981), p. 659.
25. C. Klingshirn and H. Haug, *Physics Reports* 70:315 (1981).
26. S.D. Kramer and N. Bloembergen, *Phys. Rev. B.* 14:4654 (1976).
27. S. Schmitt-Rink, D.S. Chemla, and D.A.B. Miller, *Adv. in Phys.* 38:89 (1989).
28. D.S. Chemla, *Physics Today.* (June, 1993) p. 46.
29. D.S. Chemla, *Physics Today.* (May, 1985) p. 57.
30. "Nonlinear Optical Properties of Organic Molecules and Crystals," D.S. Chemla and J. Zyss, eds., Academic Press, New York (1988).
31. "Introduction to Nonlinear Optical Effects in Molecules and Polymers," P.N. Prasad and D.J. Williams, eds., Wiley (1991).
32. "Molecular Nonlinear Optics," J. Zyss, ed., Academic Press, New York (1994).
33. N. Bloembergen, *Intl. J. Nonlinear Opt. Phys.* 3:439 (1994).
34. T. Kobayashi, in reference 32, p. 47.
35. S. Etemad, G.L. Baker, and Z.G. Zoos, in reference 32, p. 433.
36. B.I. Greene, J. Orenstein, and S. Schmitt-Rink, *Science* 247:679 (1990).
37. S. Mukamel, in reference 32, p. 1.
38. S. Mukamel, A. Takahashi, H.X. Wang, and G. Chen, *Science* 266:250 (1994).
39. A.J. Heeger, S. Kivelson, J.R. Schrieffer, and W.P. Su, *Rev. Mod. Phys* 60:781 (1988).
40. L. Rothberg, T.M. Jedju, S. Etemad, and G.L. Baker, *IEEE J. Quant. Elec.* 24:311 (1988).
41. S. Taekuchu, M. Yoshizawa, T. Masuda, T. Higashimura, and T. Kobayashi, *IEEE J. Quant. Elec.* 28:2508 (1992).
42. T. Kobayashi, in reference 32, p. 47.

FUNDAMENTALS OF SPECTROSCOPY OF COLLECTIVE EXCITATION IN SOLIDS

B. Di Bartolo

Department of Physics
Boston College
Chestnut Hill, Massachusetts 02167, USA

ABSTRACT

Collective excitations are the intrinsic excitations of solids. A study of their physical properties is necessary for an adequate understanding of solid state systems. The purpose of this article is to examine the basic aspects of collective excitations and of their spectroscopic properties; it will introduce the fundamental concepts and will lay down the background material useful for the treatment of the subject of this book.

In the first section we shall deal with a simple two-level system and we shall examine the effect of a time-dependent perturbation on such a system.

In the second section we shall introduce the concept of collective excitation, keeping the treatment as general as possible.

In the third section we shall treat the problem of the interaction of electromagnetic radiation with collective excitations.

In the fourth section we shall examine the conditions for the propagation of electromagnetic radiation in a dispersive linear medium.

Finally in the fifth section we shall apply the knowledge acquired in the previous sections to the treatment of different types of collective excitations.

I. INTERACTION IN A TWO-LEVEL SYSTEM

I.A. Quantum Mechanical Resonance

Let us consider a system with a time-independent Hamiltonian H_0. The time-dependent Schrödinger equation is written as follows

$$H_0 \psi = i\hbar \frac{\partial \psi}{\partial t} \qquad (1)$$

If the system is in a stationary state labeled i

$$\psi(t) = \psi_i(t) = \psi_i(0)e^{-i(E_i/\hbar)t} \tag{2}$$

where the energy values are given by

$$H_0\psi_i(0) = E_i\psi_i(0) \tag{3}$$

We shall assume that the wavefunctions $\psi_i(t)$ are orthonormal.

Let us now suppose that the system is subjected to a time-dependent perturbation represented by $H'(t)$. The system will then be represented by a wavefunction $\psi(t)$ such that

$$H\psi(t) = (H_0 + H')\psi(t) = i\hbar\frac{\partial\psi}{\partial t} \tag{4}$$

We can expand $\psi(t)$ in terms of the complete set $\psi_i(t)$

$$\psi(t) = \sum_i c_i(t)\psi_i(t) \tag{5}$$

If $H' = 0$, the coefficients c_i's are time-independent. Replacing Eq.(5) in Eq.(4),

$$(H_0 + H')\sum_i c_i(t)\psi_i(t) = i\hbar\left[\sum_i c_i(t)\frac{\partial\psi_i}{\partial t} + \sum_i \frac{\partial c_i(t)}{\partial t}\psi_i(t)\right] \tag{6}$$

Then

$$\sum_i c_i(t)H'\psi_i(t) = i\hbar\sum_i \frac{\partial c_i(t)}{\partial t}\psi_i(t) \tag{7}$$

where we have taken Eqs.(2) and (3) into account. Multiplying by $\psi_k^*(t)$ and integrating over space coordinates we obtain

$$i\hbar\sum_i \frac{\partial c_k(t)}{\partial t} = \sum_i c_i(t)\langle\psi_k(t)|H'|\psi_i(t)\rangle = \sum_i c_i(t)M_{ki}e^{i\omega_{ki}t} \tag{8}$$

where

$$\omega_{ki} = \frac{E_k - E_i}{\hbar}; \qquad M_{ki} = \langle\psi_k(0)|H'|\psi_i(0)\rangle \tag{9}$$

We shall now make the following simplifying assumptions:
  (a) The system has only two energy levels, say 1 and 2;
  (b) the diagonal matrix elements of H' are zero; and
  (c) the perturbation H' is constant, but is turned on at time $t = 0$.

The coupled equations (8) become

$$\begin{cases} \dot{c}_2(t) = -\frac{i}{\hbar}c_1(t)e^{i\omega_{21}t}M_{21} \\ \dot{c}_1(t) = -\frac{i}{\hbar}c_2(t)e^{-i\omega_{21}t}M_{21}^* \end{cases} \tag{10}$$

Differentiating the first equation above and using the second equation, we obtain for $t > 0$

$$\ddot{c}_2(t) - i\omega_{21}\dot{c}_2(t) + \frac{|M_{21}|^2}{\hbar^2}c_2(t) = 0 \qquad (11)$$

We expect $c_2(t)$ to be of the form

$$c_2(t) = Ae^{\alpha_1 t} + Be^{\alpha_2 t} \qquad (12)$$

where α_1 and α_2 are solutions of the equation

$$\alpha^2 - i\omega_{21}\alpha + \frac{|M_{21}|^2}{\hbar^2} = 0 \qquad (13)$$

or

$$\alpha_{1,2} = \frac{i}{2}\left[\omega_{21} \pm \left(\omega_{21}^2 + 4\frac{|M_{21}|^2}{\hbar^2}\right)^{1/2}\right] = \frac{i\omega_{21}}{2} \pm ia \qquad (14)$$

where

$$a = \frac{1}{2}\left(\omega_{21}^2 + 4\frac{|M_{21}|^2}{\hbar^2}\right)^{1/2} \qquad (15)$$

Let us study the time evolution of such a system starting from the following initial conditions

$$c_2(0) = 0, \quad c_1(0) = 1 \qquad (16)$$

which indicate that the system is in the state 1 at time t=0. The initial condition on c_2 gives $A = -B$; then Eq.(12) becomes

$$c_2(t) = A\left(e^{\alpha_1 t} - e^{\alpha_2 t}\right) = A\left(e^{\frac{i\omega_{21}t}{2}+iat} - e^{\frac{i\omega_{21}t}{2}-iat}\right) = Ce^{\frac{i\omega_{21}t}{2}}\sin(at) \qquad (17)$$

where $\qquad C = 2iA \qquad (18)$

The expression for $c_1(t)$ can be derived from the first of Eq.(10) and the expression (17) for $c_2(t)$:

$$c_0(t) = \frac{i\hbar e^{-i\omega_{21}t}}{M_{21}}\dot{c}_2(t) = \frac{i\hbar e^{-i\omega_{21}t/2}C}{M_{21}}\left(i\frac{\omega_{21}}{2}\sin(at) + a\cos(at)\right) \qquad (19)$$

The initial condition for $c_1(t)$ gives $C = M_{21}/(i\hbar a)$. Therefore, c_2 becomes

$$c_2(t) = Ce^{i\omega_{21}t/2}\sin(at) = \frac{M_{21}}{i\hbar a}e^{i\omega_{21}t/2}\sin(at) \qquad (20)$$

and

$$|c_2(t)|^2 = \frac{|M_{21}|^2}{\hbar^2 a^2}\sin^2(at) \qquad (21)$$

On the other hand, we can now write

$$c_1(t) = e^{-i\omega_{21}t/2}\left(\cos(at) + i\frac{\omega_{21}}{2a}\sin(at)\right) \qquad (22)$$

and

$$|c_1(t)|^2 = \left(\cos^2(at) + \frac{\omega_{21}^2}{4a^2}\sin^2(at)\right) \qquad (23)$$

We can verify that

$$|c_1(t)|^2 + |c_2(t)|^2 = 1 \qquad (24)$$

Considering the expressions for $|c_1(t)|^2$ and $|c_2(t)|^2$, we note that at time t=0, $|c_1(0)|^2 = 1$ and $|c_2(0)|^2 = 0$, as expected. Following the time t=0, the two probabilities have oscillatory behaviors; at time $t = \pi/2a$, $|c_2|^2$ and $|c_1|^2$ take their maximum and minimum values, respectively:

$$|c_2|^2_{max} = \frac{|M_{21}|^2}{\hbar^2 a^2} = \frac{4|M_{21}|^2}{\hbar^2\omega_{21}^2 + 4|M_{21}|^2} \xrightarrow{\omega_{21}\to 0} 1 \qquad (25)$$

$$|c_1|^2_{min} = \frac{\omega_{21}^2}{4a^2} = \frac{\hbar^2\omega_{21}^2}{\hbar^2\omega_{21}^2 + 4|M_{21}|^2} = 1 - \frac{4|M_{21}|^2}{\hbar^2\omega_{21}^2 + 4|M_{21}|^2} = 1 - |c_2|^2_{max} \xrightarrow{\omega_{21}\to 0} 0 \qquad (26)$$

The condition $\omega_{21} = 0$ (i.e., levels 1 and 2 degenerate) corresponds to a "quantum mechanical resonance". In this particular case

$$a = |M_{21}|^2/\hbar \qquad (27)$$

and

$$|c_2(t)|^2 = \sin^2 at = \sin^2\frac{|M_{21}|^2}{\hbar}t \qquad (28)$$

$$|c_1(t)|^2 = \cos^2 at = \cos^2\frac{|M_{21}|^2}{\hbar}t \qquad (29)$$

At the time t=0, $\qquad |c_2(0)|^2 = 0, \qquad |c_1(0)|^2 = 1$

After a time

$$t = \pi/2a = \hbar/2|M_{21}| \qquad (30)$$

the system shifts from the state 1 to the state 2:

$$|c_2(\tfrac{\pi}{2a})|^2 = 1, \quad |c_1(\tfrac{\pi}{2a})|^2 = 0$$

I.B. Static Effects of a Perturbation

Let us consider a system with the Hamiltonian

$$H = H_0 + H' \qquad (31)$$

where both H_0 and H' are independent of time. Let us call ψ_l the orthonormal eigenfunctions of H_0, and ψ'_λ the eigenfunctions of H:

$$H_0\psi_l = E_l\psi_l \qquad (32)$$

$$H\psi'_\lambda = E'_\lambda \psi'_\lambda \tag{33}$$

We can expand the eigenfunctions of H as follows:

$$\psi'_\lambda = \sum_l a_{\lambda l} \psi_l \tag{34}$$

where l ranges over all the eigenfunctions of H_0. Replacing Eq.(34) in Eq.(33),

$$H_0 \sum_l a_{\lambda l} \psi_l + H' \sum_l a_{\lambda l} \psi_l = E'_\lambda \sum_l a_{\lambda l} \psi_l \tag{35}$$

or, because of Eq.(32)

$$\sum_l a_{\lambda l}(E'_\lambda - E_l)\psi_l = \sum_l a_{\lambda l} H' \psi_l \tag{36}$$

Multiplying by ψ^*_m and integrating over the space coordinates we find

$$a_{\lambda l}(E'_\lambda - E_l) = \sum_l a_{\lambda l} H'_{ml} \tag{37}$$

where

$$H'_{ml} = \langle \psi_m | H' | \psi_l \rangle \tag{38}$$

We shall assume at this point that the energy level scheme of H_0 consists of two degenerate levels; the equations (37) can then be written as follows:

$$\begin{cases} a_{\lambda 1}(E'_\lambda - E_1) = a_{\lambda 1} H'_{11} + a_{\lambda 2} H'_{12} \\ a_{\lambda 2}(E'_\lambda - E_2) = a_{\lambda 1} H'_{21} + a_{\lambda 2} H'_{22} \end{cases} \tag{39}$$

Setting $E_1 = E_2 = E$, we get

$$\begin{cases} (E'_\lambda - E - H'_{11})a_{\lambda 1} + (-H'_{12})a_{\lambda 2} = 0 \\ (-H'_{21})a_{\lambda 1} + (E'_\lambda - E - H'_{22})a_{\lambda 2} = 0 \end{cases} \tag{40}$$

which implies that

$$\begin{vmatrix} (E'_\lambda - E - H'_{11}) & (-H'_{12}) \\ (-H'_{21}) & (E'_\lambda - E - H'_{22}) \end{vmatrix} = 0 \tag{41}$$

If we assume for simplicity that $H'_{11} = H'_{22}$, we get

$$(E'_\lambda - E - H'_{11})^2 - |H'_{12}|^2 = 0 \tag{42}$$

The equation above gives us the two eigenvalues

$$E'_1 = E + H'_{11} + |H'_{12}| \tag{43}$$

$$E'_2 = E + H'_{11} - |H'_{12}| \tag{44}$$

We have still to find the coefficients $a_{\lambda m}$. If we take $\lambda = 1$ and replace E_1' by the expression above in the relations (40), we obtain $a_{11} = a_{12}$ and

$$\psi_1' = \frac{1}{\sqrt{2}}(\psi_1 + \psi_2)$$ (45)

Taking $\lambda = 2$ and replacing the expression for E_2' in Eqs.(40), we obtain $a_{21} = -a_{22}$ and

$$\psi_2' = \frac{1}{\sqrt{2}}(\psi_1 - \psi_2)$$ (46)

It is easy to verify that

$$\langle \psi_1' | H | \psi_1' \rangle = E + H_{11}' + |H_{12}'|$$ (47)

$$\langle \psi_2' | H | \psi_2' \rangle = E + H_{11}' - |H_{12}'|$$ (48)

and

$$\langle \psi_1' | H | \psi_2' \rangle = 0$$ (49)

Let us relate now these findings to the results of the previous Section I.A:

Since in the present case we consider two degenerate states we are clearly dealing with a quantum mechanical resonance. If, in addition, we assume, as we did in Section I.A, that the diagonal matrix elements of H' are zero, we can represent the situation as in Figure 1.

We can now make the following observations:

(1) The same perturbation that is responsible for the change in time from the state 1 to state 2 is giving us a splitting of levels equal to $2|M_{21}|$.
(2) Consider this splitting to be 1 cm⁻¹ ($2|M| = hc$). Then the time it takes the system to switch from state 1 to state 2 is

$$t = \frac{\pi}{2a} = \frac{\pi\hbar}{2|M|} = \frac{1}{2c} = 1.67 \times 10^{-11} \, \text{sec}^{-1}$$ (50)

This is a very short time indeed. This case provides an illustrative example of how a perturbation that produces negligible (1 cm⁻¹) static effects may cause dramatic $(1.67 \times 10^{-11} \, \text{sec})$ dynamical effects.

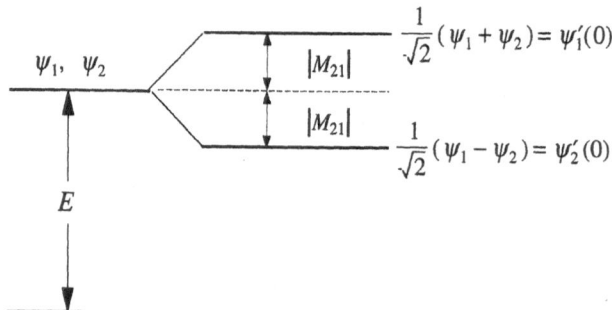

Figure 1. Static effect of a perturbation on a system consisting of two degenerate states.

The time-dependent wavefunctions of the perturbed states are given by

$$
\begin{cases}
\psi_1' = \dfrac{1}{\sqrt{2}}\left(\psi_1 + \psi_2\right)e^{-i(E_0+|M|)t/\hbar} \\[2mm]
\psi_2' = \dfrac{1}{\sqrt{2}}\left(\psi_1 - \psi_2\right)e^{-i(E_0-|M|)t/\hbar}
\end{cases}
\tag{51}
$$

At time t=0, these wavefunctions become

$$
\begin{cases}
\psi_1' = \dfrac{1}{\sqrt{2}}\left(\psi_1 + \psi_2\right) \\[2mm]
\psi_2' = \dfrac{1}{\sqrt{2}}\left(\psi_1 - \psi_2\right)
\end{cases}
\tag{52}
$$

and conversely

$$
\begin{cases}
\psi_1 = \dfrac{1}{\sqrt{2}}\left[\psi_1'(0) + \psi_2'(0)\right] \\[2mm]
\psi_2 = \dfrac{1}{\sqrt{2}}\left[\psi_1'(0) - \psi_2'(0)\right]
\end{cases}
\tag{53}
$$

Let us assume that at time t=0 the system is in the state ψ_1:

$$
\psi(t=0) = \psi_1 = \frac{1}{\sqrt{2}}\left[\psi_1'(0) + \psi_2'(0)\right]
\tag{54}
$$

The wavefunction of the system at time t is then

$$
\psi(t) = \frac{1}{\sqrt{2}}\left[\frac{1}{\sqrt{2}}\left(\psi_1 + \psi_2\right)e^{-i(E+|M|)t/\hbar} + \frac{1}{\sqrt{2}}\left(\psi_1 - \psi_2\right)e^{-i(E-|M|)t/\hbar}\right]
$$

$$
= e^{-iEt/\hbar}\left(\psi_1 \cos\frac{|M|}{\hbar}t - i\psi_2 \sin\frac{|M|}{\hbar}t\right)
\tag{55}
$$

If at the time t=0, as we assumed, the system is in the state ψ_1, at the time $t = \pi\hbar/(2|M|)$ the wavefunction of the system is

$$
\psi\left(t = \frac{\pi\hbar}{2|M|}\right) = e^{-(E/\hbar)(\pi\hbar/2M)}\left(-i\psi_2\right)
\tag{56}
$$

namely, the system is in the state ψ_2. After a time $t = 2\pi\hbar/(2|M|)$ the system goes back to the state ψ_1 and so on.

We note here that the rate at which the system changes state is proportional to the coupling energy $|M|$.

We can make the following observations on these results:

(a) When the system is in a stationary state the wavefunction, as seen in Eq.(51), includes both ψ_1 and ψ_2. There is an equal probability of finding the system in state ψ_1 or in state ψ_2.

(b) In order to put the system in a state ψ_1, it is necessary to include the two wavefunctions ψ_1' and ψ_2' which have different energies (ψ_1 is not a stationary state of $H = H_0 + H'$).

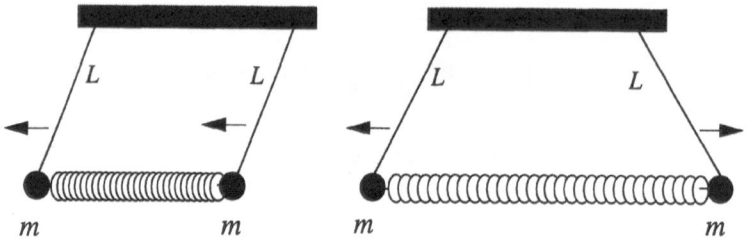

Figure 2. Normal modes of two coupled pendula.

The quantum mechanical system that we have examined is analogous to the mechanical system consisting of two equal pendula coupled by a spring. This system presents two normal modes whose patterns are presented in the following Figure 2.

If the pendula oscillate according to one of the two normal modes the system is in a stationary state and the same amount energy is associated to each pendulum. If, however, we start from the initial conditions in which one pendulum is in the rest position and the other pendulum is deflected, an exchange of energy between the two pendula takes place.

We note also that the closer the coupling, i.e. the stronger is the spring, the greater is the energy transfer rate.

II. COLLECTIVE EXCITATIONS

II.A. Setting of the Problem

Consider a linear crystal consisting a chain of N identical atoms with a unit cell of length a.

Assume that the Hamiltonian of the system has the following form

$$H = H_0 + H' \tag{57}$$

where

$$H_0 = \sum_{s}^{N} H_s \tag{58}$$

$$H' = \frac{1}{2} \sum_{\substack{s,s' \\ s \neq s'}} V_{ss'} \tag{59}$$

The H_s are one-atom Hamiltonians and the $H_{ss'}$ are terms representing the interactions between the different atoms. The ground state wavefunction of H_0 is given by

$$\psi_g = u_1 u_2 \dots u_{N-1} u_N \tag{60}$$

The first excited state of H_0 corresponds to the following N degenerate wavefunctions

$$\begin{cases} \psi_1 = v_1 u_2 \dots u_N \\ \psi_2 = u_1 v_2 \dots u_N \\ \dots \dots \dots \dots \\ \psi_N = u_1 u_2 \dots v_N \end{cases} \tag{61}$$

In these wavefunctions u_i represents a localized one-atom ground state and v_i a localized one-atom excited state. Both u_i and v_i are assumed to be nondegenerate.

The ground state of H_0 is nondegenerate; the first excited state of H_0 is N-fold degenerate, because there are N distinct ways to excited only one atom. For all these wavefunctions the energy is E_0:

$$H_0 \psi_l = E_0 \psi_l \qquad (l = 1, 2, ...N) \tag{62}$$

The eigenfunctions of H are given by

$$H\psi'_\lambda = (H_0 + H')\psi'_\lambda = E'_\lambda \psi'_\lambda \qquad (\lambda = 1, 2, ...N) \tag{63}$$

The excited-state eigenfunctions of H can be expressed as follows:

$$\psi'_\lambda = \sum_{l=1}^{N} a_{\lambda l} \psi_l \tag{64}$$

Replacing this in Eq.(63) we find

$$\sum_l a_{\lambda l}(E_0 - E'_\lambda)\psi_l + \sum_l a_{\lambda l} H' \psi_l = 0 \tag{65}$$

Multiplying by ψ_m^* and integrating

$$a_{\lambda l}(E'_\lambda - E_0 - H'_{mm}) - \sum_{l \neq m} a_{\lambda l} H'_{ml} = 0 \tag{66}$$

We note the following:

E'_λ = total energy of the system,

$E_0 + H'_{mm}$ energy of the chain if the excitation cannot move,

$E'_\lambda - (E_0 + H'_{mm}) = \varepsilon_\lambda$ energy associated with the motion of the excitation.

We see from Eq.(66) that the motion of the excitation is related to the off-diagonal terms of the H'.

II.B. Eigenfunctions

In order to look for functions that approximate well the wavefunction ψ'_λ that satisfy the eigenvalue equation (63) we want to take advantage of the symmetry of the system. We shall assume periodic boundary conditions to take care of "surface" effects and translational symmetry for the chain

$$H(x + la) = H(x) \tag{67}$$

It seems appropriate at this time to digress from the present sequence of derivations and obtain a general result for the three-dimensional case.

Let us assume that an electron is in a periodic field of force and its Hamiltonian is such that

25

$$H\left(\vec{r} + \vec{R}_n\right) = H(\vec{r}) \tag{68}$$

The eigenfunctions of H are given by the equation

$$H\psi(\vec{r}) = E\psi(\vec{r}) \tag{69}$$

We shall assume these functions to be orthonormal. Let us introduce a translation operation $T_{\vec{n}}$ that displaces the origin by $\vec{R}_n$, $\vec{R}_n$ being a lattice vector:

$$T_{\vec{n}}\psi(\vec{r}) = \psi(\vec{r} + \vec{R}_n) \tag{70}$$

Dropping for simplicity the subscript of the operation T we can write

$$TH\psi(\vec{r}) = ET\psi(\vec{r}) \tag{71}$$

or

$$H[T\psi(\vec{r})] = E[T\psi(\vec{r})] \tag{72}$$

Therefore, $T\psi(\vec{r})$ is an eigenfunction of H belonging, as $\psi(\vec{r})$, to the eigenvalue E.

If E is nondegenerate, then

$$T\psi(\vec{r}) = c\psi(\vec{r}) \tag{73}$$

with c=const. $c\psi(\vec{r})$ differs from $\psi(\vec{r})$ only because a change in the origin. Therefore

$$\int \left|c\psi(\vec{r})\right|^2 d\tau = |c|^2 \int |\psi(\vec{r})|^2 d\tau = |c|^2 = 1 \tag{74}$$

If E is an m-degenerate eigenvalue of the energy, then

$$H\left[T\psi_\mu(\vec{r})\right] = E\left[T\psi_\mu(\vec{r})\right] \tag{75}$$

with

$$T\psi_\mu(\vec{r}) = \psi_\mu\left(\vec{r} + \vec{R}_n\right) = \sum_{l=1}^{m} c_{\mu l}\psi_l(\vec{r}) \tag{76}$$

with $\mu = 1, 2, \ldots\ldots, m$. As said before, the functions ψ are orthonormal:

$$\int \psi_\mu^*(\vec{r})\psi_\lambda(\vec{r})d\tau = \delta_{\mu\lambda} \tag{77}$$

These integrals are not affected by a change of origin; therefore, the same is true of the new functions

$$\int \psi_\mu^*\left(\vec{r} + \vec{R}_n\right)\psi_\lambda\left(\vec{r} + \vec{R}_n\right)d\tau = \int \sum_{l=1}^{m} c_{\mu l}^* \psi_l^* \sum_{l'=1}^{m} c_{\lambda l'}\psi_{l'}d\tau$$

$$= \sum_{l=1}^{m}\sum_{l'=1}^{m} c_{\mu l}^* c_{\lambda l'} \int \psi_l^* \psi_{l'}d\tau = \sum_{l=1}^{m}\sum_{l'=1}^{m} c_{\mu l}^* c_{\lambda l'}\delta_{ll'} = \sum_{l} c_{\mu l}^* c_{\lambda l} = \delta_{\mu\lambda} \tag{78}$$

We observe then the following:

(1) The $m \times m$ matrix $\hat{C}$ of the coefficient c is unitary:

$$\hat{C}\hat{C}^{+} = \hat{1} \tag{79}$$

(2) Any unitary matrix can be brought to diagonal form by a unitary transformation.

(3) There exists an orthonormal set of linear combinations of the functions ψ_l which are related to a diagonal matrix $\hat{C}$:

$$T\begin{pmatrix} \psi_1 \\ \psi_2 \\ : \\ : \\ \psi_m \end{pmatrix} = \begin{pmatrix} c_{11} & 0 & \cdots & 0 \\ 0 & c_{22} & \cdots & 0 \\ 0 & 0 & \cdots & 0 \\ 0 & 0 & \cdots & 0 \\ 0 & 0 & \cdots & c_{mm} \end{pmatrix} \begin{pmatrix} \psi_1 \\ \psi_2 \\ : \\ : \\ \psi_m \end{pmatrix} \tag{80}$$

This is equivalent to

$$\psi(\vec{r} + \vec{R}_n) = c\psi(\vec{r}) \tag{81}$$

being valid for each function ψ_μ.

(4) Having made the matrix $\hat{C}$ diagonal for a particular translation $\vec{R}_n$, we can do likewise for any other translation because the translation operations commute. The matrices $\hat{C}$ related to any two lattice vector translations commute, and a set of commuting matrices can always be diagonalized simultaneously.

On the basis of the above we can assume that all the eigenfunctions can be made to satisfy the condition

$$\psi(\vec{r} + \vec{R}_n) = c\psi(\vec{r}) \tag{82}$$

for any lattice vector displacement. Therefore, we have

$$\psi(\vec{r} + \vec{R}_n) = c(\vec{R}_n)\psi(\vec{r}) \tag{83}$$

$$\psi(\vec{r} + \vec{R}_m) = c(\vec{R}_m)\psi(\vec{r}) \tag{84}$$

and

$$\psi(\vec{r} + \vec{R}_n + \vec{R}_m) = c(\vec{R}_m)\psi(\vec{r} + \vec{R}_n) = c(\vec{R}_n)c(\vec{R}_m)\psi(\vec{r}) = c(\vec{R}_n + \vec{R}_m)\psi(\vec{r}) \tag{85}$$

The coefficients c must respect two conditions:

(1) $\left| c(\vec{R}_n) \right|^2 = 1$

(2) $c(\vec{R}_n + \vec{R}_m) = c(\vec{R}_n)c(\vec{R}_m)$

Therefore we set

$$c(\vec{R}_n) = e^{i\vec{k}\cdot\vec{R}_n} \tag{86}$$

and

$$\psi(\vec{r} + \vec{R}_m) = \psi(\vec{r})e^{i\vec{k}\cdot\vec{R}_n} \tag{87}$$

The vector $\vec{k}$ is defined, apart from any reciprocal lattice vector $\vec{K}_s$, since for any such vector

$$e^{i\vec{K}_s \cdot \vec{R}_n} = 1 \tag{88}$$

for any $\vec{R}_n$. This gives us the possibility of keeping $\vec{k}$ within the first Brillouin zone.

The value of $\vec{k}$ are determined by the periodic boundary conditions that we impose on the wavefunctions.

We apply now these considerations to our linear crystal. As we noted in Eq.(64), the desired wavefunctions have the form

$$\psi'_k = \sum_{l=1}^{N} a_{kl} \psi_l \tag{89}$$

where we have changed the wavefunction index from v to k. In the expressions above k runs over N possible values and we note that the wavefunction ψ'_k may involve some degeneracy.

As we established in Eqs.(61), the first excited state of H_0 is related to the wavefunctions

$$\begin{cases} \psi_1(x) = v_1(x)u_2(x-a)u_3(x-2a)......u_N[x-(N-1)a] \\ \psi_2(x) = u_1(x)v_2(x-a)u_3(x-2a)......u_N[x-(N-1)a] \\ .... \\ \psi_N(x) = u_1(x)u_2(x-a)u_3(x-2a)......v_N[x-(N-1)a] \end{cases} \tag{90}$$

Let us consider the effect of a shift of the origin in the -x direction by an amount a:

$$\psi_i(x) = \psi_i(x-a) \tag{91}$$

and

$$\psi_1(x-a) = v_1(x-a)u_2(x-2a)u_3(x-3a).......u_N(x-Na)$$

$$= v_1(x-a)u_2(x-2a)u_3(x-3a).......u_N(x) = \psi_2(x) \tag{92}$$

We have made use of the periodic boundary conditions in the intermediate step. In general it may be seen that

$$\psi_i(x-a) = \psi_{i+1}(x) \tag{93}$$

Let us now, examine the effect of this shift of origin on the wavefunctions ψ'_k. First, according to the result (87)

$$\psi'_k(x-a) = e^{-ika}\psi'_k(x) = e^{-ika}\left(a_{k1}\psi_1 + a_{k2}\psi_2 + ...a_{kN}\psi_N\right) \tag{94}$$

On the other hand, we may also write

$$\psi'_k(x-a) = a_{k1}\psi_2 + a_{k2}\psi_3 + ...a_{kN}\psi_1 \tag{95}$$

The comparison of the coefficients in Eqs.(94) and (95) gives us

$$a_{kl} = e^{i(l-1)ka}a_{k1} \tag{96}$$

The desired wavefunctions can now be written

28

$$\psi'_k = a_{k1}\left(\psi_1 + e^{ika}\psi_2 + e^{i2ka}\psi_3 + \ldots\right) \tag{97}$$

The normalization of ψ'_k yields

$$|a_{k1}|^2 = \frac{1}{N} \tag{98}$$

and we may choose the phase of a_{k1} in such a way that

$$a_{k1} = \frac{1}{\sqrt{N}} e^{ika} \tag{99}$$

resulting in

$$a_{kl} = \frac{1}{\sqrt{N}} e^{ilka} \tag{100}$$

Therefore the approximate wavefunction of H are

ground state

$$\psi_g = |u_1 u_2 \ldots u_N\rangle \tag{101}$$

first excited state

$$\psi'_k = \frac{1}{\sqrt{N}} \sum_{l=1}^{N} e^{ilka} |u_1 u_2 \ldots v_l \ldots u_N\rangle \tag{102}$$

The above wavefunctions are approximate for the following reasons:

(1) We treated the localized electronic wavefunctions u_j as non-overlapping. In fact, any functions used to represent them do overlap.

(2) We did not take explicitly into account the interaction terms $V_{ss'}$ in generating the ψ_g and ψ'_k.

Finally, we consider the allowed values of k which, as we said, are determined by the boundary conditions that we impose. If we choose periodic boundary conditions, $\psi'_k(x + Na) = \psi'_k(x)$ and

$$a_{k,N+1} = a_{k1} \tag{103}$$

or

$$e^{ik(N+1)a} = e^{ika} \tag{104}$$

This implies

$$e^{ikNa} = 1 \tag{105}$$

with the solutions

$$k = \frac{2\pi n}{Na} \tag{106}$$

If we take $-\frac{N}{2} < n \leq \frac{N}{2}$, the range for n is compatible with k being in the first Brillouin zone.

II.C. Dispersion Relations

We have already seen that

$$a_{kl} = \frac{1}{\sqrt{N}} e^{ilka} \tag{107}$$

These coefficients appear in Eq.(66) as follows

$$a_{km}\varepsilon_k - \sum_{l \neq m} a_{kl} H'_{ml} = 0 \qquad (108)$$

where

$$\varepsilon_k = E'_k - E_0 - H'_{mm} \qquad (109)$$

Substituting Eq.(107) in Eq.(108), we obtain a relationship between ε_k and k

$$\frac{1}{\sqrt{N}} e^{imka} \varepsilon_k - \sum_{l \neq m} \frac{1}{\sqrt{N}} e^{ilka} H'_{ml} = 0$$

and

$$\varepsilon_k = \sum_{l \neq m} e^{i(l-m)ka} H'_{ml} \qquad (110)$$

This <u>dispersion relation</u> is independent of m; in addition, the N different values of k generate N values for ε_k. We may recall that

$$H'_{ml} = \langle \psi_m | H' | \psi_l \rangle \qquad (111)$$

which in the present case reduces to

$$H'_{ml} = \langle v_m u_l | H' | u_m v_l \rangle \qquad (112)$$

In many cases only nearest neighbor interactions are non-negligible; for such systems the matrix elements may be written

$$H'_{ml} = M \delta_{l,m\pm1} \qquad (113)$$

where M is the strength of the interaction. The dispersion relation represented by Eq.(110) then becomes

$$\varepsilon_k = M e^{ika} + M e^{-ika} = 2M \cos ka \qquad (114)$$

The total energy of the system is then of the form

$$E'_k = E_0 + H'_{mm} + \varepsilon_k = E_0 + H'_{mm} + 2M \cos ka \qquad (115)$$

where we have taken into account the expression (109). we observe that, as a result of the perturbation H', the N-fold degenerate state has become a band of N states (recall that we have N allowed values of k).

The dispersion relation is sketched in Figure 3.

II.D. Effective Mass

These collective excitations can be treated as <u>quasi-particles</u> with an effective mass. By forming wave packets with a spread $\Delta \vec{k}$ about $\vec{k}$ we can define a particle velocity as the group velocity

$$v_k = \frac{d\omega}{dk} = \frac{1}{\hbar} \frac{\partial \varepsilon_k}{\partial k} \qquad (116)$$

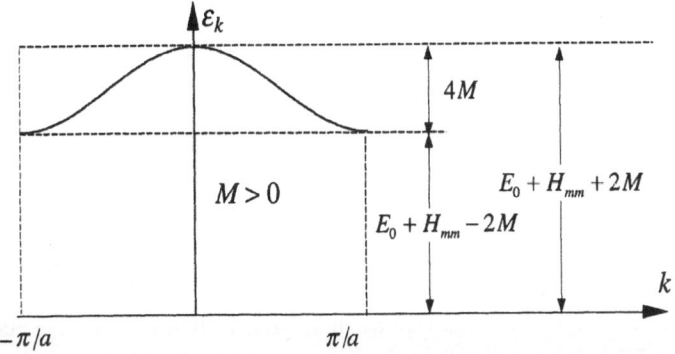

Figure 3. Dispersion relation for collective excitations of a linear chain.

We note that this formula is also valid for electromagnetic waves.

Since, for a free particle

$$\varepsilon_k = \frac{\hbar^2 k^2}{m} \tag{117}$$

we can associated to our quasi-particle the mass

$$m^* = \hbar^2 \Big/ \frac{\partial^2 \varepsilon}{\partial k^2} \tag{118}$$

In the present case

$$\varepsilon_k = 2M \cos ka \tag{119}$$

and

$$\frac{d\varepsilon_k}{dk} = -2Ma \sin ka \tag{120}$$

$$\frac{d^2 \varepsilon_k}{dk^2} = -2Ma^2 \cos ka \tag{121}$$

Therefore

$$v_k = \frac{1}{\hbar} \frac{\partial \varepsilon_k}{\partial k} = -\frac{2Ma \sin ka}{\hbar} \tag{122}$$

and

$$m^* = \hbar^2 \Big/ \frac{\partial^2 \varepsilon}{\partial k^2} = -\frac{\hbar^2}{2Ma^2 \cos ka} \tag{123}$$

ε_k, v_k and m^* are represented in Figure 4. We note that for small k

$$v_k = -\frac{2Mka^2}{\hbar} \tag{124}$$

$$m^* = -\frac{\hbar^2}{2Ma^2} \tag{125}$$

(1) The velocity is linearly dependent on both M and k. As the strength of the interaction increases, the speed of excitation propagation also increases (all other factors being equal). The dependence of v on the interatomic distance a is more subtle, since a portion of this dependence is "hidden" in Mk. As a typical example, let us consider the electric dipole-electric dipole interaction between nearest neighbors. This interaction goes as a^{-3}[1]. In this case, we have $v \propto a^{-2}$, since $k \propto a^{-1}$. We therefore see that the velocity decreases (possible dramatically) as a increases.

(2) The effective mass is inversely proportional to M. A strong interaction between nearest neighbors would then result in a relatively small effective mass (all other factors being equal). This reinforces our general notion that a strong interaction enhances the delocalization of excitation energy in the system. As for the velocity, the dependence of m^* on a requires some specification of the interaction M. For the electric dipole-electric dipole case given above, it is easy to verify that $m^* \propto a^1$. This results in the physically reasonable behavior that an increased separation distance hampers the movement of excitation energy.

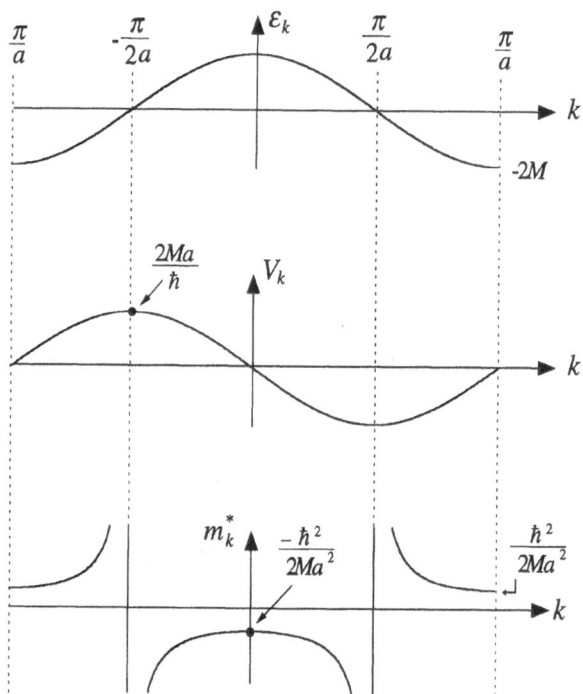

Figure 4. Energy, velocity and equivalent mass of collective excitations.

II.E. Generalization to Three Dimensions

We generalized our treatment to an ordered solid in three dimensions. Let $\vec{a}_1$, $\vec{a}_2$ and $\vec{a}_3$ be the primitive basic lattice vectors and $N = N_1 \times N_2 \times N_3$ the total number of atoms. We introduce the notion of reciprocal lattice, a geometrical construction which consists of an array of points. The primitive basic vectors of the reciprocal lattice are given by

$$\begin{cases} \vec{b}_1 = 2\pi \dfrac{\vec{a}_2 \times \vec{a}_3}{\vec{a}_1 \cdot \vec{a}_2 \times \vec{a}_3} \\[2mm] \vec{b}_2 = 2\pi \dfrac{\vec{a}_3 \times \vec{a}_1}{\vec{a}_1 \cdot \vec{a}_2 \times \vec{a}_3} \qquad (\vec{b}_i \cdot \vec{a}_j = 2\pi \delta_{ij}) \\[2mm] \vec{b}_3 = 2\pi \dfrac{\vec{a}_1 \times \vec{a}_2}{\vec{a}_1 \cdot \vec{a}_2 \times \vec{a}_3} \end{cases} \qquad (126)$$

In the reciprocal lattice we carve out the Brillouin zone, another geometrical construction in $\vec{k}$ space. It is constructed by starting from a lattice point of the reciprocal lattice, drawing lines to the nearest neighbor points and cutting these lines halfway with perpendicular planes; the smallest volume enclosed by these planes is the (first) Brillouin Zone.

The wavefunction of the crystal corresponding to the first excited state is now designated as follows:

$$\psi_k = \frac{1}{\sqrt{N}} \sum_l e^{i\vec{k}\cdot\vec{R}_l} |u_1 u_2 \dots v_l \dots u_N\rangle \qquad (127)$$

The allowed values of $\vec{k}$ are determined by the periodic boundary conditions.

The dispersion relation is given by

$$\varepsilon_{\vec{k}} = \sum_{l,l \neq m} e^{i\vec{k}\cdot(\vec{R}_l - \vec{R}_m)} H'_{ml} \qquad (128)$$

The group velocity is given by

$$\vec{v}_{\vec{k}} = \frac{1}{\hbar} \vec{\nabla}_{\vec{k}} \varepsilon_{\vec{k}} \qquad (129)$$

and the mass tensor by

$$m^* = \frac{\hbar^2}{\vec{\nabla}_{\vec{k}} \vec{\nabla}_{\vec{k}} \varepsilon_{\vec{k}}} \qquad (130)$$

II.F. Periodic Boundary Conditions and Density of States

Consider a crystal with
N_1 cells in the $\vec{a}_1$ direction,
N_2 cells in the $\vec{a}_2$ direction, and
N_3 cells in the $\vec{a}_3$ direction.
Call

$$\vec{R}_N = N_1 \vec{a}_1 + N_2 \vec{a}_2 + N_3 \vec{a}_3 \qquad (131)$$

If we impose periodic boundary conditions (PBC), we obtain the allowed values of $\vec{k}$:

33

$$\psi_{\vec{k}}\left(\vec{r} + \vec{R}_N\right) = \psi_{\vec{k}}(\vec{r}) \tag{132}$$

namely,

$$e^{i\vec{k}\cdot\vec{R}_N} = 1 \tag{133}$$

This equations determines the allowed values of $\vec{k}$. Equation (133) implies that

$$\vec{k}\cdot\vec{R}_N = 2\pi s \quad (s = \text{integer}) \tag{134}$$

The wave vector can be expressed as follows

$$\vec{k} = k_1\hat{b}_1 + k_2\hat{b}_2 + k_3\hat{b}_3 \tag{135}$$

where $\hat{b}_i$=unit vector in the $\vec{b}_i$ direction.

Therefore

$$\vec{k}\cdot\vec{R}_N = (k_1\hat{b}_1 + k_2\hat{b}_2 + k_3\hat{b}_3)\cdot(N_1\vec{a}_1 + N_2\vec{a}_2 + N_3\vec{a}_3)$$

$$= (N_1 k_1\vec{a}_1\cdot\hat{b}_1 + N_2 k_2\vec{a}_2\cdot\hat{b}_2 + N_3 k_3\vec{a}_3\cdot\hat{b}_3) = 2\pi s \tag{136}$$

This relation is satisfied if

$$\begin{cases} k_1 = 2\pi\dfrac{2\pi s_1}{N_1\vec{a}_1\cdot\hat{b}_1} \\[2mm] k_2 = 2\pi\dfrac{2\pi s_2}{N_2\vec{a}_2\cdot\hat{b}_2} \\[2mm] k_3 = 2\pi\dfrac{2\pi s_3}{N_3\vec{a}_3\cdot\hat{b}_3} \end{cases} \tag{137}$$

<u>Claim</u>

The values of s_1 greater than N_1,
the values of s_2 greater than N_2, and
the values of s_3 greater than N_3.
are redundant.

<u>Proof</u>
Let us assume that $s_i < N_i$ and let us replace s_i by $s_i + N_i$.
Then

$$k_1 = \frac{2\pi s_1}{N_1\vec{a}_1\cdot\hat{b}_1} + \frac{2\pi N_1}{N_1\vec{a}_1\cdot\hat{b}_1} = \frac{2\pi s_1}{N_1\vec{a}_1\cdot\hat{b}_1} + \frac{2\pi|\vec{b}_1|}{\vec{a}_1\cdot\hat{b}_1} = \frac{2\pi s_1}{N_1\vec{a}_1\cdot\hat{b}_1} + |\vec{b}_1|$$

$$k_2 = \frac{2\pi s_2}{N_2\vec{a}_2\cdot\hat{b}_2} + |\vec{b}_2|$$

$$k_3 = \frac{2\pi s_3}{N_3\vec{a}_3\cdot\hat{b}_3} + |\vec{b}_3|$$

and

$$k = \frac{2\pi s_1}{N_1\vec{a}_1\cdot\hat{b}_1}\hat{b}_1 + \frac{2\pi s_2}{N_2\vec{a}_2\cdot\hat{b}_2}\hat{b}_2 + \frac{2\pi s_3}{N_3\vec{a}_3\cdot\hat{b}_3}\hat{b}_3 + \vec{b}_1 + \vec{b}_2 + \vec{b}_3$$

But

$$e^{i\vec{k}\cdot(\vec{b}_1+\vec{b}_2+\vec{b}_3)} = 1$$

Therefore the values of $s_1 > N_1$, $s_2 > N_2$ and $s_3 > N_3$ are redundant.   Q.E.D.

We can then limit k_1, k_2 and k_3 to the following values:

$$k_1 = \frac{2\pi s_1}{N_1(\vec{a}_1 \cdot \hat{b}_1)} \qquad \begin{array}{l} s_1 = 1,2,.......,N_1 \\[4pt] or \pm 1, \pm 2,....,\pm\frac{N_1}{2} \end{array}$$

$$k_2 = \frac{2\pi s_2}{N_2(\vec{a}_2 \cdot \hat{b}_2)} \qquad \begin{array}{l} s_2 = 1,2,.......,N_2 \\[4pt] or \pm 1, \pm 2,....,\pm\frac{N_2}{2} \end{array}$$

$$k_3 = \frac{2\pi s_3}{N_3(\vec{a}_3 \cdot \hat{b}_3)} \qquad \begin{array}{l} s_3 = 1,2,.......,N_3 \\[4pt] or \pm 1, \pm 2,....,\pm\frac{N_3}{2} \end{array}$$

The number of states with k_1 in $(k_1 + dk_1)$, k_2 in $(k_2 + dk_2)$ and k_3 in $(k_3 + dk_3)$ is given by

$$ds_1 ds_2 ds_3 = \frac{N_1 N_2 N_3}{(2\pi)^3}\left[(\vec{a}_1 \cdot \hat{b}_1)(\vec{a}_2 \cdot \hat{b}_2)(\vec{a}_3 \cdot \hat{b}_3)\right] \times dk_1 dk_2 dk_3 \tag{138}$$

The infinitesimal volume element in $\vec{k}$ is given by

$$d^3\vec{k} = dk_1 dk_2 dk_3 \left(\hat{b}_1 \cdot \hat{b}_2 \times \hat{b}_3\right) \tag{139}$$

Therefore

$$ds_1 ds_2 ds_3 = \frac{N_1 N_2 N_3}{8\pi^3} d^3\vec{k}\frac{(\vec{a}_1 \cdot \hat{b}_1)(\vec{a}_2 \cdot \hat{b}_2)(\vec{a}_3 \cdot \hat{b}_3)}{\hat{b}_1 \cdot \hat{b}_2 \times \hat{b}_3} \tag{140}$$

But

$$\frac{(\vec{a}_1 \cdot \hat{b}_1)(\vec{a}_2 \cdot \hat{b}_2)(\vec{a}_3 \cdot \hat{b}_3)}{\hat{b}_1 \cdot \hat{b}_2 \times \hat{b}_3} = \frac{(\vec{a}_1 \cdot \vec{b}_1)(\vec{a}_2 \cdot \vec{b}_2)(\vec{a}_3 \cdot \vec{b}_3)}{\vec{b}_1 \cdot \vec{b}_2 \times \vec{b}_3} \tag{141}$$

We know that

$$\vec{a}_i \cdot \vec{b}_j = 2\pi\delta_{ij}$$

and

$$\vec{b}_1 \cdot \vec{b}_2 \times \vec{b}_3 = \text{volume of unit cell of the reciprocal lattice} = 8\pi^2/\Omega_a$$

where

Ω_a = unit cell of the <u>direct lattice</u>

Therefore

$$\frac{(\vec{a}_1 \cdot \vec{b}_1)(\vec{a}_2 \cdot \vec{b}_2)(\vec{a}_3 \cdot \vec{b}_3)}{\vec{b}_1 \cdot \vec{b}_2 \times \vec{b}_3} = \frac{8\pi^3}{8\pi^3/\Omega_a} = \Omega_a \tag{142}$$

and

$$ds_1 ds_2 ds_3 = \frac{N_1 N_2 N_3}{8\pi^3} d^3\vec{k}\Omega_a = \frac{V}{8\pi^3} d^3\vec{k} \tag{143}$$

because

$$N_1 N_2 N_3 \Omega_a = V = \text{volume of the crystal}$$

The volume of the (first) Brillouin Zone is also $8\pi^3/\Omega_a$. The number of allowed $\vec{k}$ values in this zone is, as expected,

$$\frac{8\pi^3}{\Omega_a}\frac{V}{8\pi^3} = \frac{N\Omega_a}{\Omega_a} = N \tag{144}$$

where $\qquad N = N_1 N_2 N_3$.

III. INTERACTION OF RADIATION WITH COLLECTIVE EXCITATION

III.A. The Radiation Field

1. The Classical Radiation Field. In a region free of currents and charges, a radiation field is defined by a vector potential $\vec{A}(\vec{r},t)$ [2,3]

$$\nabla^2\vec{A}(\vec{r},t) - \frac{1}{c^2}\frac{\partial^2\vec{A}(\vec{r},t)}{\partial t^2} = 0 \tag{145}$$

$$\vec{\nabla}\cdot\vec{A}(\vec{r},t) = 0 \tag{146}$$

$$\vec{E}(\vec{r},t) = -\frac{1}{c}\frac{\partial\vec{A}(\vec{r},t)}{\partial r} \tag{147}$$

$$\vec{B}(\vec{r},t) = \vec{\nabla}\times\vec{A}(\vec{r},t) \tag{148}$$

Equation (145) is the so-called field equation, Eq.(16) indicates that we have adopted the Coulomb gauge, and Eqs.(147) and (148) give the electric field and the magnetic field in terms of $\vec{A}(\vec{r},t)$, respectively.

A typical solution of the field equation is given by

$$A(\vec{r})q(t) = \vec{\pi}\left(\frac{4\pi c^2}{V}\right)^{1/2}e^{i\vec{k}\cdot\vec{r}}|q|e^{-i\omega t} \tag{149}$$

where $\omega = kc$. We note that Eq.(146) implies

$$\vec{\pi}\cdot\vec{k} = 0 \tag{150}$$

namely, that the polarization of the wave is perpendicular to the direction of the wave vector. The allowed values of $\vec{k}$ are determined by the boundary conditions of the problem.

Summing over all wave vectors ($\vec{k}$) and all polarizations (σ), we obtain the general solution of the field equation

$$\vec{A}(\vec{r},t) = \sum_\alpha\sum_\sigma\left[q_\alpha^\sigma(t)\vec{A}_\alpha^\sigma(r) + q_\alpha^{\sigma*}(t)\vec{A}_\alpha^{\sigma*}(r)\right] \tag{151}$$

where

$$q_\alpha^\sigma(t) = |q_\alpha^\sigma|e^{-i\omega_\alpha t} \tag{152}$$

$$\vec{A}_\alpha^\sigma(r) = \vec{\pi}_\alpha^\sigma\left(\frac{4\pi c^2}{V}\right)^{1/2}e^{i\vec{k}_\alpha\cdot\vec{r}} \tag{153}$$

The Hamiltonian of radiation field can be derived from the expression for the energy of the field

$$\frac{1}{8\pi}\int\left[\left(\vec{E}\right)^2+\left(\vec{B}\right)^2\right]d^3\vec{r} \tag{154}$$

and the relations (147) and (148) for $\vec{E}$ and $\vec{B}$, respectively. Because of the orthogonality of the various Fourier components of the field, cross terms with subscripts $\alpha\alpha'$ $(\alpha \neq \alpha')$ drop and we are left with (see [2]):

$$H = \sum_\alpha \sum_\sigma \omega_\alpha^2\left(q_\alpha^\sigma q_\alpha^{\sigma*}+q_\alpha^{\sigma*}q_\alpha^\sigma\right) \tag{155}$$

We note the followings:

(1) The Hamiltonian H of the radiation field is the sum of independent terms

$$H_\alpha^\sigma = \omega_\alpha^2\left(q_\alpha^\sigma q_\alpha^{\sigma*}+q_\alpha^{\sigma*}q_\alpha^\sigma\right) \tag{156}$$

(2) The coordinates q_α^σ represent the <u>normal coordinates</u> of the field.
(3) No approximation has been made.
(4) We use the form $\omega_\alpha^2\left(q_\alpha q_\alpha^*+q_\alpha^* q_\alpha\right)$ rather than the form $2\omega_\alpha^2 q_\alpha q_\alpha^*$ in preparation for our move into quantum mechanics: in the quantum mechanical treatment q and q^* become non-commuting operators.

The values of $\vec{k}$, as we said before, are determined by the boundary conditions, When, as it is the case here, the wavelength of the radiation is much smaller than the dimension of the spatial region under consideration, any sum over $\vec{k}$ is in effect an integral and the relevant information is the density of states, i.e., the number of states with $\vec{k}$ in $\left(\vec{k},\vec{k}+d\vec{k}\right)$. In order to find this quantity, periodic boundary conditions can be used; these conditions give for the possible values of the three components of $\vec{k}$

$$\begin{cases} k_x = n_x \dfrac{2\pi}{L_x} \\ k_y = n_y \dfrac{2\pi}{L_y} \\ k_z = n_z \dfrac{2\pi}{L_z} \end{cases} \tag{157}$$

where the volume has been taken as a parallelopiped of sides L_x, L_y and L_z and where

$$n_x, n_y, n_z = 0, \pm1, \pm2, \pm3, \ldots\ldots \tag{158}$$

The subscript α used before stands for a particular choice of n_x, n_y, and n_z.

The number of modes with $\vec{k}$ in $\left(\vec{k},\vec{k}+d\vec{k}\right)$ is given by

37

$$\frac{L_x dk_x}{2\pi} \frac{L_y dk_y}{2\pi} \frac{L_z dk_z}{2\pi} = \frac{L_x L_y L_z}{8\pi^3} dk_x dk_y dk_z$$

$$= \frac{V}{8\pi^3} k^2 dk \sin\theta d\theta d\phi = \frac{V}{8\pi^3} \frac{\omega^2}{c^3} d\omega d\Omega \tag{159}$$

where $\quad d\Omega = \sin\theta d\theta d\phi$

2. The Quantum Radiation Field. Consider one term of the Hamiltonian (155)

$$H_\alpha = \omega_\alpha^2 \left(q_\alpha q_\alpha^* + q_\alpha^* q_\alpha \right) \tag{160}$$

where we have dropped for convenience the superscript σ. We introduce two new real variables for each α

$$\begin{cases} Q_\alpha = q_\alpha + q_\alpha^* \\ P_\alpha = -i\omega_\alpha \left(q_\alpha - q_\alpha^* \right) = \dot{Q}_\alpha \end{cases} \tag{161}$$

The Hamiltonian H_α, when written in terms of Q_α and P_α, takes the form

$$H_\alpha = \tfrac{1}{2} \omega_\alpha^2 Q_\alpha^2 + \tfrac{1}{2} P_\alpha^2 \tag{162}$$

Q_α and P_α are real variables that satisfy Hamiltonian's equations; their Poisson brackets are given by

$$\begin{cases} \{Q_\alpha, P_{\alpha'}\} = \sum_i \left(\dfrac{\partial Q_\alpha}{\partial Q_i} \dfrac{\partial P_{\alpha'}}{\partial P_i} - \dfrac{\partial Q_\alpha}{\partial P_i} \dfrac{\partial P_{\alpha'}}{\partial Q_i} \right) = \delta_{\alpha\alpha'} \\ \{Q_\alpha, Q_{\alpha'}\} = \{P_\alpha, P_{\alpha'}\} = 0 \end{cases} \tag{163}$$

The prescription for moving over from a classical to a quantum mechanical treatment is simple. In the latter treatment Q_α and P_α become Hermitian operators and their commutator is obtained by replacing the Poisson brackets as follows:

$$\{Q_\alpha, P_{\alpha'}\} \rightarrow \frac{1}{i\hbar} [Q_\alpha, P_{\alpha'}] \tag{164}$$

Then we obtain

$$\begin{cases} [Q_\alpha, Q_{\alpha'}] = [P_\alpha, P_{\alpha'}] = 0 \\ [Q_\alpha, P_{\alpha'}] = i\hbar \delta_{\alpha\alpha'} \end{cases} \tag{165}$$

q_α and q_α^*, which are related to Q_α and P_α by the relations (161), become two (non-Hermitian) operators which we shall call q_α and q_α^+, respectively. The commutation relation of these two operators are easily derived:

$$\begin{cases} [q_\alpha, q_{\alpha'}] = [q_\alpha^+, q_{\alpha'}^+] = 0 \\ [q_\alpha, q_{\alpha'}^+] = \dfrac{\hbar}{2\omega_\alpha} \delta_{\alpha\alpha'} \end{cases} \tag{166}$$

We may replace q_α and q_α^+ by the dimentionless operators

$$a_\alpha = \left(\frac{2\omega_\alpha}{\hbar}\right)^{1/2} q_\alpha, \qquad a_\alpha^+ = \left(\frac{2\omega_\alpha}{\hbar}\right)^{1/2} q_\alpha^+ \tag{167}$$

and find

$$\begin{cases} [a_\alpha, a_{\alpha'}] = [a_\alpha^+, a_{\alpha'}^+] = 0 \\ [a_\alpha, a_{\alpha'}^+] = \delta_{\alpha\alpha'} \end{cases} \tag{168}$$

The Hamiltonian of the radiation field can now be written

$$H = \sum_\alpha \omega_\alpha^2 \left(q_\alpha q_\alpha^+ + q_\alpha^+ q_\alpha\right) = \sum_\alpha \hbar\omega_\alpha \left(a_\alpha^+ a_\alpha + \tfrac{1}{2}\right) \tag{169}$$

Reintroducing the polarization index σ

$$H = \sum_\alpha \sum_\sigma \hbar\omega_\alpha \left(a_\alpha^{\sigma+} a_\alpha^\sigma + \tfrac{1}{2}\right) \tag{170}$$

The Hamiltonian

$$H_\alpha^\sigma = \hbar\omega_\alpha \left(a_\alpha^{\sigma+} a_\alpha^\sigma + \tfrac{1}{2}\right) \tag{171}$$

has the energy eigenvalues

$$E_\alpha^\sigma = \hbar\omega_\alpha \left(n_\alpha^\sigma + \tfrac{1}{2}\right) \tag{172}$$

where $n_\alpha^\sigma = 0,1,2,\ldots\ldots$ The eigenvalues of H_α^σ are simply given by the kets $\left| n_\alpha^\sigma \right\rangle$.

The Hamiltonian, eigenvalues and eigenfunctions of the radiation field are now listed:

$$H = \sum_\alpha \sum_\sigma H_\alpha^\sigma = \sum_\alpha \sum_\sigma \hbar\omega_\alpha \left(a_\alpha^{\sigma+} a_\alpha^\sigma + \tfrac{1}{2}\right) \tag{173}$$

$$E_{n_1^{\sigma_1} n_1^{\sigma_2} n_2^{\sigma_1} \ldots\ldots} = \sum_\alpha \sum_\sigma \hbar\omega_\alpha \left(n_\alpha^\sigma + \tfrac{1}{2}\right) \tag{174}$$

$$\psi_{n_1^{\sigma_1} n_1^{\sigma_2} n_2^{\sigma_1} \ldots\ldots} = \prod_\alpha \prod_\sigma \left| n_\alpha^\sigma \right\rangle \tag{175}$$

One can see from the above relations that the radiation field may be thought of as a collection of an infinite number of harmonic oscillators, one for each (α, σ) component, with different degrees of excitation n_α^σ. Alternatively, the radiation field may be thought of as an ensemble of photons: n_α^σ is the number of photons present for each wavevector $\vec{k}_\alpha$ and polarization α.

In the quantum-mechanical treatment the vector potential represents an operator which can be expressed as follows:

$$\vec{A} = \sum_\alpha \sum_\sigma \left[\vec{A}_\alpha^\sigma q_\alpha^\sigma + \vec{A}_\alpha^{\sigma *} a_\alpha^{\sigma +} \right] = \sum_\alpha \sum_\sigma \left(\frac{4\pi c^2}{V} \right)^{1/2} \left(\frac{\hbar}{2\omega_\alpha} \right)^{1/2} \vec{\pi}_\alpha^\sigma \left(e^{i\vec{k}_\alpha \cdot \vec{r}} a_\alpha^\sigma + e^{-i\vec{k}_\alpha \cdot \vec{r}} a_\alpha^{\sigma +} \right)$$

$$= \sum_\alpha \sum_\sigma \left(\frac{hc^2}{\omega_\alpha V} \right)^{1/2} \vec{\pi}_\alpha^\sigma \left(e^{i\vec{k}_\alpha \cdot \vec{r}} a_\alpha^\sigma + e^{-i\vec{k}_\alpha \cdot \vec{r}} a_\alpha^{\sigma +} \right) \tag{176}$$

We note that the operators a_α^σ and $a_\alpha^{\sigma +}$ operate as follows:

$$\begin{cases} a_\alpha^{\sigma +} \left| n_\alpha^\sigma \right\rangle = \sqrt{n_\alpha^\sigma + 1} \left| n_\alpha^\sigma + 1 \right\rangle \\ a_\alpha^\sigma \left| n_\alpha^\sigma \right\rangle = \sqrt{n_\alpha^\sigma} \left| n_\alpha^\sigma - 1 \right\rangle \end{cases} \tag{177}$$

III.B. The Form of the Interaction

Consider a particle of mass m and charge q under the action of a radiation field $\vec{A}(\vec{r},t)$ and of a potential $\phi(\vec{r},t)$.

The equation of motion of such a particle is given by the Hamiltonian

$$H = \frac{\left(\vec{p} - \frac{q}{c} \vec{A} \right)^2}{2m} + q\phi \tag{178}$$

where $\vec{p}$ = linear momentum of the particle. This can be easily justified by considering the Hamilton's equation, which give the correct expression for the (Lorentz) force acting on the particle.

The Hamiltonian H can be written as follows:

$$H = \frac{(\vec{p})^2}{2m} - \frac{q}{2mc} \left(\vec{p} \cdot \vec{A} + \vec{A} \cdot \vec{p} \right) + \frac{q^2}{2mc^2} \left(\vec{A} \right)^2 + q\phi$$

$$= \frac{(\vec{p})^2}{2m} - \frac{q}{mc} \left(\vec{p} \cdot \vec{A} \right) + \frac{q^2}{2mc^2} \left(\vec{A} \right)^2 + q\phi \tag{179}$$

since $\left[\vec{p}, \vec{A} \right] = 0$, because of the Coulomb gauge. The interaction term, which is linear in the field, is relevant here. This term is

$$H_1 = -\frac{q}{mc} \vec{p} \cdot \vec{A} = -\frac{q}{m} \sum_\alpha \sum_\sigma \left(\frac{h}{\omega_\alpha V} \right)^{1/2} \left(a_\alpha^\sigma e^{i\vec{k}_\alpha \cdot \vec{r}} + a_\alpha^{\sigma +} e^{-i\vec{k}_\alpha \cdot \vec{r}} \right) \vec{\pi}_\alpha^\sigma \cdot \vec{p} \tag{180}$$

In the case of several particles

$$H_1 = -\left\{ \sum_\alpha \sum_\sigma \left(\frac{h}{\omega_\alpha V} \right)^{1/2} \sum_i \left[\frac{q_i}{m_i} \left(a_\alpha^\sigma e^{i\vec{k}_\alpha \cdot \vec{r}} + a_\alpha^{\sigma +} e^{-i\vec{k}_\alpha \cdot \vec{r}} \right) \left(\vec{\pi}_\alpha^\sigma \cdot \vec{p}_i \right) \right] \right\} \tag{181}$$

III.C. Absorption and Emission Processes

Let us continue with the case of a charged particle under the action of a potential ϕ and of a radiation field $\vec{A}$. the Hamiltonian of the system which consists of the particle and the radiation field is given by

$$H = \frac{1}{2m}\left(\vec{p} - \frac{q}{c}\vec{A}\right)^2 + q\phi + \frac{1}{8\pi}\int\left[\left(\vec{E}\right)^2 + \left(\vec{B}\right)^2\right]d^3\vec{r}$$

$$= \left(\frac{\left(\vec{p}\right)^2}{2m} + q\phi\right) + \frac{1}{8\pi}\int\left[\left(\vec{E}\right)^2 + \left(\vec{B}\right)^2\right]d^3\vec{r} - \frac{q}{2mc}\left(\vec{p}\cdot\vec{A}\right) + \frac{q^2}{2mc^2}\left(\vec{A}\right)^2 \quad (182)$$

We can express H as follows:

$$H = H_0 + H_1 + H_2 \quad (183)$$

where

$$H_0 = \frac{1}{2m}\left(\vec{p} - \frac{q}{c}\vec{A}\right)^2 + q\phi + \frac{1}{8\pi}\int\left[\left(\vec{E}\right)^2 + \left(\vec{B}\right)^2\right]d^3\vec{r}$$

$$= -\frac{\hbar^2}{2m}\nabla^2 + q\phi + \sum_\alpha\sum_\sigma \hbar\omega_\alpha\left(a_\alpha^{\sigma+}a_\alpha^\sigma + \tfrac{1}{2}\right) \quad (184)$$

and H_1 and H_2 are the terms linear and quadratic in the field, respectively.

The method to be applied here consists in considering H_0 as the Hamiltonian of the "unperturbed" system, given simply by the sum of the Hamiltonian of the particle and the Hamiltonian of the radiation field and taking H_1 and H_2 as time-dependent perturbations of the system which may induce transitions between the different eigenstates of H_0. These eigenstates are given by

$$\psi_{e;n_1^{\sigma_1}n_1^{\sigma_2}......} = \psi^e\prod_\alpha\prod_\sigma\left|n_\alpha^\sigma\right\rangle \quad (185)$$

where ψ^e = eigenfunction of the particle and $\left|n_\alpha^\sigma\right\rangle$ eigenfunction of the (α,σ) radiation oscillator. The energies of these states are given by

$$E_{e;n_1^{\sigma_1}n_1^{\sigma_2}......} = E^e + \sum_\alpha\sum_\sigma \hbar\omega_\alpha\left(n_\alpha^\sigma + \tfrac{1}{2}\right) \quad (186)$$

where E^e = energy of the particles and the sum over α and σ gives the energy of the radiation field.

In the case of one photon absorption, the initial and the final states are given by

$$\psi_i = \left|\psi_i^e\right\rangle\left|n_\alpha^\sigma\right\rangle\left|n_{\alpha'}^{\sigma'}\right\rangle...... \quad (187)$$

$$\psi_f = \left|\psi_f^e\right\rangle\left|n_\alpha^\sigma - 1\right\rangle\left|n_{\alpha'}^{\sigma'}\right\rangle...... \quad (188)$$

respectively. Since

$$\left\langle n_\alpha^\sigma - 1\right|a_\alpha^\sigma\left|n_\alpha^\sigma\right\rangle = \sqrt{n_\alpha^\sigma} \quad (189)$$

41

the relevant matrix element for the process of absorption of one photon is

$$\left\langle \psi_f^e; n_\alpha^\sigma - 1 \middle| H_1 \middle| \psi_i^e; n_\alpha^\sigma \right\rangle = -\frac{q}{m}\left(\frac{h}{\omega_\alpha V}\right)^{1/2} \left\langle \psi_f^e \middle| e^{i\vec{k}_\alpha \cdot \vec{r}} \, \vec{\pi}_\alpha^\sigma \cdot \vec{p} \middle| \psi_i^e \right\rangle \sqrt{n_\alpha^\sigma} \quad (190)$$

In the case of one photon emission, the initial and the final states are given by

$$\psi_i = \left| \psi_i^e \right\rangle \middle| n_\alpha^\sigma \right\rangle \middle| n_{\alpha'}^{\sigma'} \right\rangle \cdots \cdots \quad (191)$$

$$\psi_f = \left| \psi_f^e \right\rangle \middle| n_\alpha^\sigma + 1 \right\rangle \middle| n_{\alpha'}^{\sigma'} \right\rangle \cdots \cdots \quad (192)$$

respectively. Since

$$\left\langle n_\alpha^\sigma + 1 \middle| a_\alpha^\sigma \middle| n_\alpha^\sigma \right\rangle = \sqrt{n_\alpha^\sigma + 1} \quad (193)$$

the relevant matrix element for the process of emission of one photon is

$$\left\langle \psi_f^e; n_\alpha^\sigma + 1 \middle| H_1 \middle| \psi_i^e; n_\alpha^\sigma \right\rangle = -\frac{q}{m}\left(\frac{h}{\omega_\alpha V}\right)^{1/2} \left\langle \psi_f^e \middle| e^{-i\vec{k}_\alpha \cdot \vec{r}} \, \vec{\pi}_\alpha^\sigma \cdot \vec{p} \middle| \psi_i^e \right\rangle \sqrt{n_\alpha^\sigma + 1} \quad (194)$$

Since the radiation field has a continuous density of states, both absorption and emission processes are associated with a probability per unit time. By applying the Fermi Golden Rule, we derive the probability per unit time of finding the system (particle + radiation field) with one less or one more photon of energy $h\omega_\alpha$ and polarization $\vec{\pi}_\alpha^\sigma$ in the solid angle $\left(\Omega_\alpha, \Omega_\alpha + d\Omega_\alpha\right)$; it is given by

$$P_\alpha^\sigma d\Omega_\alpha = \frac{2\pi}{\hbar^2} \left| M_\alpha^\sigma \right|^2 g(\omega_\alpha) \quad (195)$$

where

$$g(\omega_\alpha) = \frac{V\omega_\alpha^2}{8\pi^3 c^3} d\Omega_\alpha \quad (196)$$

and

$$\left| M_\alpha^\sigma \right|^2 = \begin{cases} \dfrac{q^2}{m^2}\dfrac{h}{\omega_\alpha V} \left\langle \psi_f^e \middle| e^{i\vec{k}_\alpha \cdot \vec{r}} \, \vec{\pi}_\alpha^\sigma \cdot \vec{p} \middle| \psi_i^e \right\rangle n_\alpha^\sigma \\[12pt] \dfrac{q^2}{m^2}\dfrac{h}{\omega_\alpha V} \left\langle \psi_f^e \middle| e^{-i\vec{k}_\alpha \cdot \vec{r}} \, \vec{\pi}_\alpha^\sigma \cdot \vec{p} \middle| \psi_i^e \right\rangle \left(n_\alpha^\sigma + 1\right) \end{cases} \quad (197)$$

In the above formula the upper (lower) row corresponds to the process of absorption (emission) of one photon. Replacing Eq.(197) in Eq.(195) and taking Eq.(196) into account, we find:

$$P_\alpha^\sigma d\Omega_\alpha = \frac{\omega_\alpha q^2}{hc^3 m^2} \left| \left\langle \psi_f^e \middle| \begin{matrix} e^{i\vec{k}_\alpha \cdot \vec{r}} \, \vec{\pi}_\alpha^\sigma \cdot \vec{p} \\ e^{-i\vec{k}_\alpha \cdot \vec{r}} \, \vec{\pi}_\alpha^\sigma \cdot \vec{p} \end{matrix} \middle| \psi_i^e \right\rangle \right|^2 \left(\begin{matrix} n_\alpha^\sigma \\ n_\alpha^\sigma + 1 \end{matrix}\right) d\Omega_\alpha \quad (198)$$

Let us consider two quantum states of the particle ψ_l^e (l stands for lower) and ψ_u^e (u stands for lower) with energies E_l and E_u, respectively and $E_l < E_u$. It is possible to show that

$$\left|\left\langle \psi_l^e \left| e^{-i\vec{k}_\alpha \cdot \vec{r}} \left(\vec{\pi} \cdot \vec{p} \right) \right| \psi_u^e \right\rangle\right|^2 = \left|\left\langle \psi_u^e \left| e^{i\vec{k}_\alpha \cdot \vec{r}} \left(\vec{\pi} \cdot \vec{p} \right) \right| \psi_l^e \right\rangle\right|^2 \tag{199}$$

The last squared matrix element is the one that would enter the transition probability for an $l \rightarrow u$ (absorption) process.

On the basis of the above result, Eq.(198) becomes

$$P_\alpha^\sigma d\Omega_\alpha = \frac{\omega_\alpha q^2}{hc^3 m^2} \left|\left\langle \psi_u^e \left| e^{i\vec{k}_\alpha \cdot \vec{r}} \vec{\pi}_\alpha^\sigma \cdot \vec{p} \right| \psi_l^e \right\rangle\right|^2 \binom{n_\alpha^\sigma}{n_\alpha^\sigma + 1} d\Omega_\alpha \tag{200}$$

The transition probability for absorption is always proportional to the number of photon n_α^σ present; the transition probability for emission consists of one part, called induced emission, which is proportional to n_α^σ and of another part, called spontaneous emission, which is present even when $n_\alpha^\sigma = 0$. We note here that the transition probability of absorption and the transition probability for induce emission between two states are equal. If two or more charged particles are present

$$P_\alpha^\sigma d\Omega_\alpha = \frac{\omega_\alpha}{hc^3} \left|\left\langle \psi_u \left| \sum_i \frac{q_i}{m_i} e^{i\vec{k}_\alpha \cdot \vec{r}_i} \left(\vec{\pi}_\alpha^\sigma \cdot \vec{p}_i \right) \right| \psi_l \right\rangle\right|^2 \binom{n_\alpha^\sigma}{n_\alpha^\sigma + 1} d\Omega_\alpha \tag{201}$$

III.D. Interaction of Photons with Collective Excitations

The relevant quantity in the creation or annihilation of a quantum of collective excitation via the absorption or emission of a photon is

$$\left\langle \psi_u \left| \sum_i \frac{q_i}{m_i} e^{i\vec{k}_\alpha \cdot \vec{r}_i} \left(\vec{\pi}_\alpha^\sigma \cdot \vec{p}_i \right) \right| \psi_l \right\rangle \tag{202}$$

where $\vec{k}_\alpha$ = wave vector of the photon.

In our case

$$\psi_l = \left| u_1 u_2 \ldots \ldots u_N \right\rangle = \text{ground state} \tag{203}$$

$$\psi_u = \frac{1}{\sqrt{N}} \sum_s e^{i\vec{k} \cdot \vec{R}_s} \left| u_1 u_2 \ldots v_s \ldots u_N \right\rangle = \text{excited state} \tag{204}$$

We can now write

$$\left\langle \psi_u \left| \sum_i \frac{q_i}{m_i} e^{i\vec{k}_\alpha \cdot \vec{r}_i} \left(\vec{\pi}_\alpha^\sigma \cdot \vec{p}_i \right) \right| \psi_l \right\rangle = \left\langle \psi_u \left| \sum_i C_i e^{i\vec{k}_\alpha \cdot \vec{r}_i} \right| \psi_l \right\rangle$$

$$= \frac{1}{\sqrt{N}} \sum_s e^{i\vec{k} \cdot \vec{R}_s} \left\langle u_1 u_2 \ldots v_s \ldots u_N \left| \sum_i C_i e^{i\vec{k}_\alpha \cdot \vec{r}_i} \left(\vec{\pi}_\alpha^\sigma \cdot \vec{p}_i \right) \right| u_1 u_2 \ldots \ldots u_N \right\rangle$$

$$= \frac{1}{\sqrt{N}} \sum_i \sum_s e^{-i\vec{k} \cdot \vec{R}_s} \left\langle v_s \left| C_i e^{i\vec{k}_\alpha \cdot \vec{R}_i} e^{i\vec{k}_\alpha \cdot \vec{r}_i} \right| u_s \right\rangle \delta_{is}$$

$$= \frac{1}{\sqrt{N}} \sum_s e^{i\left(\vec{k}_\alpha - \vec{k}\right)\cdot\vec{R}_s} \langle v|Ce^{i\vec{k}_\alpha \cdot \vec{r}_i'}|u\rangle$$

$$= \sqrt{N}\langle v|Ce^{i\vec{k}_\alpha \cdot \vec{r}_i'}|u\rangle \delta_{\vec{k}_\alpha, \vec{k}+\vec{K}_s} \tag{205}$$

where $C = \dfrac{q}{m}\vec{\pi}_\alpha^\sigma \cdot \vec{p}$ and $\vec{r}_i = \vec{R}_i + \vec{r}_i'$.

Setting $\vec{K}_s = 0$ (no umklapp processes)

$$\left\langle \psi_u \middle| \sum_i C_i e^{i\vec{k}_\alpha \cdot \vec{r}_i} \middle| \psi_l \right\rangle = \sqrt{N}\langle v|Ce^{i\vec{k}_\alpha \cdot \vec{r}_i'}|u\rangle \delta_{\vec{k},\vec{k}_\alpha} \tag{206}$$

This selection rule is illustrated in Figure 5 which indicates that only excitations with $\vec{k} = \vec{k}_\alpha \cong 0$ can be created in absorption and only excitations with $\vec{k} = \vec{k}_\alpha \cong 0$ can produce the emission of a photon. We observe also that no dispersion effect can be seen when these processes occur, because the dispersion curve of the collective excitations and that of the others cross at <u>one</u> point.

The $\vec{k} = \vec{k}_\alpha$ rule is relaxed if more than one collective excitation is involved in the radiative process. For example, in absorption

$$\vec{k}_\alpha = \vec{k}_1 + \vec{k}_2 \tag{207}$$

would correspond to the creation of a collective excitation of wave vector $\vec{k}_1$ and of a collective excitation of wave vector $\vec{k}_2$;

$$\vec{k}_\alpha = \vec{k}_1 - \vec{k}_2 \tag{208}$$

would correspond to the creation of a collective excitation of wave vector $\vec{k}_1$ and the annihilation of s collective excitation of wave vector $\vec{k}_2$.

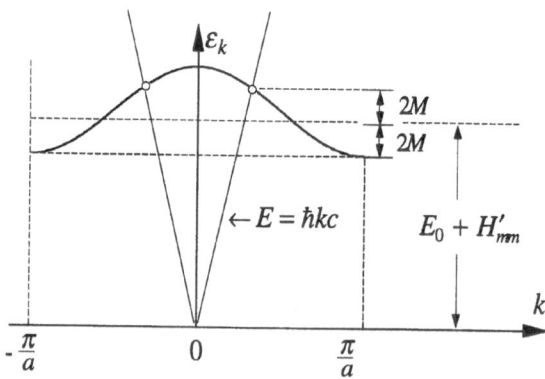

Figure 5. Radiative processes and collective excitations.

IV. PROPAGATION OF RADIATION IN A DISPERSIVE MEDIUM

IV.A. Introduction

Let us consider an ensemble of N particles of mass m and negative charge e, each elastically bound to an equilibrium position where a positive charge $|e|$ resides. Let $n_0 = N/V$ be the density of these particles, ω_0 their natural angular frequency of oscillation and γ the damping constant. Let also

$$\vec{E}(\vec{r},t) = \vec{E}_0(\vec{r})e^{i\omega t} \tag{209}$$

represent an electromagnetic (plane) wave present in the medium.

The equation of motion of the individual particle is given by

$$\ddot{z} = -\frac{K}{m}z + \frac{e}{m}\vec{E} - \gamma\dot{z} = -\omega_0^2 z + \frac{e}{m}\vec{E} - \gamma\dot{z} \tag{210}$$

where $\vec{z}(\vec{r},t)$ = displacement of the individual charged particle from its equilibrium position $\vec{r}$. We shall make the assumption that these displacements are always much smaller than the wavelength of the radiation.

The steady-state solution of (210) is

$$\vec{z}(\vec{r},t) = \vec{z}_0(\vec{r})e^{i\omega t} \tag{211}$$

where

$$\vec{z}_0(\vec{r}) = \frac{e/m}{\omega_0^2 - \omega^2 + i\gamma\omega}\vec{E}_0(\vec{r}) \tag{212}$$

The induced dipole moment is

$$e\vec{z}(\vec{r},t) = \frac{e^2/m}{\omega_0^2 - \omega^2 + i\gamma\omega}\vec{E}(\vec{r},t) = \alpha(\omega)\vec{E}(\vec{r},t) \tag{213}$$

where the polarizability is given by

$$\alpha(\omega) = \frac{e^2/m}{\omega_0^2 - \omega^2 + i\gamma\omega} = \operatorname{Re}\alpha(\omega) + i\operatorname{Im}\alpha(\omega) \tag{214}$$

and

$$\operatorname{Re}\alpha(\omega) = \frac{e^2}{m}\frac{\left(\omega_0^2 - \omega^2\right)}{\left(\omega_0^2 - \omega^2\right)^2 + \gamma^2\omega^2} \tag{215}$$

$$\operatorname{Im}\alpha(\omega) = -\frac{e^2}{m}\frac{\gamma\omega}{\left(\omega_0^2 - \omega^2\right)^2 + \gamma^2\omega^2} \tag{216}$$

The real part of the induced dipole moment is

$$\operatorname{Re}\left[e\vec{z}(\vec{r},t)\right] = \operatorname{Re}\left\{\left[\operatorname{Re}\alpha(\omega) + i\operatorname{Im}\alpha(\omega)\right]\left[\vec{E}_0\cos\omega t + i\vec{E}_0\sin\omega t\right]\right\}$$

$$= \vec{E}_0 \left\{ \left[\mathrm{Re}\, \alpha(\omega) \right] \cos \omega t - \left[\mathrm{Im}\, \alpha(\omega) \right] \sin \omega t \right\}$$

or

$$\mathrm{Re}\left[e\vec{z}(\vec{r},t) \right] = \frac{e^2}{m} \frac{\vec{E}_0(\vec{r})}{\sqrt{\omega_0^2 - \omega^2 + i\gamma\omega}} = \cos(\omega t - \phi) \tag{217}$$

where

$$\tan \phi = \frac{\gamma\omega}{\omega_0^2 - \omega^2} \tag{218}$$

The values of ϕ for $\omega = 0$, ω_0, and ∞ are $\phi = 0$, $\pi/2$, and π, respectively.

The magnetic force acting on the individual particle is equal to $\sim v/c \times$ electric force. If we let $v = v(oscillations) \cong \omega z_0$,

$$\frac{v}{c} = \frac{\omega z_0}{c} = \frac{2\pi z_0}{\lambda}$$

Taking $z_0 = 10^{-8}$ cm and $\lambda = 1{,}000$ A, v/c is on the order of 10^{-3}; therefore the magnetic force is negligible.

IV.B. Dielectric Constant

We shall introduce at this point some important definitions.

Polarization = dipole/unit volume:
$$\vec{P} = n_0 \alpha(\omega) \vec{E} \tag{219}$$

where

$\vec{E}$ = "average" applied field.

Dielectric susceptibility

$$\chi(\omega) = \frac{\vec{p}}{\vec{E}} = \frac{dipole\ /\ unit\ \ volume}{applied\ \ field} = n_0 \alpha(\omega) \tag{220}$$

Electric displacement

$$\vec{D} = \vec{E} + 4\pi\vec{P} = \vec{E} + 4\pi\chi\vec{E} = (1 + 4\pi\chi)\vec{E} = K\vec{E} \tag{221}$$

Dielectric constant

$$K(\omega) = 1 + 4\pi\chi(\omega) = n^2(\omega) \tag{222}$$

where $n(\omega)$ = index of refraction. Note that

$$\chi(\omega) = \frac{K(\omega) - 1}{4\pi} \tag{223}$$

For the system introduced in the previous section:

$$K(\omega) = 1 + 4\pi\chi(\omega) = 1 + 4\pi n_0 \alpha(\omega) = 1 + 4\pi n_0 \frac{e^2/m}{\omega_0^2 - \omega^2 + i\gamma\omega} \tag{224}$$

If more than one type of oscillator is present in the system

46

$$K(\omega) = 1 + 4\pi \sum_s \frac{n_s e_s^2/m_s}{\omega_s^2 - \omega^2 + i\gamma_s\omega} = n^2(\omega) \qquad (225)$$

For very low ω each oscillator type adds a constant contribution $\left[4\pi n_s e_s^2/\omega_s^2 m_s\right]$ to the static dielectric constant. For $\omega \gg \omega_s$ the contribution of the s-type oscillator becomes negligible.

For $\omega \cong \omega_0$

$$K(\omega) = n^2(\omega) = A + \frac{4\pi n_0 e^2/m}{\omega_0^2 - \omega^2 + i\gamma\omega} \qquad (225)$$

where

$$A = 1 + \sum_{s>0} \frac{4\pi n_s e_s/m_s}{\omega_s^2} \qquad (227)$$

A=1 if there is only one oscillator type.

We can express the complex dielectric constant as follows

$$K(\omega) = K_r(\omega) + iK_i(\omega)$$

For $\omega \cong \omega_0$[4]

$$K_r(\omega) = A + \frac{e^2}{m} \frac{4\pi n_0 e^2/m}{\left(\omega_0^2 - \omega^2\right)^2 + \gamma^2\omega^2}\left(\omega_0^2 - \omega^2\right) \approx A + \frac{B\Delta\omega}{\left(\Delta\omega\right)^2 + \left(\gamma/2\right)^2} \qquad (228)$$

$$K_i(\omega) = -\frac{\omega\gamma\left(4\pi n_0 e^2/m\right)}{\left(\omega_0^2 - \omega^2\right)^2 + \gamma^2\omega^2} \approx -\frac{B\gamma/2}{\left(\Delta\omega\right)^2 + \left(\gamma/2\right)^2} \qquad (229)$$

where $B = \left(4\pi n_0 e^2/m\right)/2\omega_0$

The approximate expressions for and are represented in Figure 6.

The condition $K_r(\omega) \leq 0$ does not allow the propagation of radiation in the medium. This condition, in the limit $\gamma \to 0$, corresponds to the "forbidden" frequency region

$$\omega_0 \leq \omega \leq \omega_L \qquad (230)$$

where

$$\omega_L = \omega_0 \sqrt{1 + \frac{4\pi n_0 e^2/m}{A\omega_0^2}} \qquad (231)$$

Still in the limit $\gamma \to 0$:

$$K(\omega_L) = 0, \qquad K(\omega_0) = \infty,$$

$$K(\infty) = A, \qquad K(0) = A + \frac{4\pi n_0 e^2/m}{\omega_0^2}$$

and

$$\frac{K(\infty)}{K(0)} = \frac{\omega_0^2}{\omega_L^2} \qquad (232)$$

The last relation is called the Lyddane-Sachs-Teller relation.

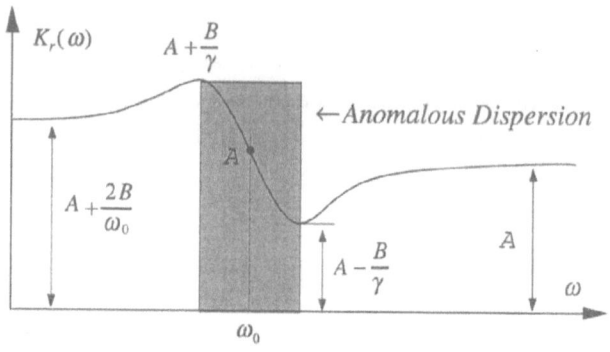

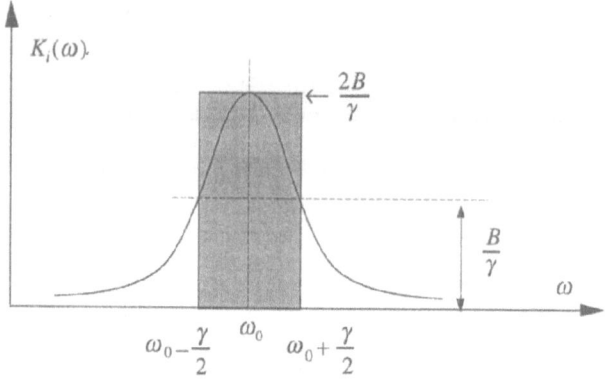

Figure 6. Real and imaginary part of the dielectric constant of an ensemble of oscillating charges.

IV.C. Propagation of an Electromagnetic Wave in a Dispersive Medium

The Maxwell equations can be expressed in general as follows:

$$
\begin{cases}
\vec{\nabla} \cdot \vec{B} = 0 \\
\vec{\nabla} \times \vec{E} + \dfrac{1}{c}\dfrac{\partial \vec{B}}{\partial t} = 0 \\
\vec{\nabla} \cdot \vec{E} = 4\pi\rho \\
\vec{\nabla} \times \vec{B} - \dfrac{1}{c}\dfrac{\partial \vec{E}}{\partial t} = \dfrac{4\pi}{c}\vec{j}
\end{cases}
\tag{233}
$$

Consider now a medium which is uncharged $(\rho = 0)$, polarizable (polarization = $\vec{P}$), and non-magnetic $(\mu = 1, \ \vec{B} = \vec{H})$. We have in this case

$$
\rho = \rho_{pol} = -\vec{\nabla} \cdot \vec{P}
\tag{234}
$$

$$
\vec{j} = \vec{j}_{pol} + \vec{j}_{true} = \frac{\partial \vec{P}}{\partial t} + \sigma \vec{E}
\tag{235}
$$

where $\sigma=$ conductivity of the medium. Now we can write

$$\vec{\nabla} \cdot \vec{E} = 4\pi \rho_{pol} = -4\pi \vec{\nabla} \cdot \vec{P} \tag{236}$$

$$\vec{\nabla} \cdot \left(\vec{E} + 4\pi \vec{p}\right) = \vec{\nabla} \cdot K\vec{E} = 0 \tag{237}$$

Also

$$\vec{\nabla} \times \vec{B} = \frac{1}{c}\frac{\partial \vec{E}}{\partial t} + \frac{4\pi}{c}\vec{j} = \frac{1}{c}\frac{\partial \vec{E}}{\partial t} + \frac{4\pi}{c}\left(\vec{j}_{pol} + \vec{j}_{true}\right)$$

$$= \frac{1}{c}\frac{\partial \vec{E}}{\partial t} + \frac{4\pi}{c}\left(\frac{\partial \vec{P}}{\partial t} + \sigma\vec{E}\right)$$

$$= \frac{1}{c}\frac{\partial}{\partial t}\left(\vec{E} + 4\pi\vec{P}\right) + \frac{4\pi}{c}\sigma\vec{E}$$

$$= \frac{1}{c}\left(K\frac{\partial \vec{E}}{\partial t} + 4\pi\sigma\vec{E}\right) \tag{238}$$

Therefore the Maxwell equation for the medium can be expressed as follows:

$$\begin{cases} \vec{\nabla} \cdot \vec{H} = 0 \\ \vec{\nabla} \times \vec{E} + \dfrac{1}{c}\dfrac{\partial \vec{H}}{\partial t} = 0 \\ \vec{\nabla} \cdot K\vec{E} = 4\pi\rho \\ \vec{\nabla} \times \vec{H} = \dfrac{1}{c}\left(K\dfrac{\partial \vec{E}}{\partial t} + 4\pi\sigma\vec{E}\right) \end{cases} \tag{239}$$

We look for wave-like solutions

$$\begin{cases} \vec{E}(\vec{r},t) = \vec{E}_0 e^{-i\left(\vec{\tau}\cdot\vec{r}-\omega t\right)} \\ \vec{H}(\vec{r},t) = \vec{H}_0 e^{-i\left(\vec{\tau}\cdot\vec{r}-\omega t\right)} \end{cases} \tag{240}$$

Using these expressions in the Maxwell equations (239) we obtain

$$\begin{cases} \vec{\tau} \cdot \vec{H}_0 = 0 \\ \vec{\tau} \times \vec{E}_0 = \dfrac{\omega}{c}\vec{H}_0 \\ K\vec{\tau} \cdot \vec{E}_0 = 0 \\ -\vec{\tau} \times \vec{H}_0 = \dfrac{1}{c}(\omega K - 4\pi\sigma i)\vec{E}_0 \end{cases} \tag{241}$$

We shall consider first the case

$$K = 0, \qquad \vec{\tau} \cdot \vec{E} \neq 0 \tag{242}$$

In such a case

$$\vec{D} = K\vec{E} = \vec{E} + 4\pi\vec{P} = 0 \tag{243}$$

and

49

$$\vec{E} = -4\pi\vec{P} \tag{244}$$

If we neglect damping $(\sigma \approx 0)$

$$\vec{H}_0 = 0 \qquad \text{and} \qquad \vec{\tau} \times \vec{E} = 0 \tag{245}$$

Therefor in this case $\vec{E}$ represents a <u>longitudinal</u> wave with the wave vector $\vec{\tau}$ parallel to $\vec{E}$ and $\vec{P}$.

We consider next the case $K \neq 0$,

that gives

$$\begin{cases} \vec{\tau} \cdot \vec{H}_0 = 0 \\ \vec{\tau} \times \vec{E}_0 = \dfrac{\omega}{c}\vec{H}_0 \\ \vec{\tau} \cdot \vec{E}_0 = 0 \\ -\vec{\tau} \times \vec{H}_0 = \left(\dfrac{\omega K}{c} - \dfrac{4\pi\sigma i}{c} \right)\vec{E}_0 \end{cases} \tag{246}$$

In this case $\vec{\tau}$ is perpendicular to both $\vec{E}$ and $\vec{H}$. We can take the y-axis in the $\vec{E}$ direction, the z-axis in the $\vec{H}$ direction, and the x-axis in the $\vec{\tau}$ direction. We obtain from Eqs.(246)

$$\begin{cases} \tau \ E_0 = \dfrac{\omega}{c}H_0 \\ \tau \ H_0 = \left(\dfrac{\omega K}{c} - \dfrac{4\pi\sigma i}{c} \right)E_0 \end{cases} \tag{247}$$

or, combining the two equations above

$$\left(\frac{c\tau}{\omega} \right)^2 = K - \frac{4\pi\sigma}{\omega}i = K_r + iK_i \tag{248}$$

Therefore

$$\begin{cases} K_r = K \\ K_i = -\dfrac{4\pi\sigma}{\omega} \end{cases} \tag{249}$$

The index of refraction is also complex

$$\frac{c\tau}{\omega} = n = n_r + in_i = \sqrt{K_r + iK_i} \tag{250}$$

The real and imaginary parts of the complex index of refraction and of the complex dielectric constant are related as follows:

$$\begin{cases} n_r^2 - n_i^2 = K_r \\ 2n_r \ n_i = K_i \end{cases} \tag{251}$$

Now we can write

$$\vec{E}(\vec{r},t) = \vec{j}E_0 e^{-i\tau x + i\omega t} = \vec{j}E_0 e^{-i(\omega/c)nx + i\omega t}$$

$$= \vec{j}E_0 e^{-i(\omega/c)(n_r + in_i)x + i\omega t} = \vec{j}E_0 e^{-i[(\omega/c)n_r x + \omega t]} e^{(\omega/c)n_i x} \quad (252)$$

$|\vec{E}|^2$ drops as

$$e^{2(\omega/c)n_i x} = e^{-\eta(\omega)x} \quad (253)$$

where $\eta(\omega)$ is defined as the <u>absorption coefficient</u> and is given by

$$\eta(\omega) = -2\frac{\omega}{c}n_i = -\frac{\omega K_i}{cn_r} \quad (254)$$

Note that if $\sigma = 0$, $K_i = 0$ and $\eta(\omega) = 0$ (no absorption).

Let us now relate these findings to the system of N-charged particles introduced in Section IV.A. We know that

$$K(\omega) = K_r(\omega) + iK_i(\omega) \quad (255)$$

where now, setting $K = 1$,

$$K_r(\omega) = 1 + \frac{4\pi n_0 e^2/m}{\left(\omega_0^2 - \omega^2\right)^2 + \gamma^2\omega^2}\left(\omega_0^2 - \omega^2\right) \xrightarrow{\gamma \to 0} 1 + \frac{4\pi n_0 e^2/m}{\omega_0^2 - \omega^2} \quad (256)$$

$$K_i(\omega) = \frac{-4\pi n_0 e^2/m}{\left(\omega_0^2 - \omega^2\right)^2 + \gamma^2\omega^2}\left(\gamma\omega\right) = \frac{4\pi\sigma}{\omega} \quad (257)$$

or

$$\sigma(\omega) = \frac{\omega^2\gamma\, n_0 e^2/m}{\left(\omega_0^2 - \omega^2\right)^2 + \gamma^2\omega^2} \xrightarrow{\gamma \to 0} 0 \quad (258)$$

Consider now the case $K = 0$. Neglecting losses this corresponds to

$$K_r(\omega) = 1 + \frac{4\pi n_0 e^2/m}{\omega_0^2 - \omega^2} = 0 \quad (259)$$

The frequency for which $K_r = 0$ is

$$\omega_L = \omega_0\sqrt{1 + \frac{4\pi n_0 e^2/m}{m\omega_0^2}} \quad (260)$$

Also, since

$$\vec{P} = n_0\alpha(\omega)\vec{E} = n_0\frac{e^2/m}{\omega_0^2 - \omega^2}\vec{E} \quad (261)$$

we have

$$\vec{D} = K\vec{E} = \vec{E} + 4\pi\vec{P} = \left(1 + 4\pi n_0\alpha\right)\vec{E} = \left[1 + \frac{4\pi n_0 e^2/m}{\omega_0^2 - \omega^2}\right]\vec{E} = 0 \quad (262)$$

as expected, in agreement with Eq.(243). $\vec{E}$ represents here a longitudinal wave with the wave vector $\vec{\tau}$ parallel to the induced dipole moments. Also $\omega_L > \omega_0$, due to the fact that the longitudinal electric field has the effect of increasing the force constant of the oscillators.

If $K \neq 0$ we have from Eqs.(248) and (256), neglecting losses,

$$\left(\frac{c\tau}{\omega}\right)^2 = K_r = 1 + \frac{4\pi n_0 e^2/m}{\omega_0^2 - \omega^2} = n_r^2 \tag{263}$$

and

$$\tau = \frac{\omega}{c}\sqrt{1 + \frac{4\pi n_0 e^2/m}{\omega_0^2 - \omega^2}} \tag{264}$$

The dispersion curves are represented in Figure 7.

We can now summarize our finding as follows:[5,6]

(1) For $\omega = \omega_L$, $K = 0$ and only longitudinal waves can propagate in the medium; the dispersion curve for longitudinal modes is a straight line with ω independent of τ.

(2 For values of ω such that $\omega_0 < \omega < \omega_L$, $K(\omega)$ is negative and the index of refraction is imaginary. No propagation is possible and the incident radiation is reflected. An incoming wave with a frequency in the above interval would be transmitted only through a thin slab of thickness $1/|\tau|$.

(3) When dealing with the transverse modes we cannot think of "photons" and "oscillators" independently; in fact, we have a completely new system of "modes" due to the coupling of the radiation with the oscillating charges. We can eliminate the effect of this coupling by letting $e \to 0$ or $m \to \infty$; in this case $\omega = \tau c$ (photons have vacuum dispersion) and $\omega_L = \omega_0$ (oscillators oscillate at the frequency ω_0).

(4) For the lower branch of the T modes the closer ω is to ω_0, namely the larger is τ, less electromagnetic energy and more mechanical energy is present in the wave. The contrary is true for small τ and ω, but as $\tau, \omega \to 0$ some residual mechanical energy (which represents the energy of static polarization) is present.[7]

(5) For the upper T branch, the larger is τ, less mechanical energy and more electromagnetic energy is present in the wave.[7]

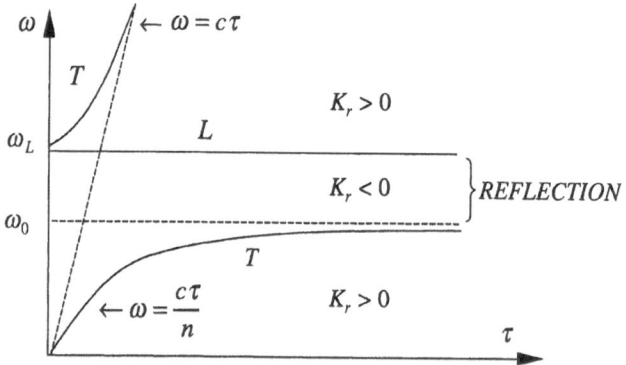

Figure 7. Dispersion relations for L (longitudinal) and T (transverse) modes.

This general behavior has been looked at in the limiting case of $\gamma = \sigma = 0$ (no losses). The losses of the mechanical oscillators will make a transverse wave (which is in part mechanical and electromagnetic) lose energy. This may give us a mechanism or model for the losses experienced by an electromagnetic wave propagating through the system.

The incoming radiation couples with a certain number of transverse modes of the system. These modes are of mixed (mechanical and electromagnetic) nature; the mechanical part of these modes represents the sink of the energy. The excitation of these modes is restored by the incoming radiation. An electromagnetic wave of high ω (upper branch) will not lose any energy and will pass unattenuated through the system.

V. EXAMPLES OF COLLECTIVE EXCITATIONS

V.A. Phonons

1. Summary of Properties

(a) In a molecule or solid, the electrons provides the potential in which the atoms perform their vibrational motions. These motions in solids are often called lattice vibrations.

(b) These motions can be thought of as a superposition of normal modes of vibration which are 3N in number, if N = number of atoms in the solid.(More precisely, since the solid has also three translational and three rotational degrees of freedom, the number of normal modes of vibration is 3N-6).

(c) A normal mode of vibration is a pattern of motion in which all the atoms of the solid participate. There may, however, be localized modes which involve a relatively small number of atoms.

(d) In treating the vibration of solids the harmonic approximation is used. This approximation implies that there is no exchange of energy between the normal modes.

(e) Each normal mode is equivalent to a harmonic oscillator. The vibrations of a solid are then equivalent to a collection of 3N harmonic oscillations. If an oscillator of frequency ω is in its nth excited state, this fact is also expressed by saying that n phonons, each of energy $\hbar\omega$, are present in the solid.

(f) The different normal modes are not completely isolated, but are, rather, in speaking terms, due to their anharmonicity: this provides the mechanism by which the solid reaches thermal equilibrium.

(g) When considering the thermal vibrations of a solid, it may be interesting to have an idea of how many phonons may be present in a solid at, say, room temperature. The number of phonons in a frequency interval $(\omega, \omega + d\omega)$ is given by $\bar{n}(\omega)\rho(\omega)d\omega$, where:

$$\bar{n}(\omega) = \left(e^{\hbar\omega/kT} - 1\right)^{-1}, \text{ and } \rho(\omega) = density\ of\ phonon\ states \propto \omega^2/c_s^3$$

It is the value of c_s =velocity of sound in solids $\sim 5\times10^5$ cm/sec (versus c = velocity of light $= 3\times10^{10}$ cm/sec) that makes the number of phonons extremely large. For the sake of comparison, the total number of phonons/cm³ in black body radiation at T= 300 K is ~ 6.4×10^8, the number of photons/cm³ in a typical laser medium (λ = 6,300 A, 1 W/cm²) is 10^8, the number of phonons/cm³ in a solid (with a Debye temperature T_D= 1,000 K) at T= 300 K is ~ 3.5×10^{23}. The sheer number of phonons may give us an idea of their importance in affecting the spectral characteristics of ions in solids.

(h) If more than one atom is present in the unit cell, the dispersion curves of the vibrations include optical branches.

2. Infrared Absorption by Ionic Solids. In this case the "oscillators" interacting with the electromagnetic radiation are the optical modes of lattice vibrations with small $\vec{k}$ (intra-molecular vibrations).

The splitting of optical phonon branches into longitudinal and transverse takes place in accord with the Lyddane-Sachs-Teller relation (232). This is experimentally confirmed when comparing ω_L/ω_T obtained by inelastic neutron scattering[8,9] with the experimental values of $[K(0)/K(\infty)]^{1/2}$:[10, p.185]

	NaI	KBr	GaAs
ω_L/ω_T	1.44 ± 0.05	1.39 ± 0.02	1.07± 0.02
$[K(0)/K(\infty)]^{1/2}$	1.45 ± 0.03	1.38 ± 0.03	1.08

The process of infrared absorption can be related to the diagram in Figure 8. If we consider "the system" as encompassing the vibrating solid and the electromagnetic radiation, a state of the system with a photon of wave vector, say, $\vec{k}_A$ may be represented as follows:

$$\underbrace{\left|1_{\vec{k}_A}\right\rangle|0\rangle|0\rangle|0\rangle}_{photon\ part}......\underbrace{|0\rangle|1\rangle|n\rangle|n'\rangle}_{phonon\ part}...... \tag{265}$$

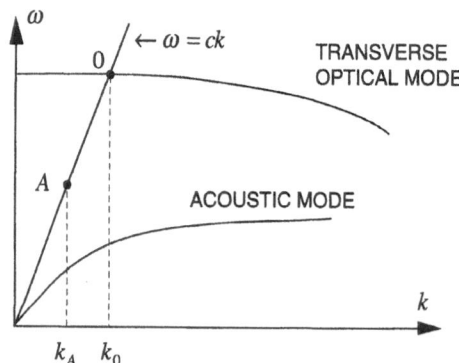

Figure 8. The coupling of electromagnetic radiation with transverse optical phonon modes.

The two states

$$\left|1_{\vec{k}_0}\right\rangle\left|0_{\vec{f}_0}\right\rangle = 1 \text{ photon of wave vector } \vec{k}_0, \text{ no phonon of wave vector } \vec{f}_0 = \vec{k}_0$$

$$\left|0_{\vec{k}_0}\right\rangle\left|1_{\vec{f}_0}\right\rangle = \text{no photon of wave vector } \vec{k}_0, \text{ 1 phonon of wave vector } \vec{f}_0 = \vec{k}_0$$

are in resonance and may be split by the ion-photon interaction. The matrix element of this interaction taken among the two states above may be zero to first-order, but may be different from zero to second-order, In any case, a splitting of the two degenerate states will follow, and the situation around the crossing point 0 of Figure 8 will be distorted as in Figure 9. The real stationary states are a mixture of phonons and photons; they are what we call <u>polariton states</u>.

Polariton states occupy a small part of the Brillouin zone: $\left|\vec{k}_0\right| \approx 10^4 \ cm^{-1}$ whereas the dimension of the Brillouin zone is ~ 10^8 cm⁻¹.

The various points in Figure 9 represent the following:
A = photon in the solid, P_0 = pure photon, P_2 = phonon, P_1 = polariton.
For large $\vec{k}$ (small λ) we have pure photons and pure phonons.

The mechanism for absorption consists of the following:

(1) The incoming beam maintains a certain excitation of the T (transverse) normal modes.
(2) These T-modes decay because their mechanical part is damped.
(3) The excitation is continuously restored by the incoming beam.

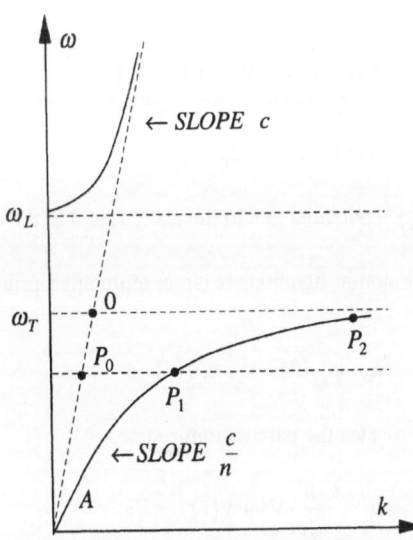

Figure 9. Polariton states due to the interaction of transverse optical phonons with electromagnetic radiation.

55

V.B. Excitons

1. <u>General Theory</u>. Consider a system of N atoms in an ordered array. The "electronic" Hamiltonian is given by

$$H = H_0 + H' \tag{266}$$

where

$$H_0 = -\sum_i \frac{\hbar^2 \nabla_i^2}{2m} + \sum_i V(\vec{r}_i) \tag{267}$$

$$H' = \sum_{i<j} \frac{e^2}{r_{ij}} \tag{268}$$

We shall assume the following:
(1)   The nuclei are at rest in their equilibrium position, and
(2)   The electrons are extra-core, one per atom.

The approximate ground state wavefunction of the system can be written as follows:

$$A\big(a_{01}a_{02}\ldots\ldots a_{0N}\big) = \frac{1}{\sqrt{N!}}
\begin{vmatrix}
a_{01}(1) & a_{01}(2) & \ldots\ldots & a_{01}(N) \\
a_{02}(1) & a_{02}(2) & \ldots\ldots & a_{02}(N) \\
\multicolumn{4}{c}{\ldots\ldots\ldots\ldots\ldots\ldots\ldots\ldots} \\
\multicolumn{4}{c}{\ldots\ldots\ldots\ldots\ldots\ldots\ldots\ldots} \\
a_{0N}(1) & a_{0N}(2) & \ldots\ldots & a_{0N}(N)
\end{vmatrix} \tag{269}$$

where $a_{0i}(j) = a_0\big(\vec{r}_j - \vec{R}_i\big)$ = ground state "atomic" wavefunction

A = antisymmetryzing operator

For each atom

$$h_i(\vec{r}) = -\frac{\hbar^2}{2m}\nabla^2 + V\big(\vec{r} - \vec{R}_i\big) \tag{270}$$

and

$$h_i(\vec{r})\ a_{ni}(\vec{r}) = \varepsilon_n\ a_{ni}(\vec{r}) \tag{271}$$

where $V\big(\vec{r} - \vec{R}_i\big)$ = potential due to the nucleus and to the core electrons.

If the overlap of atomic functions is large, more appropriate one-electron states are the eigensolutions of

$$h_0\ \psi_{n\vec{k}}(\vec{r}) = \varepsilon_{n\vec{k}}\ \psi_{n\vec{k}}(\vec{r}) \tag{272}$$

where the subscript n indicates the parent atomic state,

$$h_0 = -\frac{\hbar^2}{2m}\nabla^2 + V(\vec{r}) \tag{273}$$

and $V(\vec{r})$ = periodic crystal potential.

56

A solution of Eq.(272) is a Block-type wavefunction

$$\psi_{n\vec{k}}(\vec{r}) = e^{i\vec{k}\cdot\vec{r}} u_{n\vec{k}}(\vec{r}) \qquad (274)$$

These functions have the usual properties of orthonormality and closure

$$\begin{cases} \psi_{n\vec{k}}^*(\vec{r})\psi_{n\vec{k}'}(\vec{r})d^3\vec{r} = \delta_{\vec{k}\vec{k}'}, \\[2mm] \sum_{\vec{k}} \psi_{n\vec{k}}^*(\vec{r})\psi_{n\vec{k}'}(\vec{r}) = \delta(\vec{r}-\vec{r}') \end{cases} \qquad (275)$$

The Block functions $\psi_{n\vec{k}}(\vec{r})$ are completely delocalized. We can construct localized states known as <u>Wannier function</u>, which we define as follows:

$$a_{ni}(\vec{r}) = a_n(\vec{r}-\vec{R}_i) = \frac{1}{\sqrt{N}}\sum_{\vec{k}} e^{-i\vec{k}\cdot\vec{R}_i}\psi_{n\vec{k}}(\vec{r}) = \frac{1}{\sqrt{N}}\sum_{\vec{k}} e^{-i\vec{k}\cdot(\vec{r}-\vec{R}_i)} u_{n\vec{k}}(\vec{r}) \qquad (276)$$

Unlike the Block functions which depend on both $\vec{k}$ and $\vec{r}$, the Wannier functions depend on $(\vec{r}-\vec{R}_i)$ only and are localized; in fact, the phases for the different terms are $\sim$ zero when $\vec{r} \approx \vec{R}_i$, and the functions $u_{n\vec{k}}$ adds constructively.

The Wannier functions are orthonormal:

$$\int a_{ni}^*(\vec{r})a_{nj}(\vec{r})d^3\vec{r} = \frac{1}{N}\int \sum_{\vec{k}} e^{i\vec{k}\cdot\vec{R}_i}\psi_{n\vec{k}}^*(\vec{r})\sum_{\vec{k}'} e^{-i\vec{k}'\cdot\vec{R}_j}\psi_{n\vec{k}'}(\vec{r})d^3\vec{r}$$

$$= \frac{1}{N}\sum_{\vec{k}}\sum_{\vec{k}'} e^{i\vec{k}\cdot\vec{R}_i}e^{-i\vec{k}'\cdot\vec{R}_j}\int \psi_{n\vec{k}}^*(\vec{r})\psi_{n\vec{k}'}(\vec{r})d^3\vec{r}$$

$$= \frac{1}{N}\sum_{\vec{k}} e^{i\vec{k}\cdot(\vec{R}_i-\vec{R}_j)} = \delta_{ij} \qquad (277)$$

and have the closure property

$$\sum_i a_{ni}^*(\vec{r})a_{ni}(\vec{r}') = \frac{1}{N}\sum_{\vec{k}}\sum_{\vec{k}'} \psi_{n\vec{k}}^*(\vec{r})\psi_{n\vec{k}'}(\vec{r}')\sum_i e^{i(\vec{k}-\vec{k}')\cdot\vec{R}_i}$$

$$= \frac{1}{N}\sum_{\vec{k}}\sum_{\vec{k}'} \psi_{n\vec{k}}^*(\vec{r})\psi_{n\vec{k}'}(\vec{r}')N\delta_{\vec{k}\vec{k}'}$$

$$= \sum_{\vec{k}} \psi_{n\vec{k}}^*(\vec{r})\psi_{n\vec{k}}(\vec{r}') = \delta(\vec{r}-\vec{r}') \qquad (278)$$

We note that for each atomic state n, there exists a set of N Block wavefunctions characterized by the N possible values of $\vec{k}$. Such being the case, if n=0 (ground state) the zero-order ground state of the system is given by

$$\psi_G = A\left| \psi_{0\vec{k}_1}(\vec{r}_1)\psi_{0\vec{k}_2}(\vec{r}_2)\ldots\ldots\psi_{0\vec{k}_N}(\vec{r}_N)\right\rangle \qquad (279)$$

It can be shown[6] that this function, apart sign, is the same as

$$\psi_G = A \left| a_{01}(\vec{r}_1) a_{02}(\vec{r}_2) \ldots\ldots a_{0N}(\vec{r}_N) \right\rangle \tag{280}$$

In general the determinantal functions

$$A \left| \psi_{n\vec{k}_1}(\vec{r}_1) \psi_{n\vec{k}_2}(\vec{r}_2) \ldots\ldots \psi_{n\vec{k}_N}(\vec{r}_N) \right\rangle \tag{281}$$

and

$$A \left| a_{n\vec{R}_1}(\vec{r}_1) a_{n\vec{R}_2}(\vec{r}_2) \ldots\ldots a_{n\vec{R}_N}(\vec{r}_N) \right\rangle \tag{282}$$

are identical if the state they refer to can be represented by a single determinant (closed-shell situation). The a-based function is equal to the ψ-based functions times a unitary matrix whose elements are

$$\frac{1}{\sqrt{N}} e^{-i\vec{k}_s \cdot \vec{R}_t} \qquad (s, \ t = 1,2,3,\ldots\ldots,N)$$

and whose determinant is equal to 1.

We want now to describe an excited state of the system. The simplest excitation is obtained by raising an electron from a ground state orbital a_{0h} to an orbital a_{1e} corresponding to a higher atomic state. This situation is represented by the wavefunction

$$\psi_1 = A \left| a_{01} a_{02} \ldots\ldots a_{0,h-1} a_{1e} a_{0,h+1} \ldots\ldots a_{0N} \right\rangle \tag{283}$$

and corresponds to a "hole" in the "atom" at position $\vec{R}_h$ and an extra electron in the "atom" at position $\vec{R}_e$. States of the type ψ_1 above are called <u>electron transfer states</u>.

Let us call

$$\beta = \vec{R}_e - \vec{R}_h \tag{284}$$

For each $\vec{\beta}$, there will be several states of the type ψ_1. In the case of a linear crystal with four atoms we get the following excited states:

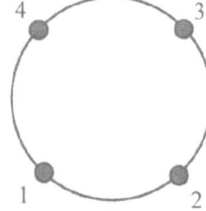

$$
\begin{cases}
\beta = 0: & \begin{matrix} |1 \quad 0\rangle \\ R_1 \quad R_1 \end{matrix} & \begin{matrix} |2 \quad 0\rangle \\ R_2 \quad R_2 \end{matrix} & \begin{matrix} |3 \quad 0\rangle \\ R_3 \quad R_3 \end{matrix} & \begin{matrix} |4 \quad 0\rangle \\ R_4 \quad R_4 \end{matrix} \\[2ex]
\beta = 1: & \begin{matrix} |1 \quad 1\rangle \\ R_1 \quad R_2 \end{matrix} & \begin{matrix} |2 \quad 1\rangle \\ R_2 \quad R_3 \end{matrix} & \begin{matrix} |3 \quad 1\rangle \\ R_3 \quad R_4 \end{matrix} & \begin{matrix} |4 \quad 1\rangle \\ R_4 \quad R_1 \end{matrix} \\[2ex]
\beta = -1: & \begin{matrix} |1 \ -1\rangle \\ R_1 \quad R_4 \end{matrix} & \begin{matrix} |2 \ -1\rangle \\ R_2 \quad R_1 \end{matrix} & \begin{matrix} |3 \ -1\rangle \\ R_3 \quad R_2 \end{matrix} & \begin{matrix} |4 \ -1\rangle \\ R_4 \quad R_3 \end{matrix} \\[2ex]
\beta = 2: & \begin{matrix} |1 \quad 2\rangle \\ R_1 \quad R_3 \end{matrix} & \begin{matrix} |2 \quad 2\rangle \\ R_2 \quad R_4 \end{matrix} & \begin{matrix} |3 \quad 2\rangle \\ R_3 \quad R_1 \end{matrix} & \begin{matrix} |4 \quad 2\rangle \\ R_4 \quad R_2 \end{matrix}
\end{cases}
$$

We have a total of 16 $\psi_1\left(\vec{R}_h,\vec{\beta}\right)$ states and the matrix of the Hamiltonian is 16×16. In the designation above R_i R_j indicates that the hole is at position R_i and electron at position R_j.

We note that in general the diagonal terms of the Hamiltonian $\left\langle \psi_1\left(\vec{R}_h,\vec{\beta}\right)\left|H\right|\psi_1\left(\vec{R}_h,\vec{\beta}\right)\right\rangle$ are the same for a certain value of $\vec{\beta}$ and for all $\vec{R}_h$, because of the translational symmetry of the system.

At this point we construct states in the exciton representation:

$$\psi_1\left(\vec{k},\vec{\beta}\right) = \frac{1}{\sqrt{N}}\sum_{\vec{R}_h} e^{i\vec{k}\cdot\vec{R}_h}\,\psi_1\left(\vec{R}_h,\vec{\beta}\right) \tag{283}$$

These states allow a diagonalization with respect to $\vec{R}_h$, but not $\vec{\beta}$:

$$\left\langle \psi_1\left(\vec{k},\vec{\beta}\right)\left|H\right|\psi_1\left(\vec{k}',\vec{\beta}'\right)\right\rangle = \frac{1}{N}\sum_{\vec{R}_h}\sum_{\vec{R}_{h'}} e^{-i\vec{k}\cdot\vec{R}_h} e^{i\vec{k}'\cdot\vec{R}_{h'}}\left\langle \psi_1\left(\vec{R}_h,\vec{\beta}\right)\left|H\right|\psi_1\left(\vec{R}_{h'},\vec{\beta}'\right)\right\rangle \delta_{\vec{k}\vec{k}'}$$

$$= \frac{1}{N}\sum_{\vec{R}_h}\sum_{\vec{R}_{h'}} e^{-i\vec{k}\cdot\left(\vec{R}_h - \vec{R}_{h'}\right)}\left\langle \psi_1\left(\vec{R}_h,\vec{\beta}\right)\left|H\right|\psi_1\left(\vec{R}_{h'},\vec{\beta}'\right)\right\rangle \delta_{\vec{k}\vec{k}'}$$

$$= \sum_{\vec{R}_{h'}} e^{-i\vec{k}\cdot\left(\vec{R}_h - \vec{R}_{h'}\right)}\left\langle \psi_1\left(\vec{R}_h,\vec{\beta}\right)\left|H\right|\psi_1\left(\vec{R}_{h'},\vec{\beta}'\right)\right\rangle \delta_{\vec{k}\vec{k}'}$$

$$= \sum_{\vec{R}_{h'}} e^{i\vec{k}\cdot\vec{R}_{h'}}\left\langle \psi_1\left(0,\vec{\beta}\right)\left|H\right|\psi_1\left(\vec{R}_{h'},\vec{\beta}'\right)\right\rangle \delta_{\vec{k}\vec{k}'} = H_{\vec{\beta}\vec{\beta}'}\delta_{\vec{k}\vec{k}'} \tag{284}$$

where

$$H_{\vec{\beta}\vec{\beta}'} = \sum_{\vec{R}_h} e^{i\vec{k}\cdot\vec{R}_h}\left\langle \psi_1\left(0,\vec{\beta}\right)\left|H\right|\psi_1\left(\vec{R}_h,\vec{\beta}\right)\right\rangle \tag{285}$$

2. The Frenkel Exciton. In this case $\vec{\beta} = 0$ and the electron making the transition does not leave the "cell" where it resides. The exciton is in this case called a Frenkel exciton, a packet of waves which are linear combinations of atomic excited wavefunctions[11]. No $\vec{\beta}$ diagonalization is necessary.

The ground state is given by

$$\psi_G = A\left|a_0(\vec{r}-\vec{R}_1)a_0(\vec{r}-\vec{R}_2)......a_0(\vec{r}-\vec{R}_N)\right\rangle \tag{286}$$

The electron transfer state is given by

$$\psi_1(\vec{R}_h) = A\left|a_0(\vec{r}-\vec{R}_1)a_0(\vec{r}-\vec{R}_2)......a_1(\vec{r}-\vec{R}_h)......a_0(\vec{r}-\vec{R}_N)\right\rangle \tag{287}$$

and the generic matrix element of Hamiltonian is

$$\left\langle \psi_1\left(\vec{R}_h\right)\left|H\right|\psi_1\left(\vec{R}_{h'}\right)\right\rangle \tag{288}$$

We construct states in the exciton representation as follows:

$$\psi_1(\vec{k}) = \frac{1}{\sqrt{N}} \sum_{\vec{R}_h} e^{i\vec{k}\cdot\vec{R}_h} \psi_1(\vec{R}_h)$$ (289)

The ground state energy of the system is

$$\langle \psi_G | H | \psi_G \rangle = \langle \psi_G | H_0 | \psi_G \rangle + \langle \psi_G | H' | \psi_G \rangle = N\varepsilon_0 + \Delta E_g$$ (290)

where ε_0 = energy of the ground atomic state, and

$$\Delta E_g = \langle \psi_G | H' | \psi_G \rangle$$

The expectation value of the energy in the first excited state is

$$\langle \psi_1(\vec{k}) | H | \psi_1(\vec{k}) \rangle = \frac{1}{N} \sum_{\vec{R}_h} \sum_{\vec{R}_{h'}} e^{-i\vec{k}\cdot(\vec{R}_h - \vec{R}_{h'})} \langle \psi_1(\vec{R}_h) | H | \psi_1(\vec{R}_{h'}) \rangle$$

$$= \sum_{\vec{R}_{h'}} e^{-i\vec{k}\cdot(\vec{R}_h - \vec{R}_{h'})} \langle \psi_1(\vec{R}_h) | H | \psi_1(\vec{R}_{h'}) \rangle$$

$$= \sum_{\vec{R}_h} e^{i\vec{k}\cdot\vec{R}_h} \langle \psi_1(0) | H | \psi_1(\vec{R}_h) \rangle$$

$$= \langle \psi_1(0) | H | \psi_1(0) \rangle + \sum_{\vec{R}_h \neq 0} e^{i\vec{k}\cdot\vec{R}_h} \langle \psi_1(0) | H | \psi_1(\vec{R}_h) \rangle$$

$$= \varepsilon_1 + (N-1)\varepsilon_0 + \langle \psi_1(0) | H' | \psi_1(0) \rangle + \sum_{\vec{R}_h \neq 0} e^{i\vec{k}\cdot\vec{R}_h} \langle \psi_1(0) | H' | \psi_1(\vec{R}_h) \rangle$$

$$= \varepsilon_1 + (N-1)\varepsilon_0 + H'_{\vec{R}_h \vec{R}_h} + \sum_{\vec{R}_h \neq 0} e^{i\vec{k}\cdot\vec{R}_h} H'_{0\vec{R}_h}$$ (291)

where ε_1 = energy of the excited atomic state.

The matrix elements of H taken between two ψ_1 states with different $\vec{k}$ are zero.

The total change in energy going from the ground state to an excited state is

$$\Delta E = \left[\varepsilon_1 + (N-1)\varepsilon_0 + H'_{\vec{R}_h \vec{R}_h} + \sum_{\vec{R}_h \neq 0} e^{i\vec{k}\cdot\vec{R}_h} H'_{0\vec{R}_h} \right] - \left[N\varepsilon_0 + \Delta E_g \right]$$

$$= (\varepsilon_1 - \varepsilon_0) + (H'_{\vec{R}_h \vec{R}_h} - \Delta E_g) + \sum_{\vec{R}_h \neq 0} e^{i\vec{k}\cdot\vec{R}_h} H'_{0\vec{R}_h}$$

$$= \Delta E \ (atomic) + \Delta V \ = \ \text{(change in interaction energy)} + \varepsilon(\vec{k})$$ (292)

where

$$\Delta E \ (atomic) = (\varepsilon_1 - \varepsilon_0) \approx 3eV$$ (293)

$$\Delta V = (H'_{\bar{R}_h\bar{R}_h} - \Delta E_g) \tag{294}$$

and finally

$$\varepsilon\left(\vec{k}\right) = \sum_{\bar{R}_h \neq 0} e^{i\vec{k}\cdot\bar{R}_h} H'_{0\bar{R}_h} \tag{295}$$

gives the <u>dispersion relation</u>.

We can make at this point the following observations:

(1) If $H' = 0$ the excited state is N-fold degenerate. As a result of the perturbation H' we have a band of states. (See Figure 10)

(2) The matrix elements $H'_{\bar{R}_h\bar{R}_{h'}}$ are responsible for the fact that the excitation energy does not stay in one atom, but moves along.

(3) Lattice vibrations and other periodicity-destroying entities such as defects, destroy the coherence of the exciton states and result in the scattering of excitons.

(4) If we consider only nearest neighbor interactions and set $H'_{0\bar{R}_h} = M$ we obtain for a linear lattice

$$\varepsilon(\vec{k}) = 2M\cos ka \tag{296}$$

and for a cubic lattice

$$\varepsilon(\vec{k}) = 2M\sum_i \cos k_i a \tag{297}$$

(5) Finally, if $\varepsilon(\vec{k})$ is known we can define

$$\vec{v}_g = \frac{1}{\hbar}\vec{\nabla}_{\vec{k}}\varepsilon\left(\vec{k}\right) \tag{298}$$

and

$$\frac{1}{m^*} = \frac{1}{\hbar^2}\vec{\nabla}_{\vec{k}}\vec{\nabla}_{\vec{k}}\varepsilon\left(\vec{k}\right) \tag{299}$$

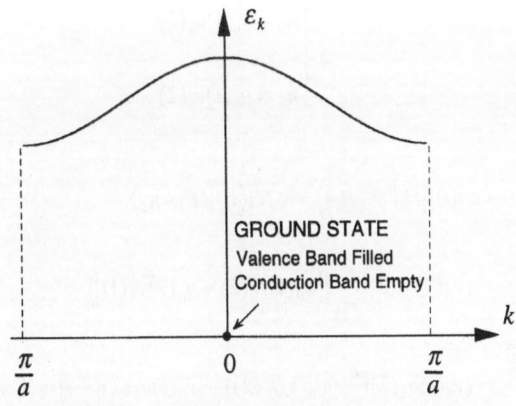

Figure 10. Frenkel exciton states.

We want now to direct our attention to the matrix elements responsible for the transfer of energy

$$H'_{\vec{R}_h \vec{R}_{h'}} = \left\langle \psi_1(\vec{R}_h) \middle| H' \middle| \psi_1(\vec{R}_{h'}) \right\rangle \tag{300}$$

We define the <u>product states</u> as follows:

$$\pi_h = \left| a_{01} a_{02} \ldots\ldots a_{1h} \ldots\ldots a_{0N} \right\rangle \tag{301}$$

$$\pi_{h'} = \left| a_{01} a_{02} \ldots\ldots a_{1h'} \ldots\ldots a_{0N} \right\rangle \tag{302}$$

Then

$$\psi_1(\vec{R}_h) = A\pi_h \tag{303}$$

$$\psi_1(\vec{R}_{h'}) = A\pi_{h'} \tag{304}$$

It is easy to show (see [12], p.165) that

$$A(A\pi) = A^2 \pi = \sqrt{N!} A\pi \tag{305}$$

and

$$\left\langle \psi_1(\vec{R}_h) \middle| H' \middle| \psi_1(\vec{R}_{h'}) \right\rangle = \left\langle A\pi_h \middle| H' \middle| A\pi_{h'} \right\rangle$$

$$= \left\langle \pi_h \middle| H' \middle| A^2 \pi_{h'} \right\rangle = \sqrt{N!} \left\langle \pi_h \middle| H' \middle| A\pi_{h'} \right\rangle \tag{306}$$

In order to relate the present notation with that used in Section II we shall rename the a functions as follows:

a_0 = ground state function u

$$a_0\left(\vec{r}_i - \vec{R}_h\right) = u_h(i) \tag{307}$$

a_1 = ground state function v

$$a_1\left(\vec{r}_i - \vec{R}_h\right) = v_h(i) \tag{308}$$

Then

$$\begin{cases} \pi_0 = v_0(1)u_h(2)\ldots\ldots \\[2mm] \pi_h = u_0(1)v_h(2)\ldots\ldots \end{cases} \tag{309}$$

and

$$\left\langle \psi_1(0) \middle| H' \middle| \psi_1(\vec{R}_h) \right\rangle = \sqrt{N!} \left\langle \pi_0 \middle| H' \middle| A\pi_h \right\rangle$$

$$= \left\langle v_0(1)u_h(2) \middle| \frac{e^2}{r_{12}} \middle| \left[u_0(1)v_h(2) - u_0(2)v_h(1) \right] \right\rangle$$

$$= \left\langle v_0(1)u_h(2) \middle| \frac{e^2}{r_{12}} \middle| u_0(1)v_h(2) \right\rangle - \left\langle v_0(1)u_h(2) \middle| \frac{e^2}{r_{12}} \middle| v_h(1)u_0(2) \right\rangle \tag{310}$$

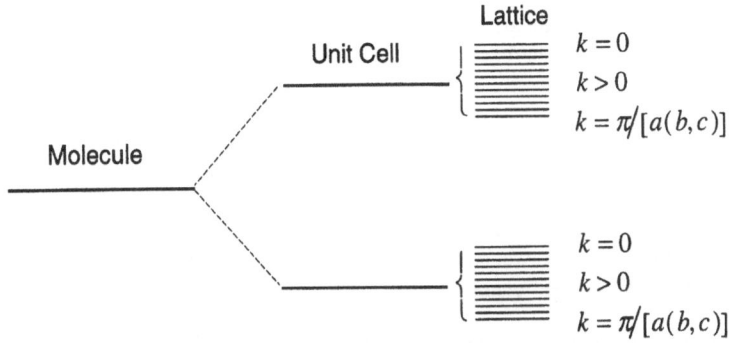

Figure 11. Davydov splitting due to the presence of two interacting molecules in the unit cell.

The first term represents the interaction between two charge distributions $v_0^*(1)u_0(1)$ located at position 0 and $u_h^*(2)v_h(2)$ located at position $\vec{R}_h$; this <u>direct interaction</u> is Coulombic and long-range in nature; this term is the same that appears in Eq.(112). The other term represents an <u>exchange interaction</u>; its range is short.

If each atom contains two electrons, rather than one, the formula (310) can be written as follows:

$$\langle v_0(1,2)u_h(3,4)|H'|u_0(1,2)v_h(3,4)\rangle - \langle v_0(1,2)u_h(3,4)|H'|v_h(1,2)u_0(3,4)\rangle \qquad (311)$$

where the $u(i,j)$ and $v(k,l)$ functions represents the ground and the excited atomic states, respectively. The ground state is generally a singlet , the excited state can be a singlet or a triplet. The direct interaction term vanishes if the excited state is a triplet; if the excited state is a singlet this term can be written as follows:

$$2\langle v_0(1)u_h(3)|\frac{e^2}{r_{13}}|u_0(1)v_h(3)\rangle \qquad (312)$$

where $u(i)$ and $v(j)$ are ground state and excited state electron orbital, respectively.

Up to now we have considered only one atom or one molecule per unit cell. If there are several, say n, identical, but differently oriented molecules per unit cell, the interactions among these molecules produce a splitting into n levels and , in essence, n exciton states for each $\vec{k}$. The stronger is the interaction among different molecules in the unit cell, the greater is the splitting of exciton states for a certain $\vec{k}$. The splitting for the case n=2 is reported in Figure 11.[13] The additional splitting that is present even if the excited state of the molecule is non-degenerate is called the <u>Davydov splitting</u>. It is obvious that this splitting is restricted to excited states and it cannot be referred to <u>one</u> molecule in the crystal. It is also clear that such splitting does not exist for molecules in solution.

The Davydov splitting has been observed in the spectra of organic molecules. It may range from a few hundred to several thousand cm^{-1}.[13] It may also occur in inorganic crystals such as MnF_2; for this system there are two Mn^{2+} ions per unit cell; they are distinguished from each other by the orientation of the F^- ions. The interaction among the

Mn^{2+} ions is of antiferromagnetic nature. However, since this interaction is small, the Davydov splitting is not experimentally measurable.[14]

3. The Wannier Exciton. The general matrix element $\left\langle \psi_1\left(\vec{k},\vec{\beta}\right)\middle| H \middle| \psi_1\left(\vec{k}',\vec{\beta}'\right)\right\rangle$ was used by Wannier to develop his effective mass theory[15]. We can arrive at the same conclusions by making the following observations. Certain semiconductors such as Germanium and Silicon have large dielectric constant, large bands and small gaps. In these materials the interaction between electron and hole is weak and in fact electron and hole maintain much of their "free" character.

The Hamiltonian of the Wannier exciton can be written as follows:

$$H = -\frac{\hbar^2}{2m_e}\nabla_e^2 - \frac{\hbar^2}{2m_h}\nabla_h^2 - \frac{e^2}{K\left|\vec{r}_e - \vec{r}_h\right|} \tag{313}$$

where m_e and m_h are the effective masses of the electron and of the hole, respectively, $\vec{r}_e$ and $\vec{r}_h$ are the coordinates of the electron and of the hole, respectively, and K is the dielectric constant of the medium. The following equivalence can be established.

Hydrogen atom	Wannier exciton
e^2	e^2/K
m = mass of the electron	m_e
M = mass of the nucleus	m_h
$\mu = \dfrac{mM}{m+M} \approx m$	$\mu' = \dfrac{m_e m_h}{m_e + m_h}$
Energy $= \dfrac{\hbar^2 K^2}{2(m+M)} + E_n$ (314)	Energy $= \varepsilon_c + \dfrac{\hbar^2 K^2}{2(m_e + m_h)} + E_n$ (315) (ε_c=energy of bottom of conduction band)
$E_n = -\dfrac{me^4}{2\hbar^2}\dfrac{1}{n^2} = -\dfrac{e^2}{2a_0}\dfrac{1}{n^2}$ $= -\dfrac{13.6}{n^2}$ eV (316)	$E_n = -\dfrac{\mu' e^4}{2\hbar^2 K^2}\dfrac{1}{n^2} = -\dfrac{e^2}{2a_0'}\dfrac{1}{Kn^2}$ $= -\dfrac{13.6}{n^2}\left(\dfrac{\mu'}{mK^2}\right)$ eV (317)
$a_0 = \dfrac{\hbar^2}{me^2} \approx 0.5$ A (318)	$a_0' = \dfrac{\hbar^2 K}{\mu' e^2} = a_0\left(\dfrac{mK}{\mu'}\right)$ (319)
$a_n = n^2 a_0$ (320)	$a_n' = n^2 a_0' = n^2 a_0\left(\dfrac{mK}{\mu'}\right)$ (321)

Example

Energy gap between the valence and the conduction bands = 0.3 eV

$$m_e = 0.02 \ m, \qquad m_h = 0.4 \ m, \qquad K = 12$$

We find

$$E_n = -\frac{13.6}{n^2}\left(\frac{\mu'}{mK^2}\right) = -\frac{0.0018}{n^2} \ eV$$

$$a_0' = a_0\left(\frac{mK}{\mu'}\right) = 0.5\frac{12}{0.019} \approx 316 \ A$$

$$a_n' = 316n^2 \ A$$

The binding energy of the exciton is 0.0018 eV, while $kT \approx (1/40) \ eV$ at room temperature; en exciton created in this system at room temperature would be thermally unstable and would quickly decay into an electron and a hole.

We note the following:

(1) In general an effective mass may not be defined. Rather, because of the directionality of the dependence of the one-electron energy in $\vec{k}$ space, only an effective mass tensor has meaning. In this case it is not possible to make the center of mass transformation which is necessary to define the simple exciton model. The valence energy band (for example, the 2p band) may also be degenerate at the top along a certain $\vec{k}$ direction and not along another $\vec{k}$ direction; likewise for the conduction band at its bottom. This could contribute to the anisotropy of the problem and to additional degeneracy of the exciton levels.

(2) The 1s eigenfunction of the H atom is given by

$$\psi_{100} = \frac{1}{\sqrt{\pi}} a_o^{-3/2} e^{-(r/a_0)}$$

and is different than zero for r=0, namely at the position of the nucleus. This fact leads to the so-called quantum-defects in the spectra of hydrogen-like atoms.

A similar situation arises when an exciton is in the 1s state in that the probability of finding both electron and hole in the same cell is not negligible. This may produce deviation from the Wannier-type situation and, consequently, a correction to the "ground state" energy of the exciton may be necessary.

(3) Consider an exciton of reduced mass μ and radius a. Its angular momentum being, say, $\hbar$, is given by

$$\hbar = \mu v a = \mu a^2 \omega$$

where $v = a\omega$ and ω is the angular frequency of the rotational motion of the exciton. We find

$$\omega = \frac{\hbar}{\mu a^2} \quad \text{and} \quad a = \sqrt{\frac{\hbar}{\mu\omega}}$$

We call <u>critical radius</u> the one that we obtain by putting $\omega = \omega_0$ = lowest resonance frequency of the system = frequency of the "optical" lattice vibration branch:

$$a_{crit} = \sqrt{\frac{\hbar}{\mu\omega_0}}$$

By setting $\omega_0 \approx 3 \times 10^{-13}$ sec^{-1}, $\mu = 0.5m$

we obtain $a_{crit} = \sqrt{\frac{\hbar}{m\omega_0}} = \sqrt{\frac{10^{-27}}{0.5 \times 9.1 \times 10^{-28} \times 3 \times 10^{13}}} \approx 27$ A

If $a < a_{crit}$ the electron-hole pair revolves at frequencies greater than ω_0 and the high-frequency dielectric constant has to be used. If $a << a_{crit}$ the electron-hole pair revolves at frequencies smaller then ω_0 and the low-frequency dielectric constant is the one that has to be used.

For many semiconductors the low-frequency dielectric constant has been used to explain the experimental spectral data.

4. The Intermediate Case. We have examined the two limiting cases: (i) exciton with zero radius (Frenkel exciton) and (ii) exciton with large radius (Wannier exciton). By "large" radius we mean a radius large in comparison to the lattice constant.

The can be seen by considering the following argument. When electron and hole are close to each other the limitation of the spatial dimension of the pair makes it necessary, as dictated by the uncertainty principle, the use of a large number of waves (namely a large portion of the $\vec{k}$ space) in order to describe the exciton. The range of $\vec{k}$ values necessary may actually involve regions of the $\vec{k}$ space where the $\varepsilon(\vec{k})$ dependence for both the valence and the conduction band is not parabolic. For small radii the Wannier model breaks down.

On the other hand, the Frenkel model breaks down for large radii, namely for a solid in which the electrons are not tightly bound to the parent atoms or molecules. As a consequence of the overlap of atomic or molecular orbital, a Frenkel-type of diagonalization of the matrix of the Hamiltonian is not possible and consequently waves of the Frenkel-type do not represent adequately the situation.

Even if the two extreme simplifications are not applicable to the intermediate case, we note that the "general theory" presented in Section V.B.1 is adequately set to handle in principle any case. In particular, Frenkel (atomic-like) and Wannier (non-atomic-like) excitons could occur in the <u>same</u> solid. We may cite the case of solid xenon where the n=1 is a Frenkel exciton and the n=2,3 are Wannier excitons.[5]

An interesting "intermediate" case is the one presented by simple ionic crystals like the alkali halides. The basic excitation due to the absorption of a photon is the transfer of an electron from a halogen atom to a neighboring alkali atom. In this case it is clear that the Frenkel model cannot be used because of the electron transfer and the Wannier model cannot be used because of the small exciton radius. For more details on this subject we refer the reader to [6].

5. The Photon-Exciton System. In analogy with the phonon-photon case we may say the following:

(1) Photons couple with long-wavelength exciton states.

(2) Exciton states are "modes", namely express "resonance" of the system and we associate "oscillators" to them and apply to them the general considerations of Section IV.

(3) There may be "longitudinal" and "transverse" exciton states. This is the case when the ground atomic state is of the 1s type and the excited atomic state is of the 2p type. The relevant quantity, according to Eq.(206) is

$$\vec{p}(\vec{k}_\alpha) = \langle v | \vec{p} e^{i\vec{k}_\alpha \cdot \vec{r}'} | u \rangle \approx \langle v | \vec{p} | u \rangle \tag{322}$$

which has three components proportional to x, y and z. If the polarization vector of the radiation field is in the z-direction only the z-component of $\vec{p}(\vec{k}_\alpha)$ couple with the field.

(4) Finally the photon-transverse exciton interaction creates polaritons, as depicted in Figure 12, where

A = photon in the field, P_0 = pure photon, P_2 = exciton, P_1 = polariton.

6. The Photon-Exciton-Phonon System. In order to have a more complete model of the interactions of the radiation field with a solid we have to include in our treatment both the exciton system and the phonon system. Figure 13 illustrates the situation. We shall assume that the photon energies are such that the switch "infrared absorption" is open and the switch "absorption" is closed. The process of absorption takes place as follows:

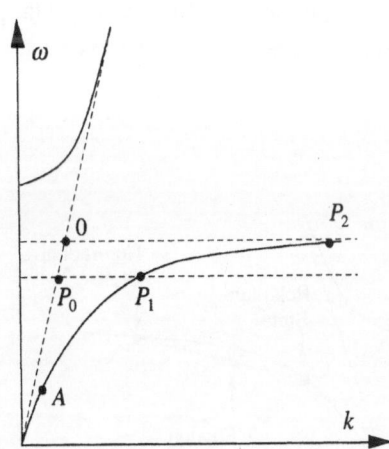

Figure 12. Polariton states due to the interaction of transverse exciton states with electromagnetic radiation.

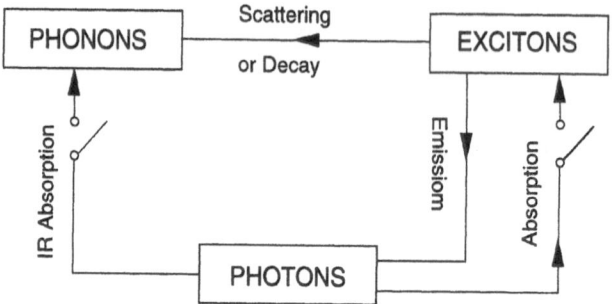

Figure 13. The photon-exciton-phonon system.

(1) An incoming photon excites the solid to a polariton state (or creates a polariton wave packet).

(2) The polariton travels undisturbed and recreates the photon at the surface of solid, or the polariton is scattered by a phonon or a point defect into a nonradiative state. (See Figure 14.)

(3) In the words of Pekar:[16]

"*Absorption becomes possible only as result of exciton decay accompanied by a thermal transition of the system to excited states of another type.*"

7. Indirect Transitions. The presence of lattice vibrations makes it possible to observe phonon-assisted transitions. The selection rules are simple determined by the conservation of linear momentum and of energy. We shall call

$\vec{k}_e$ = wave vector of the exciton, E_e = energy of the exciton

$\vec{q}$ = wave vector of the phonon, $\hbar\omega_q$ = energy of the phonon

$\vec{k}_\alpha$ = wave vector of the photon

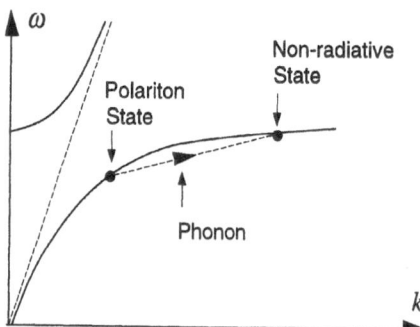

Figure 14. Scattering of a polariton into a nonradiative state.

and here are the possible cases:

(a) Creation of an exciton, accompanied by the creation of a phonon

$$\begin{cases} \vec{k}_e = -\vec{q} \\ \hbar c k_\alpha = E_e + \hbar \omega_q \end{cases} \tag{323}$$

(b) Creation of an exciton, accompanied by the annihilation of a phonon

$$\begin{cases} \vec{k}_e = \vec{q} \\ \hbar c k_\alpha = E_e - \hbar \omega_q \end{cases} \tag{324}$$

This process does not take place at very low temperature (when there are no phonons around).

(c) Annihilation of an exciton, accompanied by the creation of a phonon

$$\begin{cases} \vec{k}_e = \vec{q} \\ \hbar c k_\alpha = E_e - \hbar \omega_q \end{cases} \tag{325}$$

(d) Annihilation of an exciton, accompanied by the annihilation of a phonon

$$\begin{cases} \vec{k}_e = -\vec{q} \\ \hbar c k_\alpha = E_e + \hbar \omega_q \end{cases} \tag{326}$$

This process does not take place at very low temperature.

The cases (a) and (b) are illustrated in Figure 15. Since we are dealing with phonons near the zone boundary, the phonon energy $\hbar \omega_q$ is nearly independent of $\vec{q}$.

Near T=0, only phonon creation is relevant. By considering exciton states with energy E_e, it is evident that a band exists in absorption for these temperature. The absorption edge starts at $E_e + \hbar \omega_q$ and extends to higher energies. As temperature is raised, a new absorption edge appears at $E_e - \hbar \omega_q$ corresponding to phonon absorption. As a result, the absorption spectrum is now comprised of two absorption edges separated by $2\hbar \omega_q$.

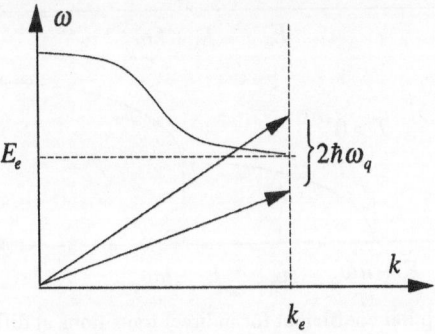

Figure 15. Indirect transitions.

One can therefore discern the presence of such indirect transitions by examining the temperature dependence and separation of the absorption edges. A thorough review of both the theoretical and experimental aspects of indirect transitions is given by Mclean[17]. The effects of temperature on the absorption coefficient are illustrated in Figure 16.

V.C.   Magnons

1. Setting of the Problem. Consider a system consisting of N spins coupled ferromagnetically. This system may be realized by N atoms with orbital angular momentum zero and spin angular momentum S=1/2. The ground state of this system consists of all spins lining in the same direction, say z.

An excitation of this system may be produced by raising the temperature by a small amount so that one spin is flipped, Since each spin has equal probability of being flipped, a spin wave perturbation will travel throughout the system.

If the temperature is increased further, two spins may be flipped and two spin waves will travel with different velocities, will meet and will scatter. If the flipped spins are adjacent we have what is called a spin complex and a situation in which the energy is lower than that when the flipped spins are separated. If more than two spins are flipped, more scattering may be present and larger spin complexes may be formed.

The basic approximation that is made in spin wave theory is that spin waves are independent of each other; this approximation is clearly valid for $T << T_f$. T_f, called the ferromagnetic Curie temperature, is the temperature above which all ferromagnetic order is destroyed, and the system behaves as a paramagnet.

Within the limits of this approximation the total energy of the spin waves is equal to the sum of the energies of the individual spin waves. Dyson has shown that the error in the calculation of the magnetization when the above approximation is used is less than 5% for T=0.5 T_f.[18]

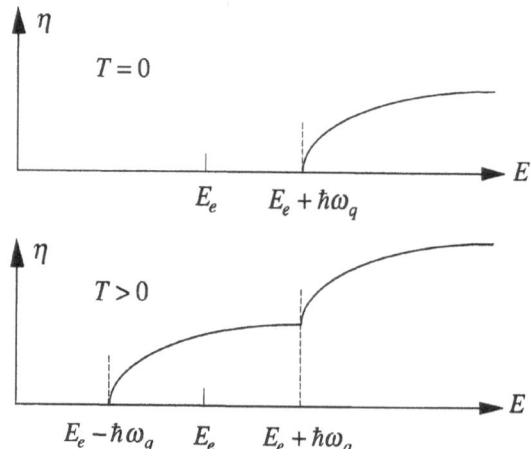

Figure 16. Absorption coefficient for indirect transitions at different temperatures.

2. Hamiltonian and Eigenstates.

The relevant Hamiltonian of the coupled spin system is given by

$$H = -2J \sum_{l,m}' \vec{S}_l \cdot \vec{S}_m \tag{327}$$

where the prime in the sum excludes the term with l=m, and where J>0. J=230k for Ni and 205k for Fe (k=Boltzmann constant). $\vec{S}$ are operators represented by the <u>Pauli matrices</u>:

$$\begin{cases} S^x = \dfrac{1}{2}\begin{pmatrix} 0 & 1 \\ 1 & 0 \end{pmatrix} \\[3mm] S^y = \dfrac{1}{2}\begin{pmatrix} 0 & -i \\ i & 0 \end{pmatrix} \\[3mm] S^z = \dfrac{1}{2}\begin{pmatrix} 1 & 0 \\ 0 & -1 \end{pmatrix} \end{cases} \tag{328}$$

The eigenfunctions of S^2 and S^z are

$$\alpha = \begin{pmatrix} 1 \\ 0 \end{pmatrix}; \qquad S^2\alpha = \frac{3}{4}\alpha, \qquad S_z\alpha = \frac{1}{2}\alpha \tag{329}$$

$$\beta = \begin{pmatrix} 0 \\ 1 \end{pmatrix}; \qquad S^2\beta = \frac{3}{4}\beta, \qquad S_z\beta = -\frac{1}{2}\beta \tag{330}$$

We define

$$\begin{cases} S^+ = S^x + iS^y = \begin{pmatrix} 0 & 1 \\ 0 & 0 \end{pmatrix} \\[3mm] S^- = S^x - iS^y = \begin{pmatrix} 0 & 0 \\ 1 & 0 \end{pmatrix} \end{cases} \tag{331}$$

It is

$$\begin{cases} S^+\beta = \alpha \\ S^-\alpha = \beta \end{cases} \tag{332}$$

Note also that

$$S_l^x S_m^x + S_l^y S_m^y = \frac{1}{2}\left(S_l^+ S_m^- + S_l^- S_m^+\right) \tag{333}$$

Therefore we can write

$$H = -2J \sum_{l,m}' \vec{S}_l \cdot \vec{S}_m = -2J \sum_{l,m}' \left[S_l^x S_m^x + S_l^y S_m^y + S_l^z S_m^z\right]$$

$$= -J \sum_{l,m}' \left[S_l^+ S_m^- + S_l^- S_m^+ + 2S_l^z S_m^z\right] \tag{334}$$

The ground state of the system is

$$\psi_G = |\alpha(1)\alpha(2)\alpha(3)\ldots\ldots\alpha(N)\rangle \tag{335}$$

71

When we operate on this state with H we obtain

$$H\psi_G = -J\sum_{l,m}{}' \left[S_l^+ S_m^- + S_l^- S_m^+ + 2S_l^z S_m^z \right] \; |\alpha(1)\alpha(2)\alpha(3)......\alpha(N)\rangle$$

$$= -J\frac{1}{4}nN\psi_G = E_0\psi_G \tag{336}$$

$E_0 = -J(1/4)nN$ is the energy of the ground state, n is the number of nearest neighbors. Also, we have counted only once the interaction between two atoms; from this the extra factor of 1/2. The state of the system when one spin is flipped is described by the wavefunction

$$\psi_1(p) = |\alpha(1)\alpha(2)......\beta(p)......\alpha(N)\rangle \tag{337}$$

When we operate on this state with H we obtain

$$H\psi_1(1) = -J\sum_{l,m}{}' \left[S_l^+ S_m^- + S_l^- S_m^+ + 2S_l^z S_m^z \right] \; |\alpha(1)\alpha(2)......\beta(p)......\alpha(N)\rangle$$

$$= -J\left[\sum_{\substack{l,m \\ l\neq p, m\neq p}}{}' S_l^z S_m^z + \sum_{r=1}^{n} S_p^+ S_{p+r}^- + \sum_{r=1}^{n} S_p^z S_{p+r}^z \right] \; |\alpha(1)\alpha(2)......\beta(p)......\alpha(N)\rangle$$

$$= \left[-\frac{1}{4}n(N-2)J + \frac{1}{2}nJ \right]\psi_1(p) - J\sum_{r=1}^{n}\psi_1(p+r)$$

$$= \left(-\frac{1}{4}nN + n \right)J\psi_1(p) - J\sum_{r=1}^{n}\psi_1(p+r) \tag{338}$$

or

$$H\psi_1(p) = A\psi_1(p) - J\sum_{r=1}^{n}\psi_1(p+r) \tag{339}$$

where

$$A = nJ\left(1 - \frac{N}{4} \right) \tag{340}$$

The N functions $\psi_1(p)$ $(p=1,2,3,......N)$ are <u>not</u> eigenfunctions of H.

We can form linear combinations of $\psi_1(p)$ functions as follows:

$$\psi_1(\vec{k}) = \sum_{p=1}^{N} e^{i\vec{k}\cdot\vec{R}_p}\psi_1(p) \tag{341}$$

where $\vec{k}$ is determined by the usual periodic boundary conditions.

Now we obtain

$$H\psi_1(\vec{k}) = \sum_{p=1}^{N} e^{i\vec{k}\cdot\vec{R}_p}H\psi_1(p) = \sum_{p=1}^{N} e^{i\vec{k}\cdot\vec{R}_p}\left[A\psi_1(p) - J\sum_{r=1}^{n}\psi_1(p+r) \right]$$

$$= A\psi_1(\vec{k}) - J\sum_{r=1}^{n}\sum_{p=1}^{N} e^{i\vec{k}\cdot\vec{R}_p}\psi_1(p+r) \tag{342}$$

But

$$\sum_{p=1}^{N} e^{i\vec{k}\cdot\vec{R}_p}\psi_1(p+r) = \sum_{p=1}^{N} e^{i\vec{k}\cdot\vec{R}_{p+r}}\psi_1(p+r)e^{i\vec{k}\cdot(\vec{R}_p-\vec{R}_{p-r})}$$

$$= \sum_{p=1}^{N} e^{i\vec{k}\cdot\vec{R}_p}\psi_1(p)e^{i\vec{k}\cdot(\vec{R}_{p+r}-\vec{R}_p)} = \psi_1(\vec{k})e^{i\vec{k}\cdot\vec{r}} \tag{343}$$

where $\vec{r}$ = vectorial distance from the given atom to a nearest neighbor. Therefore

$$H\psi_1(\vec{k}) = A\psi_1(\vec{k}) - J\left(\sum_r e^{i\vec{k}\cdot\vec{r}}\right)\psi_1(\vec{k}) = \left(-n\frac{NJ}{4} + nJ - J\sum_r e^{i\vec{k}\cdot\vec{r}}\right)\psi_1(\vec{k})$$

$$= \left[E_0 + J\left(n - \sum_r e^{i\vec{k}\cdot\vec{r}}\right)\right]\psi_1(\vec{k}) = E_{\vec{k}}\psi_1(\vec{k}) \tag{344}$$

where

$$E_{\vec{k}} = E_0 + J\left(n - \sum_r e^{i\vec{k}\cdot\vec{r}}\right) \tag{345}$$

is the dispersion relation. For small values of $\vec{k}$

$$e^{i\vec{k}\cdot\vec{r}} = 1 + i\vec{k}\cdot\vec{r} - \frac{(\vec{k}\cdot\vec{r})^2}{2} \tag{346}$$

$$\sum_r e^{i\vec{k}\cdot\vec{r}} = n - \frac{1}{2}\sum_r (\vec{k}\cdot\vec{r})^2 \tag{347}$$

and

$$E_{\vec{k}} - E_0 = \frac{1}{2}J\sum_r (\vec{k}\cdot\vec{r})^2 \tag{348}$$

where $|\vec{k}\cdot\vec{r}| \ll 1$. For cubic symmetry

$$\sum_r (\vec{k}\cdot\vec{r})^2 = \frac{1}{3}k^2\sum_r r^2 \tag{349}$$

and

$$E_{\vec{k}} - E_0 = \frac{1}{6}Jk^2\sum_r r^2 = Jk^2 a^2 \tag{350}$$

where a = lattice constant.

The system of spin waves can be thought of as a collection of harmonic oscillators. Second quantization techniques can be applied and can allow us to express the dispersion relation (345) as follows:

$$\hbar\omega = \left(n - \sum_r e^{i\vec{k}\cdot\vec{r}}\right)J \tag{351}$$

The z-component of the total spin is easily obtained by operating on the wavefunction of the spin system by means of the operator

$$S_z = \sum_{i=1}^{N} S_i^z$$

By operating with S_z on ψ_G of Eq.(335) we find that the z-component of the total spin when the system is in the ground state is

$$M_s = N\frac{1}{2} = NS$$

The z-component of the total spin when a spin wave vector $\vec{k}$ is excited is obtained as follows:

$$S_z \psi_1(\vec{k}) = \sum_i S_i^z \sum_p e^{i\vec{k}\cdot\vec{R}_p} \psi_1(p) = \sum_p e^{i\vec{k}\cdot\vec{R}_p} \sum_i S_i^z \psi_1(p)$$

$$= \sum_p e^{i\vec{k}\cdot\vec{R}_p} (N-2)\frac{1}{2}\psi_1(p) = \left(N\frac{1}{2}-1\right)\psi_1(\vec{k}) \qquad (352)$$

Therefore when a spin wave is excited

$$M_s = NS - 1 \qquad (353)$$

and the z-component of each spin will be (S-1/N).

If $n_{\vec{k}}$ spin waves of wave vector $\vec{k}$ are excited the z-component of the total spin is

$$M_s = NS - n_{\vec{k}} \qquad (354)$$

and the z-component of each spin will be

$$\frac{M_s}{N} = S - \frac{n_{\vec{k}}}{N} \qquad (355)$$

3. <u>Semiclassical Treatment</u>[10,19]. In this approximation each spin is a vector of length $[S(S+1)]^{1/2}$ precessing about the z-axis. If a number of spin waves $n_{\vec{k}}$ of wave vector $\vec{k}$ are excited then each spin precesses about the z-axis describing a circles of amplitude u as in Figure 17.

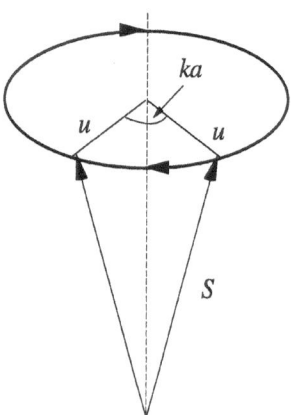

Figure 17. Precession of a spin about the z-axis.

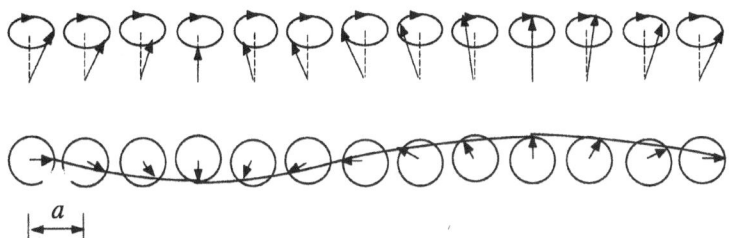

Figure 18. Precession of a linear string of spins[10,19].

By approximating $[S(S+1)]^{1/2}$ with S the z-component of a spin for u << S is given by

$$\sqrt{S-u^2} \approx S - \frac{u^2}{2S} \qquad (356)$$

We have to set this equal to the expression (354):

$$S - \frac{u^2}{2S} = S - \frac{n_{\bar{k}}}{N} \qquad (357)$$

and we find, now calling u by a new name, $u_{\bar{k}}$,

$$u_{\bar{k}}^2 = \frac{2Sn_{\bar{k}}}{N} \qquad (358)$$

The radius $u_{\bar{k}}$ is the same for all spins. The precession of a linear string of spins is now illustrated in Figure 18.

4. Thermodynamics of Magnons. The energy of the spin-wave system is given by

$$E_{n_{\bar{k}_1} \, n_{\bar{k}_2} \, n_{\bar{k}_3} \, \cdots \cdots \, n_{\bar{k}_N}} = \sum_{\bar{k}} n_{\bar{k}} E_{\bar{k}} \qquad (359)$$

We simplify the notation as follows:

$$n_{\bar{k}} = n_i \qquad (360)$$
$$E_{\bar{k}} = E_i \qquad (361)$$

Then we can write the partition sum of the system as follows:

$$Z = \sum_r e^{-\beta E_r} = \sum_{n_1 n_2 \ldots n_N} e^{-\beta(n_1 E_1 + n_2 E_2 \ldots \ldots n_N E_N)}$$

$$= \sum_{n_1=0}^{\infty} e^{-\beta n_1 E_1} \sum_{n_2=0}^{\infty} e^{-\beta n_2 E_2} \ldots \ldots \ldots \sum_{n_N=0}^{\infty} e^{-\beta n_N E_N}$$

$$= \frac{1}{1-e^{-\beta E_1}} \frac{1}{1-e^{-\beta E_2}} \ldots \ldots \ldots \frac{1}{1-e^{-\beta E_N}} \qquad (362)$$

and

$$\ln Z = -\sum_{i=1}^{N} \ln\left(1 - e^{-\beta E_i}\right) \qquad (363)$$

75

where $\beta = 1/kT$. Therefore the internal energy of the system is given by

$$\overline{E} = -\frac{\partial \ln Z}{\partial \beta} = \sum_{\vec{k}} \frac{E_{\vec{k}}}{e^{\beta E_{\vec{k}}} - 1} = \sum_{\vec{k}} \overline{n}(E_{\vec{k}}) E_{\vec{k}} \qquad (364)$$

where

$$\overline{n}(E_{\vec{k}}) = \frac{1}{e^{E_{\vec{k}}/kT} - 1} \qquad (365)$$

represents the average number of spin waves. We can also think of a spin wave of wave vector $\vec{k}$ as a particle called __magnon__ of linear momentum $\hbar\vec{k}$ and energy $E_{\vec{k}}$; $\overline{n}(E_{\vec{k}})$ is then the average number of magnons of energy $E_{\vec{k}}$ present at temperature T.

Magnons are clearly bosons since the interchange of two reversed spins does not change the state of the system. Indeed, Eq.(365) expresses the statistics one would expect for bosons whose total number is not fixed.

The number of allowed $\vec{k}$ values in $(\vec{k}, \vec{k} + d\vec{k})$ is given by

$$g(\vec{k})d^3\vec{k} = \frac{V}{8\pi^3} d^3\vec{k} \qquad (366)$$

The number of allowed $\vec{k}$ values with $k = |\vec{k}|$ in $(k, k + dk)$ is given by

$$g(k)dk = 4\pi k^2 \frac{V}{8\pi^3} dk = \frac{Vk^2}{2\pi^2} dk \qquad (367)$$

The number of magnons with energy in $(E, E + dE)$ is given by

$$g(E)dE = g(k)dk = g(k)\frac{dk}{dE} dE$$

$$= \frac{Vk^2}{2\pi^2} \frac{dk}{d(Ja^2k^2)} dE = \frac{V}{2\pi^2} \frac{k}{2Ja^2} dE = \frac{V}{4\pi^2} \left(\frac{1}{Ja^2}\right)^{3/2} E^{1/2} dE \qquad (368)$$

where we have used the dispersion relation (350).

The total number of magnons is given by

$$\sum_{\vec{k}} \overline{n}(E_{\vec{k}}) = \int_0^{E_{max}} \overline{n}(E)g(E)dE = \frac{V}{4\pi^2} \left(\frac{1}{Ja^2}\right)^{3/2} \int_0^{E_{max}} \frac{E^{1/2}}{e^{E/kT} - 1} dE$$

$$= \frac{V}{4\pi^2} \left(\frac{1}{Ja^2}\right)^{3/2} (kT)^{3/2} \int_0^{E_{max}/kT} \frac{x^{1/2}}{x^{E/kT} - 1} dx \qquad (369)$$

For $T \ll E_{max}/k$ we have

$$\sum_{\vec{k}} \overline{n}(E_{\vec{k}}) = \frac{V}{4\pi^2} \left(\frac{1}{Ja^2}\right)^{3/2} (kT)^{3/2} \int_0^{\infty} \frac{x^{1/2}}{x^{E/kT} - 1} dx$$

$$= 0.1174 V \left(\frac{kT}{Ja^2}\right)^{3/2} \qquad (370)$$

Example

$$J = 205k; \qquad a = 2 \times 10^{-8} cm; \qquad T = 20^{\circ}K$$

$$\frac{\sum \bar{n}}{V} = 0.1174 \left(\frac{kT}{Ja^2} \right)^{3/2} = \left(0.1174 \frac{k \times 20}{205k \times \left(2 \times 10^{-8} \right)^2} \right)^{3/2} \approx 4.5 \times 10^{20} cm^{-3}$$

The change of the spontaneous magnetization due to the excitation of spin waves is

$$\frac{\Delta M}{M(0)} = \frac{M(0) - M(T)}{M(0)} = \frac{\sum \bar{n}}{N} = 0.1174 V \left(\frac{kT}{Ja^2} \right)^{3/2} \frac{V}{N} \tag{371}$$

But

$$N = V \frac{Q}{a^3} \tag{372}$$

where Q = 1, 2, and 4 for simple cubic, body-centered cubic and face-centered cubic lattice, respectively. Then

$$\frac{\Delta M}{M(0)} = \frac{\sum \bar{n}}{N} = 0.1174 \left(\frac{kT}{Ja^2} \right)^{3/2} \frac{a^3}{Q} = \frac{0.1174}{Q} \left(\frac{kT}{J} \right)^{3/2} = A T^{3/2} \tag{373}$$

where $A = [(0.1174 / Q) \times (k / J)^{3/2}]$. The above formula expresses the so-called Block $T^{3/2}$ law.

The internal energy of the system is given by

$$\sum_{\vec{k}} \bar{n}(E_{\vec{k}}) E_{\vec{k}} = \int_0^{E_{max}} \bar{n}(E) E g(E) dE = \frac{V}{4\pi^2} \left(\frac{1}{Ja^2} \right)^{3/2} \int_0^{E_{max}} \frac{E^{3/2}}{e^{E/kT} - 1} dE$$

$$= \frac{V}{4\pi^2} \left(\frac{1}{Ja^2} \right)^{3/2} (kT)^{5/2} \int_0^{E_{max}/kT} \frac{x^{3/2}}{e^x - 1} dE \tag{374}$$

For $T \ll E_{max}/k$ we have

$$\sum_{\vec{k}} \bar{n}(E_{\vec{k}}) E_{\vec{k}} \approx \frac{V}{4\pi^2} \left(\frac{1}{Ja^2} \right)^{3/2} (kT)^{5/2} \int_0^\infty \frac{x^{3/2}}{e^x - 1} dE = \frac{1.7844}{4\pi^2} \frac{V}{Ja^2} (kT)^{5/2} \tag{375}$$

and the specific heat per unit volume of the magnon system is

$$\frac{1.7844}{4\pi^2} \left(\frac{1}{Ja^2} \right)^{3/2} k^{5/2} \frac{5}{2} T^{3/2} = 0.113 \left(\frac{T}{Ja^2} \right)^{3/2} k^{5/2} \tag{376}$$

V.D. Plasmons

1. Dielectric Response of an Ensemble of Oscillators. It is appropriate at this point to summarize some of the results obtained in Section IV. We dealt then with a system of N particles of mass m and negative charge e, each elastically bound to an equilibrium position where a positive charge $|e|$ resides; the angular frequency of oscillation is ω_0 and γ is the damping constant. If the system is under the action of a plane wave

$$\vec{E}(\vec{r}, t) = \vec{E}_0(\vec{r}) e^{i\omega t} \tag{377}$$

then the particles oscillate displacing themselves from their equilibrium positions by the amount

$$\vec{z}(\vec{r},t) = \frac{e/m}{\omega_0^2 - \omega^2 + i\gamma\omega} \vec{E}_0(\vec{r})e^{i\omega t} \qquad (378)$$

The induced dipole moment is

$$e\vec{z}(\vec{r},t) = \alpha(\omega)\vec{E}(\vec{r},t) \qquad (379)$$

where

$$\alpha(\omega) = polarizability = \frac{e^2/m}{\omega_0^2 - \omega^2 + i\gamma\omega} \qquad (380)$$

The real part of the induced dipole moment is

$$\mathrm{Re}\left[e\vec{z}(\vec{r},t)\right] = \frac{e^2}{m} \frac{\vec{E}_0(\vec{r})}{\sqrt{\left(\omega_0^2 - \omega^2\right)^2 + \gamma^2\omega^2}} \cos(\omega t - \phi) \qquad (381)$$

where

$$\tan\phi = \frac{\gamma\omega}{\omega_0^2 - \omega^2} \qquad (382)$$

$\phi = 0, \ \pi/2 \ and \ \pi$ for $\omega = 0, \ \omega_0 \ and \ \infty$, respectively.

The polarization is given by

$$\vec{P}(\vec{r},t) = n_0\alpha(\omega)\vec{E}(\vec{r},t) = \chi(\omega)\vec{E}(\vec{r},t) \qquad (383)$$

where

$$\chi(\omega) = dielectric \ susceptibility = n_0\alpha(\omega) \qquad (384)$$

$$n_0 = N/V \qquad (385)$$

The electric displacement is given by

$$\vec{D}(\vec{r},t) = \vec{E}(\vec{r},t) + 4\pi\vec{P}(\vec{r},t) = \left[1 + 4\pi\chi(\omega)\right]\vec{E}(\vec{r},t) = K(\omega)\vec{E}(\vec{r},t) \qquad (386)$$

where the dielectric constant $K(\omega)$ can be written

$$K(\omega) = 1 + \frac{4\pi n_0 e^2/m}{\omega_0^2 - \omega^2 + i\gamma\omega} \qquad (387)$$

In the limit of $\gamma = 0$

$$K(\omega) = 1 + \frac{4\pi n_0 e^2/m}{\omega_0^2 - \omega^2} \qquad (388)$$

This function is represented in Figure 19. We note that

$$K(\omega) < 0 \quad for \quad \omega_0 < \omega < \omega_L \qquad (389)$$

where

$$\omega_L = \omega_0\sqrt{1 + \frac{4\pi n_0 e^2/m}{\omega_0^2}} \qquad (390)$$

and

$$K(\omega_L) = 0 \qquad (391)$$
$$K(\infty) = 1 \qquad (392)$$
$$K(0) - K(\infty) = \frac{4\pi n_0 e^2/m}{\omega_0^2} \qquad (393)$$

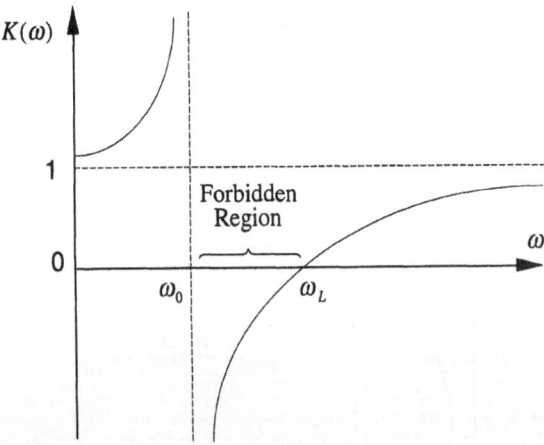

Figure 19. Dielectric constant of an ensemble of oscillators.

2. Dielectric Response of a Free Electron Gas. In order to handle this case we set in the expressions we reported

$$\omega_0 = 0, \quad \gamma = 0 \tag{394}$$

and find that, in this case, under the action of the field $\vec{E}(\vec{r},t)$ given in Eq.(377)

$$e\vec{z}(\vec{r},t) = \alpha(\omega)\vec{E}(\vec{r},t) \tag{395}$$

where

$$\alpha(\omega) = -\frac{e^2}{m\omega} \quad (\phi = 0) \tag{396}$$

The polarization is given by

$$\vec{P}(\vec{r},t) = n_0 \alpha(\omega)\vec{E}(\vec{r},t) = \chi(\omega)\vec{E}(\vec{r},t) \tag{397}$$

where

$$\chi(\omega) = n_0 \alpha(\omega) = -\frac{n_0 e^2}{m\omega^2} \tag{398}$$

The electric displacement is

$$\vec{D}(\vec{r},t) = K(\omega)\vec{E}(\vec{r},t) = \left[1 + 4\pi\chi(\omega)\right]\vec{E}(\vec{r},t) \tag{399}$$

where the dielectric constant $K(\omega)$ is

$$K(\omega) = 1 - \frac{4\pi n_0 e^2 / m}{\omega_0^2} \tag{400}$$

This function is represented in Figure 20. We note that

$$K(\omega_L) = 0 \tag{401}$$

where

$$\omega_L = \sqrt{\frac{4\pi n_0 e^2}{m}} = \omega_p = plasma\ frequency \tag{402}$$

and $K(\omega)$ can be written

$$K(\omega) = 1 - \frac{\omega_L^2}{\omega^2} \tag{403}$$

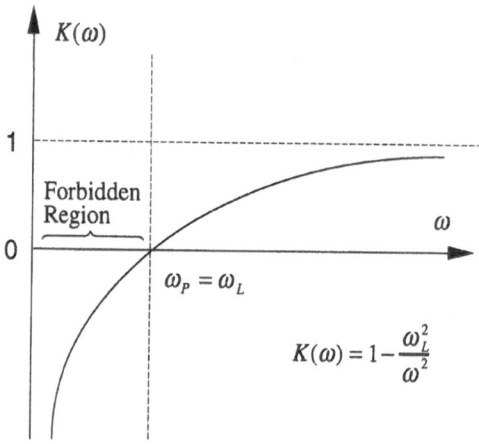

Figure 20. Dielectric constant of a free electron gas.

3. <u>Transverse Optical Modes in a Plasma</u>. A <u>plasma</u> can be defined as a medium with equal concentration of positive and negative charges, of which at least one charge type is mobile.[10] In solids such a medium may be realized by the conduction electrons whose negative charges are balanced by the positive charges of the ion cores.

The dispersion relation for transverse electromagnetic waves in a plasma is given by

$$\omega = \frac{ck}{\sqrt{K}} = \frac{ck}{\sqrt{1 - \frac{\omega_p^2}{\omega^2}}} \tag{404}$$

or

$$\omega^2 = \omega_p^2 + c^2 k^2 \tag{405}$$

or

$$\frac{\omega}{\omega_p} = \sqrt{1 + \left(\frac{ck}{\omega_p}\right)^2} \tag{406}$$

The last expression is represented in Figure 21. ω cannot be smaller than ω_p for transverse electromagnetic waves; when such waves have $\omega < \omega_p$ and are incident on the plasma they are reflected. If $\omega > \omega_p$, $K(\omega) < 0$, as in Figure 20.

If we use for e and m the values of the charge and mass of the electron, respectively, we find

$$\omega_p = \omega_L = \sqrt{\frac{4\pi n_0 e^2}{m}} \approx 5.7 \times 10^4 \sqrt{n_0} \tag{407}$$

and

$$\lambda_p = \frac{2\pi c}{\omega_p} \approx \frac{3.32 \times 10^6}{\sqrt{n_0}} \tag{408}$$

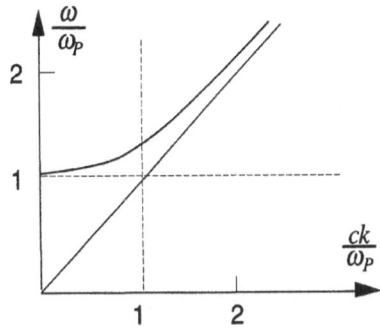

Figure 21. Dispersion relation for transverse electromagnetic waves in a plasma[10].

Example of ω_p and λ_p are given in the following table.

$n_0 \ (cm^{-3})$	10^{22}	10^{18}	10^{14}	10^{10}
$\omega_p \ (sec^{-1})$	5.7×10^{15}	5.7×10^{13}	5.7×10^{11}	5.7×10^{9}
$\lambda_p \ (cm)$	3.3×10^{-5}	3.3×10^{-3}	0.33	33

λ_p is the cut-off wavelength; no radiation with $\lambda < \lambda_p$ can be transmitted through the plasma.

   This result is in accord with the experimental data; for alkali metals it is found that a metal film is transparent only if $\lambda < \lambda_p$. We report in the following table, from [10], the relevant information.

	Li	Na	K	Rb	Cs
$\lambda_p \ (A)$ (calculated)	1550	2090	2870	3220	3620
$\lambda_p \ (A)$ (observed)	1550	2100	3150	3400	-

The optical properties of metals are reviewed in [20].

   We note also that the reflection of light from metals is similar to the reflection of radio waves from the ionosphere.

   4. <u>Longitudinal Optical Modes in a Plasma</u>. The electron gas undergoes a free (longitudinal) oscillation at $\omega = \omega_p$, the cut-off frequency of the transverse electromagnetic waves.

The electron move with respect to the positive background:

$$\vec{P} = -n_0 e \vec{z} \qquad (409)$$

But $K(\omega_p) = 0$, then

$$\vec{D} = K\vec{E} = \vec{E} + 4\pi\vec{P} = 0 \qquad (410)$$

and

$$\vec{E} = -4\pi\vec{P} = 4\pi n_0 e \vec{z} \qquad (411)$$

The equation of motion of the electrons in the unit volume is

$$n_0 m \frac{d^2\vec{z}}{dt^2} = -n_0 e\vec{E} = -4\pi n_0^2 e^2 \vec{z} \qquad (412)$$

or

$$\frac{d^2\vec{z}}{dt^2} + \frac{4\pi n_0 e^2}{m}\vec{z} = 0 \qquad (413)$$

or

$$\frac{d^2\vec{z}}{dt^2} + \omega_p^2\vec{z} = 0 \qquad (414)$$

where ω_p is given by Eq.(402).

A plasma oscillation of small wavevector has $\omega \approx \omega_p$. The dispersion relation for longitudinal oscillators will not be derived here; the reader is referred to Pines[21] who gives

$$\omega = \omega_p \left(1 + \frac{3v_F^2}{10\omega_p^2}k^2 + \ldots\ldots\ldots\ldots \right) \qquad (415)$$

where v_F = Fermi velocity.

Plasma oscillations in metals are collective longitudinal excitations of the gas which consists of the conduction electrons. A <u>plasmon</u> is a quantum of plasma oscillation. The typical energy of a plasmon can be obtained by using for n_0 the typical metallic density of $n_0 \approx 10^{23}$. We obtain then

$$\omega_p = 5.7 \times 10^4 \ \sqrt{n} = 5.7 \times 10^4 \ \sqrt{10^{23}} \approx 2 \times 10^{16} \ \ sec^{-1}$$
$$\hbar\omega_p = 10^{-27} \times 2 \times 10^{16} = 2 \times 10^{11} \ \ ergs \approx 12eV$$

Therefore no thermal excitation of plasmons can take place.

Plasmons can be produced by passing an electron through a thin metallic film or reflecting an electron from a film. The charge of the electron couples with the fluctuations of the $\vec{E}$ field. The transmitted or reflected electron shows an energy loss equal to $n(\hbar\omega_p)$ where n = number of plasmons created.

We note also that there may be "surface" plasmons, different in ω_p from "volume" plasmons.

Finally, plasma oscillations can be excited in dielectric film of , say, Si, Ge, InSb. In a dielectric all the valence band electrons oscillate with respect to the ion cores.

The following table presents some typical values of plasmon energies.

Materials	Observed $\hbar\omega_p$ (in eV)	Calculated $\hbar\omega_p$ (in eV)
Li	7.12	7.96
Na	5.71	5.58
K	3.72	3.86
Mg	10.6	10.9
Al	15.3	15.8
Dielectrics		
Si	16.4 - 16.9	16
Ge	16 - 16.4	16
InSb	12 - 13	12

ACKNOWLEDGMENT

The author would like to thank Mr. Gang Lei for helpful discussions on the subject of collective excitations and for the accurate typing of this article and for preparation of the figures.

REFERENCES

1. D. L. Dexter, J. Chem. Phys., **21**, 836(1953).

2. B. Di Bartolo, *Spectroscopy of the Excited State,* B Di Bartolo, ed., Plenum Press, New York, 1976.

3. B. Di Bartolo, *Luminescence of Inorganic Solids*, B Di Bartolo, ed., Plenum Press, New York, 1978.

4. A. von Hippel, *Dielectrics and Waves*, Wiley, New York, 1954.

5. D. L. Dexter and R. S. Knox, *Excitons*, Wiley Interscience, New York, 1965.

6. R. S. Knox, *Theory of Excitons*, Academy Press, New York and London, 1963.

7. M.Born and K.Huang, *Dynamical Theory of Crystal Lattices*, Oxford, London, 1954.

8. A. D. B. Woods, et al., Phys. Rev., **131**, 1025(1963).

9. J. L. T. Waugh and G. Dolling, Phys. Rev., **132**, 2410(1963).

10. C. Kittel, *Introduction to Solid State Physics*, Wiley, New York, 1971.

11. J. Frenkel, Phys. Rev., **37**, 17(1931); Phys. Rev., **37**, 1276(1931).

12. B. Di Bartolo, *Optical Interactions in Solids*, Wiley, New York, 1968.

13. H. C. Wolf, *Solid State Physics*, **9**, 1(1959).

14. J. Hegarty, Ph.D. Thesis, University College, Galway, Ireland(1976), unpublished.

15. G. H. Wannier, Phys. Rev., **52**, 191(1937).

16. S. I. Pekar, Soviet Physics JETP, **11**, 1286(1960).

17. T. P. Mclean, *Progress in Semiconductors* A.F. Gibson et al., eds., Vol.5, Wiley, New York, 1961.

18. F. J. Dyson, Phys. Rev., **102**, 1217(1956); and Phys. Rev., **102**, 1230(1956).

19. A. H. Morrish, *The Physical Principles of Magnetism*, Wiley, New York, 1965.

20. M. P. Givens, *Solid State Physics*, **6**, 313(1958).

21. D. Pines, *Elementary Excitations in Solids*, Benjamin, New York, 1963.

LIGHT-MATTER INTERACTION · EXPERIMENTAL ASPECTS

Claus F. Klingshirn

Institut für Angewandte Physik
der Universität Karlsruhe
Kaiserstraße 12
D-76128 Karlsruhe
Germany

ABSTRACT

After the introduction of the concept of collective excitations in the preceeding contribution we stress in this part various aspects of light matter interaction in semiconductors, essentially from an experimental point of view. In doing so, we consider first the optical properties of a classical ensemble of oscillators and proceed then to quantum mechanics. We treat the weak and strong coupling approaches between light and the collective excitations of matter and shall see, how the second approach leads to the concept of polaritons. We outline some possibilities to measure the dispersion relation of these polaritons directly. In the next parts we concentrate on the electronic system of semiconductors and present the luminescence processes under increasing excitation, introducing in this context the so-called many-particle effects.

In the last part we outline how the light emission from highly excited systems can be used to produce light emitting devices like laser diodes. We restrict ourselves mainly to quasistationary excitation conditions, while experimental and theoretical aspects of ultrafast spectroscopy will be treated in some of the following contributions.

I. INTRODUCTION

In the preceeding contribution to this school and to his book, B. Di Bartolo has introduced the concept of collective excitations in solids [1]. The quanta of these collective excitations are e.g. the phonons, plasmons, magnons or excitons. They are called quasiparticles, because they exhibit particle and wave character like photons, electrons or other elementary particles in vacuum. The prefix "quasi" has been introduced because the quanta of the elementary excitations considered in this school exist only in a (generally crystalline) solid, but not in vacuum in contrast to the "real" particles.

Spectroscopy and Dynamics of Collective Excitations in Solids
Edited by Di Bartolo, Plenum Press, New York, 1997

Another reason is, that the quantitiy $\hbar\mathbf{k}$ which appears in the dispersion relation $E\,(\mathbf{k})$ is a quasi-momentum, which is conserved only modulo integer multiples of the vectors of the reciprocal lattice. Furthermore the eigenfunctions of the elementary excitations contain usually a plane wave factor exp $\{i\mathbf{k}\mathbf{r}\}$, but are usually no eigenfunctions of the momentum operator $-i\hbar\,\vec{\nabla}$. An example are the Bloch functions in a periodic lattice, which fulfill the Ewald-Bloch theorem:

$$\alpha_k\,(\mathbf{r}) = e^{i\mathbf{k}\mathbf{r}}\,u_k\,(\mathbf{r}) \text{ with } u_k\,(\mathbf{r}) = u_k\,(\mathbf{r}+\mathbf{R})$$

where $\mathbf{R}$ is a translation vector of the periodic lattice. More information on the concept of quasiparticles can be found in textbooks on solid state physics like [2], in the proceedings of an earlier school of this series [3] and in the references given therein.

In this contribution we concentrate from an experimental point of view on the interaction of light with the elementary excitations of the electronic system of semiconductors. We shall start with the classical model of an ensemble of (un-)coupled oscillators and deduce from this model the frequency and $\mathbf{k}$-dependence of the complex dielectric function $\epsilon(\omega)$, of the complex index of refraction $\tilde{n}\,(\omega)$, and absorption (or better extinction) $\alpha\,(\omega)$.

In a next step we elucidate the corrections, which have to be added to these results, if quantum mechanics is applied, stressing the concepts of weak and of strong coupling between the electromagnetic radiation field and the elementary excitations in the semiconductor. The second approach will lead us to the concept of polartions and we shall present several experimental techniques which allow to determine the dispersion relations of these polaritons experimentally.

In section III we concentrate on the light emission from excited semiconductors with emphasis on incoherent emission i.e. luminescence. We give first an overview of the various interaction processes which occur in semiconductors with increasing density of electron-hole pairs, and present then some selected examples including semiconductor structures of reduced dimensionality like quantum-wells or -dots.

In section IV we shall use these knowledges to go into an aspect of applied semiconductor sciences discussing the way of operation of light-emitting devices like laserdiodes.

If not stated otherwise, we concentrate on direct-gap semiconductors because their optical properties are more exciting than the ones of indirect gap materials, at least according to the subjective feeling of the author. To keep this contribution of a finite length we shall deal essentially with intrinsic properties of semiconductors and we assume if not stated otherwise that the duration of the excitation pulses is longer than the lifetime T_1 or at least the phase coherence time T_2 of the excited species so that a quasi-stationary distribution of the collective excitations can develop. In doing so, we are lying the ground for the contributions to this school by Hvam and by Zimmermann who will concentrate on the experimental [4] and theoretical [5] description of ultrafast and coherent phenomena in semiconductor optics.

For further reading of various aspects of semiconductor optics we recommend e.g. [1, 6 - 18] and references given therein.

Since the number of publications on the topic chosen here is extremely large, we can give only a very limited selection of references and we apologize for this shortcoming.

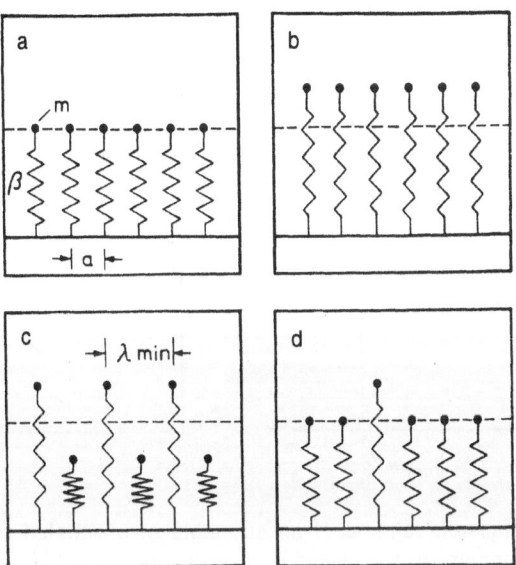

Figure 1. An ensemble of uncoupled oscillators (a), elongated in phase (b) or anti-phase (c) and a wave-packet (d).

II. POLARITONS

In this section we advance from the classical model of an ensemble of oscillators to the polariton concept and its experimental verification.

II. A. Optical properties of an ensemble of oscillators

We consider a three dimensional ensemble of uncoupled oscillators attached to their "lattice points" as sketched for one dimension in Fig 1. These oscillators are characterized by their eigenfrequency ω_0' which is given without damping by

$$\omega_0' = (\frac{\beta}{m})^{\frac{1}{2}} \tag{1}$$

where ß is the spring constant and m the mass.

The uncoupled oscillators will oscillate with this frequency if they are all elongated in phase or in antiphase. The first case corresponds to infinite wavelength or $k=0$ the second one to the shortest, physically meaningful wavelength λ_{min} corresponding to a wavevector

$$k_{max} = 2\pi/\lambda_{min} = \pi/a \tag{2}$$

This is just the boundary of the first Brilloninzone in the reciprocal space of a simple cubic lattice. For more details on this topic see e.g. [2, 15].

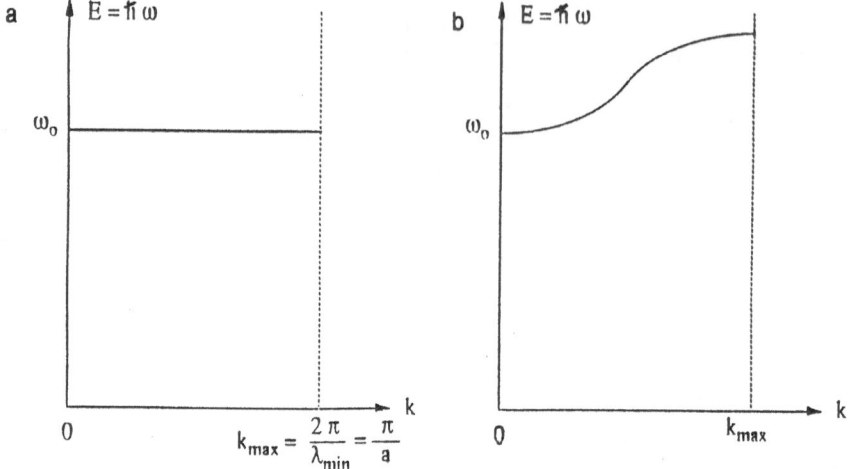

Figures 2 The dispersion relation of an ensemble of uncoupled (a) and of coupled osciallators (b). Compare to Figs 1 and 6, respectively.

Relation (1) holds also for all wavelengths and -vectors between $k = 0$ and k_{max} so that the dispersion relation ω (**k**) is just a horizontal line as shown in Fig 2a. The group velocity of a wave paket v_g is given by (3b)

$$v_{phase} = \frac{\omega}{k} \tag{3a}$$

$$v_{group} = \frac{d\omega}{dk} \tag{3b}$$

and an inspection of Fig 1d or of Fig 2a shows immediately that a wave paket will not move. This is a special example of the general rule

$$\text{zero coupling} \Leftrightarrow \text{zero band width} \Leftrightarrow \text{zero group velocity} \tag{4a}$$
$$\text{finite coupling} \Leftrightarrow \text{finite band width} \Leftrightarrow \text{finite group velocity} \tag{4b}$$

We now want to couple the oscillators to the electric field of an incident plane electromagnetic wave, polarized in z direction and propagating in x direction

$$\mathbf{E} = \mathbf{E}_0 \, e^{i\,(\mathbf{kr}-\omega t)} = (0, 0, E_0) \, e^{i\,(k_x x-\omega t)} \tag{5}$$

To this end, we give every mass a small electric charge e and we place for neutrality reasons the opposit charge in the equilibrium position of every oscillator.

We obtain thus for the equation of motion of the elongation u(t) for every oscillator the expression (7). There we introduced a small damping term $m\gamma u$ and assumed that the spacing or lattice constant a of the oscillators is much smaller than the wavelength λ of the electromagnetic field so that we can neglect the phase shift of **E** from one oscillator to the next. This assumption is equivalent to the dipole approximation in which one expands the term $e^{i\mathbf{kr}}$ in a power series and truncates after the constant term

$$e^{ikr} = 1 + kr + \frac{(kr)^2}{2!} + \dots \approx 1 \tag{6}$$

resulting in

$$m\ddot{u} - \gamma m\dot{u} + \beta u = e\, E_0\, e^{-i\omega t} \tag{7}$$

The general solution of (7) is given in (8)

$$u(t) = u_0 \exp\{i(\omega_0'^2 - \gamma^2/4)^{1/2}t\} \exp\{-t\gamma/2\} + u_p\, e^{-i\omega t} \tag{8}$$

The first term on the right hand side describes a transient feature, namely for weak damping an oscillation with the damped eigenfrequency $(\omega_0^2 - \gamma^2/4)^{1/2}$ which decays to zero for $t\gamma \gg 1$. Then we are left with the forced or driven oscillation with frequency ω which is given by the second term on the r.h.s. The amplitude u_p of the latter can be calculated by inserting the ansatz (8) into (7) resulting in

$$u_p = \frac{eE_0}{m} \frac{1}{\omega_0'^2 - \omega^2 - i\omega\gamma} \tag{9}$$

This is the well known expression for the resonance of a driven harmonic oscillator. The amplitude u_p is connected with a dipole moment p_0 of every oscillator

$$p_0 = eu_p \tag{10}$$

and a polarization P_0 given by (11) where N is the number of oscillators per unit of volume

$$P_0 = Np_0 = \frac{Ne^2}{m} \frac{1}{\omega_0'^2 - \omega^2 - i\omega\gamma} E_0 \tag{11}$$

or a polarizability α of every oscillator

$$\hat{\alpha} = \frac{eu_p}{E_0} = \frac{e^2}{m} \frac{1}{\omega_0'^2 - \omega^2 - i\omega\gamma} \tag{12}$$

We can consider the term on the r.h.s. of (11) in front of E_0 as a linear response function because it describes an answer **P** which depends linearly on the stimulus **E**. Consequently all what follows from this approach is called linear optics. The Maxwell equation, which connects the electric field **E**, the polarization density **P** and the electric displacement **D** reads in SI units

$$\mathbf{D} = \epsilon_0 \mathbf{E} + \mathbf{P} \tag{13}$$

Inserting (11) in (13) yields

$$D = \epsilon_0 \left[1 + \frac{Ne^2}{m} \frac{1}{\omega_0'^2 - \omega^2 - i\omega\gamma}\right] E \qquad (14)$$

By comparison with the usual linear approach

$$\mathbf{D} = \epsilon_0 \, \epsilon \, (\omega) \, \mathbf{E} \qquad (15)$$

or

$$\frac{1}{\epsilon_0} P = (\epsilon \, (\omega) - 1) \, E = \chi(\omega) E \qquad (15b)$$

where $\epsilon(\omega)$ is the dielectric function and $\chi(\omega)$ the susceptibility we obtain

$$\epsilon \, (\omega) = 1 + \frac{Ne^2}{m\epsilon_0} \frac{1}{\omega_0'^2 - \omega^2 - i\omega\gamma} = \chi \, (\omega) + 1 \qquad (16)$$

We introduce now a first correction or refinement to the dielectric function $\epsilon(\omega)$. We considered until now only the external electric field E acting on the oscillators. This approach is good for dilute systems like gases. For a dense system like a solid we should include in the local field $\mathbf{E}_{loc}$ the external one E plus all the electric field contributions from the other dipoles. Taking into account this effect leads for e.g. cubic materials to the so-called Clausius-Mosotti or Lorenz-Lorentz equation

$$\frac{\epsilon(\omega) - 1}{\epsilon(\omega) + 2} = \frac{1}{3} N \frac{e^2}{m} \frac{1}{\omega_0'^2 - \omega^2 - i\omega\gamma} \qquad (17)$$

We can recover from (17) the simple form of the dielectric function in (16) however with a shifted eigenfrequency

$$\omega_0^2 = \omega_0'^2 - \frac{Ne^2}{3m\epsilon_0} \qquad (18)$$

This eigenfrequency ω_0 is the only experimentally accessible one for dense systems and we write from now on

$$\epsilon(\omega) = 1 + \frac{Ne^2}{m} \frac{1}{\omega_0^2 - \omega^2 - i\omega\gamma} \qquad (19)$$

If we have a system which contains oscillators of various eigenfrequencies ω_{0j} and "oscillatorstrength" $f_j = \left(\frac{Ne^2}{m}\right)_j$ we obtain in linear response theory

$$\epsilon(\omega) = 1 + \sum_j \frac{f_j}{\omega_{0j}^2 - \omega^2 - i\omega\gamma_j} \qquad (20)$$

This is essentially the so-called Helmholtz-Ketteler formula or Kramers-Heisenberg dielectric function. We note further, that the contribution of one term in the sum of

(20) tends to zero for $\omega >> \omega_{0j}$ and to a constant value, namely $\dfrac{f_j}{\omega_{0j}^2}$ for $\omega << \omega_{0j}$.

Consequently we can neglect for an isolated resonance, i.e. one for which the other resonance are "sufficiently" far away, all lower lying resonances and sum up the contributions of all higher lying resonances in a so-called back ground dielctric constant ϵ_b resulting in

$$\epsilon(\omega) = \epsilon_b + \frac{f_0}{\omega_0^2 - \omega^2 - i\omega\gamma_0} \tag{21}$$

Next we should like to mention the following aspect: in the absence of free space charges ϱ one of Maxwell's equations read

$$\text{div } \mathbf{D} = \vec{\nabla} \cdot \mathbf{D} = \varrho = 0 \tag{22a}$$

In vacuum $\mathbf{D} = \epsilon_0 \mathbf{E}$ holds and (22) is generally used to argue that electromagnetic waves are transversal waves with $\mathbf{D} \parallel \mathbf{E} \perp \mathbf{k}$ and with $\vec{\nabla} \times \mathbf{E} = -\dot{\mathbf{B}}$ so that $\mathbf{E} \perp \mathbf{B} \perp \mathbf{k}$. This argument holds also in matter where we have

$$\text{div } \mathbf{D} = \text{div } \epsilon_0 \, \epsilon(\omega) \, \mathbf{E} = \varrho = 0 \tag{22b}$$

But apart from the solution $\mathbf{E} \perp \mathbf{k}$ there exists for special frequencies the solution $\epsilon(\omega) = 0$. For these frequencies also a longitudinal wave is possible with vanishing $\mathbf{D}$ (and $\mathbf{B}$) and antiparallel $\mathbf{E}$ and $\mathbf{P}$. Therefore we call these special frequencies longitudinal eigenfrequencies ω_L and note that we have for vanishing damping the simple relations

$$\omega_L^2 - \omega_0^2 = \frac{f}{\epsilon_b} \tag{23a}$$

$$\frac{\omega_L^2}{\omega_0^2} = \frac{\epsilon_b + f}{\epsilon_b} = \frac{\epsilon_s}{\epsilon_b} \tag{23b}$$

or in words:

$$\text{finite } f \Leftrightarrow \text{finite } \Delta_{LT} = \omega_L - \omega_0 \tag{23c}$$

with Δ_{LT} denoting the longitudinal-transvers splitting. The equation (23b) is the so-called Lyddane-Sachs-Teller relation, where ϵ_s is called static dielectric constant since it describes $\epsilon(\omega)$ well below the isolated resonance at ω_0.

In Figs 3 to 5 we give for zero and small damping the real and imaginary parts of $\epsilon(\omega)$, of the complex index of refraction $\tilde{n}(\omega) = n(\omega) + i\kappa(\omega)$ which is connected with $\epsilon(\omega)$ simply by

$$\tilde{n}^2(\Omega) = \epsilon(\omega) = (n(\omega) + i\kappa(\omega))^2 \tag{24}$$

and the reflectivity $R(\omega)$ of a single surface of a medium containing the model oscillators against vacuum. $R(\omega)$ is given for the intensities and normal incidence by

$$R(\omega) = \frac{(n(\omega)-1)^2 + \kappa^2(\omega)}{(n(\omega)+1)^2 + \kappa^2(\omega)} \tag{25}$$

The term small damping used already several times means $\gamma < \Delta_{LT}$.

We see in Figs 3 to 5 nicely the resonance behaviour in $\epsilon(\omega)$ and $\tilde{n}(\omega)$ and the so-called stop- or reststrahlband in the reflectivity between ω_0 and ω_L. This is a frequency intervall in which no oscillatory solution exists for $\gamma = 0$ and an incident wave is totally reflected. The absorption spectrum $\alpha(\omega)$ is connected with κ by

$$\alpha(\omega) = 2\frac{\omega}{c}\kappa(\omega) \tag{26}$$

This relation follows from replacing $k_v = \dfrac{2\pi}{\lambda_v}$ for an electro magnetic wave in vacuum by the following expression in matter

$$k_M = \tilde{n}(\omega)k_v = \tilde{n}(\omega)\frac{\omega}{c} \tag{27}$$

and from remembering that the intensity i.e. the energy flux density of a harmonic wave is always proportional to its amplitude squared.

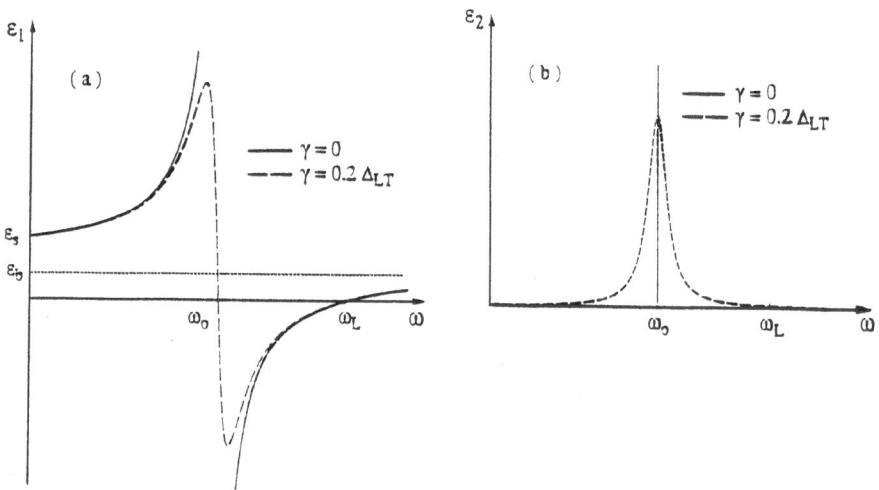

Figure 3 The real and imaginary parts of $\epsilon(\omega)$ for vanishing and small damping.

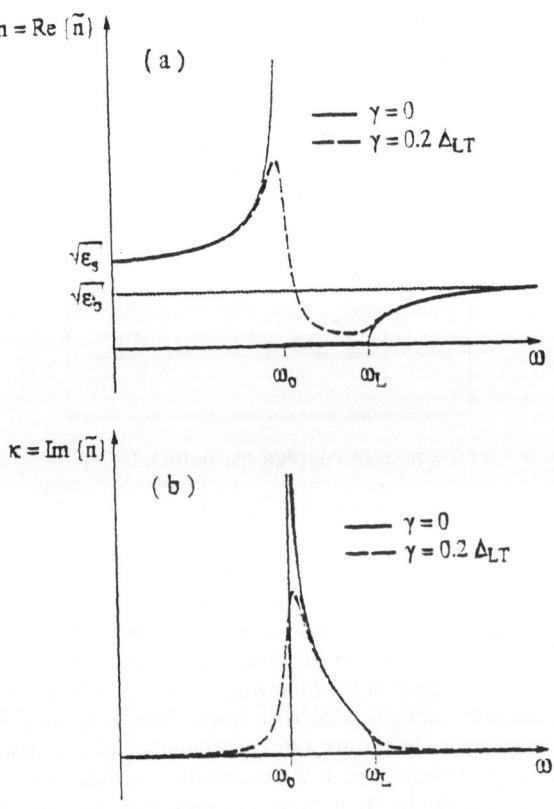

Figure 4 The real and imaginary parts of ñ for vanishing and small damping

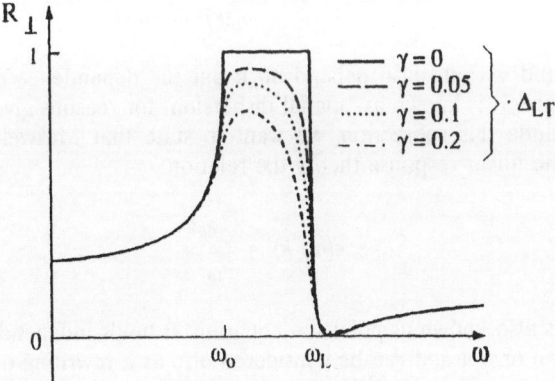

Figure 5 The intensity reflection spectrum for normal incidence and vanishing or small damping.

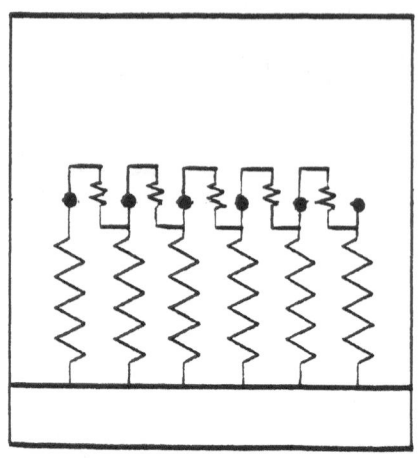

Figure 6 An ensemble of coupled oscillators. Compare to Fig 1 a

As can be seen from (15a) the dielectric function connects two vector fields namely **D** and **E**. Consequently $\epsilon(\omega)$ can have tensor character and for anisotropic materials $\epsilon(\omega)$ is indeed a tensor. This gives rise to the rich variety of phenomena known as birefringence and dichroism. The description of them goes beyond the scope of this contribution. We assume in the following always that $\epsilon(\omega)$ is a scalar function if not stated otherwise and refer the reader for birefringence e.g. to [15, 19, 20].

Now we want to study shortly the consequences which arise, if we skip the assumption of uncoupled oscillators. If we couple the oscillators to each other e.g. in the way shown schematically in Fig 6, their eigenfrequency becomes **k** dependent, i.e. ω_0 (**k**), in the way shown schematically in Fig 2b since the coupling springs are not elougated for $\lambda \rightarrow \infty$ but contribute as an additional restoring force for finite λ. As a consequence the dielectric function of e.g. (21) will be now a function of the two independent variables ω and **k**

$$\epsilon(\omega, k) = \epsilon_b + \frac{f}{\omega_0^2(k) - \omega^2 - i\omega\gamma} \tag{28}$$

In principle f and γ could also depend on **k** but the dependence of ω_0 is the most prominent feature. It is known as "spatial dispersion" for reasons given below.

To conclude this subsection, we want to state that Maxwells equations give together with the linear response theory the relation

$$\epsilon(\omega, k) = \frac{c^2 k^2}{\omega^2} \tag{29}$$

This equation is also known as polariton equation. It holds independent if we include spatial dispersion or not and can be considered also as a rewriting of (24) and (27) if we remember that $k_v = \frac{2\pi}{\lambda_v} = \frac{\omega}{c}$.

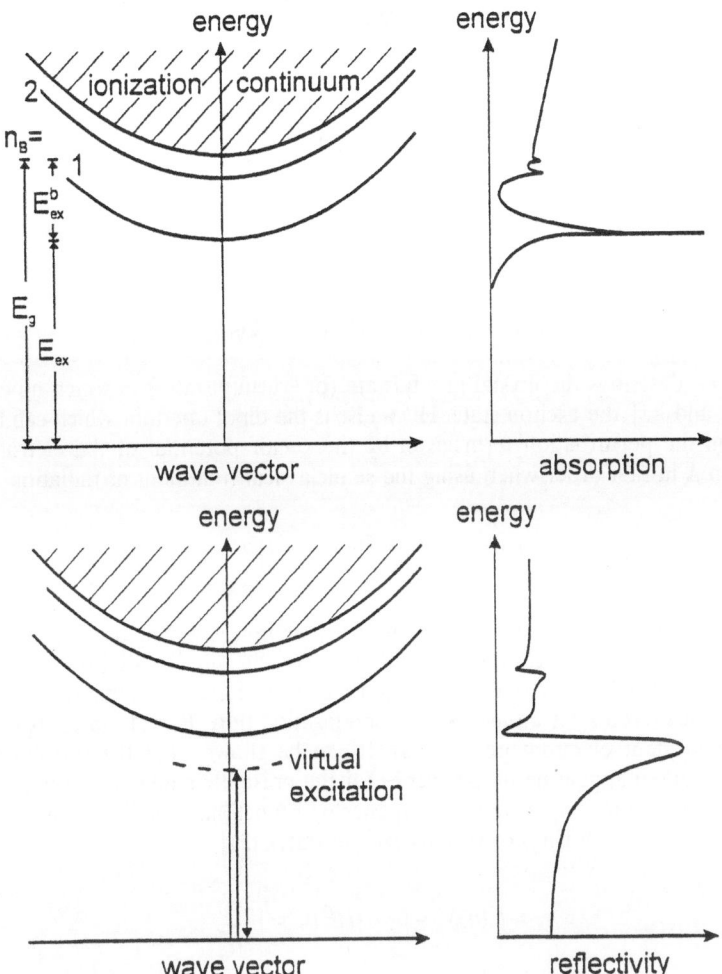

Figure 7 The dispersion curves of the exciton and a virutal excitation (left column) and the resulting absorption and reflection spectra (right column) (schematic drawings)

II. B. The weak and the strong coupling approach and the concept of polaritons

We leave now the system of model oscillators and replace them by the collective excitations in solids the quanta of which are the phonons, plasmons, excitons or magnons. The oscillator strengths f which was in eq. (19, 20) just an abreviation of some "mechanical" parameters, contains from now on the quantum mechanical transition matrix element, which describes e.g. in dipol approximation the coupling between the electromagnetic wave, the quanta of which are the photons and the collective excitations in the solid, the quanta of which are the above mentioned quasiparticles.

To elucidate this concept of weak coupling, we show in Fig 7 the series of exciton states in a direct gap semiconductor with dipole allowed band-to band transition and simple parabolic bands, followed by the states of the ionization contiuum, the onset of which coincides with the energy of the forbidden gap E_g.

An absorption process can be described in the way illustrated in Fig 7a by the solid arrow. A photon with an energy equal to the exciton energy comes in and is absorbed under the simultaneous creation of an exciton. The energy conservation reads

$$\hbar\omega = E_{ex} = E_g - E^b_x \tag{30}$$

where E^b_x denotes the binding energy of the exciton.

The transition probability is proportional to the matrix element squared, e.g. in Fermi's golden rule

$$W_{if} \sim |<f|H^D|i>|^2\varrho(E) \tag{31}$$

where $|i>$ describes the crystal groundstate (or "vacuumstate") in which no exciton is present, and $<f|$ the exciton state. $H^D \propto eEr$ is the dipol operator which can be deduced from the perturbation term given by the vector potential of the elctromagnetic radiation **A** in first order when using the semiclassical treatment of radiation

$$H^{(1)} = -\frac{e}{m}\,A\,\frac{\hbar}{i}\,\vec{\nabla} \;\rightarrow\; H^D \propto er \tag{32}$$

In dipole approxiamtion the momentum of the photon is assumed to be zero (see eq (6)) so that momentum conservation allows to reach only the exciton and continum states at $\mathbf{k} = 0$. To have a real absorption process, we have to assume further, that the exciton is scattered after an average phaserelaxation time T_2 and looses its coherence with the incident electromagnetic field. It can be shown, that the transition matrix element squared appearing in (31) varies for the presently made assumptions with the main quantum numer n_B of the exciton envelope function. This finding is reflected in the oscillator strength $f(n_B)$ of various exciton states [21]

$$f(n_B) \propto |<c|H^D|v>|^2\frac{1}{n_B^3} \tag{33}$$

which varies like n_B^{-3} and is otherwise proportional to the band-to-band transition matrix element squared.

The transitions into the continum states do not start with a square-root dependence as expected in effective mass approximation from (31) but are enhanced at and above E_g by the so-called Sommerfeld enhancement factor which has in three dimensional systems a square root singularity at E_g, so that the continum transitions start at E_g with a finite value and vary only sligtly with higher photon energies [21-23].

The above statements allow us already to understand the excitons, sketched in Fig 7b. The spectrum of the refractive index can be deduced either from the absorption spectrum with the help of the Kramers-Kronig relations [24] or by considering virutal excitations in the sense sketched in Fig 7c. A virtual transition means the creation of a state (e.g. an exciton) with a photon of an energy which does not fulfil (30). This is possible in quantum mechanics for a time Δt which is connected with the energy mismatch ΔE by (34)

$$\Delta E \cdot \Delta t \approx \hbar \tag{34}$$
$$\text{with}$$

$$\Delta E = |\hbar\omega - E_{ex}|$$

At latest after the time Δt has elapsed, the virtually excited state has to disappear again, in simpliest case by emitting a photon which has a certain phase-delay to the incident one, but is otherwise identical to it.

At this stage, we can already understand qualitatively, how to describe the increase of the refractive index $n(\omega)$ when approaching a resonance from lower photon energies (see Fig 4) since Δt and thus the phase delay increase with decreasing ΔE. On the other hand the phase velocity of light in a medium v_{ph} is connected with the phase velocity in vacuum c by

$$V_{ph} = \frac{\omega}{k} = \frac{c}{n(\omega)} \qquad (35)$$

See also (3). With the knowledge of the spectra of absorption α (ω) or of the imaginary part $\kappa(\omega)$ of $\tilde{n}(\omega)$ and of its real part $n(\omega)$ we can calculate via (25) also the spectrum of reflection as shown schematically in Fig 7d.

The description of the interaction of light with the quasi particles of solids, which we have outlined above for the case of excitons, relies evidently on a perturbative treatment. This is the so-called weak coupling approach. It allows to understand qualitatively and to some extend quantitatively the optical spectra of phonons, excitons or plasmons.

If we look close to the optical spectra, the weak coupling approach is often not sufficient for a quantitative description and the strong coupling approach has to be used. This statement is especially true for high quality samples at low temperatures where both the phonon- and defect- contributions to the damping γ are weak.

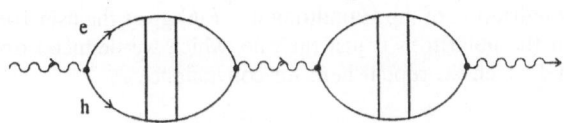

Figure 8 Sketch of the exciton polariton

We shall elucidate the strong coupling or polariton concept from various sides starting with a first attempt in prose, and using the exciton-photon coupling as an example.

We assume that a photon is transmitted through the interface from vacuum into a semiconductor. There it excites (virtually) an exciton state which has a finite oscillator strength. Just because of this finite osciallator strength and the finite polarization **P** connected with it (see eq (12)) this virtually created exciton radiates a photon, which after a while creates again an exciton and so on.

Consequently what propagates as "light" in matter is a mixture of the elctromagnetic radiation field and a polarization wave at the medium. This polarization wave in the medium is the analogon to the forced or driven oscillation in the classical picture of the model oscillators mentioned above.

If this mixed electro-magnetic and polarization wave is quantized, we find e.g. from the (internal) photoeffect, that the energy of the quanta is still $\hbar\omega$. The name of the quanta of this new mixed collective excitation is polariton. If the polarization wave is predominantly a phonon, plasmon or exciton one we speak about phonon-, plasmon- or exciton polaritons, respectively.

In a diagrammatic presentation the above explanation can be sketched like in Fig 8. A photon comes in and creates an electron-hole pair. The two carriers interact by their Coulomb-attraction (symbolized by the two vertical lines in the bubbles) to form the exciton, and recombine to give a photon etc.

This process can be written in a formula like (36) in the framework of second quantization (for more details see e.g. [9])

$$ H = \sum_k \hbar\omega_k\, a_k^+\, a_k + \sum_{k'} E\,(k')\, B_k^+\,,\, B_{k'} + i\hbar \sum g_k\, (B_k^+\, a_k + h.c.) \qquad (36) $$

The total Hamiltonion consist of the Hamiltonion for photons with their number operator $a_{k+} a_k$ and of the excitons $B_k^+\, B_k$ and the interaction term, which describes e.g. the annihilation of a photon a_k and the creation of an exciton B^+_k under momentum conservation. The g_k contain the transition matrix elements of the type discussed above e.g. in connection with eq (31).

If we treat this third term as a small perturbation, we are back in the weak coupling limit. If, however, we diagonalize the Hamiltonion of (36) by introducing suitable linear combinations of the creation and annihilation operators for photons and excitons we get the operators for the quanta of the mixed state, the polaritons.

Before we continue to discuss the dispersion relation of polaritons we state, that light propagating through matter is in the form of polaritons, whenever the refractive index n deviates from 1. This means for example, that whenever we look out of the window e.g. during a boaring lecture, we use the polariton concept for the time during which the light quanta travel through the glass.

The diagonalization of the Hamiltonion in (36) gives the user-friendly result, that the dispersion of the polaritons is just the one, which we deduced from the polariton equation (29) and which we repeat here for convenience

$$ \frac{c^2 k^2}{\omega^2} = \epsilon(\omega) = \epsilon_b + \frac{1}{\omega_o^2(k) - \omega^2 - i\omega\gamma} \qquad (37) $$

The two outer parts of (37) are an implicit representation of the dispersion relation of the polaritons $\hbar\omega = E = f(\mathbf{k})$.

Alternatively the polariton dispersion can be deduced from the dispersion of photons and of excitons (or phonons or plasmons) and the quantum mechanical non-crossing rule. The latter one says, that a crossing point between two eigenenergies as a function of some parameter, here the wavevector $\mathbf{k}$, disappears when the two states interact. The splitting in this avoided crossing is the larger, the stronger the coupling is.

We elucidate this concept in Fig 9. On the left hand side we give the dispersion relation of photons by a steep straight line indicating their small but non-vanishing momentum $\hbar\mathbf{k}$ and of the quanta of the collective excitation in the solid namely from top to bottom of phonons, plasmons and the $n_B = 1$ exciton.

We note that the dispersion relations of phonons is for the range of wave-vectors considered here a horizontal line, resembling thus the one in Fig 2a without spatial dispersion. For the plasmons we remember that the transverse eigenfrequency ω_0 is zero for $\mathbf{k} = 0$ since a gas, including a gas of free carriers, has a vanishing shear stiffness G so that the transverse eigenfrequency

$$\omega_0 = (G/\varrho)^{1/2} \qquad (38)$$

vanishes; ϱ is the mass density. The plasmon frequency ω_{pl}

$$\omega_{pl} = (\frac{e^2 n}{m_e \epsilon \epsilon_0})^{1/2} \qquad (39)$$

(with n = carrier density and m_e = effective mass of carriers) is the longitudinal eigenfrequency. We do not indicate orders of magnitude for plasmons, since $\hbar \omega_{pl}$ is situated in the range of a few tens of meV for highly doped or highly excited semi-conductors ($n \approx 10^{18} cm^{-3}$) and in the five to ten eV regime for metals ($n \approx 10^{23} cm^{-3}$). For excitons we have evidently the case of spatial dispersion discussed in context with Fig 2b and 6.

If the elementary excitation does not couple to the radiation field, this is all, and if the coupling is weak, this is still a good approximation.

In the strong coupling picture, we obtain the dispersion of phonon-, plasmon- and exciton-polaritons on the right hand side.

For the cases a and b the dispersion starts at the origin with a steep slope. These parts are said to be "photon-like". Then they bend over to a phonon- or exciton-like behaviour. This whole part of the dispersion is know as the lower polariton branch (LPB). There is a finite transvers-longitudinal splitting Δ_{LT} which is proportional to the oscillator strength (see(23c)), a longitudinal eigenstate and the upper polariton branch (UPB) both starting at ω_L. For the plasmons the LPB coincides essentially with the abscissa for the reasons given above.

The spatial dispersion of the excitons has among others the following consequences. There is for every frequency at least one propagating mode. As a consequence, there is no true stop-band between $\hbar \omega_0$ and $\hbar \omega_L$ and the reflectivity reaches even for vanishing damping only values considerably below 1 as shown schematically already in Fig 7d. Above $\hbar \omega_L$ there are even two propagating modes for one energy with different wave vectors and the same polarization. An incident beam will therefore be split in matter into two which propagate on the LPB and UPB. For oblique incidence they propagate in two different directions and two short pulses will be separated even for normal incidence into two which propagate with different group velocities according to (3). Both effects are the reason, why the dependence ω_0 (k) is also known as spatial dispersion. It should be noted that the above phenomena occur already in isotropic materials and have nothing to do with birefringence, which describes a spatial separation of beams with orthogonal polarisations.

The fact, that there are two propagating modes in the vicinity of a resonance with spatial dispersion, is not covered by the boundary conditions which follow from Maxwell's equations and lead e.g. to Fresnel's formulae. An additional boundary condition (abc) must be introduced which gives the "branching ratio" i.e. which says which fraction of the incident light propagates on the LPB and on the UPB, respective-ly. As a rule of thumb one can state, that light propagates essentially on the LPB for energies which a few or more Δ_{LT} below ω_0 and on the UPB a few Δ_{LT} above. In the resonance region both fractions can be comparable. For more details of this topic we refer the reader e.g. to [6, 7, 15, 20] and original references given therein. The disper-sion relation E(k) or ω(k) can be obtained also from Fig 4, if we exchange the ordinate and the abscissa and multiply the n axis with ω/c since the wave vector in the medium is just $n(\omega)\dfrac{\omega}{c}$ (27).

For the sake of completeness we shortly note that between ω_0 and ω_L so-called surface-polariton modes may exist, which propagate only along the interface between two different media e.g. vacuum and a semiconductor and which decay exponentially to both sides. For details see e.g. [15, 25] and references therein.

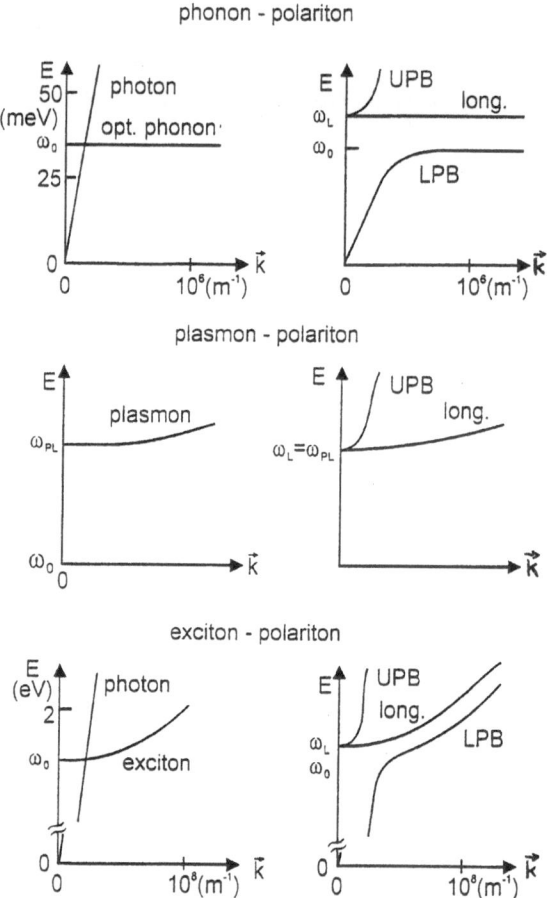

Figure 9 The dispersion relation of some elementary excitations of solids and of photons in the weak coupling picture (left column) for phonons (a) plasmons (b) and excitons (c) and the corresponding polariton dispersion (right column) in the strong coupling or polariton picture. For clarity vanishing damping is assumed and only the real part of k is given.

II. C Selected experimental methods to measure the dispersion of polaritons

In this subsection we present some of the numerous possibilities of **k** space spectroscopy, which allow to measure the dispersion relation of polaritons directly. We first mention an example for the phonon polariton, but concentrate then on the exciton polariton.

Raman scattering is usually explained as a scattering between light (generally in the spectral range $\hbar\omega$ between the phonon and the exciton resonances $\hbar\omega_{Ph} < \hbar\omega < E_{ex}$) and phonons. In the weak coupling picture one would state, that a photon propagates in the semiconductor or insulator and virtually excites an exciton. This virtually excited exciton emitts or absorbes an optical phonon and emitts a Raman-photon. In this picture one would expect to see as Stokes or anti-Stokes shift the energy of the optical phonons independent of the scattering geometry since the phonon dispersion is essentially flat. This is experimentally only true for scattering geometries close to backward scattering (see Fig 10a) since the wave vector of the created (or annihilated) phonon is so large that the state is situated on the phonon-like part of the dispersion relation as seen from a comparison of Figs 9a and c.

In geometries close to forward scattering however, the wave vector of the created particle can be very small (see Fig 10b) and this is the regime in which the dispersion of the phonon polariton is measured and where the experimental findings can be explained only in the polariton picture since the Raman shift starts with zero energy and momentum transfer for exact forward scattering and scans the LPB of the phonon for increasing scattering angle. In addition the longitudinal and the UPB may be observed. The correct description is then, that a photon-like exciton polariton in created in the sample by the incident (laser-) light beam, which is scattered under emission or absorption of a phonon polariton on the LPB (or for small k_{Phon} also on the UPB) or of a longitudinal phonon. The scattered photon-like exciton polariton leaves the sample as a Raman photon, when it is transmitted through the surface of the semiconductor or insulator.

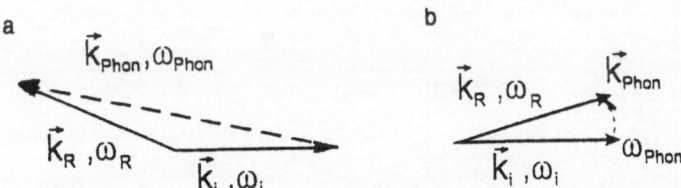

Figure 10 Geometries for Raman-scattering; backword geometry (a) forward geometry (b). The indices i, R and Phon stand for the incident photon (like exciton polariton), the created or annihilated phonon (-like polariton) and the Raman emission.

Beautiful experiments have been performed with this technique which demonstrate clearly the concept of phonon polaritons. Examples are found in [15, 26, 27] and references therein.

Now we concentrate on exciton polaritons. A basically simple method to determine the spectrum of the real part of the refractive index $n(\omega)$ and with (27) the dispersion relation is the investigation of the refraction of light in a prism. Since the absorption (or better extinction) coefficient can reach in the exciton resonance values well above 10^4 cm^{-1}, extremely thin prisms are necessary in this spectral range to keep the product of $\alpha(\omega) \cdot d$ in the experimentally accesible range ≤ 5. Thin prisms have necessaryly a small angle. Fortunately some CdS platelets grow in the desired shape. In Fig 11 we give two examples for CdS showing nicely the birefringence and dichroism around the A exciton resonance, and the spatial dispersion of the exciton like part of the LPB. It should be noted that values of $n(\omega)$ up to 25 have been observed corresponding to $\epsilon_1(\omega) = 625$ or $k = 3 \cdot 10^8$ m^{-1} i.e. a few percent of the first Brillouin zone.

Another possibility of $\mathbf{k}$ (or momentum-) space spectroscopy is the investigation of the Faby-Perot modes of a resonator formed by the semiconductor under investigation and using e.g. the natural reflectivity of the plan-parallel surfaces which some of the as grown platelet type samples have.

Transmission maxima occur, when an integer number of half-waves fits into the resonator. Consequently the polariton energies of the transmission maxima are equally spaced on the k axis with $\Delta k = \dfrac{\pi}{d}$ where d is the geometrical sample thickness, resulting in

$$k_m = \frac{m\pi}{d} \text{ or } \hbar\omega_m = E\left(m\,\frac{\pi}{d}\right) \qquad (40)$$

with m = 1, 2, 3, ...

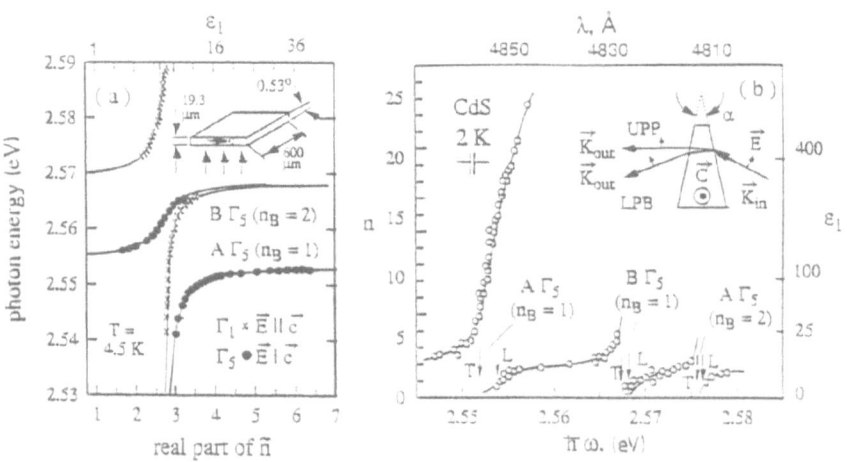

Figure 11 The real part of the refractive index as a function of photon energy in CdS. According to [28, 29].

If the order m of one transmission maximum (or reflexion minimum) is known, or the refractive index n, it is possible to reconstruct the whole dispersion curve from (40). In Fig 12 we give an example for CdSe. Below the transverse eigenfrequency one sees the rapidly decreasing spectral distance between the reflection minima, which correspond to the transmission maxima, due to the increase of $n(\omega_m)$ and above the narrow modulation due to the LPB superimposed on the wide one resulting from the UPB. The decrease of the modulation degree with increasing photon energy reflects the decreasing fraction of light, which travels on the LPB. Wave vectors up to $4 \cdot 10^8$ m^{-1} have between reached and the curvature of the exciton - like part of the LPB is nicely seen. Data for other materials are found in [15, 31].

For the investigation of the exciton polariton by resonant Brillouin scattering, by two- and three-photon absorption or by two-photon Raman scattering we refer the reader to [15, 32], to contributions to preceeding schools of this series [33] and the references given therein. Instead we proceed to two examples which are based on the propagation of short laser pulses i.e. pulses shorter than T_2.

A pulse of short duration (typical in the ps regime) and a spectral width (which is limited by the duration) much smaller than Δ_{LT} propagates in the sample with the group velocity v_g given in (3b). By measuring the time of flight through a sample of a thickness d of typical a few μm, one can determine v_g and thus the slope of the dispersion relation. By varying the incident photon energy it is possible to deduce the dispersion relation. An example is given in Fig 13. The calculated and measured energy dependences of the group velocity coincide perfectly. It should be noted that values of v_g as low as $5 \cdot 10^{-5}$ c have been observed. Examples for the "time-of-flight" method for other semiconductor materials are given e.g. in [35, 36].

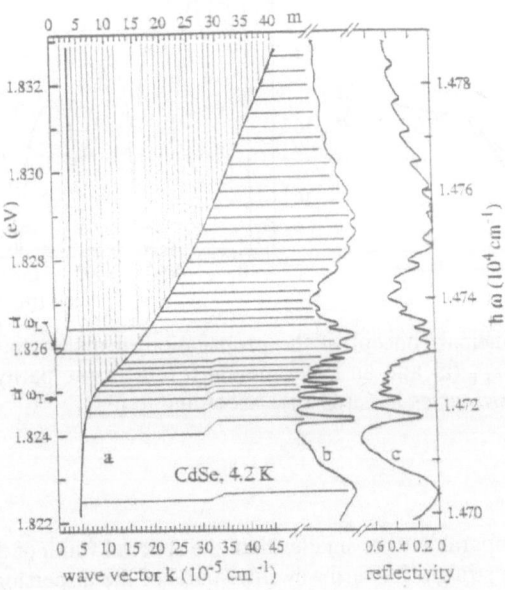

Figure 12 The dispersion of the $A\Gamma_5$ -exciton plariton in CdSe with a measured and a calculated reflection spectrum according to [30]

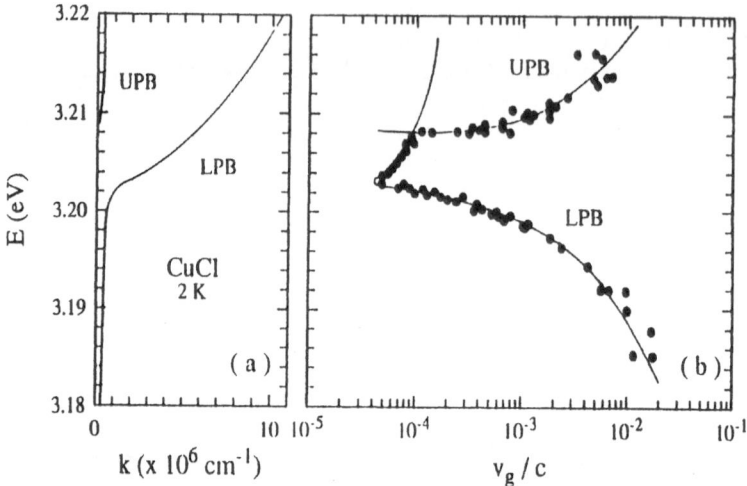

Figure 13 The dispersion relation of the exciton polariton in CuCl, and the calculated and measured group velocities. According to [34].

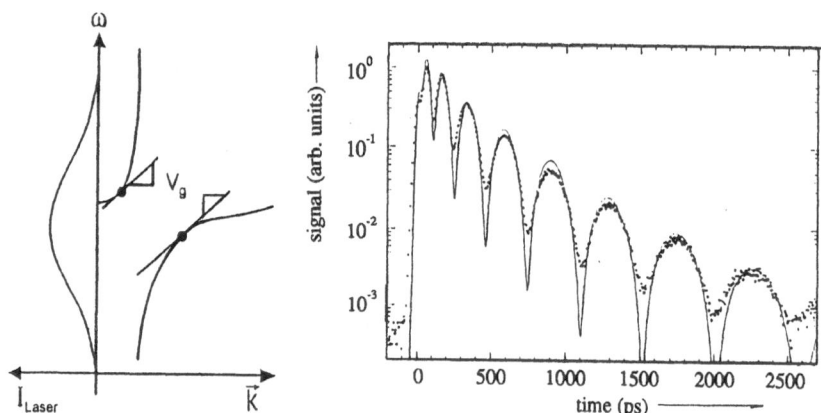

Figure 14 The principal concept of the propagation of a short pulse with a spectral width larger than Δ_{LT} (a) and an experimental result for the parity forbidden $n_B = 1$ exciton of the yellow series in Cu_2O (b). According to [37].

    If Δ_{LT} is comparable to or smaller than the spectral width of the pulse, the above method fails in the straight forward way since parts of the dispersion relation with very different group velocities are excited simmultaneously. In this case, a more elaborate technique works, however. We explain the principle idea in Fig 14a. When the incident pulse hits the surface of the sample, it excites polariton states both on the LPB and the UPB around the resonance. States which have equal slope propagate with the same group velocity through the sample and reach simultaneously the opposite side of the sample. Due to their frequency difference, a beating occurs and is observed in the transmitted intensity. The temporal period of this beating increases with time, since the pairs on the polariton dispersion which have high group velocities are spectrally further separated then the ones with low group velocities, as follows from an inspection of Fig 14a.

In Fig 14b experimental and calculated curves are shown for the parity forbidden $n_B = 1$ exciton resonance of the yellow series in Cu_2O [37]. The above outlined features are clearly visible. In addition the whole structure decays with time indicating a phase relaxation time T_2 of about 200 ps. This value is even longer than the T_2 time of intrinsic excitons in direct gap semiconductors with dipole allowed band to band transitions and which of the order of 10 ps at low temperatures and densities. For a collection of data see e.g. [15]. It should be noted that the value of Δ_{LT} deduced from the fit of the data in [37] is only 5 μeV, compared to typical values of allowed exciton transitions ranging in semiconductors from 0.5 to 10 meV. This shows clearly, that the polariton concept works and has to be used for a detailed quantitative understanding not only for allowed transitions treated e.g. in Figs 11 and 12, but even for transitions with weak oscillator-strength. The application of a technique similar to the one of Fig 13a to allowed transitions has been reported in [38]. Both above experiments show, that the polariton concept works also for short pulses. Deviation may be expected only for even shorter pulses in the 10 fs regime or below.

The concept of polaritons is presently being worked out for semiconductor structures of reduced dimensionality. For details see e.g. [39, 40].

III. LIGHT EMISSION FROM SEMICONDUCTORS

In this section we review the intrinsic light emission processes for direct gap semiconductors with increasing electron-hole pair density. As stated already in the introduction, we concentrate mainly on quasistationary conditions.

Light emission means in this context luminescence, optical gain or related phenomena. It seems important to note at this state, that the luminescence yield η_{Lum} is usually considerably below unity even for the direct gap semiconductors. Only for some selected quantum well structures values of η_{Lum} and for some laser diodes differential efficiencies close to 1 have been reported [41].

III. A The general scenario with increasing excitation

In all pure, high quality semiconductors the scenario of Fig 15 of many particle processes has been observed with increasing electron-hole pair density. At low excitation densities (and temperatures) the optical properties of semiconductors are governed in the spectral vicinity of the gap by excitonic features including their continum states. At low temperatures mainly the bottle neck region of the lowest polariton resonance is populated, togehter with dipole forbidden triplet states and bound exciton complexes not treated here [15]. At higher temperatures the excitons are increasingly ionized, expecially if their binding energy is smaller than the LO phonon energy and if the chemical binding of the semiconductor has a sufficiently strong polar character, so that the ionization process of the exciton to the continum states can be accomplished by the absorption of one LO phonon. We shall see examples for free exciton luminescence later.

With increasing density the so-called intermediate density regime is reached. This regime is characterized by the fact, that excitons are still good quasi particles, but their density is sufficient that they interact during their lifetime.

There are e.g. elastic or inelastic scattering processes between excitons, and preferentially at higher temperatures, between excitons and free carriers. At low temperatures, two excitons may be bound together to form an excitonic molecule or biexciton.

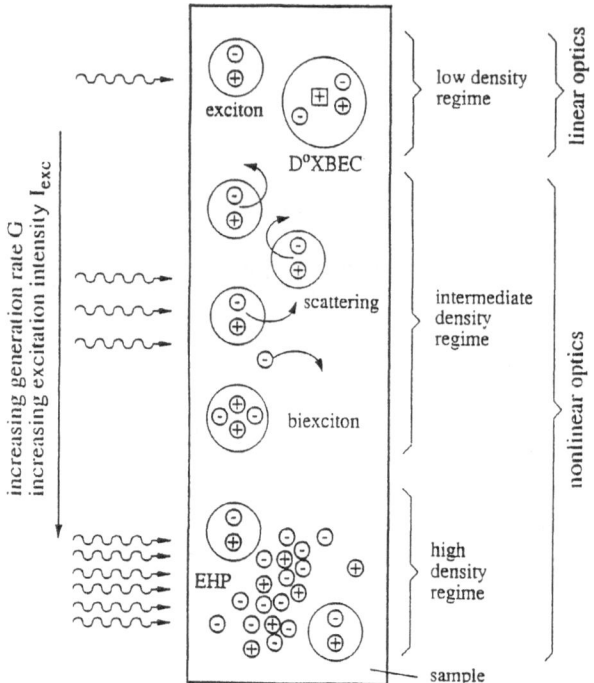

Figure 15 The general scenario of intrinsic interaction processes in the electronic system of semiconductors with increasing excitation density.

The above mentioned processes give rise to a rich variety of new recombination processes, which lead to new luminescence bands, to optical amplification (gain) or to excitation induced increase or bleading of absorption. We shall present later some examples for this density regime.

At the highest excitation densities, the average distance between electrons and holes is, at least in a fraction of the excited volume, comparable to the excitonic Bohr radius. Under these conditions, the excitons cease to exist as individual quasiparticles and a new collective phase is formed, the so-called electron-hole plasma. The formation of the plasma is connected with a reduction of the band-gap due to exchange and correlation effects among the carriers, the disappearance of the exciton resonances from the optical spectra due to screening of the Coulomb interaction and due to phase-space filling. If the chemical potential μ of the electron-hole pair system, which is defined as the enegetic distance between the quasi-Fermi levels of electrons and holes (41), falls above the reduced band gap $E'_g (n_p)$ (42) then the band-to-band transition is inverted, resulting again in optical gain, essentially in the spectral region between μ and $E'_g(n_p)$

$$\mu (n_P, T_P) = E^e_F (n_P, T_P) - E^h_F (n_P, T_P) \qquad (41)$$

$$\mu (n_P, T_P) \geq E'_g (n_P) \qquad (42)$$

where n_P and T_P give the density and temperature of carrier pairs, respectively.

Quasiequilibrium thermodynamics predict below a certain critical temperature T_c for the EHP a first order phase transition into a liquid like phase (EHL) and a low density vapour phase consisting of excitons and free carriers. This phase separation can develop especially in indirect gap semiconductors due to the long carrier lifetime. It has been shown, that the EHL is in the form of small droplets (EHD) and the phase diagram and the coexistence region are very similar to the ones of a real or van der Waals gas. For more details see [15, 42-44] and references therein.

III. B Semiconductor structures of reduced dimensionality

Since we want to present in the following also examples for semiconductor structures of reduced dimensionality, we introduce here shortly the relevant concepts. For further reading see e.g. [8 - 15] or [45 - 51] and references therein.

We show in Fig 16a a single quantum well. A material with a smaller gap E_g (e.g. GaAs) and a thickness of a few nm is sandwiched between a material with higher bandgap (e.g. $Ga_{1-y}Al_yAs$). The motion of the carriers is quantized in growth- (or z-) direction, while they can still move as free particles in the two-dimensional x-y plane. In (b) we show a multiple quantum well. In these structures the barriers are so thick, that the electron and hole states in adjacent wells do not interact. A superlattices has in contrast narrower barriers so that the exponential tails of the carrier wave functions in the barriers overlap, giving rise to a "miniband" for the motion of the carriers in z-direction. See also relation (4).

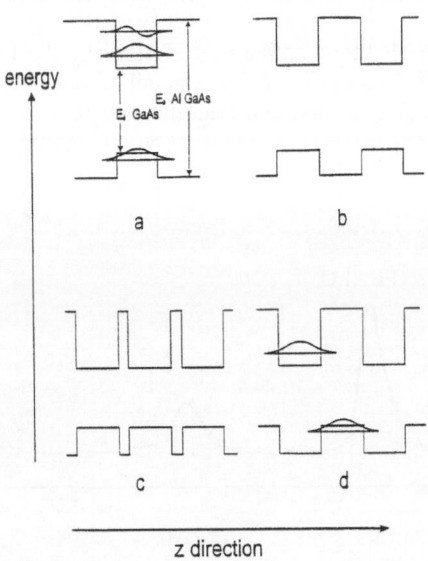

Figure 16 Sketch of a single (a) and a multiple quantum well (b) and of a superlattice of type I (c) and of type II (d).

The cases in Fig 16a-c are said to be of type I, since the electrons and holes are confined in the same layer. Depending on the band allignment, one can find also the situation of Fig 16d, which is known as a type II superlattice. In this case the different carriers are confined in the differnt layers and a transition between them is said to be indirect in real space, in analogy the transitions which are indirect in momentum space. The system CdS/CdSe is of type II and the system GaAs/AlAs becomes of type II for layers of both materials, which are only a few atomic layers thick.

Quasi one-dimensional structures, so-called quantum wires can be formed by lateral structuring of quantum wells, by growth of the well on V-groove shaped substrates or ridges or on highly indexed, corrugates surfaces. The latter case leads to the so-called quantum well wire superlattices (QWWSL). For details see e.g. [15, 44, 47-49].

The transition to quasi zero-dimensional quantum dots can be achieved either by lateral structuring of quantum wells in two dimensions, leading generally to disc-shaped structures [48, 49] or by growing roughly spherical semiconductor nano-crystals in some insulating matrix like glass, organic solvents or sol-gel systems or even in the pores of crystals like zeolithes [50, 51].

III. C Selected experimental results

We present now some selected experimental results for the various density regimes mentioned above and for structures of various (quasi-)dimensions. Even with the restrictions mentioned in the introduction of this contribution the amount of results is almost unlimited. The choice which we present here is therefore to a large extend subjective.

We start with two examples of the low temperature and low density luminescence of semiconductors.

First we consider ZnO as an example for bulk materials. There, the luminescence yield of free excitons is generally rather low, due to the limited transmission through the interface, which is governend by the reflectivity and the angle of total internal reflection. The latter aspect is a consequence of the k-conservation parallel to the interface.

In polar materials the coupling to preferentially the LO phonons gives rise to satellites as shown for ZnO in Fig 17. More details about this process and on the zero-phonon line found in [15, 52 - 54] and references therein.

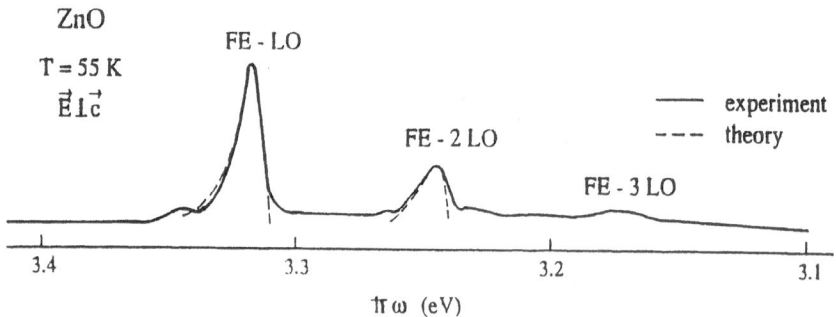

Figure 17 The first three phonon satellites of the exciton luminescence, in ZnO. According to [52].

It should be noted that bound exciton complexes, donor-acceptor pairs and other extrinsic luminescence channels have often much higher luminescence efficiencies than the intrinsic ones, especially at low temperautes and even for high quality samples. More information on these processes in ZnO is found e.g. in [54].

As next example we show in Fig 18 the luminescence and absorption spectra of two AlGaAs MQW samples of different qualities. The vanishing shift between the absorption maximum of the hh exciton resonance and of the luminescence when reabsorption is considered and the narrow with of both bands around 1 meV prove the high quality of the upper sample. Defects in the well and interface (as they appear as a consequence of growth interruption [56]) and fluctuations of the well width [57] may increase these quantities to the 10 meV range [58] as shown in the lower example of Fig 18.

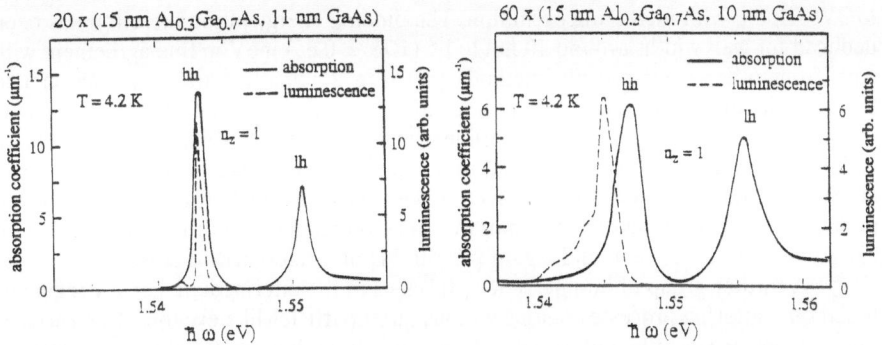

Figure 18 The absorption and luminescence spectra of two AlGaAs MQW samples. According to [55].

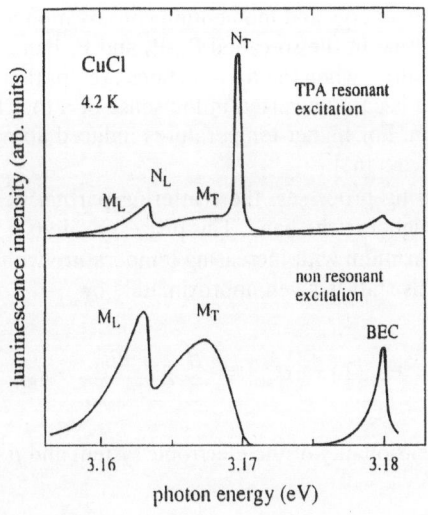

Figure 19 The biexciton luminescence in CuCl. According to [7].

Now we enter the intermediate density regime, giving some examples for the radiative decay of biexcitons. Biexcitons or excitonic molecules are entities, which consist of two electron-hole pairs. The ground state of the biexciton is usually bound, i.e. it is energetically lower (at k = 0) by the binding energy E^b_{biex} compared to the energy of two free excitons. E^b_{biex} is generally in the range from below 1 meV up to about 20 meV depending on the material. As a consequence, low temperatures of the lattice and of the electron system are a pre-requiste to observe biexcitons. The simpliest radiative decay process of a biexciton is in a free exciton and a photon (- like polariton). For details see e.g. [6, 7, 9, 15] and references therein. In Figs 19 to 21 we give examples for systems of different dimenisonality, starting in Fig 19 with bulk CuCl. The two broad bands result form decay channels leaving a transverse exciton like polariton (M_T) or a longitudinal exciton (M_L) in the sample. The lineshape, which is essentially an inverted Boltzmann distribution, is governed by the dispersion of excitons and biexcitons and their (thermal) population. The two narrow bands N_T and N_L observed additionally under resonant two-photon excitation are due to a cold gas of biexcitons. In Fig 20 we show the normalized exciton and biexciton luminescence in an AlGaAs MQW sample with increasing excitation. The binding energy of the biexciton has been deduced for well widths around 10 nm to be (1.75 ± 0.05) meV in fine agreement with data from other spectroscopic methods [59]. Fig 21 finally shows the exciton and biexciton luminescence for different excitation intensities from CuBr quantum dots embedded in a glass matrix. The superlinear increase of the low energy emission band with increasing pump-power compared to the high energy exciton band, its faster decay after ps pulsed excitation and its spectral position as a function of the average dot radius confirm the biexcitonic origin of the emission band [60]. Furhter information on biexcitons in QD can be found e.g. in [12, 50, 51] and references therein.

A further group of luminescence processes in the intermediate density regime is based on scattering processes among various quasi particles like excitons, free carriers, LO-phonons or states bound at some defects. For a review see [6] and the references therein. In the so-called inelastic scattering processes at least one exciton - like polariton is scattered on to the photon - like branch of the dispersion curve and appears as luminescence quantum while another particle is scattered into a higher state under energy and - if applicable - momentum conservation.

In the so-called inelastic exciton - exciton scattering between two exciton - like polaritons in the n_B = 1 state, one is scattered onto the photon - like LPB while the other one reaches under energy and momentum conservation a higher state n_B = 2,3,... or the continuum resulting in the so-called P_2, P_3 and P_∞ bands.

At low temperatures, when the n_B > 1 states are not thermally populated i.e. for $k_B T < Ry^*$ this process is easily inverted in the sense of a four level laser system [6, 7] resulting in optical gain. For higher temperatures induced absorption can be observed due to the reverse process [61].

In other scattering processes, the scattering partner is a free carrier which is scattered into a state higher in the band. This process exhibits a rather pronounced red-shift of its emission maximum with increasing temperature with respect to the position of the free exciton. This shift is given approximately by

$$E_{ex}(T) - \hbar \omega_{max} = \frac{d}{2} \beta \frac{m_{ex}}{m_{carrier}} \cdot k_B T \tag{43}$$

where d gives the dimensionality of the electronic system and β is of the order of unity.

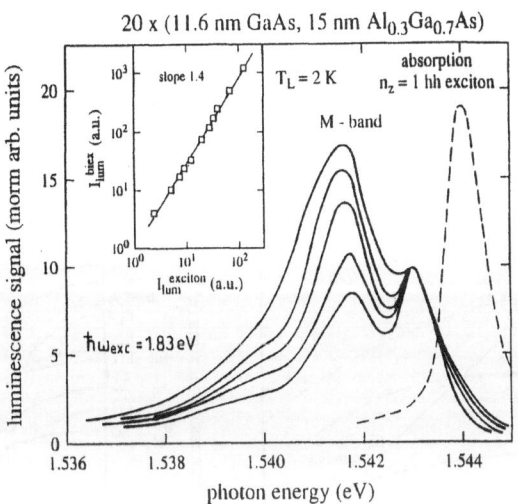

Figure 20 The normalized luminescence of biexcitons in AlGaAs MQW with increasing excitation. According to [55].

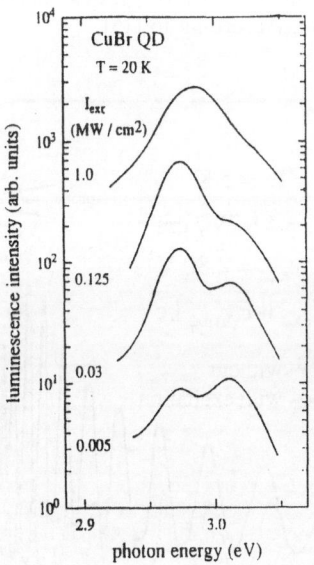

Figure 21 The luminescence of biexcitons in CuBr quantum dots with increasing excitation. According to [60].

In a third example the scattering partner is a LO phonon which is emitted leading to the LO-phonon replica discussed already in connection with Fig 17.

In Fig 22 we give an overview of the temperature dependence of the spectral positions of various emission bands in the low and medium density regimes as a function of temperature for bulk ZnO and CdS. These processes are known since almost three decades and obtained new interest recently in connection with the laser processes thin of II-VI epitaxial layers and superlattices. We come back to this aspect in section IV.

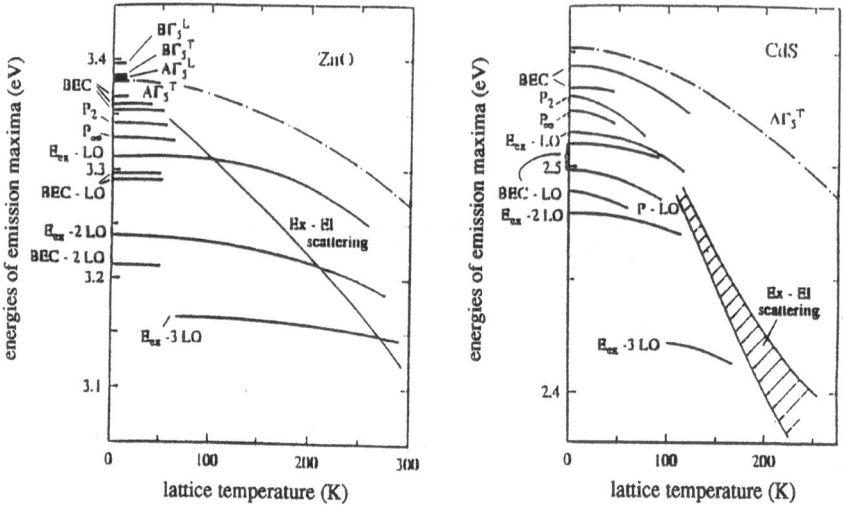

Figure 22 The temperature dependence of various emission maxima in ZnO and CdS as a function of temperature. According to [15].

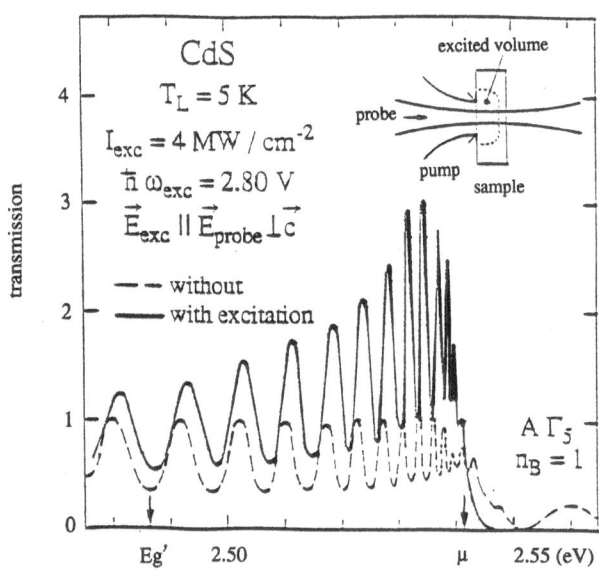

Figure 23 Pump-and-probe beam of the transmission spectra of a 3d CdS platelet at low temperature. According to [62].

As examples for the high-density or electron-hole plasma (EHP) regime we present two low-temperature spectra in Figs 22 and 23. Fig 22 shows the transmission of a CdS platelet at low temperature without and with additional pump beam on. Without pump the sample is transparent below the exciton energy. Actually it exhibits clear Fabry-Perot modes in this range because its plan-parallel, as-grown surfaces. Then we see to higher energies the transmission dip due to the lowest free exciton, a transparent window above followed by higher exciton states. With the pump-beam on, the excitonic features disappear from the transmission and reflection spectra [6, 15] indicating the formation of an EHP. Depending on the density reached and on the temperature of the carrier system, the band-to-band transition can be inverted. This is in simpliest approximation the case when the chemical potential of the electron-hole pair system is situated above the normalized gap mentioned with eq. (41, 42).

In the case of Fig 22 the gap is renormalized from a value of 2.58 eV without excitation to 2.49 eV. The chemical potential μ is situated at 2.54 eV indicating a binding energy of the EHP with respect to a gas of free excitons of about 10 meV. A careful analysis of the shape of the optical amplification, occuring in Fig 23 evidently between 2.48 eV and 2.54 eV, allows to determine the values of n_P, T_P, E'_g (n_P) and μ (n_P, T_P). Here n_P is around $3 \cdot 10^{18}$ cm^{-3} and $T_P \approx 35$ K. Similar data have been found for many direct gap II-VI and III-V compounds. See e.g. [6, 9, 11, 13, 15] and references therein.

Additionally, optical gain may result from inelastic scattering processes in the EHP even for densities below the ones defined by (42).

In indirect gap semiconductors with long carrier lifetime and for quantum wells grown on corrugated surfaces the transition to a liquid - like plasma state has been observed [9, 15, 44, 64]. In direct gap materials, the carrier lifetime is usually too short for the development of a spatial phase separation in an EHL and a surrounding gas below T_C [65].

Fig 24 shows in contrast to Fig 22 the optical denity $\alpha(\omega)$d and not the transmission spectra of an AlGaAs MQW for various excitation conditions. Apart from some finer details not discussed here, we see for this quasi-two-dimensional system a similar behaviour as in Fig 23 for a bulk sample, namely the disappearence of the excitonic features with increasing excitation density, the appearance of gain and a reduction of the band gap of the lowest quantized structures here from about 1.53 eV without excitation to density dependent values around 1.47 eV.

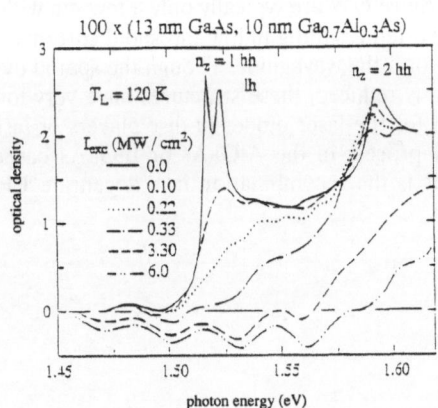

Figure 24 Pump-and-probe beam spectra of the optical density a quasi 2d AlGaAs MQW sample. According to [63].

IV. LIGHT EMITTING SEMICONDUCTOR DEVICES

In this section we address some aspects of applied research resulting from the (stimulated) light emission of semiconductors. See also [66, 67] and references therein. While (resonant) optical pumping is the method of choice for basic research, especially for materials which cannot be doped ambipolar at all or not sufficiently to reach the high injection levels in a forward biased diode to fulfil the conditions of e.g. (41, 42), electrical puming is an undispensable prerequiste of widespread applications of light-emitting diodes (LED) and of laser diodes.

IV. A Commercial laser diodes

In Fig 25 we show from a to c the spatial band structure of increasingly sophisti-cated laser-diodes, biased in forward direction. Part a gives a simple homojunction. Away from the junction the distribution of electrons and holes is given by the (tempera-ture dependent) Fermi energy E_F. In the transition region there is a splitting into two quasi Fermi levels for electrons and holes $E_F^{e, h}$ to describe the nonequilibrium situa-tion. To fulfill the condition (41, 42) the doping level in the p and / or n region must be so high, that at least in one part a degenerate gas of carriers is formed i.e. that the Fermi level is in the band. This type of structures has very high laser threshold currents and can be operated only at reduced temperatures and in a pulsed mode.

A major breakthrough for laser diodes was the formation of double heterostruc-tures as shown in part b. An intrinisic region with lower band gap is sandwiched between two doped semiconductors with higher gap e.g. p $Al_{1-y}Ga_yAs$ / i GaAs/ n $Al_{1-y}Ga_yAs$. As can be seen, the doping level can be lower and the decrease of the refracti-ve index in the barriers compared to the well at the emission wave length of the well can be used simultaneously to form a wave-guiding structure in one direction. Fur-thermore one has a perfect spatial overlap between the envelope of the electromagnetic radiation field and the inverted part of the semiconductor if one obtains also a confine-ment of the (stimulated) emission in a second direction. This can be achieved by stripe like contracts or by the formation of a ridge structure as shown in Fig 26. Such structu-res allow already cw operation. Even lower threshold currents of a few mA can be obtained using single or multiple quantum wells in graded index separate confinement hetero-structures (GRINSCH) shown schematically in part c of Fig 25. The injected carriers are fed by the funnel-shaped bandstructure very efficiently into the QW in the center of the i-region. Since QW are typically only a few nm wide, only extremely small volumes have to be inverted resulting in low threshold currents. The wider, outer parts of the intrinsic region form the waveguide. Though the spatial overlap between carriers and the radiation field is reduced, these structures have very low laser thresholds and are presently used e.g. for the laser diodes in disc players or laser printers.

The basic laser process in the AlGaAs or InGaAs based structures operating from the red to the IR is the recombination in a degenrate EHP as discussed in the previous section.

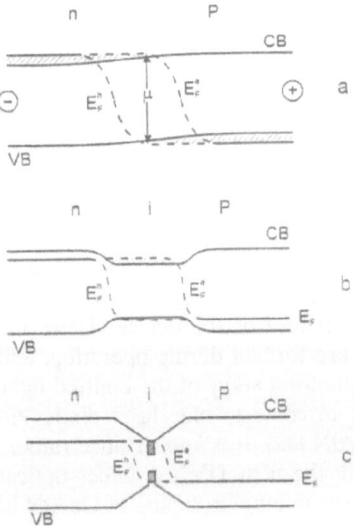

Figure 25 The spatial band-alignment of a forward biased simple diode (a) a double-heterostructure (b) and a GRINSCH diode (c).

IV. B Research activities

Many research efforts aim presently towards low laser thresholds, e.g. for fiber optics communication, high output power e.g. as pump sources for solid state lasers, one dimensional arrays of edge emitters or even two dimensional arrays of surface emitting structures for parallel electro-optic data handling and / or image processing.

Another very active field of research concerns the shift of the range of available emission wavelenths through the whole visible spectrum to the near UV for display purposes or for increased packing density in optical data storage.

The first blue-green laser diodes were based on ZnSe/ZnCdSe structures [68]. Nowadays ZnSSe/ZnSe or ZnCdMgTe structures are discussed to extend the range of available wavelength. The development of this field including the attempts of epitaxial growth and of ambipolar doping can be nicely followed in the series of International [69] and European [70] II-VI conferences.

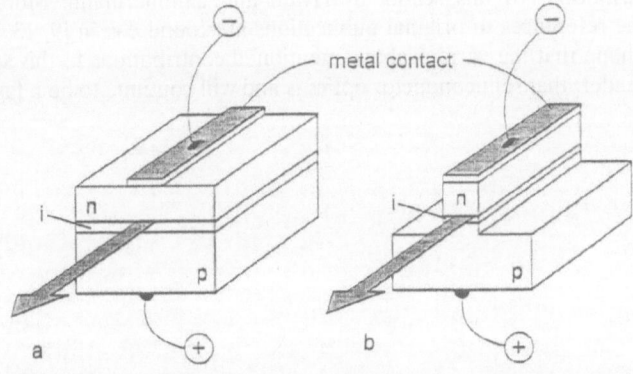

Figure 26 Laterale confinement of the photon field by gain-guiding with a stripe-like contact (a) or by a ridge structure (b).

Since the binding energy of excitons tends to increase with increasing band gap [9, 15] and the excitonic Bohr radius decreases, excitons are more stable at high densities in wide gap semiconductors. As a consequence, excitonic processes like the inelastic scattering processes presented in section III, emission from localized excitons states or biexcitons are discussed as physical origine of the stimulated emission in addition to the EHP. For some recent publications on this field see e.g. [69, 70, 71] and references therein.

A crucial problem of II-VI laserdiodes is still their lifetime. Progress has been made from values around 1 s to about 1 h but another four decades have to be covered to come to widely applicable devices. The lifetime is presently limited by the formation of nonradiative centers, so-called dark spot or dark line defects, which grow under laser-operation up to the failure of the device. These dark lines are most probably dislocations loops which are formed during operation, and which are caused - in the worst case - by the high photon energy of the emitted light itself.

Another approach to come to blue laser diodes are wide gap III-V materials especially the system GaAlN [72]. It is known since rather long time, that GaN shows gain and laser emission in the near UV e.g. under optical pumping [73]. Recently it became possible to produce ambipolar doping of GaAlN based structures. Since these materials are considerably harder than the II-VI compounds, dislocation formation is a less stringent problem. On the other hand, only spontaneous but no stimulated light emission has been obtained so far.

So an exciting completition can be expected for the future between these two groups of materials or rather between the scientists devoting their work to the development of technically useful devices. To this end, not only lifetime, thershold currents and emission wavelength will be important, but also the compatibility with excisting GaAs or even better Si technologies.

V. CONCLUSION AND OUTLOOK

As mentioned in the introduction, we concentrated here mainly on the optical properties of semiconductors under stationary or pulsed quasi-stationary excitation conditions. A rather rapidly developing field of research concentrates in recent years on the coherent interaction of light with the electronic system of semiconductors. The keywords are e.g. photon-echos, quantum-beats, the ac- or optical Stark effect, the THz emission from coherent optical phonon pakets from or from beating electronic excitations or memory (non-Markovian) processes. Some of the above aspects will be treated in the contributions to this school by Hvam and Zimmermann. More information including the references to original publications are found e.g. in [9, 13, 15, 16].

We hope that the various above mentioned contributions to this school demonstrate the reader, that semiconductor optics is and will continue to be a fascinating field of research.

REFERENCES

[1]   B. Di Bartolo, this volume

[2]   Ch. Kittel, Einführung in die Festkörperphysik, 7. Aufl., Oldenbourg, München (1988) and Introduction to Solid State Physics, 6th ed., John Wiley and Sons, New York (1986)
      H. Ibach. H. Lüth, Festkörperphysik, 3. Aufl., Springer, Berlin (1991)
      N.W. Ashcraft, N.D. Mermin, Solid State Physics Holt, Rinehard and Winston, New York (1976)
      K. Kopitzki, Einführung in die Festkörperphysik, Teubner, Stuttgart (1989)
      O. Madelung, Introduction to Solid-State Theory, Springer Series in Solid-State Sciences, 2nd ed, Springer, Berlin (1981)

[3]   Collective Excitations in Solids, B. Di Bartolo ed. NATO, ASI Series B 88, Plenum Press, New York (1983)

[4]   J.M. Hvam, this volume

[5]   R. Zimmermann, this volume

[6]   C. Klingshirn and H. Haug, Physics Reports 70, 315 (1981)

[7]   B. Hönerlage, R. Levy, J.B. Grun, C. Klingshirn, K. Bohnert, Physics Reports 124, 161 (1985)

[8]   Optical Switching in Low-Dimensional Systems, H. Haug and L. Banyai eds., NATO ASI Series B 194, Plenum, New York (1989)

[9]   H. Haug and S.W. Koch, Quantum Theory of the Optical and Electronic Proper-ties of Semiconductors, 3rd ed., World Scientific, Singapore (1995)
      N. Peyghambarian, S.W. Koch, A. Mysyrowicz, Introduction to Semiconductor Optics, Prentice Hall, Englewood Cliffs, New York (1993)
      Optics of Semiconductor Nanostructures, F. Henneberger, S. Schmitt-Rink and E.O. Göbel eds., Akademie Verlag VCH, Weinheim (1993)

[10] H. Kuzmany, Festkörperspektroskopie, Springer, Berlin (1991)

[11] Weng W. Chow, S.W. Koch and M. Sargent III, Semiconductor Laser Physics, Springer, Berlin (1994)

[12] L. Banyai and S.W. Koch, Semiconductor Quantum Dots, World Scientific Series on Atomic, Molecular and Optical Physics 2, World Scientific, Singapore (1993)

[13] R. Zimmermann, Many Particle Theory of Highly Excited Semiconductors, Teubner Texte Physik 18, Teubner Leipzig (1988)

[14] P.T. Landsberg, Recombination in Semiconductors, Cambridge University, Press, Cambridge (1991)

[15] C. Klingshirn, Semiconductor Optics, Springer, Berlin (1995)

[16]   M. Wegener and W. Schäfer, Semiconductor Optics and Transport, Springer, Berlin, in press

[17]   Proceedings of the International Workshops on "Nonlinear Optics and Excitation Kinetics in Semiconductors"
NOEKS I, phys. stat. sol. b 146, 311-391, and ibid. 147, 699-756, K65-69
NOEKS II, phys. stat. sol. b 159, 9-484 (1990)
NOEKS III, phys. stat. sol. b 173, 9-477 (1992)
NOEKS IV, phys. stat. sol. b 188, 9-587 (1995)

[18]   Contributions to previous proceedings of this school e.g. by B. Di Bartolo, J.M. Hvam, D. Fröhlich and C. Klingshirn, like NATO ASI series B, Vols. 114, 249, 301, 339

[19]   W. Kleber, Einführung in die Kristallographie, 8th ed., VEB Verlag Technik, Berlin (1965)

[20]   V.M. Agranovich and V.L. Ginzburg, Crystal Optics with Spatial Dispersion and Excitons, Springer Series in Solid State Sci. 42, 2nd ed., Springer, Berlin (1984)
R. Claus, L. Merten and J. Brandmüller, Light Scattering by Phonon Polaritons, Springer Tracts in Mod. Phys. 75, Springer, Berlin (1975)

[21]   R.J. Elliot, Phys. Rev. 108, 1384 (1957)
S. Nikitine, Progr. Semicond. 6, 233 and 269 (1962)
R.S. Knox, Theory of Excitons, Solid State Physics. Suppl. 5, Academic Press, New York (1963)
Polarons and Excitons, C.G. Kuper and G.D. Whitfield eds., Plenum Press, New York (1963)
M.M. Denisov and V.P. Makarov, phys. stat. sol. b 56, 9 (1973)

[22]   See [21] and
M. Shinada and S. Sugano, J. Phys. Soc. Japan 21, 1936 (1966) and references therein

[23]   B. Langen, H. Leiderer, W. Limmer, W. Gebhardt, M. Ruff and U. Rössler, J. Crystal Growth 101, 718 (1990)

[24]   F. Wooten, Optical Properties of Solids, Academic Press, New York (1972)
H. Kuzmany, Festkörperspektroskopie, Springer, Berlin (1990)
abc Physik, Brockhaus, Leipzig (1972)

[25]   Modern Problems in Condensed Matter Sciences, North Holland, Amsterdam, Vol 1, Surface Polaritons, V.M. Agranovich and D.L. Mills eds.
Vol 9, Surface Excitations, V.M. Agranovich and R. London eds. and references therein

[26]   R. Claus, L. Merten and J. Brandmüller, Light Scattering by Phonon Plaritons, Springer Tracts in Mod. Phys. 75, Springer, Berlin (1975)

[27]   Ch. Henry and J.J. Hopfield, Phys. Rev. Lett. 15, 964 (1965)
N. Marshall and B. Fischer, Phys. Rev. Lett 28, 811 (1972)

[28] I. Broser, R. Broser, E. Beckmann and E. Birkicht, Solid State Commun. 38, 1209 (1981)

[29] M.V. Lebedev, M.I. Strashnikova, V.B. Timofeev and V.V. Chernyi, JETP Lett. 39, 366 (1984)

[30] V.A. Kiselev, B.S. Razbirin and I.N. Uraltsev, phys. stat. sol. b 72, 161 (1975) and references therein

[31] I.V. Makarenko, I.N. Uraltsev and V.A. Kiselev, phys. stat. sol. b 98, 773 (1980)
      T. Mita and N. Nagasawa, Solid State Commun. 44, 1003 (1984)

[32] B. Hönerlage, R. Lévy, J.B. Grund, C. Klingshirn and K. Bohnert, Physics Reports 124, 161 (1985)

[33] C. Klingshirn NATO ASI Series B 301, 119 (1992)
      D. Fröhlich, NATO ASI Series B 339, 289 (1995)

[34] Y. Masumoto, Y. Unuma, Y. Tanaky and S. Shinoya, J. Phys. Soc. Japan 47, 1844 (1979)

[35] R.G. Ulbrich and G.W. Fehrenbach, Phys. Rev. Lett. 43, 963 (1979)

[36] T. Itho, P. Lavallard, J. Reydellet and C. Benoit à la Guillaume, Solid State Commun. 37, 925 (1981)
      Y. Segawa, Y. Aoyagi and S. Namba, J. Phys. Soc. Japan 52, 3664 (1983)

[37] D. Fröhlich, A. Kulik, B. Uebbing, A. Mysyrowicz, V. Langer, H. Stolz and W. von der Osten, Phys. Rev. Lett. 67, 2343 (1991) and phys. stat. sol. b 173, 31 (1992)

[38] K.-H. Pankte, P. Schillak, J. Erland, V.G. Lyssenko, B.S. Razbirin and J.M. Hvam, phys. stat. sol. b 173, 91 (1992)
      K.-H. Pankte and J.M. Hvam, Intern. Journal of Modern Physics B 8, 73 (1994)

[39] T. Rappen, G. Mohs and M. Wegener, phys. stat. sol. b 173, 77 (1992)

[40] F. Tassone and F. Bassani, Il Nuovo Cimento 14D, 1241 (1992)
      L.C. Andreani, phys. stat. sol. b 188, 29 (1995) and D.S. Citrin, ibid. p 43
      N.A. Gippius, E.A. Mulyarov and S.G. Tikhodeev, ibid. p 57

[41] D. Bimberg, T. Wolf and J. Böhrer, NATO ASI Series B 249, 577 (1991)
      R. Ulbrich, ibid. p 197
      K. Wasa, K. Tsubouchi and N. Mikoshiba, Japan. J. of Appl. Phys. 19, L653 (1980)

[42] R. Zimmermann, Many Particle Effects of Highly Excited Semiconductors, Teubner Texte Physik 18, Teubner, Leipzig (1988)

[43] See e.g. [9, 14] and references therein

[44]   H. Kalt, Optical Properties of III-V Semiconductors: the Influence of the Multi-Valley Bandstructure, Springer Series in Solid State Sciences, Springer, Berlin (1995)

[45]   Excitonic Optical Nonlinearities, D.S. Chemla ed. JOSA B2, 1135-1243 (1985)

[46]   Semiconductor Quantum Wells and Superlattices: Physics and Applications, D.S. Chemla and A. Pinczuk eds., IEEE J. QE 22, 1609-1921 (1986)
Quantum Well Heterostructures and Superlattices, J.J. Coleman ed., IEEE J. QE 24, 1579-1798 (1988)

[47]   S. Schmitt-Rink, D.A.B. Miller and D.S. Chemla, Adv. Phys. 38, 9 (1989)
C. Klingshirn, Semiconductor, Science and Technology 5, 457 (1990)
E.O. Göbel and K. Ploog, Progr. Quant. Electron. 14, 289 (1990)
R. Congolani and K. Ploog, Adv. Phys. 40, 535 (1991)
R. Congolani, Physical Scripta T 49B, 470 (1993)

[48]   Ch. Gréus, L. Butov, F. Daiminger, A. Forchel, P. Knipp and T.L. Reinecke, Phys. Rev. B 47, 7626 (1993)
Ch. Gréus, R. Spiegel, P.A. Knipp, T.L. Reinecke, F. Faller and A. Forchel, ibid. B 49, 5753 (1994) and
P. Ils, A. Forchel, K.-H. Wang, Ph. Pagnod-Rossiaux and L. Goldstein, ibid. B 50, 11-46 (1994)
P. Ils, Ch. Gréus, A. Forchel, V.D. Kulakovskii, N.A. Gippius and S.G. Tikhodeev, ibid. B 51, 4273 (1995)
M. Illing, G. Bacher, T. Kümmell, A. Forchel, T.G. Andersson, D. Hommel, B. Jobst and G. Landwehr, Appl. Phys. Lett (1995), in press
and references given therein

[49]   O. Schilling, A. Forchel, A. Kohl and S. Brittner, J. Vac. Sci. Technol. (1993) in press and references given therein

[50]   U. Woggon, Optical Properties of Semiconductor Quantum Dots, Habilitation Thesis, Kaiserslautern (1995) to be published in Springer Tracts in Mod. Physics (1996)
S. Gaponenko and U. Woggon, phys. stat. sol. b (1995) in press and referneces therein.

[51]   R. Reisfeld, NATO ASI Series B 301, 601 (1992) and ibid. 339, 491 (1995) and refernces given therein

[52]   C. Klingshirn, phys. stat. sol. b 71, 547 (1975)

[53]   S. Permogarov in Excitons, E.I. Rashba and M. Sturge eds., Mod. Probl. Cond. Mat. Sci. 2, 177 (1982)

[54]   R. Kuhnert, R. Helbig and K. Hümmer, phys. stat. sol. b 107, 83 (1981)
R. Kuhnert and R. Helbig, J. Luminesc. 26, 203 (1981)

[55]   D. Oberhauser, K.-H. Pantke, W. Langbein, V.G. Lyssenko, H. Kalt, J.M. Hvam, G. Weimann and C. Klingshirn, phys. stat. sol. b 173, 53 (1992)

[56] M.A. Herman, D. Bimberg and J. Christen, J. Appl. Phys. 70, R1 (1991)
      M. Gurioli, A. Vinattieri, M. Colocci, C. Dapris, J. Massies, G. Neu, A. Bosacchi
      and S. Franchi, Phys. Rev. B 44, 3115 (1991)
      H. Yu, P.B. Mookherjee, R. Murray and A. Yoshinaga, J. Appl. Phys. 77, 1217
      (1995)

[57] J.A. Gaj, W. Grieshaber, C. Bodin-Deshayes, J. Cibert, G. Fenillet, Y. Merle
      d'Aubigné and A. Wasiela, Phys. Rev. B 50, 5512 (1994) and references given
      therein

[58] Fang Yang, M. Wilkinson, E.J. Austin and K.P. O'Donnell, Phys. Rev. Lett 70,
      323 (1993) and ibid. 72, 1945 (1994)

[59] K.-H. Pantke, D. Oberhauser, V.G. Lyssenko, J.M. Hvam and G. Weimann,
      Phys. Rev. B 47, 2413 (1993)

[60] U. Woggon, O. Wind, W. Langbein, O. Gogolin and C. Klingshirn, J. Luminesc.
      59, 135 (1994)

[61] S.W. Koch, H. Haug, G. Schmieder, K. Bohnert and C. Klingshirn, phys. stat. sol.
      b 89, 431 (1978)
      R. Lindwurm and H. Haug, Z. Physik B 53, 281 (1983) and references therein

[62] H.E. Swoboda, F.A. Majumder, V.G. Lyssenko, C. Klingshirn and L. Banyai, Z.
      Physik B 70, 341 (1988)

[63] K.-H. Schlaad, Ch. Weber, D.S. Chemla, J. Cunningham, C.V. Hoof, G. Borghs,
      G. Weimann, W. Schlapp, H. Nickel and C. Klingshirn, phys. stat. sol. b 159, 173
      (1990) and Phys. Rev. B. 43, 4268 (1991)

[64] H. Kalt, Festkörperprobleme / Advances in Solid State Physics 32, 145 (1992)

[65] K. Bohnert, M. Anselment, G. Kobbe, C. Klingshirn, H. Haug, S.W. Koch, S.
      Schmitt-Rink and F.F. Abraham, Z. Physik B 42, 1 (1991)

[66] K.J. Ebeling, Integrierte Optoelektronik, 2nd ed, Springer, Berlin (1992)

[67] See e.g. S.M. Sze, Physics of Semiconductor Devices, 2nd ed., Wiley-Interscience,
      New York (1981) or [11] and references therein

[68] M.A. Haase, J. Qiu, J.M. De Puydt and H. Cheng, Appl. Phys. Lett 59, 1272
      (1991)
      H. Jeon, J. Ding, W. Patterson, A.V. Nurmikko, W. Xie, D.C. Grillo, M. Kobay-
      ashi and R.L. Gunshor, ibid., p 3619

[69] The proceedings of the Intern. Conf. on II-VI Semiconductors have been published in
      J. Crystal Growth <u>59</u> (1982)
      J. Crystal Growth <u>72</u> (1985)
      J. Crystal Growth <u>86</u> (1988)
      J. Crystal Growth <u>101</u> (1990)
      J. Crystal Growth <u>117</u> (1992)
      J. Crystal Growth <u>138</u> (1994)

[70] The proceeding of some other, preferentially European Workshops on II-VI Compounds can be found e.g. in NATO ASI Series B <u>200</u> (1989)
      Semicond. Sci. and Technol. <u>6</u>, 9A (1991)
      Advanced Materials for Optics and Electronics <u>3</u>, 1-308 (1991)
      Materials Science Forum (1995) in press

[71] J. Ding et al., Phys. Rev. B <u>47</u>,10528 (1993)
      C. Klingshirn et al., J. Crystal Growth, <u>138</u>, 787 (1994)
      J. Gutowski et al., phys. stat. sol. b. <u>187</u>, 423 (1995)
      M. Lowisch, F. Kreller, J. Puls and F. Henneberger, Japn. J. Appl. Phys. <u>34</u>, 83 (1995)
      and references given therein

[72] S.T. Kim, H. Amano, I. Akasaki and N. Koide, Appl. Phys. Lett. <u>64</u>, 1535 (1994) and references given therein
      S. Nakamura, T. Mukai and M. Senoh, ibid., p 1687
      T. Matsuoka, phys. stat. sol. b <u>187</u>,471 (1995)
      S. Strite, Festkörperprobleme / Advances in Solid State Physics <u>34</u>, 79 (1995)

[73] R. Dingle, K.L. Shaklee, R.F. Leheny and R.B. Zetterstrom, Appl. Phys. Lett. <u>19</u> 5 (1971)
      J.M. Hvam and E. Ejder, J. Luminesc. <u>12/13</u>, 611 (1976)
      I.M. Catalano, A. Cingolani, M. Ferrara, M. Lugarà and A. Minafra, Solid State Commun. <u>25</u>, 349 (1978)
      R. Dai, W. Zhuang, K. Bohnert and C. Klingshirn, Z. Physik B <u>46</u>, 189 (1982)

THEORETICAL DESCRIPTION OF COLLECTIVE EXCITATIONS: BLOCH EQUATIONS AND RELAXATION MECHANISMS

R. Zimmermann

Max Planck Group "Semiconductor Theory"
at Humboldt University Berlin
Hausvogteiplatz 5-7, D-10117 Berlin

ABSTRACT

Using second quantization and the Bloch representation, the equation of motion for the density matrix in a solid is derived. The decoupling on the Hartree Fock level leads to the semiconductor Bloch equations. Excitons are introduced as collective electron-hole excitations. The optical Stark effect and electron relaxation by phonon scattering are presented as application of the density matrix technique.

I. INTRODUCTION

I. A. Maxwell's equation and optical response

In optical transmission and reflection experiments only one light beam is used, and the absorption and refraction coefficient can be determined. Nonlinear optical experiments have typically two beams applied (pump and probe). Examples are the induced or nonlinear response, the optical Stark effect, and four wave mixing.

Whereas here a classical description of the light field suffices, the understanding of luminescence (spontaneous and stimulated) or lasing needs a quantized treatment (photons). In the present article the so-called semiclassical level is used, which means to treat the light field classically, but interacting with the quantized material system. More details on the theory of semiconductor optics along these lines can be found in the monography by Haug and Koch [1].

After eliminating the magnetic field, Maxwells equation condense into the wave equation for the electric field

$$\Delta_{\mathbf{r}}\mathbf{E} - \frac{1}{c^2}\partial_{tt}\mathbf{E} = \frac{4\pi e}{c^2}\partial_t\mathbf{j} \tag{1}$$

where the electronic current **j** acts as source of the induced field. Introducing electrodynamic potentials via $\mathbf{E}(\mathbf{r},t) = -\partial_t \mathbf{A}(\mathbf{r},t) + \nabla_{\mathbf{r}}\phi(\mathbf{r},t)$ and confining to transverse fields ($\phi = 0$) we get as wave equation for the vector potential

$$\Delta_{\mathbf{r}}\mathbf{A} - \frac{1}{c^2}\partial_{tt}\mathbf{A} = -\frac{4\pi e}{c^2}\mathbf{j} . \tag{2}$$

Fourier transformation according to

$$\mathbf{A}(\mathbf{r}t) = \frac{1}{(2\pi)^3}\int d\mathbf{q}\, \frac{1}{2\pi}\int d\omega\, e^{i(\mathbf{q}\mathbf{r}-\omega t)}\,\mathbf{A}(\mathbf{q}\omega) \tag{3}$$

gives

$$\left(-q^2 + \frac{\omega^2}{c^2}\right)\mathbf{A}(\mathbf{q}\omega) = -\frac{4\pi e}{c^2}\mathbf{j}(\mathbf{q}\omega) . \tag{4}$$

The current term can be incorporated into the left hand side defining the optical dielectric function $\epsilon(\mathbf{q}\omega)$

$$\epsilon(\mathbf{q}\omega) = 1 + \frac{4\pi e}{\omega^2}\frac{\mathbf{j}(\mathbf{q}\omega)}{\mathbf{A}(\mathbf{q}\omega)} \tag{5}$$

which modifies the light propagation in the usual way. For simplicity, we have dropped the tensor character of ϵ.

For typical semiconductor applications, the photon momentum $q = 2\pi/\lambda$ is a negligible quantity and can be put equal to zero.

I. B. Interacting electrons and second quantization

The Hamiltonian for the electrons interacting with the transverse electromagnetic field (Coulomb gauge, $div\mathbf{A} = 0$) reads

$$\mathcal{H} = \sum_{j=1}^{N}\left[\frac{1}{2m_0}(\mathbf{p}_j - e\mathbf{A}(\mathbf{r}_j t))^2 + V_c(\mathbf{r}_j)\right] + \frac{1}{2}\sum_{i\neq j}^{N}\frac{e^2}{|\mathbf{r}_i - \mathbf{r}_j|} . \tag{6}$$

The electrons move in the crystal potential $V_c(\mathbf{r})$ due to the ionic lattice, and the last term takes into account the mutual Coulomb interaction of the electrons.

The total N-particle wave function must be antisymmetric since electrons obey Fermi statistics,

$$\Psi(\mathbf{r}_1 s_1, \cdots \mathbf{r}_j s_j, \cdots \mathbf{r}_l s_l, \cdots \mathbf{r}_N s_N) = -\Psi(\mathbf{r}_1 s_1, \cdots \mathbf{r}_l s_l, \cdots \mathbf{r}_j s_j, \cdots \mathbf{r}_N s_N) . \tag{7}$$

For non-interacting particles, this is achieved by a determinant

$$(\mathbf{r}_1 s_1 \cdots \mathbf{r}_N s_N \,|\, q_1 \cdots q_n) = \frac{1}{\sqrt{N!}}\begin{vmatrix} \psi_{q_1}(\mathbf{r}_1 s_1)\,\psi_{q_1}(\mathbf{r}_2 s_2)\cdots \psi_{q_1}(\mathbf{r}_N s_N) \\ \cdots \\ \psi_{q_N}(\mathbf{r}_1 s_1)\,\psi_{q_N}(\mathbf{r}_2 s_2)\cdots \psi_{q_N}(\mathbf{r}_N s_N) \end{vmatrix} \tag{8}$$

formed with one-particle wave functions (q_j are quantum numbers). The clumsy handling of determinants can be rationalized introducing the concept of second quantization. Then,

$$(\mathbf{r}_1 s_1 \cdots \mathbf{r}_N s_N \,|\, q_1 \cdots q_n) = (\mathbf{r}_1 s_1 \cdots \mathbf{r}_N s_N \,|\, \hat{\psi}_{q_1}^{\dagger} \cdots \hat{\psi}_{q_N}^{\dagger} \,|\, O) \tag{9}$$

where $\hat{\psi}_q^{\dagger}$ is the creation operator which puts an electron into state q. The annihilation operator $\hat{\psi}_q$ takes an electron out of state q. A nice by-product is that the Hilbert space for fixed particle number N can be extended to the Fock space with variable particle number. The determinant calculus is reflected by the following (Fermion) commutation rules

$$\hat{\psi}_q^\dagger \hat{\psi}_{q'} + \hat{\psi}_{q'} \hat{\psi}_q^\dagger = [\hat{\psi}_q^\dagger, \hat{\psi}_{q'}]_+ = \delta_{qq'} \tag{10}$$
$$\hat{\psi}_q \hat{\psi}_{q'} + \hat{\psi}_{q'} \hat{\psi}_q = [\hat{\psi}_q, \hat{\psi}_{q'}]_+ = 0 \ .$$

This was the anticommutator in contrast to the usual (Boson) commutator:

$$\hat{A}\hat{B} - \hat{B}\hat{A} = [\hat{A}, \hat{B}] \ . \tag{11}$$

Now we are equipped to introduce the density operator $\hat{N}_q = \hat{\psi}_q^\dagger \hat{\psi}_q$ whose eigenvalues can be "1" (state q is occupied) and "0" (state q is empty). In contrast, for Bosons any occupation number $n = 0, 1, 2, \cdots$ is possible. The expectation value

$$(q_1 \cdots q_N | \hat{\psi}_q^\dagger \hat{\psi}_q | q_1 \cdots q_N) \equiv \langle \hat{\psi}_q^\dagger \hat{\psi}_q \rangle^0 = 1 \text{ or } 0 \tag{12}$$

tests if state q is present in $q_1 \cdots q_N$ or not.

The extension from a non-interacting wave function to the full one and to a statistical ensemble (see below) leads to the density matrix defined as

$$\rho_{qq'}(t) = \langle \hat{\psi}_{q'}^\dagger \hat{\psi}_q \rangle \ . \tag{13}$$

Using space and spin $(\mathbf{r}, s)$ as "quantum numbers" we have

$$[\hat{\psi}_s^\dagger(\mathbf{r}), \hat{\psi}_{s'}(\mathbf{r}')]_+ = \delta_{ss'}\delta(\mathbf{r} - \mathbf{r}') \tag{14}$$

and can apply the replacement

$$F(\mathbf{r}_j s_j) \Rightarrow \sum_s \int d\mathbf{r} \, \hat{\psi}_s^\dagger(\mathbf{r}) \, F(\mathbf{r}s) \, \hat{\psi}_s(\mathbf{r}) \tag{15}$$

to rewrite the Hamiltonian (6) into second quantized form

$$\hat{\mathcal{H}} = \sum_s \int d\mathbf{r} \, \hat{\psi}_s^\dagger(\mathbf{r}) \left[\frac{1}{2m_0} (-i\hbar\nabla_{\mathbf{r}} - e\mathbf{A}(\mathbf{r}t))^2 + V_c(\mathbf{r}) \right] \hat{\psi}_s(\mathbf{r}) \tag{16}$$
$$+ \frac{1}{2} \sum_{ss'} \int d\mathbf{r} \, d\mathbf{r}' \, \hat{\psi}_s^\dagger(\mathbf{r}) \hat{\psi}_{s'}^\dagger(\mathbf{r}') \frac{e^2}{|\mathbf{r}_i - \mathbf{r}_j|} \hat{\psi}_{s'}(\mathbf{r}') \hat{\psi}_s(\mathbf{r}) \ .$$

I. C. Equation of motion and density matrix

We start with the N-particle Schrödinger equation and form the expectation value of the operator $\hat{B}$

$$i\hbar\partial_t \Psi(t) = \hat{\mathcal{H}}\Psi(t) \qquad \langle \hat{B} \rangle_t = (\Psi(t), \hat{B}\Psi(t)) \ . \tag{17}$$

The time derivative gives

$$i\hbar\partial_t \langle \hat{B} \rangle_t = (-\hat{\mathcal{H}}\Psi, \hat{B}\Psi(t)) + (\Psi, \hat{B}\hat{\mathcal{H}}\Psi(t)) \tag{18}$$
$$= (\Psi, (\hat{B}\hat{\mathcal{H}} - \hat{\mathcal{H}}\hat{B})\Psi) = \langle [\hat{B}, \hat{\mathcal{H}}] \rangle \ .$$

Generalizing the pure state $\Psi(t)$ to a statistical operator $\hat{W}(t)$,

$$\langle \hat{B} \rangle_t = Tr(\hat{W}(t)\hat{B}) \ , \tag{19}$$

allows to rewrite

$$i\hbar\partial_t\langle\hat{B}\rangle_t = Tr(\hat{W}(t)[\hat{B},\hat{\mathcal{H}}]) = Tr([\hat{\mathcal{H}},\hat{W}(t)]\hat{B}) \tag{20}$$

which can be understood as equation of motion for the statistical operator

$$i\hbar\partial_t\hat{W}(t) = [\hat{\mathcal{H}},\hat{W}(t)] \ . \tag{21}$$

Introducing the time evolution operator $\hat{U}(t)$ we have as solution

$$i\hbar\partial_t\hat{U}(t) = \hat{\mathcal{H}}\hat{U}(t) \ ; \quad \hat{W}(t) = \hat{U}(t-t_0)\,\hat{W}(t_0)\,\hat{U}^\dagger(t-t_0) \tag{22}$$

and find

$$\langle\hat{B}\rangle_t = Tr(\hat{W}(t_0)\,\hat{U}^\dagger(t-t_0)\,\hat{B}\,\hat{U}(t-t_0)) \overset{\text{def}}{=} Tr(\hat{W}(t_0)\,\hat{B}(t)) \ . \tag{23}$$

By moving the time dependence from $\hat{W}$ to $\hat{B}$ we have just switched to the Heisenberg picture

$$-i\hbar\partial_t\hat{B}(t) = [\hat{\mathcal{H}},\hat{B}(t)] \ . \tag{24}$$

Note that the commutation rules hold for equal times only!

Now we can define the density matrix with Heisenberg operators as

$$\rho_{ss'}(\mathbf{rr}'t) = Tr\left(\hat{W}(t_0)\,\hat{\psi}_{s'}^\dagger(\mathbf{r}'t)\,\hat{\psi}_s(\mathbf{r}t)\right) \equiv \langle\hat{\psi}_{s'}^\dagger(\mathbf{r}'t)\,\hat{\psi}_s(\mathbf{r}t)\rangle \ . \tag{25}$$

Within first quantization the particle current reads

$$\mathbf{j} = \sum_{j=1}^{N}\partial_t\mathbf{r}_j = \frac{1}{m_0}\sum_{j=1}^{N}(\mathbf{p}_j - e\mathbf{A}(\mathbf{r}_jt)) \ . \tag{26}$$

The local current is given by the symmetric expression

$$\mathbf{j}(\mathbf{r}) = \frac{1}{2m_0}\sum_{j=1}^{N}[(\mathbf{p}_j - e\mathbf{A}(\mathbf{r}_jt))\delta(\mathbf{r}-\mathbf{r}_j) + \delta(\mathbf{r}-\mathbf{r}_j)(\mathbf{p}_j - e\mathbf{A}(\mathbf{r}_jt))] \tag{27}$$

which transforms into second quantization and Heisenberg picture as

$$\hat{\mathbf{j}}(\mathbf{r}t) = \frac{-i\hbar}{2m_0}\sum_s\left(\hat{\psi}_s^\dagger(\mathbf{r}t)\nabla_\mathbf{r}\hat{\psi}_s(\mathbf{r}t) - \left[\nabla_\mathbf{r}\hat{\psi}_s^\dagger(\mathbf{r}t)\right]\hat{\psi}_s(\mathbf{r}t)\right) - \frac{e}{m_0}\mathbf{A}(\mathbf{r}t)\sum_s\hat{\psi}_s^\dagger(\mathbf{r}t)\hat{\psi}_s(\mathbf{r}t) \ . \tag{28}$$

Now, the current expectation value can be expressed as

$$\mathbf{j}(\mathbf{r}t) = \frac{-i\hbar}{2m_0}\sum_s(\nabla_\mathbf{r} - \nabla_{\mathbf{r}'})\,\rho_{ss}(\mathbf{rr}'t)\bigg|_{\mathbf{r}=\mathbf{r}'} - \frac{e}{m_0}\mathbf{A}(\mathbf{r}t)\sum_s\rho_{ss}(\mathbf{rr}t) \tag{29}$$

pointing to the importance of spatially non-diagonal elements in the density matrix.

I. D. Bloch representation

For a periodic solid it is advantageous to work in reciprocal space. Bloch functions $\phi_{\mathbf{k}\nu}$ and band structure $E_{\mathbf{k}\nu}$ are defined as

$$\left[-\frac{\hbar^2}{2m_0}\Delta_\mathbf{r} + V_c(\mathbf{r})\right]\phi_{\mathbf{k}\nu}(\mathbf{r}s) = E_{\mathbf{k}\nu}\,\phi_{\mathbf{k}\nu}(\mathbf{r}s) \tag{30}$$

where the band index ν is understood to contain the spin degeneracy. The periodicity of the crystal potential leads to the Bloch property

$$\phi_{\mathbf{k}\nu}(\mathbf{r} + \mathbf{R}, s) = e^{i\mathbf{k}\mathbf{R}} \phi_{\mathbf{k}\nu}(\mathbf{r}, s) \tag{31}$$

where $\mathbf{R}$ are the lattice vectors. Bloch creation operators are introduce via

$$\hat{\psi}_s^\dagger(\mathbf{r}t) = \sum_{\mathbf{k}\nu} c_{\mathbf{k}\nu}^\dagger \phi_{\mathbf{k}\nu}^*(\mathbf{r}, s) \tag{32}$$

(for simplicity we write $c_{\mathbf{k}\nu}^\dagger$ instead of $\hat{c}_{\mathbf{k}\nu}^\dagger$).

Neglecting the spatial dependence of $\mathbf{A}(\mathbf{r}t)$ which is valid in the longwave limit, we transform the Hamiltonian (16) into the Bloch basis. The kinetic part gives

$$\hat{\mathcal{H}}^{kin} = \sum_{\mathbf{k}\nu} \left(E_{\mathbf{k}\nu} + \frac{e^2 A^2(t)}{2m_0} \right) c_{\mathbf{k}\nu}^\dagger c_{\mathbf{k}\nu} - \frac{eA(t)}{m_0} \sum_{\mathbf{k}\nu\nu'} \mathbf{p}_{\mathbf{k}\nu'\nu} c_{\mathbf{k}\nu'}^\dagger c_{\mathbf{k}\nu} \tag{33}$$

with the momentum matrix element

$$\mathbf{p}_{\mathbf{k}\nu'\nu} = \int d\mathbf{r}\, \phi_{\mathbf{k}\nu'}^*(\mathbf{r})(-i\hbar\nabla_{\mathbf{r}})\phi_{\mathbf{k}\nu}(\mathbf{r}) . \tag{34}$$

The band diagonal term is simply $\hbar\mathbf{p}_{\mathbf{k}\nu\nu} = m_0\nabla_{\mathbf{k}}E_{\mathbf{k}\nu}$. Taking the Bloch overlap integrals at $\mathbf{k} = 0$ which is a reliable approximation for direct gap semiconductors, the Coulomb interaction attains the simple form

$$\hat{\mathcal{H}}^{Coul} = \frac{1}{2} \sum_{\mathbf{k}\mathbf{k}'\mathbf{q}\nu\nu'} v_{\mathbf{q}}\, c_{\mathbf{k}+\mathbf{q}\nu}^\dagger c_{\mathbf{k}'-\mathbf{q}\nu'}^\dagger c_{\mathbf{k}'\nu'} c_{\mathbf{k}\nu} \tag{35}$$

with the Fourier transform of the Coulomb potential, $v_{\mathbf{q}} = 4\pi e^2/\Omega q^2$. The current (averaged over the normalization volume Ω) is

$$\mathbf{j}(t) = \frac{1}{m_0\Omega} \sum_{\mathbf{k}\nu\nu'} \rho_{\mathbf{k}\nu\nu'}(t)\left(\mathbf{p}_{\mathbf{k}\nu'\nu} - e\mathbf{A}(t)\delta_{\nu\nu'}\right) \tag{36}$$

with the density matrix in Bloch representation

$$\rho_{\mathbf{k}\nu\nu'}(t) = Tr\left(\hat{W}(t_0) c_{\mathbf{k}\nu'}^\dagger(t) c_{\mathbf{k}\nu}(t)\right) \equiv \langle c_{\mathbf{k}\nu'}^\dagger(t) c_{\mathbf{k}\nu}(t) \rangle . \tag{37}$$

Its equation of motion reads

$$-i\hbar\partial_t\rho_{\mathbf{k}\nu\nu'} = \langle[\hat{\mathcal{H}}, c_{\mathbf{k}\nu'}^\dagger]c_{\mathbf{k}\nu}\rangle + \langle c_{\mathbf{k}\nu'}^\dagger[\hat{\mathcal{H}}, c_{\mathbf{k}\nu}]\rangle \tag{38}$$

and is evaluated using the commutator rules

$$[c_1^\dagger c_2, c_3^\dagger] = c_1^\dagger c_2 c_3^\dagger - c_3^\dagger c_1^\dagger c_2 = c_1^\dagger c_2 c_3^\dagger + c_1^\dagger c_3^\dagger c_2 = c_1^\dagger[c_2, c_3^\dagger]_+ = c_1^\dagger \delta_{2,3} \tag{39}$$

with the result

$$[\hat{\mathcal{H}}, c_{\mathbf{k}\nu'}^\dagger] = \left(E_{\mathbf{k}\nu'} + \frac{e^2 A^2(t)}{2m_0} \right) c_{\mathbf{k}\nu'}^\dagger - \frac{eA(t)}{m_0} \sum_\nu \mathbf{p}_{\mathbf{k}\nu'\nu} c_{\mathbf{k}\nu}^\dagger + \sum_{\mathbf{k}'\mathbf{q}\eta} v_{\mathbf{q}}\, c_{\mathbf{k}+\mathbf{q}\nu'}^\dagger c_{\mathbf{k}'-\mathbf{q}\eta}^\dagger c_{\mathbf{k}'\eta} . \tag{40}$$

Together with the Hermitian conjugate for $-[\hat{\mathcal{H}}, c_{\mathbf{k}\nu}]$ we obtain

$$-i\hbar\partial_t\rho_{\mathbf{k}\nu\nu'} = (E_{\mathbf{k}\nu'} - E_{\mathbf{k}\nu})\rho_{\mathbf{k}\nu\nu'} - \frac{eA(t)}{m_0} \sum_\eta \left(\rho_{\mathbf{k}\nu\eta}\, \mathbf{p}_{\mathbf{k}\eta\nu'} - \mathbf{p}_{\mathbf{k}\nu\eta}\, \rho_{\mathbf{k}\eta\nu'} \right) \tag{41}$$

$$+ \sum_{\mathbf{k}'\mathbf{q}\eta} v_{\mathbf{q}} \left(\langle c_{\mathbf{k}+\mathbf{q}\nu'}^\dagger c_{\mathbf{k}'-\mathbf{q}\eta}^\dagger c_{\mathbf{k}'\eta} c_{\mathbf{k}\nu}\rangle - \langle c_{\mathbf{k}\nu'}^\dagger c_{\mathbf{k}'\eta}^\dagger c_{\mathbf{k}'-\mathbf{q}\eta} c_{\mathbf{k}+\mathbf{q}\nu}\rangle \right) .$$

Note that the A^2 terms have dropped out completely. Since the two-operator expectation value (density matrix) couples to a four-operator one, we face here a typical hierarchy problem. The simplest decoupling which leads to a closed set of equations is discussed in the next chapter.

127

II. THE SEMICONDUCTOR BLOCH EQUATIONS

II. A. Response of independent particles

For the time being, we drop the Coulomb interaction and linearize with respect to **A**. The equilibrium density matrix ($\mathbf{A} = 0$) is given by

$$\rho^{eq}_{\mathbf{k}\nu\nu'} = \delta_{\nu\nu'} f(E_{\mathbf{k}\nu}); \quad f(E) = \frac{1}{e^{(E-\mu)/k_B T} + 1} \tag{42}$$

where the Fermi function $f(E)$ results from $W^{eq} = \exp(-(\hat{\mathcal{H}} - \mu\hat{N})/k_B T)$, the statistical operator in equilibrium (μ – chemical potential). Eq. (41) reduces then to

$$- i\hbar \partial_t \rho_{\mathbf{k}\nu\nu'} = (E_{\mathbf{k}\nu'} - E_{\mathbf{k}\nu}) \rho_{\mathbf{k}\nu\nu'} - \frac{e\mathbf{A}(t)}{m_0} (f(E_{\mathbf{k}\nu}) - f(E_{\mathbf{k}\nu'})) \mathbf{p}_{\mathbf{k}\nu\nu'} \tag{43}$$

or Fourier transformed ($\mathbf{A}(t) = \mathbf{A}(\omega) \exp(-i\omega t)$)

$$(E_{\mathbf{k}\nu} - E_{\mathbf{k}\nu'} - \hbar\omega) \rho_{\mathbf{k}\nu\nu'}(\omega) = -\frac{e\mathbf{A}(\omega)}{m_0} (f(E_{\mathbf{k}\nu}) - f(E_{\mathbf{k}\nu'})) \mathbf{p}_{\mathbf{k}\nu\nu'} . \tag{44}$$

Insertion into the current

$$\mathbf{j}(\omega) = \frac{1}{m_0 \Omega} \sum_{\mathbf{k}\nu\nu'} \rho_{\mathbf{k}\nu\nu'}(\omega) \mathbf{p}_{\mathbf{k}\nu'\nu} - \frac{e\mathbf{A}(\omega)}{m_0 \Omega} \sum_{\mathbf{k}\nu} f(E_{\mathbf{k}\nu}) \tag{45}$$

gives for the optical dielectric function (5)

$$\epsilon(\omega) = 1 - \frac{4\pi e^2}{m_0 \omega^2} \left(\frac{1}{\Omega} \sum_{\mathbf{k}\nu\nu'} \frac{|\mathbf{p}_{\mathbf{k}\nu\nu'}|^2}{m_0} \frac{f(E_{\mathbf{k}\nu}) - f(E_{\mathbf{k}\nu'})}{E_{\mathbf{k}\nu} - E_{\mathbf{k}\nu'} - \hbar\omega} + \frac{1}{\Omega} \sum_{\mathbf{k}\nu} f(E_{\mathbf{k}\nu}) \right) \tag{46}$$

which is called Ehrenreich-Cohen expression. The last term is the bare plasma response $-\omega_{pl}^2/\omega^2$ with $\omega_{pl}^2 = 4\pi n_0 e^2/m_0$ (n_0 is the total electron density).

The imaginary part is closely related to the absorption coefficient

$$\mathrm{Im}\,\epsilon(\omega) = \frac{4\pi^2 e^2 \hbar^2}{m_0^2} \frac{1}{\Omega} \sum_{\mathbf{k}\nu\nu'} \frac{|\mathbf{p}_{\mathbf{k}\nu\nu'}|^2}{(E_{\mathbf{k}\nu} - E_{\mathbf{k}\nu'})^2} (f(E_{\mathbf{k}\nu'}) - f(E_{\mathbf{k}\nu})) \delta(E_{\mathbf{k}\nu} - E_{\mathbf{k}\nu'} - \hbar\omega) \tag{47}$$

and describes direct interband transitions between occupied and empty bands. At zero temperature, theses are valence ($f(E_{\mathbf{k}\nu'}) = 1$) and conduction bands ($f(E_{\mathbf{k}\nu}) = 0$).

II. B. Hartree-Fock decoupling

When including the Coulomb interaction, we have to deal with 4-operator expectation values as $\langle c_1^\dagger c_2^\dagger c_3 c_4 \rangle$. The Hartree-Fock (HF) approximation keeps only correlations on the statistical level. We show that this is correct for any non-interacting state

$$|Z\rangle = c_{z1}^\dagger c_{z2}^\dagger \cdots c_{zN}^\dagger |0\rangle \tag{48}$$

where N different quantum states $z1 \cdots zN$ are occupied.
The single-particle density is here

$$\langle c_1^\dagger c_2 \rangle \equiv (Z | c_1^\dagger c_2 | Z) = \delta_{1,2} \cdot \delta_{2 \in Z} \tag{49}$$

and one realizes the rule: What has been annihilated must be created again! In

$$\langle c_1^\dagger c_2^\dagger c_3 c_4 \rangle \equiv (Z | c_1^\dagger c_2^\dagger c_3 c_4 | Z) \tag{50}$$

3 and 4 are annihilated and must therefore occur within $|Z\rangle$. But there are two possibilities to create 3 and 4 again:

$$2 = 3 \quad \text{and} \quad 1 = 4 : \quad +\delta_{2,3} \cdot \delta_{1,4} \cdot \delta_{3 \in Z} \cdot \delta_{4 \in Z} \tag{51}$$

and a "crossed" or "exchange term" with one additional commutation (sign change):

$$2 = 4 \quad \text{and} \quad 1 = 3 : \quad -\delta_{2,4} \cdot \delta_{1,3} \cdot \delta_{3 \in Z} \cdot \delta_{4 \in Z} \; . \tag{52}$$

This is combined into ($3 \neq 4$ holds automatically)

$$(Z | c_1^\dagger c_2^\dagger c_3 c_4 | Z) = (\delta_{2,3} \cdot \delta_{1,4} - \delta_{2,4} \cdot \delta_{1,3}) \cdot \delta_{3 \in Z} \cdot \delta_{4 \in Z} \tag{53}$$

and can be rewritten in terms of single-particle densities

$$\langle c_1^\dagger c_2^\dagger c_3 c_4 \rangle \Rightarrow \langle c_1^\dagger c_4 \rangle \langle c_2^\dagger c_3 \rangle - \langle c_1^\dagger c_3 \rangle \langle c_2^\dagger c_4 \rangle \; . \tag{54}$$

This is the Hartree-Fock decoupling or "mean field approximation" which is applied even if one-particle densities are non-diagonal (unrestricted HF).

In the HF decoupling of eq. (41) we start with

$$\langle c_{\mathbf{k}+\mathbf{q}\nu'}^\dagger c_{\mathbf{k}'-\mathbf{q}\eta}^\dagger c_{\mathbf{k}'\eta} c_{\mathbf{k}\nu} \rangle \Rightarrow \delta_{\mathbf{q},0} \langle c_{\mathbf{k}\nu'}^\dagger c_{\mathbf{k}\nu} \rangle \langle c_{\mathbf{k}'\eta}^\dagger c_{\mathbf{k}'\eta} \rangle - \delta_{\mathbf{k}+\mathbf{q},\mathbf{k}'} \langle c_{\mathbf{k}'\nu'}^\dagger c_{\mathbf{k}'\eta} \rangle \langle c_{\mathbf{k}\eta}^\dagger c_{\mathbf{k}\nu} \rangle \tag{55}$$

(note that the density matrix is diagonal in $\mathbf{k}$) and obtain

$$\sum_\mathbf{q} v_\mathbf{q} \langle c_{\mathbf{k}+\mathbf{q}\nu'}^\dagger c_{\mathbf{k}'-\mathbf{q}\eta}^\dagger c_{\mathbf{k}'\eta} c_{\mathbf{k}\nu} \rangle \Rightarrow \sum_{\mathbf{k}'} \left(v_{\mathbf{q}=0}\, \rho_{\mathbf{k}\nu\nu'}\, \rho_{\mathbf{k}'\eta\eta} - v_{\mathbf{k}-\mathbf{k}'}\, \rho_{\mathbf{k}'\eta\nu'}\, \rho_{\mathbf{k}\nu\eta} \right) \; . \tag{56}$$

The first term is the direct or Hartree term, the second one the exchange or Fock term. In the present case, the Hartree term is cancelled when adding the other 4-operator decoupling:

$$-i\hbar \partial_t \rho_{\mathbf{k}\nu\nu'} = (E_{\mathbf{k}\nu'} - E_{\mathbf{k}\nu}) \rho_{\mathbf{k}\nu\nu'} - \frac{e\mathbf{A}(t)}{m_0} \sum_\eta \left(\rho_{\mathbf{k}\nu\eta}\, \mathbf{P}_{\mathbf{k}\eta\nu'} - \mathbf{P}_{\mathbf{k}\nu\eta}\, \rho_{\mathbf{k}\eta\nu'} \right) \tag{57}$$

$$- \sum_{\mathbf{k}'\eta} v_{\mathbf{k}-\mathbf{k}'} \left[\rho_{\mathbf{k}\nu\eta} \rho_{\mathbf{k}'\eta\nu'} - \rho_{\mathbf{k}'\nu\eta} \rho_{\mathbf{k}\eta\nu'} \right] \; .$$

These are the "semiconductor Bloch equations" for the density matrix, sometimes written with the "effective" (that is Coulomb modified) Rabi frequency

$$\Omega_{\mathbf{k}\eta\nu'}^{\text{eff}}(t) = \frac{e\mathbf{A}(t)}{m_0} \mathbf{P}_{\mathbf{k}\eta\nu'} + \sum_{\mathbf{k}'} v_{\mathbf{k}-\mathbf{k}'} \rho_{\mathbf{k}'\eta\nu'} \tag{58}$$

as

$$-i\hbar \partial_t \rho_{\mathbf{k}\nu\nu'} = (E_{\mathbf{k}\nu'} - E_{\mathbf{k}\nu}) \rho_{\mathbf{k}\nu\nu'} - \sum_\eta \left(\rho_{\mathbf{k}\nu\eta}\, \Omega_{\mathbf{k}\eta\nu'}^{\text{eff}}(t) - \Omega_{\mathbf{k}\nu\eta}^{\text{eff}}(t)\, \rho_{\mathbf{k}\eta\nu'} \right) \; . \tag{59}$$

Let us restrict to a two-band situation ($\nu = c, v$) and introduce

polarization	$P_\mathbf{k} = \rho_{\mathbf{k}cv}$	$P_k^* = \rho_{\mathbf{k}vc}$	(60)
distribution	$f_{\mathbf{k}c} = \rho_{\mathbf{k}cc}$	$f_{\mathbf{k}v} = \rho_{\mathbf{k}vv}$.	(61)

The polarization equation reads

$$-i\hbar\partial_t P_{\mathbf{k}} = \left(-E_{\mathbf{k}cv} - \sum_{\mathbf{k}'} v_{\mathbf{k}-\mathbf{k}'}(f_{\mathbf{k}'v} - 1 - f_{\mathbf{k}'c}) + \frac{e\mathbf{A}(t)}{\hbar}\nabla_{\mathbf{k}}E_{\mathbf{k}cv}\right) P_{\mathbf{k}} \quad (62)$$

$$-(f_{\mathbf{k}c} - f_{\mathbf{k}v})\left(\frac{e\mathbf{A}(t)}{m_0}\mathbf{p}_{\mathbf{k}cv} + \sum_{\mathbf{k}'} v_{\mathbf{k}-\mathbf{k}'}P_{\mathbf{k}'}\right).$$

The interband energy $E_{\mathbf{k}cv} = E_{\mathbf{k}c} - E_{\mathbf{k}v}$ is corrected by HF self energies (and light field) in the first line. Note that $E_{\mathbf{k}v}$ contains already the HF shift of the filled valence band, thus the "1" has to be subtracted. The second line gives the Pauli blocking of the effective Rabi coupling (or gap function)

$$\Delta_{\mathbf{k}} = \frac{e\mathbf{A}(t)}{m_0}\mathbf{p}_{\mathbf{k}cv} + \sum_{\mathbf{k}'} v_{\mathbf{k}-\mathbf{k}'}P_{\mathbf{k}'}. \quad (63)$$

In the equations for the distributions all energy and self-energy terms cancel, and the Coulomb interaction gives $-\rho_{\mathbf{k}cv}\rho_{\mathbf{k}'vc} + \rho_{\mathbf{k}'cv}\rho_{\mathbf{k}vc}$ resulting in

$$-i\hbar\partial_t f_{\mathbf{k}c} = \Delta_{\mathbf{k}}P_{\mathbf{k}}^* - P_{\mathbf{k}}\Delta_{\mathbf{k}}^*. \quad (64)$$

For the conduction band distribution follows therefore the kinetic equation

$$\partial_t f_{\mathbf{k}c} = (2/\hbar)\mathrm{Im}(P_{\mathbf{k}}\Delta_{\mathbf{k}}^*); \quad \partial_t f_{\mathbf{k}v} = -(2/\hbar)\mathrm{Im}(P_{\mathbf{k}}\Delta_{\mathbf{k}}^*) \quad (65)$$

which has been complemented by a similar one for the valence band.

II. C. Linear response of excitons

Electrons and holes can form bound states in close analogy to the Hydrogen atom. The corresponding Schrödinger equation is

$$\left(E_g - \frac{\hbar^2}{2m_e}\Delta_{\mathbf{r}_e} - \frac{\hbar^2}{2m_h}\Delta_{\mathbf{r}_h} - \frac{e^2}{\epsilon_0|\mathbf{r}_e - \mathbf{r}_h|}\right)\Phi_n(\mathbf{r}_e\mathbf{r}_h) = \mathcal{E}_n\Phi_n(\mathbf{r}_e\mathbf{r}_h). \quad (66)$$

Direct optical transitions have nearly zero momentum transfer, and only the relative motion of the exciton is important, $\Phi_n(\mathbf{r}_e\mathbf{r}_h) = \phi_n(\mathbf{r} = \mathbf{r}_e - \mathbf{r}_h)$. Therefore,

$$\left(E_g - \frac{\hbar^2}{2\mu}\Delta_{\mathbf{r}} - \frac{e^2}{\epsilon_0 r}\right)\phi_n(\mathbf{r}) = \mathcal{E}_n\phi_n(\mathbf{r}) \quad (67)$$

suffices to calculate the optical properties ($1/\mu = 1/m_e + 1/m_h$ defines the reduced effective mass). The absorption is proportional to

$$\mathrm{Im}\,\epsilon(\omega) \sim \sum_n |\phi_n(\mathbf{r} = 0)|^2\delta(\hbar\omega - \mathcal{E}_n), \quad (68)$$

and oviously only s-states contribute. The bound states have energies

$$\mathcal{E}_n = E_g - \frac{Ry^*}{n^2} \quad (69)$$

with the effective Rydberg $Ry^* = 13.6\,\mathrm{eV}\cdot\mu/(m_0\epsilon_0^2)$ and the effective Bohr radius $a^* = 0.053\,\mathrm{nm}\cdot\epsilon_0 m_0/\mu$. For bulk GaAs, these values are $Ry^* = 4\,\mathrm{meV}$ and $a^* = 11\,\mathrm{nm}$.

Instead of the square-root interband density of states for non-interacting electron-hole pairs, we end up with a series of sharp lines at bound states and a Sommerfeld enhanced continuum

$$Im\,\epsilon(\omega) \sim \sum_{n=1}^{\infty} \frac{2}{n^3} \delta(E + 1/n^2) + \frac{\Theta(E)}{1 - \exp(-2\pi/\sqrt{E})} \qquad (70)$$

with reduced energy $E = (\hbar\omega - E_g)/Ry^*$.

But where are excitons in the density matrix theory? The polarization equation in HF was

$$-i\hbar\partial_t P_{\mathbf{k}} = \left(-E_{\mathbf{k}cv} - \sum_{\mathbf{k'}} v_{\mathbf{k}-\mathbf{k'}}(f_{\mathbf{k'}v} - 1 - f_{\mathbf{k'}c}) - \frac{e\mathbf{A}(t)}{\hbar}\nabla_{\mathbf{k}}(E_{\mathbf{k}v} - E_{\mathbf{k}c})\right) P_{\mathbf{k}} \quad (71)$$

$$-(f_{\mathbf{k}c} - f_{\mathbf{k}v})\left(\frac{e\mathbf{A}(t)}{m_0}\mathbf{p}_{\mathbf{k}cv} + \sum_{\mathbf{k'}} v_{\mathbf{k}-\mathbf{k'}}P_{\mathbf{k'}}\right) \,.$$

When linearizing according to $P \sim A$, the distributions reduce to Fermi functions. After Fourier transformation,

$$\left(E_{\mathbf{k}cv} - \sum_{\mathbf{k'}} v_{\mathbf{k}-\mathbf{k'}}(f(E_{\mathbf{k'}c}) + 1 - f(E_{\mathbf{k'}v})) - \hbar\omega\right) P_{\mathbf{k}}(\omega) \qquad (72)$$

$$- (f(E_{\mathbf{k}v}) - f(E_{\mathbf{k}c})) \sum_{\mathbf{k'}} v_{\mathbf{k}-\mathbf{k'}}P_{\mathbf{k'}}(\omega) = \frac{e\mathbf{A}(\omega)}{m_0}\mathbf{p}_{\mathbf{k}cv}(f(E_{\mathbf{k}v}) - f(E_{\mathbf{k}c}))$$

holds showing HF self energy, Pauli blocking of the interaction, and phase space occupation of the source term. For a non-excited semiconductor ($f_c = 0, f_v = 1$), this reduces to

$$(E_{\mathbf{k}cv} - \hbar\omega) P_{\mathbf{k}}(\omega) - \sum_{\mathbf{k'}} v_{\mathbf{k}-\mathbf{k'}}P_{\mathbf{k'}}(\omega) = \frac{e\mathbf{A}(\omega)}{m_0}\mathbf{p}_{\mathbf{k}cv} \qquad (73)$$

which is the inhomogeneous exciton equation in **k**-space. Its homogeneous form is

$$(E_{\mathbf{k}cv} - \mathcal{E}_n)\,\phi_n(\mathbf{k}) - \sum_{\mathbf{k'}} v_{\mathbf{k}-\mathbf{k'}}\phi_n(\mathbf{k'}) = 0 \,. \qquad (74)$$

Assuming a parabolic band dispersion $E_{\mathbf{k}cv} = E_g + \hbar^2 k^2/2\mu$ this is just the Fourier transform of the real-space exciton equation (compare eq. (67))

$$\left(E_g - \frac{\hbar^2}{2\mu}\Delta_{\mathbf{r}} - v(\mathbf{r})\right)\phi_n(\mathbf{r}) = \mathcal{E}_n\phi_n(\mathbf{r}) \,. \qquad (75)$$

Let us expand the polarization into the exciton states ϕ_n,

$$P_{\mathbf{k}}(\omega) = \sum_n \phi_n(\mathbf{k})C_n(\omega) \,, \qquad (76)$$

with the result

$$\sum_n (\mathcal{E}_n - \hbar\omega)\,\phi_n(\mathbf{k})C_n(\omega) = \frac{e\mathbf{A}(\omega)}{m_0}\mathbf{p}_{\mathbf{k}cv} \,. \qquad (77)$$

Using the wave function orthogonality, the polarization comes out to be

$$P_{\mathbf{k}}(\omega) = \sum_{\mathbf{k'}} G(\mathbf{k}\mathbf{k'}\omega) \frac{e\mathbf{A}(\omega)}{m_0}\mathbf{p}_{\mathbf{k'}cv} \qquad (78)$$

whith the resolvent (or excitonic Green's function) in k-space

$$G(\mathbf{k}\mathbf{k}'\omega) = \sum_n \frac{\phi_n(\mathbf{k})\,\phi_n^*(\mathbf{k}')}{\mathcal{E}_n - \hbar\omega} \,. \tag{79}$$

The current contains only interband terms (factor 2 for spin),

$$\mathbf{j}(\omega) = \frac{2}{m_0\Omega} \sum_{\mathbf{k}} \left(\rho_{\mathbf{k}cv}(\omega)\mathbf{p}_{\mathbf{k}vc} + \rho_{\mathbf{k}vc}(\omega)\mathbf{p}_{\mathbf{k}cv}\right) - \frac{e\mathbf{A}(\omega)}{m_0}n_0 \,. \tag{80}$$

By band interchange, $\rho_{\mathbf{k}cv}(\omega) = P_{\mathbf{k}}(\omega)$, $\rho_{\mathbf{k}vc}(\omega) = P_{\mathbf{k}}^*(-\omega)$ can be derived, and the optical dielectric function is cast into

$$\epsilon(\omega) = 1 + 2\frac{4\pi e^2}{m_0^2\omega^2}\sum_n \left(\frac{|M_{ncv}|^2}{\mathcal{E}_n - \hbar\omega} + \frac{|M_{ncv}|^2}{\mathcal{E}_n + \hbar\omega}\right) - \frac{4\pi n_0 e^2}{m_0\omega^2} \tag{81}$$

with the excitonic matrix element

$$M_{ncv} = \frac{1}{\Omega}\sum_{\mathbf{k}} \phi_n(\mathbf{k})\mathbf{p}_{\mathbf{k}cv} \approx p_{cv}\phi_n(\mathbf{r}=0) \tag{82}$$

for allowed interband transitions, $\mathbf{p}_{\mathbf{k}cv} = $ const. Isolating the (leading) $1s$-part gives a one-oscillator expression

$$\epsilon(\omega) = \epsilon_b + 2\frac{4\pi e^2\hbar^2 p_{cv}^2}{m_0^2\,\mathcal{E}_{1s}^2}\frac{1}{\pi a^{*3}}\frac{2\mathcal{E}_{1s}}{\mathcal{E}_{1s}^2 - \hbar\omega^2} = \epsilon_b + \frac{f_{1s}}{1 - (\hbar\omega/\mathcal{E}_{1s})^2} \tag{83}$$

where the ground-state wave function

$$\phi_{1s}(\mathbf{r}) = \frac{1}{\sqrt{\pi a^{*3}}}\,e^{-r/a^*} \tag{84}$$

has been used. The imaginary part of eq. (81) reads in full accordance with the real-space derivation eq. (68)

$$Im\,\epsilon(\omega) = 2\frac{4\pi^2 e^2\hbar^2 p_{cv}^2}{m_0^2 E_g^2}\sum_n |\phi_n(0)|^2\left(\delta(\hbar\omega - \mathcal{E}_n) - \delta(\hbar\omega + \mathcal{E}_n)\right) \,. \tag{85}$$

II. D. Semiconductor Bloch equations (rotating wave)

For a not too short pump pulse, we write

$$\mathbf{A}(t) = \frac{\mathbf{F}_p(t)}{i\omega_p}\left(e^{-i\omega_p t} - e^{i\omega_p t}\right) \tag{86}$$

with a field amplitude $\mathbf{F}_p(t)$ slowly varying on the scale of $1/\omega_p$. If $\hbar\omega_p \approx E_{\mathbf{k}cv}$ the corresponding polarization decomposition gives

$$P_{\mathbf{k}}(t) = e^{-i\omega_p t}\cdot\mathcal{O}(1/(\hbar\omega_p - E_{\mathbf{k}cv})) + e^{i\omega_p t}\cdot\mathcal{O}(1/(\hbar\omega_p + E_{\mathbf{k}cv})) \,. \tag{87}$$

In the rotating wave approximation (RWA) only the first dominant term is kept, $P_{\mathbf{k}}(t) \sim e^{-i\omega_p t}$. Consequently, $P_{\mathbf{k}}(t)$ is driven by the first term of eq. (86) only (interband source), and the intraband terms can be dropped since they oscillate with $\pm 2\omega_p$ or zero frequency. If we add a phenomenological dephasing $\Gamma = 1/T_2$, the polarization equation reads

$$- i\hbar \partial_t P_{\mathbf{k}} = \left(-E_{\mathbf{k}cv}^{HF} + i\Gamma \right) P_{\mathbf{k}} - \left(f_{\mathbf{k}c} - f_{\mathbf{k}v} \right) \Delta_{\mathbf{k}} \tag{88}$$

with the RWA gap function

$$\Delta_{\mathbf{k}} = \frac{e\mathbf{A}^+(t)}{m_0} \mathbf{p}_{\mathbf{k}cv} + \sum_{\mathbf{k}'} v_{\mathbf{k}-\mathbf{k}'} P_{\mathbf{k}'} . \tag{89}$$

The distributions have in leading order a slow variation, and on defining the inversion as $I_{\mathbf{k}} = (f_{\mathbf{k}c} - f_{\mathbf{k}v})/2$ we get from eq. (65)

$$\hbar \partial_t I_{\mathbf{k}} = 2 \mathrm{Im} \left(P_{\mathbf{k}} \Delta_{\mathbf{k}}^* \right) \equiv 2 (\mathrm{Im}\, P_{\mathbf{k}} \cdot \mathrm{Re}\, \Delta_{\mathbf{k}} - \mathrm{Re}\, P_{\mathbf{k}} \cdot \mathrm{Im}\, \Delta_{\mathbf{k}}) . \tag{90}$$

Introducing a three-component vector notation at given $\mathbf{k}$,

$$\vec{S} = \begin{pmatrix} \mathrm{Re}\, P \\ \mathrm{Im}\, P \\ I \end{pmatrix} \qquad \vec{\Gamma} = \begin{pmatrix} \Gamma \\ \Gamma \\ 0 \end{pmatrix} \qquad \vec{H} = \begin{pmatrix} 2\mathrm{Re}\, \Delta \\ 2\mathrm{Im}\, \Delta \\ E^{HF} \end{pmatrix} \tag{91}$$

the equations for polarization and inversion can be cast into the compact form called semiconductor Bloch equations

$$\left(\hbar \partial_t + \vec{\Gamma} \right) \vec{S} = \vec{H} \times \vec{S} . \tag{92}$$

This is similar to the spin Bloch equations where spin precession with damping is described. With $\Gamma = 0$, the length of $\vec{S}$ is conserved which means strict coherence or the conservation law

$$(\mathrm{Re}\, P)^2 + (\mathrm{Im}\, P)^2 + I^2 = |P_{\mathbf{k}}|^2 + \frac{1}{4} \left(f_{\mathbf{k}c} - f_{\mathbf{k}v} \right)^2 = const. = \frac{1}{4} . \tag{93}$$

Further, the total electron number at given $\mathbf{k}$ is conserved, too:

$$f_{\mathbf{k}c} + f_{\mathbf{k}v} = const. = 1 . \tag{94}$$

Eliminating the valence distribution gives

$$|P_{\mathbf{k}}|^2 + f_{\mathbf{k}c}^2 - f_{\mathbf{k}c} = 0 , \tag{95}$$

a fixed relation between polarization and distribution. The solution of eq. (95), however, leaves one sign open,

$$f_{\mathbf{k}c} = \frac{1}{2} \pm \sqrt{\frac{1}{4} - |P_{\mathbf{k}}|^2} . \tag{96}$$

Time-dependent calculations have shown that the initial sign "-1" holds permanently for pumping below resonance, $\hbar \omega_p < \mathcal{E}_n$.

After splitting off the leading time dependence, $P_{\mathbf{k}}(t) \Rightarrow P_{\mathbf{k}}(t) \exp(-i\omega_p t)$, the polarization equation reads

$$\left(-i\hbar \partial_t + E_{\mathbf{k}cv} - \sum_{\mathbf{k}'} v_{\mathbf{k}-\mathbf{k}'} \left(1 - \sqrt{1 - 4|P_{\mathbf{k}'}|^2} \right) - \hbar \omega_p \right) P_{\mathbf{k}} - \tag{97}$$

$$-\sqrt{1 - 4|P_{\mathbf{k}}|^2} \sum_{\mathbf{k}'} v_{\mathbf{k}-\mathbf{k}'} P_{\mathbf{k}'} = \mu\, F_p(t) \sqrt{1 - 4|P_{\mathbf{k}}|^2} \tag{98}$$

with field amplitude $F_p(t)$ and effective dipole matrix element $\mu = (e p_{cv})/(i m_0 \omega_p)$.

133

II. E. Real and virtual density

How does a finite dephasing $\Gamma > 0$ destroy the coherence? In the conduction band density

$$n_c(t) = 2 \sum_{\mathbf{k}} f_{\mathbf{k}c} \qquad (99)$$

the Coulomb terms cancel,

$$\hbar \partial_t n_c = 4 \mathrm{Im} \, \mu^* F_p(t) \sum_{\mathbf{k}} P_{\mathbf{k}} \; . \qquad (100)$$

Now let us consider low excitation and determine the density up to order F_p^2. Obviously, a linear expansion of $P_{\mathbf{k}}$ is sufficient

$$\left(-i\hbar \partial_t + E_{\mathbf{k}cv} - i\Gamma - \hbar \omega_p\right) P_{\mathbf{k}} - \sum_{\mathbf{k}'} v_{\mathbf{k}-\mathbf{k}'} P_{\mathbf{k}'} = \mu \, F_p(t) \; . \qquad (101)$$

Fourier transformation and exciton representation gives as before

$$P_{\mathbf{k}}(\omega) = \mu \, F_p(\omega) \sum_{n} \frac{\phi_n(\mathbf{k}) \, \phi_n^*(\mathbf{r} = 0)}{\mathcal{E}_n - \hbar \omega_p - \hbar \omega - i\Gamma} \; . \qquad (102)$$

If summing over $\mathbf{k}$ in n_c the linear dielectric function $\epsilon(\omega)$ appears,

$$\hbar \partial_t n_c = F_p(t) \, \mathrm{Im} \, \frac{1}{4\pi^2} \int d\omega \, e^{-i\omega t} \, F_p(\omega) \, \epsilon(\omega_p + \omega) \; , \qquad (103)$$

and the full integration over time gives the final density

$$n_c(t \to \infty) = \frac{1}{4\pi^2 \hbar} \int d\omega \, |F_p(\omega)|^2 \, \mathrm{Im} \, \epsilon(\omega_p + \omega) \qquad (104)$$

expressed as a frequency integral over the pump intensity, weighted with the absorptive part of ϵ. What happens at finite times? If the pump amplitude varies only slowly, a Taylor expansion of $\epsilon(\omega)$ around ω_p can be accepted, giving

$$
\begin{aligned}
\partial_t n_c &= F_p(t) \, \mathrm{Im} \, \frac{1}{4\pi^2 \hbar} \int d\omega \, e^{-i\omega t} \, F_p(\omega) \left(\epsilon(\omega_p) + \omega \, \frac{d\epsilon(\omega_p)}{d\omega_p} \right) \qquad (105) \\
&= F_p(t) \, \mathrm{Im} \, \frac{1}{2\pi \hbar} \left(F_p(t) \, \epsilon(\omega_p) + i\partial_t F_p(t) \, \frac{d\epsilon(\omega_p)}{d\omega_p} \right) \\
&= \frac{1}{2\pi \hbar} \left(F_p^2(t) \, \mathrm{Im} \, \epsilon(\omega_p) + F_p(t) \, \partial_t F_p(t) \, \frac{d\mathrm{Re} \, \epsilon(\omega_p)}{d\omega_p} \right) \; .
\end{aligned}
$$

The time integration is straightforward and yields

$$n_c(t) = \frac{1}{2\pi \hbar} \, \mathrm{Im} \, \epsilon(\omega_p) \int_{-\infty}^{t} dt' \, F_p^2(t') + \frac{1}{4\pi} \, \frac{d\mathrm{Re} \, \epsilon(\omega_p)}{d\omega_p} \, F_p^2(t) \; . \qquad (106)$$

The first term integrates over the pulse and adds up to the final value as expected from a "real density". The second term is called "virtual density" since it increases and decreases synchronously with the pulse. However, the given decomposition does not mean any difference in the physical action of the density. For small dephasing Γ, $\mathrm{Im} \, \epsilon(\omega_p) \ll d\mathrm{Re} \, \epsilon(\omega_p)/d\omega_p$ holds, and the virtual density dominates which is the signature of coherence.

134

III. THE OPTICAL STARK EFFECT

III. A. Atomic case

The optical Stark effect is a typical pump-probe nonlinear effect, where an optical transition is shifted under the action of an intense pump beam. Before we investigate the semiconductor specifics we show the essentials in an atomic two-level system. The Hamiltonian reads

$$H = \omega_1 |1\rangle\langle 1| + \omega_0 |0\rangle\langle 0| + \mu F_p e^{-i\omega_p t} |1\rangle\langle 0| + h.c. \tag{107}$$

or within a rotating frame

$$H = \omega_1 |1\rangle\langle 1| + (\omega_0 + \omega_p) |0\rangle\langle 0| + \mu F_p |1\rangle\langle 0| + h.c. \tag{108}$$

The solution is searched within the basis of excited and ground state, $|1\rangle$ and $|0\rangle$ respectively.

$$\begin{vmatrix} \omega_1 - E & \mu F_p \\ \mu F_p & \omega_0 + \omega_p - E \end{vmatrix} = 0 \tag{109}$$

gives the eigenvalues

$$E_\pm = \frac{1}{2}\left(\omega_1 + \omega_0 + \omega_p \pm \sqrt{(\omega_1 - \omega_0 - \omega_p)^2 + (2\mu F_p)^2}\right), \tag{110}$$

and the transition energies are (going back to the original frame)

$$\Omega_\pm = \omega_p \pm \sqrt{(\omega_1 - \omega_0 - \omega_p)^2 + (2\mu F_p)^2}. \tag{111}$$

The square root expression is often called Rabi frequency. Expanding in leading order in the field strength gives

$$\Omega_+ = \omega_1 - \omega_0 + \frac{2\mu^2 F_p^2}{\omega_1 - \omega_0 - \omega_p}, \tag{112}$$

and one realizes that the Stark shift is proportional to intensity F_p^2 and inversely proportional to the detuning $\delta = \omega_1 - \omega_0 - \omega_p$.

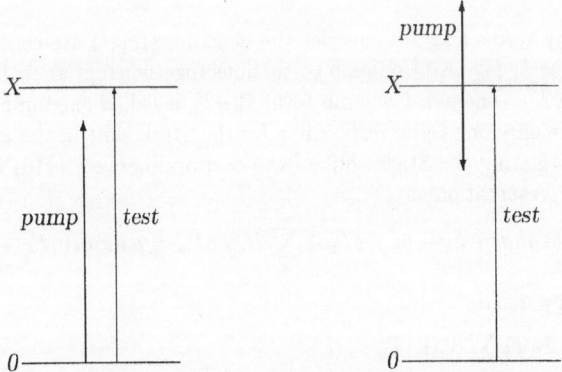

Fig. 1. Schematic view of the interband (left) and intraband (right) optical Stark effect

III. B. Interband optical Stark effect in semiconductors

In semiconductors, the two-level system is replaced by the ground state exciton. But more important, the Coulomb interaction introduces additional nonlinearities which modify the Stark effect via the exciton-exciton interaction, as has been pointed out in the pioneering papers of Schmitt-Rink, Chemla, and Haug [2].

Experimentally, the optical Stark effect in semiconductors is detected as a shift of the exciton lines in a test beam transmission or reflection, if at the same time a nonresonant pump laser hits the sample. Depending on the detuning, we can distinguish between the interband case (pump frequency close to the exciton transition) and the intraband case as shown in Fig. 1.

To begin with the interband case, we separate pump and probe driven contributions in the field strength and the polarization, $F_p(t) \Rightarrow F_p(t) + \delta F(t)$, $P_\mathbf{k}(t) \Rightarrow P_\mathbf{k}(t) + \delta P_\mathbf{k}(t)$. Upon linearizing of the full coherent equation

$$\left(-i\hbar\partial_t + E_{kcv} - \sum_{\mathbf{k}'} v_{\mathbf{k}-\mathbf{k}'} \left(1 - \sqrt{1 - 4|P_{\mathbf{k}'} + \delta P_{\mathbf{k}'}|^2} \right) - \hbar\omega_p \right) (P_\mathbf{k} + \delta P_\mathbf{k}) - \quad (113)$$

$$-\sqrt{1 - 4|P_\mathbf{k} + \delta P_\mathbf{k}|^2} \sum_{\mathbf{k}'} v_{\mathbf{k}-\mathbf{k}'} (P_{\mathbf{k}'} + \delta P_{\mathbf{k}'}) = \mu \left(F_p(t) + \delta F(t) \right) \sqrt{1 - 4|P_\mathbf{k} + \delta P_\mathbf{k}|^2}$$

with respect to δF and $\delta P_\mathbf{k}$ we get the result $(W_\mathbf{k} = \sqrt{1 - 4|P_\mathbf{k}|^2})$

$$\left(-i\hbar\partial_t + E_{kcv} - \sum_{\mathbf{k}'} v_{\mathbf{k}-\mathbf{k}'} (1 - W_{\mathbf{k}'}) - \hbar\omega_p \right) \delta P_\mathbf{k} - W_\mathbf{k} \left(\mu\,\delta F + \sum_{\mathbf{k}'} v_{\mathbf{k}-\mathbf{k}'}\,\delta P_{\mathbf{k}'} \right) - \quad (114)$$

$$-2P_\mathbf{k} \sum_{\mathbf{k}'} v_{\mathbf{k}-\mathbf{k}'} \frac{P_{\mathbf{k}'}\,\delta P_{\mathbf{k}'}^* + P_{\mathbf{k}'}^*\,\delta P_{\mathbf{k}'}}{W_{\mathbf{k}'}} + 2\frac{P_\mathbf{k}\,\delta P_\mathbf{k}^* + P_\mathbf{k}^*\,\delta P_\mathbf{k}}{W_\mathbf{k}} \left(\mu\,F_p + \sum_{\mathbf{k}'} v_{\mathbf{k}-\mathbf{k}'}\,P_{\mathbf{k}'} \right) = 0 \,.$$

A coupled system for $\delta P_\mathbf{k}$ and $\delta P_\mathbf{k}^*$ has been obtained. The coupling is of order $|F_p|^4$ and gives rise to the gain feature at $\hbar\omega = 2\hbar\omega_p - \mathcal{E}_n$ which is below the pump frequency. However, in lowest order the Stark effect is proportional to $|F_p|^2 \delta F$ (χ_3 level), and the coupling can be neglected. Now retaining only terms in χ_3 quality we have

$$\left(-i\hbar\partial_t + E_{kcv} - \sum_{\mathbf{k}'} v_{\mathbf{k}-\mathbf{k}'} \left(2|P_{\mathbf{k}'}|^2 - 2P_{\mathbf{k}'}^* P_{\mathbf{k}'} \right) + 2P_\mathbf{k}^* \mu\,F_p - \hbar\omega_p \right) \delta P_\mathbf{k} - \quad (115)$$

$$-\sum_{\mathbf{k}'} v_{\mathbf{k}-\mathbf{k}'} \left(1 - 2|P_\mathbf{k}|^2 + 2P_\mathbf{k} P_\mathbf{k}^* \right) \delta P_{\mathbf{k}'} = \left(1 - 2|P_\mathbf{k}|^2 \right) \mu\,\delta F \,.$$

In the blocking terms and self energies the densities $(|P_\mathbf{k}|^2)$ are complemented by polarisation terms $P_\mathbf{k}^* P_{\mathbf{k}'}$. All Coulomb corrections together represent the exciton-exciton interaction (XX). The non-Coulomb term $P_\mathbf{k}^* \mu\,F_p$ is called exciton-photon interaction (XP), it is the only one to be responsible for the Stark shift in the atomic case.

For investigating the Stark shift of the exciton lines, eq. (115) is rewritten using the exciton representation as

$$(-i\hbar\partial_t + \mathcal{E}_n - \hbar\omega_p)\,\delta P_n + \sum_n H_{nm}\,\delta P_m = \mu\,\delta F\,\phi(0)\,(1 - O_n) \quad (116)$$

with the energy matrix

$$H_{nm} = 2\mu\,F_p \sum_\mathbf{k} \phi_n^*(\mathbf{k})\,P_\mathbf{k}^*\,\phi_m(\mathbf{k}) \quad (117)$$

$$+ 2\sum_{\mathbf{k}\mathbf{k}'} v_{\mathbf{k}-\mathbf{k}'}\,\phi_n^*(\mathbf{k}) \left[\left(|P_\mathbf{k}|^2 - P_\mathbf{k} P_{\mathbf{k}'}^* \right) \phi_m(\mathbf{k}') - \left(|P_{\mathbf{k}'}|^2 - P_\mathbf{k}^* P_{\mathbf{k}'} \right) \phi_m(\mathbf{k}) \right]$$

and the oscillator strength

$$O_n = \frac{2}{\phi_n(0)} \sum_{\mathbf{k}} \phi_n^*(\mathbf{k}) |P_{\mathbf{k}}|^2 . \tag{118}$$

The Stark shift $\Delta\mathcal{E}_n \equiv H_{nn}$ of the exciton energy $\mathcal{E}_n$ is given by

$$\Delta\mathcal{E}_n = 2\mu\, F_p \sum_{\mathbf{k}} |\phi_n(\mathbf{k})|^2 \cdot P_{\mathbf{k}}^* + 2 \sum_{\mathbf{k}\mathbf{k}'} v_{\mathbf{k}-\mathbf{k}'} \left(\phi_n^*(\mathbf{k}) - \phi_n^*(\mathbf{k}') \right) \cdot P_{\mathbf{k}} \cdot \left(P_{\mathbf{k}}^* - P_{\mathbf{k}'}^* \right) \cdot \phi_n(\mathbf{k}') \tag{119}$$

where $P_{\mathbf{k}}$ has to be taken linear in F_p. For stationary excitation it is useful to express it via the exciton Green's function

$$P_{\mathbf{k}} = \mu\, F_p \sum_{n} \frac{\phi_n(\mathbf{k})\, \phi_n^*(\mathbf{r}=0)}{\mathcal{E}_n - \hbar\omega_p - i0} = \mu\, F_p \sum_{\mathbf{k}'} G(\mathbf{k}\mathbf{k}'\omega_p) \equiv \mu\, F_p\, G_p(\mathbf{k}) \tag{120}$$

getting

$$\Delta\mathcal{E}_n = 2(\mu F_p)^2 \left\{ \sum_{\mathbf{k}} \phi_n^2(\mathbf{k})\, G_p(\mathbf{k}) \right. \tag{121}$$

$$\left. + \sum_{\mathbf{k}\mathbf{k}'} v_{\mathbf{k}-\mathbf{k}'} \left(\phi_n(\mathbf{k}) - \phi_n(\mathbf{k}') \right) G_p(\mathbf{k}) \left(G_p^*(\mathbf{k}) - G_p^*(\mathbf{k}') \right) \phi_n(\mathbf{k}') \right\} .$$

The first line stems from the exciton-photon interaction, whereas the second one comprises the exciton-exciton interaction. Near to the resonance, the Coulomb Green's function is inversely proportional to the detuning $\delta = \mathcal{E}_{1s} - \hbar\omega_p$, and the general structure is

$$\Delta\mathcal{E}_n = 2\mu^2 F_p^2 \left\{ \frac{XP}{\delta} + \frac{XX}{\delta^2} \right\} . \tag{122}$$

Far away from the resonance, the atomic limit holds, but close to the resonance the exciton-exciton interaction dominates. Numerical results for the Stark shift as well as the change in oscillator strength (eq. 118) have been obtained e.g. in [3]. A comparison with experimental data is shown in Fig. 2.

Note however that a full treatment of the optical Stark effect within the χ_3 level has to include biexciton interactions which are outside the Hartree Fock level studied here.

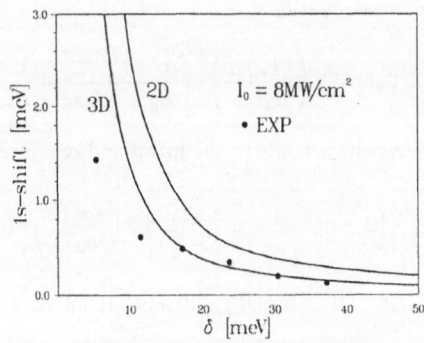

Fig. 2. Measured Stark shift of excitons in dependence on detuning below the exciton resonance. Experimental data (dots, AlGaAs quantum well) are compared with calculations (full curves) for the ideal three- and two-dimensional case. From [4]

III. C. Intraband optical Stark effect

Here, no resonance with internal transitions can be used to simplify the treatment. In the Bloch representation and dropping for the moment the Coulomb interaction, we have as Hamiltonian

$$\hat{\mathcal{H}} = \sum_{\mathbf{k}\nu} \left(E_{\mathbf{k}\nu} + \frac{e^2 A(t)^2}{2m_0} \right) c_{\mathbf{k}\nu}^\dagger c_{\mathbf{k}\nu} - \frac{eA(t)}{m_0} \sum_{\mathbf{k}\nu\nu'} \mathbf{p}_{\mathbf{k}\nu\nu'} c_{\mathbf{k}\nu}^\dagger c_{\mathbf{k}\nu'} \qquad (123)$$

which contains linear and quadratic terms in $\mathbf{A}(t)$. The band-nondiagonal density matrix $\rho_{\mathbf{k}\nu\nu'} = \langle c_{\mathbf{k}\nu'}^\dagger c_{\mathbf{k}\nu} \rangle$ obeys

$$(E_{\mathbf{k}\nu} - E_{\mathbf{k}\nu'} - i\hbar\partial_t)\,\rho_{\mathbf{k}\nu\nu'} + \frac{e\mathbf{A}}{m_0} \sum_{\eta} (\rho_{\mathbf{k}\nu\eta}\,\mathbf{p}_{\mathbf{k}\eta\nu'} - \mathbf{p}_{\mathbf{k}\nu\eta}\,\rho_{\mathbf{k}\eta\nu'}) = 0 \ . \qquad (124)$$

Due to the energy difference, the A^2 term has dropped out. But we expect an A^2 term within effective mass approximation! To see what happens we consider two bands only. With $\mathbf{p}_{\mathbf{k}cv} \sim \mathbf{p}_{cv}$ and $\mathbf{p}_{\mathbf{k}\nu\nu} = (m_0/m_\nu)\hbar\mathbf{k} + O(k^2)$ we have

$$\left(E_{\mathbf{k}cv} - \frac{e\mathbf{A}}{\mu}\hbar\mathbf{k} - i\hbar\partial_t \right) \rho_{\mathbf{k}cv} = -\frac{e\mathbf{A}}{m_0}\mathbf{p}_{cv}\,(\rho_{\mathbf{k}cc} - \rho_{\mathbf{k}vv}) \ . \qquad (125)$$

Instead of solving for $\rho_{\mathbf{k}\nu\nu}$ we apply the conservation law

$$\rho_{\mathbf{k}cc} - \rho_{\mathbf{k}vv} = -\sqrt{1 - 4\rho_{\mathbf{k}cv}\,\rho_{\mathbf{k}vc}} \qquad (126)$$

for the inversion. In the intraband case, $|\rho_{\mathbf{k}cv}|^2$ is expected to be small, but since p_{cv} is large we have to keep track of this term which has been called virtual density above.

Expansion into pump and test, $\mathbf{A} = \mathbf{A} + \delta\mathbf{A}$, $\rho_{\mathbf{k}cv} = P_{\mathbf{k}} + \delta P_{\mathbf{k}}$, gives

$$\left(E_{\mathbf{k}cv} - \frac{e\mathbf{A}}{\mu}\hbar\mathbf{k} - i\hbar\partial_t \right) \delta P_{\mathbf{k}} + \frac{e\mathbf{A}\mathbf{p}_{cv}}{m_0} 2P_{\mathbf{k}}^*\,\delta P_{\mathbf{k}} = -\frac{e\mathbf{p}_{cv}}{m_0}\sqrt{1 - 4\,|\,P_{\mathbf{k}}\,|^2}\,\delta\mathbf{A} \qquad (127)$$

linear in the test contribution. For the pump field, we take $\mathbf{F}(t) = 2\mathbf{F}_p\cos(\omega_p t)$ resp. $\mathbf{A}(t) = -2(\mathbf{F}_p/\omega_p)\sin(\omega_p t)$. The pump-driven polarization is then

$$P_{\mathbf{k}}(t) = \frac{eF_p\,p_{cv}}{i\omega_p m_0} \left(\frac{e^{-i\omega_p t}}{E_{\mathbf{k}cv} - \hbar\omega_p} - \frac{e^{+i\omega_p t}}{E_{\mathbf{k}cv} + \hbar\omega_p} \right) \qquad (128)$$

and gives in the test-driven equation

$$\frac{eA(t)\,p_{cv}}{m_0} 2P_{\mathbf{k}}^*(t) = 2\left(\frac{eF_p\,p_{cv}}{\omega_p m_0} \right)^2 \left(\frac{1 - e^{+2i\omega_p t}}{E_{\mathbf{k}cv} - \hbar\omega_p} - \frac{1 - e^{-2i\omega_p t}}{E_{\mathbf{k}cv} + \hbar\omega_p} \right) \ . \qquad (129)$$

Due to $E_{\mathbf{k}cv} \approx E_g \gg \hbar\omega_p$ which holds in the intraband case, we continue with

$$\cdots = 2\left(\frac{eF_p\,p_{cv}}{\omega_p m_0} \right)^2 \frac{1}{E_g}\left(2 - e^{+2i\omega_p t} - e^{-2i\omega_p t} \right) = 2\left(\frac{p_{cv}}{m_0} \right)^2 \frac{1}{E_g}(eA(t))^2 = \frac{1}{2\mu}(eA(t))^2 \qquad (130)$$

since $4p_{cv}^2/m_0 E_g = m_0/\mu$. The desired result is that an A^2 term has been recovered with reduced effective mass,

$$\left(E_g + \frac{1}{2\mu}(\hbar\mathbf{k} - e\mathbf{A}(t))^2 - i\hbar\partial_t \right) \delta P_{\mathbf{k}}(t) - \sum_{\mathbf{k}'} v_{\mathbf{k}-\mathbf{k}'}\,\delta P_{\mathbf{k}'}(t) = \frac{e\mathbf{p}_{cv}}{m_0}\delta\mathbf{A}(t) \ . \qquad (131)$$

The Coulomb interaction has been restored here (for details see [5]). In real space, this reads

$$\left(E_g + \frac{1}{2\mu}(-i\hbar\nabla_\mathbf{r} - e\mathbf{A}(t))^2 - i\hbar\partial_t - \frac{e^2}{\epsilon r}\right)\delta P(\mathbf{r},t) = \delta(\mathbf{r})\frac{e\mathbf{p}_{cv}}{m_0}\delta\mathbf{A}(t) \qquad (132)$$

and can be visualized as the exciton equation in an oscillating frame. Although $\hbar\omega_p \ll E_g$ holds, we still have a slow motion of the exciton, $E_B \ll \hbar\omega_p$. As a first approximation, we can therefore take the temporal average of the kinetic energy. The term linear in A vanishes, but the quadratic one gives rise to

$$\Delta_1 = \frac{1}{2\mu}\overline{(e\mathbf{A}(t))^2} = \frac{1}{\mu}\left(\frac{eF_p}{\omega_p}\right)^2 \qquad (133)$$

which is the Stark shift of the intraband case. As a numerical example, we consider GaAs ($\mu = 0.044$, $N = 3.3$). An intense CO_2 laser ($\hbar\omega_p = 116.5$ meV) with $I_p = 100$ MW/cm^2 gives a field strength of $F_p = 6.23 \cdot 10^4$ V/cm in the sample, which results in a Stark shift of $\Delta_1 = 5.0$ meV. Nevertheless $|P_k| = \sqrt{\Delta_1/E_g} = 0.06$ is small as anticipated.

Fig. 3 shows that a nearly rigid blue shift of the exciton spectrum is indeed the dominating feature. The obvious broadening of the Stark-shifted exciton line is due to multiple reflections inside the sample. Well described by the calculation is even the weak polarization dependence. It is related to the participation of heavy- and light-hole valence bands, extending the two-band model studied here.

The full calculation outside the simple time average shows that eq. (132) contains a weak satellite structure at $E_X - \hbar\omega_p$, which is nothing else than the two-photon absorption "CO_2 photon + Dye photon → exciton".

It is instructive to look at the intraband Stark effect of free electron-hole pairs since here an analytical solution is possible. Splitting off the test field by

$$\delta P_\mathbf{k}(t) = \frac{e p_{cv}}{m_0}\delta A\, e^{-i\omega t}\delta p_\mathbf{k}(t) \qquad (134)$$

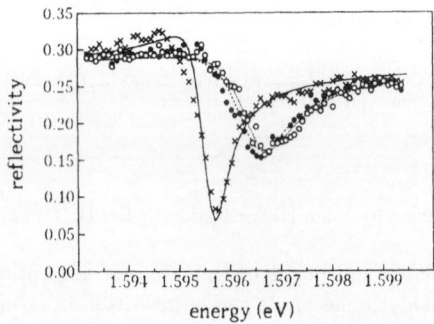

Fig. 3. Intraband optical Stark effect in CdTe. Measured reflection spectra without CO_2 laser (crosses) and with a CO_2 pump laser of 40 MW/cm^2. Open circles: CO_2 laser and dye laser polarization parallel; full dots: CO_2 and dye laser polarization perpendicular. Calculations are shown as solid line (linear spectrum), dotted line (parallel configuration), and dashed line (perpendicular configuration). From [6]

and dropping the Coulomb term we have

$$\left(-i\hbar\partial_t + E_g + \frac{1}{2\mu}(\hbar\mathbf{k} - e\mathbf{A}(t))^2 - \hbar\omega - i\Gamma\right)\delta p_{\mathbf{k}}(t) = 1 . \tag{135}$$

The test response is given by

$$\epsilon(\omega) = \epsilon_b + \frac{4\pi e^2 p_{cv}^2}{m_0^2\omega^2}\sum_{\mathbf{k}}\overline{\delta p_{\mathbf{k}}(t)} \tag{136}$$

with the time average over one pump field cycle. Straightforward integration of eq. (135) gives

$$\delta p_{\mathbf{k}}(t) = \frac{i}{\hbar}\int_{-\infty}^{t}dt'\exp\left\{\frac{i}{\hbar}\int_{t}^{t'}d\tau\left[\frac{1}{2\mu}(\hbar\mathbf{k} - e\mathbf{A}(\tau))^2 + E_g - \hbar\omega - i\Gamma\right]\right\} . \tag{137}$$

With the explicit form of $\mathbf{A}(t)$ the integral over τ can be done at once, followed by the integration over a shifted $\mathbf{k}$ variable. After some algebra follows ($E = (\hbar\omega - E_g)/\hbar\omega_p$)

$$\epsilon(\omega) = \epsilon_b + C\cdot\sqrt{\omega_p}\,i\int_{0}^{\infty}\frac{ds}{(is+0)^{3/2}}\exp\left[(iE-\Gamma)s - iA(s) - iB(s)\cos(\omega_p t)\right] \tag{138}$$

with

$$A(s) = \frac{\Delta_1}{\hbar\omega_p}\left[s - \frac{2}{s}(1 - \cos s)\right] ; \quad B(s) = \frac{\Delta_1}{\hbar\omega_p}\left[\frac{2}{s}(1 - \cos s) - \sin s\right] . \tag{139}$$

The temporal average produces the Bessel function $J_0(B(s))$. A careful pole treatment results in

$$Im\,\epsilon(\omega) = C\cdot\sqrt{\omega_p}\int_{0}^{\infty}\frac{ds}{s^{3/2}}\left[e^{-\Gamma s}\sin\left(E\,s - \pi/4 - A(s)\right)J_0(B(s)) + 1/\sqrt{2}\right] . \tag{140}$$

For large s, $A(s) = s\Delta_1$ holds, and the Stark shift $+\Delta_1$ is obvious.

Even the limit of a static electric field can be extracted: $s \equiv \omega_p t$ tends to zero, and with

$$B(s) = A(s) \Rightarrow \frac{\Delta_1}{\hbar\omega_p}\frac{s^3}{12} = W^3\frac{t^3}{24} ; \quad W^3 = \frac{\hbar^2(2eF)^2}{\mu} \tag{141}$$

(W is the field energy) we obtain

$$\epsilon(\omega) = \epsilon_b + C\cdot i\int_{0}^{\infty}\frac{dt}{(it+0)^{3/2}}\exp\left[(i\omega - iE_g - \Gamma)t - iW^3t^3/12\right] , \tag{142}$$

respectively

$$Im\,\epsilon(\omega) = C\cdot\int_{0}^{\infty}\frac{dt}{t^{3/2}}\left[e^{-\Gamma t}\sin\left((\omega - E_g)t - \pi/4 - W^3t^3/12\right) + 1/\sqrt{2}\right] . \tag{143}$$

This is just the modulated and broadened square root absorption found in a constant electric field (Franz-Keldysh effect). It can be rewritten in terms of Airy functions as usual,

$$Im\,\epsilon(\omega) = C\cdot\pi\sqrt{W}\left[x\,Ai^2(-x) + (Ai'(-x))^2\right] , \tag{144}$$

with $x = (\hbar\omega - E_g)/W$.

Therefore, the formalism presented here is able to span the range from high frequencies (interband Stark effect) to medium ones (intraband case) down to the static electric field.

IV. ELECTRON-PHONON INTERACTION AND RELAXATION

IV. A. Electron-phonon kinetics

A description of optically excited carriers in solids would be incomplete without inclusion of scattering. In a phenomenological way, these are often introduced "by hand" as longitudinal (T_1) and transverse (T_2) relaxation times, as we did in the foregoing chapters. A microscopic theory couples the electrons to scattering partners – these could be lattice vibrations (phonons) or density fluctuations (plasmons). Since on short time scales energy conservation does not hold, a real-time description is preferable. Among the possible methods (projection technique, non-equilibrium Green's functions, density matrix) we concentrate on the last one, in accordance with the general frame of the article. This method allows a rather physical interpretation and is not complicated by the (formal) introduction of imaginary times or double-time contours. A central point will be memory effects (non-Markoff scattering) which at the same time are crucial to understand realistic (non-Lorentzian) optical lineshapes. A complete derivation can be found in [7].

We focus on phonon scattering and extend the Hamiltonian in Bloch representation

$$
\hat{\mathcal{H}} = \sum_{\mathbf{k}\nu} E_{\mathbf{k}\nu}\, c_{\mathbf{k}\nu}^{\dagger}\, c_{\mathbf{k}\nu} - \frac{e\mathbf{A}(t)}{m_0} \sum_{\mathbf{k}\nu\nu'} \mathbf{p}_{\mathbf{k}\nu\nu'}\, c_{\mathbf{k}\nu}^{\dagger}\, c_{\mathbf{k}\nu'} + \frac{1}{2} \sum_{\mathbf{q}\mathbf{k}\nu\mathbf{k}'\nu'} v_{\mathbf{q}}\, c_{\mathbf{k}+\mathbf{q}\,\nu}^{\dagger}\, c_{\mathbf{k}'\nu'}^{\dagger}\, c_{\mathbf{k}\nu'}\, c_{\mathbf{k}'+\mathbf{q}}
$$
$$
+ \sum_{\mathbf{q}} \hbar\omega_{\mathbf{q}}\left(a_{\mathbf{q}}^{\dagger} a_{\mathbf{q}} + \frac{1}{2}\right) + \sum_{\mathbf{k}\mathbf{q}\nu} M_{\mathbf{q}}(a_{\mathbf{q}} + a_{-\mathbf{q}}^{\dagger})c_{\mathbf{k}+\mathbf{q}\,\nu}^{\dagger}\, c_{\mathbf{k}\nu}\,. \tag{145}
$$

Apart from electron operators $c_{\mathbf{k}\nu}^{\dagger}, c_{\mathbf{k}\nu}$ phonon creation and annihilation operators $a_{\mathbf{q}}^{\dagger}, a_{\mathbf{q}}$ appear which obey Boson commutation rules

$$
[a_{\mathbf{q}}, a_{\mathbf{q}'}^{\dagger}] \equiv a_{\mathbf{q}}\, a_{\mathbf{q}'}^{\dagger} - a_{\mathbf{q}'}^{\dagger}\, a_{\mathbf{q}} = \delta_{\mathbf{q},\mathbf{q}'}\,. \tag{146}
$$

The interband light field $\mathbf{A}(t)$ is treated in rotating wave approximation which allowed to drop the A^2 term. As new quantities the phonon distribution

$$
N_{\mathbf{q}} = \langle a_{\mathbf{q}}^{\dagger} a_{\mathbf{q}} \rangle \tag{147}
$$

and the so-called phonon-assisted density matrix are introduced

$$
T_{\nu\nu'}^{+}(\mathbf{k}\mathbf{q}) = M_{\mathbf{q}} \langle a_{\mathbf{q}}^{\dagger}\, c_{\mathbf{k}-\mathbf{q}\nu'}^{\dagger}\, c_{\mathbf{k}\nu} \rangle \;; \qquad T_{\nu\nu'}^{-}(\mathbf{k}\mathbf{q}) = M_{\mathbf{q}} \langle a_{-\mathbf{q}}\, c_{\mathbf{k}-\mathbf{q}\nu'}^{\dagger}\, c_{\mathbf{k}\nu} \rangle\,. \tag{148}
$$

Within the Hartree-Fock decoupling the equation of motion reads

$$
-i\hbar\partial_t \rho_{\mathbf{k}\nu\nu'} = (E_{\mathbf{k}\nu'} - E_{\mathbf{k}\nu})\,\rho_{\mathbf{k}\nu\nu'} - \frac{e\mathbf{A}(t)}{m_0} \sum_{\eta} (\rho_{\mathbf{k}\nu\eta}\, \mathbf{p}_{\mathbf{k}\eta\nu'} - \mathbf{p}_{\mathbf{k}\nu\eta}\, \rho_{\mathbf{k}\eta\nu'}) \tag{149}
$$
$$
+ \sum_{\mathbf{q}} [T_{\nu\nu'}^{-}(\mathbf{k}\mathbf{q}) + T_{\nu\nu'}^{+}(\mathbf{k}\mathbf{q}) - T_{\nu\nu'}^{-}(\mathbf{k}+\mathbf{q},\mathbf{q}) - T_{\nu\nu'}^{+}(\mathbf{k}+\mathbf{q},\mathbf{q})]
$$
$$
- \sum_{\mathbf{k}'\eta} v_{\mathbf{k}-\mathbf{k}'}[\rho_{\mathbf{k}\nu\eta}\, \tilde{\rho}_{\mathbf{k}'\eta\nu'} - \tilde{\rho}_{\mathbf{k}'\nu\eta}\, \rho_{\mathbf{k}\eta\nu'}]\,.
$$

$E_{\mathbf{k}\nu}$ includes the (unexcited) HF self energy, and $\tilde{\rho}_{\mathbf{k}\nu\nu'} \equiv \rho_{\mathbf{k}\nu\nu'} - \delta_{\nu\nu'} \cdot \delta_{\nu\in\upsilon}$. Again a conservation law holds by summing over $\mathbf{k}$: The total density matrix is not changed via phonon (T) and Coulomb terms (v). We continue to the next hierarchy level and derive an equation of motion for the phonon-assisted density matrix

$$
\begin{aligned}
-i\hbar\partial_t T^{\pm}_{\nu\nu'}(\mathbf{kq}) =\ & (E_{\mathbf{k-q}\nu} - E_{\mathbf{k}\nu} \pm \hbar\omega_{\mathbf{q}})\, T^{\pm}_{\nu\nu'}(\mathbf{kq}) \\
& - \frac{e\mathbf{A}(t)}{m_0} \sum_{\eta} \left[T^{\pm}_{\nu\eta}(\mathbf{kq})\, \mathbf{p}_{\mathbf{k-q}\eta\nu'} - \mathbf{p}_{\mathbf{k}\nu\eta}\, T^{\pm}_{\eta\nu'}(\mathbf{kq}) \right] \\
& + M_{\mathbf{q}}^2 \left[(N_{\mathbf{q}} + 0/1)(\rho_{\mathbf{k}\nu\nu'} - \rho_{\mathbf{k-q}\nu\nu'}) \pm \sum_{\eta} \rho_{\mathbf{k}\nu\eta}(\delta_{\eta\nu'} - \rho_{\mathbf{k-q}\eta\nu'}) \right] \\
& + v_{\mathbf{q}}\, (\rho_{\mathbf{k}\nu\nu'} - \rho_{\mathbf{k-q}\nu\nu'}) \sum_{\mathbf{k'}\eta} T^{\pm}_{\eta\eta}(\mathbf{k'q})
\end{aligned}
\tag{150}
$$

$$
- \sum_{\mathbf{k'}\eta} v_{\mathbf{k-k'}} \left[\rho_{\mathbf{k}\nu\eta} T^{\pm}_{\eta\nu'}(\mathbf{k'q}) - \underline{\rho_{\mathbf{k-q}\eta\nu'} T^{\pm}_{\nu\eta}(\mathbf{k'q})} - \tilde{\rho}_{\mathbf{k'}\nu\eta} T^{\pm}_{\eta\nu'}(\mathbf{kq}) + \tilde{\rho}_{\mathbf{k'-q}\eta\nu'} T^{\pm}_{\nu\eta}(\mathbf{kq}) \right] .
$$

Higher order expectation values have been decoupled according to

$$
\langle c^{\dagger} c\, a^{\dagger} a \rangle \Rightarrow \langle c^{\dagger} c \rangle \langle a^{\dagger} a \rangle = \rho\, N
\tag{151}
$$

which is equivalent to the usual Born approximation (expansion of the scattering integral up to $M_{\mathbf{q}}^2$). We could follow the non-equilibrium phonon distribution $N_{\mathbf{q}}$ by deriving its equation of motion, too. For the sake of simplicity, however, we treat here the phonons in equilibrium, as relevant for a strong heat bath coupling.

IV. B. Exciton dephasing and lineshape

Looking for the linear response ($\rho_{\mathbf{k}cc} = 0, \rho_{\mathbf{k}vv} = 1$) only the underlined Coulomb term in (150) contributes

$$
\begin{aligned}
-i\hbar\partial_t T^{\pm}_{cv}(\mathbf{kq}) =\ & (E_{\mathbf{k-q}v} - E_{\mathbf{k}c} \pm \hbar\omega_{\mathbf{q}})\, T^{\pm}_{cv}(\mathbf{kq}) + \sum_{\mathbf{k'}} v_{\mathbf{k-k'}} T^{\pm}_{cv}(\mathbf{k'q}) \\
& + M_{\mathbf{q}}^2 (N_{\mathbf{q}} + 0/1)(P_{\mathbf{k}} - P_{\mathbf{k-q}}) .
\end{aligned}
\tag{152}
$$

Here, an exciton representation with nonzero c.o.m momentum $\mathbf{q}$ is needed

$$
(E_{\mathbf{k}c} - E_{\mathbf{k-q}v} - \mathcal{E}_{n\mathbf{q}})\, \phi_n(\mathbf{kq}) - \sum_{\mathbf{k'}} v_{\mathbf{k-k'}}\, \phi_n(\mathbf{k'q}) = 0 .
\tag{153}
$$

For parabolic bands

$$
\mathcal{E}_{n\mathbf{q}} = \mathcal{E}_n + \frac{\hbar^2 q^2}{2M} ; \quad \phi_n(\mathbf{kq}) = \phi_n(\mathbf{k} - \frac{m_e}{M}\mathbf{q})
\tag{154}
$$

holds ($M = m_e + m_h$), and the expansion according to

$$
T^{\pm}_{cv}(\mathbf{kq}) = \sum_n \phi_n(\mathbf{k} - \frac{m_e}{M}\mathbf{q})\, T^{\pm}_{n\mathbf{q}} ; \quad P_{\mathbf{k}} = \sum_n \phi_n(\mathbf{k})\, P_n
\tag{155}
$$

gives

$$
-i\hbar\partial_t T^{\pm}_{n\mathbf{q}} = (-\mathcal{E}_{n\mathbf{q}} \pm \hbar\omega_{\mathbf{q}})\, T^{\pm}_{n\mathbf{q}} + M_{\mathbf{q}}^2 (N_{\mathbf{q}} + 0/1) \sum_{\mathbf{k}} \phi_n^*(\mathbf{k} - \frac{m_e}{M}\mathbf{q}) (P_{\mathbf{k}} - P_{\mathbf{k-q}}) .
\tag{156}
$$

Further, defining the exciton charge density matrix element

$$
\sum_{\mathbf{k}} \phi_n^*(\mathbf{k} - \frac{m_e}{M}\mathbf{q}) (\phi_l(\mathbf{k}) - \phi_l(\mathbf{k} - \mathbf{q})) = \langle n | e^{i\mathbf{qr}m_e/M} - e^{-i\mathbf{qr}m_h/M} | l \rangle \equiv Q_{nl}
\tag{157}
$$

142

we obtain for the phonon-assisted density matrix

$$(\mathcal{E}_{nq} \mp \hbar\omega_q - i\hbar\partial_t) T_{nq}^{\pm} = M_q^2 (N_q + 0/1) \sum_l Q_{nl} P_l \tag{158}$$

and the polarization correspondingly

$$(\mathcal{E}_{m0} - i\hbar\partial_t) P_m = \sum_k \phi_m^*(k) \frac{eA(t)}{m_0} p_{cv} + \sum_l Q_{nm}^* \left(T_{nq}^- + T_{nq}^+\right) . \tag{159}$$

Time integration gives

$$T_{nq}^{\pm}(t) = \frac{i}{\hbar} \int_{-\infty}^t dt' \, e^{-i(\mathcal{E}_{nq} \mp \omega_q)(t-t')} M_q^2 (N_q + 0/1) \sum_l Q_{nl} P_l(t') , \tag{160}$$

and insertion into eq. (159)

$$(\mathcal{E}_{m0} - i\hbar\partial_t) P_m(t) = \frac{eA(t)}{m_0} p_{cv} \, \phi_m^*(\mathbf{r}=0) + i \int_{-\infty}^t dt' \sum_l \sigma_{ml}(t-t') P_l(t') \tag{161}$$

with the "memory kernel"

$$\sigma_{ml}(t-t') = \frac{1}{\hbar} \sum_{qn} M_q^2 Q_{nm}^* Q_{nl} \left[(N_q+1) e^{-i(\mathcal{E}_{nq}+\omega_q)(t-t')} + N_q e^{-i(\mathcal{E}_{nq}-\omega_q)(t-t')}\right] . \tag{162}$$

Dephasing can be seen best in the stationary regime (after splitting off $\exp(-i\omega t)$)

$$(\mathcal{E}_{m0} - \hbar\omega) P_m = \frac{ep_{cv}A_0}{m_0} \phi_m^*(0) - \sum_l \Sigma_{ml}(\omega) P_l \tag{163}$$

with the exciton-phonon self energy

$$\Sigma_{ml}(\omega) = -i \int_0^\infty d\tau \, \sigma_{ml}(\tau) \, e^{i(\omega+i0)} . \tag{164}$$

Explicitly, this reads

$$\Sigma_{ml}(\omega) = -\sum_{qn} M_q^2 Q_{nm}^* Q_{nl} \left(\frac{N_q+1}{\mathcal{E}_{nq} + \hbar\omega_q - \hbar\omega - i0} + \frac{N_q}{\mathcal{E}_{nq} - \hbar\omega_q - \hbar\omega - i0} \right) \tag{165}$$

and describes phonon emission (including stimulated emission, first term) and absorption (second term).

If diagonal terms are assumed to dominate, the solution is simply

$$P_m = \frac{ep_{cv}A_0}{m_0} \cdot \frac{\phi_m^*(0)}{\mathcal{E}_{m0} + \Sigma_{mm} - \hbar\omega} \tag{166}$$

and can be added up to get the optical dielectric function:

$$\epsilon(\omega) = \epsilon_b + \frac{8\pi e^2}{m_0^2 \omega^2} \sum_n \frac{|p_{cv}\phi_m(0)|^2}{\mathcal{E}_{m0} + \text{Re}\,\Sigma_{mm}(\omega) - i\Gamma_{mm}(\omega) - \hbar\omega} . \tag{167}$$

Note that the dephasing $\Gamma_{mm}(\omega)$ depends on frequency which leads to a non-Lorentzian lineshape. The well-known exponential Urbach tail can be traced back to this dependence, but needs a multi-phonon treatment. If one improves over the diagonal assumption the damping is somewhat reduced [7].

IV. C. Electron kinetics with memory

Here we drop all Coulomb terms and write the optical excitation via the polarization as a source term

$$S_{\mathbf{k}}(t) = \left(\frac{ep_{cv}}{\hbar m_0}\right)^2 A_0(t) \int_{-\infty}^{t} dt' \, A_0(t') \, e^{-(t-t')\Gamma} \, 2\cos((t-t')(E_{kcv} - \omega)) \tag{168}$$

which holds for low excitation. Using the symmetry of the intraband assisted density matrix T_{cc} we obtain ($\rho_{\mathbf{k}cc} \equiv f_{\mathbf{k}}$)

$$\partial_t f_{\mathbf{k}} = S_{\mathbf{k}}(t) - \frac{2}{\hbar} \sum_{\mathbf{q}} Im\left[T_{cc}^{-}(\mathbf{kq}) + T_{cc}^{+}(\mathbf{kq})\right] \tag{169}$$

and

$$-i\hbar\partial_t T_{cc}^{\pm}(\mathbf{kq}) = (E_{\mathbf{k-q}} - E_{\mathbf{k}} \pm \hbar\omega_{\mathbf{q}}) T_{cc}^{\pm}(\mathbf{kq}) \tag{170}$$
$$+ M_{\mathbf{q}}^2\left[(N_{\mathbf{q}} + 0/1)(f_{\mathbf{k}} - f_{\mathbf{k-q}}) \pm f_{\mathbf{k}}(1 - f_{\mathbf{k-q}})\right] .$$

Integrating the differential equation gives (t_0 is a time before the pulse arrives)

$$T_{cc}^{+}(\mathbf{kq}, t) = \frac{i}{\hbar} \int_{t_0}^{t} dt' \, e^{i(E_{\mathbf{k-q}} - E_{\mathbf{k}} + \omega\mathbf{q})(t-t')} M_{\mathbf{q}}^2 \left[N_{\mathbf{q}}(f_{\mathbf{k}} - f_{\mathbf{k-q}}) + f_{\mathbf{k}}(1 - f_{\mathbf{k-q}})\right]_{t'} \tag{171}$$

which is inserted into (169)

$$\partial_t f_{\mathbf{k}} = S_{\mathbf{k}}(t) + \frac{2}{\hbar^2} \sum_{\mathbf{k'}} M_{\mathbf{k-k'}}^2 \int_{t_0}^{t} dt' \tag{172}$$
$$\{ \quad \cos(E_{\mathbf{k'}} - E_{\mathbf{k}} - \omega_{\mathbf{k-k'}})(t-t') \left[N_{\mathbf{k-k'}}(f_{\mathbf{k'}} - f_{\mathbf{k}}) + f_{\mathbf{k'}}(1 - f_{\mathbf{k}})\right]_{t'}$$
$$- \quad \cos(E_{\mathbf{k'}} - E_{\mathbf{k}} + \omega_{\mathbf{k-k'}})(t-t') \left[N_{\mathbf{k-k'}}(f_{\mathbf{k}} - f_{\mathbf{k'}}) + f_{\mathbf{k}}(1 - f_{\mathbf{k'}})\right]_{t'} \} .$$

We have obtained a kinetic equation with memory where the scattering term depends on the whole "history" of the system. To get rid of the memory two assumptions are necessary: i) the distribution varies little with time, $f_{\mathbf{k}}(t') \Rightarrow f_{\mathbf{k}}(t)$, ii) the time integration result is replaced by a delta function

$$\int_{t_0}^{t} dt' \cos(\omega(t - t')) = \frac{\sin(\omega(t - t_0))}{\omega} \Rightarrow \pi\delta(\omega) . \tag{173}$$

Then, eq. (172) reduces to the standard Boltzmann equation

$$\partial_t f_{\mathbf{k}} = S_{\mathbf{k}}(t) + \frac{2\pi}{\hbar} \sum_{\mathbf{k'}} M_{\mathbf{k-k'}}^2 \cdot \tag{174}$$
$$\{ \quad \delta(E_{\mathbf{k'}} - E_{\mathbf{k}} - \hbar\omega_{\mathbf{k-k'}}) \left[N_{\mathbf{k-k'}}(f_{\mathbf{k'}} - f_{\mathbf{k}}) + f_{\mathbf{k'}}(1 - f_{\mathbf{k}})\right]$$
$$- \quad \delta(E_{\mathbf{k'}} - E_{\mathbf{k}} + \hbar\omega_{\mathbf{k-k'}}) \left[N_{\mathbf{k-k'}}(f_{\mathbf{k}} - f_{\mathbf{k'}}) + f_{\mathbf{k}}(1 - f_{\mathbf{k'}})\right] \} .$$

There is no inherent memory time in the Cosine of eq. (172). However, an effective memory depth is set by the smoothness of the distribution as a function of momentum. In the numerical calculations (Fig. 4) we have used bulk GaAs parameters with a simplified LO phonon scattering. The wavy features in the distribution are related to the phonon oscillation time $2\pi/\omega_{LO} = 115$ fs and are a clear fingerprint of scattering with memory. Time-resolved reflection measurements have revealed such type of oscillations [8], but in this case the phonons have modulated the distribution via a surface potential.

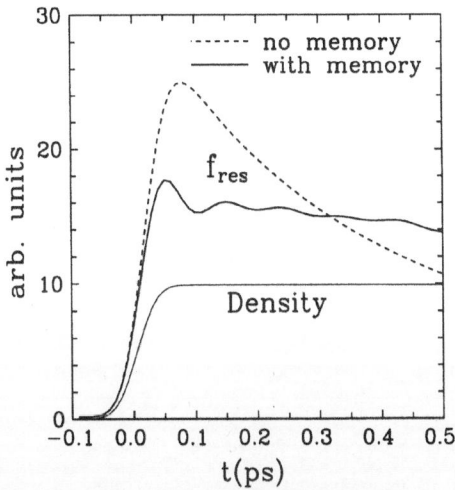

Fig. 4. Temporal evolution of the distribution function at resonance f_{res} following an optical pulse (60 fs FWHM, 40 meV excess energy). The carriers relax via phonon scattering into equilibrium ($kT_0 = 15$ meV). Thin curve - total density. From [9]

REFERENCES

1. H. Haug and St.W. Koch, "Quantum Theory of the Optical and Electronic Properties of Semiconductors", World Scientific, Singapore 1993.

2. S. Schmitt-Rink and D.S. Chemla, Phys. Rev. Letters **57**, 2752 (1986);
   S. Schmitt-Rink, D.S. Chemla, and H. Haug, Phys. Rev. **B37**, 941 (1988).

3. R. Zimmermann, Adv. Solid State Phys. (Festkörperprobleme) **30**, 295 (1990).

4. W. Schäfer, Adv. Solid State Phys. (Festkörperprobleme) **28**, 63 (1988).

5. R. Zimmermann and D. Fröhlich, in "Optics of Semiconductor Nanostructures", Eds. F. Henneberger, S. Schmitt-Rink, and E.O. Göbel, Akademieverlag, Berlin 1993, p. 20.

6. D. Fröhlich, B. Uebbing, T. Willms, and R. Zimmermann, J. Luminescence **58**, 227 (1994).

7. R. Zimmermann, phys. stat. sol. (b) **159**, 317 (1990).

8. G.C. Cho, W. Kütt, and H. Kurz, Phys. Rev. Letters **65**, 764 (1990).

9. R. Zimmermann, J. Luminescence **53**, 187 (1992).

LINEAR AND NONLINEAR OPTICAL SPECTROSCOPY: SPECTRAL, TEMPORAL AND SPATIAL RESOLUTION

Jørn M. Hvam

Mikroelektronik Centret
The Technical University of Denmark
DK-2800 Lyngby, Denmark

ABSTRACT

Selected linear and nonlinear optical spectroscopies are being described with special emphasis on the possibility of obtaining simultaneous spectral, temporal and spatial resolution. The potential of various experimental techniques is being demonstrated by specific examples mostly taken from investigations of the electronic, and opto-electronic, properties of semiconductor nanostructures.

I. INTRODUCTION

Optical spectroscopy has been very instrumental for the development of science and technology during the last century. In particular, physics, chemistry and materials science have benefited greatly from the very detailed and precise information about the elementary excitations of atoms, molecules and solids, that has been obtained from optical spectra. Energy level diagrams as well as dynamical properties (state lifetimes) have been retrieved. With the development of laser sources with ever shorter pulse lengths, from nanoseconds to down below 10 femtoseconds, more and more of the dynamical features are studied directly in the time domain.[1]

It is relevant to ask i) if one can obtain the same information from optical spectra (in the frequency domain) as from direct time resolution of the optical response, and ii) how one can combine the two to perform optical experiments with the maximum information allowed by the uncertainty principles (tranform limited spectroscopy). In transform limited spectroscopy, informations obtained from the frequency domain and from the time domain are only equivalent if the systems are equally prepared. This requires that the optical excitations are perturbative, i.e. that the system response is linear and that excited state populations, and population interactions, are negligible.

Due to the wave nature of light, the spatial resolution of the optical information retrieved in spectroscopy has usually been limited to about half the wavelength of the light applied. This, however, is not a fundamental limitation imposed by the uncertainty relations, but rather the Abbe diffraction limit valid for propagating light modes. The spatial resolution can be enhanced dramatically by making use of non-propagating, evanescent light modes. In this case, the spatial resolution is determined by the size of a small aperture placed in close proximity to the investigated surface or object. With the advent of scanning probe microscopy, various techniques of controlling the distance between a scanning probe and a sample surface, with almost atomic precision, have been developed. Hence, also scanning near-field optical microscopes (SNOM's) have been constructed, combining optical spectroscopy with spectral, and possibly temporal, resolution with optical microscopy featuring nanometer spatial resolution.

In the present article, I shall discuss various aspects of linear and nonlinear optical spectroscopy with special emphasis on the possibilities of combining spectral, temporal and spatial resolution. In the next section, I shall briefly discuss various types of linear optical spectroscopy, including modulation spectroscopy as a precursor for the discussion of nonlinear optical spectroscopies in section III. Quantum beat spectroscopy, as a special case of dynamic spectroscopy in the time domain, is discussed separately in section IV. Section V is devoted to spatial resolution and near-field microscopy. Concluding remarks are made in section VI.

II. LINEAR SPECTROSCOPY

The basic optical response of materials is governed by Maxwell's equations for the electric field $\mathbf{E}$, displacement $\mathbf{D}$, the magnetic field $\mathbf{H}$, and induction $\mathbf{B}$. For nonmagnetic materials, $\mathbf{B} = \mu_0\mathbf{H}$, where μ_0 is the magnetic permeability, and $\mathbf{D} = \epsilon_0\epsilon\mathbf{E} = \epsilon_0\mathbf{E} + \mathbf{P}$, where (ϵ_0) ϵ is the (vacuum) dielectric funtion and $\mathbf{P}$ is the macroscopic polarization field of the medium. Usually the optical properties are described by the polarization field[2]

$$P = \chi{:}E = \epsilon_0(\epsilon-1){:}E \tag{1}$$

where the susceptibility χ is, in general, a tensor. In media without free charge, plane wave solutions of the type

$$E = E_0 e^{i(q \cdot r - \omega t)} \tag{2}$$

with complex wavevector $\mathbf{q} = \mathbf{q}_1 + i\mathbf{q}_2$, are solutions to the wave equation, provided $\epsilon = 0$ for longitudinal waves ($\mathbf{q} \parallel \mathbf{E}$), or for transverse waves ($\mathbf{q} \perp \mathbf{E}$):

$$q \cdot q = \epsilon_0\mu_0\epsilon\omega^2 = \frac{\epsilon\omega^2}{c^2} \tag{3}$$

where $\epsilon = \epsilon_1 + i\epsilon_2$ is the complex dielectric function describing the dielectric, and hence the optical, properties of the medium. Alternatively, is used the complex refractive index

$$N = n + ik = \sqrt{\epsilon} = \sqrt{\epsilon_1 + i\epsilon_2} \tag{4}$$

where n is the real part of the refractive index and k is the extinction coefficient. From the inverse relations

$$\epsilon_1 = n^2 - k^2 \quad ; \quad \epsilon_2 = 2nk \tag{5}$$

it is obvious that ϵ_1 essentially describes the refractive properties and ϵ_2 the absorptive properties of the medium.

For linear media, there are fundamental relations between the real part and the imaginary part of the response function, namely the Kramer-Kronig relations [2],[3]

$$\epsilon_1(\omega) = 1 + \frac{2}{\pi} P \int_0^\infty \frac{\omega' \epsilon_2(\omega') d\omega'}{\omega'^2 - \omega^2} \tag{6}$$

and

$$\epsilon_2(\omega) = - \frac{2\omega}{\pi} P \int_0^\infty \frac{\epsilon_1(\omega') d\omega'}{\omega'^2 - \omega^2} \tag{7}$$

where P indicates the principle value of the integrals. Hence, if the absorption over the full spectral range is known, one can determine the refractive index from a Kramer-Kronig relation and vice versa. One has to remember, though, that it is only strictly valid for linear response, and that experimentally the absorption, or the refraction, is usually only determined over a limited spectral range.

It is instructive to relate the optical parameters (ϵ_1, ϵ_2 or n,k) to simple experiments af absorption, transmission and reflection. For ordinary wave propagation (Eq. (2)), the wave is damped in the direction of propagtion, i.e. $q_1 \parallel q_2$. Then the dispersion relation, expressed in Eq. (3) requires

$$q_1 = \frac{n\omega}{c} \quad ; \quad q_2 = \frac{k\omega}{c} \tag{8}$$

which for a plane wave propagating in the z-direction gives

$$E = E_0 e^{-\frac{k\omega z}{c}} e^{i(\frac{n\omega z}{c} - \omega t)} \tag{9}$$

This wave propagates with the phase velocity $v_f = c/n$ and the damping of the intensity in the direction of propagation is given by

$$I(z) = E^* E = E_0^2 e^{-\frac{2k\omega z}{c}} \equiv E_0^2 e^{-\alpha z} \tag{10}$$

with the absorption coefficient $\alpha = 2k\omega/c$.

At normal incidence, the intensity reflection coefficient in vacuum (or air) from the surface of a medium with optical parameters n and k is

$$R = \frac{(n-1)^2 + k^2}{(n+1)^2 + k^2} \tag{11}$$

and the transmission through a slab of the same material is given by the intensity transmission coefficient [4]

149

$$T = \frac{(1-R)^2\, e^{-\alpha d}(1 + \frac{k^2}{n^2})}{1 - R^2 e^{-2\alpha d}} \qquad (12)$$

In Eq. (12), multiple reflections at the slab surfaces are taken into account, and interference effects are averaged out (rough surfaces). If absorption is not too strong ($k^2 \ll n^2$) and the sample is not too thin ($\alpha d \gg 1$), a simpler expression for the transmission coefficient is valid

$$T = (1 - R)^2\, e^{-\alpha d} \qquad (13)$$

From Eqs. (11)-(13), it is seen that in order to determine the optical constants, two experiments have to be performed. Either a transmission and a reflection experiment or two reflection experiments at different incident angles. In the latter case, the polarization state of the light also has to be controlled, since there is a different reflection coefficient for p-polarized light (**E** parallel to the plane of incidence) and for s-polarized light (**E** perpendicular to the plane of incidence). If a transmission or reflection experiment is done over a sufficiently large spectral range, one can supplement a single experiment with the Kramer-Kronig relations to obtain both optical constants (functions of frequency).

Transmission, or absorption, spectroscopy is widely applied to characterize materials like semiconductors both in the bulk and in lower-dimensional nanostructures, e.g. quasi two-dimensinal layer structures that may be artificially grown by molecular beam epitaxy (MBE).[5] In absorption, one essentially measures the product of the optical density of states and the optical transition matrix elements.

Absorption of photons in a semiconductor leads to the formation of electron-hole pairs with a significant final-state Coulomb interaction. At low temperatures, this may lead to the formation of hydrogen-like bound states, excitons, that significantly alters the absorption spectra with a hydrogen-like series of discrete lines below the fundamental absorption edge. Even at temperatures where the lifetime of the bound states is negligibly short and no sharp exciton lines are observed, the free electron-hole pair, or continuum state, absorption is significantly enhanced by Coulomb correlations.

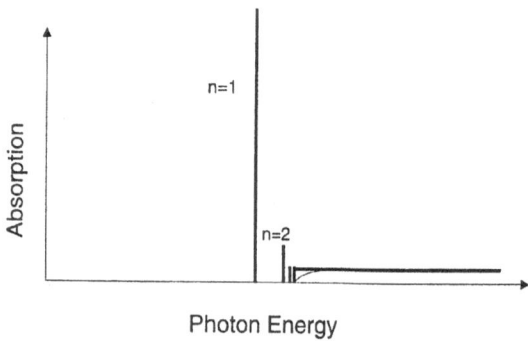

Fig. 1 Absorption in semiconductors according to the 3D Elliot formula

In bulk, or three-dimensional (3D), semiconductors the absorption coefficient is given by the Elliot formula [6]:

$$\alpha(\omega) = \alpha_0^{3D} \frac{\hbar\omega}{E_0} [\sum_{n=1}^{\infty} \frac{4\pi}{n^3} \delta(\Delta + \frac{1}{n^2}) + \Theta(\Delta) \frac{\pi e^{\pi/\sqrt{\Delta}}}{\sinh(\pi/\sqrt{\Delta})}] \qquad (14)$$

where $\Delta = (\hbar\omega - E_g)/E_0$, $\hbar\omega$ is the photon energy, E_g is the energy gap, E_0 is the 3D exciton binding energy, n is the principle exciton quantum number and the free carrier absorption coefficient $\alpha_0^{3D} \propto |p_{cv}|^2 \Theta(\Delta) \sqrt{\Delta}$, where p_{cv} is the interband dipole matrix element, and $\Theta(\Delta)$ is the Heaviside step function. In Eq. (14), damping has been neglected. The sum represents the exciton series and the last term in the bracket gives the Coulomb enhancement of the continuum states, the so-called Sommerfeld enhancement factor, as sketched in Fig. 1.

In MBE-grown quantum wells, or other types of layered structures where electrons, holes and excitons are confined to move freely only in the layers, a two-dimensional (2D) Elliot formula is valid [6]

$$\alpha(\omega) = \alpha_0^{2D} \frac{\hbar\omega}{E_0} [\sum_{n=1}^{\infty} \frac{4}{(n-\frac{1}{2})^3} \delta(\Delta + \frac{1}{(n-\frac{1}{2})^2}) + \Theta(\Delta) \frac{e^{\pi/\sqrt{\Delta}}}{\cosh(\pi/\sqrt{\Delta}}] \qquad (15)$$

where now $\Delta = (\hbar\omega - E_{rs})/E_0$ and E_{rs} is the band edge corresponding to the r'th electron and s'th hole subbands. The 2D free carrier absorption coefficient $\alpha_0^{2D} \propto |p_{cv}|^2 \Theta(\Delta)$ is now essentially a step function. The 2D exciton binding energy is enhanced by a factor of 4 over the 3D exciton binding energy, but the higher excited states are less dominant than in the 3D case. Again there is a Coulomb enhancement around the e-h continuum band edges, as sketched also in Fig. 2.

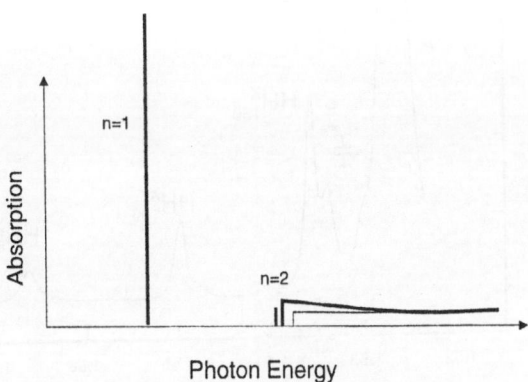

Fig. 2 Sketch of the absorption in a semiconductor quantum well according to the 2D Elliot formula.

For obtaining information about population effects and population dynamics, photo-luminescence (PL) experiments are very useful, particularly for the lowest lying and longest living energy states.[5] Figure 3 shows an example of the characterization of a sample that is MBE-grown with GaAs/AlGaAs multiple quantum wells of different well widths.[7] The upper trace shows absorption and the lower trace luminescence. From the spectral positions of the different lines one can observe for different well widths the splitting of heavy hole (HH) excitons and light hole (LH) excitons, the confinement energies and the exciton binding energies. In luminescence, one observes HH exciton luminescence as well as HH biexciton luminescence from which biexciton binding energies can be determined.[8] Biexcitons are bound states of two excitons forming an excitonic molecule. From the narrow linewidths and the small shift between absorptions and emission lines (Stoke's shift) one can conclude that the sample quality is good with little inhomogeneous broadening and localization effects.[5]

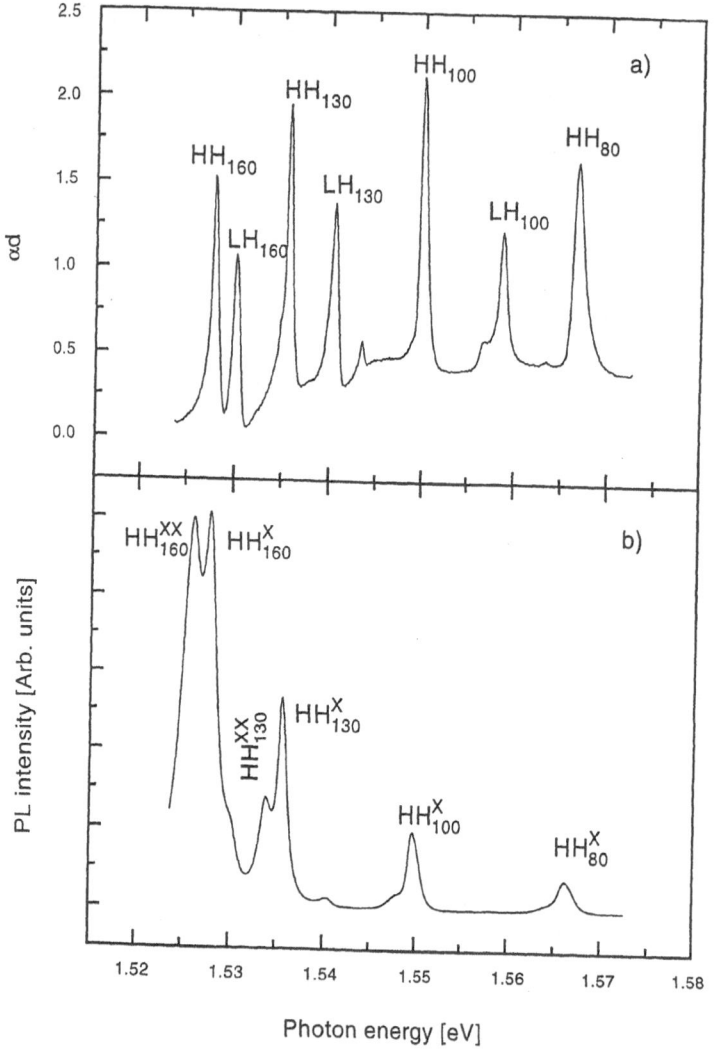

Fig. 3 Optical characterization by absorption (a) and photoluminescence (b) of a GaAs/AlGaAs multiple quantum well sample with well widths 80, 100, 130, and 160 Å.

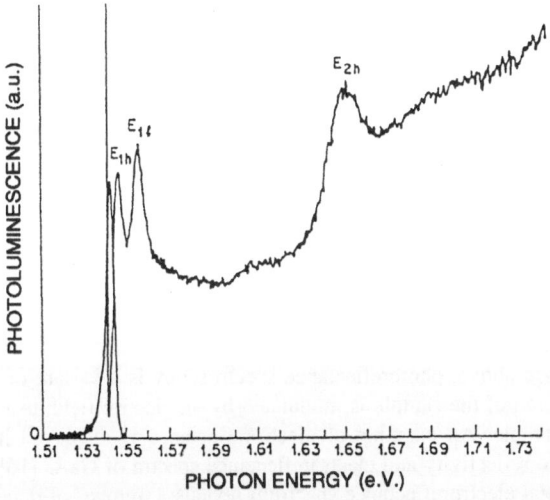

Fig. 4 Photoluminescence and excitation spectra at 5K for a p-type modulation doped multiple quantum well sample with 115 Å and 150 Å barriers. From Ref. 9.

In samples where transmission experiments are difficult or impossible to perform, e.g. epitaxially grown layer on a non-transparent substrate, one can still get information about the spectral dependence of the absorption coefficient by performing a photoluminescence excitation (PLE) experiment, i.e. the luminescence of a low-lying state is studied as a function of the excitation photon energy. This is usually a very sensitive method, and an example [9] is shown in Fig. 4. If the sample is photoconducting, similar information can be obtained by measuring photocurrent as a function of the excitation photon energy.

II.A. Modulation Spectroscopy

Bound excitonic states in semiconductors appear as distinct spectral lines in linear absorption and emission experiments as in atomic and molecular spectroscopy. The characteristic features like spectral positions, oscillator strengths and damping are therefore readily observable. In solid dielectrics and semiconductors in general, however, the optical spectra are continuous in energy and the distinct features to be retrieved are critical point and singularities in the optical density of states, that may appear as weak pumps and kinks in the optical spectra.[3],[4]

To enhance the visibility of weak spectral features, and improve the signal to noise ratio, various types of modulation spectroscopy have been developed over the years.[10],[11] The general idea of modulation spectroscopy is to modulate some physical parameter that influences the optical response and then detect the optical signal in synchronous with the modulation (lock-in technique). The result is a differential spectrum with respect to the parameter that is being modulated. the simplest form of modulation spectroscopy is wavelength modulation where the wavelenth response of the spectrometer is being modulated, e.g. by a vibrating mirror or a glass plate.[12] In this case, spectra are just being differentiated with respect to the wavelength, sharpening up the spectral features.

In most other modulation spectroscopies, some external field, or potential, applied to the sample is being modulated. The external agent may be temperature, stress, magnetic or electric fields. In this case the spectra are differentiated with respect to the external field, and in addition one gets information about the coupling between the extenal field and the optical properties as represented by the thermo-optical, the piezo-optical, the magneto-optical and the electro-optical coefficients. Electro-optical effects, have been widely used for modulation spectroscopy particularly in the form of electroreflectance. Traditionally, the electric field has been applied externally, e.g. in an electrolytic liquid [13] or in a Schottky barrier [14]. In semiconductors, there are often build-in electric fields (surface or interface fields) that can be modulated externally by photoexcitation.[15] This has given rise to the powerful technique of photoreflectance, that I shall briefly describe in the next subsection.

II.B. Photoreflectance Spectroscopy

As mentioned above, photoreflectance spectroscopy is a variant of electroreflectance where the reflection of the sample is modulated by an electric field, usually normal to the surface of the sample. An example of electroreflectance is displayed in Fig. 5 showing room-temperature reflectivity and electroreflectance spectra of GaAs.[16] It is clearly seen that the differential electroreflectance spectrum reveals a number of details that are barely seen in the reflectivity spectrum and hence enhances the sensitivity of the optical spectroscopy by many orders of magnitude. The sharpness of the electroreflectance spectra as compared to the reflectivity spectrum itself was first explained by Aspnes [17] who showed that, writing the dielectric function $\varepsilon(E,F,\gamma)$ as a function of photon energy E, electric field F, and damping γ, the changes in the dielectric function for small fields could be expressed as the third derivative with respect to the energy: $\Delta\varepsilon \sim d^3(E^2\varepsilon\,(E,0,\gamma))/dE^3$.

Similarly, Fig. 6 shows photoreflectance spectra for a GaAs/AlGaAs heterojunction and multiple quantum well samples with different well widths L_z.[15] A large number of HH and LH interband transitions are clearly seen in the spectra, and the confined energy levels for different well widths can be precisely determined.

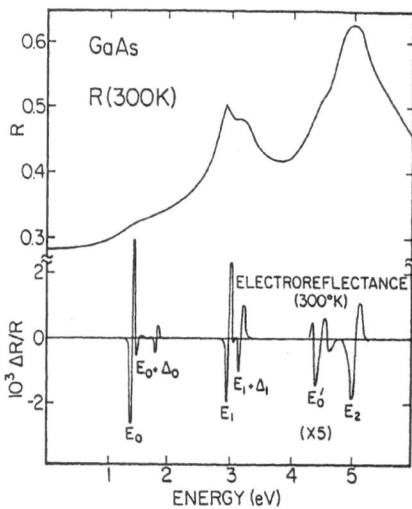

Fig. 5 Room temperature reflectivity (top) and electroreflectance (bottom) spectra of GaAs. From Ref. 16.

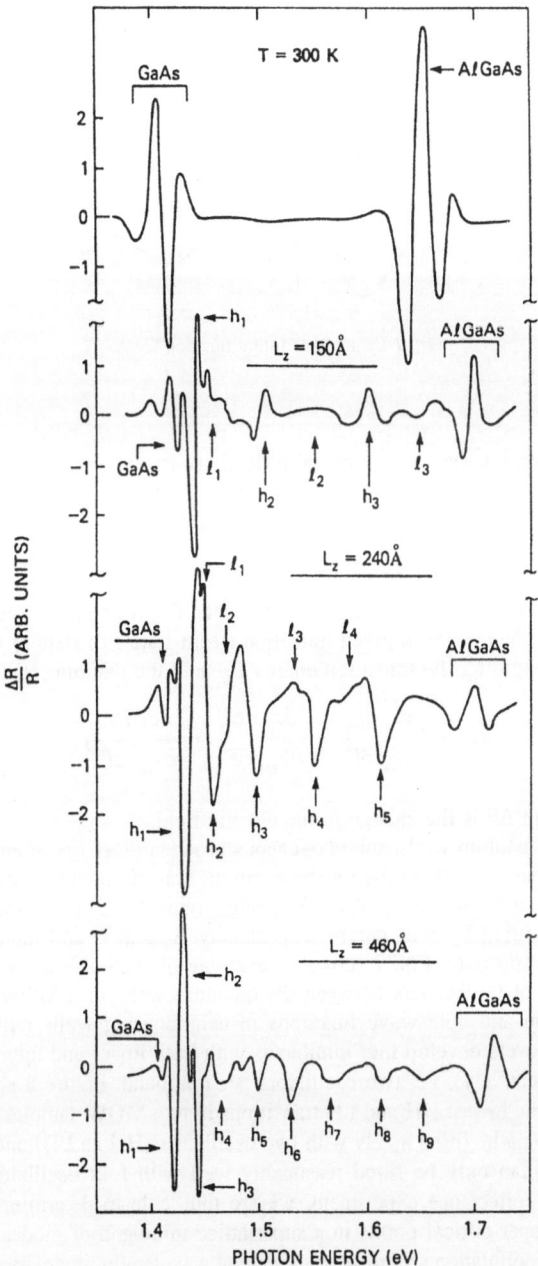

Fig. 6 Photoreflectance spectra for a GaAs/AlGaAs heterojunction (top) and multiple quantum well samples with well widths, L_z. The labels h_n, l_n, n=1,2... represent parity-allowed light and heavy hole interband transitions. From Ref. 15.

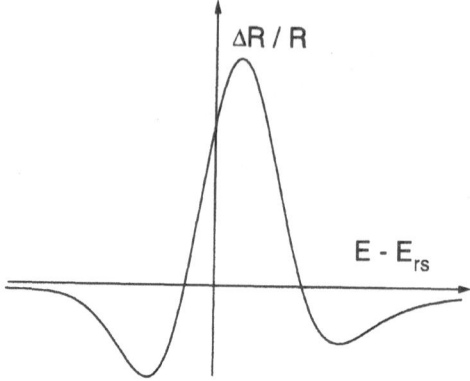

Fig. 7 Example of fitting function for ΔR/R.

The measured relative change of the reflectivity is often expressed by[18]

$$\frac{\Delta R}{R} = \alpha \Delta \epsilon_1 + \beta \Delta \epsilon_2 \tag{16}$$

where α and β are the so-called Seraphin coefficients. For fitting purposes, one may express the field induced changes for a given transition (from state r to state s) via the changes in the oscillator strength f_{rs}, the transition energy E_{rs}, and the damping γ_{rs} by

$$\Delta \epsilon_i = [\frac{\partial \epsilon_i}{\partial f_{rs}}\frac{\partial f_{rs}}{\partial F} + \frac{\partial \epsilon_i}{\partial E_{rs}}\frac{\partial E_{rs}}{\partial F} + \frac{\partial \epsilon_i}{\partial \gamma_{rs}}\frac{\partial \gamma_{rs}}{\partial F}] \cdot \Delta F \tag{17}$$

where $i = 1, 2$ and ΔF is the change in the electric field.

In multiple quantum wells, inhomogeneous broadening (Γ) is often important giving Gaussian lineshapes of $\epsilon_2(E-E_{rs},\Gamma)$ for the excitonic transitions. From $\epsilon_2(E-E_{rs},\Gamma)$, the real part $\epsilon_1(E-E_{rs},\Gamma)$ can be found by Kramer-Kronig transformation, so that ΔR/R for each oscillator can be fitted by four parameters, namely E_{rs} and Γ and their derivatives with respect to the electric field. Fig. 7 shows an example of such a fitting function for ΔR/R.

If the width of the barriers between the quantum wells in a MQW sample becomes small, the electron and hole wave functions in neighbouring wells will overlap and the confined energy levels develop into minibands with both lower and upper extremal points (M_0 and M_1, respectively), i.e. two oscillators per miniband. Figure 8 shows photoreflectance spectra for the lowest HH and LH transitions in two MQW samples. One with barrier width 100 Å that can be fitted nicely with two oscillators (HH an LH) and one with barrier width 30 Å, that can only be fitted reasonably well with four oscillators. From a room temperature photoreflectance experiment, we are thus able to determine the existence of both lower and upper critical points in a superlattice miniband of modest width.

The various modulation spectroscopies, except wavelength modulation, rely on the fact that the optical properties depend, at least perturbatively, on an applied external field. In photoreflectance, the applied field is again light (changing a built-in electric field). This means that one is no longer strictly speaking in the linear regime, because the optical properties depend on the light field itself. As such, photoreflectance is a kind of precursor for nonlinear optical spectroscopy which is the topic of the next section.

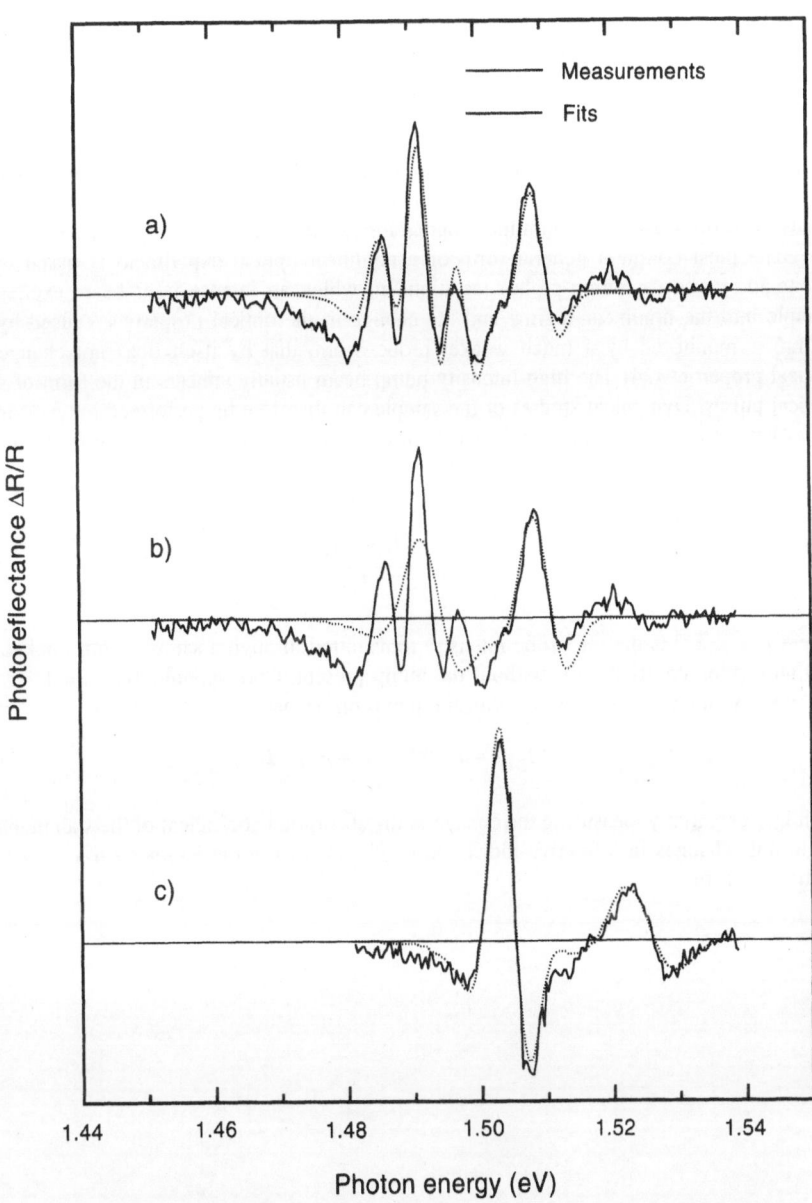

Fig. 8 Photoreflectance in different multiple quantum well samples. (a) 30Å barriers fitted with 2 oscillators; (b) 30Å barriers fitted with 4 oscillators; (c) 100Å barriers fitted with 2 oscillators.

III. NONLINEAR SPECTROSCOPY

There are several reasons for the increased activities in nonlinear optics during the last few decades. First of all, the technological development of new lasers[19] has made it possible to achieve a variety of high-quality, high-intensity laser sources for exciting optical media well into the nonlinear regime. The nonlinear optical properties of materials have by themselves attracted a lot of attention because of the potential interests for optical and optoelectronic devices for the increasing market of optical signal processing.[20] Finally, a number of nonlinear optical spectroscopies [21],[22],[23] have emerged that utilize nonlinear optical processes to gain new information about elementary excitations of materials, particularly final state interactions and dynamical features. In this section, I shall focus on various aspects of nonlinear optics for spectroscopy.

The most common general form of a nonlinear optical experiment is based on an excite-and-probe (or pump-probe) technique in which an intense laser beam excites the sample into the nonlinear regime and the change in the optical properties, caused by the pump, is monitored by a much weaker probe beam, that by itself does not change the optical properties.[24] The high-intensity pump beam usually appears in the form of short optical pulses. Dynamical studies of the sample can therefore be performed by varying the delay between the pump pulse and a weak probe pulse. Such an excite-and-probe technique is sketched in Fig. 9, as a differential transmission (DTS) experiment.

The DTS signal is experimentally determined as

$$I_{DTS} = \frac{I_{PP} - I_P}{I_P} \tag{18}$$

where $I_P = I_0\, e^{-\alpha d}$ is the the probe intensity transmitted through a sample, with thickness d and absorption coefficient α, without the pump present. Correspondingly, $I_{PP} = I_0\, e^{-(\Delta\alpha + \alpha)d}$ is the transmitted probe intensity with the pump on. Hence,

$$I_{DTS} = e^{-\Delta\alpha d} - 1 \approx -\Delta\alpha d \tag{19}$$

with I_{DTS} essentially measuring the change in the absorption coefficient of the slab medium. From this, changes in refractive index can be obtained from the Kramer Kronig relations with some care.

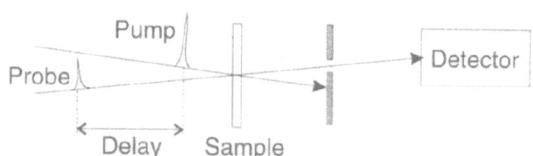

Fig. 9 Transient pump-probe experiment: Differential transmission.

In semiconductors, significant changes occur in the optical properties around the band edge when the material is excited at, or above, the bandgap, i.e. when a large number of photocarriers are created. At moderate carrier densities, a band-filling nonlinearity occurs which will shift the absorption edge towards higher energies (blue shift). At higher carrier densities, many-body effects give rise to band gap renormalizations resulting in a red shift counteracting, and eventually gaining over, the band-filling effect. At the same time, significant broadening of the band edge will take place due to carrier-carrier collisions. In Fig. 10 is sketched a band-filling nonlinearity with the changes in the absorption coefficient and the corresponding changes in the index of refraction.[3] If excitonic features are dominating the near-gap optical properties, the typical nonlinearities at low and medium intensities are, as discussed above for photoreflectance, excitonic bleaching (loss of oscillator strength), shifts (optical Stark shift) and broadening (collisions). Eventually, the excitons will be completely screened by other excitons and free carriers, and the band edge nonlinearities will prevail.

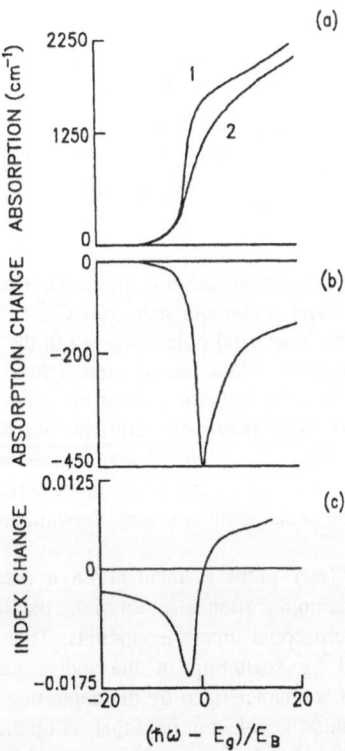

Fig. 10 Bandfilling nonlinearity in a semiconductor. (a) The absorption spectra at low (1) and high (2) carrier densities, (b) The change in absorption coefficient, and (c) the corresponding change in the refractive index, as calculated from (b) by a Kramers-Kronig transformation. From Ref. 3.

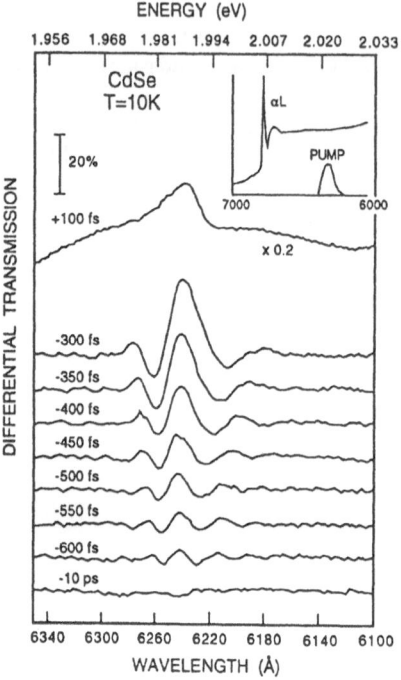

Fig. 11 Spectral hole burning in CdSe with femtosecond time resolution. From Ref. 27.

In Fig 11, a slightly different situation is illustrated. Here, the intense laser pump pulse (100 fs) is tuned into the band continuum states of a CdSe crystal held at low temperatures. The DTS spectra show that a spectral hole is burned in the absorption spectrum, shown in the inset.[25],[26] For positive delays, the spectral hole is broadening and shifting down towards the band edge, giving information about the scattering and the energy relaxation of the photexcited carriers. What may seem surprising at a first glance, are the DTS spectra showing distinct interferences at negative delays, exceeding significantly the laser pulse width, i.e. there is a pump induced change in the probe signal, even when the probe arrives well before the pump. The explanation is straight forward and is a typical coherence effect, as illustrated in Fig. 12.

When an ultrafast laser pulse is incident on a medium, in resonance with real excitations, a macroscopic polarization is set up in the medium as a coherent superposition of all the individual microscopic dipole excitations. This macroscopic polarization will eventually be destroyed by scattering of the individual dipoles. The decay of the macroscopic polarization is characterized by the dephasing time T_2. Within this time, the polarization may reemit into the coherent field that set up the polarization in the first place. Hence the transmitted (probe) pulse is stretched on the output side of the medium, with a tail extending well over the dephasing time of the optical resonance. If an intense pump pulse arrives in the medium within this dephasing time it is therefore able to modify the overall transmitted probe pulse, as illustrated in Fig. 12. The spectral fringes in the DTS spectra for negative delays are just the Fourier transforms of the temporal modification of the probe signal, as sketched in Fig. 12.

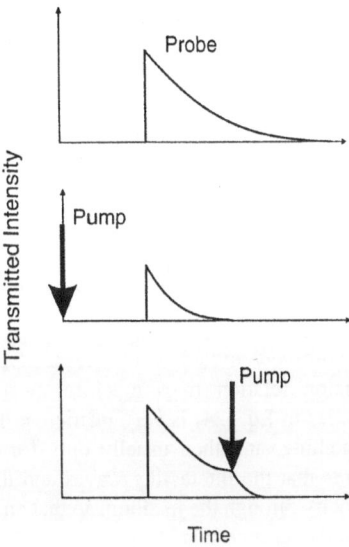

Fig. 12 Transient coherent transmission without pump (upper figure), and with pump for positive (center figure) and for negative (lower figure) delay.

In Sec. II, the optical response of materials was phenomenologically described by expressing the relation between the electric field **E** of the electromagnetic radiation and the resulting polarization **P** of the medium. In the nonlinear case, the polarization is traditionally expanded in powers of **E** [21],[27]

$$P(r,t) = \chi^{(1)} \cdot E + \chi^{(2)} \cdot EE + \chi^{(3)} \cdot EEE + \cdots$$
$$= P^{(1)}(r,t) + P^{(2)}(r,t) + P^{(3)}(r,t) + \cdots \tag{20}$$

where the terms on the right hand side are in a short-hand vector notation, and in reality are multiple space and time integrals.

If the electromagnetic field is composed of a group of monochromatic plane waves

$$E(r,t) = \sum_i E_i(k_p\omega_i) = \sum_i (E_{0i} e^{i(k_i \tau - \omega_i t)} + c.c.) \tag{21}$$

the polarization can likewise be expanded in plane waves

$$P^{(n)}(r,t) = \sum_j P_j^{(n)}(k_p\omega_j) = \sum_j (P_{0j}^{(n)} e^{i(k_j \tau - \omega_i t)} + c.c.) \tag{22}$$

where the different components can be expressed by

161

$$P^{(n)}(k,\omega) = \chi^{(n)}(k,\omega) \cdot E_1(k_1,\omega_1) E_2(k_2,\omega_2) \cdots E_n(k_n,\omega_n) \qquad (23)$$

The n'th order nonlinear susceptibility $\chi^{(n)}(k,\omega)$ is the Fourier transform of the corresponding response function $\chi^{(n)}(\mathbf{r}-\mathbf{r}_1,t-t_1,....\mathbf{r}-\mathbf{r}_n,t-t_n)$ in Eq. 20. Furthermore, the only components that survive the space and time integrations are those that conserve energy (for times $t-t_i \gg 2\pi/\omega_i$) and wavevector:

$$\omega = \sum_i \pm\omega_i \quad \wedge \quad k = \sum_i \pm k_i \qquad (24)$$

where +/- are entered for absorbed/emitted waves, respectively. Due to the fixed, and generally nonlinear, dispersion relation $\omega = \omega(k)$ in most media, the simultaneous fulfillment of the two equalities in Eq. (24) is very restrictive. It is only met under special conditions, called phase-matching, and then usually only for one particular polarization wave. Phase-matching insures that the interacting waves and the nonlinear polarization(s) propagate with the same velocity through the medium, so that an appreciable energy transfer between waves can take place over a certain length.

For steady-state conditions, the form in Eq. (23) is very convenient. In the usual dipole approximation, or local response theory, the susceptibility is furthermore independent of wavevector, so that the medium is fully described by the frequency dependent, but wavevector independent, nonlinear susceptibilities

$$\chi^{(n)}(k,\omega) = \chi^{(n)}(\omega) = \chi^{(n)}(\omega;\pm\omega_1,\pm\omega_2,...,\omega_n)) \qquad (25)$$

where in the last expression we have adopted the usual notation for nonlinear susceptibilities, [21] indicating the applied frequencies and the resulting frequency from Eq. (24).

It is easily seen that in systems with inversion symmetry there can be no second order nonlinearity, i.e. $\chi^{(2)} \equiv 0$. However, in systems without inversion symmetry such as $LiNbO_3$ KDP, BBO, GaAs, or any semiconductor surface, the lowest order optical nonlinearities will be of second order as expressed by

$$P^{(2)}(k,\omega) = \chi^{(2)}(\omega;\omega_1,\omega_2) \cdot E_1(k_1,\omega_1) E_2(k_2,\omega_2) \qquad (26)$$

where the input fields may come from the same beam or from two different coherent beams. From the energy conservation in Eq. (24), this term is responsible for sum and difference frequency generations, as well as for parametric generation where one intense pump wave with frequency ω creates signal (ω_1) and idler (ω_2).

Parametric generation and amplification is gaining increasing importance in optical spectroscopy providing tunable laser sources, both c.w. and pulsed, in broad spectral ranges (in the visible and infrared). Sum and difference generation is used very conveniently to transform a signal into another wavelength range where one may have more sensitive or faster detectors (up- or down-conversion). With ultrafast laser pulses, sum and difference generation has become the standard technique to obtain ultrafast time resolution by correlation (auto or cross). In these situations, phase-matching is essential in order to obtain a reasonably efficient conversion. The standard technique for phase-matching is to use birefringent crystals for the nonlinear mixing and let the fundamental and the up (down) converted waves propagate as ordinary and extraordinary waves, respectively. Accurate phase-matching is then achieved by varying the direction of propagation with respect to the optical axis of the crystal, e.g. KDP.[21]

In materials with inversion symmetry ($\chi^{(2)} \equiv 0$), the lowest order nonlinearities are due to third order effects of the form

$$P^{(3)}(k,\omega) = \chi^{(3)}(\omega;\omega_1,\omega_2,\omega_3) \cdot E_1(k_1,\omega_1)E_2(k_2,\omega_2)E_3(k_3,\omega_3) \qquad (27)$$

If only one monochromatic light beam is present in the medium, then $\omega_1=\omega_2=\omega_3=\omega'$ and $k_1=k_2=k_3=k'$ in Eq. (27), which from Eq. (24) offers two possibilities: 1) Third harmonic generation with $\omega=3\omega'$ and $k=3k'$, where phase-matching is generally difficult to achieve, and 2) degenerate four-wave mixing (DFWM) within the same beam with $\omega=\omega'$ and $k=k'$. This situation is always perfectly phase-matched and the nonlinear polarization

$$P^{(3)}(k,\omega) = \chi^{(3)}(\omega;\omega,-\omega,\omega)\ |E(k,\omega)|^2 E(k,\omega) \qquad (28)$$

expresses the situation of intensity dependent refractive index $n(I)$ and absorption coefficient $\alpha(I)$. With $n(I) = n_0 + n_2 I$ and $\alpha(I) = \alpha_0 + \alpha_2 I$, we find n_2 and α_2 from the real and imaginary part of $\chi^{(3)}$

$$n_2 = \frac{Re\chi^{(3)}}{\epsilon_0 n_0^2 c} \quad \wedge \quad \alpha_2 = \frac{\omega Im\chi^{(3)}}{\epsilon_0 n_0^2 c^2} \qquad (29)$$

The intensity dependent refractive index is responsible for self-focussing (or defocussing) and self-phase modulation. These phenomena are very important for solid state ultrafast lasers and for the propagation of intense light pulses (solitons) in optical fibers. The intensity dependent absorption coefficient describes saturable absorption or induced absorption, depending on the sign of α_2.

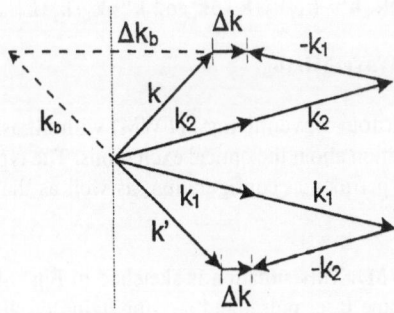

Fig. 13 Wavevector conservation for DFWM with two beams, k_1, k_2.

With two light beams $\omega_1=\omega_2=\omega$, $k_1=k_2=k$, and $\cos\theta = \mathbf{k}_1 \cdot \mathbf{k}_2 / k^2 < 1$, as in Fig. 13, there are two possibilities with perfect phase-match: $\mathbf{k}=\mathbf{k}_1$ and $\mathbf{k}=\mathbf{k}_2$. As in the case with only one beam, the nonlinear signals are generated in a direction with a large linear signal and may therefore be difficult to detect at moderate intensities. There are also two possibilities with near phase-match: $\mathbf{k} = 2\mathbf{k}_2\text{-}\mathbf{k}_1\text{-}\Delta\mathbf{k}$ and $\mathbf{k}'= 2\mathbf{k}_1\text{-}\mathbf{k}_2\text{-}\Delta\mathbf{k}$, where $\Delta\mathbf{k}$ is the wave vector mismatch perpendicular to the sample plane. This geometry (see Fig. 13) has the advantage that the nonlinear signal is a collimated beam generated in a direction where there is no linear signal. It is therefore very well suited for the detection of even very small nonlinear signals. The linear background can to a high degree be eliminated by simple spatial filtering (see Fig.14).

DFWM can also be viewed upon as a case of light-induced gratings.[28] The two incident beams interfere and create stationary polarization gratings in the nonlinear medium with scattering vectors $\pm(\mathbf{k}_2\text{-}\mathbf{k}_1)$ and a grating constant $\Lambda=2\pi/|\mathbf{k}_2\text{-}\mathbf{k}_1|$. These gratings in turn scatter the incident beams $\mathbf{k}_1$ and $\mathbf{k}_2$ into the directions $\mathbf{k}'$ and $\mathbf{k}$, respectively, as also discussed above (see Fig. 13). The signal intensity in the scattered direction is[28]

$$I_s \propto d^2 \; \frac{\sin^2(\frac{\Delta k \cdot d}{2})}{(\frac{\Delta k \cdot d}{2})^2} \tag{30}$$

where d is the sample thickness, or nonlinear interaction length in general. Near-phase-match requires $\Delta kd \ll 1$, i.e. thin sample geometry. If the sample is very thin ($kd \ll 1$), the grating is essentially two-dimensional, and back scattering (with $\Delta k_b \approx 2k$) will occur with about the same intensity as in the forward direction (see Fig. 13). Note, however, that in both directions the signal will be weak, because $I_s \propto d^2$. Therefore, strong resonance enhancement will normally be necessary to observe a back scattered signal. On the other hand, it does open up the possibility to observe the nonlinear interaction in a resonance where strong linear absorption prevents transmission even through a thin sample[29].

With three incident light beams, more freedom is at hand to perform different types of four-wave mixing experiments. One advantage is, that two coherent light beams at an angle are sufficient to set up a light-induced grating in the medium, which can then be probed independently by the third beam that need no coherent relation to the first two beams. One can thus set up a nonlinear grating by exciting resonantly with the two beams and then probe the grating in the transparent region. In the degenerate case $\omega_1=\omega_2=\omega_3$, and with three different beam directions, $\mathbf{k}_1$, $\mathbf{k}_2$ and $\mathbf{k}_3$, a certain phase mismatch $\Delta\mathbf{k}$ again requires thin samples, $d \ll 1/\Delta k$, and gives rise to first order scattering ($\chi^{(3)}$) in the three directions $\mathbf{k}=\mathbf{k}_1\text{-}\mathbf{k}_2+\mathbf{k}_3\text{-}\Delta\mathbf{k}$, $\mathbf{k}'=\text{-}\mathbf{k}_1+\mathbf{k}_2+\mathbf{k}_3\text{-}\Delta\mathbf{k}$ and $\mathbf{k}''=\mathbf{k}_1+\mathbf{k}_2+\mathbf{k}_3\text{-}\Delta\mathbf{k}$.

III.A. Transient Four-Wave Mixing

Performing transient four-wave mixing (TFWM) with ultrashort laser pulses, one can obtain dynamical information about the optical excitations. The type of information obtained depends on the actual experimental configuration, as well as the character of the samples investigated.

1. Two-Beam TFWM. This situation is sketched in Fig. 14. The two incident laser pulses are split off the same laser pulse, and are impinging on the sample with a variable optical delay between them. In order for the two laser pulses to interact coherently in the nonlinear medium, for example by setting up a polarization grating, the delay between them should not exceed the dephasing time of the nonlinear polarization in the medium, caused by the first laser pulse. The nonlinear TFWM signal is then self-diffraction of the second pulse in the grating set up by the coherent overlap between the polarizations from the first and the second pulse. For pulse #1 arriving first ($\tau_{12} > 0$), as in Fig. 14, a signal will be emitted in the direction $2\mathbf{k}_2 - \mathbf{k}_1$ as indicated. For pulse #2 arriving first ($\tau_{12} < 0$), the signal is emitted in the direction $2\mathbf{k}_1 - \mathbf{k}_2$ (dashed in Fig.14).

This result is obtained by solving the two-level optical Bloch equations to third order, which for δ-pulses yields [30],[31]

$$P^{(3)}(r,t,\tau_{12}>0) \propto \Theta(t)\mu_{eg}^4 \, e^{i[(2k_2-k_1)\tau - \Omega_{eg}t + \Omega_{eg}^*\tau_{12}]} \tag{31}$$

where t is the time after the arrival of the second pulse, $\Theta(t)$ is the Heaviside step function and $\Omega_{eg} = \omega_{eg} - i\gamma_{eg}$ is the complex transition frequency with the damping $\gamma_{eg} = 1/T_2 + 1/2T_1$. T_1 is the longitudinal (population) lifetime and T_2 is the transverse (polarization) lifetime, or dephasing time. The intensity of the TFWM signal in the direction $2\mathbf{k}_2 - \mathbf{k}_1$ is then

$$I_{DFWM} \propto |P^{(3)}(t,\tau_{12}>0)|^2 \propto \Theta(t)\mu_{eg}^8 \, e^{-2\gamma_{eg}(t+\tau_{12})} \tag{32}$$

decaying exponentially with time as well as with delay between the two pulses. If the TFWM signal is spectrally resolved and integrated with a slow detector, it corresponds to detecting the absolute square of the Fourier transform of Eq. (31), which is

$$I_{DFWM} \propto |P^{(3)}(\omega,\tau_{12}>0)|^2 \propto \frac{\mu_{eg}^8}{|\Omega_{eg} - \omega|^2} \, e^{-2\gamma_{eg}\tau_{12}} \tag{33}$$

where ω is the optical frequency of the signal detected. Hence, the whole Lorentzian spectrum decays exponentially with increasing delay. The correlation traces of the time-integrated TFWM signal contain information about the dephasing rate $\gamma_{eg} = 1/T_2 + 1/2T_1$, as well as the resonance enhancements of the nonlinear signal.[32]

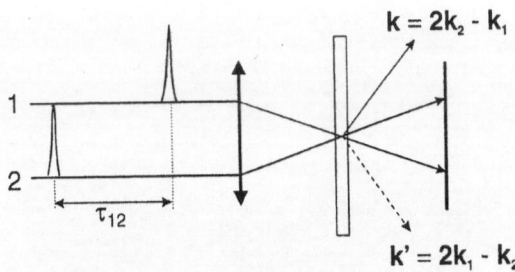

Fig. 14 Transient DFWM with two beams.

If the two-level system is inhomogeneously broadened, an integration has to be made over the distribution of transition frequencies, e.g. $g(\omega_{eg}) \propto \exp\{-(\omega_{eg}-\omega_0)^2/(\Delta\omega_{eg})^2\}$

$$P_{inhom}^{(3)}(t) = \int_0^{\infty} P^{(3)}(t)g(\omega_{eg})d\omega_{eg} \tag{34}$$

In this case, destructive and constructive interferences between the different transition frequencies play a significant role, and the signal in the $2\mathbf{k}_2-\mathbf{k}_1$ direction appears as a photon echo at the time $t = \tau_{12}$ [30],[31]

$$I_{PE} \propto |P_{inhom}^{(3)}(t,\tau_{12}>0)|^2 \propto \mu_{eg}^8\, e^{-\frac{\Delta\omega_{eg}^2(t-\tau_{12})^2}{2}}\, e^{-2\gamma_{eg}(t+\tau_{12})} \tag{35}$$

The time integrated Fourier transformed photon echo then takes the form (for $\tau_{12} \gg 1/\Delta\omega_{eg}$)

$$I_{PE} \propto |P^{(3)}(\omega,\tau_{12}>0)|^2 \propto \mu_{eg}^8\, e^{-\frac{2|\Omega_{eg}-\omega|^2}{(\Delta\omega_{eg})^2}}\, e^{-4\gamma_{eg}\tau_{12}} \tag{36}$$

which for the same homogeneous damping rate decays twice as fast as the free polarization decay in Eq. (33). From a comparison of the observed TFWM decay with the linewidth, as observed in absorption or emission, it can be decided whether the transition is homogeneously or inhomogeneously broadened. Knowing that, the dephasing time, and thereby the homogeneous linewidth, can be determined from either Eq. (33) or Eq. (36).[30],[31]

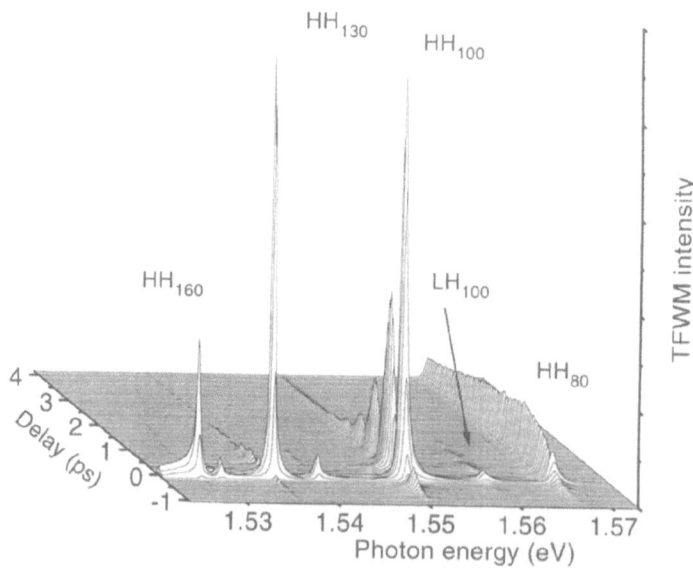

Fig. 15 TFWM spectra as a function of delay in a sample with quantum wells of different thicknesses

An example of a TFWM experiment is displayed in Fig. 15 showing FWM spectra as a function of delay between the incident pulses. The sample is specially prepared and contains several series of multiple quantum wells (MQW) of GaAs/AlGaAs with the different well widths 80, 100, 130, and 160 Å. It is excited with 100 fs laser pulses spectrally centered at the heavy-hole (HH) exciton resonance of the 100 Å quantum wells, but covering the whole spectral range shown. The TFWM signal shows pronounced resonance enhancements, as in Eqs. (33) and (36), around all the HH and light-hole (LH) exciton resonances. The temporal behaviour of the different resonances, however, is distinctly different for the different well widths. The heavy-hole exciton resonance of the 80 Å quantum wells (HH_{80}) peaks for a finite delay τ and has a rather slow decay. This is typical for a photon echo (PE) type signal (Eq. (36)) where the decay reveals the dephasing time of the relatively low density of (localized) HH excitons.

The other quantum wells that are located in the center, or at the low energy side, of the laser profile show a more prompt signal, typical for a free polarization decay (FPD), with a faster decay. The faster decay is not accompanied by a spectral broadening, and can therefore not be due to an increased dephasing rate caused by scattering with the continuum states, simultaneously excited by the laser pulse. It is rather due to destructive interference within the wavepacket formed by the spectrally broad excitation of the wider quantum wells. Another type of interference manifests itself by the oscillatory modulation of the decay of the FPD of the HH_{100} and the LH_{100} resonances. This modulation is due to a quantum beat interference between the coherently excited HH and LH excitons of the 100 Å quantum wells. Quantum beat spectroscopy will be treated separately in Sec. IV.

From PE and FPD experiments, coherence effects and scattering rates and mechanisms have been studied in a variety of semiconducting materials and structures over the past decade. Particularly in nanostructures, the information gained about the homogeneous and inhomogenous character of the broadening is of interest in order to characterize the quality of the heterostructure interfaces. Figure 16 shows TFWM spectra around the HH exciton resonance in a GaAs/AlGaAs MQW sample. Two peaks with FPD character are seen for short delays. The splitting of the lines corresponds to the difference of the exciton energy in quantum wells of widths differing by one monolayer. These lines have Lorentzian shapes,

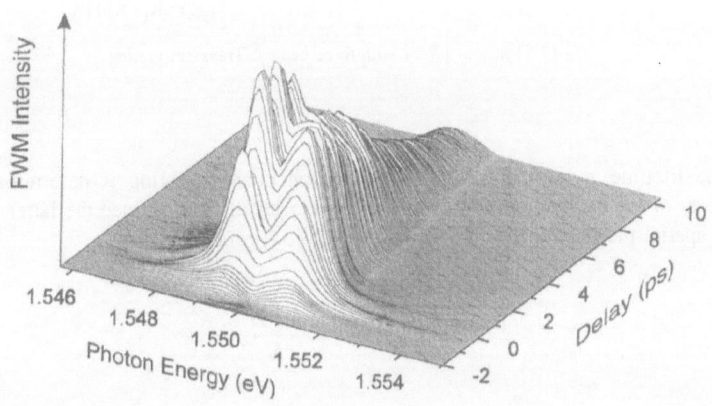

Fig. 16 TFWM spectra from GaAs/AlGaAs sample with monolayer interface fluctuations.

as in Eq. (33). For longer delays, one line is seen decaying like a PE. This line is centered between the two initial lines and it has a Gaussian spectral shape as in Eq. (36). The two FPD lines are coming from monolayer flat regions of the quantum wells with widths differing by one monolayer.[33] These regions are (much) larger in extent than the exciton diameter (150 Å). The PE signal stems from regions of the quantum wells where monolayer interface fluctuations occur on a lengths scale smaller than the exciton diameter, so that the exciton effectively averages over the corresponding fluctuating potential. This gives rise to inhomogenous broadening and localization effects, as revealed by the TFWM-PE experiment.

2. Three-beam TFWM. This situation is sketched in Fig. 17. The two first pulses arrive simultaneously, or well within the dephasing time of the material ($\tau_{12} \ll T_2$), and interfere coherently to set up a nonlinear grating in the medium. This grating can then be detected at variable time delays, τ_{13}, by diffraction of the third pulse. If also $\tau_{13} \ll T_2$, then pulse #3 will diffract off a coherent polarization grating set up by pulse #1 and pulse #2, as in the self-diffraction case above. If, however, ω is in resonance with an electronic excitation in the material, a real excitation density grating may persist long after the coherent polarization grating has disappeared by dephasing. This grating, however, still bears the fingerprint of the coherent overlap between the two first pulses, and can even give rise to a photon echo signal, stimulated by the third pulse, but dependent on the delay τ_{12}. If, on the other hand, the scattered signal is recorded as a function of delay τ_{13} of the third pulse, the decay of the incoherent excitation population is being monitored. This type of experiments is therefore well suited to separate the purely coherent contribution to the optical nonlinearities from the more long-lived incoherent contributions from a high density of excited carriers in the medium.

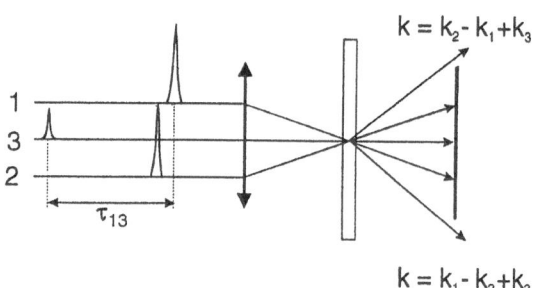

Fig.17 Transient FWM with three beams: Transient grating.

The lifetime τ_G of the incoherent excitation density grating is determined by the lifetime τ_R of the excited carriers as well as by carrier diffusion, since the latter will wash out the spatial modulation of the carrier density.[34]

$$\tau_G^{-1} = \tau_R^{-1} + \frac{4\pi^2 D}{\Lambda} \qquad (37)$$

where D is the carrier diffusion coefficient and Λ is the grating constant, as determined by the wavelength λ of the exciting light and the angle θ between the two interfering beams ($\mathbf{k}_1$ and $\mathbf{k}_2$)

$$\Lambda = \frac{2\pi}{|\mathbf{k}_1 - \mathbf{k}_2|} = \frac{\lambda}{2\sin(\frac{\theta}{2})} \tag{38}$$

The grating lifetime is determined from the decay of the integrated intensity of the scattered test signal as a function of the delay τ_{13} of the test pulse (#3). Hence, the carrier lifetime and the diffusion coefficient can be determined separately from Eq. (37) by performing such transient grating experiments at different angles θ.

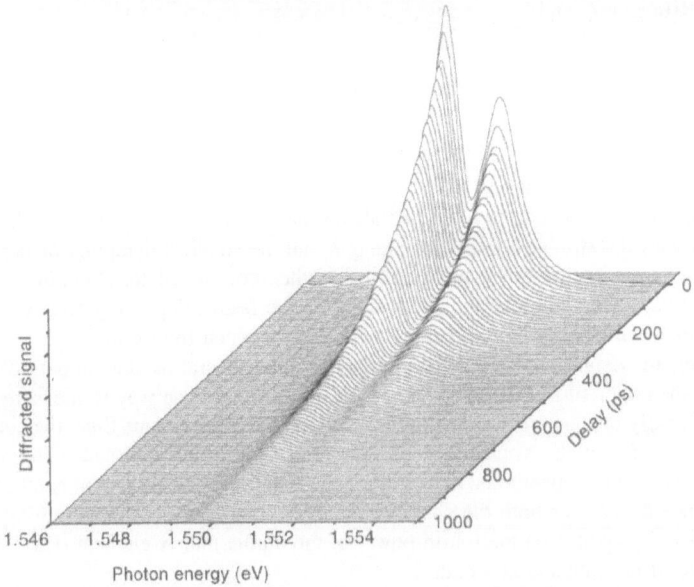

Fig. 18 Decay of the transient grating spectra of the monolayer split HH excitons in Fig. 16

Figure 18 shows transient grating decays for the monolayer split HH excitons in Fig. 16. An analysis of such grating decays for different grating constants yield an average diffusion coefficient $D_x = 2$ cm^2/s, a recombination lifetime $\tau_R = 365$ ps and a transfer rate of excitons from the higher energy regions to the lower energy regions of $\beta = 0.33$ ns^{-1}.[35] The exciton diffusion coefficient is about an order of magnitude slower than previously reported for GaAs/AlGaAs quantum wells.[36] This is explained by the trapping (localization) of the excitons in the border regions between the monolayer-flat regions, which is consistent with the slow transfer of excitons between these regions.

IV. QUANTUM BEAT SPECTROSCOPY

Quantum beat spectroscopy (QBS) was developed already in the sixties as a high resolution spectroscopy operating in the time domain.[37] When two or more close-lying atomic levels are resonantly excited by the same coherent light pulse, the atom is prepared in a mixed state that is a coherent superposition of pure states. This will give rise to an interference term in the wave function that will manifest itself, e.g. by an oscillation of the resonance fluorescence decay observed immediately after the short-pulse excitation.

The situation is illustrated by the nearly degenerate two-level system in Fig. 19, where a simultaneous excitation from the ground state $|0\rangle$ to the close-lying excited levels $|1\rangle$ and $|2\rangle$ is provided by the same coherent laser pulse at t = 0. After the excitation, the fluorescence decay can be expressed by[38]

$$I(t) = \Theta(t) \, I_0 \, [\mu_{10}^4 e^{-2\gamma_{10}t} + \mu_{20}^4 e^{-2\gamma_{20}t} + 2\mu_{10}^2\mu_{20}^2 e^{-(\gamma_{10} + \gamma_{20})t}\cos\omega_{21}t \,] \qquad (39)$$

where I_0 is proportional to the integrated intensity of the exciting laser pulse and the square of the interaction length, μ_{i0} is the dipole matrix element for the transition $|0\rangle \rightarrow |i\rangle$ with the damping coefficient γ_{i0} and

$$\omega_{21} = \frac{E_2 - E_1}{\hbar} \qquad (40)$$

where E_i is the energy of level i.

From Eq. (39), it is seen that the overall resonance fluorescence decay is dominated by the level with the strongest oscillator strength and the smallest damping, or dephasing, rate. The interference term, however, is strongest when both oscillator strengths are large, and the damping of the modulations is governed by the fastest dephasing rate. The modulation frequency directly reveals the energy splitting between the excited states. Thus, in systems where the dephasing time is long enough, so that several oscillation periods can be observed in the coherent fluorescence decay, this is a very precise way to measure energy splittings of nearly degenerate systems in the time domain. At the same time, the coherence and dephasing time of the system can be measured in a simple resonance fluorescence experiment. It should be mentioned that a resonance fluorescence experiment is intrinsically nonlinear since it involves both the absorption and the subsequent coherent emission of a photon. Hence, it depends on the fourth power of the dipole matrix element (Eq. (39)) and on the square of the interaction length.

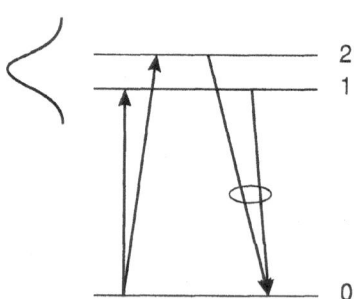

Fig. 19 Resonance fluorescence in three-level system

Interference and beat phenomena may also appear in higher order nonlinear optical signals and spectroscopies. The TFWM decays, or correlation traces, of Eqs. (33) and (36) assumed the sytems to be two-level like. If, on the other hand, there is a multiplicity of nearly degenerate levels, the decay of the nonlinear signal will be affected much the same way as in the resonance fluorescence, described above. However, the quantum interferences, or beats, will in this case appear as a function of real time t as well as a function of delay τ between the incident pulses. Assuming equal dipole matrix elements of the two transitions, the following expression is obtained [38]

$$I_{FWM} \propto 5(e^{-2\gamma_{10}t} + e^{-2\gamma_{20}t}) + 2e^{-(\gamma_{10}+\gamma_{20})t}\cos(\omega_{21}(t+\tau)) + 4(e^{-2\gamma_{10}t} + e^{-2\gamma_{20}t})\cos(\omega_{21}\tau) \quad (41)$$
$$+ 8e^{-(\gamma_{10}+\gamma_{20})t}\cos(\omega_{21}t) + 8e^{-(\gamma_{10}+\gamma_{20})t}\cos(\omega_{21}(t-\tau))$$

The oscillatory behaviour of the signal as a function of t and τ is clearly revealed, as shown in Fig. 20. Notice that also for $t = \tau$, distinct oscillations can be observed.

In an experiment where real time resolution is not achieved, the maximum information is obtained by spectrally resolving the time integrated nonlinear signal. In this case the nonlinear signal is found from the Fourier transform of the polarization. If the beating resonances are spectrally well separated, the signal near one resonance can be expressed [38]

$$I_{FWM} \propto \frac{4\mu_{10}^8 e^{-2\gamma_{10}\tau} + \mu_{10}^4\mu_{20}^4 e^{-\gamma_{20}\tau} + 4\mu_{10}^6\mu_{20}^2 e^{-(\gamma_{10}+\gamma_{20})\tau}\cos(\omega_{21}\tau)}{(\omega_{10} - \omega)^2 + \gamma_{10}^2} \quad (42)$$

revealing the modulation as a function of τ.

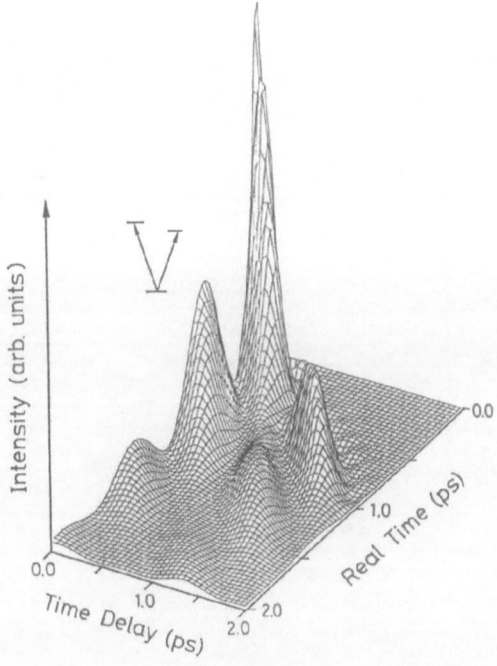

Fig. 20 TFWM signal from a three-level system featuring quantum beats.

In the above is assumed a homogeneously broadened system with a free polarization decay, but a similar modulation of the correlation trace is obtained with inhomogeneous broadening, i.e. in a photon echo experiment, as first shown by Lampert, Compaan and Abella[39],[32]. If, however, the observed modulations are due to classical polarization interferences occuring in the detector, then they will be smeared out by inhomogeneous broadening. In a homogeneously broadened system it may be difficult to distinguish between true quantum beats, e.g. from a three-level system, and polarization interferences from independent two-level systems.[40]

For two independent two-level systems with the complex transition frequencies $\Omega_{j0} = \omega_{j0} - i\gamma_{j0}$, the TFWM signal is similarly found as [41]

$$I_{FWM} \propto |P^{(3)}(t,\tau>0)|^2 \propto \Theta(t)[e^{-2\gamma_{10}t} + e^{-2\gamma_{20}t} + 2e^{-(\gamma_{10}+\gamma_{20})t}\cos(\omega_{21}(t-\tau))] \quad (43)$$

which is shown in Fig. 21. Note that for $t = \tau$ there is no modulation of the signal. This is a fingerprint of classical polarization interference from noninteracting two-level systems. In frequency space, the signal near the resonance ω_{10} has the following form

$$I_{FWM} \propto \frac{N_1^2 e^{-2\gamma_{10}\tau}}{(\omega_{10} - \omega)^2 + \gamma_{10}^2} + \frac{2N_1 N_2 e^{-\gamma_{20}\tau}\cos(\omega_{21}\tau)}{(\omega_{10} - \omega)(\omega_{20} - \omega) + \gamma_{10}\gamma_{20}} \quad (44)$$

The important thing to notice here is that the modulation of the nonlinear signal as a function of τ will undergo a phase shift of π, when passing through either one of the resonances. This is in contrast to the true quantum beat in Eq. (42) and an observation of the phase of the modulation as a function of detuning around the resonance therefore offers an experimental possibility of distinguishing the two phenomena.[41]

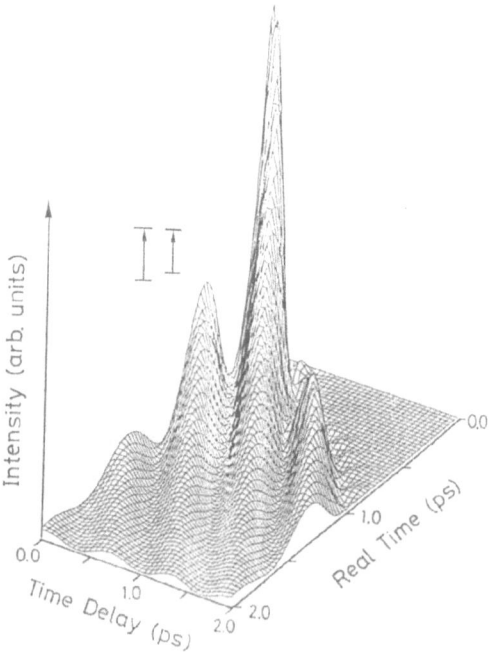

Fig. 21 TFWM signal from two independent two-level systems featuring polarization interference

As already mentioned, the quantum beat spectroscopy was first used on independent single atom systems where the dephasing times are in the microsecond or even millisecond range, allowing for coherent transient measurements without the need of ultrafast laser sources and detectors. Correspondingly, the energy splittings that can be investigated are in the megahertz to gigahertz range, i.e. ultrahigh spectral resolution from an optical point of view. Quantum beat spectroscopy is typically used to measure hyperfine splittings and Zeeman effects in atomic excitations in various environments. Since such nearly degenerate atomic levels can be rather complex with a number of optically active levels, it may be necessary to Fourier transform the fluorescence decay before an identification of the various beat levels can take place.

The reason why QBS for a long time was restricted to noninteracting atomic systems lies in the fundamental conditions for observing quantum beats

$$\gamma < \omega_{21} < \Delta f \tag{45}$$

where $\Delta f = 1/\tau_L$ is the band width of the exciting laser with pulse length τ_L. In isolated atoms, dampings are in the megaherz range, so that only moderately short laser pulses are required. In liquids or solids, e.g. semiconductors, electronic damping is in the terahertz range, so that picosecond or subpicosecond laser pulses are required to study possible quantum beats. However, such lasers are now readily available.[1]

Another aspect of strong interatomic interactions in solids is the formation of electronic energy bands that will impede the observation of interferences between nearly degenerate discrete energy levels. In semiconductors, however, the Coulomb interaction between photoexcited electron-hole pairs form hydrogen-like bound states, excitons, that can be either freely moving in the crystal or bound to local impurities (free or bound excitons). These excitonic states are the lowest elementary electronic excitations of semiconductors and introduce narrow resonant optical transitions just below the fundamental band gap. Excitons are very important for the optical properties, and in particular the nonlinear optical properties, of semiconductors in this spectral range.[42] Thus, recently a number of observations of quantum beats have been made on excitons in semiconductors, [43],[44] particularly in GaAs quantum well structures[45],[46],[47],[48],[49],[50],[51]. An example was already shown in Fig. 15.

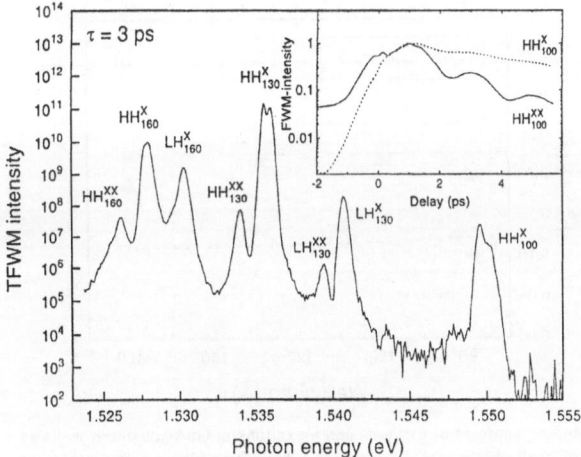

Fig. 22 TFWM spectrum from the multiple quantum well sample in Figs. 3 and 15. HH^{xx} and LH^{xx} indicate biexcitons associated with the HH and the LH valence bands, respectively. Inset shows quantum beats between excitons and biexcitons.

The formation of excitonic molecules, or biexcitons, in GaAs/AlGaAs quantum wells has been studied extensively by quantum beats in differential transmission and TFWM experiments. This has provided an unambiguous identification of the biexciton state by applying polarization selection rules in the TFWM experiments, and from the observation of spectrally resolved TFWM signals with exciton-biexciton quantum beats it has been possible to determine the biexciton binding energy.[52],[53],[54] The HH biexcitons in GaAs quantum wells were seen previously with less certainty in luminescence spectra[55],[56],[57]. An example of HH biexciton luminescence was shown already in Fig. 3b, and a TFWM spectrum of the same sample is shown in Fig. 22. The HH biexcitons are observed for all the well widths and even the light hole (LH) biexciton is observed for the 130 Å quantum wells. In the inset of Fig. 22 is shown the quantum beat modulation of the TFWM signal in the biexciton resonance of the 100 Å quantum wells. From the quantum beat periods as well as from the TFWM spectra, the binding energy of the HH biexcitons could be determined as a function of the well width. The result is shown in Fig. 23 together with calculated results by Kleinmann[58] and from our own group.[8] The experimental results indicate that the ratio between the binding energy of the biexciton and the binding energy of the exciton is about 0.2 independently of the well width. This is in agreement with our theoretical calculation based on a fractional dimension model.[59]

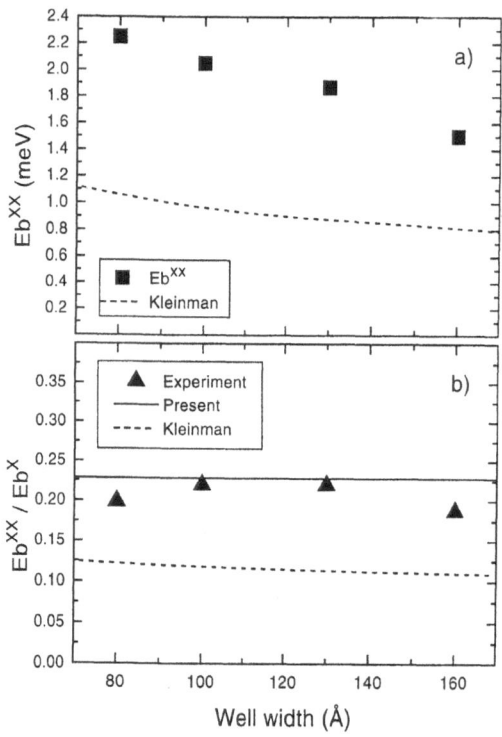

Fig. 23 Binding energies of excitons and biexcitons in GaAs quantum wells as a function of well width. (a) Experimental biexciton binding energy (points) compared with a theoretical calculation by Kleinmann (Reference 58). (b) Ratio between the biexciton binding energy and the exciton binding energy. Points are experimental, full curve is our calculation and dashed curve is calculation by Kleinmann (Ref. 58).

V. SPATIAL RESOLUTION - MICROSCOPY

In the linear and nonlinear optical spectroscopies discussed so far, all the information obtained has to be considered as spatial averages over the excitation volume of the sample investigated. Hence, the optical parameters are macroscopic values with little information about spatial variations on the microscopic atomic level. In coherent light scattering experiments some information about spatial correlations can be obtained from the validity and the use of wavevector conservaion rules in optical transition and scattering experiments in media with translational symmetries. This is of course well known from x-ray scattering, but also in optical (Raman) spectroscopy spatial correlation effects on a sub-wavelength scale have been applied to yield information about the structure of solids, e.g. semiconductor alloys.[60]

In optical microscopy, the spatial resolution is normally observed to be worse than $\lambda/2$, where λ is the wavelength of the light applied. Since this resolution in many textbook on quantum mechanics has been used to illustrate the Heisenberg uncertainty principles, it is often stated that the resolution of an optical microscope is limited to $\lambda/2$ by the uncertainty principles, and therefore ultimate. This is not true. The resolution of an optical microscope is set by the Abbe diffraction limit, valid for propagating light field modes active in the far field. For the optical near-field in a nonpropagating evanescent field mode, the spatial resolution can be much higher and only limited by geometry, i.e. by the size of an aperture and its distance from the object to be studied, as sketched in Fig. 24 [61],[62].

In a distance z above a surface, extending over the x-y plane, the electric field can be expressed by an integral over the surface of a scattering function s(x,y) containing the spatial information[63]

$$E(0,0,z) = \int_{-\infty}^{\infty} s(x,y) \, e^{i(-k_x x - k_y y + k_z z - \omega t)} dx dy = S(k_x, k_y) \, e^{i(k_z z - \omega t)} \tag{46}$$

where $k_z = (k^2 - k_x^2 - k_y^2)^{1/2}$ and $k = 2\pi/\lambda$. The Fourier transform $S(k_x, k_y)$ expresses the spatial frequency components of the scattering function. The far-field only contains propagating waves (k_z real) and the spatial frequencies are limited by $k_x^2 + k_y^2 < k^2$. The near-field, on the other hand, contains both propagating and evanescent waves including the spatial frequencies $k_x^2 + k_y^2 \geq k^2$.[64]

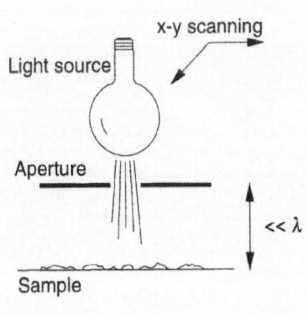

Fig. 24 Principle behind SNOM

Hence, the scattering function can be expressed by

$$s(x,y) = \int_{-\infty}^{\infty} S(k_x,k_y)e^{i(k_x x+k_y y)}dk_x dk_y$$

$$= \int_{k_x^2+k_y^2<k^2} S(k_x,k_y)e^{i(k_x x+k_y y)}dk_x dk_y + \int_{k_x^2+k_y^2\geq k^2} S(k_x,k_y)e^{i(k_x x+k_y y)}dk_x dk_y$$

(47)

where the first term on the right hand side of Eq. (47) contains the propagating modes and the last term contains only the non-propagating, or evanescent, modes. The propagating wave intensity contains only spatial frequencies $< 2k$, yielding the well-known spatial resolution of $\Delta x \geq \lambda/2$. The evanescent field modes, however, contain spatial frequencies $> 2k$, so that the near field contains all spatial frequencies.

If one wants a spatial resolution of $\Delta x \approx \lambda/10$, then spatial frequencies $k_x \approx 10\times 2\pi/\lambda$ are required. Thus, the intensity of the evanescent wave decays very rapidly as a function of the distance z from the surface, like $\exp\{-20\pi z / \lambda\}$. For a wavelength of 500 nm this means that one has to be within a distance of about 8 nm from the surface to detect this field mode. A crucial point in near-field optical microscopy (and spectroscopy) is therefore to approach and to control very small objects to near-atomic distances from the surface to be investigated. With the invention of the scanning tunneling microscope (STM) in 1982 [65] and the atomic force microscope in 1986 [66], new techniques were developed by which one is able to scan small microprobes at small and well-controlled distances from a surface. This development of scanning probe microscopes (SPM) also includes the scanning near-field optical microscope (SNOM)[67][68][69][70], where the the probe tip plays the role of a small optical aperture, acting as a light source, or as a detector, or as both. Accordingly, four different SNOM configurations are illustrated in Fig. 25.

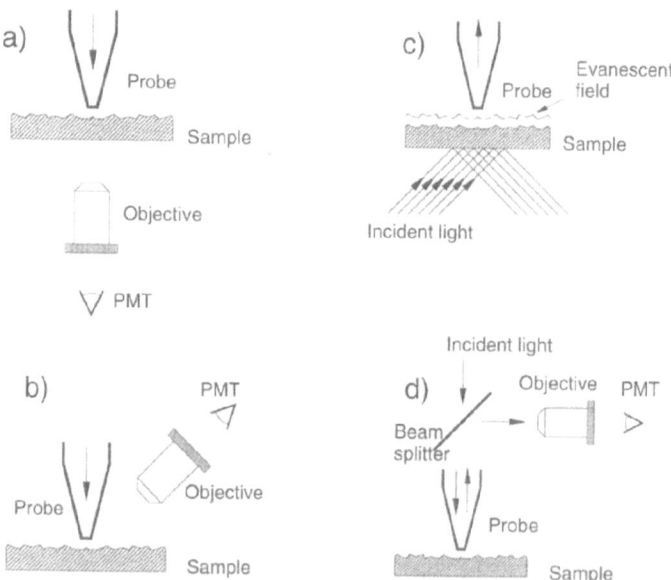

Fig. 25 Four different SNOM configurations. a) Transmission SNOM, b) reflection SNOM, c) photon STM, and d) external reflection SNOM.

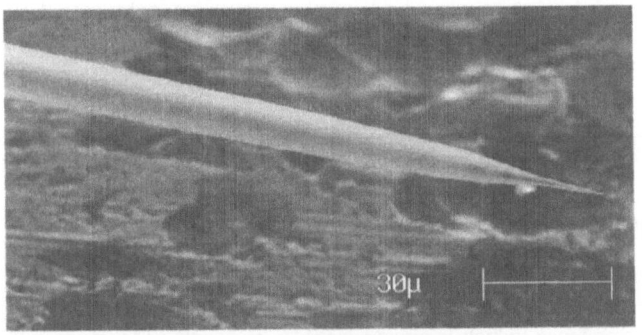

Fig. 26 SEM picture of pulled fibre probe

There are various probe fabrications that have been have been applied so far. One type is a microfabricated SiN probe tip where laser light is focused on the bottom facet of a pyramidal tip. Another type is based on a tapered micropipette formed by a heating/pulling process and filled with a fluorescent dye acting as a point light source.[71] The most succesful and widespread SNOM probe until now is a tapered fibre[72] pulled, like the micropipette, to very small dimensions. An example of such a pulled fibre tip is shown in Fig. 26, where the radius of curvature of the very end of the tip is about 30 nm. A metal coating applied from the side under rotation will leave a very small aperture hole at the fibre tip that has been shown to give a resolution better than 20 nm.[72] Even without metal coating of the fibre tip, we have obtained an optical image with a resolution of the same order of magnitude.[73]

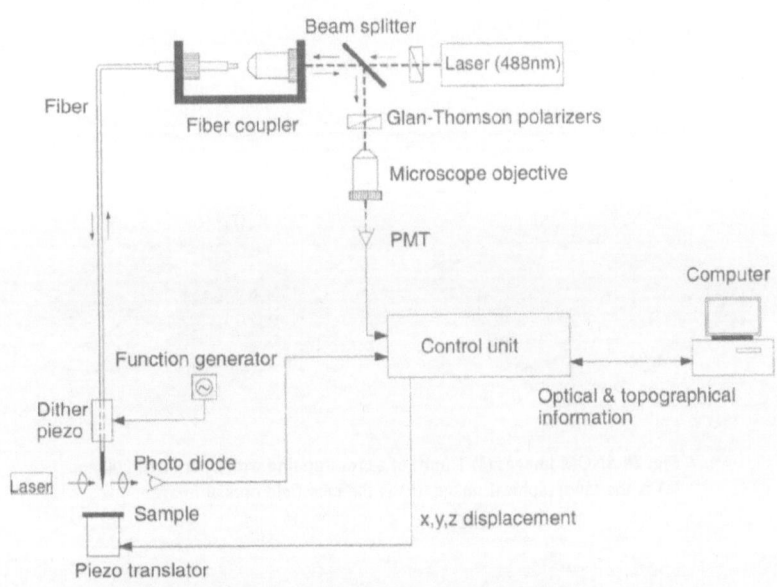

Fig. 27 Combined AFM/SNOM microscope configuration.

The regulation of the probe-sample distance is vital for the spatial resolution in a SNOM. With metal coated probes and a conductive sample the distance can be controlled via the tunnel current as in an STM.[65],[74] In this configuration simultaneous STM and SNOM images are obtained. Since an evanescent optical light field varies exponentially with distance just as a tunnel current, distance regulation can also be achieved via the near-field optical signal feed-back. In this case, however, the images will contain a mixture of optical and topographical information that may not be easily interpreted. Distance regulation can also be performed via atomic forces as in an AFM [75][76]. Figure 27 shows such a combined AFM/SNOM configuration.

The cantilever has been removed from a standard non-contact AFM and replaced by a pulled tapered fiber tip mounted on a piezo element that oscillates the tip laterally near its resonance frequency, which is around 250 kHz and with a Q-factor of about 200. The dither motion of the fiber probe is detected via a laser diode and a photo detector (see Fig. 27). The detected signal consists of a small AC component on top of a background DC signal. The amplitude of the AC signal is approximately proportional to the oscillation amplitude of the fiber probe. The rectified AC part of the detected signal is used as a feed-back signal to keep the probe-sample distance constant while scanning. In this way, topographical and near-field optical images are recorded simultaneously. Fig. 28 shows the 1x1 μm^2 images of a cross-grating with a period of 200 nm. The two-dimensional periodicity is clearly resolved both in the topographical image Fig. 28a and in the near-field optical image Fig. 28b.

The advantage of SNOM over the other SPM's is that light is the probing agent with a large spectral bandwidth (parallelism, multiplexing, etc.) allowing for identification and characterization via optical spectroscopy and also for modifications via selective photophysics or photochemistry. Magnetic recording via SNOM opens an alley to optical data storage of extremely high densities[77] and SNOM-based lithography could open up for future nanoelectronics.[78] Meanwhile nanoscale (fluorescence) spectroscopy offers itself as a tool for studying single molecules, cell membranes and DNA in the fields of chemistry and biochemistry and atomic clusters and semiconductor nanostructures on the physics and the technology side.[79]

a)   b)

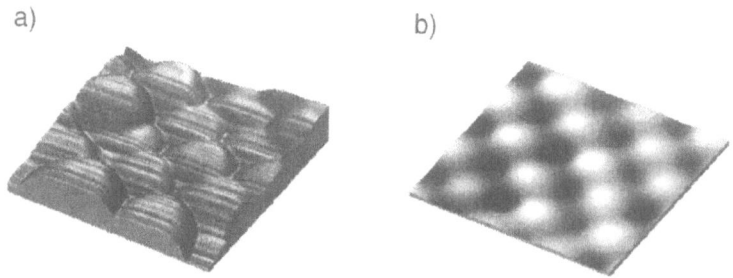

Fig. 28 SNOM images (1x1 μm^2) of a cross-grating with a period of 200 nm.
(a) is the topographical image, (b) is the near.field optical image.

VI. CONCLUDING REMARKS

In this article I have tried to give a flavour of various aspects of linear and nonlinear optical spectroscopy as a tool for studying the dynamics of collective excitations in solids. The examples are almost exclusively taken from excitons in semiconductors, because this is the field with which I am most familiar. Even so, the article does not give credit to the vast amount of work that has been done in linear and nonlinear optical spectroscopy of semiconductors over the last decade or more.

My aim with this work has not been to review the field, but merely to illustrate the power and the potential of optical spectroscopy with spectral, temporal and spatial resolution. The topics, that are considered, are strongly biased towards my own interests and I have in many cases taken examples from the work of my own group. I hope to have demonstrated a number of optical spectroscopies and techniques that have been succesfully applied to semiconductors in recent years. One major reason for this development is the progress in laser technology, particularly ultrafast lasers, and another driving force is the recent developments in micro technology, allowing the new SPM techniques to add spatial resolution to the conventional optical techniques.

ACKNOWLEDGEMENTS

I am deeply grateful to the large number of collegues, students and guests in my group for contributing, directly and indirectly, to my general understanding of optical spectroscopy and its results. Particularly, I would like to thank D. Birkedal, V. G. Lyssenko, J. Erland, K. El-Sayed, S. Madsen, and H. Gislason for their significant contributions to this work, which was supported by the Danish Natural Science Research Council and CNAST - Center for Nanostructures. Particularly, I want to thank C.B. Sørensen, III-V NANOLAB for preparing many of the samples used in this work.

REFERENCES

1. See e.g. *Ultrafast Phenomena IX*, P.F. Barbara, W.H. Knox, G.A. Mourou, and A.H. Zewail, Eds. (Springer-Verlag, Berlin Heidelberg, 1994) and previous Proceedings of the same series.
2. C.F. Klingshirn, *Semiconductor Optics* (Springer-Verlaf Berlin Heidelberg, 1995).
3. N. Peyghambarian, S.W. Koch, and A. Mysyrowicz, *Introduction to Semiconductor Optics* (Prentice Hall Inc., Englewood Cliffs, 1993).
4. D.L. Greenaway and G. Harbeke, *Optical Properties and Band Structure of Semiconductors* (Pergamon Press, Oxford, 1968).
5. E.O. Göbel and K. Ploog, Prog. Quant. Electr. **14**, 289 (1990).
6. H. Haug and S.W. Koch, *Quantum Theory of the Optical and Electronic Properties of Semiconductors* (World Scientific, Singapore, 1993).
7. J. Erland, D. Birkedal, V.G. Lyssenko, and J.M. Hvam, JOSA B special issue on *Radiative Processes and Dephasing in Semiconductors*, edited by J. Kuhl and D.S. Citrin (in press).
8. D. Birkedal, J. Singh, V.G. Lyssenko, J. Erland, and J.M. Hvam, Phys. Rev. Lett. **76** (1996).
9. R.C. Miller and D.A. Kleinman, J. Luminescence **30**, 520 (1985).
10. M. Cardona, in *Modulation Spectroscopy* (Academic Press, New York, 1969)
11. B.O. Seraphin, in *Semiconductors and Semimetals vol. 9* (Academic Press, New York, 1972) p. 1
12. I. Balslev, Phys. Rev. **143**,636 (1966).
13. E.E. Mendez,L.L. Chang, G. Ladgren, R. Ludeke, L. Esaki, and F.H. Pollak, Phys. Rev. Lett. **46**, 1230 (1981).
14. M. Erman, J.B. Theetan, P. Frijlink, S. Gaillard, F.J. Hia, and C. Alibert, J. Appl. Phys. **56**, 3241 (1984).
15. O.J. Glembocki, B.V. Shanabrook, N. Bottka, W.T. Beard, and J. Comas, Appl. Phys. Lett. **46**, 970 (1985).

16. H.R. Phillip and H. Ehrenreich, Phys. Rev. **129**, 1550 (1963).
17. D.E. Aspnes, Solid State Commun. **8**, 267 (1973).
18. B.O. Seraphin and N. Bottka, Phys. Rev. **148**, 628 (1966).
19. A.I. Ferguson, in NATO ASI Series B **339**, *Nonlinear Spectroscopy of Solids; Advances and Applications*, edited by B. DiBartolo (Plenum Press, New York, 1994), p. 225
20. S.D. Smith and R.F. Neale (Eds.), *Optical Information Technology; State-of-the-Art Report* (Springer Verlag, Berlin, 1993)
21. Y.R. Shen, *The Principles of Nonlinear Optics* (Wiley & Sons, New York, 1984).
22. R.W. Boyd, *Nonlinear Optics* (Academic Press, San Diego, 1992).
23. J.M. Hvam, in NATO ASI Series B **339**, *Nonlinear Spectroscopy of Solids; Advances and Applications*, edited by B. DiBartolo (Plenum Press, New York, 1994), p. 91
24. P. Meystre and M. Sargent III, *Elements of Quantum Optics* (Springer-Verlag, Berlin Heidelberg, 1990).
25. B. Fluegel, N. Peyghambarian, M. Lindberg, S.W. Koch, M. Joffre, D. Hulin, A. Migus, and A. Antonetti, Phys. Rev. Lett. **59**, 2588 (1987).
26. S.W. Koch, N. Peyghambarian, and M. Lindberg, J. Phys. C **21**, 5229 (1988).
27. N. Bloembergen, *Nonlinear Optics* (W.A. Benjamin, Inc., New York, 1965).
28. A. Maruani and D.S. Chemla, Phys. Rev. B **23**, 841 (1981).
29. J.M. Hvam and C. Dörnfeld, in *Optical Switching in Low-Dimensional Systems*, H. Haug and L. Banyai, eds. (Plenum Publishing Corp., New York, 1989) pp. 233-241.
30. T. Yajima and Y. Taira, J. Phys. Soc. Japan **47**, 1620 (1979).
31. J. Erland and I. Balslev, Phys. Rev. A **48**, 1765 (1994).
32. J. Erland, K.-H. Pantke, V. Mizeikis, V.G. Lyssenko, and J.M. Hvam, Phys. Rev. B **50**, 15047 (1994).
33. D. Birkedal, V.G. Lyssenko, K.-H. Pantke, J. Erland, and J.M. Hvam, Phys. Rev. B **51**, 7977 (1995).
34. S.C. Moss, J.R. Lindle, H.J. Mackey, and A. Smirl, Appl. Phys. Lett. **39**, 227 (1981).
35. D. Birkedal, unpublished results.
36. D. Oberhauser, K.-H. Pantke, J.M. Hvam, G. Weimann, and C. Klingshirn, Phys. Rev. B **47**, 6827 (1993).
37. S. Haroche, in *High Resolution Laser Spectroscopy*, K. Shimoda, ed. (Springer, Berlin, Heidelberg, 1976) pp. 253-313.
38. K.-H. Pantke and J.M. Hvam, Int. J. Modern Phys. B **8**,73 (1994).
39. L.Q. Lampert, A. Compaan, and I.D. Abella, Phys. Rev. A **4**, 2022 (1971).
40. M. Koch, J. Feldmann, G. von Plessen, E.O. Göbel, P. Thomas, and K. Köhler, Phys. Rev. Lett. **69**, 3633 (1992).
41. V.G. Lyssenko, J. Erland, I. Balslev, K.-H. Pantke, B.S. Razbirin, and J.M. Hvam, Phys. Rev. B **48**, 5720 (1993).
42. D.S. Chemla and D.A.B. Miller, J. Opt. Soc. Am. B **2**, 1155 (1985).
43. H. Stolz, V. Langer, E. Schreiber, S. Permogorov, and W. von der Osten, Phys. Rev. Lett. **67**, 679 (1991).
44. V. Langer, H. Stolz, and W. von der Osten, Phys. Rev. Lett. **64**, 854 (1990).
45. K. Leo, T.C. Damen, J. Shah, E.O. Göbel, and K. Köhler, Appl. Phys. Lett. **57**, 19 (1990).
46. B.F. Feuerbacher, J. Kuhl, R. Eccleston, and K. Ploog, Solid State Commun. **74**, 1279 (1990).
47. K. Leo, G.O. Göbel, T.C. Damen, J. Shah, S. Schmitt-Rink, W. Schäfer, J.F. Müler, K. Köhler, and P. Ganser, Phys. Rev. B **44**, 5726 (1991).
48. D. Oberhauser, K.-H. Pantke, W. Langbein, V.G. Lyssenko, H. Kalt, J.M. Hvam, G. Weiman, and C. Klingshirn, phys. stat. sol. (b) **173**, 53 (1992).
49. S. Schmitt-Rink, D. Bennhardt, V. Heuckeroth, P. Thomas, P. Haring, G. Maidorn, H. Bakker, K. Leo, D.-S. Kim, J. Shah, and K. Köhler, Phys. Rev. B **46**, 10460 (1992).
50. R. Eccleston, J. Kuhl, D. Bennhardt, and P. Thomas, Solid State Commun. **86**, 93 (1993).
51. J. Feldmann, T. Meier, G. von Plessen, M. Koch, E.O. Göbel, P. Thomas, G. Bacher, C. Hartmann, H. Schweizer, W. Schäfer, and N. Nickel, Phys. Rev. Lett. **70**, 3027 (1993).
52. S. Bar-Ad and I. Bar-Joseph, Phys. Rev. Lett. **68**, 349 (1992).
53. K.-H. Pantke, D. Oberhauser, V. G. Lyssenko, J.M. Hvam, and G. Weimann, Phys. Rev. B **47**, 2413 (1993).
54. D.J. Lovering, R.T. Phillips, G.J. Denton, and G.W. Smith, Phys. Rev. Let. **68**, 1880 (1992).
55. R.C. Miller, D.A. Kleinmann, A.C. Gossard, and O. Munteanu, Phys. Rev. B **25**, 6545 (1982).
56. S. Charbonneau, T. Steiner, M.L.W. Thewalt, E.S. Koteles, J.Y. Chi, and B. Elman, Phys. Rev. B **38**, 3583 (1988).
57. R.T. Phillips, D.J. Lovering, G.J. Denton, and G.W. Smith, Phys. Rev. B **45**, 4308, (1992).

58. D.A. Kleinman, Phys. Rev. B **28**, 871 (1983).
59. J.Singh, D. Birkedal, V.G. Lyssenko, and J.M. Hvam, Phys. Rev. B (to be published).
60. J.A. Kash, J.M. Hvam, J.C. Tsang, and T.F. Kuech, Phys Rev. B **38**, 5776 (1988).
61. J.A. O'Keefe, J. Opt. Soc. Am. **46**, 359 (1956).
62. E.A. Ash and G. Nicholls, Nature London **237**, 510 (1972).
63. G.A. Massey, Appl. Opt. **23**, 658 (1984).
64. D. Courjon, J.-M. Vigoureux, M. Spajer, K. Sarayeddine, and S. Leblanc, Appl. Opt. **29**, 3734 (1990).
65. G. Binnig, H. Rohrer, C. Gerber, and E. Weibel, Phys. Rev. Lett. **49**, 57 (1982).
66. G. Binnig, C.F. Quate, and C. Gerber, Phys. Rev. Lett. **56**, 930 (1986).
67. D.W. Pohl, W. Denk, and M. Lanz, Appl Phys. Lett. **44**, 651 (1984).
68. U. Dührig, D.W. Pohl, and F. Rohner, J. Appl. Phys. **59** 3318 (1986).
69. E. Betzig and J. Trautman, Science **257**, 189 (1992).
70. E. Betzig, A. Lewis, A. Harootunian, M. Isaacson, and E. Kratschmer, Biophys. J. **49**, 269 (1986).
71. E. Betzig, M. Isaacson, and A. Lewis, Appl. Phys. Lett. **51**, 2088 (1987).
72. E. Betzig, J. Trautman, T.D. Harris, J.S. Weiner,and R.L. Kostelak, Science **251**, 1468 (1991).
73. S. Madsen, T. Olesen, and J.M. Hvam, submitted to Applied Optics
74. K. Lieberman and A. Lewis, Appl. Phys. Lett. **62**, 1335 (1993).
75. R. Toledo-crow, P.C. Yang, Y. Chen, and M. Vaez-Iravani, Appl.Phys. Lett. **60**, 2957 (1992).
76. E. Betzig, P.L Finn, and S.J. Weiner, Appl. Phys. Lett. **60**, 2484 (1992).
77. E. Betzig, J.K. Trautman, R. Wolfe, E.M. Gyorgy, and P.L. Finn, Appl. Phys. Lett. **61**, 142 (1992).
78. M. Rudman, A. Lewis, A. Mallul, V. Haviv, I. Turovets, A. Schemelinin, and I. Nebenzahl, J. Appl. Phys. **72**, 4379 (1992).
79. H.F. Hess, E. Betzig, T.D. Harris, L.N. Pfeiffer, and K.W. West, Science **264**, 1740 (1994).

THE STUDY OF COLLECTIVE EXCITATIONS IN SOLIDS BY INELASTIC NEUTRON SCATTERING

Tormod Riste

Institutt for energiteknikk
N-2007 Kjeller, Norway

ABSTRACT

This paper reviews some of the properties which make neutrons a unique and versatile probe for dynamics of solids, and gives examples of results that have been obtained in some of the most central problem areas. Introductory chapters also treat aspects of neutron sources and the most important types of instrumentation for inelastic scattering.

1. INTRODUCTION

I. A. Neutron Properties and Neutron Scattering [1]

Neutrons have several uniquely valuable properties for probing collective excitations in solids. Thermal neutrons have a wavelength (λ) distribution with a maximum around 1.8Å and, by definition, an energy $\hbar\omega$ (about .025 eV) corresponding to thermal excitations in solids. This is in contrast to X-rays and electrons which at 1 Å have energies 12 keV and 3.5 eV, respectively. The momentum $\hbar k$, associated with the wavenumber k ($=2\pi/\lambda$), and energy $\hbar\omega$ of thermal neutrons are therefore perfectly matched for measurement of the disperson relations of the excitations. Neutrons interact with atoms in two ways: through a strong, short-range interaction with the nucleus, and through a coupling between the magnetic moment of the neutron and the atomic magnetization. Consequently both phonon-like and spinwave-like excitations are accessible through neutron scattering.

The interaction of neutrons with matter is relatively weak. The attenuation of a thermal neutron beam in aluminium is therefore only one percent per millimeter, compared with 99 percent or more for X-rays. The advantage of the high penetration power is that the sample can conveniently be kept in containers for changes of pressure and temperature, and that bulk properties of the sample are measured. The disadvantage is the weak intensity of the scattered beam, and the difficulty in measuring surface properties. The disadvantage is compounded by the low fluxes of neutron sources compared with e.g. synchrotron sources, the ratio being ~10^{-15} for the same energy band width.

Neutron scattering is thus a signal-limited technique and a relatively expensive tool. It is, however, widely used because it in many cases provides information on the structure of materials that cannot be obtained by other means. This is, in particular, true for the dynamic structure, the excitations.

Information on the excitations is obtained through inelastic scattering. The total energy and momentum is conserved in the scattering process: the energy lost (gained) by the neutron is gained (lost) by the sample, and the momentum absorbed (emitted) by the neutron is emitted (absorbed) by the sample. When scattered the neutron is transferred from an incident wave vector $\bar{k}$ to a final wave vector $\bar{k}'$. In a scattering experiment one determines

the momentum transfer $\quad \bar{Q} = \bar{k} - \bar{k}'$ $\qquad\qquad\qquad$ (1)

the energy transfer $\quad \Delta E = \dfrac{\hbar^2 k^2}{2m} - \dfrac{\hbar^2 k'^2}{2m} = \hbar\omega$ $\qquad\qquad$ (2)

A useful construction is the scattering triangle of Fig. 1. In elastic scattering $|\bar{k}| = |\bar{k}'|$ and $Q = 2k\,\sin\theta = (4\pi\,\sin\theta)/\lambda$. In elastic scattering ($\Delta E = 0$) from a periodic lattice $\bar{Q} = \bar{G}_{hkl}$, a reciprocal lattice vector. In inelastic scattering $\bar{Q} = \bar{G} + \bar{q}$ where $\bar{q}$ is the additional momentum transfer from the excitation. Simultaneous measurement of $\bar{Q}$ and ΔE thus gives access to $\omega(q)$, the dispersion relation.

Inelastic Scattering ($k' \neq k$)

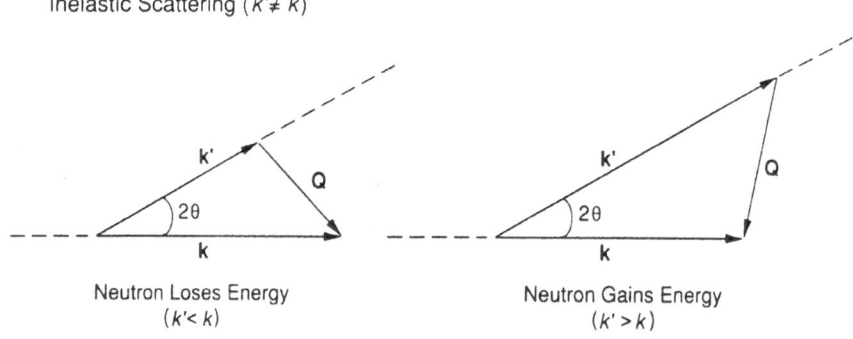

Neutron Loses Energy $\qquad\qquad\qquad$ Neutron Gains Energy
$(k' < k)$ $\qquad\qquad\qquad\qquad\qquad\qquad$ $(k' > k)$

Fig. 1. The scattering triangle in inelastic neutron scattering.

184

I. B. Neutron Sources [2]

Present-day neutron sources are of two types: reactors and spallation sources. Reactors produce neutrons by nuclear fission, while spallation sources produce them by striking heavy metal targets with high-energy protons from an accelerator. In both types of sources, neutrons are slowed down to energies required for scattering experiments by appropriate moderator-reflector assemblies, whose configuration and temperature are optimised for the type of experiment that uses them.

The average energy of neutrons from a water moderator at ambient temperature is about 25 thousandths of an electron volt (25 meV), and the average energy from a liquid-hydrogen moderator at 20 kelvins is around 5 meV. The wavelength of a 25-meV neutron is 1.8 angstroms (1.8 x 10^{-10} meter), which is of the same order as typical interatomic distances and, therefore, is quite suitable for diffraction experiments.

Mostly, reactors operate in a continuous mode and produce high integrated fluxes of neutrons of cold and thermal energies for both scattering experiments and isotope production. Spallation sources are most effectively operated in a pulsed mode (10 - 100 Hz) and give high peak fluxes of cold and thermal neutrons, as well as large quantities of epithermal neutrons (~0.1 - 10 eV) for time-of-flight experiments.

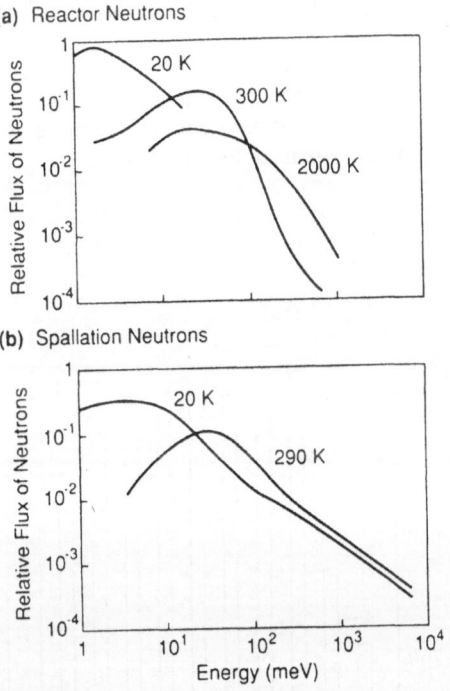

Fig. 2. Reactor and spallation neutrons.
a)The relative flux of neutrons as a function of energy for the high-flux reactor at the Institut Laue-Langevin in Grenoble, France. The curves show the distribution of neutrons emerging from moderators at temperatures of 20, 300, and 2000 kelvins.
b) Similar distribution curves for neutrons generated at the spallation spurce at Los Alamos (LANSCE) by moderators at temperatures of 20 and 290 kelvins.

With few exceptions, research reactors have been built as multi-purpose units with equally high, or higher, priority for irradiation services (such as materials testing of aircraft and energy production components), isotope production, and transmutational doping of semiconductors. The irradiation services usually require uniform fluxes over extensive volumes, and may involve loop installations that are expensive to operate. In recent years, the interest in materials testing in reactors has decreased, and the economic basis for operating these reactors is being questioned. Ageing is another reason why many research reactors will soon have to be shut down. The lifetime of a reactor depends both on the neutron flux and on the accessibility of vital components, but 40 years is a reasonable maximum estimate. The great majority of research reactors in OECD countries were commissioned between 1957 and 1967. It appears then that less than ten out of the about 30 existing reactors might still be in operation after 2005.

The tendency is now going away from multi-purpose reactors. The first reactor to be built as a dedicated neutron beam source was the HFBR reactor at Brookhaven. Since then, dedicated reactors have been built in Grenoble (Institut Laue-Langevin) and in Saclay (ORPHÉE). These reactors have compact cores with a high power density and are probably prototypes for future research reactors.

The total capacity of neutron instruments within OECD countries also decreases after year 2000, but less drastically than the number of reactors. Figure 3 plots the instrument capacity, defined as the number of instruments weighted by the source flux. Spallation sources are not included.

So far, only four spallation sources have been placed in operation: two in the United States, one in the United Kingdom and one in Japan. A spallation source in Switzerland will be operative from 1996. The total measuring capacity of the spallation sources will then be 250, on the scale of Fig. 3.

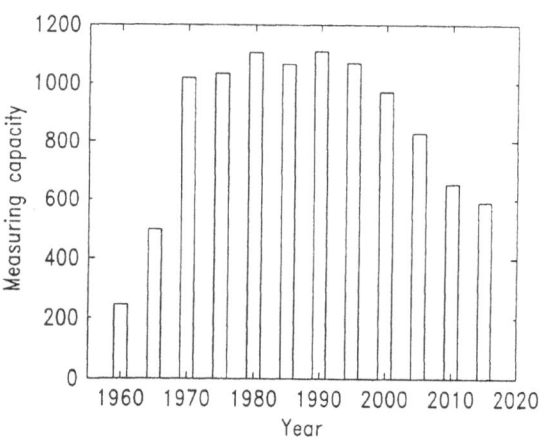

Fig. 3. Measuring capacity, i.e. number of instruments weighted by the source flux, for neutron reactors in OECD countries.

I. C. Instrumentation for Inelastic Scattering

In inelastic scattering we have to determine the neutron wave vector before and after the scattering event, see Fig. 1. The most commonly used instrument for this is the triple axis spectrometer, see Fig. 4, which was first introduced by Brockhouse [3]. A first crystal X_1, the mono-chromator selects neutrons of a suitable energy from the white reactor spectrum and directs them towards the sample S. The sample scatters these neutrons in various directions. A second crystal X_2, the analyzer, Bragg - reflects neutrons of a particular energy into a detector. The angles at the monochromator (θ_M), sample (ϕ) and analyzer (θ_A) can be set and varied in a programmed manner. At one parameter setting a measurement is made for a single scattering vector $\bar{Q}$ and energy transfer ΔE. B.N. Brockhouse, who shared the 1994 Nobel prize in physics with C.G. Shull, introduced some efficient spectrometer scans for mapping out a dispersion relation. In Fig. 5 the neutron probe, for reasons given below, is designated by an ellipse. By successive settings of the three axes one can move the neutron probe in (ω,q)-space in two rational ways: in a constant -q scan ("vertically"), and in constant -ω scan (horizontally"), and record the corresponding intensities $I(\omega)$ and $I(q)$.

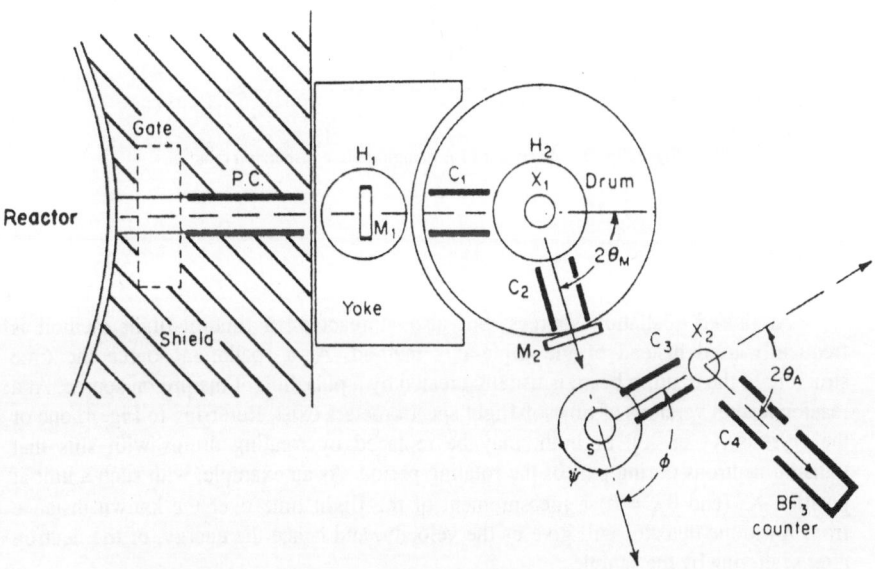

Fig. 4. Schematic drawing of the Chalk River triple axis spectrometer (Brockhouse, 1961).

The volume of the resolution ellipsoid of the neutron probe is determined by the crystalline perfections of X_1, S and X_2, and of the angular widths of the collimators C_1 to C_4. As usual, any improvement of the resolution implies a reduction of the intensity.

The raw data have to be resolution-corrected. Knowing the resolution function $R(\bar{Q},\omega)$, i.e. the ellipsoid, one could in principle de-convolute the data to obtain the corrected set of (ω,q) coordinates and the line-width of the excitation. The common, practical procedure is, however to start from a plausible assumption about the scattering law (see the Appendix), convolute it with the resolution function and compare it with the observed data. This procedure is repeated with an adjusted assumption until the best agreement with the data is obtained.

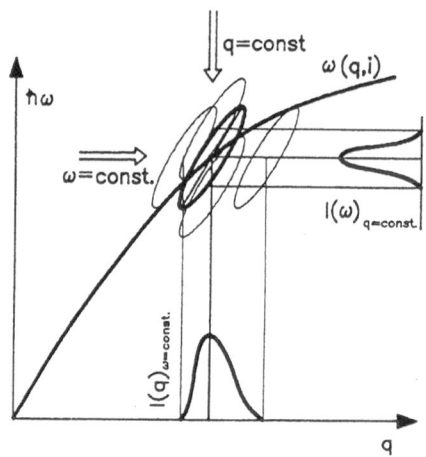

Fig. 5. Spectrometer scans for mapping out a dispersion relation.

At pulsed spallation sources, but also at reactors, a time-of-flight method is frequently used instead of the triple-axis method. At a spallation source the time structure of the neutron beam is usually created by a pulsation of the proton source. At a reactor several versions of time-of-flight spectrometers exist. Referring to Fig. 4, one of the crystals X_1 or X_2, or both, may be replaced by rotating drums with slits that transmit neutrons during part of the rotating period. As an example, with such a unit at position X_2 (and $\theta_A = 0$) a measurement of the flight time over the known distance from X_2 to the detector will give us the velocity, and hence the energy, of the neutron after scattering by the sample.

Two experimental methods exist for measuring excitations of very low energies, i.e. small values of ΔE, the back-scattering method [4] and the spin-echo method [5], which can measure values as low as ~μeV and ~neV, respectively. The high resolution achieved with these methods requires the use of cold, low-energy neutrons.

In the back-scattering spectrometer the analyzer part has several crystals operating with $2\theta_A$ close to 180^0. From the Bragg relation, which gives $(dk'/k') \sim \cot g\,\theta$, we see that the energy resolution then is very good, in practise $\sim 10^{-4}$. The analyzer is set to work at a fixed outgoing energy. In order to measure a small energy transfer one must then be able to vary the incoming energy with the same precision. One way of doing that is to vary the lattice constant of the monochromator crystal in small steps, e.g. by varying the temperature.

A spin-echo spectrometer works with polarized neutrons, i.e. neutrons whose spin (or magnetic moment) all point along a common direction. When the neutrons pass through a magnetic field of strength H, the spin processes with a Larmor frequency $\omega_L \sim H$, independent of the neutron velocity. The number of precessions over a given distance depends on the neutron velocity, and provides us with a very sensitive method for an analysis of the neutron energy after scattering. A spin-echo spectrometer has two consecutive precession coils of identical length, but with opposite precession directions. Any dephasing of the spins in the first coil, due to the velocity spread in the incident beam, is then restored in the second coil. A sketch of a spin-echo spectrometer is shown in Fig. 6.

Most modern neutron spectroscopy methods work in (q,ω)-space, one exception being the spin-echo method which works in (q,t)-space. Still another method is a

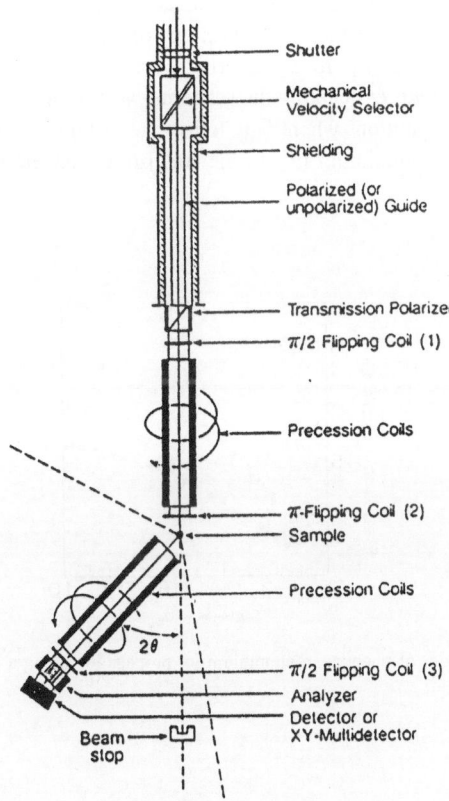

Fig. 6. Schematic layout of the spin echo spectrometer i operation at ORPHEE in Saclay.

correlation method [6], working in (r,t)-space, which we have used in the measurement of macro- or mesoscopic phenomena in liquid crystals [7]. This method only works for anisotropic scatterers, such as liquid crystals with their elongated molecules. Dynamic phenomena on a time scale extending to hours have been measured in this way.

2. STRUCTURAL EXCITATIONS

Atoms in a crystal are not frozen to their position in the lattice. Thermal energy causes the lattice to distort, and the atoms to oscillate about their lattice sites, and this motion can be described by a superposition of waves moving through the lattice. These waves are the phonons. The frequency of a phonon depends on the wavenumber of the distortion, the masses of the atoms and of the binding forces ("springs") betwen them. Phonon frequencies are typically a few times 10^{12}Hz, corresponding to a few meV, and the scattering by a phonon gives an appreciable, and easily measurable, fractional energy change of thermal neutrons.

II. A. Phonon Dispersion Relations

A phonon is characterized by a frequency ω, a wavevector $\bar{q}$ and a polarization vector $\bar{e}$. For each wavevector there may be a number of phonon branches, depending on the crystal structure. In a monoatomic cubic lattice, for a wavevector along a high-symmetry direction, there are three branches: one of longitudinal polarization and two of transverse polarization [8]. In a scattering experiment the cross section formula contains a term $\bar{Q}.\bar{e}$, with $\bar{Q}$ denoting the scattering vector, as before, and it is possible to choose scattering conditions where only transverse or longitudinal modes contribute. Fig. 7 shows data for aluminium [9]. In the [00ζ] direction the two lower transverse modes are degenerate.

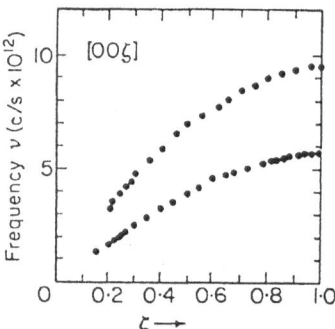

Fig. 7. Dispersion relations of phonons in aluminium for propagation along a cube edge. Experiments by Yarnell et al. [9]

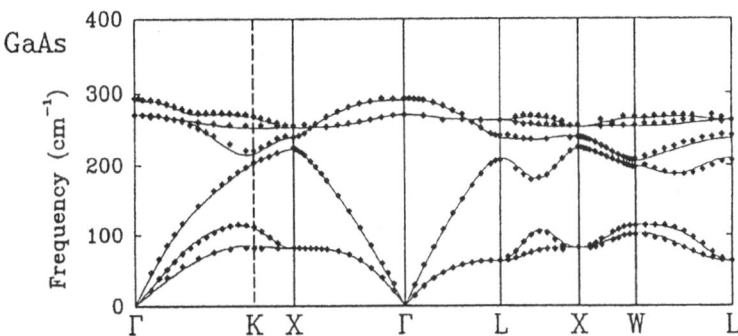

Fig. 8. Phonon dispersion curves for GaAs. The experimental data at T = 12K from Strauch and Dorner [10] are given by the points; the experimental uncertainty is typically 0.02 THz (less than the heigth of the points). The lines give the results of ab initio model calculations by Giannozzi et al. [11]. The letters below the figure give the notation for the symmetry directions or points.

In a crystal with two atoms per unit cell there are six branches, three acoustic and three optic ones. The important semiconductor GaAs has fcc structure and has two atoms per unit cell. The symmetry is still high enough that the branches in high-symmetry directions are purely longitudinal or transverse with respect to q. Some of the six branches are degenerate. Experimental data by Strauch and Dorner [10] taken at 12K, are shown in Fig. 8. These data are well reproduced by ab initio band model calculations [11]. Basic ingredients in the model are the electron distribution in the bonds, their movements relative to the ions, and the Coulomb interactions.

In general there are 3 n branches for a lattice with n atoms per unit cell. Hence, for GaAs, there are six branches in low-symmetry directions. In sapphire (Al_2O_3) there are two formula units per unit cell, and the less symmetric, rhombohedral structure implies that there are 15 non-degenerate phonon branches [12].

II. B. Structural Phase Transitions and Soft Modes

Inelastic neutron scattering has been a central tool in elucidating the mechanisms for structural phase transitions. The canonic example is the soft-mode phase transition in $SrTiO_3$. At 105K the lattice transforms continuously from a high-temperature cubic to a low-temperature tetragonal phase. In 1960 Cochran and Anderson [13] suggested that the phase transition in certain ferroelectrics might result from an instability in a normal vibrational mode of the lattice. In $SrTiO_3$ and other perovskites the relevant mode is connected with a coupled rotation of oxygen octahedra. As $T^+ \rightarrow 105K$, the rotational motion slows down, as seen from the lowering phonon energy at a zone boundary, see Fig. 9.

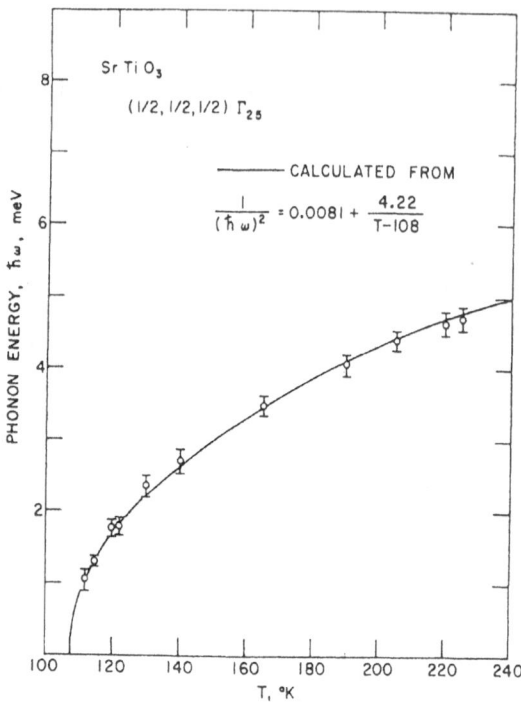

Fig. 9. Soft phonon energy approaching the 105K structural transition in $SrTiO_3$. [14]

The rotation stops at the transition, and the lattice "freezes" to a lower symmetry. Actually, the soft mode does not give an exhaustive description of the transition. Scattering experiments have shown that at the transition much of the response resides in a central, zero-energy mode. [15,16,17]

II. C. Electron-Phonon Interactions

Electron-phonon interactions are fundamental to the properties of metals in general, and to superconductors in particular. Phonon linewidth for the strongly coupled superconductors Nb and Nb_3Sn have been studied in considerable detail by neutron scattering by the Brookhaven group [18,19]. For phonons with energies below the superconducting energy gap, the electron-phonon interaction vanishes in the superconducting state because these phonons do not have sufficient energy to break the Cooper pairs. The changes observed in the linewidths of these phonons as one lowers the temperature through Tc provides a direct measure of the electron-phonon interaction, see Fig. 10.

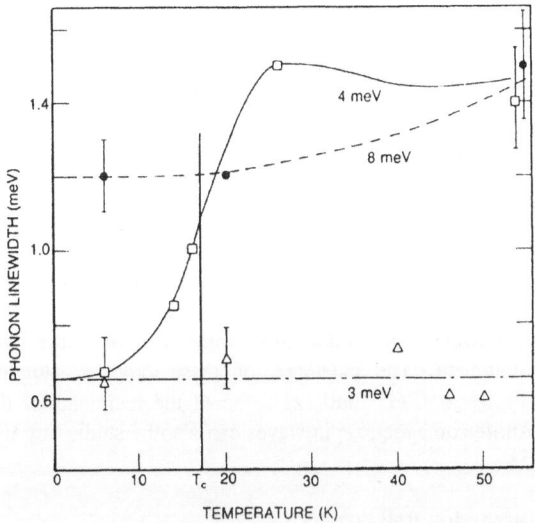

Fig. 10. Electron-phonon interactions [19]. The unusually large linewidth of the phonon with wavevector 0.18[110] (hω = 4 meV) in Nb_3Sn is suppressed if the superconducting gap energy, $2\Delta(T)$ rises above the phonon energy. The gap energy rises from 0 (above T_c at 18.3 K) to 7 meV at T = O. The phonon with wavevector 0.3[110] (hω = 8 meV) is unaffected. The temperature dependence of the linewidth of the 0.2[100] (hω = 3 meV) is shown for comparison.

The electron-phonon interaction manifests itself also in anomalies in the phonon dispersion relation. This Kohn anomaly occurs when the phonon wavevector matches a caliper of the Fermi surface. The phonon decays into an electron-hole pair, and the interaction creates a dip in the phonon energy, the Kohn anomaly. For this process to be important, there should be a large density of states at points on the Fermi surface separated by the phonon wavevector. Clearly this condition is fulfilled for low-dimensional systems where the Fermi surface has large, flat sections separated by $2k_F$. An example [20] is KCP, see Fig. 11.

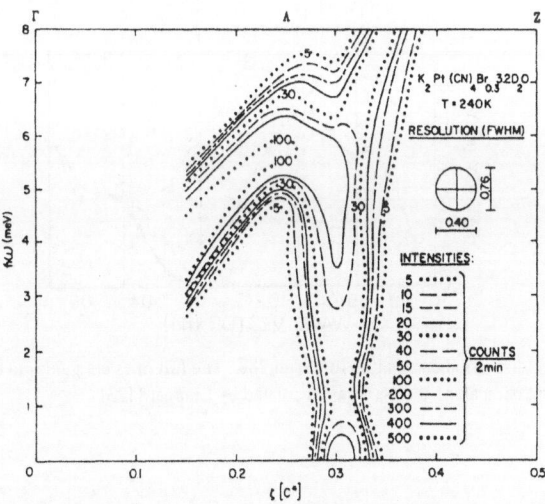

Fig. 11. Inelastic scattering intensity. Contours for KCP showing a sharp Kohn anomaly in the acoustic branch near $\zeta \approx 0.3c^*$.

3. MAGNETIC EXCITATIONS

Neutrons have a magnetic moment, and can be scattered from solids whose atoms have a net magnetic moment, i.e. unpaired electron spins. Ferromagnetic materials are magnetic because the magnetic moments at different atoms align spontaneously. The moments at different sites are coupled, and thermal excitation in the form of orientational waves can pass through the lattice of moments. These spin waves or magnons are the magnetic analogue of the phonons, and have dispersion relations that can be measured by inelastic neutron scattering.

Antiferromagnetic materials have an equal number of "up" and "down" orientations of the magnetic moments, and are therefore macroscopically nonmagnetic. The lattice may, however, be subdivided into two or more sublattices with "ferromagnetic" alignment. The existence of these ordered sublattices was first demonstrated by Professor C.G. Shull [21], one of the recipients of the 1994 Nobel prize in physics. Antiferromagnetic spin waves can also be studied by inelastic neutron scattering.

III. A. Magnon Dispersion Relations

In analogy with phonons, the magnon dispersion relation is determined by the lattice geometry and the force constants. The relevant lattice is now the magnetic lattice, i.e. the configuration (position, direction and size) of magnetic moments. The dominant interaction is the exchange interaction between spins $H = -2J\Sigma S_i S_j$, where J denotes the exchange integral. It is positive for ferromagnetic interaction and negative for antiferromagnetic interaction.

Ferromagnetic gadolinium has two atoms per unit cell, so we expect two branches, and this is verified in the experiments by Koehler et al. [22]. In Gd the exchange interaction between localized spins is mediated by the conduction electrons. The exchange interaction is then q-dependent and depends on the geometry of the Fermi surface.

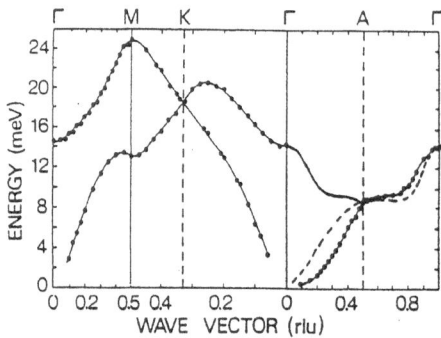

Fig. 12. Spin-wave dispersion curves [22] for Gd at 78K. The full lines are guides to the eye; the dashed lines show the dispersion along the c-axis as calculated by Lindgård [23].

Metals like Fe and Ni, remain magnetic up to high temperatures. Spin waves then extend to very high energies, and can only be measured at high-flux sources and incident neutrons of relatively high energy. To measure the dispersion relation in Ni, which has a very weak magnetic moment (0.6 µB), Mook et al. [24] used a sample of ^{60}Ni weighing 0.5 kg! The data are shown in Fig. 13. The splitting of the branch reflects an interaction with the Stoner continuum of states, and is a characteristic of the intinerant character of electrons in metals.

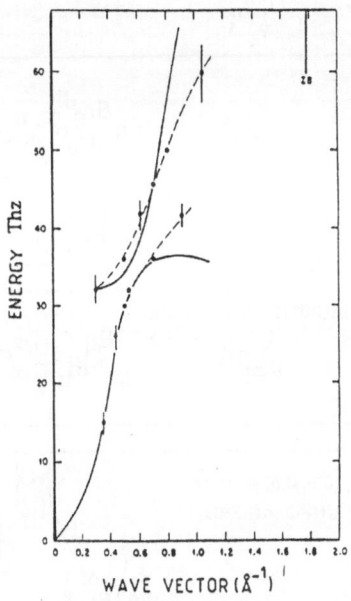

Fig. 13. Dispersion curves for Ni showing evidence for the Stoner excitation [24].

High-energy antiferromagnetic excitations may be responsible for the unusual physical properties of high-T_c super-conductors. The existence of such spin-wave like states is shown by Fig. 14 [24].

195

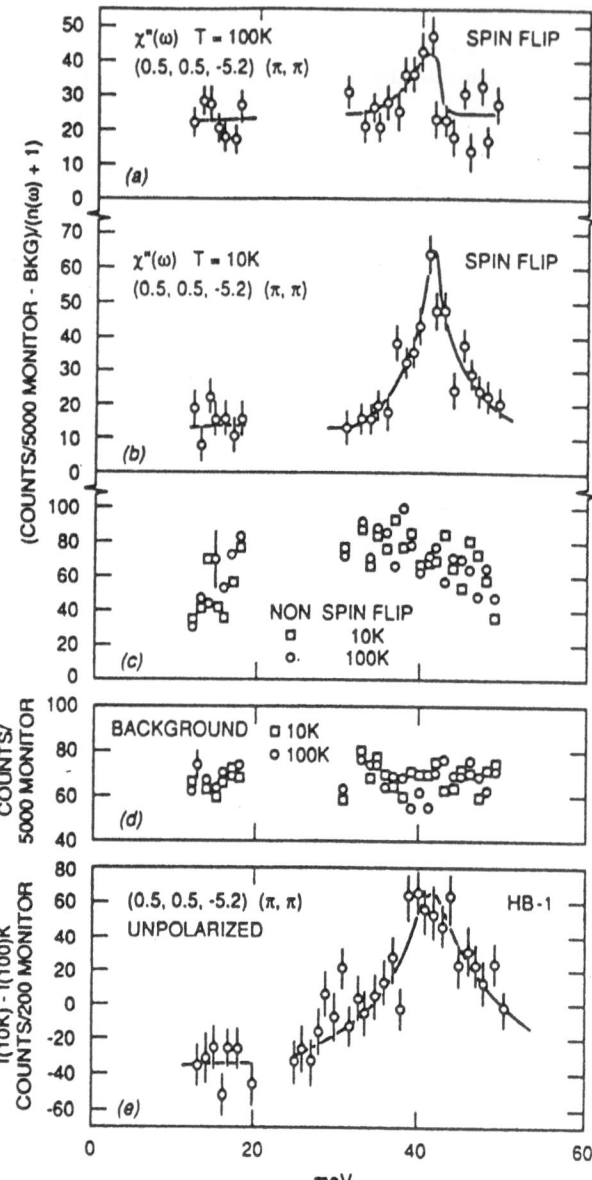

Fig. 14. Neutron scattering measurements [25] showing the scattering at (π,π) reciprocal lattice position for $YBa_2Cu_3O_7$.

The scattering intensity due to these excitations is much lower than that of phonons, and polarization analysis had to be used in order to identify their magnetic origin. A sketch of a spectrometer for polarization analysis [26] is shown in Fig. 15. This instrument measures, not only the angular and energy distribution of scattered neutrons, but also whether or not the neutrons have flipped their spins in the scattering process.

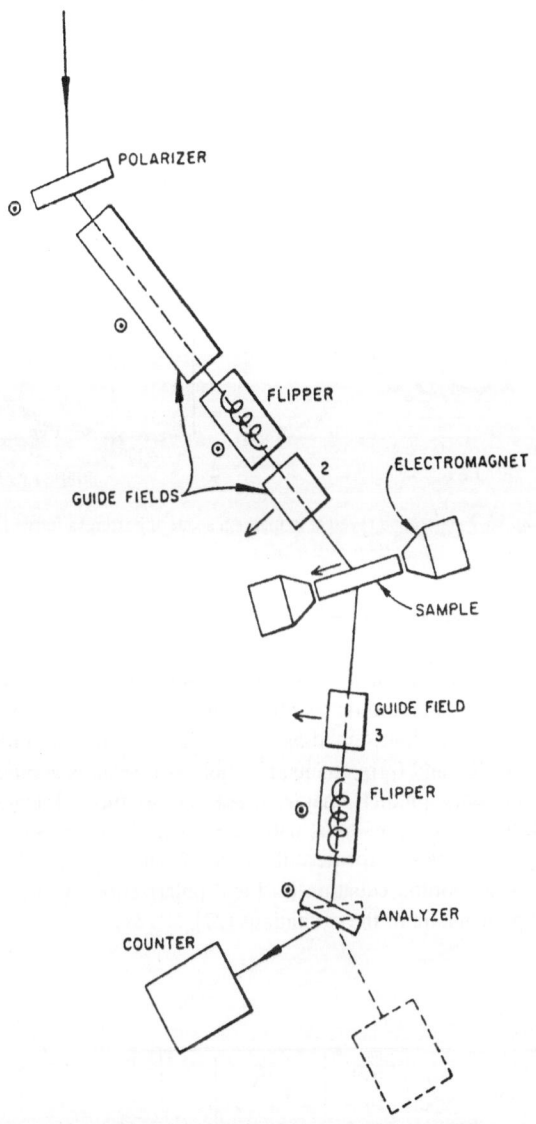

Fig. 15. Experimental arrangement for polarization analysis. Arrows adjacent to the guide fields show the direction of the magnetic field sensed by the neutrons.

The magnetic scattering of neutrons is not isotropic in nature, only the component of the sample magnetization perpendicular to the scattering vector $\bar{Q}$ is effective in the scattering process. In the field configurations of Fig. 15 only spin correlations transverse to the magnetization will be magnetically scattered. If magnons are present, they will manifest themselves as a spin-flipped component, whereas phonon scattering gives unflipped neutron spins. Fig. 16 gives an example in which magnon scattering has been identified in this way.

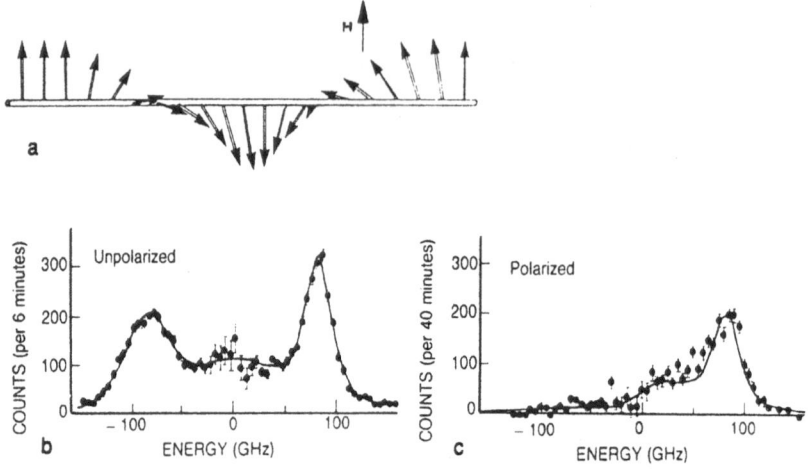

Fig. 16. Polarization reversal in magnon scattering from Li ferrite [28].

III. B. Low-Dimensional Magnetism

In low-dimensional magnets there are collective, propagative excitations at temperatures where long-range order is absent. An example of this is the 1-D magnet $CsNiF_3$. Here the spins are constrained by crystal-field anisotropy to lie in a plane, and they are weakly coupled and form chains. If a field ~ 1 Tesla is applied, perpendicular to the chain axis, theory predicts magnetic solitons to form. Magnetic solitons are regions in which the nickel spins twist about the chain direction, see Fig. 17a. Linear spin wave theory is insufficient for describing the dynamics, but solitons are solutions of the non-linear sine-Gordon equation [28] and polarization analysis was once more used in identifying the nature of the excitations [27].

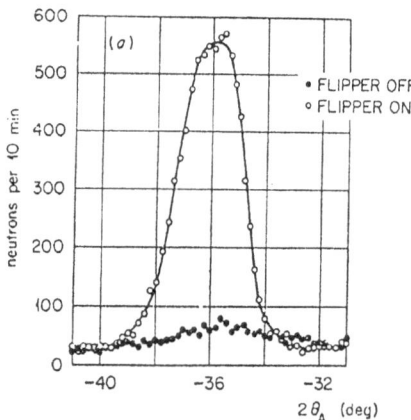

Fig. 17. Magnetic solitons [27]. a: Schematic representation of a magnetic soliton. Arrows indicate direction of spins. b: Spectrum for inelastic neutron scattering without polarization analysis. c: Spectrum with polarization analysis. Energies have been divided by Planck's constant, so that 100 GHz is 0.417 meV.

REFERENCES

1. Of the many excellent review on this topic I have made use of
   a. R. Pynn, Los Alamos Science No. 19, p. 1 (1990)
   b. J.D. Axe and R.M. Nicklow, Phys. Today, 38, No. 1, p. 26 (1985)
   c. "Lecture Notes of the First Summer School on Neutron Scattering", (A. Furrer ed.), Paul Scherrer Institut, Villigen, Switzerland (1993)
2. T. Riste, "Neutron Beams and Synchrotron Sources", p. 63. OECD Paris (1994)
3. B.N. Brockhouse, "Inelastic Scattering of Neutrons in Solids and Liquids", Vol. I, p. 259, IAEA, Vienna (1961)
4. M. Birr, A. Heidemann and B. Alefeld, Nucl. Instr. Methods 95, 435 (1971)
5. F. Mezei, Z. Physik 255, 146 (1972)
6. K. Otnes, T. Riste and H.B. Möller, Kerntechnik 44 Suppl., 753 (1984)
7. L. Dobrzynski, K. Otnes and T. Riste, J. Neutron Research 2, 155 (1994)
8. C. Kittel, "Introduction to Solid State Physics" (6th edition) p. 82, John Wiley, New York (1986
9. J.L. Yarnell, J.L. Warren and R.G. Wenzel, Phys. Rev. Lett. 13, 13 (1964)
10. D. Strauch and B. Dorner, J. Phys. Cond. Matter 2, 1457 (1990)
11. P. Giannozzi, S. de Gironcoli, P. Pavone and S. Baroni, Phys. Rev. B. 43, 7231 (1991)
12. H. Schober, D. Strauch, and B. Dorner, Z. Phys. B. Cond. Matter B 92, 273 (1993)
13. P.W. Anderson, "Fizika dielektrov" (G.I. Skanavi ed.), p. 290 Akad. Nauk SSSR, Moscow (1960); W. Cochran, Phys. Rev. Lett. 3, 412 (1959)
14. G. Shirane and Y. Yamada, Phys. Rev. 177, 858 (1969)
15. T. Riste, E.J. Samuelsen, K. Otnes and J. Feder, Solid State Commun. 9, 1455 (1971
16. S.M. Shapiro, J.D. Axe, G. Shirane, and T. Riste, Phys. Rev. B 6, 4332 (1972)
17. S.R. Andrews, Phase Transitions, 11 181 (1988)
18. S.M. Shapiro, G. Shirane and J.D. Axe, Phys. Rev. B 12, 4899 (1975)
19. J.D. Axe and G. Shirane, Phys. Rev. Lett. 30, 214 (1973)
20. K. Carneiro, G. Shirane, S.A. Werner and S. Kaiser, Phys. Rev. B 13, 4258 (1976)
21. C.G. Shull and J.S. Smart, Phys. Rev. 76, 1256 (1949)
22. W.C. Koehler, H.R. Child, R.M. Nicklow, H.G. Smith, R.M. Moon and J.W. Cable, Phys. Rev. Lett. 24, 16 (1970)
23. P.A. Lindgaard, J. de Physique C1 - 238 (1971)
24. H.A. Mook and D.McK.Paul, Phys. Rev. Lett. 54, 227 (1985)
25. H.A. Mook, M. Yethiraj, G. Aeppli, T.E. Mason and T. Armstrong, Phys. Rev. Lett. 70, 3490 (1993)
26. R.M. Moon, T. Riste and W.C. Koehler, Phys. Rev. 181, 920 (1969)
27. K. Kakurai, R. Pynn, B. Dorner and M. Steiner, J. Phys. C 17 L 123 (1984)
28. H.J. Mikeska, Phys. Rev. B 12, 2794 (1975)

APPENDIX

Formulae of Neutron Scattering

In the Born approximation the probability that an incident plane wave of wave vector $\vec{k}$ will be scattered by a weak potential $V(\vec{r})$ to become an outgoing plane wave with wave vector $\vec{k}'$ is proportional to

$$\left| \int e^{i\vec{k}\vec{r}} V(\vec{r})^{-i\vec{k}'\vec{r}} d^3r \right|^2 = \left| \int e^{i\vec{Q}\vec{r}} V(\vec{r}) d^3r \right|^2 \tag{1}$$

with $\vec{Q} = \vec{k} - \vec{k}'$, and the integration is over the volume of the sample. For scattering of neutrons by an assembly of nuclei, such as a crystal, the potential $V(\vec{r})$ is

$$V(\vec{r}) = \sum_j b_j \delta(\vec{r} - \vec{r}_j) \tag{2}$$

with b_j denoting the scattering length of neucleus j at position $\vec{r}_j$.

Using (1) and (2) Van Hove showed that the scattering law, i.e. the number of neutrons scattered per incident neutron, with a given momentum and energy change, can be written as

$$I(\vec{Q}, \omega) \frac{k'}{k} \sum_{j,k} b_j b_k \int_{-\infty}^{\infty} \langle e^{-i\vec{Q}\vec{r}_k(0)} e^{i\vec{Q}\vec{r}(t)} \rangle e^{-i\omega t} dt \tag{3}$$

For an assembly of identical atoms the expression under the summation sign can be written

$$Nb^2 \int_{-\infty}^{\infty} G(\vec{r}, t) e^{-i\vec{Q}\vec{r}} d^3r \tag{4a}$$

where

$$G(\vec{r}, t) = \frac{1}{N} \sum_{j,k} \langle \delta(\vec{r} - [\vec{r}_k(0) - \vec{r}_j(t)]) \rangle \tag{4b}$$

N is the number of atoms in the sample. The delta function is zero, except when the position of an atom k at time zero and the position of an atom j at time t are separated by the vector $\vec{r}$. The summation is thus over pairs of atoms, and the time-dependent pair-correlation function $G(\vec{r},t)$ describes how the correlation between two particles evolves with time.

The van Hove neutron scattering law can now be written

$$I(\vec{Q},\omega) = Nb^2 \frac{k'}{k} \int_{-\infty}^{\infty} G(\vec{r},t)e^{-i\vec{Q}r}e^{-i\omega t}d^3r\,dt \tag{5}$$

The physics of the scattering sample is contained in $G(\vec{r},t)$. If the time dependent states are written in the phonon approximation, the scattering law for a phonon of wave vector $\vec{q}$ in branch m, and at temperature T, is

$$I_m(\vec{Q},\omega) = \left|G_m(\vec{q},\vec{Q})\right|^2 F_{q,m}(\omega,T) \tag{6}$$

The inelastic structure factor G_m

$$G_m(q,Q) = \sum_{unitcell} b\frac{1}{\sqrt{M}}\left[\vec{Q}\vec{e}_m(\vec{q})\right]\exp\left[-W(\vec{Q})+i\vec{Q}\vec{u}_m\right] \tag{7}$$

W is the Debye-Waller factor. The displacement $\vec{u}_m$ oscillates in time with the phonon frequency $\Omega_m(\vec{q})$.

The frequency response function is

$$F_{q,m}(\omega,T) = \frac{\omega}{1-\exp(-\hbar\omega/kT)}\frac{\delta(\omega\pm\Omega_{m(q)})}{\omega^2} \tag{8}$$

The formulae for magnetic scattering are more complicated, because the interaction is anisotropic, as mentioned above. The pair correlation now also contains the correlation between spin orientations at different sites. The time-dependent states of spin correlations can be written in the spin wave approximation.

EXCITATION DYNAMICS IN ORGANIC MOLECULES, SOLIDS, FULLERENES AND POLYMERS

Paras N. Prasad

Photonics Research Laboratory
Department of Chemistry
State University of New York at Buffalo
Buffalo, NY 14260-3000

ABSTRACT

This review deals with excitation dynamics in organic molecules, solids, fullerenes and polymers and various nonlinear optical techniques to probe them. For the benefit of readers not familiar with excitations in organic materials, a brief review of nature of molecular eigenstates and collective excitations in organic solids is presented. This "Fundamental" section is followed by a description of nonlinear optical interactions. Various nonlinear optical techniques using ultrafast laser pulses are discussed under "Experimental Techniques". Then "selected results" are presented to illustrate the application of these techniques. Finally, to encourage involvements of new researchers in this rich field, the review concludes with a discussion of new opportunities for research.

I. FUNDAMENTALS

I. A. Molecular Eigenstates

Organic materials, whether in a crystalline form or in the form of an amorphous polymer, are molecular materials which consist of chemically bonded molecular units weakly interacting through Vanderwaals forces. The nature of excitations in molecular solids can be related to the molecular eigenstates which are solutions of the molecular Schrödinger equation [1]. The molecular Hamiltonian can be written as [1]

Spectroscopy and Dynamics of Collective Excitations in Solids
Edited by Di Bartolo, Plenum Press, New York, 1997

$$H_M = \frac{-\hbar^2}{2m_e} \sum_i \nabla_i^2(r_i) - \sum_{i,\alpha} \frac{Z_\alpha e^2}{r_{i\alpha}} + \sum_{i,j} \frac{e^2}{r_{ij}} + \sum_{\alpha,\beta} \frac{Z_\alpha Z_\beta e^2}{r_{\alpha\beta}} - \sum_\alpha \frac{\hbar^2}{2M_\alpha} \nabla_\alpha^2(r_\alpha) \qquad (1)$$

In the above equation, i,j label the electrons while α,β label the nuclei. Using the Born-Oppenheimer Approximation, one partitions the Hamiltonian as

$$H_M = H_e(r_i, r_\alpha) + H_N(r_\alpha) \qquad (2)$$

While $H_e(r_i, r_\alpha)$ is the electronic Hamiltonian containing all the terms, except, the last term, the nuclear kinetic contribution, of equation 1. The solution of the electronic Hamiltonian gives the electronic potential energy surface $U_e(r_\alpha)$. The solution of a Schrödinger equation using $H_N(r_\alpha) + U_e(r_\alpha)$ gives the energy states for nuclear motion. Further partitioning of energy states is obtained by a transformation of the coordinates to the center of mass based system. This leads to separation of three different kinds of nuclear motions: (i) translation, i.e., the displacement of the center of mass, (ii) rotation of the entire molecule, and (iii) vibrations which describe relative motions of atoms within the molecule. Appropriate partitioning of the nuclear Schrödinger equation using $H_N(r_\alpha) + U_e(r_\alpha)$, obtained by the coordinate transformation and their subsequent solutions, provide the translational, rotational and vibrational energy states of a molecule. The translational energy of a molecule is not quantized, while the rotational and vibrational energies are quantized.

The most widely used quantum mechanical method used to obtain the molecular electronic energies is where the molecular orbitals are constructed from a linear combination of atomic orbitals (hence LCAO-MO) [1]. The most comprehensive calculation would involve inclusion of all electrons (i,j) of the electronic Hamiltonian $H_e(r_i, r_\alpha)$. This approach is called the ab-initio method. Here one first optimizes the geometry by minimization of the electronic energy vs. the nuclear coordinates and evaluates all quantum mechanical integrals explicitly. Until recently such calculations were limited only to di- or tri-atomic molecules; but with the availability of supercomputers, the ab-initio calculations now can be carried out for a relatively large-size molecules. For large molecules and polymers, a more practical approach is a semi-empirical approach which does not consider all electrons but often only the valence (outer) electrons. Furthermore, on the basis of the nature of overlap of electronic wavefunctions in a chemical bond, a separation of σ and π electrons is made. The σ electrons are the ones involved in a single bond formation, where the overlap of the electronic atomic orbitals of the bonded atoms is along the internuclear axis. When two atoms form a multiple bond, one bond is of σ-type, while the additional bonds are of π-type; because due to geometric limitation imposed, the additional orbitals involved can overlap only laterally. The electrons involved in these π-bonds are the π-electrons. Some of the semi-empirical calculations only consider the π-electrons which give the lower energy excitations in π-bonded systems. Other semi-empirical methods consider both σ- and π-electrons. Furthermore, in a semi-empirical method, the integrals are parameterized by fitting the spectroscopic data. Also, some overlap integrals are neglected. Of special interest from the excitation and dynamics point of view are the highest occupied molecular orbitals (HOMO) and the lowest unoccupied molecular orbitals (LUMO). The lowest energy excitation involves the promotion of an electron from HOMO to LUMO. In the case of a structure containing π-bonding, the HOMO and LUMO are π (bonding) and π^* (antibonding) orbitals.

I. B. Conjugated Polymers: Solitons and Polarons

Of special interest is a conjugated organic structure which involves alternate single and double bonds [1]. In these systems, the adjacent π-orbitals overlap to give rise to π-electron delocalization which leads to a resonance splitting of π-excitations. In the ideal case of an infinite conjugated polymer, the π and π* states split into bands analogous to the valence and the conduction bands, and the polymer would ideally behavior-like a pseudo one-dimensional semiconductor.

In real one-dimensional conjugated polymers such as polyacetylene shown below,

polyacetylene

the electronic excitation is strongly coupled to the vibrations of the chain. The problem of polyacetylene has been theoretically treated by Su, Schifffer and Heeger (SSH) [2,3]. The Hamiltonian of the SSH model is

$$H = \frac{K}{2}\sum_{n}(U_n - U_{n+1})^2 + \frac{M}{2}\sum_{n}\dot{U}_n^2 - \sum_{n,s}\left\{t_0 + \alpha(U_n - U_{n+2})\right\}\left(C_{n+1,s}^+ C_{n,s} + C_{n,s}^+ C_{n+1,s}\right) \qquad (3)$$

In equation 3, U_n is the displacement of CH group along chain axis from its equilibrium position. $C_{n,s}(C_{n,s}^+)$ is the annihilation (creation) operator of the π-electron on the nth site. K is the stiffness of the σ-bond; M is mass of CH group; α is the electron-phonon coupling constant and t_0 is the hopping integral of the chain. The model predicts that optical excitation (from π- to π*-band), in polyacetylene, will lead to rapid deformation forming a pair of solitons and antisolitons (charge defects) as represented below.

The energy states of these pairs lie at the midgap state. Therefore, new spectroscopic transitions corresponding to these midgap states arise which have been spectroscopically observed by transient spectroscopy [3]. Polyacetylene represents a degenerate ground states system where the reversal of the order of single and double bonds around a soliton defect leads to energetically equivalent situation. Other conjugated polymers such as polythiophene do not represent a degenerate ground state [3]. In such a case the deformation produces gap states corresponding to polarons and bipolarons [3].

I. C. Fullerenes

Fullerenes represent a special class of materials. It can be considered as a cluster because it consists of 60, 70, etc. carbon atoms [4]. However, these carbon atoms are chemically bonded; therefore, fullerenes are molecules. The most widely studied fullerene is C_{60} which consists of 60 π-bonds and 180 σ-bonds and the carbon atoms are bonded producing hexagons and pentagons. The arrangement is like a soccer ball. Molecular excitations in fullerene molecules can be treated in the same way as that for other molecules [4]. Because of the large number of bonds involved, the most practical approach to calculate molecular electronic states has been a semi-empirical approach [4-6].

I. D. Excitations in Solid State

Next we discuss the changes which occur in molecular eigenstates when a molecule is placed in a solid environment. The most pronounced changes occur in the rotational and translational motions of a molecule. In a solid phase they become hindered and give rise to new collective excitations called phonons [7]. Although, in general, a collective phonon excitation in a molecular solid has mixed rotational (librational in solid phase) and translational characters, under certain molecular and crystal symmetries, one can have separate librational and translational phonons [7]. An example is a centrosymmetric molecule forming a crystal with one molecule per unit cell. For this system, the Hamiltonian for lattice displacement can be written as [8]:

$$H_s = \tfrac{1}{2} \sum_\ell \mu_s (\dot{U}_\ell^s)^2 + \tfrac{1}{2} \sum_\ell \phi_{\ell,\ell}^{s,s} (U_\ell^s)^2 + \sum_\ell \phi_{\ell,\ell+a}^{s,s} U_\ell^s U_{\ell+a}^s \tag{4}$$

Here S = L for libration for which U_s = I (the moment of inertia), and S = T for translation for which U_s = M (the mass). Also, ℓ represents a site and a represents lattice spacing. The ϕ terms represent the force constant terms and due to the constraints arising from the invariance of potential energy they exhibit the following relationships [7]:

$$\phi_{\ell,\ell}^{T,T} = -2\phi_{\ell,\ell+a}^{T,T} \;;\;\; \phi_{\ell,\ell}^{L,L} = 4\phi_{\ell,\ell+a}^{L,L} \tag{5}$$

One introduces the following transformations:

$$(2\phi_{\ell,\ell+a}^{L,L} I)^{\frac{1}{4}} U_\ell^L = q_\ell^L \tag{6}$$

$$(\phi_{\ell,\ell+a}^{T,T} M)^{\frac{1}{4}} U_\ell^T = q_\ell^T \tag{7}$$

$$q_\ell^s = (\tfrac{1}{2}\hbar)^{\frac{1}{2}} (b_{s,\ell} + b_{s,\ell}^+) \tag{8}$$

and a Bogolyubov transformation from the real site space to a reciprocal k-space as [8]:

$$b_{s,k} = \frac{1}{\sqrt{N}} \sum_\ell \left[\cos h\, \theta_{s,k} b_{s,\ell} + \sin h\, \theta_{s,k} b_{s,\ell}^+ \right] e^{ik\ell} \tag{9}$$

206

The Hamiltonian (4) then takes the diagonal form:

$$H_s = \sum_k \hbar\omega_{s,k}(b^+_{s,k}b_{s,k} + 1/2) \tag{10}$$

It can be seen from equation 11 that the librational phonon behaves as an optical phonon, i.e.,

$$\text{where } \omega_{L,k} = (2\phi^{L,L}_{\ell,\ell+a}/I)^{\frac{1}{2}}(2 + \cos ka)^{\frac{1}{2}} \text{ for libration} \tag{11}$$

$$\text{and } \omega_{T,k} = 2(\phi^{T,T}_{\ell,\ell+a}/M)^{\frac{1}{2}} \sin\frac{ka}{2} \text{ for translation} \tag{12}$$

nonzero frequency value for $k = 0$ which can be optically excited. The dispersion of translational phonon gives the acoustic phonon branch for which $\omega = 0$ at $k = 0$.

The effect of intermolecular interactions on the internal vibrations and electronic excitations is often described in the weak coupling limit by using perturbation theory. Under weak intermolecular interaction, they behave as Frenkel excitons (tight binding limit) [9]. For example, the Hamiltonian for a tightly bound electronic excitation (no separation of the electron-hole pair) can be written as [9,10]:

$$H_s = \sum_\ell (E_0^s + D^s)b^+_{s,\ell}b_{s,\ell} + \tfrac{1}{2}\sum_{\ell,\ell'} M^s_{\ell,\ell'}\left[b^+_{s,\ell}b_{s,\ell'} + b_{s,\ell}b^+_{s,\ell'}\right] \tag{13}$$

In the above equation, s labels the molecular excitation: E_0^s is the excitation energy of the molecule and D^s is the static shift in this energy due to intermolecular interactions. In other words $E_0^s + D^s$ is the site energy at any lattice site ℓ (assume energetically identical) $M^s_{\ell,\ell'}$ is the resonance (excitation exchange) interaction between two sites ℓ and ℓ'. If one uses the nearest neighbor interaction and the following Fourier transformations to k space [9]:

$$b_{s,k} = (N)^{-\frac{1}{2}}\sum_\ell b_{s,\ell}e^{ik\cdot\ell} \tag{14}$$

then the Hamiltonian of equation 13 transforms into a diagonal form with the eigenvalues given by [9,10]

$$b^+_{s,k} = (N)^{-\frac{1}{2}}\sum_\ell b_{s,\ell}e^{ik\cdot\ell} \tag{15}$$

$$E_k^s = E_0^s + D^s + M^s \cos k\cdot a \tag{16}$$

The $k = 0$ level of the exciton band is optically accessible. Figure 1 summarizes the correlation between the molecular states and the crystal states. The presence of positional, orientational and substitutional disorder leads to different degrees of manifestations depending

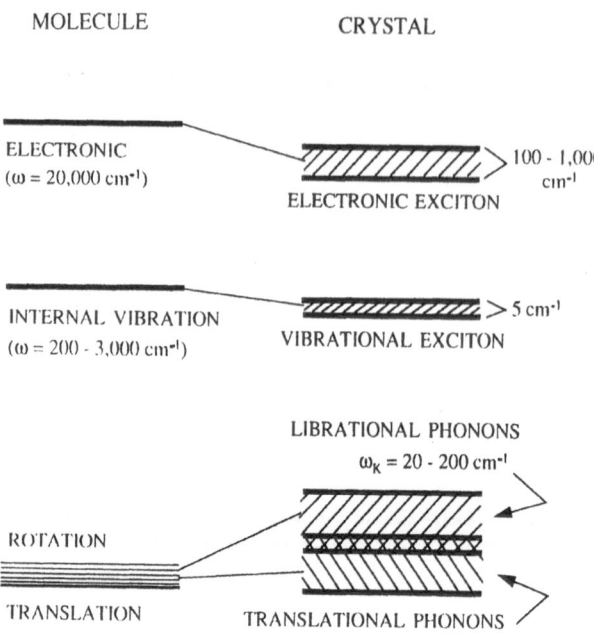

MOLECULE   CRYSTAL

ELECTRONIC
(ω = 20,000 cm^{-1})

ELECTRONIC EXCITON
100 - 1,000 cm^{-1}

INTERNAL VIBRATION
(ω = 200 - 3,000 cm^{-1})

VIBRATIONAL EXCITON
> 5 cm^{-1}

LIBRATIONAL PHONONS
ω_K = 20 - 200 cm^{-1}

ROTATION

TRANSLATION

TRANSLATIONAL PHONONS
(ACOUSTIC PHONONS)
ω_K = 0 - 200 cm^{-1}

Figure 1. Correlation Diagram between the molecular states and the crystal states of a molecular solid.

on their relative strength compared to resonance interactions for a given type of excitation. The presence of a disorder such as in polymers, which are usually amorphous, destroys the k-characterization, often localizing the excitation.

I. E. Interaction with the Radiation Field

The interaction of a molecule in the presence of an intense optical field can be described by an induced dipole moment due to a change in electronic distribution created by the electric field, E, of radiation. The induced dipole moment is expressed as a power series [11]:

$$\mu_{ind} = \alpha \cdot E + \beta : EE + \gamma : EEE + ... \tag{17}$$

The linear term involving polarizability α is generally considered for weak optical field, and it describes the linear response such as ordinary refraction and linear absorption. The higher hyperpolarizability terms β and γ describe the molecular nonlinear optical responses. One can use several quantum chemical methods to compute the terms α, β and γ [11].

An analogous power expansion is used for the induced polarization P at the bulk level [11,12]:

$$P = \chi^{(1)} \cdot E + \chi^{(2)} : EE + \chi^{(3)} \vdots EEE + \ldots = \chi_{eff} \cdot E \qquad (18)$$

In equation 18, $\chi^{(1)}$ is the linear susceptibility which is generally adequate to describe the optical response in the case of a weak optical field (ordinary light). The terms $\chi^{(2)}$ and $\chi^{(3)}$ are the second- and third-order nonlinear optical susceptibilities which describe the nonlinear optical response of the medium. Since the electric field E is oscillating at a frequency ω (in the case of an optical pulse, this frequency is the corresponding optical frequency), a more precise notation for the nonlinear susceptibilities carries frequency specification. The generalized notation for the nonlinear susceptibility, for example the third order susceptibility, in the frequency representation is $\chi^{(3)}(-\omega_4; \omega_1, \omega_2, \omega_3)$ for a process in which three input waves of frequencies ω_1, ω_2 and ω_3 interact in the medium to generate an output wave at frequency ω_4. The bulk susceptibilities can be related to corresponding microscopic coefficients α, β and γ of equation 17 if one uses the weak intermolecular coupling limit of an oriented gas model. Under this model, the bulk susceptibilities $\chi^{(n)}$ are derived from the corresponding microscopic coefficients by using simple orientationally-averaged site sums with appropriate local field correction factors which relate the applied field to the local field at a molecular site. Under this approximation [11]

$$\chi^{(2)}(-\omega_3; \omega_1, \omega_2) = F(\omega_1)F(\omega_2)F(\omega_3) \sum_n < \beta^n(\theta, \phi)> \qquad (19)$$

$$\chi^{(3)}(-\omega_4; \omega_1, \omega_2, \omega_3) = F(\omega_1)F(\omega_2)F(\omega_3)F(\omega_4) \sum < \gamma^n(\theta, \phi)> \qquad (20)$$

In the above equations, β^n and γ^n represent the microscopic coefficients at site n which are averaged over molecular orientations θ and ϕ and summed over all sites n. The terms $F(\omega_i)$ are the local field corrections for a wave of frequency ω_i. Generally, one utilizes the Lorentz approximation for the local field, in which case [11,12]

$$F(\omega_i) = n_0^2(\omega_i) \qquad (21)$$

In this equation, $n_0(\omega_i)$ is the linear refractive index of the medium at frequency ω_i.

From equation 19, it is clear that even for molecular systems with $\beta \neq 0$, the bulk second-order nonlinearity, determined by the second-order nonlinear susceptibility $\chi^{(2)}$, will be absent if the bulk structure is centrosymmetric or amorphous in which case $\sum < \beta_n(\theta, \phi) = 0$. Therefore, for a molecular system to give rise to second-order effect the conditions are that (i) $\beta \neq 0$ and (ii) the bulk structure is non-centrosymmetric.

Since γ is a fourth rank tensor, its average does not vanish even in a centrosymmetric structure. Therefore, an isotropic medium such as an amorphous polymer will exhibit third-order nonlinear optical response. However, a system may still show large differences in the $\chi^{(3)}$ value if it is oriented. An example is a conjugated polymeric structure which has the largest component of the γ tensor along the polymer chain [13]. If only this component contributes, then the largest value of $\chi^{(3)}$ will correspond to a bulk in which all the polymeric chains are oriented in the same direction. The largest component of $\chi^{(3)}$ in this case will be along the orientation direction. In contrast, the $\chi^{(3)}$ value in a truly amorphous phase of the same polymer will be reduced by a factor of five. These orientation effects have been observed in stretch-oriented polymers [13,14].

Manifestations of the second-order nonlinearity ($\chi^{(2)}$ effects) are second harmonic generation and linear electric field dependence of the refractive index of a medium. Manifestations of the third-order nonlinearity ($\chi^{(3)}$ effects) are third harmonic generation and intensity dependence of refractive index (real part of $\chi^{(3)}$ effects). The imaginary part of $\chi^{(3)}$ gives rise to intensity dependence of absorption, i.e., two-photon absorption.

Photorefractive materials have emerged [15] as an important class of nonlinear optical materials which show large optically induced refractive index changes. Even though the overall manifestation is like a third-order nonlinear optical effect (light induced refractive index change), the mechanism involves a photo-induced charge separation which sets up an electric field to produce a refractive index change by the $\chi^{(2)}$ electro-optic mechanism [15]. Therefore, a photorefractive material is actually a $\chi^{(2)}$ medium which also contains centers to produce photo-induced carriers. Furthermore, the carriers must have sufficient mobility to separate and set up a strong electric field. Inorganic ionic crystals such as $LiNbO_3$, $BaTiO_3$ and SBN have been used in the past for photorefractive effects [15]. Polymeric materials have presently emerged as a new class of photorefractive materials [16-19]. They are very attractive for a photorefractive effect because they exhibit the largest $\chi^{(2)}$. Also, polymeric materials offer the flexibility to tailor their structure and composition to control other properties necessary for photorefractive effects. Therefore, it is possible to decrease the response time, which traditionally is very slow in the inorganic ionic photorefractive crystals.

Among various third-order processes, one of the most interesting one, which provides great insight into dynamical processes, is that in which all the input frequencies and the output frequencies are the same, i.e., that described by the susceptibility $\chi^{(3)}(-\omega; \omega, -\omega, \omega)$. This combination of frequencies corresponds to phenomena derived from intensity dependence of refractive index. In this case, one can expect an enhancement of the third-order nonlinear process when the input frequency approaches a material resonance at ω_0. However, there may also be a two-photon enhancement, i.e., a contribution from a resonance at $\omega_0 \approx 2\omega$. A theoretical description to treat this third-order nonlinear optical response under resonance condition is presented here.

The resonant behavior of a material (assumed to be centrosymmetric for simplicity) can be described by expressing the material susceptibility as [20,21]

$$\chi = \chi_g^{(1)} + \chi_g^{(3)} EE + \frac{N}{N_0} (\chi_e^{(1)} - \chi_g^{(1)}) + \chi_g^{(5)} EEEE + \cdots. \tag{22}$$

The frequency representation is dropped for simplicity of notation, with an understanding that, in the present context, the nonlinear susceptibility is $\chi^{(3)}(-\omega;\omega,-\omega,\omega)$. In equation 22, $\chi_g^{(n)}$ stands for the linear and higher order susceptibilities of a material consisting of ground state molecules, and $\chi_e^{(1)}$ is the linear susceptibility corresponding to excited state molecules, their density being N. N_0 is the total density of molecules. For example, under one-photon absorption conditions and for excited species decaying with a unimolecular relaxation rate, N can be expressed as [20,21]

$$N(t) = N_0 \kappa e^{-t/\tau} \int_{-\infty}^{t} I(t')e^{t'/\tau}dt', \tag{23}$$

where κ is the absorption cross section, τ is the relaxation time and I is the light intensity. Since I is proportional to E^2, this contribution to the susceptibility is of the same order in the electric field as the coherent third-order $\chi^{(3)}$ term. The temporal behavior of N depends on the kinetics of formation of excited species and their decay. Thus, one can conceptually treat a resonant third-order process as containing two contributions: the coherent contribution (four-photon parametric mixing) coming from the true third-order susceptibility, and an incoherent contribution, due to the $(\chi_e^{(1)} - \chi_g^{(1)})$ term of equation 22, coming from the population of excited species. The latter contribution will depend on factors like the laser pulse properties, the experimental geometry and kinetic properties of the excited species [22,23]. In the case of two-photon generated species, the population N is proportional to I^2. Therefore, the incoherent contribution derived from excited state population will have the same electric field dependence as $\chi^{(5)}$. It is the incoherent contribution of nonlinear response which can be used to probe the dynamics of excitations.

Near a resonance, whether one photon or two-photon, the susceptibility $\chi^{(3)}(-\omega;\omega,-\omega,\omega)$ becomes complex with a non-negligible imaginary part resulting from absorption. A pure two-photon process is a third-order nonlinear process in which two photons are absorbed simultaneously when the combined energy of the two photons matches a resonance of the medium. Though saturation of an one-photon absorption is basically the modification of the absorption of a medium by a laser beam, it can also be treated as a third-order process since it corresponds to some degree of ground state depletion thus an intensity dependent absorption coefficient. At a one-photon resonance, therefore, the effective imaginary part of $\chi^{(3)}$ will be negative (decrease of absorption with increasing intensity) while near two-photon resonance, it would be positive [24].

II. EXPERIMENTAL TECHNIQUES

Two main nonlinear techniques to probe the dynamics of excitations in organic and polymeric materials are described below.

II. A. Degenerate Four-Wave Mixing

Under the conditions of a degenerate four-wave mixing experiment, three input beams of the same frequency are made to interact in the sample. Their interaction is often described by the formation of a grating produced by two input beams and subsequent Bragg diffraction of a third (probe) input beam [11,25]. The Bragg diffraction condition is the conservation of the wavevectors; i.e., $\Sigma k_i = 0$, where k_i stands for the wavevectors of the four interacting beams. This is the phase-matching condition. For the phase-matched interaction, the amplitude of the generated beam (labeled as beam 4) can be described as (neglecting for simplicity transient effects due to travel of pulses through the medium) [20,21].

$$\frac{d\epsilon_4(z,t)}{dz} = i\frac{2\pi\omega^2}{k_4c^2}[\chi_g^{(3)}\epsilon_1(t)\epsilon_2^*(t)\epsilon_3(t)$$

(24)

$$+(\chi_e^{(1)}-\chi_g^{(1)})\kappa e^{-t/\tau}\epsilon_3(t)\int_{-\infty}^{t}\epsilon_1(t')\epsilon_2^*(t')e^{t'/\tau}dt'],$$

The first term in the square bracket on the right-hand side describes coherent four-wave mixing due to ground state third-order nonlinearity, and the second term describes the population grating effect (the incoherent contribution) due to one-photon excited species, which is noninstantaneous, and, through the convolution integral, provides an integrating effect. Even though this contribution for a one-photon induced process will have the same electric field dependence as $\chi^{(3)}$ (in both cases, the signal is proportional to I^3), this contribution will be critically dependent on laser pulse properties. If τ is much longer than the laser pulse width, this term can give an effective nonlinearity that becomes larger for longer laser pulses due to build-up of the excited state concentration over the pulse width (integrating effect). In the case of two-photon generated species, the population grating will produce the DFWM signal, which will show an I^5 dependence.

Since the susceptibilities near a resonance are complex, equation 24 in fact describes the formation of both the phase and amplitude gratings, and both gratings can have the coherent and incoherent parts. Since the quantity measured in DFWM is the intensity of the generated (the fourth) beam, the DFWM signal will be the sum of squares of the real and imaginary contributions.

II. B. Transient Absorption and Kerr gate Involving Two-Wave Mixing

In a two-wave mixing experiment, there are just two input beams. The interaction of two beams that fulfills the phase-matching relation is, however, a modification of one beam by the presence of another. Examples of these two-wave mixing phenomena that are important for a systematic investigation of dynamics of excitations are transient absorption and the optical Kerr gate effect. In a transient absorption experiment, a strong beam (pump) is incident on a sample; the nonlinear response of the medium is monitored by the changes in transmission characteristics of a weak probe beam. In analogy with DFWM, a similar theoretical description of the two-wave mixing phenomena can also be developed where the change in the amplitude of the probe field can be given as [20,21]

$$
\frac{d\epsilon_1(z,t)}{dz} = i\frac{2\pi\omega^2}{k_1 c^2} [\chi_g^{(3)} \epsilon_2(t)\epsilon_2^*(t)\epsilon_1(t)
$$

$$
+ (\chi_e^{(1)} - \chi_g^{(1)})\kappa e^{-t/\tau}\epsilon_1(t) \int_{-\infty}^{t} \epsilon_2(t')\epsilon_2^*(t')e^{t'/\tau}dt'].
$$

(25)

In this case again, the contribution to the modification of the ϵ_1 amplitude may come from both the real and the imaginary parts of the susceptibilities. However, since the transient transmission measurement is performed against the background of the initial field amplitude $\epsilon_1(0)$, one can easily show that the contribution from the real parts of the respective susceptibilities will be negligible (since their contribution is 90° out of phase with the initial field). The transient absorption signal is derived from the imaginary part of $\chi^{(3)}$ and the imaginary part of the change in the linear susceptibility. Clearly, the contribution due to imaginary part of $\chi^{(3)}$ will give rise to two-photon absorption of the probe beam due to a bias from the pump beam, while the imaginary part of the change in the linear susceptibility may have a character of a bleaching signal (for a negative change in the imaginary part of the first-order susceptibility) or induced absorption signal (for a positive change due to absorption by excited species). This picture may be additionally complicated if one takes into account the possibility of creation of excited species by two-photon absorption.

The optical Kerr gate experiment differs from the previous two experiments in that the polarization state of the probe beam is analyzed [26]. In a homodyne version of the Kerr gate experiment, a probe beam of a given polarization interacts with the sample pumped by a strong beam [26]. As a result, a perpendicular polarization component in the probe beam is created through optically induced birefringence due to the pump beam. This component is then detected after passing the probe beam through a cross analyzer. The description of this experiment is also provided by equation 25, which, however, should be written for both polarization components of the probe beam, i.e., $\epsilon_{1,x}$ and $\epsilon_{1,y}$. Proper tensor components of the susceptibilities should be taken too. The main difference in interpreting equation 25 comes from the boundary conditions for the probe beam, since the probe beam enters the sample as the x component only, and the y component is produced by the nonlinear interaction. In effect, the Kerr gate signal is produced by the real part of the third-order susceptibility if this change is anisotropic (a good example of this effect is light-indued orientation of anisotropic molecules in liquids).

II. C. Use of Higher Order Diffraction in DFWM for Bimolecular Processes

The above description assumes that the excited state decay channel is first-order (single excitation decay). Many organic solids and polymers show biexcitonic decay which involves annihilation of two excitons simultaneously. In such a case the kinetic equation for the decay of the excited state is governed by the equation [22]:

$$\frac{dn_{ex}(r,t)}{dt} = \sigma NI(r,t) - k_1 n_{ex}(r,t) - \kappa_2 n_{ex}^2(r,t)... \tag{26}$$

where k_1 and k_2 stand for the rate constants for the first- and second-order decay, respectively. $I(r,t)$ describes the temporal and spatial profiles of the exciting light intensity, σ stands for the absorption cross-section and N is the concentration of the absorbing center. The presence of non-linear decay term $k_2 n_{ex}^2(r,t)$ makes the grating due to change in susceptibility $\Delta\chi = \chi_e^{(1)}-\chi_g^{(1)}$, formed by the two pump beams in a degenerate four-wave mixing experiment, nonsinusoidal. This change can formally be expressed as a Fourier series [22]:

$$\Delta\chi = C_0(t) + C_1(t)\cos(\Delta k \cdot r) + C_2(t)\cos(2\Delta k \cdot r) + ... \tag{27}$$

where $\Delta k = k_1 - k_2$ for the two-pump beams.

For linear decay one has a first-order diffraction proportional to $C_1(t)^2$ at a phase-matched direction of $k_z = k_3 \pm \Delta k$. But due to the nonlinear decay term $k_2 n_{ex}^2$, one also gets higher-order diffraction, such as second-order diffraction proportional to $C_2(t)^2$. The phase-matched direction for this diffraction will be $k_3 \pm 2\Delta k$. Therefore, by monitoring the time-evolution of this second-order diffraction, one can study the dynamics of biexcitonic processes [22].

II. D. Novel Approach of Phase-Sensitive Optically Heterodyned Kerr gate

The above methods do not provide the sign of the imaginary portion of $\chi^{(3)}$.

For this purpose, the phase-sensitive optically heterodyned Kerr gate method [27] was recently developed in our Photonics Research Laboratory. This method takes advantage of selective (i.e., optically phase-tuned) enhancement of either the real or the imaginary component of the nonlinear response of medium under study, followed by a polarization sensitive optical detection and is equally suited for the study of solid, liquid (solutions) as well as thick or thin film samples.

The details of this method are described elsewhere [27]. The coupled-wave equation governing the optical Kerr gate experiment may be presented in a form similar to equation 25. However, the boundary condition now is $\mathscr{E}_{1,y}(z = 0) = 0$, since the initial incident light field is assumed to be x-polarized, and the y component is produced in the sample due to the action of the proper components of the sample's third-order polarizability tensor. In general, $\chi^{(3)}$ is complex; therefore, equation 25 will contain both real and imaginary terms. The relevant component of the generated field, $\mathscr{E}_{1,y}(L)$, due to the imaginary part of $\chi^{(3)}$ is in-phase with the local oscillator (formed by a portion of the original beam), $\mathscr{E}_{1,x}$, whereas the component due to the real part of $\chi^{(3)}$ is $\pi/2$ out-of-phase. Hence, the amplitude of the y component of the Kerr signal at the sample exit can be presented as $\mathscr{E}_{1,y}(L) = i\zeta_{im}(L)$, where ζ is a function which, in the simplest case, will contain field amplitudes and components of $\chi^{(3)}$; L stands for the beams interaction path. The indices r and im indicate the real and the imaginary parts of the function ζ.

In homodyne-detection OKG one simply measures the y component of $\mathscr{E}_1$ by crossing the polarizer and the analyzer; the measured intensity contains then contributions from both ζ_r and ζ_{im}. Heterodyne detection involves mixing of the OKG signal with a given fraction of a local oscillator signal (which may be the transmitted portion of the original probe itself). In our case the analyzer is rotated by some angle, ϕ, to admit a small contribution from the x component of the field. This component, practically equal to $\mathscr{E}_{1,x}(0)$ (neglecting small Kerr-induced contribution), constitutes a local oscillator field. In the presented technique, the phase relation between the response signal and the local oscillator beam is established by the presence or absence of a phase retardation element (a properly oriented quarter-wave plate with one principal axis parallel to the probe polarization, x) in the probe beam in front of the analyzer. The quarter-wave plate imposes a fixed $\pi/2$ phase bias between the x component (local oscillator) and the y component (Kerr signal) of the field. Thus without the quarter-wave plate the field after the analyzer is [27]:

$$\mathscr{E}_1(L) \; \alpha \; \mathscr{E}_{1,x}(0) \; \sin\phi \; + \; \mathscr{E}_{1,y}(L) \; \cos\phi \; , \tag{28}$$

and in the presence of the quarter-wave plate

$$\mathscr{E}_1(L) \; \alpha \; \mathscr{E}_{1,x}(0) \; \sin\phi \; + \; i \; \mathscr{E}_{1,y}(L) \; \cos\phi \; . \tag{29}$$

The lock-in detected intensity at the chopping frequency, $I^* = (nc/2\pi)\mathscr{E}_1(L)\mathscr{E}_1^*(L)$ (n standing for the refractive index of the sample), contains, therefore, terms proportional to $\cos^2\phi$, $\sin^2\phi$ and $\sin\phi\cos\phi$. For small angles up to first order in ϕ we arrive at

$$I = \frac{nc}{2\pi} \left(\zeta_r^2(L) \; + \; \zeta_{im}^2(L) \; - \; 2 \, \mathscr{E}_{1,x}(0) \, \zeta_{im}(L) \, \phi \right) , \tag{30}$$

and

$$I = \frac{nc}{2\pi} \left(\zeta_r^2(L) \; + \; \zeta_{im}^2(L) \; - \; 2 \, \mathscr{E}_{1,x}(0) \, \zeta_r(L) \, \phi \right) , \tag{31}$$

without and with a $\pi/2$ phase bias imposed, respectively. In the former case the detection favors the imaginary component of the signal (ζ_{im}), while in the latter case the real component (ζ_r) is favored.

Now, carrying out the measurements as a function of the angle ϕ and making use of equations 30,31 one can fit the dependence of the heterodyned OKG signal with the form $I = z_1 + z_2\phi$. The coefficient z_2 is proportional either to the imaginary component, $\zeta_{im}(L)$, or to the real component, $\zeta_r(L)$. The $\zeta_{r(im)}(L)$ can be substituted with an effective third-order nonlinearity $(\phi_K^{(3)})_{r(im)}$ and the proper beam intensities. Hence we obtain

$$Z_{2,r(im)} = -C\Re_2 n^{-1/2} (\phi_K^{(3)})_{r(im)} I_2 I_1 , \tag{32}$$

where $C = 4\pi^2\omega^2L/k_1c^3$, I_1 and I_2 stand for the pump and the probe beams intensities, respectively, and R_2 represents a correction factor for attenuation of the beams in the sample due to linear absorption.

Performing the dependence of the OKG signal on the analyzer angle ϕ we obtain linear plots of the type $I = z_1 + z_2\phi$. Having the z_2 coefficients for the investigated sample, z_2^{samp}, and for the reference sample, z_2^{ref}, of known susceptibility, $\chi_{(ref)}^{(3)}$, we can readily determine the real and the imaginary parts of the third-order susceptibility of the sample, $\chi_{samp}^{(3)}$.

$$\left(\chi_{samp}^{(3)}\right)_{r(im)} = \frac{Z_{2,r(im)}^{samp}}{Z_{2,r(im)}^{ref}} \frac{1}{R_2^{samp}} \left(\chi_{2,r(im)}^{ref}\right)_{r(im)} \tag{33}$$

The use of femtosecond pulses allows us to separate the coherent (instantaneous) nonlinear optical response form the delayed incoherent response resulting either from excited state population or from orientational nonlinearity.

Some examples of the various types of resonances and excited state dynamics which may contribute to the incoherent resonant nonlinearities, yielding thus significant imaginary component of the nonlinear susceptibility, are illustrated below.

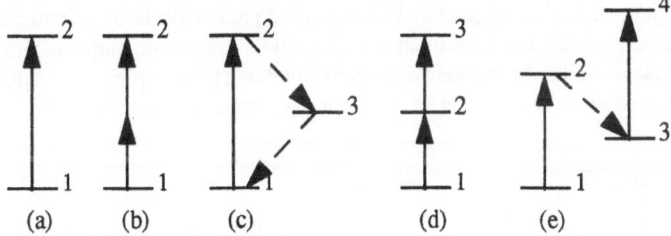

The diagrams represent two-, three- or four-level systems which are coupled by a one-photon resonance, a direct two-photon resonance or a sequential two-photon absorption. The different resonances and their excited state dynamics will manifest differently in the time domain and in the intensity dependencies of their real and imaginary parts of $\chi^{(3)}$. Also, the signs of $Im[\chi^{(3)}]$ will vary.

For example, a direct two-photon resonance (case b) will yield a nonlinear incoherent response (time-delayed component) of the homodyne optical Kerr gate signal dependent on the fifth power of light intensity. On the other hand, the phase-tuned optically heterodyned Kerr gate method will give, in this a case, a third power of the intensity dependence of the signal. This technique should also reveal here a positive sign of the imaginary part of $\chi^{(3)}$, while a supplementary transient absorption experiment will indicate an induced (two-photon) absorption with instantaneous response. In the case of one-photon dynamics (diagram a), the respective signals in the homodyne and the heterodyne OKG experiment will depend on the third and the second power of the input intensity. Transient absorption signal will show here a saturation (one-photon) absorption profile. In the case of relaxed sequential absorption (case e), incoherent nonlinear response may show a complicated temporal behavior of the heterodyne and homodyne-detected Kerr gate signals. Depending on whether the initially populated level (2), relaxed level (3), or subsequently pumped level (4) yield the largest change of linear susceptibility ($[\chi_e^{(1)} - \chi_g^{(1)}]$ in equation 22 above) with respect to the ground state, the time dependence of the signal will be different. Assuming that the transition 1 to 2 can be saturated, the $Im[\chi^{(3)}]$ will be negative for short time delay, yet the transient absorption may show time-delayed induced absorption.

A very interesting, and often yielding remarkably high nonlinearities [27], is the case (c) where not the initially reached state (2), but rather the relaxed state (3) (or a manifold of states) gives predominant contribution to the nonlinear optical response. Various modifications of this case are possible where, e.g., state (2) is populated by a two-photon excitation or subsequently populated states are of different parity.

III. SELECTED RESULTS

III. A. Excitations in Fullerenes

The excitation dynamics for fullerenes have been extensively investigated by both linear and nonlinear optical techniques. Its properties both in the solution phase and in the pure solid form have been studied [5,6,28-35]. Electronic excitations in fullerenes are primarily molecular with the excitation in the pure crystalline form described under the tight binding Frenkel limit [29]. The lowest lying excitation in the C_{60} crystal corresponds to a Frenkel excitation of energy 1.846 eV above the ground state [5,29]. C_{60} also shows a strong excited state absorption which has been shown to provide optical power limiting behavior [34,35]. This behavior is exhibited by the nonlinear transmission curve as shown in Figure 2 [35]. The output power levels off at high input power.

The excitations in C_{60} solution and in condensed phase have been investigated by the nonlinear optical techniques of third harmonic generation, degenerate four-wave mixing and optical Kerr gate as well as by transient absorption involving the pump-probe technique using femtosecond pulses [5,6,28-33]. The excited state dynamics is extremely fast.

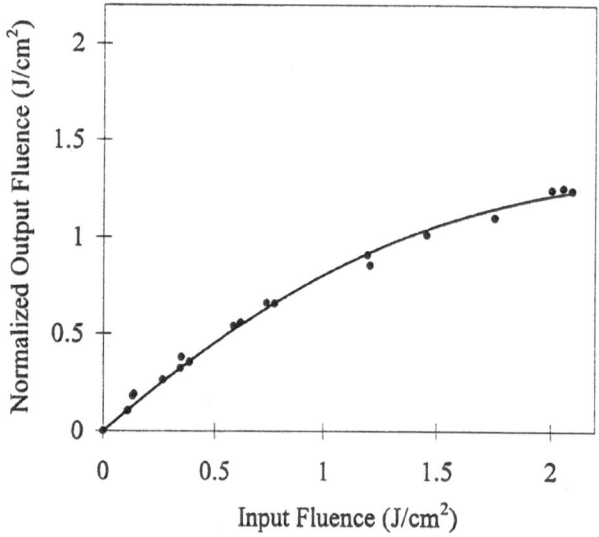

Figure 2. Nonlinear transmission behavior of C_{60} doped in a sol-gel matrix.

III. B. Excitation Dynamics in Conjugated Structures

Excitation dynamics in conjugated structures, because of their enhanced nonlinear response derived from π-conjugation, is ideally suited for investigation by nonlinear optical techniques. A good example of the dynamics probed by the Kerr effect is that for canthaxanthin carotenoid. This compound in tetrahydrofuran solution was investigated by both the homodyne and optically heterodyned Kerr gate techniques discussed in the earlier section [27]. The use of 60 femtosecond pulses at 620 nm, obtained from an amplified-colliding pulse mode locked laser system, permitted the separation of the instantaneous, coherent part from that of the time-delayed incoherent response. The result is shown in Figure 3. The power dependence of the delayed response establishes that it is due to excitations

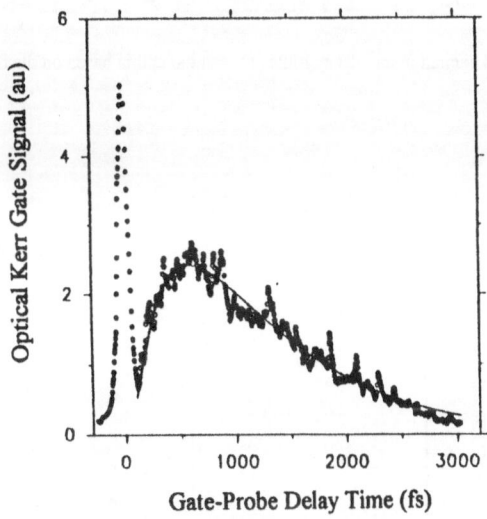

Figure 3. Femtosecond time-resolved homodyne Kerr gate signal from a solution of canthaxanthin in tetrahydrofuran solvent. The solid line is the theoretical fit using energy level scheme of Figure 4.

created by direct-two-photon absorption. Another unusual feature exhibited in Figure 3 is that the delayed incoherent response (due to the $(\chi_e^{(1)}-\chi_g^{(1)})$ term of equation 25) shows a buildup followed by a decay. This feature indicates that the initially two-photon pumped excitation does not produce any significant change of linear susceptibility, but it is a lower energy excitation (2^1Ag state) subsequently populated by nonradiative relaxation which produces a dominant change in the linear susceptibility responsible for the incoherent nonlinear optical response. The proposed excitation and decay routes are shown in Figure 4. The fit of the time-evolution of the delayed Kerr response yields a buildup time of 0.22 ps and a decay time of 2.0 ps for the intermediate state.

Figure 5 shows the optical Kerr gate signal as a function of the angle of heterodyning for the imaginary part of $\chi^{(3)}$. From the negative slope, it is clear that the sign of the $I_m\chi^{(3)}$ is positive which will correspond to a two-photon absorption [27].

III. C. Use of Higher Order Diffractions for Biexcitonic Decay

This technique discussed above has been used to investigate the excitation dynamics of many conjugated structures. The example presented here is for a conjugated molecule: a red dye called perylenetetracarboxylic dianhydride, which was vacuum deposited as a film [22]. This film was studied by degenerate four-wave mixing using 400 femtosecond pulses at

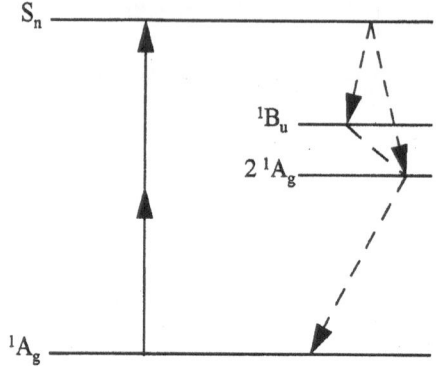

Figure 4. The proposed excitation and decay routes for canthaxanthin based on the homodyne Kerr gate data of Figure 3.

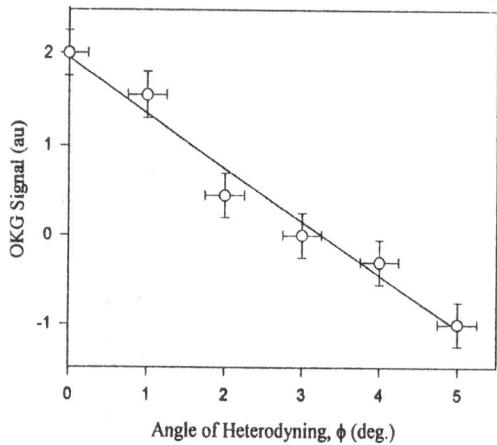

Figure 5. Optical Kerr gate signal as a function of the angle of heterodyning for the imaginary component of $\chi^{(3)}$. The system is canthaxanthin in tetrahydrofuran.

602 nm from a dye amplified Nd:Yag pumped system. The results of the time-resolved first-order and second-order diffractions are shown in Figure 6. A theoretical fit of the curve yields the value of $k_2 n_{ex}^{max}$ as changing from $2 \times 10^{10} sec^{-1}$ at low intensities to $1 \times 10^{12} sec^{-1}$ at high intensities.

III. D. Excitation Dynamics in Conjugated Polymers

As discussed above conjugated polymers exhibit a rich variety of excitations such as Frenkel excitons, polarons, and solitons [3]. They have been extensively studied using transient absorption, four wave mixing and Kerr-gate. Two examples are presented here. The first is the case of poly-p-phenylene vinylene, which exhibits Frenkel type tightly bound

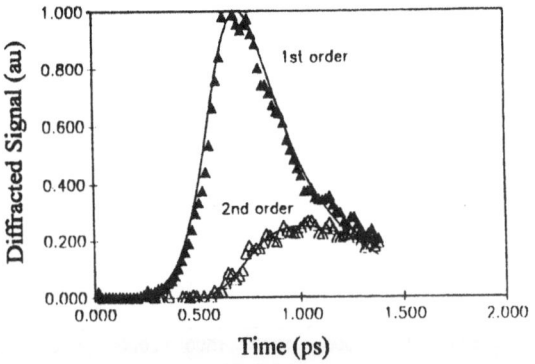

Figure 6. Time-resolved first- and second-order diffractions for the vacuum evaporated dye film of perylene tetracarboxylic dianhydride. The solid line represents the theoretical simulation curves.

excitons. It has been investigated in the form of unoriented polymer film and a stretch oriented film with both uniaxial and biaxial orientations as well as in the form of an optical composite with silica produced by sol-gel processing [23,36]. This composite exhibits enhanced optical quality. The degenerate four-wave mixing signal obtained by using 60 femtosecond pulses at 620 nm (an amplified colliding pulse mode locked system) is shown in Figure 7 for two different pump power levels. At high pump power levels a tail builds up

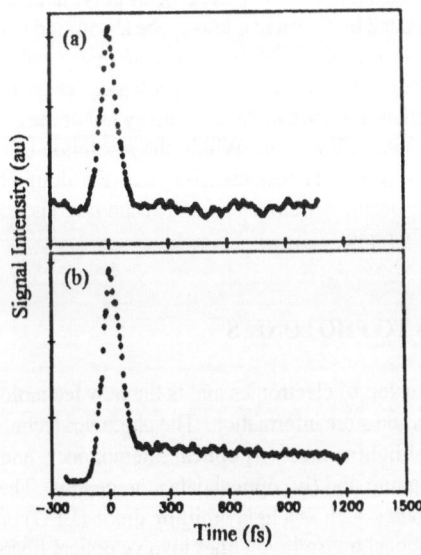

Figure 7. Time-resolved degenerate four-wave mixing signal for the poly-p-phenylene vinylene. Sol-gel glass composite film at (a) low power, (b) high power levels.

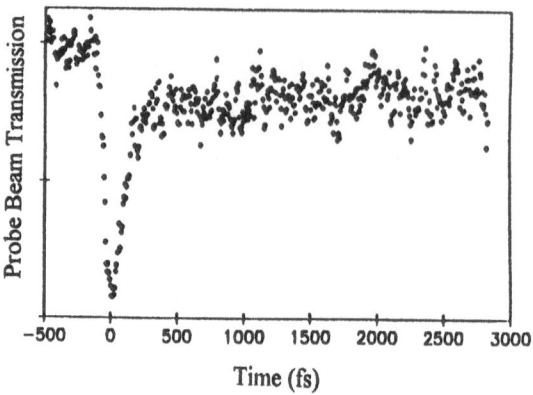

Figure 8. Time-resolved transient absorption signal for the poly-p-phenylene vinylene: sol-gel composite film.

indicating an incoherent delayed component. The power dependence of this delayed component indicates that it is due to a two-photon pumped excitation. Figure 8 shows the transient absorption results, *i.e.* time resolved induced absorption of a weak probe beam in the presence of a strong pump beam. The transient absorption exhibits a strong peak at zero delay time which corresponds to the direct two-photon absorption. In addition, it exhibits a weak long tail which arises from subsequent one-photon absorption of the probe beam by the excited species. In other words, it arises from the excited state absorption. This example nicely illustrates how two techniques can be used together to create a better understanding of excitation dynamics.

The second example is that of a conjugated polymer which involves coupled lattice deformation (conformational defect). Here the case of polythiophene is discussed which has been extensively investigated by transient pump-probe absorption [37] and by degenerate four-wave mixing [38]. Here results of the femtosecond time resolved degenerate four-wave mixing studies conducted in our laboratory are presented using 400 femtosecond pulses at 620 nm. This wavelength is absorbed in the low energy tail of the absorption band. The time-resolved response is shown in Figure 9. Within the resolution of the optical pulse, the rise time of the nonlinear response is instantaneous and the dominant decay leading to the formation of a polaron occurs within 200 femtoseconds. Therefore, the conformational deformation is extremely rapid.

IV. APPLICATIONS TO PHOTONICS

Photonics is the analog of electronics and is the new technology in which photons are used to transmit, process and store information. The photonics technology involves four major components: (i) optical light source, (ii) optical interconnects and transmission lines, (iii) optical signal processing unit and (iv) optical data storage unit. The optical light source can be an incoherent light source such as a light emitting diode (LED) or a coherent laser source. The interconnects and optical transmission lines involve optical fibers or channel waveguides. For optical signal processing, one utilizes the manifestations of nonlinear optical effects whereby an optical carrier signal is modified in phase or frequency by the application of an

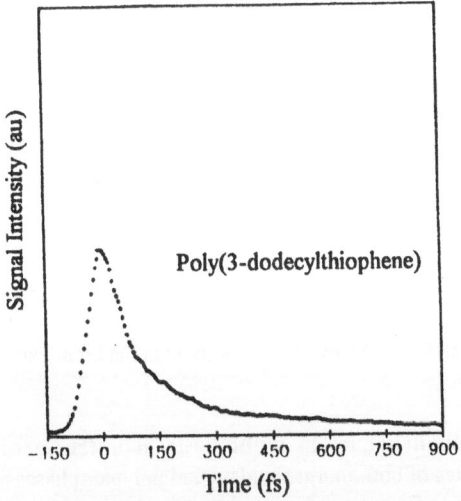

Figure 9. Time-resolved degenerate four-wave mixing signal for a polythiophene film.

electric field (electro-optic effect) or an optical field ($\chi^{(3)}$ effect). For optical data storage, particularly erasable memory, photorefractive media have emerged as an important class of materials in which the information can be stored three-dimensionally as photorefractive volume holograms.

In order to develop components both from materials engineering and device architecture points of view, a proper understanding of dynamics of optical excitations is very essential. For example, for both LED and laser optical sources, an understanding of excited state dynamics is very important. Compact and tunable solid state dye lasers are very attractive for applications to photonics. Again, an understanding of phonon-assisted relaxation is crucial for the development of highly efficient and temporally stable solid state lasers. Another exciting prospect for coherent light source is exemplified by up-conversion lasers which provide an opportunity to produce lasing in the blue region for more compact writing of information. In this regard, a significant development is that of a two-photon pumped laser.

For the fabrication of channel waveguides and optical circuitry, methods based on photobleaching have been proposed which again require studies of excited state physical and chemical processes. For optical signal processing, the ultimate time resolution is provided by a $\chi^{(3)}$ type nonlinear process which is an all optical effect (as opposed to $\chi^{(2)}$ electro-optic effect). For photorefractivity, optically excited charge-carriers provide the space-charge field which produces the required refractive index change by the electro-optic mechanism. Again, photoinduced charge carrier generation efficiency is intimately related to the excitation dynamics.

Organic and polymeric materials exhibit both the structural flexibility and the rich variety of excitation dynamics. Furthermore, because of the presence of highly deformable π-bonds, conjugated polymers exhibit enhanced nonlinear optical responses. For these reasons, organic and polymeric structures have emerged as an important class of photonic materials. They show exciting prospects in design and fabrication of all major four components discussed above. These prospects together with new opportunities for research are discussed in the next section.

V. NEW OPPORTUNITIES FOR RESEARCH

V. A. Excitation Dynamics

Some areas of research opportunities are listed below:

- **Excitation dynamics in polymers**. Conjugated polymers exhibit a rich variety of excitations such as self-trapped excitons, solitons, bipolarons, interaction dynamics between side-chain and main-chain excitations. There is a considerable opportunity for both theoretical and experimental research.

- **Exciton strings**. An exciting new topic is the bound states of multiple excitons [39].

- **Excitation dynamics in multiphasic nanostructured composites**. These composites of both inorganic glass and polymer phases are prepared by sol-gel processing. These two phases are separated only on nanometer scale. Therefore, energy transfer between the two phases can be controlled.

V. B. Solid State Dye Lasers, Polymer Lasers

The merits and issues to address and opportunities for research are listed below:

Merits: compact, suitable for integrated optics in the forms of channel waveguides and fibers; up-conversion lasing, multi wavelength lasing, chemical and bio-sensors.

Issues:

- Operational lifetime
- Quenching
- Efficiency

Opportunities:

- Energy transfer and relaxation
- Structure-property relation for two-photon excitation
- Waveguide processes in polymers
- Processing of optical fibers and channel waveguides

V. C. Polymer Light Emitting Diodes

Merits: Ease of fabrication, large area displays, relatively low cost of fabrication, ease of tunability of wavelength.

Issues:

- Low efficiency
- Short operational life-time

Needs:

- A better understanding of physics of relevant processes
- Better materials optimized both for electrical and optical properties

V. D. Nonlinear Optics of Polymers for Optical Signal Processing

Merits: ease of processing, large non-resonant nonlinearity, femtosecond all-optical ($\chi^{(3)}$) response, high-optical damage threshold

Current status:

- Electro-optic polymers for nearest term applications
- Non-resonant $\chi^{(3)}$ still too small for applications
- Harmonic generation: inorganic crystals currently have an edge

Opportunities:

- A better understanding of structure-property relation
- Beyond local field approximation
    - Role of new collective excitations
    - Role of exciton strings
    - Local field enhancement in composite materials
- Optical waveguide processes
- Cascaded higher-order nonlinearities

V. E. Photorefractive Polymers for Optical Data Storage

Merits: Ease of fabrication, tunability of spectral response, high efficiency, low cost

Issues:

- Low mobility of carriers which leads to strong electric field dependence
- Short storage time
- Materials with long time stability

Opportunities:

- Theoretical and experimental studies of space-charge field
- Kinetics of trapping and detrapping
- Modeling and design of multifunctional polymers
- Use of novel inorganic:organic composite materials

VI. ACKNOWLEDGEMENTS

The research work at the Photonics Research Laboratory has been supported by the Air Force Office of Scientific Research and the Polymer Branch of Air Force Wright Laboratory under contract number F4962093C0017. The author also wishes to thank Dr. J. D. Bhawalkar and Mr. G. Ruland of Photonics Research Laboratory for their help.

VII. REFERENCES

1. I.N. Levine, "Quantum Chemistry," Allyn and Bacon Inc., Boston (1974).
2. W.P. Su, J.R. Schrieffer, and A.J. Heeger, *Phys. Rev. Lett.* 42:1698 (1979).
3. T.A. Skotheim, "Handbook of Conducting Polymers," Vols. 1 and 2, Marcel Dekker, Inc., New York (1986).
4. H.W, Kroto, J.R. Heath, S.C. O'Brien, R.F. Curl and R. E. Smalley, *Nature* 318:162 (1985).
5. "Fullerenes and Photonics", Z.H. Kafafi, ed., SPIE Proceedings, Vol. 2284 (1994).
6. G.B. Talapatra, N. Manickam, M. Samoc, M.E. Orczyk, S.P. Karna, and P.N. Prasad, *J. Phys. Chem.* 96:5206 (1992).
7. G. Venkataraman and V.C. Sahni, *Rev. Mod. Phys.* 24:409 (1970).
8. P.N. Prasad and R. Kopelman, *J. Chem. Phys.* 58:126 (1973).
9. A.S. Davydov, "Theory of Molecular Excitons," translated by B. Dresner, Plenum, New York (1971).
10. R.M. Hochstrasser and P.N. Prasad *in:* "Excited States," Vol. 1, E.C. Lim, ed., Academic Press, New York (1974), p. 79.
11. P.N. Prasad and D.J. Williams, "Introduction to Nonlinear Optical Effects in Molecules and Polymers," Wiley Interscience, New York (1990).
12. N. Bloembergen, "Nonlinear Optics," Benjamin, New York (1965).
13. D.N. Rao, J. Swiatkiewicz, P. Chopra, S.K. Ghoshal, and P.N. Prasad, *Appl. Phys. Lett.* 48:1187 (1986).
14. B.P. Singh, P.N. Prasad, and F.E. Karasz, *Polymer* 29:1940 (1988).
15. P. Yeh, "Introduction to Photorefractive Nonlinear Optics," Wiley, New York (1993).
16. S. Ducharme, J.C. Scott, R.J. Tweeg, and W.E. Moorner, *Phys. Rev. Lett.* 66:1846 (1991).
17. Y. Zhang, Y. Cui, and P.N. Prasad, *Phys. Rev. B* 46:9900 (1992).
18. B. Kippelen, K. Tamusa, N. Peyghambarian, A.B. Padias, and H.K. Hall, Jr., *J. Appl. Phys.* 74:3617 (1993).
19. M.E. Orczyk, B. Swedek, J. Zieba, and P.N. Prasad, *J. Appl. Phys.* 76:4995 (1994).
20. Y. Cui, M.T. Zhao, G.S. He, and P.N. Prasad, *J. Phys. Chem.* 95:6842 (1991).
21. M.T. Zhao, Y. Cui, M. Samoc, P.N. Prasad, M.R. Unroe, and B.A. Reinhardt, *J. Chem. Phys.* 93:864 (1991).
22. M. Samoc and P.N. Prasad, *J. Chem. Phys.* 91:6643 (1989).
23. Y. Pang, M. Samoc, and P.N. Prasad, *J. Chem. Phys.* 94:5282 (1991).
24. M.E. Orczyk, J. Swiatkiewicz, G. Huang, and P.N. Prasad, *J. Phys. Chem.* 98:7307 (1994).
25. Y.R. Shen, "The Principles of Nonlinear Optics", Wiley, New York (1984).
26. P.P. Ho *in:* "Semiconductors Probed by Ultrafast Laser Spectroscopy", Vol. II, R.R. Alfano, ed., Academic Press, New York (1984), p. 410.
27. M.E. Orczyk, M. Samoc, J. Swiatkiewicz, and P.N. Prasad, *J. Phys. Chem.* 98:2524 (1993).
28. J.R. Linde, R.G.S. Pong, F.J. Bartoli, and Z.H. Kafafi, *Phys. Rev. B* 48: 9447 (1993).
29. R. Zamboni, M. Muccini, R. Daniels, C. Taliani, H. Mahn, W. Müller, and H.U. ter Meer *in:* "Fullerens and Photonics," SPIE Proceedings, 2284:121 (1994).
30. R.A. Cheville and N.J. Halas, *Phys. Rev. B* 45:4548 (1992).
31. S.D. Bronson, M.K. Kelly, U. Wenschuh, R. Buhleir, and J. Kuhl, *Phys. Rev. B* 46:7329 (1992)
32. F. Kajzar, C. Taliani, M. Muccini, R. Zamboni, S. Rossini, and R. Danieli, "Fullerenes and Photonics", SPIE Proceedings, Vol. 2284, (1994), p. 58.

33. P.N. Prasad, G. Huang, M.E. Orczyk, J. Swiatkiewicz, J. Zieba, M. Berrada, and S. Miyata, SPIE Proceedings, Vol. 2284, (1994), p. 228.

34. B.L. Justus, Z.H. Kafafi, and A.L. Huston, *Opt. Lett.* 18:1603 (1993).

35. R. Gvishi, G. Ruland, D. Kumar, U. Narang, and P.N. Prasad, *Chem. Mat.* (in Press).

36. J. Swiatkiewicz, P.N. Prasad, F.E. Karasz, M.A. Druy, and P. Glatkowski, *Appl. Phys. Lett.* 56:892 (1990).

37. U. Stamm, M. Taiji, M. Yoshizawa, K. Yoshin, and T. Kobayashi, *Mol. Cryst. Liq. Cryst.* 182A:147 (1990).

38. Y. Pang and P.N. Prasad, *J. Chem. Phys.* 93:2201 (1990).

39. M. Kuwata-Gonokami, N. Peyghambarian, K. Meissner, B. Fluegel, Y. Sato, K. Ema, R. Shimano, S. Mazumdar, F. Guo, T. Kokihiro, H. Ezaki, and E. Hanamura, *Nature* 367:47 (1994).

THE IR VIBRATIONAL PROPERTIES OF COMPOSITE SOLIDS AND PARTICLES: THE LYDDANE-SACHS-TELLER RELATION REVISITED

A. J. Sievers

Laboratory of Atomic and Solid State Physics, Materials Science Center
and the Center for Radiophysics and Space Research
Cornell University
Ithaca, New York 14853-2501
USA

ABSTRACT

Hot pressed zinc sulfide, dispersion hardened with diamond particles has turned out to be an ideal material in which to explore the optical complexity introduced by perhaps the simplest optical composite structure. Because the transverse vibrational mode of ZnS is IR active while that of diamond is IR inactive, the vibrational polarization properties of the resulting medium are extremely inhomogeneous at far infrared frequencies. This novel system has turned out to be extremely useful for identifying the important dynamical features underlying the optical properties of all transparent composites and complex dielectrics. One result of far ir measurements of the reststrahlen region of this system is the discovery of a generalized Lyddane-Sachs-Teller (LST) relation for solids and liquids. The characteristic dynamical frequencies are defined in terms of second moments of the relevant response functions, and very general causality arguments are used to obtain the generalized LST relation. The measured electronic excitation properties of silicon have been used to illustrate the generality of the resulting relation, which provides a useful connection between the long-wavelength dynamical behavior of nonmetallic condensed media and their static and high-frequency dielectric properties. When the characteristic vibrational frequencies of small disordered dielectric particles are described in terms of optical moments of the appropriate response functions, a generalized Fröhlich relation follows which connects the second moment of the small particle response function to the dc dielectric properties of the particle. This result is then used to obtain a new representation for the Clausius-Mossotti relation. Finally, the frequency dependent extinction cross section of an ellipsoidal particle of arbitrary size is examined from the same perspective of sum rules and optical moments. A number of general results can be found. It is demonstrated here that an extinction strength sum rule exists which directly relates the effective number of oscillators in the scattering particle to the integral over the extinction cross section spectrum, independent of particle shape. In addition, the characteristic frequency for the extinction cross section spectrum has the interesting property of being independent of particle size. It is also shown that the frequency expression characterizing the extinction properties of an ellipsoid of arbitrary size has the same form as the Fröhlich relation for small single crystal particles. The expression agrees with the generalized Fröhlich relation that relates the squared frequency characterizing the absorption behavior of an ellipsoid in the Rayleigh limit with the small particle dielectric constant. The end result is that there is a much stronger connection between the scattering and absorptive properties of a large disordered particle and the absorptive properties of a small single crystalline particle than previously recognized.

I. VIBRATIONAL POLARIZATION WAVES IN INHOMOGENEOUS MEDIA

I. A. Soft Mode Ferroelectric Crystals

The Lyddane-Sachs-Teller (LST) relation[1] connecting the static and dynamic properties of the vibrational response has played an important role in the early interpretation of displacive ferroelectricity. Fröhlich[2] and Cochran[3,4] first recognized that an IR active optic mode must show a temperature dependent frequency which drops to zero frequency at the paraelectric to ferroelectric transition temperature because the dc dielectric constant $\varepsilon(0)$ becomes extremely large at that temperature. This result can be seen most easily by examining the long wavelength dielectric response $\hat{\varepsilon}(\omega)$ of a cubic crystal with a single IR active vibrational mode, namely,

$$\frac{\hat{\varepsilon}(\omega)}{\varepsilon_\infty} = \frac{\omega_l^2 - \omega^2 - i\gamma\omega}{\omega_t^2 - \omega^2 - i\gamma\omega}, \tag{1}$$

where ω_l is the longitudinal optic mode frequency, ω_t the transverse frequency, γ is the damping constant and ε_∞ is the high frequency dielectric constant. Evaluating this relation at zero frequency gives the LST relation,

$$\varepsilon(0)/\varepsilon_\infty = \omega_l^2/\omega_t^2 . \tag{2}$$

This relation provides a simple connection between the long wavelength transverse and longitudinal optic mode frequencies, ω_t and ω_l, and the dc and optical dielectric constants, $\varepsilon(0)$ and ε_∞ in a diatomic insulating crystalline solid. [In the small damping limit where $\gamma \ll \omega_t$, ω_t and ω_l give the pole and the zero of $\hat{\varepsilon}(\omega)$.

Experimentally, the temperature dependence of the dc dielectric constant in the paraelectric phase of a ferroelectric crystal near T_C is observed to follow quite accurately a Curie - Weiss law,

$$\varepsilon(0,T) = \frac{C}{|T - T_c|} \tag{3}$$

Cochran showed that the combination of eqs. (2) and (3) dictate a specific functional form for the transverse optic mode frequency, namely, $\omega_t^2 = A(T - T_c)$ for temperatures above the transition temperature. The dynamical picture is that the TO mode becomes softer and softer as the transition temperature is approached until a vibrational instability occurs. Equation (2) has been generalized to cubic crystals with more than two atoms per unit cell[4] and also to include damping and anharmonicity.[5-7] In addition, Barker[6,7] has found that eq. (2) can be obtained from a causality argument if the response is approximated by a δ-function mode at ω_t . All of these extensions treat single crystals.

At the same time that progress was being made to better characterize the LST relation, more precise experimental far IR measurements on the polarization modes of some displacive ferroelectric crystals have tended not to agree with the LST prediction.[8-10] A number of experiments now indicate that near T_C, temperature-dependent changes in the dc dielectric constant are observed, although the soft mode frequency is found to be temperature independent over the same interval.[9,10] The classic ferroelectric BaTiO₃ is a case in point since the soft mode frequency remains unchanged $\omega_t \sim 60$ cm⁻¹ over a 100 K temperature interval near the cubic-to-tetragonal phase transition, even though the dc dielectric constant continues to increase as T_c is approached.[9] The conclusion is that near the transition temperature an additional relaxation mechanism is dominant and provides most of the temperature dependence of the dc dielectric constant, with the end result being similar to that found in hydrogen bonded ferroelectric crystals.[6] Such additional relaxation phenomena are not included in the LST soft mode described by eqs. (2) and (3). It now appears that the LST relation as given above does not provide the necessary connection between the static and dynamic properties of all displacive ferroelectrics. However, one may ask if a more general LST relation does exist which could provide an exact connection between the static and dynamic properties of these more complex systems. Such a connection could provide a new clue as to the important temperature-dependent dynamical feature to monitor near the phase transition temperature. Clearly, if relaxation dynamics is the controlling mechanism, it makes little sense to focus attention exclusively on the temperature dependence of TO mode frequency near T_c.

Different kinds of far IR experiments on composite systems have been used to demonstrate that such a generalized LST relation does exist,[11] and this development is reviewed next. The results point directly to a general derivation of the LST which makes use of the Kramers-Kronig relations so that all linear non-conducting dielectric systems are covered, including ferroelectrics, liquids and inhomogeneous media.[12,13] The key to the generalized LST relation is in the identification of the characteristic frequencies themselves. One way to identify the characteristic frequencies has been described in Ref. [6], where a number of modes are defined, each represented by a complex frequency. However, this method does not generate a universal LST relation. Another more general way to identify the characteristic frequencies of an arbitrary system is with a moment representation and it is shown here that appropriately defined moments do provide the correct framework for uncovering the generalized LST relation.

I. B. Far Infrared Experiments on a ZnS:diamond Composite

Far IR measurements on ZnS:diamond composites provided the first experimental demonstration that the original LST relation is too specific to describe disordered systems.[11] This composite represents perhaps the simplest optical composite that one can imagine. Since the ZnS has the zinc-blende structure there are two different atoms per unit cell thus one IR active TO mode producing a reststrahlen band centered near 300 cm^{-1}. The diamond component also has two atoms per unit cell but since they are identical there is no electric dipole moment in first order so at these same frequencies only the dc contribution from the electronic degrees of freedom contributes. Hence there is only one frequency dependent component in the ZnS:diamond composite in the far IR providing a simple and nearly ideal extension away from single crystals to a more complex disordered system which has the topology of Swiss cheese.

The samples were made from high purity ZnS powder with an average particle diameter of 0.5 mm and synthetic diamond powder with particle sizes ranging from 0.01 to 1 mm. To remove impurities, the diamond powder was cleaned with HF acid. To enhance dispersion of the two powders, they were mixed in several green state processes such as ultrasonic agitation, shear mixing, and impact milling at liquid-nitrogen temperature. These mixtures were then hot pressed in an atmosphere of He + 4% H$_2$ at a pressure of 200 MPa at around 1000 °C for 30 minutes. The final densities of the samples were at least 99% of the theoretical value. Careful control of the hot pressing parameters was required in order to: (a) limit the phase transformation to a non-cubic phase of ZnS at high temperatures, and (b) maintain the small grain size of the original powder.

The reflectivity measurements on the room temperature samples were made over the frequency range of 40-900 cm^{-1} using a fourier transform spectrometer. Since the near IR indices of refraction of these two materials are nearly the same, scattering losses are greatly reduced even at high fill fractions in the percolation regime. In order to avoid coherent effects from the reflections at both surfaces, one surface of each sample was polished and the other one was roughened. A Kramers-Kronig analysis of these far IR reflectivity data on samples with a volume fill fraction of diamond from 0% up to 55% produces the required transverse loss function, $\text{Im}[\hat{\varepsilon}(\omega)]$, and the longitudinal loss function, $\text{Im}[-1/\hat{\varepsilon}(\omega)]$.

These experimental results are shown in Fig. 1(a) and 1(b). In both cases, as the diamond concentration increases, the strengths of the narrow loss peaks, located near 275 and 350 cm^{-1}, respectively, decrease in height and broaden; however, inspection of the two pictures shows that the peak positions themselves, defined as ω_t and ω_l, remain relativelyunchanged in position. Each loss function broadens in an asymmetric way with the expanding wing extending toward the frequency peak of the other kind of loss function.

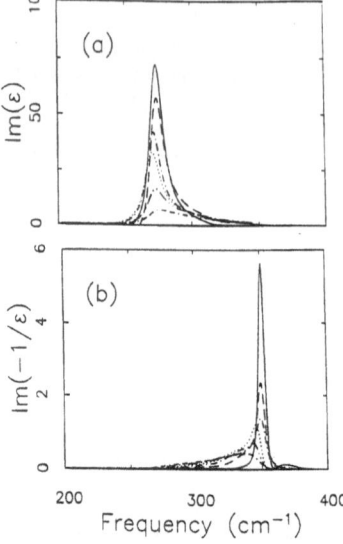

Figure 1. Frequency dependence of the imaginary part of the response functions obtained from a Kramers-Kronig analysis of the reflectivity data for ZnS:diamond composites. (a) for a transverse field. (b) for a longitudinal field. The volume fill fraction of diamond ranges from 0 to 55% in steps of 11% with the solid line $f = 0$ and the dash-dash-dot line $f = 55\%$. (After Ref. [11])

Figure 2(a) shows the measured values of ω_l^2/ω_t^2 (triangles and right hand ordinate) and $\varepsilon(0)/\varepsilon_\infty$ (squares and left hand ordinate) at various fill fractions. The values of $\varepsilon(0)/\varepsilon_\infty$, which are measured by fitting the general shape of measured reflectivity in the long wavelength limit with a single Lorentz oscillator model, displays a systematic decease as f increases, while the values of ω_l^2/ω_t^2 are nearly independent of f. Clearly, the LST relation connecting $\varepsilon(0)/\varepsilon_\infty$, and ω_l^2/ω_t^2 in this way is not valid for composite materials. Figure 2(a) also shows that even if one uses a standard effective medium theory (EMA) for composites,[14] it produces the same discrepancy between the calculated ω_l^2/ω_t^2 (dashed line) and the calculated $\varepsilon(0)/\varepsilon_\infty$ (solid line) for these ZnS:diamond composites. Since at higher fill fraction the spectral weight is greatly shifted in Fig. 1 perhaps some moment involving the response function is a more appropriate way to determine the characteristic frequency of the system rather than the peak value? By trial and error it has been determined that a specific second moment for the frequencies of interest produces much better agreement between the ac and dc sets of data. These experimental results are shown in Fig. 2(b). The moment that produces this good agreement has an interesting form. Let $\hat{\chi}_i(\omega)$ be the response function for the particular experimental probe (i) where $i = 1,t$ then

$$\left\langle \omega^2 \right\rangle_i = \int_0^\infty d(\ln \omega) \, \omega^2 \, \mathrm{Im}[\hat{\chi}_i(\omega)] \bigg/ \int_0^\infty d(\ln \omega) \, \mathrm{Im}[\hat{\chi}_i(\omega)]. \tag{4}$$

With the second moment integrals given by eq. (4), $<\omega^2>_l$ and $<\omega^2>_t$ are then calculated and the ratios $<\omega^2>_l/<\omega^2>_t$ are found (circles). Again the results are compared with the

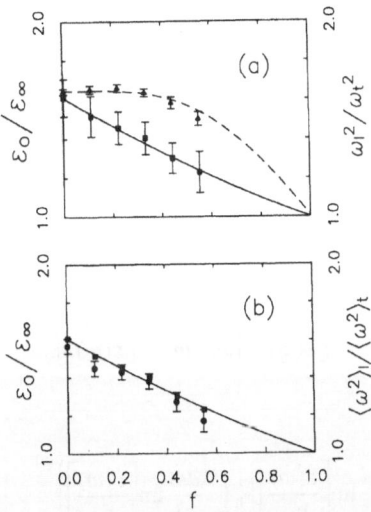

Figure 2. Fill fraction dependence of $\varepsilon_0/\varepsilon_\infty$ and the ratio of the squares of the characteristic loss function frequencies for ZnS:diamond composites. The f values are the same as in Fig. 1. (a) Comparison between measured $\varepsilon_0/\varepsilon_\infty$ (squares) and ω_l^2/ω_t^2 (triangles), where ω_l and ω_t are the respective frequencies at which the longitudinal and transverse loss functions are a maximum. The EMA predictions for $\varepsilon_0/\varepsilon_\infty$ (solid line) and ω_l^2/ω_t^2 (dashed line) are in good agreement with experiments. (b) Comparison between the measured $\varepsilon_0/\varepsilon_\infty$ (squares) and $<\omega^2>_l/<\omega^2>_t$ (circles), where $<\omega^2>_l$ and $<\omega^2>_t$ are the characteristic longitudinal and transverse squared frequencies based on the second momentum representation. The agreement between these quantities validates the generalized LST relation for inhomogeneous ZnS:diamond composites. The EMA prediction is shown as a solid line. (After Ref. [11])

measured values of $\varepsilon(0)/\varepsilon_\infty$ (squares) at various fill fractions in Fig. 2(b). The excellent agreement between $\varepsilon(0)/\varepsilon_\infty$ and $<\omega^2>_l/<\omega^2>_t$ determined in this fashion demonstrates that within the experimental uncertainties an LST-like relation holds for these ZnS:diamond composites. But is this connection an accident? To show that it is not we examine the eq. (4) in more detail.

I. C. The Generalized LST Relation for Bulk Media

Before one can make contact with the moment representation, it is important to define the appropriate macroscopic response functions of the general infinite system. For a bulk nonmagnetic linear isotropic dielectric, let the external susceptibility be defined as

$$\vec{P}_i = \hat{\chi}_i^{ext} \, \vec{E}_i^{ext} . \tag{5}$$

where i identifies transverse or longitudinal. Only two kinds of response functions can occur since, by Helmholtz's theorem, there are only two kinds of vector polarization fields, transverse (solenoidal) and longitudinal (irrotational), in the bulk system. It can be readily shown that in terms of the macroscopic dielectric function $\hat{\varepsilon}(\omega)$, the transverse response is[13]

$$\hat{\chi}_t^{ext} = [\hat{\varepsilon}(\omega) - 1]/4\pi , \tag{6}$$

and the longitudinal response is

$$\hat{\chi}_l^{ext}(\omega) = [1 - 1/\hat{\varepsilon}(\omega)]/4\pi. \tag{7}$$

It can be shown that the loss associated with the response in eqs. (6) or (7) is simply $\mathrm{Im}[\hat{\chi}_l^{ext}(\omega)]$ hence the loss functions for the transverse and longitudinal responses are $(4\pi)^{-1}\mathrm{Im}[\hat{\varepsilon}(\omega)/\varepsilon_\infty]$ and $(4\pi)^{-1}\mathrm{Im}[-\varepsilon_\infty/\hat{\varepsilon}(\omega)]$, respectively.

With the LST-like data described in Section I. B. in mind, the two characteristic frequencies are defined in terms of weighted second moments of the longitudinal and transverse loss by[13]

$$\left\langle \omega^2 \right\rangle_l \equiv \left\{ \int_0^\infty \frac{d\omega}{\omega} \omega^2 \mathrm{Im}\left[\frac{-\varepsilon_\infty}{\hat{\varepsilon}(\omega)}\right] \right\} \left\{ \int_0^\infty \frac{d\omega}{\omega} \mathrm{Im}\left[\frac{-\varepsilon_\infty}{\hat{\varepsilon}(\omega)}\right] \right\}^{-1} \tag{8}$$

and

$$\left\langle \omega^2 \right\rangle_t \equiv \left\{ \int_0^\infty \frac{d\omega}{\omega} \omega^2 \mathrm{Im}\left[\frac{\hat{\varepsilon}(\omega)}{\varepsilon_\infty}\right] \right\} \left\{ \int_0^\infty \frac{d\omega}{\omega} \mathrm{Im}\left[\frac{\hat{\varepsilon}(\omega)}{\varepsilon_\infty}\right] \right\}^{-1}. \tag{9}$$

To simplify the ratio of eq. (8) to eq. (9) we make use of the oscillator strength sum rule to define another characteristic frequency associated with these degrees of freedom, the plasma frequency. The oscillator strength sum rule can be given in terms of either the transverse or longitudinal probe as [15]

$$\int_0^\infty d\omega\omega \mathrm{Im}[\hat{\varepsilon}(\omega)] = \frac{\pi}{2}\omega_p^2 = \int_0^\infty d\omega\omega \mathrm{Im}\left[\frac{-1}{\hat{\varepsilon}(\omega)}\right]. \tag{10}$$

No matter which probe is used to examine the dynamics of this system, the spectral strength is the same. With some effort it can also be shown that [16]

$$\int_0^\infty d\omega\omega \mathrm{Im}\left[\frac{\hat{\varepsilon}(\omega)}{\varepsilon_\infty}\right] = \int_0^\infty d\omega\omega \mathrm{Im}\left[\frac{-\varepsilon_\infty}{\hat{\varepsilon}(\omega)}\right]. \tag{11}$$

This expression follows from the Kramers-Kronig relations for $\hat{\varepsilon}(\omega)/\varepsilon_\infty$ and $-\varepsilon_\infty/\hat{\varepsilon}(\omega)$ and the free particle response at high frequencies [$\hat{\varepsilon}(\omega)/\varepsilon_\infty \to 1 - \omega_p^2/\omega^2$]. Therefore, the numerator in eq. (8) above is equal to the numerator in eq. (9) and the ratio of eq. (8)/eq. (9) simplifies to

$$\frac{\left\langle \omega^2 \right\rangle_l}{\left\langle \omega^2 \right\rangle_t} = \frac{\int_0^\infty (d\omega/\omega)\mathrm{Im}[\hat{\varepsilon}(\omega)/\varepsilon_\infty]}{\int_0^\infty (d\omega/\omega)\mathrm{Im}[-\varepsilon_\infty/\hat{\varepsilon}(\omega)]}. \tag{12}$$

A Kramers-Kronig relation connecting the real and imaginary parts of the transverse response function is [17]

$$\mathrm{Re}\left[\frac{\hat{\varepsilon}(\omega)}{\varepsilon_\infty}\right] - 1 = \frac{2}{\pi}\int_0^\infty dx \frac{x\,\mathrm{Im}[\hat{\varepsilon}(x)/\varepsilon_\infty]}{x^2 - \omega^2}. \tag{13}$$

Evaluating this expression at $\omega = 0$ produces the well know dc susceptibility sum rule, namely,

$$\frac{\varepsilon(0)}{\varepsilon_\infty} - 1 = \frac{2}{\pi} \int_0^\infty \frac{dx}{x} \mathrm{Im} \left[\frac{\hat{\varepsilon}(x)}{\varepsilon_\infty} \right]. \tag{14}$$

In a similar manner one can evaluate at $\omega = 0$ the corresponding Kramers-Kronig relation for the longitudinal response and find

$$1 - \frac{\varepsilon_\infty}{\varepsilon(0)} = \frac{2}{\pi} \int_0^\infty \frac{dx}{x} \mathrm{Im} \left[\frac{-\varepsilon_\infty}{\hat{\varepsilon}(x)} \right]. \tag{15}$$

Since equations (14) and (15) are the numerator and denominator of eq. (12), combining these three relations gives the generalized LST relation:

$$\frac{\langle \omega^2 \rangle_l}{\langle \omega^2 \rangle_t} = \frac{\varepsilon(0)/\varepsilon_\infty - 1}{1 - \varepsilon_\infty / \varepsilon(0)} = \frac{\varepsilon(0)}{\varepsilon_\infty}. \tag{16}$$

This generalized LST relation provides an exact fundamental connection between the long-wavelength dynamical behavior of a nonmetallic condensed phase and its static and high frequency dielectric properties.

I. D. The LST Relation for Bulk Silicon

To demonstrate the power of this relation we need not restrict ourselves to the vibrational properties of solids, the electronic degrees of freedom will work just as well. The measured optical properties of silicon provide a good test. The published optical constants for single crystal Si are available up to 2000 eV.[18] By calculating the squared frequencies for the longitudinal (eq. 8) and transverse (eq. 9) loss as a function of the upper limit of the integrals (in each case replace ∞ by ϖ), the number of effective silicon electrons in each moment up to that frequency becomes apparent. The ratio of these two second moments plotted as a function of frequency ϖ then gives the LST relation and brings out the contribution of the electronic components in different frequency regions. Recall that there is no IR active lattice mode for Si so only electronic degrees of freedom will appear here.

The LST results for silicon are shown in Fig. 3. Starting from low frequencies, the ratio has the value 1 until the interband transition region is passed at about 10 eV where the LST ratio grows rapidly to reach a plateau. This constant plateau value identifies the LST relation for the valance band transitions. The ratio remains constant with increasing frequency until the silicon inner shell L-band transitions are excited at about 100 eV, at which point the ratio grows to a second plateau value not very different from the valance band transitions. The lowest frequency excitations produce the largest contributions to the LST relation. Equation (16) applies to those frequency regions where the ratio has reached constant values, i.e., frequency regions between spectral features. It is worth noting that this ratio of the two optical moments actually puts additional constraints on the measured optical constants.

How is it that such a general relation as given by eq. (16) has been overlooked until recently? A variety of optical moments of the imaginary part of the dielectric function have been introduced over the years to characterize the broad spectra of ionic and covalent bonded crystals and glasses. Much of the effort in classifying electronic spectra was motivated by the work of Phillips and Van Vechten in which an "average energy gap" was identified in each dielectric material.[19,20] Other optical moments have been used to characterize the same classes of materials empirically with a "dispersion energy" parameter.[21] A different moment expression has been created to relate the optical properties to the charge distribution in a unit cell [22] and yet another to estimate the dc dielectric constant.[17] Moments have been used to derive mode frequencies for model antiferromagnets.[23] Finally, moments have been introduced to characterize the spectrum of "harmonic" glasses.[24] With such a variety of possible moment expressions, one difficulty has been to decide which expressions are true in general. For example, the dispersion energy parameter and the charge

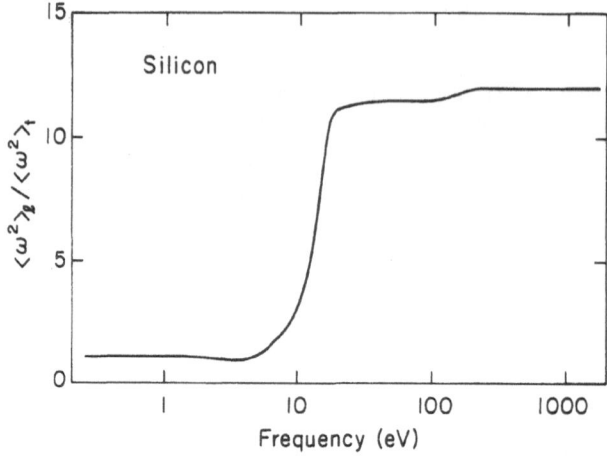

Figure 3. The generalized Lyddane-Sachs-Teller relation as a function of frequency for silicon. The measured optical constants of silicon are used to calculate the weighted second moment frequencies of the longitudinal and transverse loss as a function of frequency. By calculating the LST ratio as a function of the upper limit of the integrals in Eqs (8) and (9), the number of effective silicon electrons in the LST relation at any frequency becomes apparent. Starting from low frequencies, the two moments in eq. (16) have the same value until the interband transition appears at about 4 eV. In this frequency region the transverse second moment reaches a plateau while the longitudinal moment continues to grow until the longitudinal strength contribution from the valence electrons is exhausted. The result is the knee in the figure at about 12 eV. The next step in the LST relation to larger values at about 100 eV is produced by the inner L-state electronic transitions. (After Ref. [25])

distribution cases diverge, if one does not truncate the range of integration in some specific way, while for glasses, many of the proposed moments[24] diverge if the harmonic restriction is removed. Note that these last authors did produce an LST relation for the special case of a classical harmonic disordered solid. But it is important to recognize that the moments defined by their equation (64) with n > 0 are infinite for a general system with anharmonicity.

II. CHARACTERISTIC VIBRATIONAL FREQUENCIES OF A DISORDERED SOLID OF RESTRICTED SIZE

II. A. The Fröhlich Sphere Resonance Mode in a Cubic Diatomic Crystal

The resonance frequency of a simple diatomic cubic crystal in the form of a small but macroscopic sphere in vacuum can be obtained by noting that in the long wavelength (electrostatic) limit, the electric field inside the sample is related to the field outside by the factor $3/(\hat{\varepsilon}(\omega)+2)$. The denominator tends toward zero at the resonance condition $\hat{\varepsilon}(\omega_s)+2 \cong 0$, where a large E-field response inside the crystal is produced by a small applied external field. To identify this frequency we assume that the sphere resonance response can be represented by a single Lorentz oscillator and that the damping term can be ignored. The resulting expression at the resonance frequency is

$$\varepsilon(\omega_s) = 1 + \frac{\omega_p^2}{\left(\omega_t^2 - \omega_s^2\right)} = -2. \tag{17}$$

Solving equation (17) for the sphere resonance frequency gives

234

$$\frac{\omega_s^2}{\omega_t^2} = 1 + \frac{\omega_p^2}{3\omega_t^2}. \tag{18}$$

We can also evaluate the Lorentz oscillator response at zero frequency to obtain

$$\varepsilon(0) = 1 + \frac{\omega_p^2}{\omega_t^2}. \tag{19}$$

Adding +2 to both sides of eq. (19) and then dividing through both sides of the equation by $(\varepsilon_\infty + 2)$ gives the same quantity as found in eq. (18) so that

$$\frac{\varepsilon(0) + 2}{3} = 1 + \frac{\omega_p^2}{3\omega_t^2} = \frac{\omega_s^2}{\omega_t^2}. \tag{20}$$

This is the Fröhlich sphere resonance condition which connects the vibrational resonance of the sphere to the dc dielectric constant. In his 1949 book[2] Fröhlich showed that although both Coulomb and short range forces occur in the ionic crystal, the sphere resonance frequency ω_s only depends on the short range forces. [For a detailed derivation of this result see Born and Huang ([26])].

Equation (20) can be used to characterize the Clausius-Mossotti relation in a different way. For the bulk crystal there are two relations connecting the LO and TO modes, namely,

$$\omega_l^2 = \omega_t^2 + \omega_p^2 \text{ and } \omega_l^2 / \omega_t^2 = \varepsilon(0). \tag{21}$$

Combining these two relations gives

$$\varepsilon(0) - 1 = \omega_p^2 / \omega_t^2. \tag{22}$$

This expression can be rewritten in terms of the short range forces by inserting eq. (20) into eq. (22) to eliminate the TO frequency. The result is

$$\frac{\varepsilon(0) - 1}{\varepsilon(0) + 2} = \left(\frac{1}{3}\right)\frac{\omega_p^2}{\omega_s^2}. \tag{23}$$

Equation (23) clearly displays the ratio: (oscillator strength through ω_p^2 which depends on the number density of oscillators) / (short range forces through the second moment of the sphere resonance ω_s^2). The left side of eq. (23) is identical to the Clausius-Mossotti relation which relates the dc dielectric properties to the number density and polarizability of the ions involved.[27] Since ω_s^2 is independent of the number density while ω_p^2 is proportional to it, the ratio on the right hand side of eq. (23) expresses that same Clausius-Mossotti relation directly in terms of the spectroscopic properties of the sphere.

II. B. Generalized Fröhlich Relation for Small Disordered Particles

The generalized LST idea described in Section I can also be applied to particles small compared to the wavelength.[28] Consider the susceptibility of a linear nonmagnetic ellipsoidal dielectric particle in vacuum. Assume a uniform electric field is applied along one of the principal axes of the ellipsoid characterized by the depolarization factor N_i, where $N_1 +$

$N_2 + N_3 = 4\pi$. In the limit where the wavelength is much larger than the optical dimensions of the particle, the polarization density of the particle in the i-th principal direction is

$$\bar{P}_i = \hat{\chi}_i^{app} \, \bar{E}_i^{app}, \tag{24}$$

with

$$\hat{\chi}_i^{app} = \frac{1}{N_i} \frac{\hat{\eta}_i(\omega) - \eta_{i\infty}}{\hat{\eta}_i(\omega)}. \tag{25}$$

Here

$$\hat{\eta}_i(\omega) = \hat{\varepsilon}(\omega) + \left(\frac{4\pi}{N_i} - 1\right), \tag{26}$$

where $\hat{\varepsilon}(\omega)$ is the bulk complex dielectric function of the particle medium. According to eq. (4) the second moment expression for the ellipsoid is

$$\left\langle \omega^2 \right\rangle_{Ni} = \left\{ \int_0^\infty d(\ln x) \, x^2 \, \mathrm{Im}\left[\frac{-\eta_{i\infty}}{\hat{\eta}_i(x)}\right] \right\} \left\{ \int_0^\infty d(\ln x) \, \mathrm{Im}\left[\frac{-\eta_{i\infty}}{\hat{\eta}_i(x)}\right] \right\}^{-1}. \tag{27}$$

By taking the imaginary part of eq. (26), we find that $\mathrm{Im}[\hat{\eta}_i(\omega)] = \mathrm{Im}[\hat{\varepsilon}(\omega)]$ and the transverse second moment squared frequency of the ellipsoid, $<\omega^2>_{ti}$, is the same as for a new bulk medium described by $\hat{\eta}_i(\omega)$. In addition, $<\omega^2>_{Ni}$ is formally identical to the longitudinal second moment $<\omega^2>_l$ described above by eq. (8) for a bulk medium. The analog to eq. (11) is

$$\int_0^\infty d\omega \, \omega \, \mathrm{Im}\left[\frac{-\eta_{i\infty}}{\hat{\eta}_i(\omega)}\right] = \int_0^\infty d\omega \, \omega \, \mathrm{Im}\left[\frac{\hat{\eta}_i(\omega)}{\eta_{i\infty}}\right]. \tag{28}$$

Similar arguments apply here. Since the function $\hat{\eta}_i(\omega)/\eta_{i\infty}$ has the same form at high frequencies with a shifted ω_p, it satisfies the KK relations if $\hat{\varepsilon}(\omega)$ does and its poles are still in the lower half complex plane. Applying the equivalent expressions of eqs. (8) and (9) to $\hat{\eta}_i(\omega)$ immediately gives the analog of eq. (16) for the ellipsoid:

$$\frac{\left\langle \omega^2 \right\rangle_{Ni}}{\left\langle \omega^2 \right\rangle_t} = \frac{\eta_i(0)}{\eta_{i\infty}}. \tag{29}$$

This is the generalized Fröhlich relation. It necessarily has the same formal structure as the bulk LST relation but it describes the characteristic frequencies of ellipsoidal particles. On the right hand side of eq. (29) is the same ratio of the low frequency dielectric constant of the ellipsoid to the high frequency one as originally derived by Fröhlich for diatomic lattices. The relation for a spherical particle with $\varepsilon_\infty = 1$ is presented in eq. (20) above. On the left hand side, the expression is completely different since second moment frequencies replace the Fröhlich resonance frequencies. The Fröhlich result can be recovered if one assumes that the poles of the ellipsoid and transverse response functions are represented by delta functions at ω_{Ni} and ω_t. An interesting property of eq. (29) is that the particular ellipsoid under consideration could itself be made of a composite and the transitions of interest could be

236

valence band electronic ones so that, in general, the long wavelength $\hat{\varepsilon}(\omega)$ could be extremely intricate as a function of frequency, but nevertheless, the generalized Fröhlich relation given by eq. (29) still holds.

II. C. Connection Between the Optical Moments of an Ellipsoid and its Susceptibility

Lets consider the application of optical moments to macroscopic polarizability relations previously discussed for ordered systems in Section II. A. Again we focus our attention on an ellipsoidal particle but now look at the difference between the two optical moments $<\omega^2>_{N_i}$ and $<\omega^2>_{t_i}=<\omega^2>_t$ defined by eqs. (27) and (9). Moreover, since $\text{Im}[\hat{\eta}_i(\omega)] = \text{Im}[\hat{\varepsilon}(\omega)]$ (see eq. (26), we have

$$\int_0^\infty d\omega\omega\,\text{Im}\big[\hat{\eta}_i(\omega)\big] = \int_0^\infty d\omega\omega\,\text{Im}\big[\hat{\varepsilon}(\omega)\big] \equiv \frac{\pi}{2}\omega_p^2. \tag{30}$$

This squared frequency for the small particle provides a measure of the oscillator strength of the absorption band associated with the mode under consideration. With the help of eqs. (11), (28), (30) and the KK relations satisfied by $\hat{\eta}_i(\omega)/\eta_{i\infty}$ and $-\eta_{i\infty}/\hat{\eta}_i(\omega)$, we find that the difference between the two moments reduces to

$$\big\langle\omega^2\big\rangle_{N_i} - \big\langle\omega^2\big\rangle_t = \omega_p^2\big/\eta_{i\infty}. \tag{31}$$

Dividing eq. (31) through by $<\omega^2>_t$ and applying eq. (29) gives our first relation connecting the low and high frequency dielectric constants to the optical moments and electric susceptibility, namely,

$$\eta_i(0) - \eta_{i\infty} = \omega_p^2\big/\big\langle\omega^2\big\rangle_t, \tag{32}$$

which corresponds to eq. (22) previously obtained for the cubic diatomic lattice spherical particle. In addition, the dc limit of one of the K-K relations gives

$$\eta_i(0) - \eta_{i\infty} = \omega_p^2\big/\big\langle\omega^2\big\rangle_t = \frac{2}{\pi}\int_0^\infty \frac{d\omega}{\omega}\,\text{Im}\big[\hat{\eta}_i(\omega)\big]. \tag{33}$$

Note that even though we started this derivation for a small ellipsoidal particle in the long wavelength limit eq. (33) is independent of particle shape since both $\eta_i(0) - \eta_{i\infty}$ and $\text{Im}[\hat{\eta}_i(\omega)]$ are. This equality can be put into a standard and recognizable form with simple substitutions giving

$$\varepsilon(0) - 1 = \frac{\omega_p^2}{\big\langle\omega^2\big\rangle_t} = \frac{2}{\pi}\int_0^\infty \frac{d\omega}{\omega}\,\text{Im}\big[\hat{\varepsilon}(\omega)\big]. \tag{34}$$

The integral is recognized as a transverse susceptibility (or polarizability) sum rule. The optical moment ratio, $\omega_p^2\big/\big\langle\omega^2\big\rangle_t$, presents the effective susceptibility in a somewhat different light in terms of moments of the spectral properties. In addition, eq. (34) is recognized as a general statement of the *transverse Szigeti relation*[29]:

$$(\varepsilon_0 - 1) = 4\pi\alpha_t/V, \tag{35}$$

where α_t is the effective polarizability which he derived for short transverse waves. Our result states that in the long wavelength limit as long as only the *transverse response* of our ellipsoid is measured then the shape of the object does not matter.

In a similar manner one can ask about the difference between another combination of optical moments $< \omega^2 >_l - < \omega^2 >_{N_i}$ for the small particle. After dividing through by $< \omega^2 >_l$ and some manipulation, we find that

$$\frac{\varepsilon(0) - 1}{\varepsilon_0} = \frac{\omega_p^2}{\langle \omega^2 \rangle_l} = \frac{2}{\pi} \int_0^\infty \frac{d\omega}{\omega} \text{Im} \left[\frac{-1}{\hat{\varepsilon}(\omega)} \right], \tag{36}$$

where eq. (11) has been used. Again this relation is independent of the particle shape but here the integral represents a susceptibility sum rule for a longitudinal probe. This expression makes contact with another Szigeti relation[29]:, namely,

$$(\varepsilon_0 - 1)/\varepsilon_0 = 4\pi\alpha_l/V, \tag{37}$$

where α_l is the effective polarizability for longitudinal waves. He derived this for a wavelength much smaller than the size of the macroscopic object.

Finally, it is instructive to rewrite eq. (31) directly in terms of the optical moment of the ellipsoidal response by dividing both sides of eq. (31) by $< \omega^2 >_{N_i}$ which gives[30]

$$\frac{\eta_i(0) - \eta_{i\infty}}{\eta_i(0)} = \frac{\omega_p^2}{\eta_{i\infty} \langle \omega^2 \rangle_{N_i}} = \frac{2}{\pi} \int_0^\infty \frac{d\omega}{\omega} \text{Im} \left[\frac{-\eta_{i\infty}}{\hat{\eta}(\omega)} \right], \tag{38}$$

For this optical moment the shape of the object enters directly into the results. To appreciate the significance of eq. (38) it is helpful to specialize immediately to the case of a sphere, $N_i = 4\pi/3$, then

$$\frac{\varepsilon(0) - 1}{\varepsilon(0) + 2} = \frac{\omega_p^2}{3 \langle \omega^2 \rangle_s} = \frac{2}{\pi} \int_0^\infty \frac{d\omega}{\omega} \text{Im} \left[\frac{-3}{\hat{\varepsilon}(\omega) + 2} \right]. \tag{39}$$

Equation (39) has the familiar form of the Clausius-Mossotti relation previously described in eq. (23) for a spherical particle made up from a cubic diatomic lattice. Because of the structure shown here for the ratio of the moments, it would appear that even for a general system the squared frequency in the numerator would be proportional to the number density while the squared frequency in the denominator would depend only on short range forces when calculated this way.

The prescription for finding the effective susceptibility of a sphere, $4\pi\alpha_s/3V$, from the long wavelength optical absorption spectrum of a disordered but macroscopic system is first to calculate the appropriate integrals involving $\eta_i(\omega)$ to obtain $(\omega_p^2)/3$ [eq. (30)] and $<\omega^2>_s$, the characteristic frequency squared of an isolated particle, eq. (27). Next take the ratio. This ratio is equal to the effective susceptibility as can be seen from the form of the integral in eq. (39) but this spherical susceptibility is evaluated differently from those described by eqs. (34) and (36), where the susceptibility integral is identical with that expected for either a transverse (solenoidal) or a longitudinal (irrotational) response of a bulk medium. In this case, although eq. (39) has the form of the analogous bulk longitudinal response, eq. (36), the geometry of the object enters directly.

III. EXTINCTION SUM RULES AND THE OPTICAL MOMENTS OF LARGE PARTICLES

III. A. The Extinction Cross Section of a Particle

We have seen above that the squared frequencies in a specific second moment formalism provide a rigorous and useful way to identify the important characteristic frequencies of the electromagnetic response of a complex composite medium in the limit where the component particles sizes are much smaller than the wavelength of the radiation. Our description of a particle in this Rayleigh limit has shown the connection between the electrodynamic and static dielectric properties of a glassy particle of small but macroscopic size and the characteristic second moment frequency which replaces the squared "sphere resonance" frequency associated with the one-mode single-crystal response. The last section focused attention on particles small compared to the wavelength. Here we examine large particles following Purcell's lead.[31] By application of a K-K dispersion relation, he first determined a dc polarizability sum rule for spheroidal dust grains of arbitrary size with respect to the wavelength. We shall show how the response function approaches described in the previous section can be extended to give a polarizability sum rule, an extinction sum rule and a characteristic extinction frequency for a scattering ellipsoidal particle of arbitrary size. The resulting second moment frequency is independent of both particle crystallinity and size and depends only on the oscillator strength sum rule for the bulk dielectric function and the shape of the scattering particle.

Let us begin with the small particle regime. In this limit where the wavelength is much larger than the particle size the relation between the dielectric function and the scattering function $\hat{S}(0°, \omega)$ in the forward direction is[32]

$$\hat{\chi}_i^{app}(\omega) = \frac{1}{N_i}\left(1 - \frac{\eta_{i\infty}}{\hat{\eta}_i(\omega)}\right) = \left(\frac{c}{\omega}\right)^3 \frac{i\,\hat{S}(0°, \omega)}{V}. \tag{40}$$

Here V is the volume of the ellipsoidal particle and c the velocity of light. We have already shown that when the left hand side of this equation is inserted in eq. (4) that a characteristic squared frequency can be obtained. The same result must appear if the right hand side of the equation is used, i.e., there are K-K relations for the scattering function in the forward direction since it too is a causal response function of the particle.

Now consider the scattering function for a particle of arbitrary size. In general for particles comparable to the wavelength the scattering function describes a complex interplay between scattering and absorption. However, the optical theorem[33] provides a fundamental connection between extinction cross section of the particle which includes both scattering and absorption and the scattering matrix in the forward direction. Since both scattering and absorption are now important, it is helpful in determining the optical moments to relate the complex scattering function $\hat{S}(0°, \omega)$ and the complex extinction function $\hat{C}_{ext}(\omega)$ to a generalized causal response function $\hat{\Gamma}_i(\omega)$ for the ellipsoid which one can think of as an auxiliary function. This last function is only introduced here to make contact with the results already presented previous in the small particle section. After going through the analysis once it will become quite apparent how the same results can be obtained without introducing such an auxiliary function. The connection between the three quantities is

$$\left(\frac{c}{\omega}\right)^3 \frac{i\,\hat{S}(0°, \omega)}{V} \equiv \left(\frac{c}{\omega}\right)\frac{i\,\hat{C}_{ext}(\omega)}{4\pi V} \equiv \frac{1}{N_i}\left(1 - \frac{\Gamma_{i\infty}}{\hat{\Gamma}_i(\omega)}\right), \tag{41}$$

where the last function on the right is written in such a way that formally it resembles eq. (40).

III. B. Extinction Cross Section Sum Rules

By applying one of the K-K relations to the causal response function on the right in eq. (41) and then substituting in the complex extinction cross section function, one finds a K-

K relation involving the frequency-dependent extinction cross section, $\mathrm{Re}\{\hat{C}_{ext}(\omega)\}$, namely,

$$\frac{4\pi V}{N_i c}\mathrm{Re}\left\{\frac{\hat{\Gamma}_i(\omega)-\Gamma_{i\infty}}{\hat{\Gamma}_i(\omega)}\right\} = \frac{2}{\pi}P\int_0^\infty dx\, \frac{\mathrm{Re}\{\hat{C}_{ext}(x)\}}{x^2-\omega^2}. \tag{42}$$

We can use this relation to obtain Purcell's dc polarizability sum rule by evaluating eq. (42) in the low frequency limit where the particle is necessarily smaller than the wavelength. The real part of the response function on the left side of eq. (42) becomes equal to the small particle result in this limit so

$$\lim_{\omega\to 0}\mathrm{Re}\left\{\frac{\hat{\Gamma}_i(\omega)-\Gamma_{i\infty}}{\hat{\Gamma}_i(\omega)}\right\} = \frac{\eta_i(0)-\eta_{i\infty}}{\eta_i(0)}, \tag{43}$$

since now eq. (40) does apply. Evaluating the integral at zero frequency gives the dc sum rule which was first obtained by Purcell[31], namely,

$$\frac{4\pi V}{N_i c}\left\{\frac{\eta_i(0)-\eta_{i\infty}}{\eta_i(0)}\right\} = \frac{2}{\pi}\int_0^\infty dx\, \frac{\mathrm{Re}\{\hat{C}_{ext}(x)\}}{x^2}. \tag{44}$$

Next we examine eq. (42) in the high frequency limit. At large enough frequencies the ellipsoid becomes completely transparent since its bulk dielectric constant has the asymptotic property that $\mathrm{Re}\{\hat{\varepsilon}(\omega)\}\to 1$ and $\mathrm{Im}\{\hat{\varepsilon}(\omega)\}\to 0$ as $\omega\to\infty$. As this transparent limit is approached beyond the K-shell electronic transitions, for example, the ellipsoid can be characterized by the Rayleigh-Gans scattering theory[32,34,35] since both of the necessary conditions, $|\hat{\varepsilon}(\omega)-1|\ll 1$ and $\dfrac{\omega V^{1/3}|\hat{\varepsilon}(\omega)-1|}{c}\ll 1$, are satisfied. According to this theory in the high frequency transparent limit, the real part of the auxiliary response function on the left side of eq. (42) again simplifies to eq. (40), i.e.,

$$\lim_{\omega\to\infty}\mathrm{Re}\left\{\frac{\hat{\Gamma}_i(\omega)-\Gamma_{i\infty}}{\hat{\Gamma}_i(\omega)}\right\} = \mathrm{Re}\left\{\frac{\hat{\eta}_i(\omega)-\eta_{i\infty}}{\hat{\eta}_i(\omega)}\right\} \approx \frac{\mathrm{Re}[\hat{\varepsilon}(\omega)]-1}{\eta_{i\infty}}. \tag{45}$$

For frequencies sufficiently large compared to any characteristic absorption frequency of the particle, $\mathrm{Re}\{\hat{C}_{ext}(\omega)\}$ vanishes, and the high frequency limit of the K-K relation expressed by eq. (42) reduces to

$$\frac{\mathrm{Re}\{\hat{\varepsilon}(\omega)\}-1}{\eta_{i\infty}} = -\frac{\omega_p^2}{\eta_{i\infty}\omega^2} = -\frac{\frac{2}{\pi}\int_0^\infty dx\,\mathrm{Re}\{\hat{C}_{ext}(x)\}}{\omega^2}. \tag{46}$$

The corresponding K-K relation for the bulk dielectric function, $\hat{\varepsilon}(\omega)$, is used to produce the left equality in eq. (46) with the plasma frequency defined by eq. (10). The resulting extinction cross section sum rule, which is valid for any size particle, is[36]

$$\frac{4\pi V}{N_i c}\frac{\omega_p^2}{\eta_{i\infty}} = \frac{V\omega_p^2}{c} = \frac{2}{\pi}\int_0^\infty dx\,\mathrm{Re}\{\hat{C}_{ext}(x)\}, \tag{47}$$

independent of the ellipsoid depolarization factor, since in free space $4\pi/N_i = \eta_{i\infty}$.

III. C. Characteristic Extinction Frequency

For an optical field applied along the ith axis of the ellipsoid, the characteristic squared frequency for the generalized response function described by eq. (41) is, in analogy with eq. (27), defined as

$$\left\langle \omega^2 \right\rangle_{ext} = \left\{ \left[\int_0^\infty d(\ln \omega) \, \omega^2 \, \text{Im}\left[\frac{-\Gamma_{i\infty}}{\hat{\Gamma}_i(\omega)} \right] \right] \left[\int_0^\infty d(\ln \omega) \, \text{Im}\left[\frac{-\Gamma_{i\infty}}{\hat{\Gamma}_i(\omega)} \right] \right]^{-1} \right\}. \tag{48}$$

When this expression is rewritten in terms of the extinction cross section, the numerator reduces to the oscillator strength sum rule, eq. (47), while the denominator becomes the polarizability sum rule, eq. (44), hence the ratio of these two equations gives the characteristic squared frequency $\left\langle \omega^2 \right\rangle_{ext}$ for the extinction cross section of this ellipsoid. Putting in these quantities gives

$$\left\langle \omega^2 \right\rangle_{ext} = \left(\frac{\eta_i(0)}{\eta_{i\infty}} \right) \left[\frac{\omega_p^2}{\varepsilon(0)-1} \right]. \qquad \text{(any size particle)} \tag{49}$$

There are two limits to consider, either the ellipsoid is conducting or insulating. For a conducting ellipsoid we should point out that although in the limit of high conductivity magnetic dipole scattering contributes at low frequencies and can even exceed the contribution from electric dipole scattering in some frequency range, electric dipole scattering is always dominant at low enough frequencies. Therefore, eq. (40) is also valid for a conducting ellipsoid in the dc limit, and since a conducting ellipsoid has the property that both $\eta_i(0)$ and $\varepsilon(0) \rightarrow -\infty$, eq. (49) reduces to

$$\left\langle \omega^2 \right\rangle_{ext} = \frac{\omega_p^2}{\eta_{i\infty}}. \qquad \text{(any size conducting particle)} \tag{50}$$

For dielectric particles it is instructive to proceed somewhat differently. Applying the same second moment procedures to the bulk dielectric function $\hat{\varepsilon}(\omega)$ to obtain $\left\langle \omega^2 \right\rangle_t$, the transverse second moment gives

$$\left[\frac{\omega_p^2}{\varepsilon(0)-1} \right] = \left\langle \omega^2 \right\rangle_t. \tag{51}$$

This relation, which is independent of the particle shape, identifies a squared frequency characteristic of bulk absorption when the medium is examined with a transverse probe. Combining eqs. (49) and (51) gives

$$\frac{\left\langle \omega^2 \right\rangle_{ext}}{\left\langle \omega^2 \right\rangle_t} = \frac{\eta_i(0)}{\eta_{i\infty}}. \qquad \text{(any size insulating particle)} \tag{52}$$

This frequency expression characterizing the extinction properties of an ellipsoid of arbitrary size has the same form as the Fröhlich relation for small single crystal particles, see eq. (20). It also agrees with the generalized Fröhlich relation that relates the squared frequency characterizing the absorption behavior of an ellipsoid in the Rayleigh limit with the small particle dielectric constant, see eq. (29). Hence both squared frequency ratios, eq. (52) above for large scattering objects, and eq. (29) for small nonscattering but absorbing objects obey the same generalized Fröhlich relation. Clearly the extinction and absorption spectra in

241

the large and small dielectric particle cases are very different, since the former includes absorption and scattering while the latter only includes absorption, yet remarkably $\eta_i(0)/\eta_{i\infty}$ appears in both. As long as the ellipsoids have the same shape and orientation then independent of volume the characteristic frequencies are the same. Apparently for the large particle extinction case there is as much scattering into the forward direction as out of it so that the scattering contribution must average out when it is counted in the second moment representation given by eq. (48).

Although the sum rule and second moment expressions above have been derived for a homogeneous ellipsoid, they can be readily generalized to inhomogeneous particles such as coated spheres and composite particles since the arguments are based on causality regardless of the scattering details. In the inhomogeneous case, the left hand side of the dc sum rule [eq. (44)] is replaced by $4\pi\alpha(0)/c$ where $\alpha(0)$ is the projection along the direction of the incident field of the dc polarizability tensor for the composite particle. Since it can be calculated from electrostatic theory, and the strength sum rule has a particularly simple form with the left hand numerator of eq. (47) replaced by $V\Omega_p^2 = \int d\vec{r}\,\omega_p^2(\vec{r})$ where the integral is over the volume of the particle. The characteristic squared frequency for the extinction cross section of the composite particle becomes

$$\left\langle \omega^2 \right\rangle_{ext} = \frac{V\Omega_p^2}{4\pi\,\alpha(0)}. \tag{53}$$

which is independent of the particle volume. When the characteristic squared frequency for the extinction cross section of a particle is divided by the characteristic squared frequency for the bulk longitudinal response function, the final expression still depends on the particle shape but is independent of the plasma frequency,

$$\frac{\left\langle \omega^2 \right\rangle_{ext}}{\left\langle \omega^2 \right\rangle_{\ell}} = \frac{V_m}{4\pi\alpha(0)} \frac{[\varepsilon(0)-1]}{\varepsilon(0)}. \tag{54}$$

The resulting number, which is a feature of our arbitrarily shaped particle of arbitrary size, only depends on the dc properties of the particle. If the bulk dc dielectric constant is sufficiently large then this number only depends on the mean polarization density of the particle.

III. D. Computational Tests

Since the extinction cross section of a large particle involves a subtle interference of scattering as well as absorption processes, it is perhaps surprising that two simple sum rules represented by Eqs (44) and (47) have been found which characterize the particle properties independent of the scattering contribution. Since all of these conclusions were obtained from analytical studies of homogeneous particles, there is the question of whether or not these expressions can be applied to inhomogeneous and composite particles. For example, suppose a strongly absorbing shell contains a hole at the center. If the radiation cannot reach the center to probe the hole how can the sum rules be correct? To provide additional insight numerical tests have been made on large inhomogeneous particles which have both spectral and geometrical inhomogeneities. These Mie computations of the extinction cross section spectra, sum rules and optical moment for large spheres and for a spherical shell of varying wall thickness demonstrate that the sum rule results are independent of such inhomogeneities.

To test the extinction strength sum rule [eq. (47)] and the characteristic squared frequency $\left\langle \omega^2 \right\rangle_{ext}$ [eq. (48)] for relatively large inhomogeneous objects, we compute the frequency dependent extinction cross section for a homogeneous sphere with spectral inhomogeneity by introducing a strong resonance into the response and for a hollow spherical shell with the same spectral inhomogeneity but different wall thicknesses. Mie theory provides a precise numerical solutions for these two different cases. As the bulk dielectric function is the required input, our test medium is characterized by a single Lorentzian oscillator, namely,

$$\hat{\varepsilon}(\omega) = 1 + \frac{\omega_p^2}{\omega_0^2 - \omega^2 - i\gamma\omega}, \tag{55}$$

with $\omega_p = 1.3 \times 10^4$ cm^{-1}, $\omega_0 = 1 \times 10^4$ cm^{-1}, $\gamma = 1 \times 10^3$ cm^{-1} and $\varepsilon(0) = 2.69$. The E-field skin depth $\delta_r = 0.0564$ μm at the resonance frequency ω_0.

It has already been demonstrated that second moment expressions are Kramers-Kronig exact independent of the frequency dependent properties of the bulk dielectric function so there is no loss in generality of testing the extinction strength sum rule with this simple Lorentzian response function.

The resulting volume-normalized extinction cross section $\mathrm{Re}\left\{\hat{C}_{ext}(\omega)/V_m\right\}$ is plotted in Figure 4(a) for a homogeneous sphere with radii ranging from 0.01 μm to 2 μm. Both the size parameter $x_r = \omega_0 a/c$, with a the particle radius, and the relative skin depth $\delta_r/a = 1/x_r\kappa$, where κ is the imaginary part of the bulk refractive index, can be used to characterize the nature of the extinction spectrum. The size parameter at resonance x_r for the homogeneous spheres in Figure 4(a) goes from 0.0628 for the smallest particle (0.01 μm) to a fairly large value of 12.57 for the largest particle (2 μm). Since the 0.01 μm particle has a large value for $\delta_r/a = 5.64$ and $x_r \ll 1$ at resonance, this case essentially represents the small

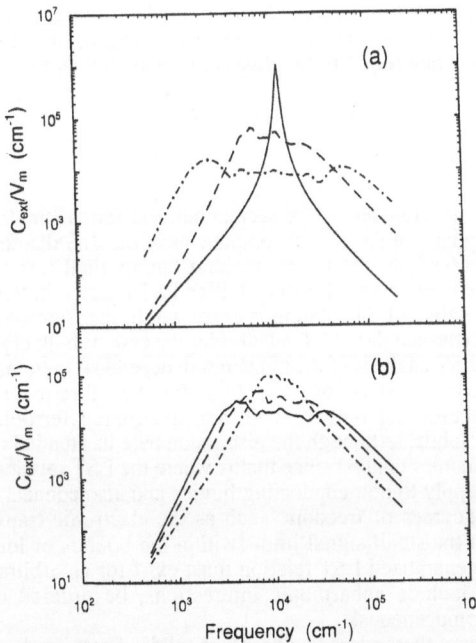

Figure 4. Mie computation of extinction spectra for particles with different kinds of inhomogeneities. (a) Calculated extinction spectrum for a homogeneous but absorbing dielectric sphere as a function of its radius. A single Lorentz oscillator with a resonance at 10,000 cm^{-1} defines the optical properties of the bulk medium. The radii are as follows: solid curve, 0.01 μm; dashed curve, 0.5 μm; and dot-dashed curve, 2.0 μm. (b) Calculated extinction spectra for a 1 μm radius hollow spherical shell as a function of its inner radius. The bulk medium is the same as (a) above. The ratio of the inner to outer radius is as follows: solid curve, q = 0; dashed curve, q = 0.75; and dot-dashed curve, q = 0.93. (After Ref. [37])

particle Rayleigh limit. Absorption is the dominant factor in the extinction cross section, see the solid curve in Fig. 4 (a), which displays the resonance structure of the bulk dielectric function; however, the 2 μm particle with $x_r \gg 1$ and $\delta_r/a \ll 1$ at resonance is in the opposite limit, now scattering is dominant and 408 partial wave coefficients are required at the highest frequency of the cross section spectrum (dot-dashed curve). Dramatic changes in the extinction spectrum occur in Fig. 4(a) as the sphere size changes between these two limits. Although the extinction spectrum is quite different in each case, the numerically determined sum rules and optical moment for all cases remains unchanged, within the accuracy of the calculation. The stronger the scatter, the higher the frequency required in the integral to exhaust the sum rule given by eq. (47). To see how the particle size enters in the sum rule calculation even though it does not appear in the final result, it's instructive to define an upper limit frequency ω_f to the integral which gives the fraction f (> 0.6) of the area counted. For small Rayleigh spheres, this frequency does not display a strong size dependence; however, for fairly large spheres (> 0.5 μm), where scattering is dominant, this normalized upper limit frequency for the integral $\omega_f/\omega_p = [3/(1-f)4\pi](\omega_p a/c)$ varies linearly with the particle radius. This simple linear dependence of the upper limit of the integral occurs because ω_f is in the Rayleigh-Gans scattering regime.

Figure 4(b) shows the volume-normalized extinction spectra calculated for hollow spherical shells with a fixed outer radius of a = 1 μm but different ratios, q, of inner to outer radius. The extinction spectra for particles of moderate q are similar to those for the homogeneous sphere since near resonance the radiation can not penetrate the shell to probe the hole inside. However, the extinction spectra for a sphere with q = 0.93 where the shell is penetrated near resonance displays a different response from that found for the other two cases. Although the extinction spectra shown in Figs. 4(a) and (b) are very different each other, the integrated extinction coefficient value is unchanged, since it is determined only by the bulk plasma frequency (eq. 30), regardless of the particle size and with or without the hole.

IV. CONCLUSIONS

The key to the development of the second moment formalism described here has been the comparative simplicity of the optical components of the ZnS:diamond composite system. The asymmetry observed for the transverse and longitudinal response functions of this composite led to a generalization of the crystalline LST relation in terms of specific optical moments. This generalized LST relation is exact within the framework of linear response theory and reduces to the standard LST relation for the case of bulk crystalline media. In the spirit of the original LST analysis of the temperature dependent optic mode dynamics of bulk displacive ferroelectric crystals, one can look for the vibrational implications of the generalized LST relation for both ordered and disordered ferroelectrics, but now the constraints are more subtle. Although the discussion here has tended to emphasize only the solid state lattice dynamics context since that is where the LST relation was discovered, the results equally well apply to non-conducting liquids and also connect the optical and static properties of other degrees of freedom, such as the electronic transitions. The general conclusion is that in the small signal limit (within the bounds of long wavelength linear response theory), a generalized LST relation must exist for an arbitrary condensed matter system which may include anharmonic interactions, be ordered or disordered, or be homogeneous or inhomogeneous.

The application of these ideas to disordered solids of restricted size has shown that the corresponding second moment frequencies of particles made from disordered materials are simply related to the "surface plasmon" modes previously calculated for single crystal particles when the characteristic frequencies are described in terms of the second moment of the appropriate response function. Next by focusing on small but macroscopic objects, we have determined in some detail how the small particle second moment results for inhomogeneous materials can be used to obtain the generalized Fröhlich relation. With these results it is fairly straightforward to construct generalized Szigeti and Clausius-Mossotti expressions. Because the dielectric function of the particle is simply the dielectric function of the bulk composite, but with the pole translated by a constant amount along the real frequency axis, the generalized Clausius-Mossotti expression in terms of optical moments describes the susceptibility of small disordered particles.

But no doubt the biggest surprise came with the success of the analysis of the extinction coefficient of large particles. The application of the low frequency polarizability sum rule [eq. (44)] and the generation of the high frequency strength sum [eq. (49)] rule are important components of the moment analysis for the extinction cross section presented here. The strength sum rule, which is surprisingly simple, directly relates the "effective" number of electrons in a grain to the integral over the extinction cross section spectrum. (Note that for a general inhomogeneous grain where the local field correction is not known, eqs. (30) and (47) still apply.) The strength sum rule result is independent of the grain aspect ratio. Although the extinction spectra for the large particle and small particle limits are very different, because the former includes absorption and scattering while the latter only includes absorption, it is noteworthy that eq. (50) and eq. (29) give the same result. When the scattering particle under analysis is exceedingly complex, the procedures described here can be used to advantage. For example, the dielectric function could be that of a liquid, a glass or a composite medium and the analysis would still hold since the characteristic second moment extinction frequency is independent of particle size and the small particle moments already have been shown to cover such diverse systems.

ACKNOWLEDGMENTS

I wish to thank T. W. Noh, J. B. Page, R. Lai, S. A. FitzGerald and R. H. Silsbee for many useful discussions and contributions. Some of this effort was supported by NSF-DMR-9312381, NASA-AAGW-2768 and ARO-DAAL03-92-G-0369. The MRL Central Facilities are supported by the NSF under Award No. DMR-9121654.

REFERENCES

1. R. H. Lyddane, R. G. Sachs, and E. Teller, Phys. Rev. **59**, 673 (1941).
2. H. Fröhlich, *Theory of Dielectrics* (Clarendon Press, Oxford, 1949).
3. W. Cochran, Phys. Rev. Lett. **3**, 521 (1959).
4. W. Cochran, Advan. Phys. **9**, 387 (1960).
5. A. S. Barker, Phys. Rev. A **136**, 1290 (1964).
6. A. S. Barker, in *Ferroelectrics*, edited by E.F. Weller (Elsevier, Amsterdam, 1967), p. 213.
7. A. S. Barker, Phys. Rev. B **12**, 4071 (1975).
8. G. Burns and B. A. Scott, Sol. State Commun. **13**, 417 (1973).
9. Y. Luspin, J. L. Servoin, and F. Gervais, J. Phys. C **13**, 3761 (1980).
10. J. P. Sokoloff, L. I. Chase, and D. Rytz, Phys. Rev. B **38**, 597 (1988).
11. S. A. FitzGerald, T. W. Noh, A. J. Sievers, L. A. Xue, and Y. Tzou, Phys. Rev. B **42**, 5469 (1990).
12. T. W. Noh and A. J. Sievers, Phys. Rev. Lett. **63**, 1800 (1989).
13. A. J. Sievers and J. B. Page, Phys. Rev. B **41**, 3455 (1990).
14. D. A. G. Bruggeman, Ann. Phys. (Leipzig) **24**, 636 (1935).
15. D. Y. Smith, in *Handbook of Optical Constants of Solids*, edited by E. Palik (Academic Press, New York, 1985), Vol. I,
16. A. J. Sievers and J. B. Page, Infrared Phys. **32**, 425 (1991).
17. L. D. Landau, E. M. Lifshitz, and L. P. Pitaevskii, *Electrodynamics of Continuous Media*, 2nd ed. (Pergamon, Oxford, 1984), p. p. 282.
18. D. F. Edwards, in *Handbook of Optical Constants of solids*, edited by E.D. Palik (Academic Press, New York, 1985), p. 554.
19. J. C. Phillips, Rev. Mod. Phys. **42**, 317 (1970).
20. J. A. Van Vechten, Pjys. Rev. **182**, 891 (1969).
21. S. H. Wemple and M. DiDomenico, Phys. Rev. B **3**, 1338 (1971).
22. J. J. Hopfield, Phys. Rev. B **2**, 973 (1970).
23. P. C. Hohenberg and W. Brinkman, Phys. Rev. B **10**, 128 (1974).
24. M. F. Thorpe and S. W. de Leeuw, Phys. Rev. B **33**, 8490 (1986).
25. A. J. Sievers, Solar Energy Mat. & Solar Cells **32**, 451 (1993).

26. M. Born and K. Huang, *Dynamical Theory of Crystal Lattices* (Oxford Univ. Press, London, 1954), p. 108.

27. N. W. Ashcroft and N. D. Mermin, *Solid State Physics* (Saunders College, Philadelphia, 1976).

28. A. J. Sievers and J. B. Page, Phys. Rev. B **41**, 12562 (1990).

29. B. Szigeti, Trans. Faraday Soc. **45**, 155 (1949).

30. A. J. Sievers, T. W. Noh, and J. B. Page, Physica A **207**, 46 (1994).

31. E. M. Purcell, Ap. J. **158**, 433 (1969).

32. C. F. Bohren and D. R. Huffman, *Absorption and Scattering of Light by Small Particles* (John Wiley & Sons, New York, 1983).

33. J. D. Jackson, *Classical Electrodynamics* (John Wiley & Sons, New York, 1975), p. 454.

34. M. Kerker, *The Scattering of Light and Other Electromagnetic Radiation* (Academic Press, New York, 1969).

35. H. C. van de Hulst, *Light Scattering by Small Particles* (Dover, New York, 1981).

36. A. J. Sievers, Optics Commun. **109**, 71 (1994).

37. R. Lai and A. J. Sievers, Optics Commun. **116**, 72 (1995).

INTRINSIC LOCALIZED MODES IN ANHARMONIC LATTICES

S. R. Bickham,[1] S. A. Kiselev[2], and A. J. Sievers

Laboratory of Atomic and Solid State Physics and Materials Science Center
Cornell University
Ithaca, NY 14853-2501

ABSTRACT

The success of the harmonic approximation in describing vibrations of condensed matter systems comes about because the plane wave amplitude at any particular site is extremely small hence it does not matter that the intermolecular potentials themselves are considered to be anharmonic in nature. Anharmonicity is generally introduced perturbatively to explain experimentally observed phenomena such as thermal expansion and phonon combination bands. Here we examine the dynamical properties of a locally excited anharmonic lattice such as might be expected to occur during an optical excitation of a coupled electron-lattice system so that the anharmonicity can play a more significant role. One development presented here is a straightforward demonstration of intrinsic localized modes in perfect one-dimensional monatomic lattices with quartic anharmonicity. These modes are similar to those associated with previously studied force constant defects, but they may be located anywhere in a perfect lattice and can actually move from site to site given the appropriate initial conditions. Associated with each localized vibrational mode in monatomic lattices with cubic and quartic anharmonicity is a localized dc distortion which forms an integral part of this new dynamical configuration. The softening of the potential with large displacements from equilibrium also produces a red-shift in the local mode frequency, but stable modes are observed in molecular dynamics simulations for a large range of anharmonicity parameters. The building blocks presented here for the development of localized modes in the one dimensional anharmonic chain can be used directly in the discussion of more complex nonlinear systems.

I. INTRODUCTION

The possibility of intrinsic localized modes in pure anharmonic crystals for sufficiently strong quartic anharmonicity [1-4] or large amplitude, first proposed in 1988, has been confirmed by numerical simulation studies [5-8] in one and two dimensions. In some ways the results are reminiscent of defect-induced local modes. For example, simulations for a 1-D diatomic anharmonic lattice show that intrinsic local modes appear both in the gap between the optic and acoustic plane wave spectrum as well as above the optic branch similar to the spectrum generated by a point defect in a harmonic diatomic lattice [9]. Although the original analytical study by Sievers and Takeno [1] focused on the odd parity vibrational mode, which in one dimension has essentially the amplitude pattern of a simple triatomic molecule, simulations [5-8] and the analytical work by Page [4] have both shown that an unusual even parity mode with the vibrational pattern of a diatomic molecule also exists. Simulations of moving even localized modes have been described briefly in the literature [5,6,11-13] .

Spectroscopy and Dynamics of Collective Excitations in Solids
Edited by Di Bartolo, Plenum Press, New York, 1997

Here we present some of the fundamental properties of intrinsic localized modes in perfect lattices. First a qualitative picture of anharmonic localization is presented in Section II, then, in Section III, the properties of stationary localized modes are examined for a monatomic lattice with quartic anharmonicity in the nearest-neighbor potential. The interesting properties of moving localized modes are introduced in Section IV, and the results of molecular dynamics simulations are compared to analytical predictions for both stationary and moving intrinsic localized modes [14]. For moving modes, we show that uniform translational motion can be produced over part of $\omega(k)$ space, with the k-range becoming more restricted as the local mode frequency increases. A gaussian-like envelope function provides a good fit to the simulated vibrational soliton, and when such an analytical form is used to characterize the motion, the vibrational frequency, the wave vector, and the group velocity can be identified. The next step in the local mode development is to include cubic anharmonicity. In the past this has only been included either as a small perturbation [15-18] or with unjustified approximations[19]. In Section V, we demonstrate that one needs a local static distortion and well as the rotating wave approximation in order to generate intrinsic localized mode solutions that agree with numerical simulations. Odd and even parity modes may then be characterized for unrestricted cubic and quartic anharmonicities [7,20]. The properties of intrinsic localized modes in monatomic lattices with cubic and quartic anharmonicity are summarized in Section VI.

II. QUALITATIVE PICTURE OF ANHARMONIC LOCALIZATION

Let us examine a 1-D monatomic lattice of N particles, with the nearest-neighbors connected by anharmonic springs. The spring potential is made up of a harmonic and a quartic term in the form

$$V_{2-4}(x) = \frac{K_2}{2} x^2 + \frac{K_4}{4} x^4 ,$$

(1)

where $K_2, K_4 > 0$ and x is the deviation of the spring's length from its equilibrium value. One effect of such a positive anharmonicity is to increase the frequency of each mode in the plane wave spectrum. Another effect is the localization and appearance of a vibrational mode above the top of the plane wave spectrum in this perfect anharmonic lattice if the amplitude is sufficiently large with respect to those of plane waves. The following qualitative argument illustrates how this may occur[1].

For a one dimensional monatomic harmonic lattice with particles of mass m connected by nearest-neighbor spring constant K_2, the normal modes are described by an orthogonal set of plane homogeneous waves which are confined to a band of frequencies with a high frequency cutoff at ω_0. The amplitude pattern of this highest frequency mode in which every atom vibrates π out of phase with its neighbor is shown in Fig. 1(a). To estimate the rms amplitude of a particular particle in a plane wave harmonic mode ω for a chain of N particles, one can use the virial theorem. For a harmonic system the mean energy is equal to two times the mean potential energy, e.g., $\hbar\omega \approx N[m\omega^2 < u_n^2 >]$ hence the rms amplitude at each site n in this plane wave mode is quite small for a long chain since $< u_n^2 >^{1/2} \approx (N)^{-1/2}(\hbar / m\omega)^{1/2}$. Next assume that a localized wave packet is constructed at $t = 0$ at a particular site with the appropriate linear superposition of these plane wave modes. In this localized excitation in which only a few atoms, N_d, are displaced from their equilibrium position the central atom would have a relatively large rms amplitude, $< u_n^2 >^{1/2} \approx (N_d)^{-1/2}(\hbar / m\omega)^{1/2}$, independent of the length of the chain; however, for $t > 0$, because of lattice dispersion, this localized excitation does not remain confined for long. For t large the excitation is uniformly distributed at all lattice sites.

For a similarly constructed one dimensional anharmonic lattice with a potential described by eq. (1), the plane wave spectrum derived in the small oscillation limit is quite similar to that found for the harmonic case. For a given plane wave mode the amplitude at each particle site is very small for a long chain, hence a Taylor's expansion of this potential in the small amplitude limit reproduces the harmonic approximation. But when a wave packet is constructed at t = 0 to represent a localized disturbance in this chain, a different situation occurs. To be specific let us assume that the wave packet is to be localized in stages. As the wave packet corresponding to Fig. 1(b) is constructed, the amplitude of the particle at the central site increases, but since the potential is no longer harmonic, so does its frequency of vibration. According to Rayleigh's theorem the frequency shift of each plane wave mode can be no larger than the frequency interval between lattice modes. As the amplitude pattern is more strongly localized around a particular site so that $N_d \rightarrow 1$, the vibrational amplitude becomes larger forcing the center frequency up through the plane wave spectrum of modes until it reaches the frequency ω_0. Since this mode is not bounded from above, it rises out of the top of the plane wave spectrum as shown in Fig. 2(a) becoming an inhomogeneous wave in the process. Given sufficient anharmonicity the end result is a true localized mode centered at the particle in question with frequency ω_ℓ. The original anharmonic lattice is now described by a set of effective harmonic oscillators, with (N-1) plane wave-like and one localized. In this example, the anharmonic system has evolved in such a way that both plane (homogeneous) and localized (inhomogeneous) waves exist simultaneously.

Another example of the anharmonic localization of lattice vibrations can be given using a 1-D diatomic chain with a negative anharmonicity in the interaction between nearest-neighbors. The potential of the springs could be chosen to be that given by eq. (1) with $K_2 > 0$ and $K_4 < 0$. Figure 2(b) illustrates the anharmonic localization of the mode that originally came from the bottom of the optic branch. In this case the localization still produces a larger amplitude at a few atoms which again increases the effective anharmonicity of the potential, but now a lower frequency results. As is the case for the monatomic lattice, the remaining modes of the lattice are plane waves with renormalized frequencies. Strictly speaking, such anharmonically driven localization of lattice vibrations is only possible when there is a frequency gap in the plane wave spectrum.

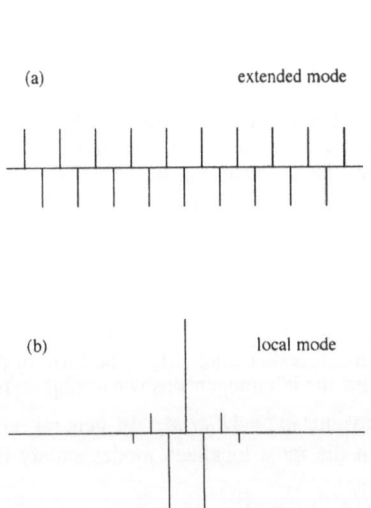

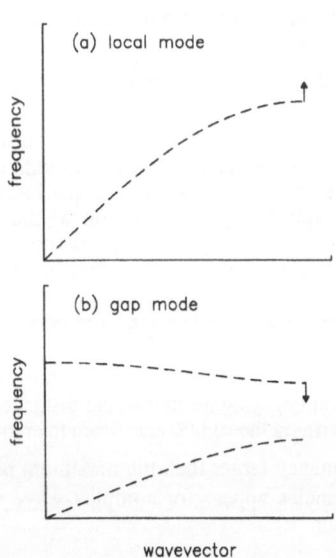

Figure 1. A schematic illustration of the eigenvectors of extended and localized modes. The extended mode (a) has a small rms amplitude at each lattice site, while the localized mode (b) has large rms amplitudes on a few sites. After Ref. (7).

Figure 2. Positions of the anharmonic mode frequencies relative to the plane wave dispersion curves. Case (a): When interparticle has positive anharmonicity, the mode at the top of the plane wave band becomes localized. Case (b): A soft interparticle potential in a diatomic lattice leads to the localization of an extended mode in the gap below the bottom of the optic band.

III. INTRINSIC LOCALIZED MODES IN MONATOMIC LATTICES WITH QUARTIC ANHARMONICITY

III. A. Derivation of Local Mode Frequencies

To bring out the similarities and differences between the eigenvectors of defect harmonic local modes and intrinsic anharmonic local modes, we demonstrate that the frequency of an intrinsic odd symmetry anharmonic local mode can be obtained to a good approximation by first using the rotating wave approximation to eliminate time from the equations of motion and then applying essentially the same formalism which has been used previously to characterize defect modes in harmonic lattices [21]. A monatomic lattice with only nearest-neighbor harmonic and anharmonic (quartic) force constants is used to identify, in the simplest manner, the dependence of the local mode frequency on anharmonicity. The potential energy is

$$U = \frac{1}{2} K_2 \sum_n (u_{n+1} - u_n)^2 + \frac{1}{4} K_4 \sum_n (u_{n+1} - u_n)^4, \tag{2}$$

where K_2 and K_4 are the harmonic and quartic potential constants, respectively and u_n identifies the displacement at site n.

The key idea which allows one to separate the coupled anharmonic problem into orthogonal homogeneous and inhomogeneous wave solutions makes use of the very different amplitudes of excursion in each kind of mode. For a plane wave mode the excursion of each atom is $\sim O(N^{-1/2})$ with N a large number, whereas the excursion of an atom in a localized mode is $\sim O(1)$ with $N \sim 1$. In the former case the homogeneous plane wave solutions can be obtained by ignoring the K_4 potential term in eq. (2) so that the quasi-harmonic dispersion relation simplifies to the standard one:

$$M \omega^2 = 2K_2[1 - \cos(qa)], \tag{3}$$

with real wave vector q and lattice constant, a.

When an anharmonic local mode exists, the full potential given by eq. (2) must be employed. However, far from the local mode site the excursion in the localized mode is again small, so in this region of the crystal the quartic term can be neglected when determining the asymptotic properties of the solution. The resulting inhomogeneous wave solution is characterized by a complex wave vector with components

$$q^* = q \pm iq' = (2j+1)\pi/a \pm iq', \tag{4}$$

where j is an integer and the sign of the second term on the right is chosen so that the solution decays with increasing distance from the local mode center [21]. The form of the first term on the right comes from the requirement that the inhomogeneous wave solution has a frequency larger than the maximum plane wave value $\omega_o^2 = 4K_2/M$. In general, new frequencies appear for complex wave vectors but the most localized modes satisfy the equation

$$\left(\frac{\omega}{\omega_o}\right)^2 = \frac{1}{2}[1 + \cosh(q'a)]. \tag{5}$$

We assume that far from the mode center all anharmonic localized modes must obey this expression. Because of the inversion symmetry of the problem the modes can be classified, in the electric dipole sense, as either odd ($u_1 = u_{-1}$, $u_0 = \alpha$) or even ($u_1 = -u_{-1} = \alpha$, $u_0 = 0$).

1. Odd Symmetry Localized Mode. To identify q' in eq. (4), we must examine the equations of motion for the central atom and its neighbors. Although we are only interested in presenting enough constraints to specify the mode, namely, the $n = 0$ and $n = 1$ equations of motion, there is some value in temporarily retaining the index n. The acceleration equation becomes

$$m \frac{d^2 u_n}{d t^2} = -K_2 \left[2u_n - u_{n+1} - u_{n-1} \right] - K_4 \left[(u_n - u_{n+1})^3 + (u_n - u_{n-1})^3 \right]. \tag{6}$$

To eliminate the time dependence, we assume that the system responds only at one local mode frequency ω and ignore the response at 3ω, 5ω, etc. which is produced by the quartic term in the potential. The logic for this approximation is that it is much more difficult for the lattice to respond at these much higher frequencies. This assumption, called the Rotating Wave Approximation (RWA), suggests solutions of the form

$$u_n(t) = \alpha \phi_n \cos(\omega t), \tag{7}$$

where α and ϕ_n are the maximum and relative amplitudes of the localized mode displacements, respectively. The time-independent equation of motion is then

$$\left(\frac{\omega}{\omega_o} \right)^2 \phi_n = \frac{1}{4} \left\{ \left[2\phi_n - \phi_{n+1} - \phi_{n-1} \right] + \frac{3}{4} \Lambda \left[(\phi_n - \phi_{n+1})^3 + (\phi_n - \phi_{n-1})^3 \right] \right\}. \tag{8}$$

For the odd symmetry conditions $\phi_0 = 1$, $\phi_n = \phi_{-n} = A(-1)^n \exp(-nq')$ for $n > 0$, and with the anharmonicity parameter defined as $\Lambda = K_4 \alpha^2 / K_2$, eq. (8) becomes

$$\left(\frac{\omega}{\omega_o} \right)^2 = \frac{1}{2} \left[(1 + Ae^{-q'}) + \frac{3\Lambda}{4} (1 + Ae^{-q'})^3 \right], \tag{9}$$

for the central atom ($n = 0$) and

$$\left(\frac{\omega}{\omega_o} \right)^2 Ae^{-q'} = \frac{1}{4} \left\{ \begin{aligned} &\left(2Ae^{-q'} + 1 + Ae^{-2q'} \right) + \\ &\frac{3\Lambda}{4} \left[\left(Ae^{-q'} + 1 \right)^3 + A^3 e^{-3q'} \left(1 + e^{-q'} \right)^3 \right] \end{aligned} \right\}, \tag{10}$$

for the nearest-neighbor one ($n = 1$). Note that $A\exp(-q')$ is the amplitude of the nearest-neighbor atom with respect to the center atom, and the amplitude of the eigenvector decreases exponentially as given by eq. (5). The maximum vibrational amplitude of the center atom, α, is defined by the particular excitation experiment. We anticipate that this grafting together of the close-in eigenvector of the anharmonic localized mode with the farther out inhomogeneous wave appropriate to a harmonic localized mode should provide a good description both for strong and weak anharmonicity but would be a less valid approximation for intermediate values. To obtain more accurate solutions for these intermediate cases, the exponential decay can be appended onto the eigenvector farther away from the center site. Although this would increase the number of equations that describe the intrinsic localized mode, modern computer algorithms [22] can easily handle as many as ten coupled nonlinear equations.

The three equations that describe our local mode example can be simplified by introducing the definitions: $\kappa = \exp(-q')$ and $\phi_1 = A\exp(-q') = A\kappa$. Equation (5) then becomes

$$\left(\frac{\omega}{\omega_o}\right)^2 = \frac{(1+\kappa)^2}{4\kappa},$$ (11)

eq. (9);

$$\left(\frac{\omega}{\omega_o}\right)^2 = \frac{1}{2}\left[(1+\phi_1) + \frac{3\Lambda}{4}(1+\phi_1)^3\right],$$ (12)

and eq. (10);

$$\left(\frac{\omega}{\omega_o}\right)^2 \phi_1 = \frac{1}{4}\left\{(2\phi_1 + 1 + \phi_1\kappa) + \frac{3\Lambda}{4}\left[(\phi_1 + 1)^3 + \phi_1^3(1+\kappa)^3\right]\right\}.$$ (13)

Hence for a given Λ, eqs. (11), (12) and (13) can be solved simultaneously for κ, ϕ_1, and $(\omega/\omega_o)^2$. The eigenvector of the odd parity mode characterized by $\Lambda = 0.9$ is plotted in the top panel of Fig. 3.

2. Even Symmetry Localized Modes. An analysis similar to that given above can be carried out for the breathing mode case where $u_1 = -u_{-1} = \alpha$ and $u_0 = 0$. This mode, plotted in the lower panel of Fig. 3, is unusual in that unlike the breathing modes associated with force constant defects, the maximum amplitude is located at the $n = \pm 2$ lattice sites. In this respect, the eigenvector more closely resembles the pocket modes that have been used to explain certain features in the far infrared spectrum of KI:Ag$^+$ [23]. Although analytic solutions are found, computer simulations show that such even modes are in fact unstable. This breathing mode breaks up into pairs of moving odd modes which are repelled from each other. These moving modes will be discussed in Section IV.

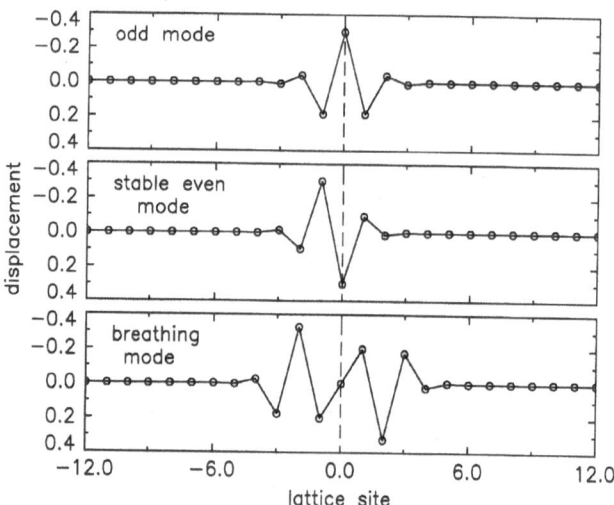

Figure 3. Eigenvectors for intrinsic localized modes of different symmetries. Although these modes are one dimensional, the displacements are plotted perpendicular to the lattice for clarity. The top panel of the figure shows an odd mode characterized by $\Lambda = 0.9$, the center panel an even mode characterized by $\Lambda = 0.9$ and the lower panel a breathing mode characterized by $\Lambda = 0.4$. The dashed line through the $n = 0$ lattice site bisects the eigenvectors of the odd and breathing modes, but not the even mode that is centered in between adjacent lattice sites. After Ref. (7).

Early simulations [6,7] and analytical work [4] have demonstrated that another type of localized vibration is stable, and although it is not compatible with the lattice symmetry, it does have even character. The eigenvector of this mode is plotted in the center panel of Fig. 3. In one dimension this strongly localized mode corresponds to the stretch mode vibration of a diatomic molecule with the next neighbors also having some amplitude. Here we find the necessary conditions for this mode, which is the analog of a force constant defect mode [24].

Following the notation used in the previous section, the atom amplitudes of the even mode are labeled $\phi_{\pm n}$, where $n > 0$ and the displacement pattern satisfies $\phi_n = -\phi_{-n}$. With $\phi_1 = 1$ and $\phi_n = A(-1)^{n-1}\exp[-(n-1)q']$ for $n > 1$, the condition for the $n = 1$ atom becomes

$$\left(\frac{\omega}{\omega_o}\right)^2 = \frac{1}{4}\left\{3 + \phi_2 + \frac{3A}{4}\left[8 + (1 + \phi_2)^3\right]\right\}. \tag{14}$$

For the $n = 2$ atom, the condition reduces to eq. (13) given above (with ϕ_1 replaced by ϕ_2), since its nearest neighbors have the same relative displacement pattern as in the odd mode case. The far field condition, eq. (5) or (11), remains unchanged for these modes.

3. Odd and Even Modes in the Limit of Weak Anharmonicity. For small values of the anharmonicity parameter A, the equation of motion may be expanded using the continuum approximation [25,26]. With the opposite-phase displacements $w_n = (-1)^n u_n$, the continuum limit of eq. (5) is

$$m\frac{\partial^2 w}{\partial t^2} + 4K_2 w + K_2 \frac{\partial^2 w}{\partial x^2} + 16K_4 w^3 = 0. \tag{15}$$

In terms of the expansion parameter $\varepsilon = \left[(\omega/\omega_o)^2 - 1\right]^{1/2} \ll 1$, the lowest order solution of eq. (15) is

$$w \approx \sqrt{2K_2/3K_4}\,\frac{\varepsilon\cos(\omega t)}{\cosh[2\varepsilon(x - x_0)]}, \tag{16}$$

where x_0 is centered on a lattice site for odd parity modes and halfway in between two sites for even parity modes. Far away from the center of the localized mode, the ratio of the nearest-neighbor displacements is

$$\frac{w_{n+1}}{w_n} \approx \frac{\cosh[2\varepsilon\,n]}{\cosh[2\varepsilon\,(n+1)]} \cong \exp(-2\varepsilon), \tag{17}$$

which is consistent with the assumption of an exponential decaying amplitude. Instead of representing the results of the continuum approximation in terms of the expansion parameter ε, it is more illuminating to write them in terms of the dimensionless anharmonicity parameter, $A = K_4 w_0^2 / K_2 = 2\varepsilon^2/3$. The frequency and relative amplitude at the nearest-neighbor site are then

$$\left(\frac{\omega}{\omega_o}\right)^2 \cong 1 + \frac{3}{2}A, \tag{18}$$

and

253

$$\phi_1 = \frac{w_1}{w_0} \cong 1 - 3\Lambda. \tag{19}$$

4. Odd and Even Modes in the Limit of Strong Anharmonicity. For the large Λ limit the relative displacement of the nearest neighbor of the odd mode can be found by combining the nonlinear terms in eqs. (12) and (13) to obtain

$$2\phi_1(1+\phi_1)^3 = (1+\phi_1)^3 + \phi_1^3. \tag{20}$$

This yields a nearest-neighbor amplitude of $\phi_1 = 0.520$, and the relationship between the local mode frequency and the anharmonicity parameter is given by

$$\left(\frac{\omega}{\omega_o}\right)^2 \cong 0.760 + 1.317\Lambda. \tag{21}$$

For the even mode, the equation for the relative displacement is

$$\phi_2\left[8 + (1+\phi_2)^3\right] = (1+\phi_2)^3 + \phi_2^3, \tag{22}$$

which yields a nearest-neighbor amplitude of $\phi_2 = 0.166$ and a local mode frequency of

$$\left(\frac{\omega}{\omega_o}\right)^2 \cong 0.792 + 1.797\Lambda. \tag{23}$$

By solving eqs. (11), (12) and (13) numerically, we can determine how the local mode frequency and the nearest-neighbor amplitude ϕ_1 vary with the anharmonicity parameter Λ for the one dimensional case. The results are shown in Fig. 4. Over most of the interval the solid line shows that $(\omega/\omega_o)^2 \propto \Lambda$ while the dashed line indicates that the nearest-neighbor

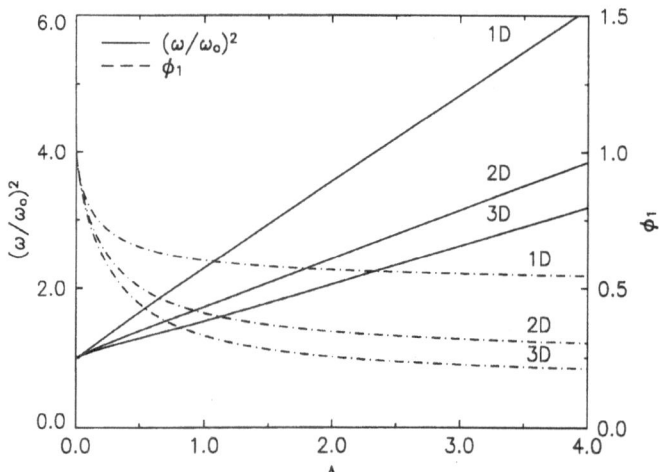

Figure 4. Local mode frequency and nearest-neighbor amplitude versus anharmonicity parameter Λ for the odd symmetry case. For one, two and three dimensions the solid curves give the calculated frequencies and the dashed curves, the relative amplitudes. After Ref. (8).

amplitude is the same as the central atom for small Λ, but approaches roughly half its value for large Λ. In the small Λ limit, the mode is delocalized with the nearest neighbors having nearly the same displacement as the central particle. For large Λ, the second neighbor has extremely small amplitude and the mode resembles a triatomic molecule with the central atom having amplitude unity and the two neighbors oppositely directed with amplitude $\approx 1/2$ [24].

5. Correction to the Rotating Wave Approximation. In developing the local mode solution, it was assumed above that the system only responded at the fundamental driving frequency. Because of the quartic potential, responses at 3ω, 5ω, etc. should also be present, and here we investigate the influence of the 3ω term on the value of the fundamental local mode frequency which we now call ω_1. To do this we generalize the RWA displacements given by eq. (7) to

$$u_n = \alpha \, \phi_n [(1-\beta)\cos(\omega_1 t) + \beta \cos(3\omega_1 t)], \tag{24}$$

using a transparent notation. Substitution of eq. (24) into the equation of motion yields

$$\left(\frac{\omega_1}{\omega_o}\right)^2 \begin{bmatrix} (1-\beta)\cos(\omega_1 t) \\ +9\beta\cos(3\omega_1 t) \end{bmatrix}$$
$$= \begin{bmatrix} (1-\beta)\cos(\omega_1 t) \\ +\beta\cos(3\omega_1 t) \end{bmatrix} f(\phi_1, \kappa) + \frac{\Lambda}{4} \begin{bmatrix} 3(1-2\beta)\cos(\omega_1 t) \\ +(1+3\beta)\cos(3\omega_1 t) \end{bmatrix} g(\phi_1, \kappa) + O(\beta^2, 5\omega_1), \tag{25}$$

where $f(\phi_1, \kappa)$ and $g(\phi_1, \kappa)$ depend on the site and parity of excitation being considered. Matching coefficients and keeping terms to order β^2 gives

$$\beta = \frac{\Lambda g(\phi_1, \kappa)}{32 f(\phi_1, \kappa) + 25 \Lambda g(\phi_1, \kappa)}, \tag{26}$$

and

$$\left(\frac{\omega_1}{\omega_o}\right)^2 = f(\phi_1, \kappa) + \frac{3}{4}\Lambda \, g(\phi_1, \kappa) \left[\frac{32 f(\phi_1, \kappa) + 23 \Lambda g(\phi_1, \kappa)}{32 f(\phi_1, \kappa) + 24 \Lambda g(\phi_1, \kappa)}\right]. \tag{27}$$

Equation (26) restricts the range of β to $0 \le \beta \le 1/25$ while the correction factor for Λ in eq. (27) is between 1 and 23/24. Since this factor is less than or equal to unity, ω_1 will be less than the uncorrected frequency ω predicted without the third harmonic term. The relative difference between the squares of these two frequencies lies in the range

$$0 \le \left(\frac{\omega}{\omega_1}\right)^2 - 1 \le \frac{1}{23}. \tag{28}$$

This difference is small, so the higher order terms in eqs. (24) and (25) usually can be safely ignored. For values of $\Lambda \ll 1$, the correction to eq. (27) is negligible and the single frequency rotating wave approximation given by eq. (7) is quite accurate.

III. B. Intrinsic Localized Modes in Two and Three Dimensions

Because of the simplicity of the potential given by eq. (2), higher dimensional cases can be treated in the same way as for the 1-D case, but with the added feature that more neighbors must be counted in the mode. We consider a monatomic lattice with nearest-neighbor

255

harmonic and anharmonic force constants that have equal transverse and longitudinal components. The separable potential energy for the three dimensional system is then

$$U = \sum_{\sigma} \sum_{l,m,n} \left\{ \frac{K_2}{2} \left[\begin{matrix} (u^{\sigma}_{l,m,n} - u^{\sigma}_{l+1,m,n})^2 \\ +(u^{\sigma}_{l,m,n} - u^{\sigma}_{l,m+1,n})^2 \\ +(u^{\sigma}_{l,m,n} - u^{\sigma}_{l,m,n+1})^2 \end{matrix} \right] + \frac{K_4}{4} \left[\begin{matrix} (u^{\sigma}_{l,m,n} - u^{\sigma}_{l+1,m,n})^4 \\ +(u^{\sigma}_{l,m,n} - u^{\sigma}_{l,m+1,n})^4 \\ +(u^{\sigma}_{l,m,n} - u^{\sigma}_{l,m,n+1})^4 \end{matrix} \right] \right\}, \tag{29}$$

where $u^{\sigma}_{l,m,n}$ identifies the σth component of the displacement at site (l,m,n). As in the one-dimensional case, the modes can be classified as either odd ($u^{\sigma}_{l,m,n} = u^{\sigma}_{-l,-m,-n}$; $u^{\sigma}_{0,0,0} = \alpha$) or even ($u^{\sigma}_{l,m,n} = -u^{\sigma}_{-l,-m,-n}$; $u^{\sigma}_{0,0,0} = 0$). The derivation and solution of the coupled nonlinear equations that describe the intrinsic localized mode is similar to the procedure outlined in Section III.A with the exponential approximation defined as $\phi_{l,m,n} = A(-1)^{l+m+n} \exp(-q' \sqrt{l^2+m^2+n^2})$. The results will be discussed in the next section.

The continuum approximation used to derive the properties of 1-D modes in the limit of weak anharmonicity cannot be easily extended to higher dimensions, but the local mode properties in the limit of strong anharmonicity may be calculated in the same manner as the 1-D results given by eqs. (21) and (23). For both odd and even parity modes, the dependence of the local mode frequency on the anharmonicity has the form

$$(\omega / \omega_o)^2 \cong a + b\Lambda, \tag{30}$$

where a and b are determined by the higher dimensional analogs of eqs. (20) and (22). The nearest-neighbor amplitudes and the coefficients in eq. (30) are given in Table 1 for one, two and three dimensions. Note that the intrinsic localized mode frequency has the strongest dependence on the anharmonicity for the one dimensional case.

The odd mode behavior versus anharmonicity for two and three dimensions is also shown in Fig. 4. One finds that to obtain a given local mode frequency, a larger anharmonicity parameter is required with increasing dimension. Although the nearest-neighbor amplitude is smaller in the higher dimensional cases, there are more neighbors, so for a fixed Λ, the mode becomes less localized as the dimension increases.

The frequency and amplitude dependencies on Λ of the even parity mode are found to be similar to those for the odd mode presented in Fig. 4. A quantitative difference is that for a given Λ, the even mode frequency is higher than the odd mode in all three cases, even though this mode involves more neighbors. This mode, which does not obey the same symmetry as the odd mode, can be viewed as a vibrational exciton. It can be constructed, approximately, by superimposing two odd modes (described in Section III.A.1) with opposite phase on nearest-neighbor sites. The phase is the analog of charge. Odd modes with opposite phase attract while those with the same phase on near neighbor sites repel.

Table 1. Local mode parameters in the strong anharmonicity limit. For the odd mode, the nearest-neighbor amplitude (defined as NN here) is $|\phi_1|$, the amplitude at the $n = \pm 1$ sites. For the even mode, this quantity is equal to $|\phi_2|$, the relative amplitude at the $n = \pm 2$ sites.

Symmetry	Dimension	NN	intercept, a	slope, b
odd	1	0.520	0.760	1.317
odd	2	0.256	0.628	0.744
odd	3	0.169	0.585	0.599
even	1	0.166	0.792	1.797
even	2	0.114	0.668	1.139
even	3	0.090	0.621	0.904

IV. MOLECULAR DYNAMICS SIMULATIONS OF STATIONARY AND MOVING INTRINSIC LOCALIZED MODES

IV. A. Translational Stability

The technique outlined in Section III for the analytical determination of the eigenfrequencies and eigenvectors of stationary intrinsic localized modes provides the starting point for simulations. Eigenvectors such as those shown in Fig. 3 can be used as initial conditions in computer simulations to test the stability of the different types of modes. One difficulty is that some local mode symmetries are not stable as stationary excitations, but instead spontaneously propagate through the lattice. For example, the odd parity mode exists as a stationary mode for fewer than one hundred periods before it begins to move between its initial position and adjacent lattice sites. Figure 5 shows the displacement versus time at two neighboring lattice sites when the initial conditions correspond to an odd parity mode characterized by $\Lambda = 0.9$. The excitation is stationary for approximately 60 vibrational periods of ω_0 before it moves from its initial position at $n = 0$. It is then briefly localized at the $n = 1$ lattice site in the time interval in between $t \approx 60$ and $t \approx 80$, but returns to the center site until $t \approx 95$. At the end of the simulation time shown in the figure, the anharmonic mode is again localized at the $n = 1$ site. Sandusky and Page have explained this phenomenon by showing analytically that this particular mode is translationally unstable to small perturbations of its eigenvector, while an even parity mode centered in between two lattice sites remains stable [27]. It is clear from these results that any systematic investigation of the properties of intrinsic localized modes in 1-D monatomic lattices needs to include the possibility of translational motion.

Another way to present time dependent simulation results is to plot the energy of the lattice at different simulation times. We have already pointed out that the breathing mode is unstable. In Fig. 6 the initial conditions are such as to correspond to an anharmonic breathing mode characterized by $\Lambda = 0.4$. The maximum amplitude is initially located at the $n = \pm 2$ lattice sites, hence there are two distinct peaks in the energy at $t = 0$. As the simulation progresses, this mode rapidly becomes translationally unstable and splits into two intrinsic localized modes of odd symmetry propagating in opposite directions. Unlike the simulation results presented for the odd mode in Fig. 5, in which the mode rocks back and forth between two lattice sites and symmetries, here it appears that a fixed group velocity can be assigned to these two moving excitations. One can speak of the breathing mode decaying but it decays into other anharmonic modes not into the plane wave spectrum.

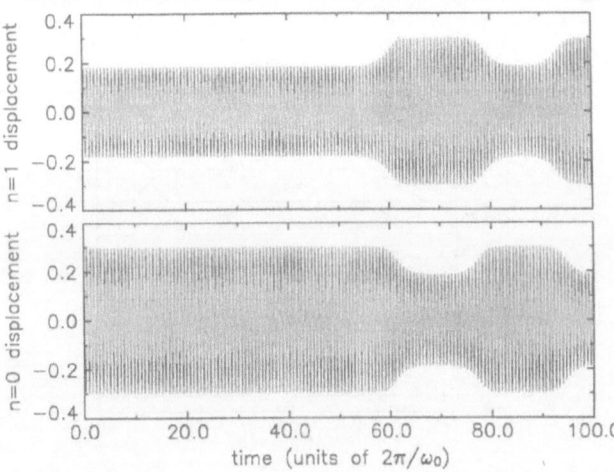

Figure 5. Simulated displacements versus time for an odd-parity intrinsic localized mode. The mode is characterized by an anharmonicity of $\Lambda = 0.9$ and is initially located the $n = 0$ lattice site. It is stable as a stationary excitation for approximately 60 periods of ω_0, but then begins to move across the lattice. At the end of the simulation time shown here, the mode is localized at the $n = 1$ site. After Ref. (28).

The simulated displacements of the other even mode are not plotted here, but they confirm that these excitations are translationally stable for at least several thousand vibrational periods. From the simulation results presented in Figs. 5 and 6, there appears to be different types of translational instabilities associated with intrinsic localized modes. The one found for odd parity modes is low order and results in a moving excitation that retains its shape, but moves slowly and aperiodically around the initial site. The translational instability found for the breathing mode is stronger, and as Fig. 6 illustrates, results in pairs of moving localized excitations that move slowly, compared to the sound velocity, in opposite directions. When isolated, these moving modes have the symmetry of an odd parity mode, but the vibrational frequencies are Doppler shifted from the frequencies of stationary ones due to the translational motion. Except where noted, further uses of the phrase "even mode" will refer to the translationally stable modes that are centered in between adjacent lattice sites.

IV. B. Stationary Intrinsic Localized Modes

The frequencies of the simulated displacements can be obtained by calculating the fast Fourier transform at each lattice site. The results should agree with the ansatz given by eq. (24). As an example, the log power spectrum calculated from the center particle displacements in Fig. 5 is plotted in Fig. 7. Only the first sixty periods of the simulation are used in the calculation to eliminate any frequency shifts associated with the translational motion. There is a very strong local mode peak in the spectrum near $\omega / \omega_0 = 1.5$, which is well above the top of the plane wave spectrum. A second peak appears at three times the fundamental frequency, but its strength is reduced by several orders of magnitude relative to the main peak. This result confirms the analytical analysis in Section III.A which showed that the correction to the RWA due to higher order harmonics is small, hence only the first term in eq. (24) is needed to accurately describe the localized vibrations.

The analytical and simulation frequencies of the local mode are plotted in Fig. 8 for several values of the quartic anharmonicity. The analytical results calculated within the RWA are represented by the solid and dashed curves for odd and even modes, respectively. There is good agreement over the entire range of Λ shown, although the negative shift of the simulation frequencies with respect to the analytical values becomes more apparent at higher anharmonicities. This red shift, caused by the distribution of a small portion of the vibrational energy into higher order harmonics of the fundamental frequency, is also stronger for the even mode than for the odd parity mode.

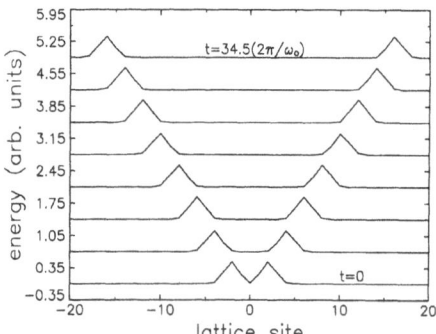

Figure 6. Energy versus lattice site at progressive times during the simulation of the dynamics of a breathing (even parity) mode. The localized mode is characterized by an anharmonicity of $\Lambda = 0.4$ and is initially centered the $n = 0$ lattice site. It rapidly separates into two odd parity modes that propagate in opposite directions with constant group velocities. In contrast, computer simulations of the even mode that is centered in between two lattice sites show that it is stable as a stationary excitation. After Ref. (28).

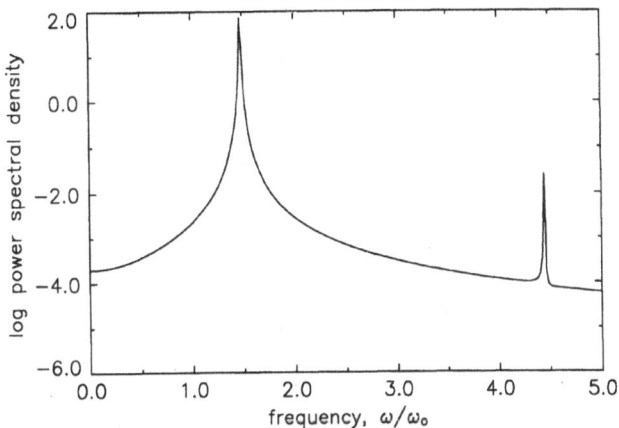

Figure 7. Log power spectrum of the odd mode displacements in Fig. 5. A fast Fourier transform was calculated over the portion of the simulation versus time plot where the mode is stationary, and this indicates that the vibrations are essentially monochromatic. A peak at the third harmonic is present, but its strength is three orders of magnitude weaker than the peak corresponding to the fundamental frequency. This particular odd mode is characterized by $\Lambda = 0.9$ and has a vibrational frequency of $\omega / \omega_0 \cong 1.5$. After Ref. (7).

IV. C. Moving Intrinsic Localized Modes

The agreement between simulation and analytical results for stationary intrinsic localized modes is encouraging; however a general theory should include the translational properties of these excitations. The idea is to explore the dispersion curve for moving even and odd localized modes with analytical and simulation tests. The satisfying results are that for both types of modes uniform translational motion can be produced by simulation over at least part

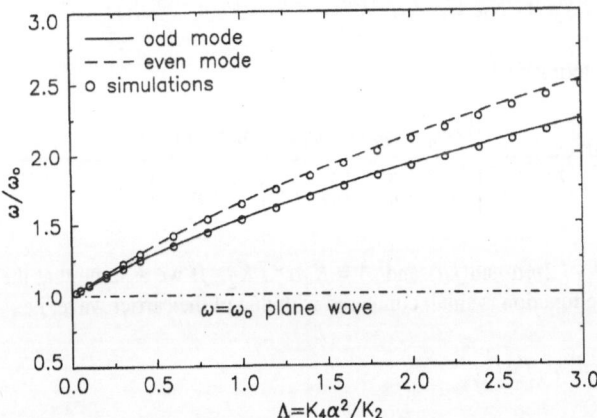

Figure 8. Simulated and analytical frequencies for odd and even parity intrinsic localized modes. The results of molecular dynamics simulations are represented by open circles, while the analytic solutions of the equations of motion are plotted by the solid and dashed curves. There is good agreement over the entire range of Λ for both parity modes, but the red shift due to higher harmonics is more pronounced with increasing frequency and quartic anharmonicity. Note that the frequencies of the odd and even parity modes become nearly degenerate at small Λ values and approach the top of the plane wave spectrum. Here ω_0 is the maximum frequency of the harmonic plane wave spectrum. After Ref. [30].

of $\omega(k)$ space with the k range becoming more restricted as the local mode frequency increases [21]. As was done for the stationary mode in Section III, we begin by considering a perfect 1-D crystal lattice in which each particle with mass m interacts with its nearest neighbors with the harmonic force constant K_2 and the anharmonic constant K_4 derived from the positive quartic anharmonic potential. The equation of motion for the displacement u_n of the nth atom from its equilibrium position is given by eq. (6).

To describe the properties of an even parity mode, it is helpful to change to a different lattice by the transformation

$$w_n = u_{n+1} - u_n.$$

(31)

Note that the symmetry type is changed from even to odd and vice versa using this new displacement field variable. Equation (6) now becomes [11]

$$\frac{d^2 w_n}{d t^2} = \frac{K_2}{m}\left(w_{n+1} + w_{n-1} - 2w_n\right) + \frac{K_4}{m}\left(w_{n+1}^3 + w_{n-1}^3 - 2w_n^3\right).$$

(32)

The maximum frequency of the plane wave spectrum is $\omega_o^2 = 4K_2/m$ since the anharmonic terms are negligible for the small amplitude limit appropriate to a plane wave spectrum [1].

An analytic solution to eq. (32) giving a moving anharmonic mode is sought by setting

$$w_n = \alpha\, \varphi_n(t)\cos(kna + \omega t),$$

(33)

where α is the maximum amplitude of the moving localized mode in a lattice with spacing a, $\varphi_n(t)$ is an envelope function, $\cos(kna + \omega t)$ is a left moving carrier wavefunction and k and ω are the wave vector and the frequency, respectively, of the carrier wave describing the localized mode. Inserting eq. (33) into eq. (32) and using the RWA, we obtain a pair of equations. Equating the sine terms gives

$$\frac{d\varphi_n}{dt} = v_k\left(\varphi_{n+1} - \varphi_{n-1}\right)\left[1 + \frac{3}{4}\Lambda\left(\varphi_{n+1}^2 + \varphi_{n+1}\varphi_{n-1} - \varphi_{n-1}^2\right)\right],$$

(34)

and the cosine terms give

$$\omega^2\varphi_n - \frac{d^2\varphi_n}{dt^2} = \frac{K_2}{m}\left\{\begin{array}{l}\left[2\varphi_n - (\varphi_{n+1} - \varphi_{n-1})\cos(ka)\right] + \\ \frac{3}{4}\Lambda\left[2\varphi_n^3 - (\varphi_{n+1}^3 + \varphi_{n-1}^3)\cos(ka)\right]\end{array}\right\},$$

(35)

where $v_k = (K_2 / 2m\omega)\sin(ka)$ and $\Lambda = K_4\alpha^2 / K_2$. If we assume that the time variation of the envelope function is small compared with that of the carrier wave, i.e.,

$$\omega^2\varphi_n \gg d^2\varphi_n / dt^2,$$

(36)

then

$$\left(\frac{\omega}{\omega_o}\right)^2\varphi_n = \frac{1}{4}\left\{\begin{array}{l}\left[2\varphi_n - (\varphi_{n+1} - \varphi_{n-1})\cos(ka)\right] + \\ \frac{3}{4}\Lambda\left[2\varphi_n^3 - (\varphi_{n+1}^3 + \varphi_{n-1}^3)\cos(ka)\right]\end{array}\right\}.$$

(37)

The simulation results presented in the following section verify that the magnitude of $d^2\varphi_n / dt^2$ is no greater than 1% of $\omega^2\varphi_n$, so the set of time-independent equations generated by eq. (37) can be used to identify the initial displacement pattern of the mode, and eq. (34) can then be used to find the initial velocity pattern. Note that for $k = 0$ and the strongly localized limit, eq. (37) reduces to eq. (14), the eigenfrequency of the stationary even mode [4,8,10].

To find the corresponding displacements and velocities for the odd mode with a particular wavevector, it is easier to work directly in u-space. Assuming a form of the solution similar to eq. (33), the differential equation corresponding to eq. (34) is

$$\frac{d\phi_n}{dt} = v_k(\phi_{n+1} - \phi_{n-1})\left\{1 + \frac{3}{4}\Lambda\left[\begin{array}{l}\phi_{n+1}^2 + \phi_{n+1}\phi_{n-1} - \phi_{n-1}^2 + \phi_n^2 \\ -2(\phi_{n+1} + \phi_{n-1})\phi_n\cos(ka)\end{array}\right]\right\}. \tag{38}$$

Once again for relatively slow time variation of the envelope function, the analog of eq. (37) is

$$\left(\frac{\omega}{\omega_o}\right)^2\phi_n = \frac{1}{4}\left\{\begin{array}{l}\left[2\phi_n - (\phi_{n+1} - \phi_{n-1})\cos(ka)\right] + \\ \frac{3}{4}\Lambda\left[\begin{array}{l}2\phi_n^3 - (\phi_{n+1}^3 + \phi_{n-1}^3)\cos(ka) \\ +(\phi_{n+1}^2 + \phi_{n-1}^2)\phi_n[1 + 2\cos^2(ka)] \\ -3(\phi_{n+1} + \phi_{n-1})\phi_n^2\cos(ka)\end{array}\right]\end{array}\right\}. \tag{39}$$

These two sets of equations [eqs. (35) and (37) or (38) and (39)] determine the initial amplitudes and velocities necessary to produce a localized mode with frequency ω and wavevector k. To truncate the series of equations in u-space, we assume that the mode is localized and introduce the following trial solution for $t = 0$:

$$\phi_0 = \alpha, \quad \phi_n = \phi_{-n}, \quad \phi_n = (-1)^n \alpha A e^{-nKa}, \tag{40}$$

for K positive [8]. For a given value of k and Λ, the time independent equations for sites $n = 0, 1$ and 2 are numerically solved for $\omega(k)/\omega_o$, A and K. Hence $\omega(k)$, the vibrational frequency, is determined and the latter two quantities give the initial amplitudes of the eigenvector. When all three quantities are substituted into either eq. (35) or (38), the initial velocities are determined. These values are used as initial conditions for the moving even and odd modes in molecular dynamics simulations. The trial solution in w-space is identical to eq. (40) with ϕ_n replaced with φ_n.

IV. D. Molecular Dynamics Simulations

To determine if these modes move according to the analytical predictions we numerically integrate the equations of motion [eq. (32)] using a fifth-order Gear predictor-corrector algorithm [29,28] for a chain of 100 particles with periodic boundary conditions. A small time step, $\Delta t = 0.01/\omega_o$, is used in order to approximately conserve energy over the time interval of the measurement.

Figure 9 presents the trajectory of the moving odd localized mode by overlaying the displacement versus time at three different sites to illustrate how the particle responds as the excitation passes and also to demonstrate that the velocity of the excitation packet is nearly constant. The farthest that we have tracked this particular moving mode is 400 lattice sites, which corresponds to a time of 1800 periods of ω_o. The ripples in the figure between the packets represent plane waves that are excited when the initial conditions do not correspond to the exact eigenfunction of the moving localized mode. The relatively small amplitudes seen here attest to the accuracy of the method used to obtain these particular initial conditions.

The calculated dispersion curves for the odd mode characterized by four different anharmonicity values are given by the dashed lines in Fig. 10. The simulation results are plotted as the open circles. In each case the solid line identifies the region of $\omega(k)$ space

261

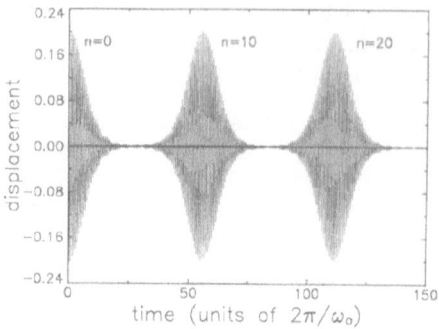

Figure 9. The localized excitation passing three different lattice sites as a function of times. The displacement amplitude of the particle at $n = 0$ decreases with increasing time. Super-imposed on top of that picture is the increasing and decreasing amplitude at sites $n = 10$ and $n = 20$ as the vibrational envelope moves along the 100 atom chain. The envelope V_g is 6% of the sound velocity. After Ref. (14).

where, according to molecular dynamics simulations, the excitation moves with a constant envelope velocity over at least 15 lattice sites. This velocity is typically smaller than 15 % of the lattice sound velocity. The absence of a solid line at small k values identifies that region where the mode either does not move or moves only a few sites before stopping. At larger k values, on the other side of the solid line, the mode moves but decelerates, making it difficult to assign a group velocity to the moving excitation. Similar results have been obtained for the even mode.

Another approach to characterize the moving excitation packet is to assume, on the basis of shape shown in Fig. 9, a specific envelope amplitude form for ϕ_n or φ_n , namely,

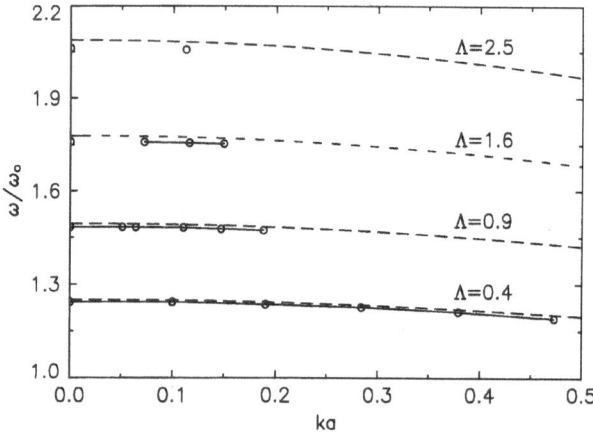

Figure 10. Dispersion curve of the odd mode for four different anharmonicity values. The values from top to bottom are $\Lambda = 2.5, 1.6, 0.9$ and 0.4. Dashed -calculated, solid-simulated. After Ref. (14).

$$w_n \text{ or } u_n = (-1)^n \alpha \exp\left\{-\left[K(na - V_g t)\right]^2\right\}\cos(kna + \omega t). \tag{41}$$

When this form is substituted directly into the equations of motion to obtain the initial conditions, the plane wave ripples in the simulation data are slightly larger than for the method described above, indicating that these initial conditions do not correspond as well to the correct eigenvector. Therefore we have fit eq. (41) to the simulation results described in the previous paragraph to obtain the dispersion curve, but did not use it to calculate the initial shape of the intrinsic localized mode. The parameters are numerically determined from the simulation amplitudes of the excitation as it travels through four consecutive lattice sites from $n = 10$ to $n = 13$. While somewhat arbitrary, these coordinates allow enough simulation time for any extra vibrational energy to dissipate into plane waves, but are close enough to the initial position at $n = 0$ to allow the moving excitations to be characterized by a uniform group velocity.

The solid line in Fig. 11 shows a typical simulation trace of displacement versus time with a gaussian function best fit represented by the dashed curve. Note that the best fit to a hyperbolic secant function, represented by the dotted curve does not agree with the simulated amplitude in the wings while the gaussian envelope matches fairly well over the entire time interval. This indicates that the eigenvector of the moving intrinsic localized mode has a shape somewhere between the hyperbolic secant function predicted by soliton theory and the exponential decay of the localized modes associated with a mass defect in a one dimensional lattice [30].

The measured dispersion curve obtained with the specific functional form given by eq. (41) is represented by the open circles in Fig. 10. For the lowest frequency modes, there is good agreement between the dashed line and the open circles. Even at the largest anharmonicity value the difference between the two curves is less than 3%, and most of this deviation is associated with the first order correction [8] to the rotating wave approximation, missing from eqs. (34), (37), (38) and (39) and hence the dashed curves.

IV. E. Discussion

Molecular dynamics simulations using a predictor-corrector algorithm are used to demonstrate the translational stability of intrinsic localized modes of different parities. The breathing even mode is found to spontaneously decompose into moving modes that are repelled from each other, while the even mode that is centered on two adjacent lattice sites is stable for thousands of vibrational periods. The odd parity mode exhibits a nearly-stable behavior, drifting from its initial site after approximately fifty periods. The vibrational

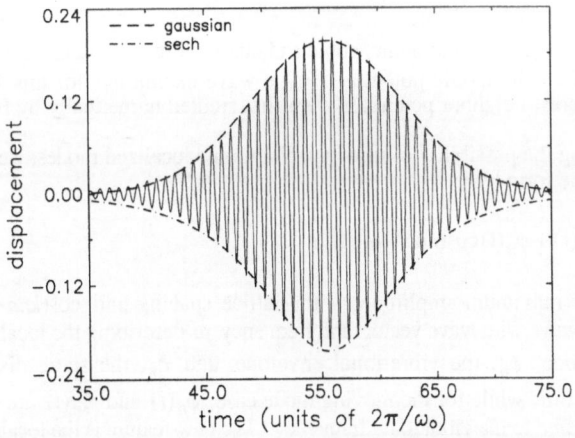

Figure 11. Displacement versus time as the vibrational excitation passes through lattice site $n = 10$. The dashed curve represents the best fit of the excitation envelope to a gaussian function lineshape as given by eq. (41). For comparison, the dot-dashed curve represents a hyperbolic secant function envelope. After Ref. (14).

frequencies of the simulated odd and exciton-like even mode are in good agreement with the values found analytically using the rotating wave approximation.

The properties of moving anharmonic localized modes in the 1-D anharmonic lattice have been derived analytically and tested using molecular dynamics simulations. The eigenvector shape is different from the hyperbolic secant function previously found for continuous systems treated in plasma physics and nonlinear optics [31-33]. Our result is also basically different from the Toda discrete lattice soliton [34], which exhibits a $sech^2$ function and contains no high frequency oscillatory behavior, and is also different from the soliton solution of the discrete Ablowitz-Ladik equation [35], which is of the form $C \sec h\left[K(na - V_g t)\right] \exp\left[i(kna - \omega t)\right]$. One fundamental difference is that eq. (32) is a typical nonlinear lattice field equation in solid state physics, while the Ablowitz-Ladik equation was originally motivated mathematically by the inverse scattering formalism in soliton theory [31] and is not derivable from a Hamiltonian. The approach used here to examine the odd vibrational modes may also be applied to two or three dimensional lattices; hence, it may identify a way to study solitons in higher dimensions, in contrast with conventional methods where the general existence of solitons has only been established in one dimension.

V. STATIONARY AND MOVING INTRINSIC LOCALIZED MODES IN LATTICES WITH CUBIC AND QUARTIC ANHARMONICITY

V. A. Solution of the Equations of Motion

In physical systems, odd order anharmonicities are present in the Taylor's expansion of the interatomic potential, so a realistic model needs to take them into account. It is just these terms which give rise to the thermal expansion of the lattice. The simplest way to examine anharmonic modes in a more realistic environment is to add cubic anharmonicity to the potential used previously. So let us now consider a perfect 1-D monatomic lattice in which each particle of mass m interacts only with its nearest neighbors through a potential that includes both cubic and quartic terms. The equation of motion for the particle at the nth site is then

$$m\frac{d^2 u_n}{dt^2} = K_2\left(u_{n+1} + u_{n-1} - 2u_n\right) + K_3\left[\left(u_{n+1} - u_n\right)^2 - \left(u_{n-1} - u_n\right)^2\right] + K_4\left[\left(u_{n+1} - u_n\right)^3 + \left(u_{n-1} - u_n\right)^3\right]$$

(42)

where K_2, K_3 and K_4 are harmonic, cubic and quartic force constants, respectively that are derived from the anharmonic potential. Plane wave excitations for this 1-D lattice with anharmonic nearest-neighbor potentials were first studied numerically by Fermi, Pasta and Ulam [15].

Anticipating the possibility of moving anharmonic localized modes, we seek a solution to Eq. (42) of the form [20]

$$u_n = \alpha\left[\xi_n(t) + \phi_n(t)\cos(kna + \omega t)\right],$$

(43)

where α is the maximum amplitude, a the lattice spacing and $\cos(kna + \omega t)$ is a left moving carrier wave with wave vector and frequency ω describing the localized mode. For a stationary mode, ϕ_n, the vibrational envelope, and ξ_n, the static displacement, are independent of time while for the moving mode case, $\phi_n(t)$ and $\xi_n(t)$ are slowly varying with time compared to the vibrational frequency. The new feature is the local static distortion which must be included to obtain analytical solutions for odd and even parity modes with unrestricted cubic anharmonicity. To obtain the solution we insert eq. (43) into eq. (42), use the RWA to remove high frequency harmonics and use the slowly varying time dependence of the envelope and static displacement terms to eliminate their second time derivatives. The results are three sets of coupled equations. To satisfy for all time the solution represented by eq. (43), the coefficients of both the in-phase and out-of-phase oscillatory terms, as well as those of the static displacement, must independently sum to zero. Equating the coefficients of the sine terms gives

$$\left(\frac{\omega}{\omega_o}\right)\left(\frac{d\phi_n}{dt}\right) = \frac{\omega_o \sin(ka)}{8}\left\{ \phi_{n+1}\begin{bmatrix} 1+\frac{3}{4}\Lambda\left(\phi_{n+1}^2 - 2\phi_{n+1}\phi_n \cos(ka) + \phi_n^2\right)+ \\ 2\Gamma(\xi_{n+1}-\xi_n)+3\Lambda(\xi_{n+1}-\xi_n)^2 \end{bmatrix} - \phi_{n-1}\begin{bmatrix} 1+\frac{3}{4}\Lambda\left(\phi_n^2 - 2\phi_n\phi_{n-1}\cos(ka) + \phi_{n-1}^2\right)+ \\ 2\Gamma(\xi_n-\xi_{n-1})+3\Lambda(\xi_n-\xi_{n-1})^2 \end{bmatrix} \right\},$$ (44)

the cosine terms,

$$4\left(\frac{\omega}{\omega_o}\right)^2 \phi_n = \left[\phi_n - \phi_{n+1}\cos(ka)\right]\begin{bmatrix} 1+\frac{3}{4}\Lambda(\phi_{n+1}^2 - 2\phi_{n+1}\phi_n \cos(ka) + \phi_n^2)+ \\ 2\Gamma(\xi_{n+1}-\xi_n)+3\Lambda(\xi_{n+1}-\xi_n)^2 \end{bmatrix}$$
$$+\left[\phi_n - \phi_{n-1}\cos(ka)\right]\begin{bmatrix} 1+\frac{3}{4}\Lambda(\phi_n^2 - 2\phi_n\phi_{n-1}\cos(ka) + \phi_{n-1}^2)+ \\ 2\Gamma(\xi_n-\xi_{n-1})+3\Lambda(\xi_n-\xi_{n-1})^2 \end{bmatrix},$$ (45)

and the static displacement terms,

$$\Lambda(\xi_{n+1}-\xi_n)^3 + \Gamma(\xi_{n+1}-\xi_n)^2 + \frac{1}{2}\Gamma(\phi_{n+1}^2 - 2\phi_{n+1}\phi_n \cos(ka) + \phi_n^2)$$
$$+(\xi_{n+1}-\xi_n)\left[1+\frac{3}{2}\Lambda(\phi_{n+1}^2 - 2\phi_{n+1}\phi_n \cos(ka) + \phi_n^2)\right] =$$
$$\Lambda(\xi_n-\xi_{n-1})^3 + \Gamma(\xi_n-\xi_{n-1})^2 + \frac{1}{2}\Gamma(\phi_n^2 - 2\phi_n\phi_{n-1}\cos(ka) + \phi_{n-1}^2)$$
$$+(\xi_n-\xi_{n-1})\left[1+\frac{3}{2}\Lambda(\phi_n^2 - 2\phi_n\phi_{n-1}\cos(ka) + \phi_{n-1}^2)\right]$$ (46)

where $\Gamma = K_3\alpha / K_2$, $\Lambda = K_4\alpha^2 / K_2$ and ω_o is again the maximum plane wave frequency. Equation (46) must be satisfied at each lattice site, therefore each side is independently equal to a constant of motion. The anharmonic localized modes we seek are characterized by exponential decays of both the vibrational amplitude, ϕ_n and the relative dc displacements, $(\xi_{n+1} - \xi_n)$, therefore all the terms in eq. (46) rapidly approach zero far from the mode center. The value of the constant of motion must then be zero for a localized mode solution, so the static equation of motion is

$$\Lambda(\xi_{n+1}-\xi_n)^3 + \Gamma(\xi_{n+1}-\xi_n)^2 + \frac{1}{2}\Gamma(\phi_{n+1}^2 - 2\phi_{n+1}\phi_n \cos(ka) + \phi_n^2)$$
$$+(\xi_{n+1}-\xi_n)\left[1+\frac{3}{2}\Lambda(\phi_{n+1}^2 - 2\phi_{n+1}\phi_n \cos(ka) + \phi_n^2)\right] = 0$$ (47)

To proceed with the analytical solution, we assume an exponential decay of the vibrational amplitudes starting with the nearest neighbor to the center particle [8]. [A better solution could be obtained by starting the exponential decay farther from the mode center, but as long as the anharmonic modes are very localized, the additional equations and variables do not significantly improve the accuracy.] With this simplifying approximation, the odd-parity eigenvector has the form,

$$\phi_0 = 1, \quad \phi_n = \phi_{-n} = (-1)^n \phi_1 \exp[-(n-1)qa] \quad \text{and} \quad \xi_0 = 0, \quad \xi_n = -\xi_{-n} \tag{48}$$

where $n > 0$. Similarly, the even mode is represented by

$$\phi_0 = 1, \quad \phi_{-1} = -1, \quad \phi_n = -\phi_{-n-1} = (-1)^n \phi_1 \exp[-(n-1)qa] \quad \text{and} \quad \xi_n = -\xi_{-n-1}. \tag{49}$$

Note that the factors of $(-1)^n$ in the expressions for ϕ_n indicate that we are measuring the wave vector from the zone boundary. For the odd mode, substitution of eq. (48) into eqs. (45) and (46) yields six coupled nonlinear equations which are numerically solved for the variables $\{\omega / \omega_o, \phi_1, \exp(-qa), \xi_1, (\xi_2 - \xi_1), (\xi_3 - \xi_2)\}$ for given values of the wave vector ka and the dimensionless anharmonicity parameters, Γ and Λ.

Similarly for the even mode, substitution of eq. (49) into eqs. (45) and (47) yields seven coupled equations which are numerically solved for the variables $\{\omega / \omega_o, \phi_1, \exp(-qa), \xi_1, (\xi_2 - \xi_1), (\xi_3 - \xi_2), (\xi_4 - \xi_3)\}$. The remaining distortions, $\xi_{n+1} - \xi_n$, are then obtained from eq. (45) for $n > 3$ and $n > 4$ for the odd and even modes, respectively, while the initial values of $d\phi_n / dt$ for moving modes are found using eq. (44).

The eigenvectors obtained using the procedure outlined above are shown in Fig. 12 for stationary odd and even modes with frequencies that are approximately 1.4 times the maximum frequency of the plane wave spectrum. The vibrational amplitudes are drawn as vectors perpendicular to the axis of motion to distinguish them from the static displacement due to the cubic anharmonicity, and the sizes of the displacements relative to the lattice constant are exaggerated for clarity (see figure caption). As the cubic anharmonicity is decreased from zero in Figs. 12(a) and (c) to significant negative values in Figs. 12(b) and (d) for the odd and even mode, respectively, the dominant effect is a expansion about the center; conversely, lattices with positive cubic anharmonicity would exhibit a compression. There is also a slight increase in the localization with the decreased cubic anharmonicity, but this effect is small for the values of the quartic anharmonicity that correspond to these particular initial conditions.

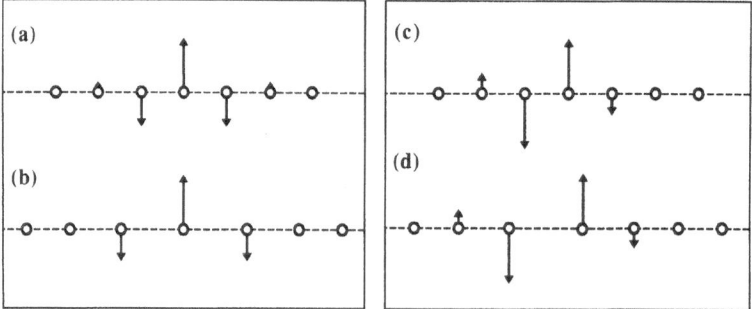

Figure 12. Vibrational amplitudes and static displacement for odd and even intrinsic localized modes. The odd mode parameters are (a) $\Lambda = 1.6$, $\Gamma = 0.0$ and (b) $\Lambda = 1.6$, $\Gamma = -2.4$ with a maximum vibrational amplitude of 0.4 and a maximum static displacement of 0.32. The lattice parameter is unity. The even mode parameters are (c) $\Lambda = 1.6$, $\Gamma = 0.0$ and (d) $\Lambda = 1.6$, $\Gamma = -1.2$ with a maximum vibrational amplitude of 0.3 and a maximum static displacement of 0.18. After Ref. (20).

V. B. Molecular Dynamics Simulations

1. Eigenvector and Stability of Stationary Localized Modes.

Eigenvectors such as those shown in Fig. 12 are obtained by solving the equations of motion and used as initial conditions in a computer simulation program [14, 20] which numerically integrates the equations of motions in a 1-D monatomic lattice with 512 sites and free end conditions. These boundary conditions are necessary because of the mismatch in the static displacements at the ends of the lattice due to the localized dc distortion. The lattice spacing, harmonic force constant and mass are all set to unity, and the simulation times are typically at least two hundred periods of the maximum plane wave frequency in order to observe the long term stability. Note that cubic anharmonicity changes the eigenvector of a stationary odd mode from $(..., -\phi_1, 1, -\phi_1,...)$ to $(..., -\phi_1-\xi_1, 1, -\phi_1+\xi_1,...)$, where ϕ_1 approaches $1/2$ for large quartic anharmonicities and the nearest-neighbor static displacement ξ_1 can be greater than unity for large cubic anharmonicity, i.e. the static displacement can be even larger than the maximum vibrational amplitude. Sandusky $et\ al.$ have reported that the odd-parity mode is unstable to this particular distortion and tends to move from its initial position, while the even mode remains stationary under the same perturbation [27]. We find that even for the large distortions used here, it takes ten to thirty periods before the odd mode moves from its initial position, while the even mode is translationally stable for at least hundreds of periods.

Figure 13 shows the simulation results for frequencies of stationary even modes as a function of the cubic anharmonicity Γ for several values of Λ, the quartic anharmonicity. These data are plotted as open circles, while the predicted frequencies from the analytical solution of the equations of motion [eqs. (45) and (47)] are represented by the different curves. There is good agreement over the entire range of cubic anharmonicities, although the red-shift of the frequency [8] due to higher order harmonics is apparent with large anharmonicities. Fourier transforms of the simulated displacements versus time curves indicate that the relative amplitude of the third harmonic is less than five percent of the strength of the fundamental, while the amplitude of the second harmonic is always completely negligible, even for the maximum value of the cubic anharmonicity shown here.

The results plotted in Fig. 13 also show that there is a limiting value of Γ for a given value of Λ. The boundary line identifying the region of stable localized modes is shown in Fig. 14. For small values of Λ, the vibrational frequency decreases with increasing Γ until it approaches the maximum frequency of the plane wave spectrum. The localized mode is

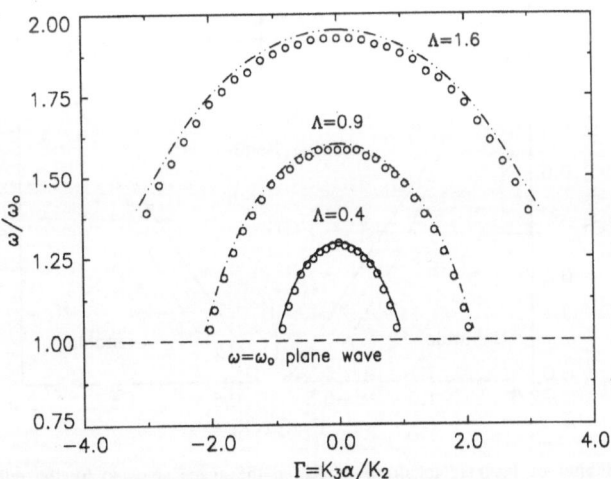

Figure 13. Frequencies of stationary even parity modes as a function of the cubic anharmonicity, Γ. The simulation and analytic results are represented by open circles and lines, respectively for several values of the quartic anharmonicity, Λ. After Ref. (20).

then observed to rapidly decay into plane waves and therefore no longer represents a stable vibrational configuration. This condition, $(\omega / \omega_o) = 1$, may be substituted into the equations of motion to obtain a stability curve in (Γ, Λ) space, which is plotted as the solid line labeled "b" in Fig. 14. This curve defines the maximum Γ value that can support intrinsic localized modes for a given value of Λ.

For larger values of Γ and Λ, the instability in the simulations occurs at frequencies well above the maximum plane wave frequency. An examination of the equations of motion shows that this different type of instability is associated with the appearance of a double minimum in the nearest-neighbor potential for sufficiently large cubic anharmonicities. With the notation $v_n = u_n - u_{n-1}$, the nearest-neighbor potential may be written

$$U_n = \frac{K_2}{2} v_n^2 + \frac{K_3}{3} v_n^3 + \frac{K_4}{4} v_n^4, \tag{50}$$

and the derivative with respect to the relative displacements is

$$\frac{d U_n}{d v_n} = K_2 v_n \left(1 + \frac{K_3}{K_2} v_n + \frac{K_4}{K_2} v_n^2 \right). \tag{51}$$

The potential energy function given by eq. (50) is plotted in Fig. 15 for three values of K_3 and fixed values of K_2 and K_4. For small cubic anharmonicities ($K_3 = 1$), U_n is a simple potential well with a single minimum at $v_n = 0$. The potential becomes increasingly asymmetric with increasing cubic anharmonicity until a second minimum appears when

$$v_n = -\frac{K_3}{2K_2} \pm \sqrt{\left(\frac{K_3}{2K_2} \right)^2 - \frac{K_4}{K_2}}. \tag{52}$$

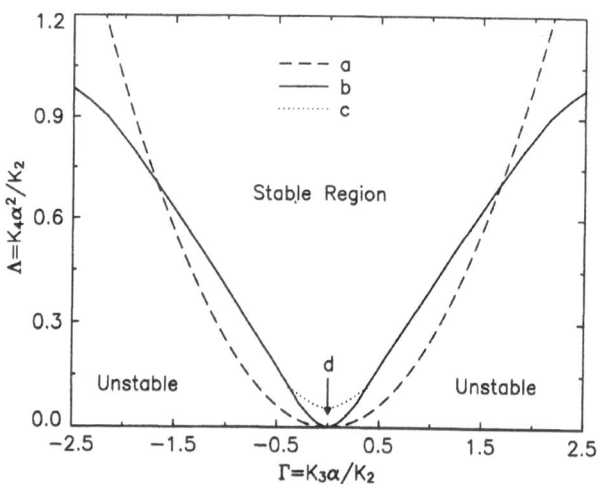

Figure 14. Habitat of intrinsic localized modes in the space spanned by the cubic and quartic anharmonicities. Curves (a) and (b) represent boundaries where the modes become unstable due to the double-well potential and the decay into plane waves, respectively. Area (d), lying between curves (b) and (c), designates the region of the continuum approximation as described in the text. After Ref. (28).

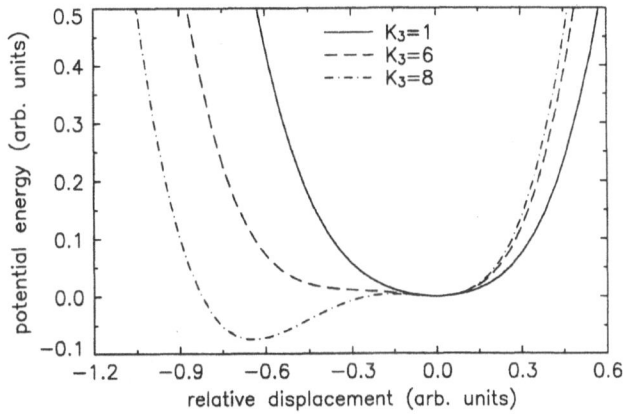

Figure 15. Nearest neighbor potential for different cubic coefficients. The coefficient of the harmonic and quartic terms are $K_2 = 1$ and $K_4 = 10$ for all three curves, which corresponds to the formation of a double well for $K_3 \geq 2\sqrt{10} \cong 6.3$. There may be stable intrinsic localized modes beyond this limit, but they are not described in the framework presented here. After Ref. (7).

For the root given by eq. (52) to be real, the potential coefficients must satisfy $K_3^2 \geq 4K_2 K_4$, or, in terms of the dimensionless anharmonicity parameters,

$$\Lambda_3^2 \geq 4\Lambda_4. \tag{53}$$

When this condition is satisfied, there is an abrupt change in the shape of the nearest-neighbor potential. For these particular parameters, Fig. 15 indicates that this occurs when the cubic coefficient lies in between $K_3 = 6$ and $K_3 = 8$.

The condition given by eq. (53) is plotted as the dashed curve labeled "a" in Fig. 14. For a particular intrinsic localized mode to be stable, not only must the vibrational frequency be above the maximum plane wave frequency, but the cubic anharmonicity must be small enough to avoid the formation of a double-well potential. This stability region is that portion of parameter space lying above both the solid and dashed curves in Fig. 14. Another type of localized mode may exist in the regime described by eq. (53); however this possibility has not yet been explored.

There is now some value in contrasting our local mode results with the earlier work by Flytzanis *et al.* on asymmetric envelope solitons within the continuum approximation [17]. The asymmetric envelope soliton solutions described there have effective widths of at least fourteen lattice sites, so to make a comparison, we solve the equations of motion for an intrinsic localized mode of approximately the same width. This solution is plotted as the dotted curve labeled "c" in Fig. 14, and the area labeled "d" then defines that region where the continuum approximation is valid. This represents only a small fraction of the area supporting stable intrinsic localized modes, therefore the local mode solutions reported here cover a much larger region of parameter space than those obtained using the continuum approximation.

2. Dispersion of Moving Intrinsic Localized Modes. The initial amplitudes and velocities for non-zero wave vectors are calculated using the same method as for stationary modes. The displacements at three equally spaced lattice sites are plotted as a

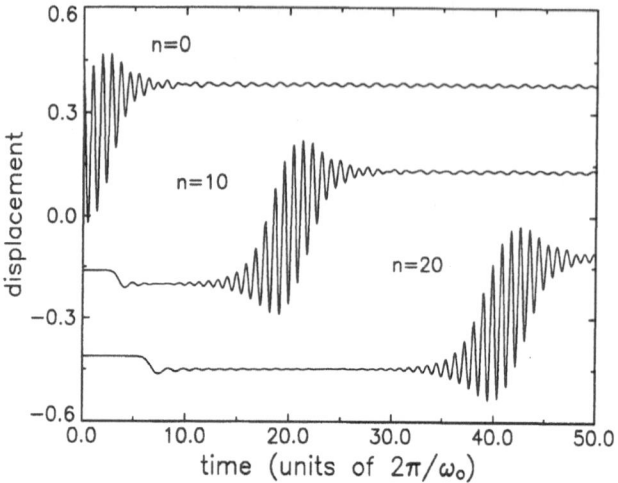

Figure 16. Displacement versus time of a moving localized mode. The response at three equally spaced lattice sites shows the motion of an intrinsic localized mode characterized by the parameters $\Lambda = 0.4$, $\Gamma = 0.8$ and $ka = 0.3$. The three curves are displaced vertically by 0.25 units for clarity. After Ref. (20).

function of time in Fig. 16 to show the static displacement and vibrational amplitude as the mode propagates through the lattice with positive cubic anharmonicity. Consider the effect of the static displacement on the particle at the $n = 10$ site in the figure. Initially, at small times, sites to the right of the mode are displaced towards the center due to the negative dc distortion. Later as the mode moves through the 10th lattice site, this displacement goes to zero and then changes sign after the mode has moved past this lattice site. The group velocity of this particular moving excitation is approximately 15% of the lattice sound velocity. Note that there is a small increase in the static displacement at the $n = 10$ site after a time of approximately three periods of the maximum plane wave frequency. This is caused by a small-amplitude, supersonic pulse that propagates in both directions away from the local mode at the beginning of the simulation. This pulse may be a long-wavelength acoustic kink soliton that is also a solution of these equations of motion [18].

Studies of moving asymmetric envelope solitons in the continuum approximation have shown that the solution is composed of the sum of two terms [17]. A cosine term modulated by a hyperbolic secant describes the high frequency vibrations, while the static displacement is represented by a hyperbolic tangent function. However the local modes studied here are not well characterized by these functions, and in fact, the offsets can only be fit to a hyperbolic tangent if this function is raised to an arbitrarily large power. Since we have not characterized the static displacement around these localized modes with an analytical function, we quantitatively describe their properties by transforming to coordinates that depend on the difference of the displacements, $v_n = u_n - u_{n-1}$.

This transformation reverses the vibrational parity, so that modes that had even parity in u-space now have odd parity in v-space and vice versa [34]. In addition, the static component now decreases exponentially with increasing distance from the center in this coordinate system. The simulated displacement versus time curves are then numerically fit using the function

$$v_n = \alpha\left[\xi_n(t) + \phi_n(t)\cos(kna + \omega t)\right], \tag{54}$$

where $\xi_n(t)$ and ϕ_n are both approximated by Gaussian functions [11]. Recall that the results in Section IV for systems with only quartic anharmonicity have shown that when determining the parameters of the carrier wave, there is some flexibility in choosing the functional form of the envelope of the moving localized mode.

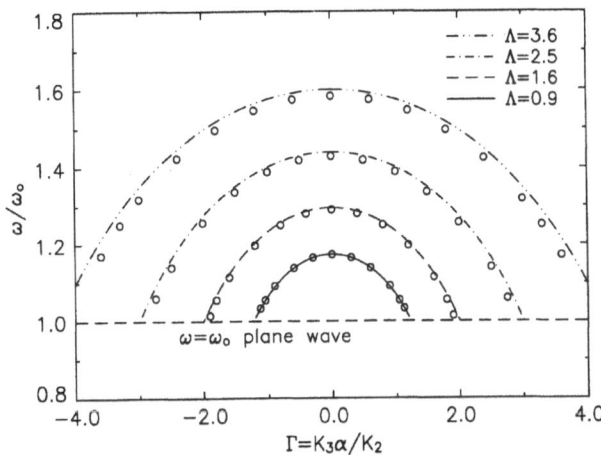

Figure 17. Frequencies of moving localized modes as a function of Γ for several values of Λ. Simulation and analytical results are represented by open circles and lines, respectively for a constant wave vector, $ka = 0.1$. After Ref. (20).

Figure 17 displays the analytic and simulation frequencies when the eigenvectors of the moving localized modes are characterized by a single value of the wave vector in the analytical calculations. The magnitude of the wave vector is $ka = 0.1$, which corresponds to moving localized modes with group velocities that are approximately 5% of the speed of sound in the monatomic chain. The simulation frequencies, represented by the open circles, are in close agreement with the frequencies that are predicted by the analytic solutions of the equations of motion, which are plotted by different lines for several values of the quartic anharmonicity. Note that this figure represents only a cross-section of the space spanned by the parameters ka, Λ and Γ and is only a small portion of the region where stable moving localized modes exist.

Figure 18 presents another way of demonstrating the simulation results. It shows the normalized frequency versus the wave vector at a particular plane crossing the Λ axis for several values of the cubic anharmonicity. The agreement between simulated and analytical frequencies is very good for small values of the wave vector and group velocity, but the simulation frequencies begin to decrease more rapidly than the analytic ones at larger wave vectors. This slowing down occurs because the initial displacements and velocities are only approximate eigenvectors, hence the mode loses amplitude and energy as the relative displacements adjust to the correct configuration. The simulation data therefore correspond to smaller values of the cubic and quartic anharmonicity than the analytical ones, and the frequencies are shifted down relative to the predicted values. The absence of simulation results in the lower right hand corner of Fig. 18 represents the region where simulations produce modes that either decay into plane waves or slow down too rapidly to be characterized by a single wave vector. The top curve, corresponding to $\Gamma = 1.8$, represents the maximum cubic anharmonicity where stationary localized modes are stable for this particular Λ value. Note that the agreement between the simulation and analytic results is better at smaller wave vectors and higher frequencies because the effect of the second derivatives that were removed from the equation of motion is minimized and the exponential approximation of the eigenvector in Eqs. (48) and (49) is more accurate.

V. C. Discussion

The intrinsic localized modes described here for a lattice with cubic and hard quartic nonlinearity in the nearest-neighbor potential provide examples of anharmonic vibrations that stabilize local static distortions. The excitation of such modes can produce either lattice

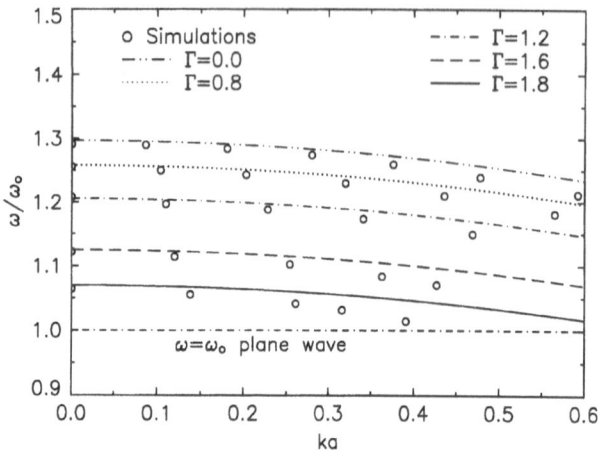

Figure 18. Moving localized mode dispersion curves for different values of Γ. Simulation and analytical results are represented by open circles and lines, respectively for $\Lambda = 1.6$. After Ref. (20).

expansion or contraction, depending on the sign of the cubic anharmonicity. For realistic potentials the cubic term is negative so localized dc expansion is associated with the generation of a localized excitation. The only restrictions on the relative sizes of the anharmonicities are that the frequency of the localized mode must be above the maximum frequency of the plane wave spectrum and the cubic anharmonicity must be small enough to avoid the formations of a double-well potential between the center particles. For realistic potentials it has been shown that the cubic term is large enough so that a local mode is not possible [36, 37]. However, for diatomic lattices which contain a gap between the optic and acoustic branches there is a new possibility: an anharmonic localized gap mode. At the bottom of the optic branch at the zone boundary, the light particle is moving π out of phase with the light atoms in the neighboring cells. Because of the form of the anharmonicity in these standard potentials, this plane wave mode can more easily convert to an anharmonic localized mode by dropping into the phonon gap while simultaneously producing a localized dc distortion of the lattice. However, an examination of localized gap modes for the nonlinear diatomic lattice is much more intricate and hence these modes are not considered here. So the local mode solutions for the (K_2, K_3, K_4) potential problem described here should be taken as simply instructive as to how one determines the new feature of the cubic potential term, i.e., the dc distortion, on the form of the solution.

VI. CONCLUSIONS

A variety of new dynamical properties have been described for one dimensional chains with sufficiently anharmonic potentials. Molecular dynamics simulations have been found to be indispensable in analyzing the stability of local mode solutions associated with these coupled nonlinear equations. When applied to test the local mode eigenvectors derived from the analytical solution of the equation of motion, very good agreement is obtained over a large range of cubic and quartic anharmonicities. In accord with Bloch's theorem, which requires translational symmetry in a perfect lattice, dispersion curves have been found for moving localized modes in the presence or absence of cubic anharmonicity, and again there is good agreement between simulation and analytical results. The most surprising general result is that large anharmonicity in a perfect 1-D lattice produces localized dynamical effects reminiscent of those previously found for defect lattices in the harmonic approximation. Spatially localized as well as delocalized vibrational modes are found in both cases. It is quite remarkable that once the anharmonic system is described in terms of such localized and delocalized modes, which turn out from the examples presented here to be only weakly interacting, the low energy vibrational excitations of the resulting system again correspond to those of a set of "effective" harmonic oscillators. The strong coupling between plane wave harmonic oscillators previously associated with large anharmonicity has disappeared because both homogeneous and inhomogeneous waves have been introduced, even though this perfect system has no boundaries or defects on which to "center" such inhomogeneous waves.

ACKNOWLEDGMENTS

We wish to thank J. B. Page and R. H. Silsbee for many useful discussions. Some of this effort was supported by NSF-DMR-9312381 and ARO-DAAL03-92-G-0369. The MRL Central Facilities are supported by the NSF under Award No. DMR-9121654.

[1] Present address: Code 6691, Naval Research Laboratories, Washington, D.C. 20375
[2] Permanent address: Institute of Spectroscopy, Russian Academy of Sciences, Troitsk, Moscow region, Russia

REFERENCES

1. A. J. Sievers and S. Takeno, Phys. Rev. Lett. **61**, 970 (1988).
2. S. Takeno and A. J. Sievers, Solid State Commun. **67**, 1023 (1988).
3. S. Takeno, K. Kisoda and A. J. Sievers, Prog. Theor. Phys. Suppl. **94**, 242 (1988).
4. J. B. Page, Phys. Rev. B **41**, 7835 (1990).
5. R. Bourbonnais and R. Maynard, Phys. Rev. Lett. **64**, 1397 (1990).
6. V. M. Burlakov, S. A. Kiselev and V. N. Pyrkov, Sol. State Comm. **74**, 327 (1990); V. M. Burlakov, S. A. Kiselev and V. N. Pyrkov, Phys. Rev. B **42**, 4921 (1990).
7. S. A. Kiselev, S. R. Bickham and A. J. Sievers, Comm. in Cond. Matt. Phys. **17**, 135 (1995).
8. S. R. Bickham and A. J. Sievers, Phys. Rev. B **43**, 2339 (1991).
9. A. S. Barker and A. J. Sievers, Rev. Mod. Phys. **47**, S1 (1975).
10. S. Takeno, J. Phys. Soc. Jpn. **59**, 1571 (1990).
11. S. Takeno and K. Hori, J. Phys. Soc. Jpn. **59**, 3037 (1990).
12. V. M. Burlakov, S. A. Kiselev and V. I. Rupasov, Phys. Lett. A **147**, 130 (1990).
13. V. M. Burlakov, S. A. Kiselev and V. I. Rupasov, JETP Lett. **51**, 544 (1990).
14. S. R. Bickham, A. J. Sievers and S. Takeno, Phys. Rev. B **45**, 10344 (1992).
15. E. Fermi, J. R. Pasta and S. M. Ulam, in *Collected Works of E. Fermi*, edited by E. Segre (University of Chicago Press, Chicago, 1955).
16. A. M. Kosevich and A. S. Kovalev, Sov. Phys.-JETP **40**, 891 (1974).
17. N. Flytzanis, S. Pnevmatikos and M. Remoissenet, J. Phys. C. **18**, 4603 (1985).
18. V. M. Burlakov and S. A. Kiselev, Sov. Phys. JETP **72**, 854 (1991).
19. S. Takeno and K. Hori, J. Phys. Soc. Jpn. **60**, 947 (1991).
20. S. R. Bickham, S. Kiselev and A. J. Sievers, Phys. Rev. B **47**, 14206 (1993).
21. A. Haug, *Theoretical Solid State Physics*, (Pergamon, Oxford, 1972), vol. 2.
22. W. H. Press, S. A. Teukolsky, W. T. Vettering and B. P. Flannery, *Numerical Recipes*, (Cambridge University Press, Cambridge, 1992).
23. K. W. Sandusky, J. B. Page, A. Rosenberg and A. J. Sievers, Phys. Rev. B **47**, 5731 (1993).
24. W. Ludwig, *Recent Developments in Lattice Theory*, G. Hoehler, Ed., Springer Tracts in Modern Physics (Springer-Verlag, Berlin, 1967), vol. 43.
25. N. Flytzanis, S. Pnevmatikos and M. Peyrard, J. Phys. A: Math. Gen. **22**, 783 (1989).
26. O. A. Chubukalo, A. S. Kovalev and O. V. Usatenko, Phys. Rev. B **47**, 3153 (1993).
27. K. W. Sandusky, J. B. Page and K. E. Schmidt, Phys. Rev. B **46**, 6161 (1992).
28. S. R. Bickham, Ph. D. thesis, Cornell University, 1995.
29. M. P. Allen and D. J. Tildesley, *Computer Simulations of Liquids*, (Clarendon Press, Oxford, 1987).
30. E. Infield and G. Rowlands, *Nonlinear Waves, Solitons and Chaos*, (Cambridge University Press, Cambridge, 1990).
31. D. L. Mills, *Nonlinear Optics*, (Springer-Verlag, New York, 1991).

32. D. L. Mills and S. E. Trullinger, Phys. Rev. B **36**, 947 (1987).
33. M. Toda, J. Phys. Soc. Jpn. **23**, 501 (1967).
34. M. Toda, *Theory of Nonlinear Lattices, 2nd Ed.*, (Springer-Verlag, New York, 1989).
35. M. J. Ablovitz and J. F. Ladik, J. Math. Phys. **17**, 1011 (1976).
36. S. A. Kiselev, S. R. Bickham and A. J. Sievers, Phys. Rev. B **48**, 13 508 (1993).
37. S. A. Kiselev, S. B. Bickham and A. J. Sievers, Phys. Rev. B **50**, 9135 (1994).

COLLECTIVE EXCITATIONS IN MAGNETIC MATERIALS

G.A. Gehring

Department of Physics and
Sheffield Centre for Advanced Magnetic Materials and Devices
University of Sheffield
Sheffield S3 7RH, UK

ABSTRACT

The collective excitations in a magnet are spin waves. These are reviewed for bulk ferromagnets and antiferromagnets. The importance of dipolar interactions for ferromagnets are discussed. There is much current interest in lower dimensional structures - layer compounds and ultra thin films for two dimensions and chain compounds for one dimensions. The form of the ground state and the excitations in lower dimensional systems are discussed for both ferromagnets and antiferromagnets. The spectroscopic methods appropriate for observing magnetic excitations are reviewed.

1. EXCITATIONS IN FERROMAGNETS

1A. General Considerations

This section is concerned with the elementary excitations in ferromagnets called *magnons* which may be excited from the ground state. Thus they may be observed experimentally at low temperature. We shall examine the magnon energies and not consider any lifetime effects which are usually very small in this regime.

A ferromagnet is an example of a co-operative system. As the temperature is raised near to a characteristic temperature the many excitations which are then excited destroy the order. In this regime the magnons are very strongly interacting and scatter off each other; this regime will not be discussed in this article.

A ferromagnet is an easily visualised example of a material in which the symmetry is lowered at the phase transition. In an insulating ferromagnet each magnetic atom has a spin s which carries a magnetic moment, below the transition the spins become aligned along some direction which is traditionally taken to be the z axis.

Although many of the basic ideas are the same the detailed theory is very different for transition metal compounds and rare earths. This is because in rare earth compounds (except those containing gadolinium for which L is zero in the ground state) the spin orbit coupling is very strong. The consequences of this are that one should not discuss ordering spins but rather the total angular momentum J and this frequently causes the exchange interaction to be anisotropic and also to produce strong magnetoelastic effects. Another difference is that the same formulation can be used to discuss the strongly magnetic rare earth metals as in these materials the f electrons are localised as in insulating materials a review of this area may be found in [1] . The rare earth compounds in which the f electrons are not fully localised show 'heavy fermion' properties and their magnetic excitation which are reviewed in [2] are also outside the scope of the article.

In a transition metal one may associate a magnetic moment with a given volume of the material and discuss the excitation energies using a Landau-Ginzburg formulation. The conclusions of this section while written in terms of an insulator actually also apply to a ferromagnetic metal provided that the wave length of the excitations are sufficiently long - this statement is clarified in section III where we discuss the effects which are specific to these metals.

The interaction responsible for the transition to the magnetic state is the exchange interaction, other important terms are the dipolar interactions (which may make a significant contribution to the ordering in rare earth crystals) and the magnetocrystalline anisotropy. The exchange Hamiltonian is given in equation (1) for convenience we will discuss the case in which the exchange interaction is only finite for nearest neighbours. In this case $R_m = R_n + \rho$.

$$H_{ex} = -\sum_{\alpha = 1n,m}^{3} \sum J^{\alpha\alpha}(R_m - R_n)S_n^{\alpha}S_m^{\alpha} \tag{1}$$

In a cubic material the values of $J^{\alpha\alpha}$ are all equal. This need not be true for a crystal of lower symmetry and in this case the z direction is chosen as the direction for which the value of $J^{\alpha\alpha}$ is the largest. The Hamiltonian above commutes with the total z component of spin,

$$\left[H_{ex}, S_{tot}^z\right] = 0 \quad where \quad S_{tot}^z = \sum_n S_n^z \tag{2}$$

A consequence of this is that all the eigenstates H are also eigenstates of S^z_{tot} and as there is only one state for which S^z_{tot} takes its maximum value of Ns (here N is the total number of magnetic atoms in the sample) the ground state is known exactly.

In three dimensions we may use mean field theory to relate J to the Curie temperature, T_c. We consider a crystal in which there is one magnetic atom in a volume Ω and the magnetic moment associated with each spin is $g\mu_B s$. Above T_c the effective field acting on any spin is given by

$$H_e = H_{ext} + \lambda M \tag{3}$$

where $\quad M = \dfrac{g\mu_B <s^z>}{\Omega} \quad$ and $\quad \lambda = \dfrac{2J^{zz}z\Omega}{g^2\mu_B^2} \tag{4}$

The number of magnetic neighbours interacting with any given spin is taken as z.

The susceptibility is given by Curie's law,

$$M = \frac{C}{T}H_{eff} \quad where \quad C = \frac{g^2\mu_B^2 S(S+1)}{3k_B T\Omega} \tag{5}$$

This leads to the Curie Weiss law and an expression for T_c in terms of J.

$$\chi = \frac{C}{T-C\lambda} \qquad T_c = C\lambda = \frac{2zs(s+1)J}{3k_B} \tag{6}$$

The Hamiltonian for magnetocrystalline anisotropy takes the following forms for (a) cubic , (b) tetragonal and (c) orthorhombic symmetry,

$$H_{anis}^{cubic} = -K_4 \sum_n S_n^{x4} + S_n^{y4} + S_n^{z4}$$

$$H_{anis}^{tet} = -K_2 \sum_n S_n^{z2} + H_{anis}^{cubic} \tag{7}$$

$$H_{anis}^{ortho} = -K_2^1 \sum_n \left(S_n^{x2} - S_n^{y2}\right) + H_{anis}^{tet}$$

1. B. Linear Spin Wave Theory for bulk crystals

At low temperatures we consider the elementary excitations away from the fully ordered ground state.

1. Exchange Dominated Regime. The elementary spin wave excitations correspond to the states in which the magnetisation has been reduced by one unit. It is convenient to define the spin raising and lowering operators and the boson spin wave operators using the Holstein -Primakoff transformation,

$$S_n^+ = S_n^x + i S_n^y = \sqrt{2s} \left(1 - \frac{a_n^+ a_n}{2s} \right)^{1/2} a_n \tag{8a}$$

$$S_n^- = S_n^x - i S_n^y = \sqrt{2s} \, a_n^+ \left(1 - \frac{a_n^+ a_n}{2s} \right)^{1/2} \tag{8b}$$

$$S_n^z = S - a_n^+ a_n \tag{8c}$$

The magnon variables b_k^+ and b_k are the fourier transforms of the a_n operators,

$$a_n = \frac{1}{\sqrt{N}} \sum_k e^{-ik_1 R_n} b_k \quad ; \quad a_n^+ = \frac{1}{\sqrt{N}} \sum_K e^{ik_1 R_n} b_k^+ \tag{8d}$$

We find the spin wave energies by writing the spin operators in the Hamiltonian in terms of the boson operators, b_K^+ and b_K , the expression below is for cubic symmetry and for an applied magnetic field B_0 along the z direction. A *magnon* is a quantum of spin reversal.

$$H_{exch} + H_{aris} + H_{zeeman} = \varepsilon_{gs} + \sum_k \left[2zsJ(1 - \gamma_k) + 4K_4 s^3 + 2gm_B B_o \right] b_k^+ b_k$$
$$+ O\left(b^+ b^+ bb \right) \tag{9a}$$

The energies of the linear spin waves are found to be,

$$\varepsilon_k = 2 Jsz(1 - \gamma_k) + 4 K_4 s^3 + 2 gm_B B_o \tag{9b}$$

278

γ_k is defined as $\gamma_k = \dfrac{1}{z}\sum_p e^{ik\rho}$

The spin wave energies are sketched in figure(1) below.

In many materials the exchange energy is much larger than the anisotropy energy, so it interesting to consider the case when both the anisotropy energy and the external field are absent.   There is then no interaction which determines the ordering, z, axis. Consequently any state in which all the spins are aligned is an equally good ground state.

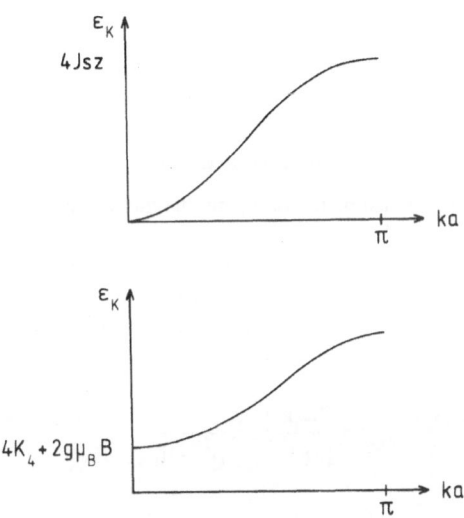

Figure (1) The upper curve shows the spin wave energy for an isotropic ferromagnet and the lower curve with the addition of anisotropy and an external magnetic field.

We may imagine the total spin being tilted through a small angle (π / N) which corresponds to the excitation of one spin wave with exactly zero k value - from our previous arguments we know that this corresponds to zero energy since both states are ground states. We should expect that the spin wave energies should be a smooth function of k and hence that the spin wave energies should tend to zero as k tends to zero. This is confirmed by the results of equation (9b), here a is the lattice spacing,

$$E_k = 2_s J(ka)^2 = D(ka)^2 \tag{10}$$

$$D = \frac{3kT_c}{z(s+1)} \tag{11}$$

This defines the 'spin wave stiffness' D. The expression given in equation (11) is *very approximate* because it relies on a mean field theory relation between J and T_c as well as linear spin wave theory

This is an example of a Goldstone Mode. Such modes exist where the ground state in the ordered phase has a broken symmetry when the interactions have full rotation symmetry and are short range. A property of such modes is that their energies vanish as the wave vector tends to zero.

The existence of the spin wave modes as derived here is most easily demonstrated by neutron scattering because the neutrons which have wavevectors which are comparable with reciprocal lattice vectors also have energies which are comparable with the energies of magnetic excitations. A neutron interacts directly with the magnetic moments and in a scattering event a magnon may be created or destroyed and both the energy and momentum changes of the neutron recorded - in this way the magnon energies may be measured over the whole Brillioun zone. Magnetic neutron scattering is reviewed in many places for example [4] and in the article by Riste in this volume.

Magnon energies can also be inferred from the thermodynamics for bulk crystals. At low temperatures excitations sufficiently few spin waves are excited that we may assume that they do not interact appreciably. In this case the deviation of the magnetisation from its saturation value and the magnetic energy may be calculated by treating the spin waves as noninteracting bosons. The results are written in terms of the temperature $T = 1/k_B\beta$.

$$M_s(o) - M_s(T) = \frac{2g\mu_B}{a^3} \sum_k \langle n_k \rangle = \frac{2\mu_B g}{(2\pi)^3} \int d^3k \frac{1}{e^{\beta\varepsilon_k} - 1} \qquad (12a)$$

$$u(T) = \sum_k \varepsilon_k < n_k > = \frac{1}{(2\pi)^3} \int \frac{d^3k \varepsilon_k}{e^{\beta\varepsilon_k} - 1} \qquad (12b)$$

At low temperatures only the very lowest energy spin waves will be excited and as these correspond to the smallest wavevectors we can use the approximation given by equation (10). It is straightforward to show that with this approximation the magnetisation deviation varies as $T^{3/2}$ and the magnetic energy as $T^{5/2}$. Because the results are well behaved it is not necessary to include the anisotropy or dipolar energy to obtain good agreement with thermodynamic quantities if J is large.

2. Dipolar Dominated Regime. In ferromagnets containing transition metal ions the dipolar interactions are weak compared with the exchange. The magnetic dipolar energy between two spins of one Bohr magneton separated by one Angstrom is equivalent to a temperature of 1.2K this is very much less than typical ordering temperatures for transition metals and their compounds which are well above room temperature. However the fact that dipolar interactions are long range and anisotropic lead to the ground state energy being shape dependent - only spheres, ellipsoids, rods and planes may be treated analytically because it is only for these shapes that the demagnetising field is uniform within the sample. The dipolar interaction is sketched below.

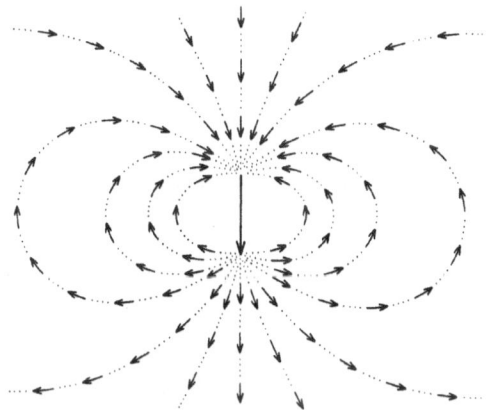

Figure (2) The magnetic dipole field from one dipole showing that dipoles like to line up parallel only when they are orientated along the line joining them.

The dipole interactions are also very important when it come to discussing the magnetic excitations for which the wave length is comparable with the size of the sample this means that they need to be included when we evaluate the spin wave modes which may be accessed by Brillioun scattering. The dipole interaction energy between two spins at R_n and R_m is given in terms of $r_{nm} = R_n - R_m$ by,

$$\frac{S_n \cdot S_m}{r_{nm}^3} - \frac{3(S_n \cdot r_{nm})(S_m \cdot r_{nm})}{r_{nm}^5} \tag{13}$$

This interaction depends on the relative position of the spins in space as well as their mutual orientation as was shown in figure (1).

The Hamiltonian including both exchange and dipolar terms may be written in terms of the spin wave operators b_k and b_k^+,

$$H = \sum_k \left[A(k) b_k^+ b_k + \frac{1}{2} B(k) \left(b_k^+ b_{-k}^+ + b_k b_{-k} \right) \right] \tag{14a}$$

where $A(\mathbf{k}) = g\mu_B \left[B_o - 4_\pi M_0 N_z \right] + Jsz(1 - \gamma_k) + 2\pi M_o \sin^2 \theta_k \tag{14b}$

and $B(\mathbf{k}) = 2\pi M_o \sin^2 \theta_k e^{-2\phi_k}$ (14c)

In these expressions M is the magnetisation , given in (4) , N_z the demagnetising factor for the particular shape of sample when the magnetisation direction is along the z axis are and θ ϕ the polar angles of the k vector relative to the direction of magnetisation (z).

This expression has new features which we discuss. The most important feature is that it contains the terms $b_k^+ b_{-k}^+$ and $b_k b_{-k}$ the presence of these terms means that the fully aligned state is no longer the ground state. The spin wave energies and the magnetisation deviation which now exists at zero temperature as well as at finite temperature are obtained following a canonical transformation,

$$\varepsilon_k^2 = A^2(\mathbf{k}) - |B(\mathbf{k})|^2$$ (15)

and

$$M(T) = \frac{g\mu_B S}{a^3} - \frac{2g\mu_B}{a^3} \sum_k \left[\frac{1}{2} \frac{(A(\mathbf{k}) - \varepsilon_k)}{\varepsilon_k} + \frac{A(k)}{\varepsilon_k} \frac{1}{e^{\beta\varepsilon_k} - 1} \right]$$ (16)

This concludes the section on linear spin wave theory in bulk ferromagnets. The work that was described briefly here is discussed in more detail in many books and reviews, for example [5,6].

1. B. Spectroscopy of Ferromagnetic Spin Waves

The only technique which is able to measure the energies of spin waves all over the Brillioun zone is neutron scattering which is discussed for example in [4] in the article by T Riste in this volume. In this subsection we examine other techniques all of which probe only the region near to $k = 0$. In a ferromagnet containing transition metal ions the exchange interaction is much larger than dipolar, anisotropy or Zeeman energy (with normal laboratory fields) but as the low energy spin wave excitations involve the spins precessing parallel to each other the exchange energy, which depends on the angle between neighbouring spins, does not come into play. The main spectroscopic techniques are,

i) Ferromagnetic resonance and the study of Damon Eshbach modes

ii) Standing spin wave resonance

iii) Parallel pumping

282

iv) Brillioun light scattering

Details of these experimental techniques may be found in [7] . Here we list only the information that may be obtained from these studies.

The ferromagnetic resonance is strictly at $k = 0$ and so as the spin precess together it must be independent of the exchange energy. The measurement is usually done as a function of magnetic field and so information is obtained both about the anisotropy energy and the g-value . The Damon Eshbach modes are the shape dependent modes which occur because of the demagnetisation energy a measurement of these modes, by either micro-wave spectroscopy or Brillioun light scattering also yields the bulk magnetisation.

Standing spin wave resonance is observed in thin films (of order 0.5 μm) of low anisotropy material such as permalloy by incorporating the thin film in a microwave cavity. In this geometry the frequency depends on anisotropy and the exchange as approximated by equation (10).

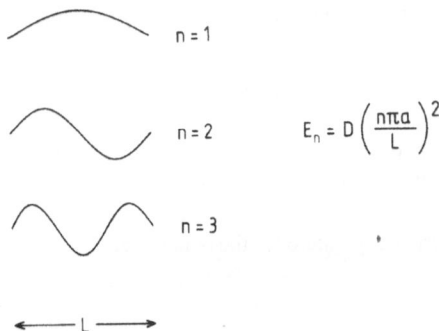

$$E_n = D \left(\frac{n \pi a}{L} \right)^2$$

Figure (3) The standing spin waves which occur in a thin film for pinned boundary conditions in a film of thickness L

A parallel pumping experiment is used to measure the inelastic lifetime of low energy spin waves in insulating materials. A spin wave is a coherent transverse precession of the magnetisation which is accompanied by a reduction of the total magnetisation in the z direction. Under suitable conditions an oscillating magnetic field at frequency 2ω parallel to the direction of magnetisation can generate two spin waves with frequency ω and equal and opposite momenta. The strength of the resonance yields the life times of the spin wave modes of frequency ω . Because these experiments are usually performed with microwaves these spin waves are usually in the regime in which their energies are dominated by anisotropy and dipolar interactions rather than the exchange.

Brillioun scattering of spin waves occur because the propagating spin wave sets up a phase grating in the material . The scattered light has a frequency shift which depends on the angle of scattering- the largest wave vector which may be excited corresponds to twice that of the incident light which is very small compared with the reciprocal lattice. Thus Brillioun scattering is an alternative technique to ferromagnetic resonance for studying the spin waves in the regime where their energies are dominated by anisotropy and dipolar effects.

1. C. Excitations in Thin Films

It has recently become possible to grow epitaxial thin films of monolayer thickness which have excited much interest recently. We first consider how thin a film should be so that it should behave qualitatively as a two dimensional object. The important point is whether in the temperature range of interest the magnetisation is substantially constant across the film thickness. The exchange interaction in a thin film may be assumed to be roughly comparable with that for bulk materials. In an insulating material where the electronic wavefunctions are localised on the atomic site the main effect of the reduced geometry is the reduced number of magnetic neighbours for the spins on the surfaces. In a metal the electronic wavefunctions may be strongly perturbed by the reduced geometry - this leads to a definite change in the value of the magnetic moment and also the exchange energy [8].

In the spin wave regime the criterion for two dimensionality is satisfied if there is insufficient energy to excite spin waves with wave length, λ, where λ is of the order of magnitude of the film thickness and is given by Na where N is the number of monolayers and a the lattice spacing. Using our approximate expression for the spin wave stiffness D, equation (11), we find that in the spin wave regime of $T < T_c/10$ (T_c corresponds to the Curie temperature for the bulk material)

$$N^2 \angle 300 / zs \tag{17}$$

Using $z = 12$ and $s = 1$ appropriate for metallic cobalt we find,

$$N \angle 5 \tag{18}$$

We consider what is special about spin waves in two dimensions. The magnetisation deviation given by equation (12) for three dimensions *diverges* in two dimensions if the spin wave energies are calculated using only the exchange energy. This occurs because if we have a spin wave energy given by equation (10) then the integral given in equation (12a), when modified for two dimensions, becomes,

$$M_s(o) - Ms(T) = \frac{2g\mu_B}{(2\pi)^z} \int \frac{d^2k}{e^{\beta D(ak)^2} - 1} \tag{19}$$

$$\sim \int \frac{dkk}{\beta D(ak^2)} \rightarrow \infty$$

The fact that this divergence occurs is related to the Mermin Wagner theorem [9] that the isotropic Heisenberg model does not order in two dimensions. Experimentally magnetism is observed in ultra thin magnetic films [4] up to and beyond room temperature. This occurs because the dipolar and anisotropy interactions exist. Although they are much smaller in magnitude than the exchange they do have the vitally important function of altering the spin wave energy in the low energy regime. From a formal point of view the Mermin Wagner theory states that there can be no long range order in two dimensions if the interactions are both short range and isotropic. The dipolar interaction is not short range and the inclusion of anisotropy clearly renders the Hamiltonian anisotropic.

The dipolar effects have the obvious effect of producing a demagnetising field which in the absence of magnetocrystalline anisotropy causes the easy direction of the spins to lie in the plane. This is in addition to the changes to the spin wave spectrum which was first considered by Maleev [10] and more recently by a number of authors [11,12].

The magnetocrystalline anisotropy is strongly affected by the film surfaces. In a metal the fact that the d electrons may only hybridise within the plane causes a large surface anisotropy [13]. In some cases, for example Co films on Au , the anisotropy is strong enough to overcome the strong demagnetising field and to force films which are sufficiently thin to become easy axis films. An easy axis film will automatically have a gap in the spin wave spectrum and this will stabilise the magnetisation.

The relative efficiency of the dipolar interaction and the magnetocrystalline anisotropy in stabilising the magnetisation was discussed by Bland et al [14] for easy plane films . The spin wave energies are given by an expression of the form of equation (15) where the coefficients A(k) and B(k) are given by the following in this case.

$$E_k^2 = \left\{ \varepsilon_k + K_p + K_2^1 + P_0 \left[f - \frac{nk}{4}(3 - \cos 2\phi_k) \right] \right\}^2$$
$$- \left\{ K_p - K_2^1 - P_0 \left[f - \frac{nk}{4}(3 - \cos 2\phi_k) \right] \right\}^2$$

(20)

In this expression the angle θ_k is the angle between the magnetisation and the k vector, ε_k is the exchange contribution to the spin wave energy $\varepsilon_k = D(ak)^2$ and P_0 is the dipolar energy which is given in terms of the saturation magnetisation per unit volume M and f is a factor to account for the effect of the discrete lattice structure of the film in the evaluation of the dipolar sums [10,11]. The anisotropy constant $K_2^{\ 1}$ is the uniaxial anisotropy causing the magnetisation to lie in the plane and K_p is defined by equation (21) below.

The anisotropy energy within the plane is given in terms of a modified cubic anisotropy constant K_4 and a uniaxial term K_2 which breaks the four-fold symmetry.

$$E(\vartheta,\phi) = K_2' \cos^2\theta + \frac{K_4'}{4}\sin^2 2\theta$$
$$+ K_2 \sin^2\theta + \frac{K_4}{4}\sin^4\theta \sin^2 2\theta \qquad (21a)$$

The magnetisation lies in plane so the equilibrium angle θ0 and the effective in-plane anisotropy energy K_p may be defined by:

$$\left.\frac{\delta E}{\delta\phi}\right|_{\varphi=\varphi_0, \theta=\frac{\pi}{2}} = 0 \qquad (21b)$$

$$2K_p = \left.\frac{\delta^2 E}{\delta\phi^2}\right|_{\varphi=\varphi_0, \theta=\frac{\pi}{2}}$$

At $k = 0$ equation (20) gives a gap energy of

$$E_o = \sqrt{4K_{2eff}^1 K_p} \qquad (22)$$

where

$$K_{2eff} = K_2^1 + P_o f$$

The spin waves defined by equation (20) contain contributions from both dipole and anisotropy energies. We now show that the anisotropy is more efficient at stabilising the magnetisation by looking at the spin wave energies at different k values. The anisotropy gives a contribution which is independent of k, the dipolar term is linear in k and the exchange term varies as k^2.

At $k = 0$ the anisotropy certainly dominates as the other two terms vanish. We ask if, as the value of k increases the energy passes directly from the anisotropy dominated regime to an exchange coupled regime or if an intermediate regime exists in which the k dependent dipole energies are dominant. We define k_d and k_e as the values of k such that the anisotropy energy is comparable to the dipolar and exchange energies respectively. If $k_d > k_e$ then at low k values the dipolar interactions are small compared with the anisotropy and at high k then the exchange dominates. Thus the k dependent dipolar terms may be neglected for all k values. The condition $k_d > k_e$ leads to a critical value of the in-plane anisotropy for which the dipolar effects may be safely ignored which is given below.

$$k_p \geq P_o^2 n^2 / 2D$$

Substituting the experimental numbers for a cobalt film [14] we find that the criterion is quite easily satisfied for ultra thin films.

There is an interesting remark that one can make about the magnetisation deviation of two dimensional films when the k dependent dipolar forces are negligible. Under these circumstances the only k dependent term comes from the exchange and it may be shown easily that the magnetisation may be integrated exactly [15].

Spin waves have been detected in ultra-thin magnetic films by Brillioun light scattering [16].

This technique probes the anisotropy and dipole dominated regime . The remarkable narrow lines which are obtained demonstrate the sample quality. This is a particularly powerful technique for use with ultra thin films as it is possible to make the measurements *in -situ*.

1D. One Dimensional ferromagnets

There are a number of excellent reviews of one dimensional magnets [17,18]. In a real physical system in which there are almost isolated spin chains there will be a temperature range over which the system will behave as if it consisted of a set of uncorrelated spin chains before long range three dimensional order sets in at very low temperatures. The experimental manifestations of this are described in [18].

Figure (4) A single reversed bond in a one dimensional magnet showing the loss of long range order

A one dimensional system may not show long range order at any non zero temperature. We can understand this by noting that as shown in figure [4] a single spin reversed bond is enough to destroy the long range order. The figure shows a sharp reversal but the long range order is lost just as effectively by a region in which the spins tilt over gradually. However at T=0 the spins will be in their ground state which may be written down exactly as the fully aligned state using the commutation relations discussed earlier.

Here we take a brief look at the collective excitations. There will be two types of excitations. The spin waves which have an energy which is calculated similar to that we found for a bulk sample including only exchange and anisotropy energies,

$$\varepsilon_k = 2Js(1 - \cos ka) + Ks \tag{23}$$

and solitons in which the magnetisation shows a 'kink' [18]. A soliton is special to one dimension. They may be created as a soliton - antisoliton pair which may then move apart to become 'free' solitons which may only be annihilated by recombining with an antisoliton. The solitons may move freely along the chain and scatter neutrons as shown by [19] and discussed by [20].

II. ANTIFERROMAGNETIC SPIN WAVES

II.A. Bulk Antiferromagnetism

We consider antiferromagnetism in three dimensions on a bipartite lattice (this is a lattice which can be partitioned into two sublattices A and B such that each A site has only B sites as its nearest neighbours and vice versa). For simplicity we first discuss the case in which there is only isotropic exchange. The Hamiltonian is given below.

$$H_{ex} = J \sum_{<nm>} \mathbf{S}_n \cdot \mathbf{S}_m \qquad \text{<nm> nearest neighbours} \tag{24}$$

As for an isotropic ferromagnet both the total component of spin along the ordering direction and the total spin commute with the Hamiltonian. The classical ground state would have each spin aligned antiparallel with its neighbours leading to a total value of the z component of spin being zero. We find the first important difference between a ferromagnet and an antiferromagnet - whereas there is a unique state of S^z_{tot} equal to Ns the number of states for which S^z_{tot} is equal to zero increases exponentially with N.

We assume that the ground state may be approximated by the classical state (Néel ground state) in which the spins are arranged so that all the spins on the A sublattice are parallel to each other and antiparallel to those on the B sublattice.

A spin deviation on the A sublattice will *reduce* the z component of spin from s to $s-1$

whereas for the B sublattice the spin deviation will *raise* the spin from $-s$ to $-s+1$.

We define the relevant operators Bose operators a and b,

Spin wave deviations:

A site s $\Rightarrow$ s to s-1 $\qquad\qquad s_n^z = s - a_n^+ a_n$

B site s $\Rightarrow$ s to $-s+1$ $\qquad\qquad s_n^z = -s + b_n^+ b_n$ $\qquad\qquad$ (25)

It is convenient to use a rotated coordinate system in which we rotate the axes on the B sublattice by π about the x -axis. This has the desired effect of changing the sign of the s^z and the s^y operators on the B sublattice so that these parts of the Hamiltonian resemble the ferromagnet that whose excitations we have already calculated. However the s^x operator is not rotated - hence we obtain the following Hamiltonian.

$$H = -J \sum_{<nm>} \left[S_n^z S_m^z + S_n^y S_m^y - S_n^x S_m^x \right] \qquad\qquad (26)$$

We transform to spin wave variables using the first terms of the transformation given by equation (8) , however we retain the notation that the operators a and b correspond to the A and B sublattices.

$$a_k^+ = \frac{1}{\sqrt{N}} \sum_n e^{-ik \cdot R_n} a_n^+ \qquad a_n^+ = \frac{1}{\sqrt{2s}} \left(s_n^x - i s_n^y \right) \qquad\qquad (27)$$

$$b_k^+ = \frac{1}{\sqrt{N}} \sum_m e^{-ik \cdot R_m} b_n^+ \qquad b_m^+ = \frac{1}{\sqrt{2s}} \left(s_m^x - i s_m^y \right) \qquad\qquad \text{as above}$$

$$H = -JNzs^2 + 2szJ \sum_k \left[a_k^+ a_k + b_k^+ b_k + \gamma_k \left(a_k^+ b_{-k}^+ + a_k b_{-k} \right) \right] \qquad\qquad (28)$$

where γ_k was defined by equation (96) in terms of the nn lattice vectors, ρ, and is given below.

$$\gamma_k = \sum_\rho e^{ik \cdot \rho}$$

The spin wave energies may be found using a Bogolubov transformation,

$$E_k^2 = (2szJ)^2 \left(1-\gamma_k^2\right) = (2szJ)^2 \left(1-\gamma_k\right)\left(1+\gamma_k\right) \tag{29}$$

The excitation spectrum is shown in figure (5). It is symmetric about $\pi/2$ as should be expected since the antiferromagnetic order has doubled the unit cell and hence halved the size of the Brillouin zone.

Inclusion of cubic magnetocrystalline anisotropy adds extra terms to the Hamiltonian which are diagonal in the spin wave operators and changes the spin wave energies to

$$E_k^2 = (2zJs)^2 \left[\left(1-\kappa \right)^2 - \gamma_k^2 \right] \qquad \text{where} \quad \kappa = 2K_4/2zJs \tag{30}$$

$$\mathrm{Lim}_{k \to 0} E_k = 2zJs \left(2\kappa + \kappa^2 \right)^{1/2}$$

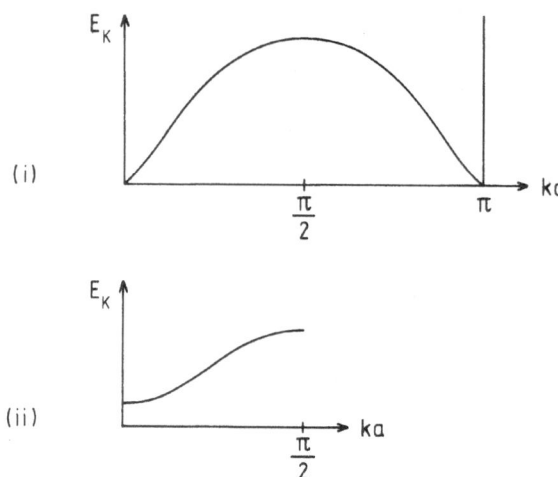

Figure (5) A sketch of the spin wave dispersion relation for (i) an antiferromagnet with exchange interactions only and (ii) including the effects of anisotropy. The figure (ii) is drawn for the reduced Brillouin zone appropriate for the antiferromagnetic structure.

We note three main differences between the ferromagnetic and antiferromagnetic spin waves: in the absence of anisotropy near to $k = 0$ the ferromagnetic spin waves vary as k^2 whereas the antiferromagnetic magnons vary as k; in a ferromagnet the anisotropy energy K_4 produces an energy gap which varies linearly as K_4 but in an antiferromagnet we see from equation (30) that the energy gap varies as $(JK_4)^{1/2}$ - as the exchange is usually much larger than the anisotropy this leads to energy gaps which are much larger in antiferromagnets than in ferromagnets; the third difference is that the zero point motion effects resulting from the necessity to perform a Bogolubov transformation in antiferromagnets can reduce the sublattice magnetisation substantially - linear spin wave theory predicts a reduction of $7.8/s$ % for an isotropic cubic antiferromagnet with spin s. All of these effects arise from the different signs for the coupling of the different spin components in equation (26). Excellent reviews of antiferromagnetic spin waves are to be found in Kittel *Quantum Theory of Solids* [6] and in Anderson *Concepts in Solids* [3].

The ground state of the antiferromagnet has total S_z component equal to zero. The one spin deviation states may have total spin of ± 1 , the energy given in equation (30) is that for the two degenerate modes. The degeneracy is split by a magnetic field. More details may be found in references [3, 6]

II.B. Spectroscopy of Antiferromagnetic Magnons

The energies of magnons may be studied over the whole Brillioun zone by neutron scattering as discussed briefly in this volume by Riste. The other methods are listed below:

i) Antiferromagnetic resonance

ii) Two magnon light scattering

iii) One magnon light scattering

As explained above the antiferromagnetic resonance (AFM) depends on the root mean square of the exchange and anisotropy. This energy is usually in the infra-red for example it is 26×10^{10} Hz for MnF_2. A measurement of the AFM may be used to obtain the anisotropy if the exchange is known. In a field the g - value is also obtained. The degeneracy of the magnon branches is broken in a field. The lower magnon branch may be driven down towards zero but before it reaches zero there is a transition to a spin flop phase in which the spins are canted with the ferromagnetic and antiferromagnetic components parallel and perpendicular to the field respectively: for further details see [21].

Two magnon light scattering occurs in an antiferromagnet because of the existence of the last two terms in equation (28). If the exchange is modulated (as by a virtual excitation to an excited state) these terms allow a Raman process to occur in which two magnons are created. Such an effect is forbidden in ferromagnets. The two magnons are created with equal and opposite wavevectors and so the observed frequency is close to the two magnon density of states with corrections due to the fact that the two magnons are created on a pair of neighbouring sites and so the interactions which occur when the higher order terms in the Holstein Primakoff transformation need to be included. [22]

One magnon Brillioun light scattering may also be observed. In this case the measurement yields information on magnons which have a low k value. The intensity is generally lower than the two magnon effect discussed above.

II .C. Low Dimensional Antiferromagnetism

An isotropic antiferromagnet is not ordered at any finite temperatures in two dimensions because of the Mermin-Wagner theorem which was discussed for ferromagnets in section Ic . However two dimensional antiferromagnetism may be stabilised by anisotropy as was found for ferromagnets. There has been an interesting question over the nature of the *ground state* for an isotropic *antiferromagnet* with a spin of 1/2 in two dimensions. As was discussed earlier the ground state wave function for an antiferromagnet is not the Néel state and the question at issue was whether the ground state does contain any long range order . The question has been settled recently in favour of a Néel state with a very large amount of zero point motion - the sublattice magnetisation was found to be reduced to   60% of its saturated value for a square lattice and 44% for the honeycomb lattice [23]. These results were obtained by Monte Carlo simulations and are in surprisingly good agreement with what would be obtained from an estimate using first order spin wave theory as described here.

The rest of this section will concentrate on one dimensional antiferromagnetic chains. There is no long range order at any temperate and so the main interest resides in understanding the nature of the ground state , including the correlation length, and the elementary excitations away from the ground state. It is found that real systems will approximate to this behaviour at sufficiently low temperatures compared with the mean field transition temperature but high enough so that interaction between chains are unimportant [18].   Antiferromagnetic spin chains are an extreme example of quantum systems and show the unusual feature that qualitatively different effects occur for integer and half integer spins [24].

The ground state and elementary excitations were found for the antiferromagnetic spin 1/2 chain many years ago and are reviewed in [25]. The ground state is a singlet [26] and has no long range order.   It is interesting to compare the exact ground state energy with that of a Néel state. For the Hamiltonian given by equation (31) the exact ground state energy is given by equation (32a) and the energy for the Néel state in equation (32b).

$$H = J \Sigma_n S_n S_{n+1} \qquad \text{where } S^z = \pm 1/2 \qquad (31)$$

$$E_0/J = 1/4 - ln2 = -0.44314... \qquad (32a)$$

$$E_{Néel}/J = -1/4 \quad = -0.25 \qquad (32b)$$

It is interesting to examine the energy for a dimerised ground state. Consider two spin 1/2 coupled together antiferromagnetically. The eigenstates of the two spin pair are a singlet and triplet , $S_T = 1$ or 0.

$$S_T = s_1 + s_2 \qquad (33a)$$

292

Using $\qquad S_T(\ S_T + 1\ ) = s_1\,(\,s_1 + 1\ )\ +\ s_2\,(\,s_2 + 1\ )\ + 2\,s_1\,.\,s_2 \qquad$ (33b)

we find the energies of the singlet and the triplet states are $-3J/4$ and $J/4$ respectively.

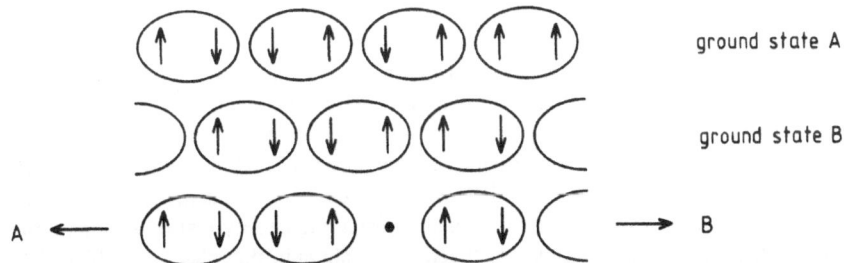

ground state A

ground state B

A $\longleftarrow$ $\qquad$ B

Figure (6) A schematic diagram of a spin dimerised chain: the spins within the ovals are coupled together to form a singlet which can then have no interactions with neighbouring singlets. The two possible ground states are shown as A and B and a defect state which is a boundary between the states A and B is also shown.

A dimerised chain is shown schematically in figure (7) - we see that half the bonds are used in dimers - the energy associated with the other bonds are exactly zero as there can be no interaction between two state of zero spin. From this we can calculate the ground state energy of a dimerised chain.

$$E_{dimer}/J \ = \ -\ 3/8 \ \ = \ -\ 0.375 \qquad\qquad (34)$$

It is interesting to note that this is much closer to the exact result than the Néel state. This is an example of a calculation of a ground state energy being reasonably good even when the wave function has serious errors. The exact ground state has algebraically decaying correlation functions and the correlations in the dimerised state are non zero only for nearest neighbours.

It has been shown [27] that the spin 1/2 chain is unstable to a Peierls distortion of the lattice which strengthens and weakens alternate bonds and stabilises a partially dimerised ground state. Dimer order may also be induced by including a next nearest neighbour antiferromagnetic exchange - this is the Mujumdar -Gosh model and is examined in [28].

The dimer state is obviously two fold degenerate as is shown in figure (6). An excited state which is a soliton is shown. It clearly requires a finite energy to create such a state but once created it is free to move . Recently an inorganic compound, $CuGeO_3$, has been found where the spin interactions approximate well to the model -this compound may also be doped . The behaviour of a dimerised spin 1/2 chain with free carrriers has also been investigated theoretically [29]. The spectroscopy of the excitations has been investigated using inelastic neutron scattering [30].

Much of the current interest in quantum spin chains dates from the Haldane conjecture [24] that the ground state of an integer spin chain has short range correlations and all the spin excitation have a finite energy. The physics may be understood by following Affleck [31] who showed that the spin one chain may be understood by considering each spin one to be made up of two spin 1/2's which are dimerised with one of the neighbours. It takes a finite energy to break a dimer and hence all states have a finite energy. But there is more than that - in this picture there is a free spin 1/2 left over at each end of the chain. This has been verified experimentally [31] and theoretically [32] as it is found that a finite chain of spin one behaves as if there is a free spin 1/2 localised near each end. Recently there has been a calculation using the density matrix renormalisation group of the energies and correlation functions of the spin one chain with biquadratic coupling [33].

II.D. Antiferromagnetism for the 1990s

This section on antiferromagnetism ends with a brief discussion on why it is that several new classes of compound have been discovered which are antiferromagnetic and several of which are part of an extended family of which some members are superconducting. The common ingredient in all the compounds is that the electrons in narrow bands and so experience very strong correlations because of their coulomb repulsion.

Materials which are currently interesting include the cuprate superconductors in which the insulating and antiferromagnetic phase of , for example, $LaCuO_3$ loses its long range antiferromagnetic order and then becomes superconducting as it is doped with barium. A similar compound, $LaMnO_3$, but for which the spin is 5/2 instead of 1/2 as for Cu is also antiferromagnetic and becomes canted and then ferromagnetic as a function of doping. The organic compounds based on BEDT-TTF also have a tendency to be either magnetic or superconducting depending on the carrier mobility [34]. The situation is different in the heavy fermion compounds such as UPt_3 , reviewed in [2,35], in this case there is a transition to an antiferromagnetic state which is associated with a very small sublattice moment followed by a transition to an unconventional superconducting state which coexists with the antiferromagnetism. Thus antiferromagnetism is associated in some way with a number of currently interesting compounds.

We address some of the questions that these observations raise :

(i)   What are 'strongly correlated electron systems' and why are they likely to be magnetic?

(ii) Why is antiferromagnetism likely ?

(iii) Why does the addition of mobile carriers destroy magnetism ?

The first question may be answered formally by noting that there is a sum rule between the electrical and magnetic fluctuations if the coulomb interactions suppress the charge fluctuations then the magnetic fluctuations are necessarily enhanced.   We can also consider a simple model system consisting of two atoms and two electrons which have antiparallel spins. There are four possibilities shown in figure (7).

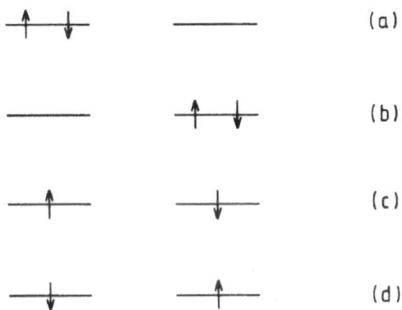

Figure (7) A sketch of the allowed states for two electrons with antiparallel spins associated with two atomic sites

If the occupation of the up spin electron is independent of that for the down spin (as would be the case if this were part of a noninteracting electron gas) then the probabilities of all four configurations would be the same and thus all equal to 1/4.   If the electrons repel strongly when they are on the same site then we should associate an additional energy , U , with the states (c) and (d). This will cause the probabilities for the states (a) and (b) to drop and the atoms are most likely to be found in their magnetic state.   Thus we see that strong electron- electron repulsion often leads to magnetic insulators.

The second issue that we wish to address is why is antiferromagnetism is frequently favoured in systems with a large value of the repulsion U compared with the energy that the electrons would gain by forming a band - this energy is just the overlap energy to transfer one electron from one site to its neighbour, it is often designated as, t . In figure (8) we show the energy of two sites with two electrons first in the case for which t is equal to zero and then to second order in t.

The energy for the parallel spin configuration is unaltered by the inclusion of finite t but the configuration with antiparallel spins is lowered. This gives a lower energy for the antiparallel configuration. (This is a very simplified verification of the well known result that the Hubbard model with one electron per site and $t/U \ll 1$ reduces to the spin 1/2 Heisenberg antiferromagnet.) Of course not all magnetic insulators are antiferromagnetic

- this depends on the geometry of the intervening ligands as expressed in the Goodenough-Kanamori rules [36] .

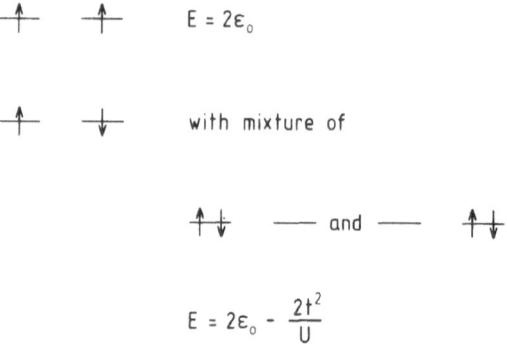

Figure (8) shows the energy of two electrons with parallel and antiparallel spins associated with two atomic sites to second order in t/U.

We now address the third question which is why is it that the presence of mobile carriers destroys antiferromagnetism. If we consider the spins to be in a Néel state then the effects of a moving hole are shown in figure (10) we see that a trail of 'wrong bonds' is left behind - of course in a Heisenberg system the antiferromagnetism may be recovered by mutual spin flips but the hole does still tend to be more localised than in a ferromagnet where it may easily form a Bloch state.

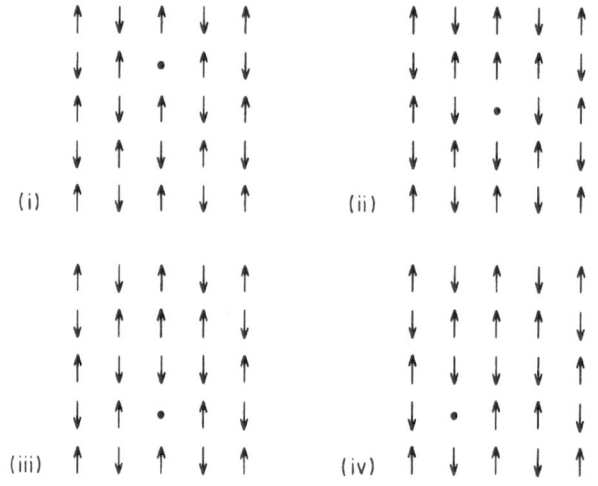

Figure (9) shows the change in the magnetic order when a hole makes three nearest neighbour moves in the sequence (i) - (ii) - (iii) - (iv) .

Thus in the Hubbard model discussed here there is a transition from the localised antiferromagnetic state to a metallic state when the band energy gained by the mobile electrons overcomes the stabilisation of the insulating antiferromagnetic state [37].

In $La_x Ba_{1-x} MnO_3$ the situation is slightly different because the manganese ion has either three or four d electrons and so one needs to consider the fourth d electon moving in a background where each Mn site has a spin of 3/2 . The coulomb interactions strongly favour the configuration in which the mobile electron has its spin parallel to the of Mn ion on which it resides. Thus the addition of mobile carriers again destroys antiferromagnetism and in this case a ferromagnetic state results [38].

III. Metallic Magnets

In this section we review the types of excitation which may occur in transition metals. At the beginning of this article we argued that even a ferromagnet should still show long wave spin waves which are substantially undamped and whose energy varies as k^2 near k equal to zero. In this section we first justify this in a little more detail and then discuss the excitations which are special to ferromagnetic metals.

Let us consider a small region centred on the point r of a ferromagnet containing several unit cells . A magnetic moment $M (r)$ equal to the value of the magnetisation averaged over the region may be associated with the point r . An energy expansion may be made in a continuum model , by symmetry the only terms which will be allowed in the energy will contain terms proportional to $M(r)$ and $\nabla M (r)$. If we consider a magnetisation deviation which varies sinusoidally,

$$M (r) = M + m \, exp (i k.r) \tag{35}$$

The presence of the gradient term ensures that the excitation energy varies as k^2. A similar argument follows for the anisotropy and hence it is common practice to use a Heisenberg model to discuss the low energy spin waves for metals and metallic films as was done earlier.

III. 1. Stoner Theory

The schematic band structure for a ferromagnetic metal is shown in figure(10) - in this simple diagram it is assumed that the majority spin band is lowered by an amount Δ relative to the minority spin band without changing its shape. (Accurate band theory calculations do show an exchange splitting but the shapes of the bands are not the same as is assumed in the simple picture.)

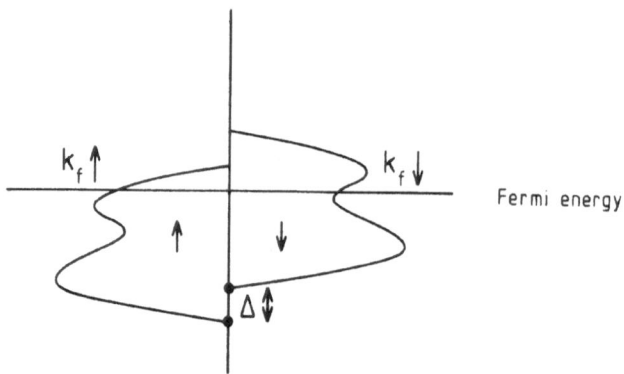

Figure (10) shows a sketch of the Stoner split bands for a transition metal.

The Fermi energy is the same for both bands and so the Fermi wave-vectors are different . A given k value that lies at the Fermi level for majority spins corresponds to an energy of Δ above the Fermi level for the minority spins. Hence the energy required to flip the spin of an electron without changing its momentum is Δ . For a process in which an electron of wave vector $k + q$ and spin up is scattered to a state with wave vector k with spin down may have an energy which is larger or smaller than Δ depending on the angles made by the wave-vectors. In order to have a process in which the energy change is zero a wave-vector change is required which is equal to the differences in the Fermi wave-vectors for the up and down spin bands. This leads to a continuum spectrum shown in figure (11) - all energies are allowed in the hatched region. Recent calculations of the spin excitations in the high energy region have been done using a first principles band structure [39] . This excitation spectrum may be compared with that for the non-magnetic case given in the paper by von Baltz in this volume.

III.2. Random Phase Approximation

This theory allows us to calculate both the spin waves and the Stoner excitations. The energy and wave-vector dependent susceptibility for a non-interacting Fermi gas is given in equation (36).

$$\chi_{\chi o}^{\uparrow\downarrow}(q,\omega) = \sum \frac{f\left(E_k^{\uparrow}\right) - f\left(E_{k+q}^{\downarrow}\right)}{E_{k+q}^{\downarrow} - E_k^{\uparrow} - \hbar\omega} \tag{36}$$

In this equation the energies $E_k^{\uparrow}$ and $E_k^{\downarrow}$ are the energies of the electron states in the Stoner model and $f(E)$ is the Fermi function. At $T = 0$ the only contributions arise from processes in which one of the f functions is unity and one is zero - this occurs if one of the energies in the denominator lies above the Fermi energy and one below.

Including the additional electron- electron repulsion on site which arises between electrons of opposite spin within the random phase approximation leads to the following result.

$$\tag{37}$$

$$\chi_{int}^{\uparrow\downarrow}\left(q\omega\right) = \frac{\chi_0^{\uparrow\downarrow}(q,\omega)}{1 - I\chi_{0\chi}^{\uparrow\downarrow}(q,\omega)}$$

This expression still has poles at the Stoner excitations but an *additional* pole arises which is the collective excitation or spin wave . (A similar result was found for plasmons in article by von Baltz in this volume .) The spin wave energy is found to be proportional to k^2 for small values of k in agreement with the discussion given in section I.. As the spin wave peak approaches the Stoner continuum it mixes and becomes broadened - an observed neutron scattering spectrum showing this feature for Ni was shown by Riste [21] and shown schematically below. The low energy spin waves may be observed using Brillioun scattering - as was discussed for ultra-thin ferromagnetic films and using ferromagnetic resonance. However the eddy current damping that exists in metals limits the usefulness of electromagnetic probes.

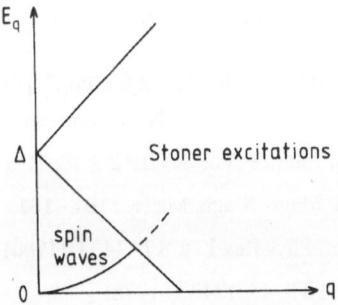

Figure (11) shows a sketch of the spin wave excitations and the Stoner continuum which may be obtained from the poles of the response function given in equation (36).

V. Conclusions

This review has focused on the elementary excitation which occur in metals and compounds of transition metal materials and their spectroscopy. The basic ideas were reviewed and the application of these to various low dimensional magnets which are of current interest were discussed.

Acknowledgements

Many thanks are due to colleagues with whom I have discussed these ideas but particularly to R Bursill and S R P Smith who helped in locating various references and also to M Ain for man˙ useful discussions on $CuGeO_3$ while we were both at Erice.

REFERENCES

1. J Jensen, A R Mackintosh *Rare Earth Magnetism; Structures and Excitations* (Oxford Univeristy Press, Oxford, 1991)

2. N Grewe, F Steglich *Handbook on the Physics and Chemistry of the Rare Earths* Vol **14** p343 Eds K A Gschneider, Jr and L Eyring (North Holland , Amsterdam, 1991)

3. P W Anderson *Concepts in solids* (W A Benjamin 1963 and Addison -Wesley reprint 1992)

4. S W Lovesey *Theory of Neutron Scattering from condensed Matter Volume 2* (Clarendon Press, Oxford 1984)

5. L R Walker *Magnetism* **I** 299 Eds G T Rado and H Suhl (Academic Press 1963)

6. C Kittel *Introduction to Solid State Physics* 6th Edition (Wiley, New York 1986) and *Quantum Theory of Solids* (Wiley New York 1963)

7. F Keffer *Encyclopedia of Physics* Vol XVIII/ 2 (1967) (Springer - Verlag, Berlin)

8. A J Freeman, R Wu , J. Magn. Magn. Mater. **104 - 107** , 1 (1992)

9. N D Mermin, H Wagner Phys Rev Lett **17**, 1133 (1966)

10. S V Maleev Sov Phys JETP **43**, 1240 (1976)

11. Y Yafet, J Kwo, E M Gyorgy Phys Rev B **33**, 6519 (1986)

12. P Bruno Phys Rev B **43**, 6015 (1991)

13. See for examples the articles by J G Gay and Richter, G H O Daalderop et al and W

J M de Jonge et al in *Ultra thin Magnetic Structures* I Eds B Heinrich and J A C Bland (Springer- Verlag, Berlin) 1994

14. J A C Bland, C Daboo, G A Gehring, B Kaplan, A J R Ives, R J Hicken and A D Johnson J Phys Cond Matt **7**, 6467 (1995)

15. J A C Bland, G A Gehring, B Kaplan, C Daboo J. Magn. Magn. Mater. **113**, 173 (1992)

16. P. Grunberg in *Light Scattering in solids* Eds V M Cardona , G Guntherodt (Springer- Verlag, Berlin 1989) and J F Cochran, B Hillebrands, G Guntherodtin *Ultra thin Magnetic Structures* II Eds B Heinrich and J A C Bland (Springer-Verlag, Berlin) 1994

17 F Keffer, H Kaplan, Y Yafet, Am. J Phys. **21**, 250 (1953)

18. L J de Jongh and A R Mediema *Experiments on Simple Magnetic Model Systems* (Taylor and Francis 1974)

19. H J Mikeska, Phys Rev B **12**, 2794 (1975)

20. K Kakurai, R Pynn, B Dorner, M Steiner, J Phys C **17** L123 (1984)

21. S Foner *Magnetism* I 383 Eds G T Rado and H Suhl (Academic Press 1963)

22. M G Cottam and D J Lockwood *Light Scattering in Magnetic Solids* (Wiley New York 1986)

23. J D Reger, J A Riera, A P Young J Phys Cond Matt **1**, 1855 (1989); J D Reger, A P Young Phys Rev B **37**, 5978 (1988).

24. F D M Haldane, Phys. Lett **93A** , 464 (1983) , Phys Rev Lett **50**, 1153 (1983).

25. Yu A Izumov and Yu N Scriabin *Statistical Mechanics of Magnetically Ordered Systems* (Consultants Bureau New York and London 1988)

26. E Lieb and D Mattis (Editors) *The Many Body Problem - An Encyclopedia of Exactly Soluble Models in One Dimension* (1993)

27. J W Bray, L V Interrante, I S Jacobs, J C Bonner in *Extended Linear Chain Compounds* , edited J S Miller (Plenum, New York 1983) **3**, 353

28. K Nomura and K Okamoto J Phys A **27**, 5773 (1994)

29. Di Tusa J F, S W Cheong, J H Park, G Aeppli, C Broholm, C T Chen Phys Rev Lett **73**, 1857 (1994)

30. I Affleck J Phys Cond Matt **1**, 3047 , Reviews in Math Phys **6**, 887 (1994)

31. M Hagiwara, K Katsumata, I Affleck, B I Halperin, J P Renard Phys Rev Lett **65**, 3181 (1990)

32. T Kennedy J Phys Cond Matt **2**, 5737 (1990)

33. T Xiang and G A Gehring , Phys Rev **B48**, 303 (1993) and R J Bursill, T Xiang and G AGehring J Phys A **28** , 2109 (1995)

34. J Singleton, J Caulfield, S Hill, S Blundell, W Lubczynski, A House, W Hayes, J Perenboom, M Kurmoo, P Day Physica B **211**, 275 (1995)

35. F. Steglich, B Buschinger, P Gengenwart, C Geibel, P Hellmann, M Lang A Link, R Modler, D Jaccard, P Link *Proceedings of International Conference on Physical Phenomena atHigh Magnetic Fields- II Talahassee* 1995 (to be published by World   Scientific)

36. J B Goodenough *Magnetism and the Chemical Bond* (Interscience 1963)

37. P W Anderson *Frontiers and Borderlines in Many Particle Physics* Ed R A Broglia and J R Schrieffer (North Holland 1988)

38. P G de Gennes Phys Rev **100**, 564 (1955)

39. J F Cooke, J A Blackman, T Morgan J Magn Magn Mater **54-57**, 1150 (1986) and J A Blackman, K N Trohidou, J F Cooke *ibid* **104**, 721 (1992)

PLASMONS AND SURFACE PLASMONS IN BULK METALS, METALLIC CLUSTERS, AND METALLIC HETEROSTRUCTURES

R. v. Baltz
Institut für Theorie der Kondensierten Materie
Universität Karlsruhe
D-76128 Karlsruhe, Germany

ABSTRACT

This article gives an introduction and survey in the theory and spectroscopy of plasmons in the bulk, as well as on boundaries of metals. First, concepts to describe the metallic state and its interaction with electromagnetic fields are summarized, then various approaches to obtain the plasmon–dispersion relations are studied. Finally, some actual questions such as correlation effects and the surface charge density profile on the plasmon–dispersion are discussed.

I. INTRODUCTION

Plasmons are quantized wave-like excitations in a plasma, i.e. a system of mobile charged particles which interact with one another via the Coulomb forces. The classical example is an ionized gas consisting of (positive) ions and (negative) electrons in a discharge tube. In metals (and in some highly doped semiconductors, too) the electrons likewise form a plasma. In contrast to the aforementioned example, however, the electrons form a degenerate Fermi-system, i.e. even at low temperature, the electrons have a large kinetic energy ($\approx$ Fermi-energy) so that (room-) temperature has little influence on the electronic excitations. The ions, on the other hand, because of their large mass have little kinetic energy and their (crystal-) structure and collective excitations (=phonons) are completely dominated by the electrons.

As a consequence of the long range nature of the Coulomb interaction the frequency of the plasma oscillations,

$$\omega_p = \sqrt{\frac{n_0 e^2}{m_0 \epsilon_0}} \tag{1}$$

is very high compared with other collective excitations like phonons. n_0 is the density and m_0 is the (free-) electron mass. For example, in Al we have $\hbar\omega_p = 15eV$, whereas, typical phonon energies are in the $10meV$ range. For a survey, see Di Bartolo's article in this book.

Collective excitations in classical plasmas were first studied by Langmuir [1]. The pioneering theoretical investigations on their quantum counterparts were carried-out by Bohm and Pines, see Pines [2]. Experimental evidence for the existence of plasmons as a well defined collective mode of the valence electrons of metals comes from characteristic energy-loss experiments. In such an experiment, one measures the energy loss–spectrum of keV electrons transmitted through a thin metallic foil, Fig. 1. The multiple excitation of this mode is also direct evidence for the quantization of the plasmon energy in units of $\hbar\omega_p$.

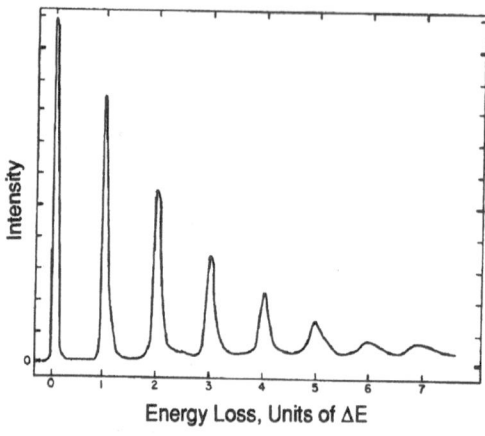

Figure 1. *Electron-energy-loss spectrum for a beam of 20keV electrons passing through an Al foil of 2580Å thickness.* $\Delta E = \hbar \omega_p = 15 eV$. *From Marton et al. [3]*

Excellent books and reviews on the theory and spectroscopy of solid state plasmas are available. In particular we recommend Pines and Nozieres [4], Platzman and Wolf [5], and DiBartolo [6]. Electron–energy–loss–plasmon spectroscopy became a major tool to study electronic excitations in solids, for surveys see Raether [7], Schnatterly [8], and Fink [9].

II. CONCEPTS TO DESCRIBE THE METALLIC STATE

II. A. The Standard Model: Jellium

To describe the characteristic metallic properties like the groundstate energy, the elementary excitations, and the interaction with electromagnetic fields, simple models have been found which are of immense value to solid state physics. For an introduction see e.g. Pines[2] or Ashcroft and Mermin [10].

The simplest quantum mechanical model of the metallic state is due to Sommerfeld. The Coulomb interactions between electrons as well as with the ions are completely neglected, yet the many particle aspect is taken into account. The single-electron states are plane-waves with wave-number **k** and energy $\epsilon_{\mathbf{k}}$ which - in accordance with the Pauli-principle - will be "filled" in **k**-space up to the radius k_F, the Fermi-wave number.

single particle states:

$| \mathbf{k} \rangle = \exp(i\mathbf{kr})$

$\epsilon(\mathbf{k}) = (\hbar\mathbf{k})^2/2m_0$

Fermi ground state:

$k_F = (3\pi^2 n)^{1/3}$

$\epsilon_F = (\hbar k_F)^2/2m_0$

$v_F = \hbar k_F/m_0$

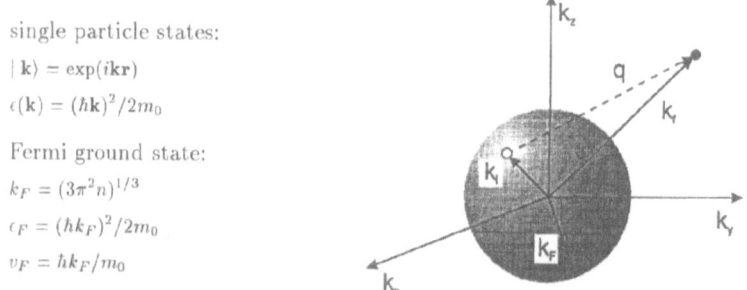

Figure 2. *Momentum distribution of the noninteracting Fermi-gas at zero temperature. In addition, a particle-hole excitation from initial state* $\mathbf{k}_i$ *to final unoccupied state* $\mathbf{k}_f$ *is shown.*

El	Z	$n/\text{Å}^3$	r_s	$k_F/\text{Å}$	ϵ_F/eV	$\hbar\omega_p/eV$	m^*/m_0	ϵ_∞	Ref
Li	1	0.0470	3.25	1.12	4.74	7.10	2.30	1.02	[7,10]
Na	1	0.0265	3.93	0.92	3.24	5.75	1.30	1.06	[10,15]
K	1	0.0140	4.86	0.75	2.12	3.80	1.20	1.15	[10,15]
Rb	1	0.0115	5.20	0.70	1.85	3.40	1.30	1.25	[10,15]
Cs	1	0.0091	5.62	0.65	1.59	2.90	1.50	1.29	[10,15]
Ag	1	0.0586	3.02	1.20	5.49	3.78	1.10		[7,10]
Be	2	0.2470	1.87	1.94	14.3	19.0	0.42	1.02	[7,10]
Mg	2	0.0861	2.66	1.36	7.08	10.5	1.30	1.01	[7,10]
Al	3	0.1810	2.07	1.75	11.7	15.0	1.40	1.11	[7,10,15]

Tab.1. *Parameters for some selected metallic elements and compounds. $\epsilon_\infty = 1 + 4\pi n\alpha$ where α is the polarizability of the ions. For Ag ϵ_∞ shows a strong dispersion peak at ω_p. Experimental data for m^* contain electron-electron and electron-phonon renormalization contributions and are, thus, larger than the bandstructure mass which is needed in (1).*

Next, we consider the influence of the Coulomb interaction between the electrons and the ions. For the "simple" metals - which include the alkalis, Al, Ga, In, Be, and Mg - the crystal potential is weak so that it is a good approximation to smear-out the ions into a positive background charge density $\rho_+ = -en_0$: Jellium. To describe the strength of the Coulomb interaction we compare the average kinetic and potential energy per electron *

$$\epsilon_{kin} = \frac{3}{5}\epsilon_F, \qquad \epsilon_{pot} \approx \frac{e^2}{4\pi\epsilon_0} \frac{1}{<r>}, \tag{2}$$

where $\langle r \rangle \approx n^{-1/3}$ is the mean electron distance. It is convenient to use "atomic units", i.e. we measure the lengths in units of Bohr-radii, energies in Rydbergs and characterize the density by the dimensionless Wigner-Seitz radius r_s.

$$\text{atomic units:} \begin{cases} \text{Bohr-radius:} & a_B = 4\pi\epsilon_0\hbar^2/m_0 e^2 = 0.529\ldots\text{Å} \\ \text{Rydberg-energy:} & R_y = e^2/8\pi\epsilon_0 a_B = 13.56\ldots eV \\ r_s\text{-parameter:} & n^{-1} = \frac{4\pi}{3}(a_B r_s)^3 \end{cases} \tag{3}$$

As $\epsilon_{pot} \propto r_s^{-1}$ but $\epsilon_{kin} \propto r_s^{-2}$, the Coulomb interaction becomes weak in the high density limit, $r_s < 1$. In most cases, however, r_s is not small but usually lies in the range 2..6, Tab. 1.

The Hamiltonian of Jellium is given by

$$\hat{H} = \sum_{j=1}^{N} \frac{\hat{p}_j^2}{2m_0} + \frac{1}{2}\sum_{i\neq j} \frac{e^2}{4\pi\epsilon_0} \frac{1}{|\hat{r}_i - \hat{r}_j|} + \hat{H}_{el-ion} + \hat{H}_{ion-ion} \tag{4}$$

The Coulomb interaction couples the states of all particles so that the exact eigenstates of $\hat{H}$ are not analytically accessible, yet reasonable approximations have been found. Owing to the uniform background charge the average field acting on one electron vanishes, so that as a first guess one might use a product of plane waves in accordance with the exclusion principle - up to k_F one electron per $\mathbf{k}$ and spin. This is the Hartree-approximation, which gives the same result for the ground state energy as the free electron model! To describe the metallic bound state one has to respect the antisymmetry of the wave function with respect to interchange of any two particles. Fortunately, in this Hartree-Fock approximation, the plane waves still provide a consistent basis, yet the single particle energies are different from the free-electron energy $\epsilon_\mathbf{k}$, Fig 2. As a result, the ground-state energy (per electron) is:

$$E_0/N = \frac{e^2}{8\pi\epsilon_0 a_B}\left[\frac{3}{5}(k_F a_B)^2 - \frac{3}{2\pi}(k_F a_B)\right] = \left[\frac{2.21}{r_s^2} - \frac{0.916}{r_s}\right]R_y \tag{5}$$

The second term in (5) is termed "exchange energy" because it results from the exchange of particles in the antisymmetrized wave-function. The difference between the exact ground state energy and the Hartree-Fock result is -by definition - the "correlation

---

* *Electron charge is $-e$, vectors in boldface, operators and tensors in boldface with hat.*

$$\hbar \mathbf{q} = \hbar \mathbf{k}_f - \hbar \mathbf{k}_i$$

$$\hbar \omega = \epsilon_f - \epsilon_i$$

$$\hbar \omega = a \hbar v_F q + \frac{(\hbar q)^2}{2m_0}$$

$$\alpha = -1 \dots 1$$

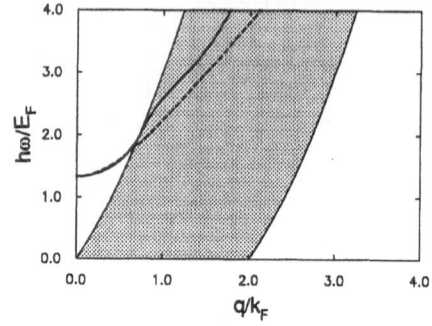

Figure 3. *Particle-hole excitation spectrum of a Fermi-gas (dotted area). The solid (dashed) lines display the plasmon-dispersion in a quantum mechanical (hydrodynamic) description.* $r_s = 2$.

energy". For Jellium a bound state is only possible for not too high densities, $r_s > 2.4$. In the low density limit, on the other hand, the electron behavior is dominated by the Coulomb interaction. For $r_s > 100$ the homogeneous state becomes unstable and crystallizes in a "Wigner - lattice", similar to the ions in a real metal [2,11].

To incorporate influences of the periodic crystal potential and the polarizability of the ions one has to replace the free-electron mass m_0 by a (bandstructure-) effective mass m^* and ϵ_0 by $\epsilon_0 \epsilon_\infty$, where the "background" dielectric constant ϵ_∞ accounts for the polarizability of the ion cores. In semimetals and heavily doped degenerate semiconductors the electron densities are quite small compared to ordinary metals, nevertheless their coupling parameter r_s may be even smaller than unity. For the ground state energy of real metals, the energy of the bottom of the band minimum and the electrostatic energy of the ions must be taken into account as well [2].

II. B. Particle - Hole Excitations

Apparently, the simplest type of excitation in a degenerate Fermi-system (at constant particle number) is by pushing one electron out of the Fermi-sea, leaving a hole behind, Fig.2. Conservation of energy and momentum requires $\hbar \mathbf{q} = \hbar \mathbf{k}_f - \hbar \mathbf{k}_i$ $\hbar \omega = \epsilon_f - \epsilon_i$ which leads to a quadratic function between ω and q. As parameter $a = k_i \cos\theta / k_F = -1 \dots 1$ the allowed region of excitations cover an entire region in the $\omega - q$ plane, Fig. 3.

II. C. Collective Excitations: Plasmons

Plasmons are collective excitations of the many electron system, i.e. all particles move coherently with a common frequency and wave-vector. In the long wave-length limit $\lambda >> 1/k_F$ this mode can be obtained quite simply. In addition we assume only a single band to contribute (one-component plasma). Consider deviations ρ_1 of the electron charge-distribution from its equilibrium value ρ_0. Then, the total charge distribution

$$\rho(\mathbf{r}, t) = \rho_+ + \rho_0 + \rho_1(\mathbf{r}, t) \tag{6}$$

will be the source of an electrical field

$$div \, \mathbf{E}(\mathbf{r}, t) = \rho_1(\mathbf{r}, t)/\epsilon_0, \qquad curl \, \mathbf{E}(\mathbf{r}, t) \approx 0. \tag{7}$$

Due to charge neutrality $\rho_0 + \rho_+ = 0$. In addition, the charge– and current density must obey the continuity equation

$$\frac{\partial \rho(\mathbf{r}, t)}{\partial t} + div \, j(\mathbf{r}, t) = 0 \tag{8}$$

where $\mathbf{j}(\mathbf{r},t) = \rho(\mathbf{r},t)\mathbf{v}(\mathbf{r},t)$ and $\mathbf{v}(\mathbf{r},t)$ denotes the mean velocity of the electrons. Finally, we need an equation to link the electron-velocity to the electrical field. In a hydrodynamic-like description this equation reads (see appendix A1):

$$m_0\, n(\mathbf{r},t)\frac{\partial \mathbf{v}}{\partial t} + grad\, P(\mathbf{r},t) = -en(\mathbf{r},t)\mathbf{E}(\mathbf{r},t). \qquad (9)$$

For low frequencies local thermodynamic equilibrium would hold and $P(\mathbf{r},t) = 2E/3V = 2\varepsilon_F n(\mathbf{r},t)/5$ is the pressure of the electron gas (at constant temperature, $T = 0$). For small amplitude oscillations, $|\rho_1| << |\rho_0|$, we obtain

$$[\frac{\partial}{\partial t} + \gamma]\mathbf{j}(\mathbf{r},t) = \frac{n_0 e^2}{m_0}\mathbf{E}(\mathbf{r},t) - \beta\, grad\, \rho_1(\mathbf{r},t) \qquad (10)$$

γ accounts for additional damping processes. Actually, plasma oscillations are a high frequency phenomenon so that the correct value for β is different from $\beta_{hyd} = v_F^2/3$. In the "random phase approximation" which is valid for $r_s < 1$ $\beta_{RPA} = 3v_F^2/5$ [2,4]. Eq. (10) is a generalization of the Drude-theory to inhomogeneous fields.

Assuming wave-like behaviour of all fields, e.g., $\rho_1(\mathbf{r},t) \propto \exp[i(\mathbf{qr} - \omega t)]$ and $\mathbf{E}, \mathbf{j} \parallel \mathbf{q}$, we obtain for the plasmon–dispersion

$$\omega_{bp}(q) = \sqrt{\omega_p^2 + \beta q^2} = \omega_p\left[1 + \frac{\beta}{2\omega_p^2}q^2 + \dots\right]. \qquad (11)$$

Bulk plasmons are, thus, longitudinally polarized charge–density waves. For neutral Fermi-systems ($\omega_p = 0$) like liquid ^3He the collective mode is a low-frequency phenomenon with sound-like dispersion $\omega = c_0 q$, $c_0 = v_F/\sqrt{3}$.

For small wave vectors, the stability of the plasmon is obvious from Fig. 3: the decay into a single electron-hole pair is forbidden by conservation of momentum and energy! Decay into several electron-hole pairs, however, is not excluded, yet as a higher order process it leads only to a small transition rate. For larger wave vectors the plasmon dispersion enters the particle-hole continuum and can decay directly into a single electron-hole pair (Landau-damping). Thus, the plasmon is expected to disappear as a well defined excitation beyond a critical wave-vector $q_c \approx k_F$.

II. D. Interaction with Electromagnetic Fields

It is clearly not feasable to work with the complete microscopic Maxwell-equations for real many particle systems. It is possible, though, to use "macroscopic fields" so that the Maxwell equations still retain their form and to construct reasonable approximations. For an introduction see Wooten [12], whereas, the state of the art of dielectric description of matter is layed down by Kirshnits et al. [13].

In a first step, one decomposes the charge and current densities into "external" and "system" or "induced" quantities.

$$\rho(\mathbf{r},t) = \rho_{ext}(\mathbf{r},t) + \rho_{ind}(\mathbf{r},t), \qquad \mathbf{j}(\mathbf{r},t) = \mathbf{j}_{ext}(\mathbf{r},t) + \mathbf{j}_{ind}(\mathbf{r},t). \qquad (12)$$

Here the adjective "external" merely refers to the control, not to the location of the sources, i.e. we assume that they are not effected by the fields induced in the system. Instead of $\rho_{ext}, \mathbf{j}_{ext}$ one often uses $\mathbf{E}_{ext}, \mathbf{B}_{ext}$, the external fields in the absence of the system charges.

For time-dependent fields, there is only but one independent "matter field", $\mathbf{j}_{ind}(\mathbf{r},t)$, - the charge density is already fixed (up to a constant) by integration of the continuity equation.

$$\frac{\partial \rho_{ind}(\mathbf{r},t)}{\partial t} + div\,\mathbf{j}_{ind}(\mathbf{r},t) = 0. \qquad (13)$$

Instead of $\mathbf{j}_{ind}(\mathbf{r},t)$ one often introduces two other fields $\mathbf{P}, \mathbf{M}$, the polarization and magnetization, by requiring

$$\mathbf{j}_{ind}(\mathbf{r},t) = \frac{\partial \mathbf{P}(\mathbf{r},t)}{\partial t} + curl\,\mathbf{M}(\mathbf{r},t), \qquad \rho_{ind}(\mathbf{r},t) = -div\,\mathbf{P}(\mathbf{r},t). \qquad (14)$$

Equations (14) automatically fulfill the continuity equation for the system charges. Such a decomposition is particularly useful for quasistatic fields for which $\mathbf{P}, \mathbf{M}$ were originally constructed. In addition, one tacitly assumes that $curl \mathbf{P} = 0$. For high frequencies, however, the decomposition of the current in terms of polarization and magnetization currents is ambiguous and one may put $\mathbf{M} = 0$ without loss of generality! In this "gauge" $\mathbf{D} = \epsilon_0 \mathbf{E} + \mathbf{P}$, $\mathbf{B} = \mu_0 \mathbf{H}$ and the Maxwell equations become particularly simple:

$$curl\, \mathbf{E}(\mathbf{r},t) = \frac{\partial \mathbf{B}(\mathbf{r},t)}{\partial t}, \qquad curl\, \mathbf{B}(\mathbf{r},t) = \mu_0 \left(\mathbf{j}_{ext}(\mathbf{r},t) + \frac{\partial \mathbf{D}(\mathbf{r},t)}{\partial t} \right),$$

$$div\, \mathbf{D}(\mathbf{r},t) = \rho_{ext}(\mathbf{r},t), \qquad div\, \mathbf{B}(\mathbf{r},t) = 0.$$

(15)

To specify the system under consideration, the Maxwell-equations must be supplemented by "material" equations, which - like (10) - link $\mathbf{j}$, or $\mathbf{P}, \mathbf{M}$ to the external (or total) fields. On a phenomenological level, this can be achieved by

$$\mathbf{D}(\mathbf{r},t) = \epsilon_0 \hat{\varepsilon} \mathbf{E}(\mathbf{r},t) = \epsilon_0 \int \int \varepsilon(\mathbf{r}, \mathbf{r}', t - t')\, \mathbf{E}(\mathbf{r}', t') d^3 \mathbf{r}'\, dt'$$

(16)

or by the inverse relation

$$\mathbf{E}(\mathbf{r},t) = \frac{1}{\epsilon_0} \hat{\varepsilon}^{-1} \mathbf{D}(\mathbf{r},t) = \frac{1}{\epsilon_0} \int \int \varepsilon^{-1}(\mathbf{r}, \mathbf{r}', t - t')\, \mathbf{D}(\mathbf{r}', t') d^3 \mathbf{r}'\, dt'$$

(17)

where $\varepsilon^{-1}(\mathbf{r}, \mathbf{r}', t - t')$ denotes the kernel of the operator $\hat{\varepsilon}^{-1}$ which is inverse to $\hat{\varepsilon}$.

For infinite, homogeneous systems, $\varepsilon(\mathbf{r}, \mathbf{r}', t - t')$ is solely a function of $\mathbf{r} - \mathbf{r}'$ and the Maxwell-equations can be simplified by Fourier-transformation.

$$\mathbf{E}(\mathbf{r},t) = \int \int \mathbf{E}(\mathbf{q}, \omega)\, e^{i(\mathbf{q}\mathbf{r} - \omega t)} \frac{d^3 q\, d\omega}{(2\pi)^4}, \qquad \mathbf{D}(\mathbf{q}, \omega) = \epsilon_0\, \varepsilon(\mathbf{q}, \omega) \mathbf{E}(\mathbf{q}, \omega),$$

$$\mathbf{E}(\mathbf{q}, \omega) = \int \int \mathbf{E}(\mathbf{r}, t)\, e^{-(i\mathbf{q}\mathbf{r} - \omega t)} d^3 \mathbf{r}\, dt, \qquad \varepsilon(\mathbf{q}, \omega) = \int \int \varepsilon(\mathbf{r}, t) e^{-i(\mathbf{q}\mathbf{r} - \omega t)}\, d^3 r dt.$$

(18)

To solve the following set of algebraic equations,

$$i\,\mathbf{q} \times \mathbf{E}(\mathbf{q}, \omega) = i\omega\, \mathbf{B}(\mathbf{q}, \omega), \qquad i\,\mathbf{q} \times \mathbf{B}(\mathbf{q}, \omega) = \mu_0 \left(\mathbf{j}_{ext}(\mathbf{q}, \omega) - i\omega\, \mathbf{D}(\mathbf{q}, \omega) \right),$$

$$i\,\mathbf{q} \cdot \mathbf{D}(\mathbf{q}, \omega) = \rho_{ext}(\mathbf{q}, \omega), \qquad i\,\mathbf{q} \cdot \mathbf{B}(\mathbf{q}, \omega) = 0,$$

(19)

it is convenient to decompose all vector-fields into longitudinal and transverse components with respect to wave vector $\mathbf{q}$, i.e.

$$\mathbf{E}(\mathbf{q}, \omega) = \mathbf{E}_\ell + \mathbf{E}_t = (\mathbf{E} \cdot \mathbf{n}_\ell) \mathbf{n}_\ell + \mathbf{n}_\ell \times (\mathbf{E} \times \mathbf{n}_\ell)$$

(20)

with unit vector $\mathbf{n}_\ell = \mathbf{q}/|\mathbf{q}|$. Even for homogeneous and isotropic systems, like Jellium, the reaction of the charged particles with respect to transverse and longitudinal fields is different, i. e. $\varepsilon(\mathbf{q}, \omega)$ is still a tensor with two different principal components $\varepsilon_\ell, \varepsilon_t$.

$$\mathbf{D}(\mathbf{q}, \omega) = \epsilon_0\, \varepsilon(\mathbf{q}, \omega)\, \mathbf{E}(\mathbf{q}, \omega) = \epsilon_0 \varepsilon_\ell(\mathbf{q}, \omega) \mathbf{E}_\ell + \epsilon_0 \varepsilon_t(\mathbf{q}, \omega) \mathbf{E}_t(\mathbf{q}, \omega)$$

(21)

$\varepsilon_\ell(0, \omega) = \varepsilon_t(0, \omega)$ is the "optical" dielectric function $\varepsilon(\omega)$. The solution of (19) is given by

$$D_\ell(\mathbf{q}, \omega) = \frac{-i}{q} \rho_{ext}(\mathbf{q}, \omega), \qquad D_t(\mathbf{q}, \omega) = \epsilon_0 \varepsilon_t(\mathbf{q}, \omega)\, \mathbf{E}_t(\mathbf{q}, \omega),$$

$$E_\ell(\mathbf{q}, \omega) = \frac{-i\, \rho_{ext}(\mathbf{q}, \omega)}{\epsilon_0\, \varepsilon_\ell(\mathbf{q}, \omega)\, q}, \qquad E_t(\mathbf{q}, \omega) = \frac{i\omega\, \mathbf{j}_{ext}(\mathbf{q}, \omega)}{\epsilon_0 [c^2 q^2 - \omega^2 \varepsilon_t(\mathbf{q}, \omega)]},$$

$$B_\ell(\mathbf{q}, \omega) = 0, \qquad B_t(\mathbf{q}, \omega) = \frac{1}{\omega} \mathbf{q} \times \mathbf{E}_t(\mathbf{q}, \omega).$$

(22)

The longitudinal component of the external current satisfies $(\mathbf{j}_{ext} - i\omega \mathbf{D})_\ell = 0$ which is equivalent to the continuity equation. Apparently, $\mathbf{B}$ is purely transverse. For non-relativistic particles, retardation effects may be neglected, hence, the electrical field is almost longitudinal. Furthermore, for slab geometries, $\mathbf{D}$ is identical with $\epsilon_0 \mathbf{E}_{ext}$, the electrical field without the system-charges.

In the hydrodynamic description the required dielectric functions are easily obtained from the Fourier-transformed equations (7,8,10)

$$\varepsilon_\ell(\mathbf{q},\omega) = 1 - \frac{\omega_p^2}{\omega(\omega + i\gamma) - \beta q^2}, \qquad \varepsilon_t(\mathbf{q},\omega) = 1 - \frac{\omega_p^2}{\omega(\omega + i\gamma)}. \tag{23}$$

Parameters γ, β can be used to fit the measured bulk losses and plasmon-dispersion. In the high density limit $r_s < 1$, $\beta_{RPA} = 3v_F^2/5$ which will be used as a reference. $\varepsilon_t(\mathbf{q},\omega)$ is identical with the standard Drude theory. Due to the singular structure of the denominator for the transverse fields in (22), this approximation will be sufficient in most cases.

The situation is more subtle for the longitudinal dielectric function. Clearly the hydrodynamic $\varepsilon_\ell(\mathbf{q},\omega)$ does not properly include particle-hole excitations which requires a microscopic theory, i.e. a kinetic or quantum treatment. Explicit analytical results were first obtained by Lindhard (see [2,4] and appendices A2,3), Fig. 4.

Causality requires that the kernel $\varepsilon(\mathbf{r}, \mathbf{r}', t-t') = 0$ if $t' > t$, i.e. the system cannot react before the pertubation is turned-on. This "trivial" property has deep consequences: In Fourier-space, $\varepsilon_t(\mathbf{q},\omega)$ is an analytic function in the complex ω–half-plane $\Im\omega > 0$ which, in turn, leads to the Kramers-Kronig relations:

$$\Re\varepsilon_t(\mathbf{q},\omega) - 1 = \frac{1}{\pi} P\!\!\int_{-\infty}^{\infty} \frac{\Im\varepsilon_t(\mathbf{q},\omega')}{\omega' - \omega}d\omega', \qquad \Im\varepsilon_t(\mathbf{q},\omega) = -\frac{1}{\pi} P\!\!\int_{-\infty}^{\infty} \frac{\Re\varepsilon_t(\mathbf{q},\omega') - 1}{\omega' - \omega}d\omega'. \tag{24}$$

Symbol "P" denotes "principal value integration", which is a prescription how to evaluate the singular integrals.

The Kramers-Kronig relations (24) are strictly valid for the transverse dielectric function only. Here, $\mathbf{D}$ is the response on the pertubation $\mathbf{E}$. For longitudinal fields the situation is opposite. ρ_{ext}, or equivalently $\mathbf{D}$, plays the role of an (arbitrarily prescribable) pertubation rather than $\mathbf{E}$. Hence, $\varepsilon^{-1}(\mathbf{r}, \mathbf{r}', t - t') = 0$ if $t' > t$ is a causal function, and $\varepsilon_\ell^{-1}(\mathbf{q},\omega)$ is analytic in $\Im\omega > 0$ and, correspondingly, the Kramers-Kronig relations read:

$$\Re\frac{1}{\varepsilon_\ell(\mathbf{q},\omega)} - 1 = \frac{1}{\pi} P\!\!\int_{-\infty}^{\infty} \frac{\Im\frac{1}{\varepsilon_\ell(\mathbf{q},\omega')}}{\omega' - \omega}d\omega', \qquad \Im\frac{1}{\varepsilon_\ell(\mathbf{q},\omega)} = -\frac{1}{\pi} P\!\!\int_{-\infty}^{\infty} \frac{\Re\frac{1}{\varepsilon_\ell(\mathbf{q},\omega')} - 1}{\omega' - \omega}d\omega'. \tag{25}$$

But, why is $\varepsilon_\ell(\mathbf{q},\omega)$ not "as good" as $\varepsilon_\ell^{-1}(\mathbf{q},\omega)$? $\varepsilon_\ell^{-1}(\mathbf{q},\omega)$ may well have a zero in $\Im\omega > 0$ so that its inverse would have a pole at just this frequency. Obviously, $\varepsilon_\ell(\mathbf{q},\omega)$ wouldn't be analytic in $\Im\omega > 0$ which, by "backshooting" kills causality between the longitudinal $\mathbf{D}$-response and $\mathbf{E}$-pertubation. Eq. (24) is correct for transverse fields whereas (25) holds for longitudinal fields! In mathematical terms: The operator $\hat{\varepsilon}$ has a zero eigenvalue (with "transverse eigenfunction") so that its inverse does not exist. Likewise for $\hat{\varepsilon}^{-1}$ and longitudinal fields. To describe the longitudinal response it is best to introduce a (longitudinal) susceptibility $\chi(\mathbf{q},\omega)$ as it is done in microscopic treatments

$$\frac{1}{\varepsilon_\ell(\mathbf{q},\omega)} = 1 + \frac{e^2}{\epsilon_0 q^2}\chi(\mathbf{q},\omega) = \frac{\rho(\mathbf{q},\omega)}{\rho_{ext}(\mathbf{q},\omega)} = \frac{\Phi(\mathbf{q},\omega)}{\Phi_{ext}(\mathbf{q},\omega)}. \tag{26}$$

$\chi(\mathbf{q},\omega)$ denotes the Fourier-transform of the density response function [2,4,12,13]

$$\chi(\mathbf{r}, \mathbf{r}', t - t') = -\frac{i}{\hbar}\theta(t - t')\langle\langle[\hat{N}(\mathbf{r}, t), \hat{N}(\mathbf{r}', t')]\rangle\rangle \tag{27}$$

which is related to the dynamic form factor $S(\mathbf{q},\omega)$ by

$$\chi(\mathbf{q},\omega) = \int_0^\infty S(\mathbf{q},\omega')\Big[\frac{1}{\hbar(\omega - \omega') + i\delta} - \frac{1}{\hbar(\omega + \omega') + i\delta}\Big]d\omega',$$
$$S(\mathbf{q},\omega) = \frac{1}{\Omega}\sum_m |\langle m \mid \hat{N}_{\mathbf{q}}^\dagger \mid 0\rangle|^2 \,\delta(\hbar\omega - E_m + E_0), \quad \omega > 0. \tag{28}$$

$|m\rangle$, E_m are the exact many-body states and energies, Ω is the volume, and $\hat{N}(\mathbf{r}, t)$ is the density operator. Clearly, the poles of $\chi(\mathbf{q},\omega)$ are identical with the excitation energies of the many body system.

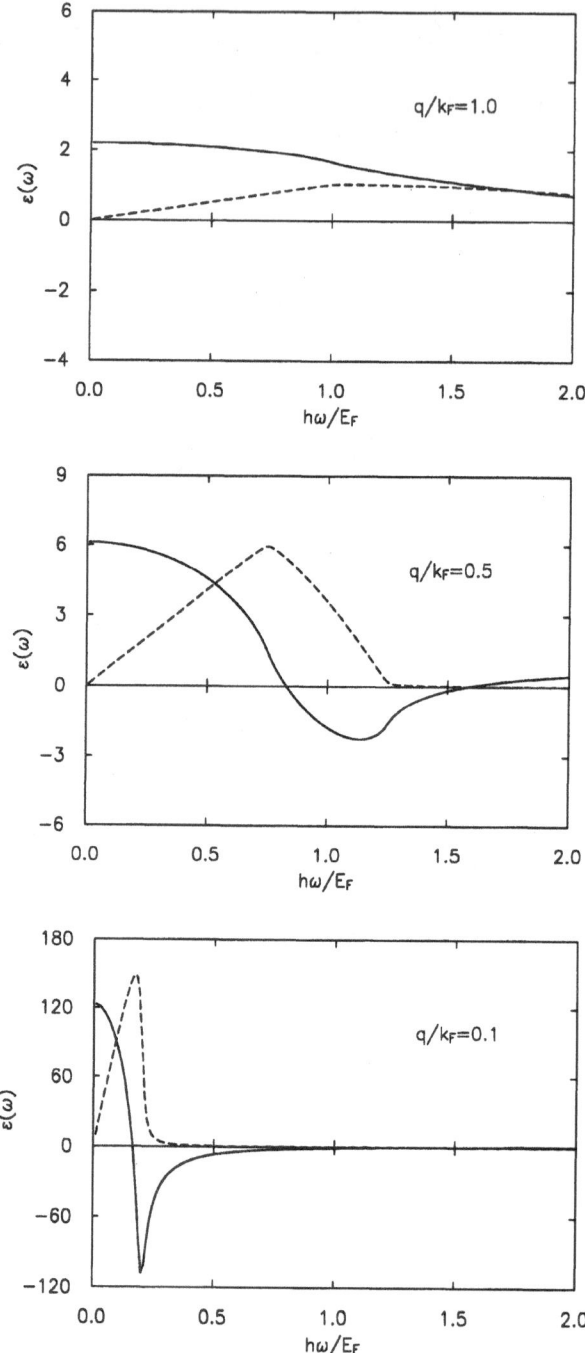

Figure 4. *Real part (solid lines) and imaginary part (dotted lines) of the Lindhard dielectric function.* $r_s = 2$.

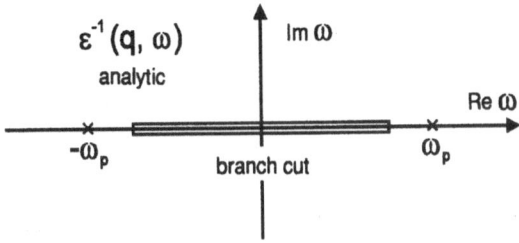

Figure 5. *Location of singularities of $\chi(\mathbf{q}, \omega)$ in RPA. The branch cut describes the particle-hole excitations whereas the poles correspond to the plasmon-mode.*

With the aid of (24,25) and the high frequency behaviour of $\varepsilon(\mathbf{q},\omega) \to 1 - (\omega_p/\omega)^2$ one can prove the following sum rules[2,4]

$$\int_0^\infty \omega \, \Im \frac{-1}{\varepsilon_\ell(\mathbf{q},\omega)} \, d\omega = \int_0^\infty \omega \, \Im \varepsilon_t(\mathbf{q},\omega) \, d\omega = \frac{\pi}{2}\omega_p^2. \tag{29}$$

Due to the translational symmetry of Jellium, the $q = 0$ limit of the exact dielectric function is identical with the Drude-result in the absence of scattering

$$\varepsilon_\ell(0,\omega) = \varepsilon_t(0,\omega) = 1 - \frac{\omega_p^2}{\omega^2}. \tag{30}$$

III. BULK PLASMONS

III. A. Dispersion, Life-Time, and Oscillator Strength

A collective excitation always corresponds to a possible oscillation of the system in the absence of an external field. Apparently, the dispersion relation $\omega_{bp}(q)$ of these modes is given by the poles of the density response function $\chi(\mathbf{q},\omega)$, or equivalently, the zeros of $\varepsilon_\ell(\mathbf{q},\omega)$:

$$\chi(\mathbf{q},\omega) = \infty, \qquad or : \varepsilon_\ell(\mathbf{q},\omega) = 0. \tag{31}$$

Bulk plasmons are purely electrical waves, $\mathbf{B} = 0$.

Causality warrants that the solutions of (31) are located on the real ω–axis or in the lower ω–half-plane when continuing $\chi(\mathbf{q},\omega)$ analytically to $\Im\omega < 0$.

$$\omega = \omega_b(q) - i\Gamma(q), \quad \Gamma(q) = \hbar/\tau > 0. \tag{32}$$

where τ is the plasmon life-time. Well defined collective modes are only those solutions with $\Gamma \ll \omega_p$, Fig. 5.

The solutions of (31) lead to peaks in the loss-function

$$P_0(\mathbf{q},\omega) = \Im \frac{-1}{\varepsilon_\ell(\mathbf{q},\omega)} \tag{33}$$

which describes the power dissipated by the external field, Fig. 6. Branch-cuts corresponding to the particle-hole excitation continuum may lead to peaks in the loss-function which will be hard to distinguish experimentally from "true" collective modes.

In the small-q limit the dispersion is parabolic and their line-width is zero for $q < q_c \approx \omega_p/v_F$ within the kinetic or RPA theory. If the pole is close to the real ω–axis we may write

$$P_0(\mathbf{q},\omega) = \frac{\Re Z(q) \, \Gamma(q) - \Im Z(q) \, [\omega - \omega_{bp}(q)]}{[\omega - \omega_{bp}(q)]^2 + \Gamma^2(q)} + P_{inc}(\mathbf{q},\omega),$$

$$Z^{-1}(q) = \frac{\partial \varepsilon(\mathbf{q},\omega)}{\partial \omega}, \quad \omega = \omega_{bp}(q), \tag{34}$$

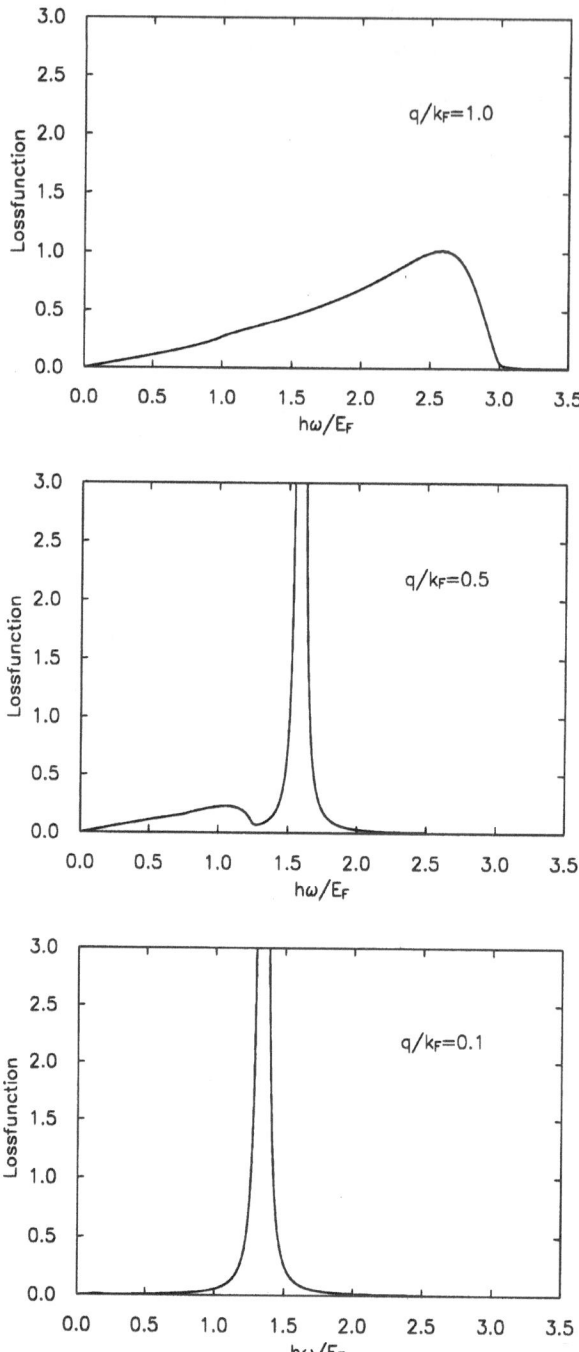

Figure 6. *Loss-function for the Lindhard dielectric function [A3]. Instrumental resolution is simulated giving ω a small imaginary part of $0.01E_F$. $r_s = 2$.*

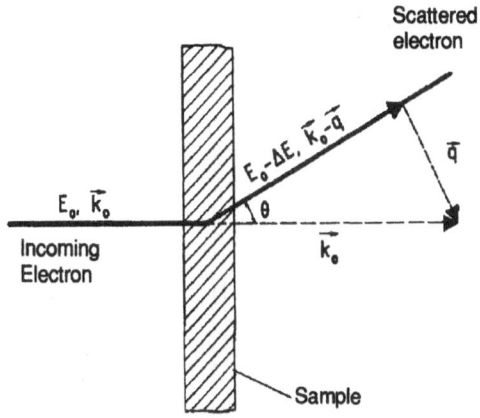

Figure 7. *Geometry of the electron–energy–loss scattering experiment [9]*

where $Z(q)$ is the residue of the pole which is the plasmon-oscillator strength. $P_{inc}(\mathbf{q}, \omega)$ describes the "incoherent" background contribution of the particle-hole excitation continuum. If $\Im Z(q) << \Re Z(q)$ the plasmon line-shape is lorenzian with width $\Gamma(q)$. Due to the sum-rule (29), $Z(q)$ is largest at $q = 0$, and decreases with increasing momentum transfer. Thus the plasmon is the dominant feature in $P(\mathbf{q}, \omega)$ at low momentum transfer.

III. B. Fundamentals of Electron-Energy-Loss-Spectroscopy (EELS)

Spectroscopy of plasmons requires the interaction with electromagnetic fields, either by radiation or with charged particles like fast electrons. As plasmons are longitudinal polarized they don't couple directly to transverse electromagnetic waves. An almost ideal tool, are fast (but nonrelativistic) electrons, which interact quasistatically via their Coulomb-field "flying" with them [2,5-9]. We follow Fink [9] on "recent developments on electron-energy-loss-spectroscopy".

The geometry of an EELS experiment is shown in Fig. 7. In the Karlsruhe-spectrometer of Dr. Fink (now at IFW Dresden) the primary energy is $E_0 = 170 keV$ which corresponds to $k_0 = 228.4 \mathring{A}^{-1}$. The scattered electrons are analyzed with respect to energy- and momentum transfer. Small q's require very small scattering angles, i.e. for $q = 1 \mathring{A}^{-1}$ the scattering angle is $4 mrad \approx 0.25°$. Decomposition of the scattering wave vector into components parallel and perpendicular to the incoming beam reveals

$$q_\parallel \approx k_0(\hbar\omega/2E_0), \qquad q_\perp \approx k_0 \sin\theta, \tag{35}$$

with $q^2 = q_\parallel^2 + q_\perp^2$. Optimum resolution of the instrument is achieved at lowest beam current ($15 nA$, beam cross section is $0.5 mm$ after passing the monochromator): $\Delta E = 80 meV$, $\Delta q = 0.04 \mathring{A}^{-1}$ (full width at half maximum).

The basic quantity measured is the differential cross-section which, as usual, can be factorized in an atomic form factor (=Rutherford cross section) and in the structure function $S(\mathbf{q}, \omega)$ which contains the dynamics of the system [2,5]. Instead of $S(\mathbf{q}, \omega)$ EELS–spectroscopists prefer the loss function $P_0(\mathbf{q}, \omega)$.

$$\frac{d^2\sigma}{d\Omega d(\hbar\omega)} = \frac{4}{a_B^2 q^4} S(\mathbf{q}, \omega) = \frac{\hbar}{(\pi a_B)^2} \frac{1}{q^2} P_0(\mathbf{q}, \omega). \tag{36}$$

The EELS cross-section decreases with increasing momentum transfer. For X-ray scattering the situation is opposite, yet EELS-resolution is much better.

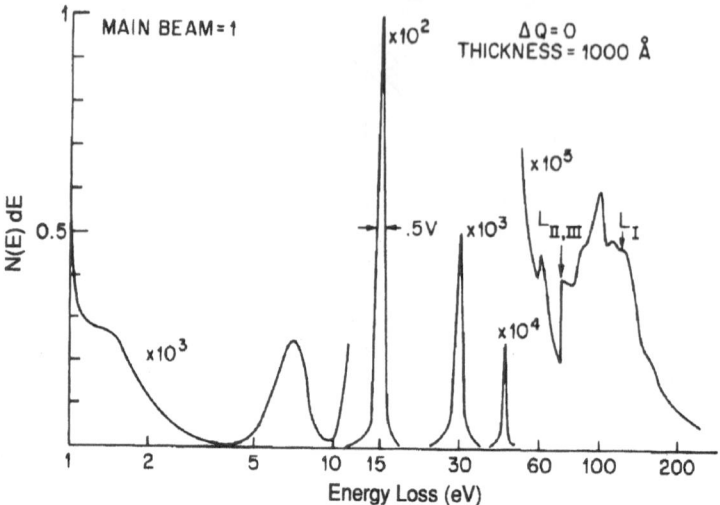

Fig.8. *Excitation spectrum of Al in the range 1-250 eV. From Schnatterly [8].*

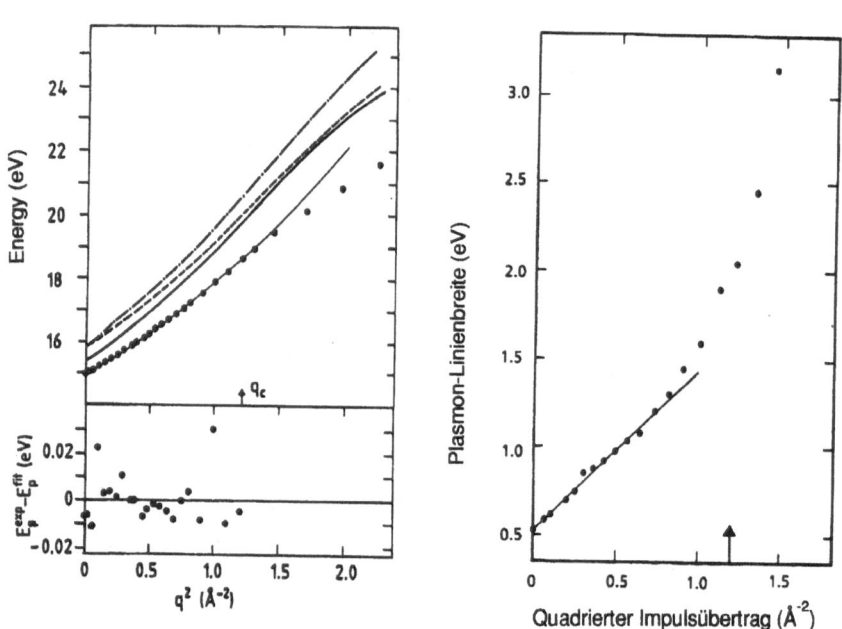

Figure 9. *Measured plasmon-dispersion and line-width in Al parallel to [100] direction compared to least square fit curve (thin line) and theories. From Sprösser-Prou et al. [14].*

III. C. Experimental Results

In the last two decades plenty of experimental and theoretical work has been performed on the dispersion of plasmons as well as on other excitations in metals. In particular, the simple metals like Al and the alkali metals (apart from Li) are regarded as nature's closest realization of Jellium. In these metals, band structure effects are expected to be small so that exchange and correlation effects may be studied. For a survey on other materials see Raether.[7].

Fig. 8 shows, as an example, the measured excitation spectrum of Al. In order of increasing energy the features of the spectrum are: an interband transition at $1.5eV$, the surface plasmon at $7eV$ (oxidized surface), the bulk plasmon at $15eV$, multiple excitation of plasmons, and the $L_{II,III}$ soft X-ray threshold at $72.5eV$ etc. With one instrument, all the elementary excitations from the near IR to the soft X-ray region can be studied [8].

Results of a high resolution ELS study of the bulk plasmon dispersion with respect to the absolute value and orientation of the transferred momentum in an Al single crystal are shown in Fig.9. The dispersion has been observed to be biquadratic in q with unique parameters over the entire q range up to the cut-off wave-vector, in contrast to earlier studies as summarized e.g. in [7].

$$\hbar\omega_{bp}(q) = \hbar\omega_p + \alpha\frac{(\hbar q)^2}{m_0} + Bq^4. \tag{37}$$

However, substantial deviations from the RPA result $\alpha_{RPA} = \frac{3}{5}\epsilon_F/\hbar\omega_p = 0.44$ have been found: $\alpha_{exp} = 0.30$. Beyond the cut-off q_c (indicated by an arrow) the plasmon still exists but with an increased line-width.

Plasmon dispersion in the alkali's seems to be even more puzzling, Fig. 10. Previous measurements of plasmon-dispersion in simple metals always showed a positive, quadratic dispersion. Rb is the first metal with almost no dispersion at all. The deviations become even more pronounced in the case of Cs which exhibits a negative dispersion. Deviations do not only occur with respect to the RPA, but to improved theories as well, Fig. 11. Particularly, the deviations increase with r_s and indicate the increasing influence of correlation effects in the metallic regime. One might interpret the negative dispersion in Cs as an incipient Wigner-crystallization of the electrons.

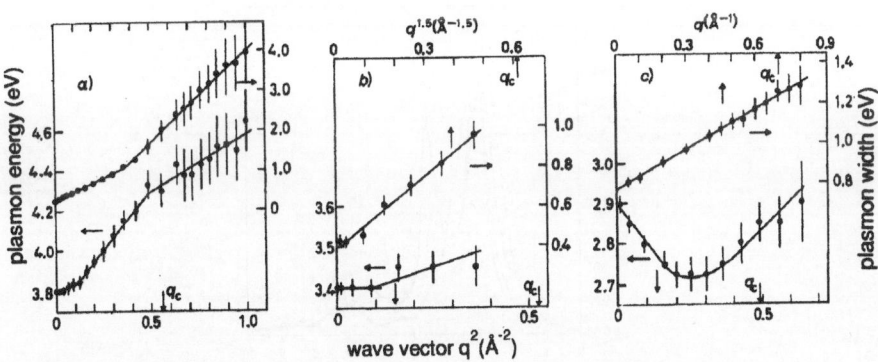

Figure 10. *Plasmon dispersion for polycrystalline K, Rb, and Cs (a-c). From v. Felde et al. [15].*

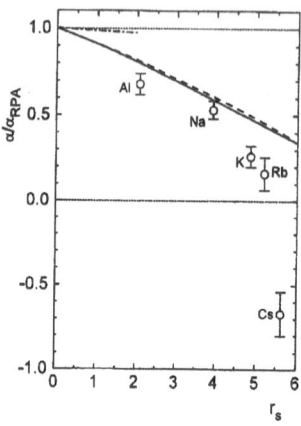

Figure 11. *Normalized plasmon dispersion coefficient as a function of the density parameter r_s. Dotted line: RPA, solid and dashed lines: theory by Vashishta and Singwi [16] and Dabrowski [17]. From v. Felde et al. [15].*

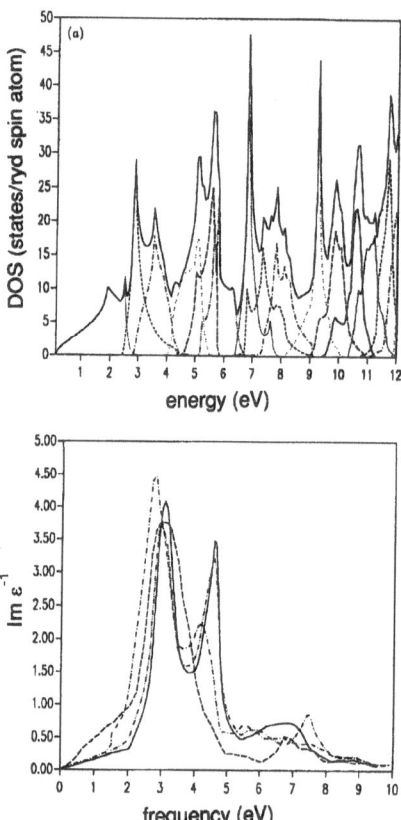

Figure 12. *Calculated excitation spectrum of Cs. Upper part: Total density of states (solid line) and contributions from individual bands (broken lines). Fermi energy is $1.88eV$. Lower part: Loss function for some wave vectors along the $(1,0,0)$ direction. $\mathbf{q} = 2\pi\kappa(1,0,0)/a$, for $\kappa = 0.1, 0.2, 0.3, 0.6$ (full, dashed-double dotted, dashed-dotted, and dashed lines). From Kollwitz and Winter [20].*

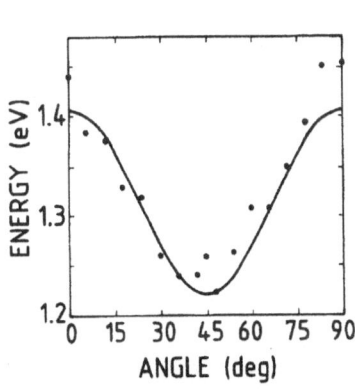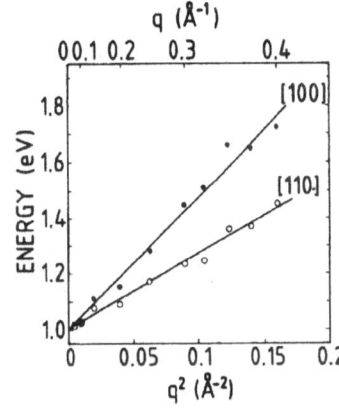

Figure 13. *Angle dependence (left) and dispersion (right) of the bulk-plasmon in $Bi_2Sr_2CaCu_2O_8$, a high T_c superconductor in the normal state. From Nücker et al. [22].*

The construction of a reasonable approximation for the interacting electron gas in the intermediate coupling range is a long standing problem. Deviations from the *RPA* arise from the neglected exchange and correlation as well as from the periodic crystal potential, which, both reduce α with respect to the *RPA* result. Up to 1986 the theoretical results, e.g. [16-17] converged to the result diplayed in Fig. 11 by the solid and dashed lines. Remarkably, the results of these ambitious many-body theories agrees very well with a kinetic theory when the many-body interactions are included by a standard density functional description, appendix A2, where

$$\alpha/\alpha_{RPA} = 1 - \left[0.092r_s + \frac{0.0034r_s^2}{1+r_s/21}\right]. \tag{38}$$

Stimulated by the experimental results by v. Felde et al. [15] several groups obtained a qualitative improvement. Taut and Sturm [18] considered the combined effect of exchange-correlation and crystal potential and Lipparini et al. [19] worked out a sum-rule approach taking multipair excitations into account. Kollwitz and Winter [20] and Aryasetiawan and Karlson [21] calculated the density response function within a *LDA* formalism. The Cs-density of states resembles those of a transition metal, Fig 12. Thus, the heavier alkalis are not free-electron–like metals!

Fig. 13 gives the "in-plane" plasmon dispersion at $300K$ in $Bi_2Sr_2CaCu_2O_8$ - one of the new high T_c superconductors [22]. The most important bands are essentially formed by the occupied orbitals in the $Cu - O$ plane. Those with the largest overlap being the $Cu\ 3d_{x^2-y^2}$ and the $O\ 2p_x$, $O\ 2p_y$ orbitals forming a quasi two-dimensional tight-binding bandstructure

$$E(\mathbf{k}) = -\frac{1}{2}t\left[cos(k_x a) + cos(k_y a)\right] \tag{39}$$

with $t \approx 1.5eV$ and $O - O$ distance $a = 3.8\overset{\circ}{A}$. Approximating the matrix elements in the *RPA* dielectric function, A3, by their free electron values ($=1$)

$$\varepsilon(\mathbf{q},\omega) = 1 - \frac{e^2}{\epsilon_0 q^2}4 \int \frac{d^3\mathbf{k}}{(2\pi)^3} f(E(\mathbf{k}))\frac{\Delta E}{(\hbar\omega + i0)^2 - (\Delta E)^2} \tag{40}$$

where $\Delta E = E(\mathbf{k}+\mathbf{q}) - E(\mathbf{k})$ the plasmon dispersion becomes in the $q \to 0$ limit [22]

$$\hbar\omega_{bp}(\mathbf{q}) = \sqrt{(\hbar\omega_p)^2 + \frac{1}{6}(ta)^2q^2\left[\frac{3}{2} + \frac{1}{2}cos4\phi\right]}, \tag{41}$$

where ϕ is the angle between $\mathbf{q}$ in the $x - y$ plane and the x axes. Other parameters are $\epsilon_\infty = 4.5$, $m^*/m_0 = 1.7$, $n_0 = 9 \times 10^{21}cm^{-3}$. There is a remarkable good agreement with experiment.

317

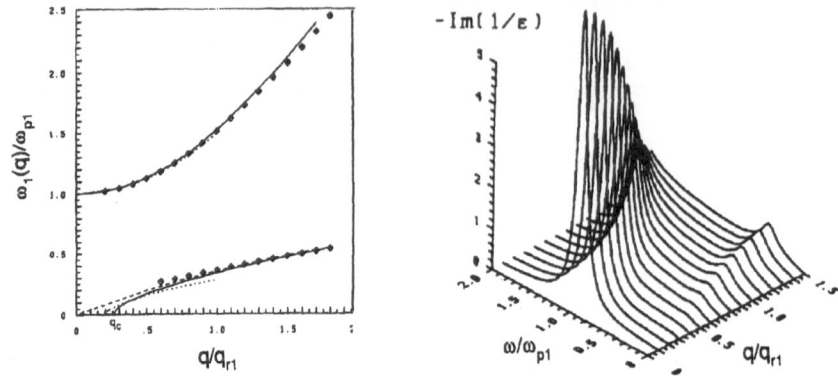

Figure 14. *Dispersion (left) and Loss-function (right) of a two component Fermi-system.* $\omega_{p2} = 0.5\omega_{p1}$, $\tau_1 = \tau_2 = 71/\omega_{p1}$. *Dashed lines:* $\tau_j = \infty$. *From [27].*

III. D. Acoustic Plasmons in a Two-Component Fermi-System

A Fermi-system with two (or more) different types of mobile charge carries e.g. a transition metal with s and d electrons exhibits two distinct collective charge density excitations: a high frequency "optical" plasmon and a sound-like "acoustic" plasmon. The dispersion of these modes, in the limit $q \to 0$ and in absence of scattering, is given by [23,24],

$$\omega_{opt}(q) = \sqrt{\omega_{p1}^2 + \omega_{p2}^2} + O(q^2),$$

$$\omega_{ac}(q) = \sqrt{\frac{\omega_{p1}^2 \beta_2 + \omega_{p2}^2 \beta_1}{\omega_{p1}^2 + \omega_{p2}^2}} \, q. \tag{42}$$

ω_{pj} and β_j denote the plasma frequencies and dispersion coefficients of the components $j = 1, 2$.

In contrast to optical plasmons, which are well-established experimentally, their acoustic counterparts have not yet been unambiguously identified, e.g. [25]. Nevertheless, there are several speculations in serious journals about their relationship to the "old high T_c" superconductors, e.g. Nb_3Sn [26].

Analogous to plasmons in a one-component Fermi-system the dispersion of acoustic and optical plasmons is obtained from

$$\varepsilon(\mathbf{q}, \omega) = 1 + \Pi_1(\mathbf{q}, \omega) + \Pi_2(\mathbf{q}, \omega) = 0, \tag{43}$$

where $\Pi_j(\mathbf{q}, \omega)$ denotes the polarization of the individual components. In the hydrodynamic description (23), we obtain

$$\varepsilon(\mathbf{q}, \omega) = 1 - \frac{\omega_{p1}^2}{\omega(\omega + i\gamma_1) - \beta_1 q^2} - \frac{\omega_{p2}^2}{\omega(\omega + i\gamma_2) - \beta_2 q^2}. \tag{44}$$

For $\gamma_j = 0$ the two branches of the solutions of (43) are given by (41). Results of a numerical study within a kinetic theory are displayed in Fig. 14. Notice that the acoustic mode is overdamped near $q = 0$ and its oscillator strength is very small.

IV. SURFACE PLASMONS

IV. A. Concept of a Surface Plasmon

The concept of surface plasmons was introduced by Ritchie[28] shortly after the discovery of bulk plasmons in metals. In recent years surface plasmons have been observed in a wide range of materials using both electron and photon spectroscopy. In addition, significant progress has been made towards a systematic use of surface plasmons as a diagnostic tool to investigate the surface charge profile of metals. For surveys see, e.g. Raether [29].

Suppose an electron gas confined to the half-space $z < 0$ with a smooth electron charge distribution near the metal-vacuum boundary, Fig. 15. If an external field acts on the plasma boundary it will induce a charge which is concentrated near the surface,

$$\rho(\mathbf{r},t) = \Re\left[\exp[i(q_x x - \omega_s t)]d(z)\right], \qquad (45)$$

where q_x is the wave number and ω_{sp} the frequency of the surface charge density wave.

For metals, the "spill-off" of the charge density across the geometrical boundary $z = 0$ is of the order of $1/k_F \approx 1\overset{\circ}{A}$, so that a macroscopic description may be sufficient for a qualitative understanding of surface charge density wave.

The charge oscillations are in turn the sources of electromagnetic fields. In particular, we are looking for "evanescent" waves, i.e. fields which decay exponentially from the surface. First, $div\,\mathbf{E}$ demands:

$$\mathbf{E}(\mathbf{r},t) = \Re\left[(\alpha A, 0, iq_x A)e^{i(q_x x - \omega t)}e^{-\alpha z}\right] \qquad (46)$$

where q_x is the same in both half-spaces but the amplitudes $A_\pm$ and decay-constants $\alpha = \pm\alpha_\pm$ are different. Second, the wave equation requires

$$\Delta\mathbf{E} + (\frac{\omega}{c})^2\epsilon_\pm(\omega)\,\mathbf{E} = 0, \quad q_x^2 - \alpha_\pm^2 = (\frac{\omega}{c})^2\epsilon_\pm(\omega), \qquad (47)$$

and, third, we have to respect the boundary condition at $z = 0$:

$$\begin{aligned}
\text{Continuity of } \mathbf{E}_x: \quad & -\alpha_+ A_+ = \alpha_- A_-, \\
\text{Continuity of } \mathbf{D}_z: \quad & q_x\epsilon_+ A_+ = q_x\epsilon_- A_-.
\end{aligned} \qquad (48)$$

Nontrivial solutions for $A_\pm$ of (48) require the determinant to vanish

$$\alpha_+\epsilon_-(\omega) + \alpha_-\epsilon_+(\omega) = 0, \qquad (49)$$

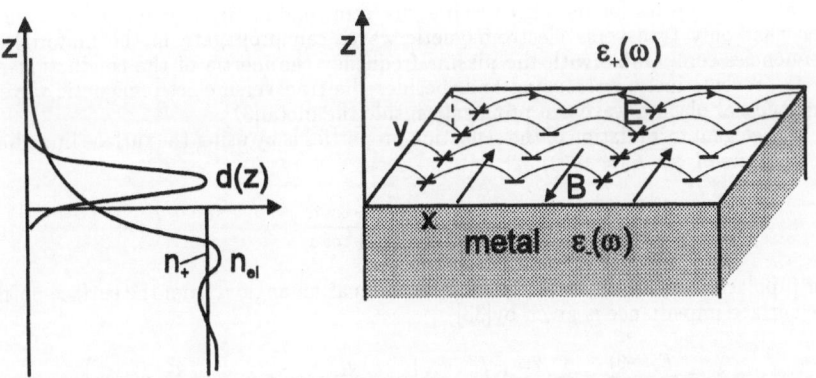

Figure 15. *Electron charge distribution (left) and electromagnetic field (right) for a plasmon at a plane surface.*

or, when using the relation between $\alpha_\pm$ and q_x:

$$\sqrt{q_x^2 - (\frac{\omega}{c})^2 \varepsilon_+(\omega)}\, \varepsilon_-(\omega) + \sqrt{q_x^2 - (\frac{\omega}{c})^2 \varepsilon_-(\omega)}\, \varepsilon_+(\omega) = 0. \tag{50}$$

This equation may be solved for q_x as a function of ω

$$q_x^2 = (\frac{\omega}{c})^2 \frac{\varepsilon_+(\omega)\varepsilon_-(\omega)}{\varepsilon_+(\omega) + \varepsilon_-(\omega)}, \tag{51}$$

which is a true solution only if $\Re\,\varepsilon_+/\varepsilon_- < 0$.

For a Drude-metal bounded by a nondispersive dielectric ε_b, we have:

$$\omega_{sp}(q_x) = \begin{cases} \dfrac{c}{\sqrt{\varepsilon_b}} q_x \left[1 - \dfrac{1}{2}(\dfrac{c}{\omega_p})^2 q_x^2 \ldots\right], & q_x \ll \omega_p/c \\[3mm] \dfrac{\omega_p}{\sqrt{1+\varepsilon_b}} \left[1 - \dfrac{1}{2}(\dfrac{\omega_p}{c})^2 (\dfrac{\varepsilon_b}{1+\varepsilon_b})^2 q_x^{-2} \ldots\right], & q_x \gg \omega_p/c. \end{cases} \tag{52}$$

Core-polarization effects in the metal can be taken into account by rescaling $\omega_p \rightarrow \omega_p/\sqrt{\varepsilon_\infty}$, $c \rightarrow c/\sqrt{\varepsilon_\infty}$, $\varepsilon_b \rightarrow \varepsilon_b/\varepsilon_\infty$.

The phase velocity of the surface-plasmon is always smaller than the speed of light in the adjacent dielectric. Thus, without participation of another system which may take-off momentum the surface-plasmon cannot decay radiatively. Even the decay into other surface-plasmons of smaller energy is not possible. Any structural feature which breaks the symmetry of the plane surface, however, may give rise to coupling: surface roughness, phonons, grating-rulings,...(For the discussion of "radiative plasmons" see Raether [29].)

The discussion of surface-plasmons can be alternatively presented as a problem of optics: Does the (inverse) reflection problem have an outgoing solution for vanishing amplitude of the incoming field?[30,31] For p-polarized light the standard Fresnel formulae for the reflection of light at a plane surface are [32]:

$$R_p = \frac{\tan(\alpha - \beta)}{\tan(\alpha + \beta)}, \qquad \frac{\sin(\alpha)}{\sin(\beta)} = n_r = \sqrt{\frac{\varepsilon_-(\omega)}{\varepsilon_+(\omega)}}. \tag{53}$$

R_p denotes the amplitude reflection-coefficient and α, β are related by Snell's diffraction law which is a consequence of the continuity of q_x at the boundary. (Formally, these relations are valid for complex α and β, too.) In the limit of vanishing incoming amplitude a nonzero outgoing wave is only possible for $R_p = \infty$, i.e. $\alpha - \beta = \pi/2$ or $\cos\alpha = -\sin\beta$. Using $\tan\alpha = q_x/q_\perp = -n_r$ and $q_\perp^2 + q_x^2 = (\omega/c)^2 \varepsilon_+$ immediately leads to (51). ($\alpha = \pi/2 - i\alpha'$). In addition, we notice that $R_p = 0$, i.e. $\alpha + \beta = \pi/2$ is just the condition for the Brewster-angle.

Standard optics for metals as well as for semiconductors is based on the assumption that only transverse electromagnetic waves can propagate in the material. At frequencies comparable with the plasma-frequency the inertia of the conduction electrons prevents instant screening and, besides the transverse electromagnetic waves, a longitudinal plasma wave can propagate inside the metal[31].

An elegant formulation of the reflection properties is by using the surface impedance Z_p (in units of $\sqrt{\mu_0/\epsilon_0} \approx 377\Omega$)[32].

$$R_p(\alpha, \omega) = \frac{Z_p - \cos\alpha}{Z_p + \cos\alpha} \tag{54}.$$

For p-polarized light, incident from the vaccuum at an angle α from the surface normal, the surface impedance is given by[33]

$$Z_p = \frac{E_x(-0)}{H_y(-0)} = \frac{1}{2\pi}(\frac{2i\omega}{c}) \int_{-\infty}^{\infty} \frac{dq_z}{q^2} \left[\frac{q_x^2}{(\omega/c)^2 \varepsilon_\ell(q,\omega)} + \frac{q_z^2}{(\omega/c)^2 \varepsilon_t(q,\omega) - q^2}\right] \tag{55}$$

where $q^2 = q_x^2 + q_z^2$, $q_x = \omega \sin\alpha/c$. Eq. (55) is valid for a sharp surface and if the electrons are scattered specularly at the surface.

The surface plasmon dispersion is obtained from the pole of (54)

$$\sqrt{q_x^2 - (\frac{\omega}{c})^2} = -\frac{2}{\pi} \int_0^\infty \frac{dq_z}{q^2} \left[\frac{q_x^2}{\varepsilon_\ell(q,\omega)} + \frac{q_z^2}{\varepsilon_t(q,\omega) - (cq/\omega)^2} \right]. \tag{56}$$

Result (56) includes many previous results for the sharp-barrier approximation. For example, in the local approximation $\varepsilon_\ell(\mathbf{q},\omega) = \varepsilon_t(\mathbf{q},\omega) = \varepsilon(\omega)$ (51) is obtained. If retardation is neglected by letting $c \to \infty$ in (56), only the term involving the longitudinal dielectric function remains,

$$-1 = \frac{2q_x}{\pi} \int_0^\infty \frac{dq_z}{q^2 \varepsilon_\ell(q,\omega)}, \tag{57}$$

which, in the hydrodynamic approximation and small wave vectors becomes Ritchie's "classic" result [28]

$$\omega_{sp}(q_x) = \frac{\omega_p}{\sqrt{2}} \left[1 + (a_1 + ia_2)q_x \right], \quad a_1 = \sqrt{\frac{3}{10}} \frac{v_F}{\omega_p}, \quad a_2 = 0. \tag{58}$$

Historic Remark: Before the discovery of the ionosphere, Zenneck and Sommerfeld [34] set-up a theory for the long distance propagation of radio-waves over (conducting) earth or sea-water of just the same type as given by (45-51), (as cited in [31]).

IV. B. Surface-Plasmon Spectroscopy

As the phase-velocity of the surface plasmon is always smaller than the velocity of light in the adjacent dielectric (or vacuum), surface plasmons cannot directly be excited by light. There are, however, two tricks to overcome this problem [29].

(a) Use of periodic surface structures. If there is a line-grating in the surface with spacing d the photon with frequency ω_0 may pick-up momentum $K_m = 2\pi m/d$, $m = 0, \pm 1, \pm 2, \ldots$ from the surface: $q_x = \omega_0 \sin \alpha/c + K_m$. If (q_x, ω_0) is on the dispersion curve, a surface plasmon may be emitted. The latter can decay and, thus, power is absorbed from the reflected beam. Radiative decay with momentum transfer K_n is also possible and lead to additional diffraction peaks which were first observed by Wood[35] as early in 1902. (As cited by Ritchie et al. [36]).

(b) ATR (attenuated total reflection) or prism method. At the boundary of a dielectric with refractive index n light with frequency ω_0 is totally reflected if the angle of incidence, α, satisfies $\sin \alpha > 1/n$, Fig. 17. In this case, there is an evanescent wave outside the dielectric propagation along the surface with wave-vector $q_x = n(\omega_0/c) \sin \alpha$ in perfect analogy with the surface plasmon. Photons in evanescent waves have $n \sin \alpha$ times larger momentum than in vaccum! The range of accesible wave vectors is $\omega_0/c < q_x < n\omega_0/c$.

Methods using periodic surface structures were first applied to metals where the plasma-frequency is in the UV region. It has not yet been possible to produce grating distances of the same order as the wavelength of light so that only but a small part of the dispersion curve near the light line has been accesible. These experiments, however, have been performed with high accuracy and even then small zone boundary gaps have been observed, Fig. 16. For doped semiconductors, on the other hand, the plasma-frequency is in the IR region and the full excitation curve has been investigated [37].

The ATR technique has two main advantages over the grating technique. (1) The surface of the sample is not destructively disturbed. (2) In the weak coupling limit, i.e. when the gap between the sample and the prism is large enough, the frequencies of the reflectivity minima directly yield the surface plasmon dispersion, Fig. 17.

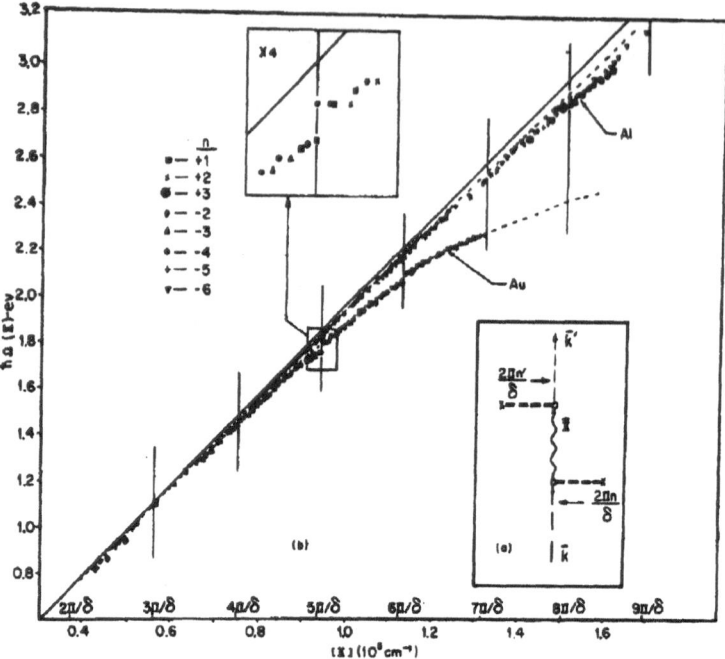

Figure 16. *Dispersion curve of surface-plasmons in Al and Au by a concave grating for varying angles between entrance and exit slits. Upper inset shows a zone-boundary gap, lower insets gives a Feynman diagram of the creation and radiative decay of the plasmon. From Ritchie et al. [36]*

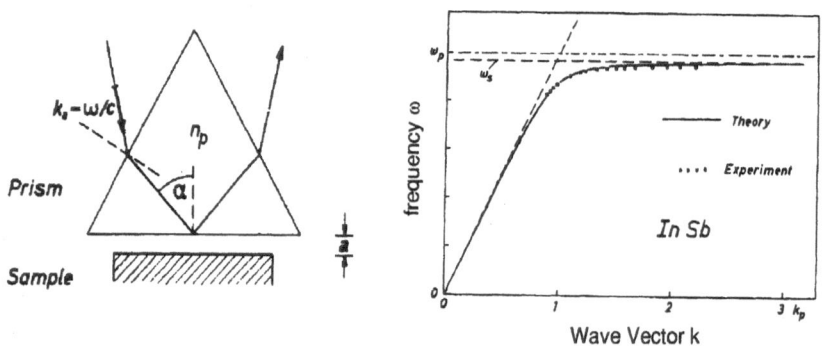

Figure 17. *Dispersion of surface plasmons in InSb as obtained from ATR-spectra. $\omega_p = 426.5 cm^{-1}$, $k_p = \omega_p/c$. $\varepsilon_\infty = 15.68$, $\gamma = 0.03\omega_p$. From Fischer et al. [37]*

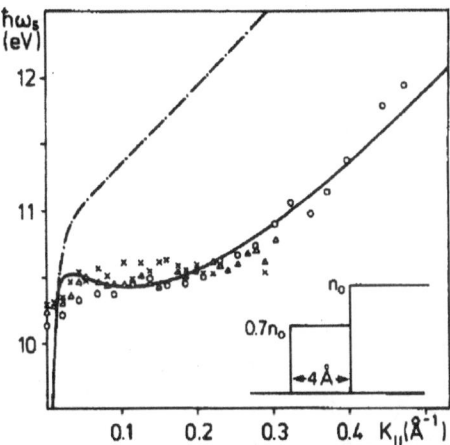

Figure 18. *Surface plasmon dispersion of Al. Experimental data from Krane and Raether[39]. The dashed–dotted and solid lines represent the result of a sharp barrier and a two-step model within the hydrodynamic description. From Forstmann and Stenschke[40].*

Passage of high energy electrons through a metal has already been discussed in connection with the spectroscopy of bulk-plasmons, however, the influence of the surfaces has been neglected. Fig.7 shows that, at normal incidence, the momentum transferred parallel to the surface is $\hbar q_x = \hbar k_0 \theta$ The angles β and θ are related by

$$\tan \beta = \theta/\theta_{\Delta E}, \qquad \theta_{\Delta E} = \Delta E/2E_0 \tag{59}$$

where E_0 is the kinetic energy of the primary electrons. If θ surpasses $\theta_{\Delta E}$ the angle β quickly approaches $90°$ so that large q_x are easily possible. However, details of the light-line requires high angular resolution. Here photons are a more suitable tool.

The loss-function for a metal foil of thickness d and area A imbedded in a dielectric with permeability ϵ_b (at normal incidence and neglecting retardation) is [29]

$$P(\mathbf{q}, \omega) \propto \Im \left\{ \frac{-1}{\epsilon(q,\omega)q^2} d + \frac{2q_\perp}{q^4} \frac{\left[\epsilon(\omega) - \epsilon_b\right]^2}{\epsilon(\omega)\epsilon_b} A \left[\frac{\sin^2(\frac{\omega d}{2v_0})}{L_+(\omega)} + \frac{\cos^2(\frac{\omega d}{2v_0})}{L_-(\omega)} \right] \right\} \tag{60}$$

with

$$L_+(\omega) = \epsilon(\omega) + \epsilon_b(\omega)\tanh(\frac{q_\perp d}{2}), \quad L_-(\omega) = \epsilon(\omega) + \epsilon_b(\omega)\coth(\frac{q_\perp d}{2}). \tag{61}$$

$q^2 = q_\perp^2 + (\omega/v_0)^2$. $(q_x = q_\perp)$. Energy losses inside the (infinite) dielectric boundaries are omitted.

With decreasing film-thickness the surface-modes at both sides become coupled and split into two modes with different frequencies. The zeros of $L_\pm$ define the coupled surface-plasmon frequencies (in the nonretarded limit).

$$\frac{\epsilon_b - \epsilon(\omega)}{\epsilon_b + \epsilon(\omega)} = \mp e^{q_x d}. \tag{62}$$

For a Drude metal imbedded in a dielectric we have for $q_x > \omega_p/c$

$$\omega_\pm(q_x) = \frac{\omega_p}{\sqrt{1 + \epsilon_b}} \sqrt{1 \pm e^{-q_x d}}. \tag{63}$$

The plus/minus sign correspond to a symmetric/antisymmetric mode in which an excess charge at one side of the surface is accompanied by an excess/deficiency of charge just opposite at the other side of the slab. The splitting into two modes can be neglected if $q_x d/2 > 1$. In the case of $50 keV$ electrons this condition means $\theta d > 10^{-2}$ (d in $\mathring{A}$ and θ in degrees).

323

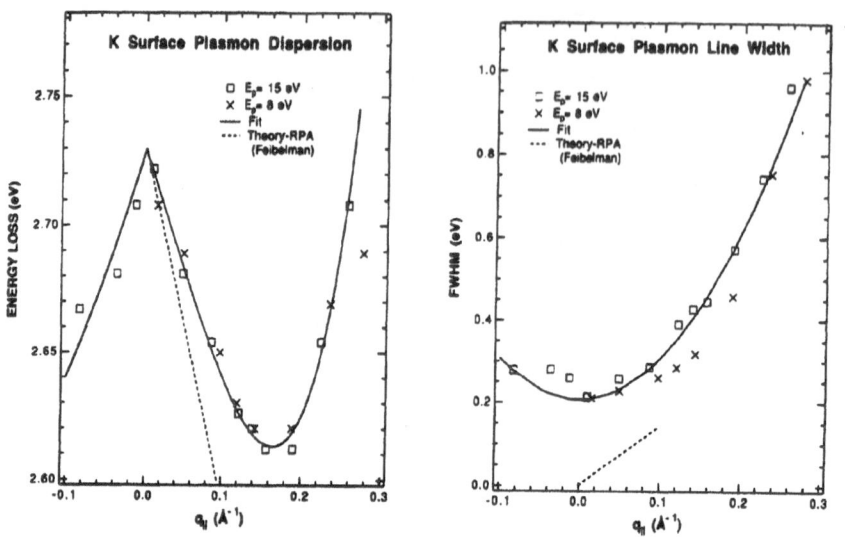

Figure 19. *Surface plasmon dispersion (left) and width (right) for Potassium. Data are shown for two different incident electron energies. Dashed Line: Feibelman's theory [44]. From Tsuei et al. [41,44].*

Experimental results for Al are displayed in Fig. 18. Contrary to the sharp barrier result (58) the expected steep increase is not observed. Bennet[38] was the first who realized that the surface plasmon-dispersion is sensitive to the surface-charge density profile. A "soft' boundary decreases the linear term in (58), which can even become negative.

There is extensive literature devoted to the study of spatial dispersion and a diffuse surface electron profile on the plasmon-dispersion curve at a metal surface. For a survey see [41]. For small wave-vectors (but still $q_x > \omega_p/c$) the surface-plasmon is of the form (58). The coefficient a_1 is the centroid of the induced surface-charge density [42-44].

$$a_1 = -\frac{\int_{-\infty}^{\infty} z \delta\rho(z)\, dz}{\int_{-\infty}^{\infty} \delta\rho(z)\, dz}. \tag{64}$$

($\delta\rho(z) = d(z)$, Fig. 15). In the absence of impurities or phonon scattering a_2 results from particle-hole excitations (Landau-damping).

Angle resolved inelastic low-energy electron reflection scattering has been used for alkali films [41,45], Fig. 19. For all alkali metals measured the dispersion coefficient $a_1 < 0$. At the frequency of the surface plasmon oscillation the induced charge is outside of the geometrical boundary, thus, the surface plasmon-dispersion is negative. Feibelman[44] has shown from microscopic considerations that, similar to the nonlocal description in the bulk in term of longitudinal and transverse dielectric functions, nonlocal surface effects can be expressed in terms of two surface response functions $d_\perp(\omega)$, $d_\parallel(\omega)$ which only depend on frequency. Remarkably, $\omega_{sp}(0)$ does not depend on the surface charge profile and, thus, is a bulk property of the metal.

For Ag, on the other hand, $a_1 > 0$ and, a strong azimutal anisotropy is observed [45-48].

V. PLASMONS IN SMALL PARTICLES AND CLUSTERS

V. A. Multipolar Plasmons

In recent years, small particles have received growing attention because of their interesting physical properties and technical importance. For an overview see e.g. [49]. Concerning the interaction of a transverse, monochromatic plane wave with a single spherical (metallic or dielectric) particle, the basic theory has been developed by Mie and Debye more than 80 years ago [50], see [32]. The analogous problem for longitudinal (electric) fields, however, received attention much later in connection with electron–energy–loss spectroscopy [51].

The system under consideration consists of a metallic sphere of radius R imbedded in a dielectric host with dielectric function $\epsilon_h(\omega)$. The metal inside the sphere will be either described by the Drude-function with $\epsilon_D(\omega)$, or within a hydrodynamic theory.

In a local dielectric description the total potential $\phi = \phi_{ind} + \phi_{ext}$ fulfills the Poisson-equation, $\Delta\phi = -\rho_{ext}/\epsilon_0\epsilon$, so that ϕ_{ind} is related to ϕ_{ext} by

$$\phi_{ind}(\mathbf{x},\omega) = \left(\frac{1}{\epsilon} - 1\right)\phi_{ext} + f(\mathbf{x},\omega). \tag{65}$$

$f(\mathbf{x},\omega)$ is a regular solution (except on the surface) of the Laplace-equation, $\Delta f = 0$. In terms of spherical harmonics, this function is represented by

$$f_{\ell m}(r,\omega) = C_1^{\ell m}(\omega)\,\theta(R-r)\,\left(\frac{r}{R}\right)^\ell + C_2^{\ell m}(\omega)\,\theta(r-R)\,\left(\frac{R}{r}\right)^{\ell+1}. \tag{66}$$

Coefficients C_1 and C_2 are determined by the requirement of continuity of the tangential component of $\mathbf{E}$ and normal component of $\mathbf{D}$ at the surface of the sphere. As a result, we obtain

$$r \leq R: \qquad \phi_{\ell m}^{ind}(r) = \frac{\ell+1}{\ell}\alpha_\ell^{c\ell}\epsilon_{met}^{-1}\,\phi_{\ell m}^{ext}(R)\left(\frac{r}{R}\right)^\ell + \left(\frac{1}{\epsilon_{met}} - 1\right)\phi_{\ell m}^{ext}(r)$$

$$r \geq R: \qquad \phi_{\ell m}^{ind}(r) = -\alpha_\ell^{c\ell}\,\epsilon_h^{-1}\,\phi_{\ell m}^{ext}(R)\left(\frac{R}{r}\right)^{\ell+1} + \left(\frac{1}{\epsilon_h} - 1\right)\phi_{\ell m}^{ext}(r). \tag{67a,b}$$

The key quantity is the classical multipolar polarizability (divided by $R^{2\ell+1}$)

$$\alpha_\ell^{c\ell}(\omega) = \frac{\epsilon_{met}(\omega) - \epsilon_h(\omega)}{\epsilon_{met}(\omega) + \frac{\ell+1}{\ell}\epsilon_h(\omega)}. \tag{68}$$

$\ell=1,2,\ldots$. For a spherical void filled with a dielectric (a noble gas "bubble") $\epsilon_{met}, \epsilon_h$ have to be interchanged.

The eigenfrequencies of the collective modes can be obtained from the poles of (68). For a metallic particle with a Drude-dielectric function, imbedded in a nondispersive host these modes are known as Mie-resonances.

$$\omega_\ell = \frac{\omega_p}{\sqrt{1 + \frac{\ell+1}{\ell}\epsilon_h}}, \tag{69}$$

As pointed out by Ekardt [52], the classical result has several deficiencies which become important for particle diameters $2R$ in the range of $2nm$ or less. For instance, according to (69) there is always a pronounced dipole resonance at $\omega_1 = \omega_p/\sqrt{1+2\epsilon_h}$, but this structure disappears in a quantum treatment for small radii. For $\ell \to \infty$, (69) approaches the surface-plasma frequency of an infinite plane metal-vacuum boundary, $\omega_\ell \to \omega_p/\sqrt{1+\epsilon_h}$, whereas in quantum theory it becomes overdamped and effectively disappears.

A qualitative similar behaviour is obtained in a hydrodynamic description, where the collective excitations are determined by the transcendental equation [53-55]:

$$\frac{j_{\ell+1}(kR)}{j_{\ell-1}(kR)} = \frac{\ell}{\ell+1}\frac{k^2}{k^2 + \kappa^2\frac{(2\ell+1)\epsilon_h}{(\ell+1)\epsilon_h+\ell}} \tag{70}$$

where $j_\ell(x)$ denotes a spherical Bessel-function and

$$k^2 = \beta^{-2}[\omega(\omega + i\gamma) - \omega_p^2], \qquad \kappa^2 = \omega_p^2/\beta. \tag{71}$$

The solutions of (70) fall into two classes:

(a) Surface-modes ($\omega_\ell < \omega_p$, $\ell = 1, 2...$).
These modes correspond to imaginary values of k so that the induced charge density is concentrated near the surface. For large spheres, $j_{\ell+1}/j_{\ell-1} \to -1$, yielding ω_ℓ of (69). With decreasing sphere-radius, ω_ℓ increases and eventually reaches ω_p at a critical radius R_ℓ and becomes a bulk mode.

(b) Bulk-modes ($\omega_\alpha > \omega_p$, $\alpha = (l, \nu)$).
k is real and the charge density oscillations are spread over the whole particle volume. For $\ell = 0$, the solution of (70) is given by $kR = x_{1,\nu}$, where $x_{\ell,\nu}$ is the νth positive root of $j_\ell(x) = 0$. From a graphical discussion, we deduce that the modes of higher multipolarity are lying in the intervals $x_{\ell+1,\nu} \leq k\,R \leq x_{\ell-1,\nu+1}$. As $k\,R$ increases from zero to infinity, the left hand side of (70) has first order poles at $x_{\ell-1,\nu}$ whereas the right hand side is positive and finite. The bulk modes are thus labelled by $\ell = 0, 1, \ldots$ and an additional index $\nu = 1, 2\ldots$. To a first approximation, we have $k\,R \approx x_{\ell-1,\nu+1}$, which (for $\ell > 0$ and $R > R_\ell$) leads to

$$\omega_{\ell,\nu} \approx -i\gamma/2 + \omega_p \sqrt{1 + \left(\frac{x_{\ell-1,\nu+1}}{\kappa R}\right)^2 - (\gamma/2\omega_p)^2}. \tag{72}$$

The ℓ, ν dependence of the collective modes is analogue to the bulk-plasmon dispersion. For $\ell = 0$, which is known as the breathing mode $x_{\ell-1,\nu+1}$ is replaced in (72) by $x_{1,\nu}$. In plasma physics the bulk modes are known as Tonks-Dattner resonances [56]. In thin metallic films they can be excited optically by p-polarized light [57] ($\ell \geq 1$).

V. B. Spectroscopy of Cluster–Plasmons and Experimental Results

At present two different experimental techniques are used to study the electronic excitations of small particles: Electron-energy loss spectroscopy (EELS) and scanning transition electron microscopy (STEM) [58]. EELS controls the momentum transfer whereas STEM controls the impact parameter.

In an EELS experiment the scattered electrons excite multipolar modes up to $\ell \approx qR$, where R is the radius of the particles so that the analysis of cluster–loss-spectrum is far more complicated than for plane surfaces. As a result, for the dielectric and hydrodynamic description the loss-functions are [55]:

$$P_{diel}(\mathbf{q}_\perp, \omega) = \frac{e^2}{\pi^2 \hbar v^2 \epsilon_o} \frac{1}{s^2} \Im \left\{ \frac{R^3}{3} \frac{-1}{\epsilon_{met}} + \frac{-1}{\epsilon_h} \int_R^\infty r^2 dr \right.$$
$$\left. + R^3 \left(\frac{1}{\epsilon_h} - \frac{1}{\epsilon_{met}}\right) \sum_{\ell=1}^\infty (2\ell+1)(\ell+1) \frac{\epsilon_{met} - \epsilon_h}{\epsilon_{met} + \frac{\ell+1}{\ell}\epsilon_h} \left[\frac{j_\ell(sR)}{sR}\right]^2 \right\}. \tag{73}$$

$s = |\mathbf{q}| = \sqrt{q_\perp^2 + (\frac{\omega}{v})^2}$. v is the electron velocity. In (73) ϵ_{met} and ϵ_h may have arbitrary frequency dependencies. An approximate result has been given before by Ashley and Ferrell [59]. For void in a metal the loss function is simply obtained from (73) by interchanging ϵ_h and ϵ_m.

$$P_{hydro}(\mathbf{q}_\perp, \omega) = \frac{e^2}{\hbar \pi^2 v^2 \epsilon_o s^2} \Im \left\{ \frac{R^3}{3} \frac{-1}{\epsilon(s, \omega)} + \frac{-1}{\epsilon_h} \int_R^\infty r^2 dr \right\} +$$
$$+ \frac{e^2}{\hbar \pi^2 v^2 \epsilon_o} \Im \left\{ \left(\frac{1}{\epsilon(s, \omega)} - 1\right)^2 \frac{kR^2}{\kappa^2 s^3} \sum_{\ell=0}^\infty (2\ell+1)^2 \times \frac{s\, j_\ell(kR)\, j_{\ell-1}(sR) - k\, j_{\ell-1}(kR)\, j_\ell(sR)}{J_\ell} \times \right.$$
$$\left. \times \left(\frac{1}{\epsilon_h}\frac{\ell}{\ell+1} j_{\ell-1}(sR) - \frac{1}{\epsilon_D} j_{\ell+1}(sR) + \frac{s^3}{k^3}\frac{\epsilon_h - 1}{\epsilon_h}\ell\frac{1}{kR} j_\ell(sR)\right) \right\} +$$
$$+ \frac{e^2}{\hbar \pi^2 v^2 \epsilon_o} \Im \left\{ \left(\frac{1}{\epsilon_h} - 1\right) \frac{R^2}{s^3} \sum_{\ell=0}^\infty (2\ell+1)^2 \frac{j_\ell'(kR)}{J_\ell} j_\ell(sR) \times \left(\frac{1}{\epsilon_h}\frac{\ell}{\ell+1} j_{\ell-1}(sR) - \frac{1}{\epsilon_D} j_{\ell+1}(sR)\right) \right\} \tag{74}$$

$\epsilon(s,\omega)$ denotes the longitudinal wave-vector dependent dielectric function of the bulk metal in hydrodynamic approximation (23). The first and second terms of (73,74) within the curly brackets represent the contributions inside and outside the particle, whereas the other terms describes surface excitations.

Results for potassium clusters in MgO are displayed in Figs. 20. (The loss functions have been divided by the prefactor of (73) and $R^3/3$ so that the result is dimensionless). Experimentally, the volume-plasmon half-width of $\hbar\gamma = 0.6$ eV [60] is considerably enhanced in comparison to its bulk value (0.24 eV) [15]. In contrast to the dielectric description, the maxima of the loss function show a considerable "blue-shift", originating from spatial dispersion for small radii which is in qualitative agreement with the experiment. The bulk-plasmon modes of higher polarity are resolved only in very small particles and for low damping.

For large radii, $R \to \infty$, the loss-functions (73,74) converge to their infinite medium results (when averaged on the incident directions): On the other side, for small radii or small momentum transfer, $sR \to 0$, we obtain:

$$P(q_\perp,\omega) = \frac{e^2}{4\pi\hbar v^2\epsilon_0}\frac{V}{s^2}\Im\left[\frac{3}{\epsilon_h}\alpha_1(\omega)\right], \qquad (75)$$

$\alpha_1(\omega)$ denotes the (electrical) dipole-polarizability.

Collective excitations on voids or noble gas bubbles in metals have also attracted experimental as well as theoretical interest, e.g. [61-64].

The size-dependence of the surface-plasmon is still an open problem. In the past there was a general agreement that the observed red-shift (with drecreasing size) is due to the spill-out of the charge density whereas theories based on sharp surfaces gave a blue-shift. For clusters imbedded in a dielectric host or voids filled with a dielectric, the spill-out is reduced by the exclusion principle so that the assumption of a fixed boundary condition seems to be well-justified. For small metallic particles in vacuum the correct surface charge density profile must be taken into account and clusters eventually require a full self-consistent quantum treatment[65].

How many metal atoms are needed to form a cluster which displays metallic behaviour? For Hg 25 atoms seem to be enough! [66-67], Fig. 21. The Hg atom has a $5d^{10}6s^2$ closed shell electronic structure so that small clusters are dominantly van der Waals bound. The width of the occupied $6s$ and empty $6p$ bands increase rougly proportional to the number of nearest neighbors. Thus the band gap decreases for increasing cluster size and becomes zero around $N = 20$.

For Na the evolution towards the bulk values of the plasmons in clusters comprising from 8 to 338 atoms has been calculated by Yannouleas et al [68] who found an increase of the Mie-plasmon energy from 2.7 to 3.2eV. The latter value is close to $\omega_p/\sqrt{3}$.

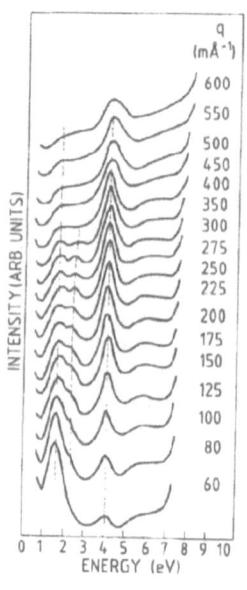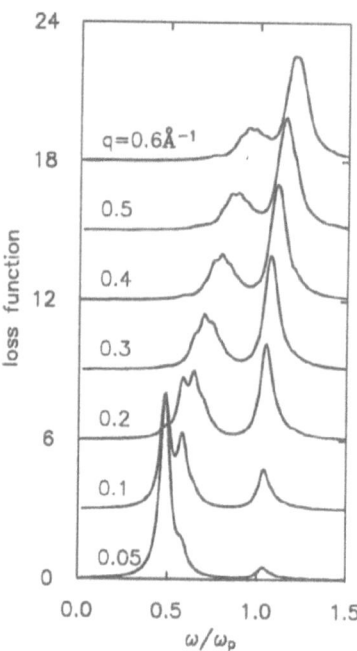

Figure 20. *Energy-loss spectrum of K-clusters in MgO [60]. (Left) experimental results, (right) hydrodynamic theory (74). Particle radius: $R = 20\text{Å}$.*

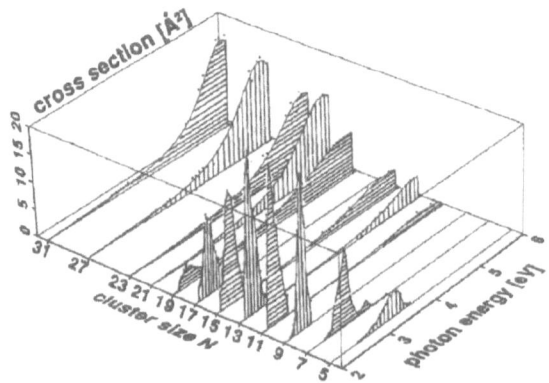

Figure 21. *Experimental Photoabsorption spectra of doubly charged Hg clusters showing an abrupt transition from atomic to collective, plasmon-like absorption as a function of cluster size. From Haberlandt et al. [66].*

VI. HETEROSTRUCTURES AND LOW-DIMENSIONAL SYSTEMS

A metallic heterostructure is an arrangement of different metals in close contact, i.e. an array of metallic sheets in which there is a negligible charge transfer between the components. In the limit of thick enough layers, one can treat the layers individually by Drude dielectric functions (23) or hydrodynamic equations (10). Typically the thickness of the individual layers lies in the range $100\ldots5000\overset{\circ}{A}$. These structures resemble semiconductor quantum-wells but little work by both theory and experiment appears to have been done on their metallic counterparts.

VI. A. Interfaces

The simplest heterostructure consists of two semi-infinite metals bounding together just in the same way as it was studied for the surface-plasmon in chapter IV. Apart from the bulk and surface plasmons in each metal there is an interface-plasmon whose dispersion is given by (51) where $\varepsilon_{\pm}(\omega)$ are both Drude-functions with bulk-plasma frequencies $\omega_{p\pm}$, Fig. 22. Neglecting retardation the frequency of the interface-plasmon is given by

$$\omega_{int} = \sqrt{\frac{\omega_{p+}^2 + \omega_{p-}^2}{2}}. \tag{76}$$

For small $q_x \to 0$ $\omega = \omega_{p-} < \omega_{p+}$. Experimental studies on interface-plasmon excitations in $Cu/RbF/GaAs$ and $Cu/Rb/Ge$ heterostructures were reported by Klauser et al. [69].

VI. B. Sandwich-Configurations

A metallic slab or foil of thickness d and dielectric function $\varepsilon(\omega)$ imbedded in a metallic host with dielectric function $\varepsilon_h(\omega)$ displays two interface modes with dispersions $\omega_{\pm}(q_x)$, Fig. 23. The plus/minus sign correspond to a symmetric/antisymmetric configuration of induced charges at the interfaces. Neglecting retardation as well as spatial dispersion these modes are determined by the zeros of $L_{\pm}(\omega)$ of (61)

$$L_+(\omega) = \varepsilon(\omega) + \varepsilon_h(\omega)\tanh(\frac{q_x d}{2}) = 0, \qquad L_-(\omega) = \varepsilon(\omega) + \varepsilon_h(\omega)\coth(\frac{q_x d}{2}) = 0. \tag{77}$$

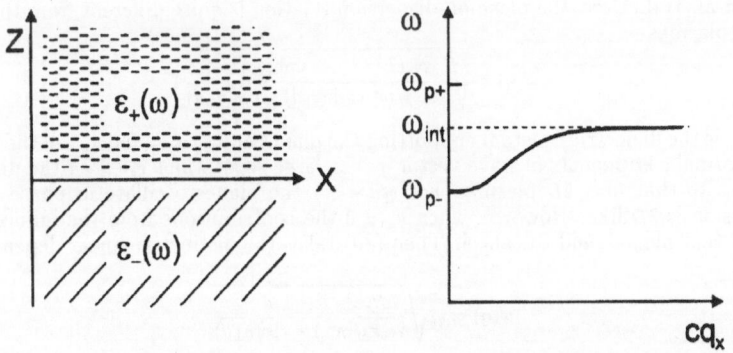

Figure 22. Geometry (left) and interface-plasmon dispersion (right) of a metal-metal contact.

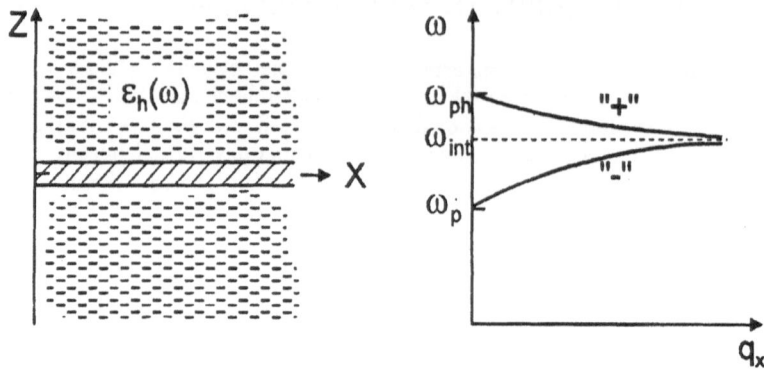

Figure 23. Geometry (left) and interface-plasmon dispersions (right) of a metallic sandwich.

VI. C. Two-Dimensional Systems

Ritchie [28] first noted that the plasmon in a thin sheet has a square-root dispersion ($d \to 0$ for the "-" mode (77)). Stern [70] later derived the explicit dispersion relation for the $2D$ plasmon (in the nonretarded limit but including spatial dispersion))

$$\omega_{2D}(q_x) = \sqrt{\frac{N_s e^2}{2m^* \epsilon_0 \bar{\varepsilon}(\mathbf{q}, \omega)}} \, q_x \, , \tag{78}$$

where N_s is the areal carrier density and $\bar{\varepsilon}$ is an effective dielectric function. For a MOS configuration consisting of a semiconductor with dielectric constant ϵ_{sc}, an (SiO_2-oxide) insulator with ϵ_{ox} and thickness d, and a perfectly screening gate

$$\bar{\varepsilon}(q_x, \omega) = \frac{1}{2}\left[\varepsilon_{sc}(\omega) + \varepsilon_{ox}(\omega) coth(dq_x)\right]. \tag{79}$$

Such $2D$-plasmons have been observed in $AlGaAs-GaAs$ heterostructures where the electrons are confined in a very narrow potential well, see e.g. Heitmann [71], or Wilkinson et al. [72].

VI. D. Two-Layer Systems

In a layered electron gas, the free charges are constrained to move on parallel planes spaced by a distance d. Such a two-layer system was studied by Olego et al. [73] and Yuh et al. [74]. Here, the plasmon-dispersion relation is quite different from that in $2D$ or $3D$ plasmas

$$\omega(\mathbf{q}) = \sqrt{\frac{N_s e^2}{2\epsilon_0 \varepsilon_M m^*} \frac{\sinh(q_{\parallel})}{\cosh(q_{\parallel} d) - \cos(q_\perp d)}} \, . \tag{80}$$

ε_M is the dielectric constant supporting the planes and $q_{\parallel} = q_x$ and $q_\perp$ are the in-plane and normal components of wave vector $\mathbf{q}$. For large separations $q_{\parallel} d \gg 1$ the dispersion reduces to that of a $2D$ plasma. For $q_{\parallel} d \ll 1$ the planes oscillate in phase and the dispersion is $3D$ like. However, when $q_\perp \neq 0$ the contributions from the induced fields in different planes tend to cancel. Then (80) takes the distinctive linear dependence

$$\omega(\mathbf{q}) = q_{\parallel}\sqrt{\frac{N_s e^2}{2\epsilon_0 \varepsilon_M m^*} \frac{d}{1 - \cos(q_\perp d)}} \, . \tag{81}$$

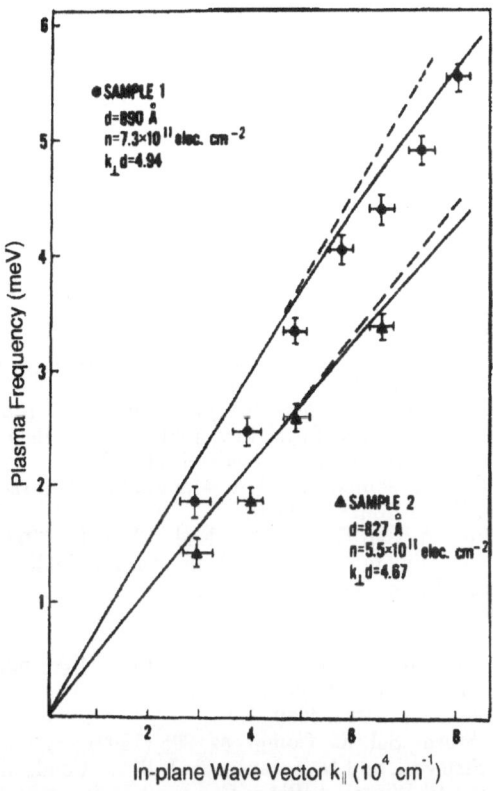

Figure 24. Dispersion relation for a two-layer plasma in $GaAs - AlGaAs$ heterostructures. Solid and dashed lines are evaluations of (80) and (81), respectively. From Olego et al. [73].

In this regime ($q_{\parallel}d \ll 1$, $q_{\perp} \neq 0$) the response is most different from that in $2D$ or $3D$ plasmas, Fig. 24.

VI. E. Superlattices

Superlattices are structures composed of alternating layers of different materials, Fig. 25. Theoretical studies on superlattice plasmons and their spectroscopy were reported e.g. by Babiker [75], Shi and Griffin [76], and Lopez-Olazagasti et al. [77].

More recently the theory of infinite metallic superlattices has found a new application in the study of high T_c superconductors. These can be viewed as periodic arrays of unit cells with a typical spacing of about $12\overset{\circ}{A}$, each of which contains up to three closely spaced CuO_2 sheets. Even at these small separations, the electronic bands are $2D$ like [78,79].

ACKNOWLEDGEMENT

I thank Prof. J. Fink for many stimulating and helpful discussions.

REFERENCES

1. I. Langmuir, Proc. Nat. Acad. Sci 14, 627 (1926).
2. D. Pines, *Elementary Excitations in Solids*, Benjamin (1964).
3. L. Marton, J.A. Simpson, H.A. Fowler, and N. Swanson, Phys. Rev. 126, 182 (1962).
4. D. Pines and Ph. Nozieres, *The Theory of Quantum Liquids*, Benjamin (1966).
5. P.M. Platzman and P.A. Wolf, Solid State Phys., Suppl. 13, Academic (1973).
6. B. DiBartolo ed., *Collective Excitation in Solids*, Nato ASI Series B, vol 88, Plenum (1981).
7. H. Raether, *Excitations of Plasmons and Interband Transitions by Electrons*, Springer Tracts in Modern Physics, Vol. 88, Springer (1980).
8. S.E. Schnatterly, Solid State Physics, 34, 275 (1979).
9. J. Fink, *Recent Developments in Energy-Loss-Spectroscopy*, Adv. in Electronics and El. Physics, Vol. 75, Academic Press, (1985).
10. N. Ashcroft and N.D. Mermin, *Solid State Physics*, Holt, Rinehart and Winston (1996).
11. D.M. Ceperley and B.J. Adler, Phys. Rev. Lett. 45, 566 (1980).
12. F. Wooten, *Optical Properties of Solids*, Academic (1972).
13. L.V. Keldysh, D.A. Kirshnits, and A.A. Maradudin eds., *The dielectric function of condensed systems*, North Holland (1989).
14. a) J. Sprösser-Prou, A. vom Felde, and J. Fink, Phys. Rev. B40, 5799 (1989).
    b) J. Sprösser-Prou, Diplom-Arbeit (1989), Institut für Nukleare Festkörperphysik, Forschungszentrum Karlsruhe (unpublished).
15. a) A. vom Felde, J. Fink, Th. Büche, B. Scheerer, and N. Nücker, Eur. Lett. 4, 1037 (1987).
    b) A. vom Felde, J. Sprösser-Prou, and J. Fink, Phys. Rev. B40, 1081 (1989).
16. P. Vashishta and K.S. Singwi, Phys. Rev. B6,875 (1972).
17. B. Dabrowski, Phys. Rev. B34, 4989 (1986).
18. M. Taut and K. Sturm, Sol. St. Comm. 82, 295 (1992).
19. E. Lipparini, S. Stringari, and K. Takayanagi, J. Phys. Cond. Mat. 6, 2025 (1994).
20. a) M. Kollwitz and H. Winter, J. Phys: Condens. Matter 7 3153 (1995).
    b) M. Kollwitz, Diplomarbeit, Institut für Nukleare Festkörperphysik, Forschungszentrum Karlsruhe (1994), unpublished.
21. F. Aryasetiawan and K. Karlson, Phys. Rev. Lett. 73, 1679 (1994).
22. N. Nücker, U. Eckern, J. Fink and P. Müller, Phys. Rev. B44, 7155 (1991).
23. D. Pines, Can. J. Phys. 34, 1379 (1956).
24. H., Fröhlich, J. Phys. C1, 544 (1968).
25. A. Pinczuk, Phys. Rev. Lett. 47, 1487 (1981).
26. J. Ruvalds, Adv. Phys. 30, 677 (1981).
27. F.C. Schaefer and R. v. Baltz, Z. Phys. B69, 251 (1987).
28. R.H. Ritchie, Phys. Rev. 106, 874 (1957).
29. H. Raether, in *Physics of Thin Films,* Vol. 9, ed. by G. Hass, M.H. Francombe, and R.W. Hoffman, Academic (1977).
30. M. Cardona, Am. J. Phys. 39, 1277 (1971).
31. F. Forstmann and R.R. Gerhardts, *Metal Optics Near the Plasma Frequency*, Springer Tracts in Modern Physics, Vol 109, Springer (1986)
32. M. Born and E. Wolf, *Principles of Optics*, Pergamon (1959).
33. R. Fuchs and K.L. Kliewer, Phys. Rev. B3, 2270 (1971).
34. J. Zenneck, Ann. Phys. (Leipzig) 23, 846 (1907); see: A. Sommerfeld, *Vorlesungen über Theoretische Physik*, VI, § 32, Leipzig (1966).
35. R.W. Wood, Phil. Mag. 4, 396 (1902).
36. R.H. Ritchie, E.T. Arakawa, J.J. Cowan, and R.N. Hamm, Phys. Rev. Lett. 21, 1530 (1968).
37. B. Fischer, N. Marschall, and H.J. Queisser, Surf. Sci. 34, 50 (1973).
38. J. Bennett, Phys. Rev. B1, 203 (1970).
39. K.J. Krane and H. Raether, Phys. Rev. Lett. 37, 1355 (1976).
40. F. Forstmann and H. Steuschke, Phys. Rev. B17, 1489 (1978).
41. K.D. Tsuei, E.W. Plummer, A. Liebsch, E. Pehlke, K. Kempa, and P. Bakshi, Surf. Sci. 247, 302 (1991).
42. J. Harris and A. Griffin, Phys. Lett. 34A, 51 (1971).
43. F. Flores and F. Garcia-Molinier, Sol. St. Comm. 11, 1295 (1972).

44. P.J. Feibelman, Phys. Rev. Lett. **30**, 975 (1973). Progress in Surface Science, Vol 12, 287 (1982).
45. Ku-Ding Tsuei, E.W. Plummer and P. Feibelman, Phys. Rev. Lett. **63**, 2256 (1989).
46. S. Suto, K.-D. Tsuei, E.W. Plummer, and E. Burstein, Phys. Rev. Lett. **63**, 2590 (1989).
47. M. Rocca, M. Lazzarino, and M. Valbusa, Phys. Rev. Lett. **67**, 3197 (1991); **69**, 2122 (1992).
48. A. Liebsch, Phys. Rev. Lett. **71**, 1451 (1993).
49. W. Halperin, Rev. Mod. Phys. **58**, 533 (1986).
50. G. Mie, Ann. Phys. (Leipzig), **25**, 377 (1908).
51. J. Crowell and R.H. Ritchie, Phys. Rev. **172**, 436 (1968).
52. W. Ekhardt, Phys. Rev. **b32**. 1961 (1985); **B33**, 8803 (1986); **B36**, 4483 (1987).
53. F. Fujimoto and K. Komaki, J. Phys. Soc. Jap. **25**, 1679 (1968).
54. M. Barberan and J. Bausells, Phys. Rev. **31**, 6354 (1985).
55. R. v. Baltz, M. Mensch, and H. Zohm, Z. Phys. **B98**,151 (1995).
56. R.B. Hall, Am. J. Phys. **31** 696 (1963).
57. A.R. Melnyk and M.J. Harrison, Phys. Rev. Lett **21** 85 (1968). Phys. Rev. **B2**, 835 (1970).
58. R.F. Egerton, *Electron Energy Loss Spectroscopy in the Electron Microscope*, Plenum (1986).
59. J.C. Ashley and T.L. Ferrell, Phys. Rev **B14**, 3277 (1976).
60. A. vom Felde, J. Fink, and W.Ekardt, Phys. Rev. Lett. **61**, 2249 (1988).
61. R. Manzke, G. Crezelius, and J. Fink, Phys. Rev. Lett **51**, 1095 (1983).
62. A. vom Felde, J. Fink, Th. Müller-Heinzerling, J. Pflüger, B. Scheer, and D. Kaletta, Phys. Rev. Lett. **53**, 922 (1984).
63. King-Sun David Wu and D.E. Beck, Phys. Rev **B36**, 998 (1987).
64. Ll. Serra, F. Garcias, J. Navarro, N. Barberan, M. Barranco, and M. Pi, Phys. Rev. **B46**, 9369 (1992).
65. M. Brack, Rev. Mod. Phys. **65**, 677 (1993).
66. H. Haberland, B. von Issendorff, Ji Yufeng, and Th. Kolar, Phys. Rev. Lett **69**, 3212 (1992).
67. K. Rademann, O. Dimopoulou-Rademann, M. Schlauf, U. Even, and F. Hensel, Phys. Rev. Lett. **69**, 3208 (1992).
68. C. Yannouleas, E. Vigezzi, and R.A. Broglia, Phys. Rev. **B47**, 9849 (1993).
69. R. Klauser et al. Surf. Sci. Lett. **255**, L557 (1991).
70. F. Stern, Phys. Rev. Lett. **18**, 546 (1967).
71. D. Heitmann, Surf. Sci. **170**, 332 (1986).
72. R.J. Wilkinson, C.D. Ager, T. Duffield, H.P. Hughes, D.G. Hasko, H. Ahmed, J.E.F. Frost, D.C. Peacock, D.A. Titchie, and G.A.C. Jones, J. Appl. Phys. **71**, 6049 (1992).
73. D. Olego, A. Pinczuk, A.C. Gossard, and W. Wiegman, Phys. Rev. **B26**, 7867 (1982).
74. E.L. Yuh, E.G. Gwinn, P.R. Pinsukanjana, W.L. Schaich, P.F. Hopkins, and A.C. Gossard, Phys. Rev. Lett. **71**, 2126 (1993).
75. M. Babiker, J. Phys. C **20**, 3321 (1987).
76. H. Shi and A. Griffin, Phys. Rev. **B44**, 11977 (1991).
77. E. Lopez Olazagasti, H. Cocoletzi, and W. Luiz Mochan, Sol. St. Comm. **78**, 9 (1991).
78. A. Griffin, Physica **C162-164**, 1427 (1989).
79. S.V. Pokrovsky and V.L. Pokrovsky, to be published
80. F. Bloch, Z. Phys. **81**, 363 (1933).
81. H. Jensen, Z. Phys. **106**, 620 (1937).
82. E,M. Lifschitz and L.P. Pitaevskii, *Physical Kinetics, Landau and Lifschitz Course on Theoretical Physics*, Vol 10, Pergamon (1981).
83. H. Ehrenreich and M.H. Cohen, Phys. Rev. **115**, 786 (1959).
84. S.L. Adler, Phys. Rev. **126**, 413 (1962).
85. N. Wiser, Phys. Rev. **129**, 62 (1963).
86. K. Sturm, Adv. Phys. **31**, 1 (1981).
87. N.D. Mermin Phys. Rev. **B1**, 2362 (1970).
88. Keun-Ho Lee and K.J. Chang, Phys. Rev. **B49**, 2362 (1994).

APPENDICES

A.1 Hydrodynamic Description

Following Bloch [80] and Jensen [81], the state of the plasma is described by the density and velocity fields $n(\mathbf{r},t)$, $\mathbf{v}(\mathbf{r},t)$, respectively. For the longitudinal response the velocity field is irrotational, $\mathbf{v}(\mathbf{r},t) = -grad\,\Psi(\mathbf{r},t)$, where $\Psi(\mathbf{r},t)$ denotes the velocity potential. The equations of motion can be derived from the action principle

$$\delta S = 0, \qquad S[n,\Psi] = \int L[n(\mathbf{r},t),\Psi(\mathbf{r},t)]\,dt \qquad (A1.1)$$

with Langrangian L and Hamiltonian H

$$L[n,\Psi] = m_0 \int n(\mathbf{r},t)\frac{\partial\Psi}{\partial t}d\mathbf{r} - H \qquad (A1.2)$$

$$H = \frac{m_0}{2}\int n(\mathbf{r})[grad\,\Psi(\mathbf{r})]^2 d\mathbf{r} - e\int \Phi_+(\mathbf{r})n(\mathbf{r})d\mathbf{r} + \frac{1}{2}\frac{e^2}{4\pi\epsilon_0}\int\int\frac{n(\mathbf{r})n(\mathbf{r}')}{|\mathbf{r}-\mathbf{r}'|}d\mathbf{r}d\mathbf{r}' + E_0[n(\mathbf{r})] \quad (A1.3)$$

where $E_0[n]$ is the (exact) ground state energy of the interacting electron gas at (local) density $n(\mathbf{r})$ and $\Phi_+(r)$ is the potential of the positive ion background.

We are interested in the small density oscillations of the plasma around it equilibrium density n_0. Correspondingly, we expand $H[n,\Psi]$ around its minimum at $n = n_0$, $\Psi = 0$. Therefore, this expansion begins with quadratic terms in $n_1 = n - n_0$:

$$H = H_0 + \frac{m_0}{2}\int n_0[grad\,\Psi(\mathbf{r})]^2 d\mathbf{r} + \frac{1}{2}\frac{e^2}{4\pi\epsilon_0}\int\int\frac{n_1(\mathbf{r})n_1(\mathbf{r}')}{|\mathbf{r}-\mathbf{r}'|}d\mathbf{r}d\mathbf{r}' + \frac{1}{2}\int\frac{\partial^2 E_0[n_0]}{\partial n_0^2}n_1^2(\mathbf{r})d\mathbf{r}\ldots \qquad (A1.4)$$

Variation with respect to $n_1(\mathbf{r},t)$ and $\Psi(\mathbf{r},t)$, leads to

$$m_0\frac{\partial\Psi(\mathbf{r},t)}{\partial t} + e\phi_1(\mathbf{r},t) - P_0 n_1(\mathbf{r},t) = 0, \qquad \frac{d}{dt}[m_0 n_1(\mathbf{r},t)] - m_0 div[n_0 grad\,\Psi(\mathbf{r},t)] = 0 \qquad (A1.5)$$

with

$$\Delta\phi_1(\mathbf{r},t) = -\frac{1}{\epsilon_0}(-e)n_1(\mathbf{r},t), \qquad P_0 = \frac{\partial^2 E_0[n_0]}{\partial n_0^2}. \qquad (A1.6)$$

In a local Hartree-Fock approximation (5)

$$E_0[n(\mathbf{r})] = \int\left[\frac{3}{5}\varepsilon_F[n(\mathbf{r})] - \frac{3}{4\pi}e^2 k_F[n(\mathbf{r})]\right]n(\mathbf{r})d\mathbf{r} \qquad (A1.7)$$

we obtain for the plasmon dispersion-coefficient

$$\beta = \frac{m_0}{n_0}P_0 = \frac{1}{3}v_F^2 - \frac{1}{3\pi}\frac{e^2 k_F}{m_0}. \qquad (A1.8)$$

As already noted in chapter 2, (A1.8) is not quantitatively correct so that β will be used as a parameter to fit the experimental plasmon–dispersion. The reason of this discrepancy lies in the roots of the hydrodynamic description itself which is correct for small q,ω, whereas, plasmons are a high frequency phenomenon. Nevertheless, the description is based on conservation laws and contains the essential physics.

A2. Kinetic Theory

In a kinetic description the state of a (one-component) plasma is described by a phase-space distribution function $f(\mathbf{r},\mathbf{p},t)$ which obeys the Boltzmann-Vlasov equation [82]

$$\frac{\partial f}{\partial t} + \mathbf{v}\frac{\partial f}{\partial \mathbf{v}} + \mathbf{F}\frac{\partial f}{\partial \mathbf{p}} = I(f). \qquad (A2.1)$$

$\mathbf{v}$ is the velocity of the particles with energy-momentum relation $\epsilon(\mathbf{p})$ and $\mathbf{F} = -e(\mathbf{E}+v\times\mathbf{B})$ is the Lorentz-force. For isotropic and elastic (impurity-) scattering the collision integral becomes

$$I(f) = \frac{1}{\tau}\Big(\langle f\rangle_\Omega - f\Big), \qquad \langle f\rangle_\Omega = \frac{1}{4\pi}\int f(\mathbf{r},\mathbf{p},t)d\Omega_p, \qquad (A2.2)$$

where τ is the scattering time and $\langle..\rangle$ denotes the angular average on the momentum directions.

Eq(A2.1) must be jointly solved with the Maxwell-equations which, in the quasistatic approximation, reduce to

$$\mathbf{E}(\mathbf{r},t) = -grad\Phi(\mathbf{r},t), \qquad \Delta\Phi(\mathbf{r},t) = -\frac{1}{\epsilon_0}[\rho_+ - en(\mathbf{r},t) + \rho_{ext}(\mathbf{r},t)], \qquad (A2.3)$$

where $n(\mathbf{r},t)$ is the electron-density

$$n(\mathbf{r},t) = \frac{2}{(2\pi\hbar)^3}\int f(\mathbf{r},\mathbf{p},t)d^3\mathbf{p}. \qquad (A2.4)$$

Next we consider small pertubations by the external field, $\Delta\Phi_{ext} = -\rho_{ext}/\epsilon_0$,

$$f = f_0 + f_1, \qquad n(\mathbf{r},t) = n_0 + n_1(\mathbf{r},t). \qquad (A2.5)$$

$f_0(\epsilon(\mathbf{p}))$ is the Fermi-function and n_0 is the equilibrium electron density. In linearized form, (A2.1) can be solved by Fourier-transformation with respect to $t,\mathbf{r}$, see e.g. [27]

$$-i\omega f_1(\mathbf{q},\mathbf{p},t) + i\mathbf{v}\mathbf{q}f_1(\mathbf{q},\mathbf{p},t) + ie\mathbf{q}\Phi\frac{\partial f_0}{\partial\epsilon_p}\mathbf{v} = I(f_1). \qquad (A2.6)$$

In particular, in the absence of collisions ($\tau = \infty$), we optain:

$$f_1(\mathbf{q},\mathbf{p},t) = -e\Phi(\mathbf{r},t)\frac{\partial f_0}{\partial\epsilon_p}\frac{\mathbf{q}\mathbf{v}_p}{\mathbf{q}\mathbf{v}_p - \omega}, \qquad \Phi(\mathbf{q},\omega) = -\frac{e}{\epsilon_0 q^2}n_1(\mathbf{q},\omega). \qquad (A2.7)$$

From (A2.5)

$$n_1(\mathbf{q},\mathbf{p},t) = \frac{em_0 p_F}{\pi^2\hbar^3}\left\{1 - \frac{\omega}{2qv_F}\ln\left[\frac{1+\left(\frac{qv_F}{\omega}\right)}{1-\left(\frac{qv_F}{\omega}\right)}\right]\right\}. \qquad (A2.8)$$

Exchange and correlation effects can be included in the same way as in appendix A1 (yet the kinetic energy has to be left-out)

$$-e\Phi \to -e\Phi + \frac{\delta^2 E_{xc}[n]}{\delta n_0^2}\,n_1. \qquad (A2.9)$$

Near $q = 0$ the plasmon dispersion is given by

$$\omega^2 = \omega_p^2 + \left[\frac{3}{5}v_F^2 + \frac{n}{m}\frac{\delta^2 E_{xc}[n]}{\delta n_0^2}\right]q^2 + \ldots \qquad (A2.10)$$

In a standard local density approximation [11]

$$E_{xc}[n(\mathbf{r})] = \int \epsilon_{xc}[n(\mathbf{r})]\,n(\mathbf{r})\,d\mathbf{r},$$

$$\epsilon_{xc}[n(\mathbf{r})] = \frac{-0.916}{r_s} - 0.045\left[(1+x^3)\ell n(1+\frac{1}{x}) + \frac{x}{2} - x^2 - \frac{1}{3}\right]. \qquad (A2.11)$$

r_s is the density parameter and $x = r_s/21$. As a result we obtain for the q^2-coefficient defined by (37), (A2.10)

$$\frac{\alpha}{\alpha_{RPA}} = 1 - \left[0.092r_s + \frac{0.0034r_s^2}{1+r_s/21}\right] \qquad (A2.12)$$

335

agrees very well with the Vashishta and Singwi' result [16]. For $r_s = 6$, $\alpha/\alpha_{RPA} = 0.35$ and $\alpha = 0$ at $r_s = 8.83$. Nevertheless, the experimental dispersion coefficient shows a much stronger r_s-dependence as given by (A2.12).

As a result we obtain for the longitudinal and transverse dielectric functions (without exchange and correlation effects but including collisions)[27,57]

$$\varepsilon_\ell(\mathbf{q}, \omega) = 1 - \frac{\omega_p^2}{\omega(\omega + i\gamma)} \frac{3}{a^2} \left(1 - \frac{\tan^{-1} a}{a}\right) \left[1 + i\frac{\gamma}{\omega}\left(1 - \frac{\tan^{-1} a}{a}\right)\right]^{-1},$$

$$\varepsilon_t(\mathbf{q}, \omega) = 1 - \frac{\omega_p^2}{\omega(\omega + i\gamma)} \frac{3}{2a^2} \left(\frac{1 + a^2}{a} \tan^{-1} a - 1\right) \tag{A2.13}$$

with abbreviations

$$a^2 = -\frac{\mathbf{q} \cdot \mathbf{q} v_F^2}{(\omega + i\gamma)^2}, \qquad \tan^{-1} z = \frac{1}{2i} \ln\left(\frac{1 + iz}{1 - iz}\right). \tag{A2.14}$$

(A2.13) hold even for complex wave-vectors, $Im\, a \geq 0$, $\ln(1) = 0$.

A3. Quantum Self-Consistent-Field-Approximation (SCFA)

In the self consistent field approximation, the response of the interacting electrons to a weak (scalar) external potential $\Phi_{ext}(\mathbf{r}, t)$ is approximated by a system of noninteracting electrons, responding to the total potential $\Phi = \Phi_{ext} + \Phi_{ind}$ [2,83].

$$\hat{\mathbf{H}} = \frac{\hat{\mathbf{p}}^2}{2m_0} + U(\mathbf{r}, t) + V(\mathbf{r}, t) \tag{A3.1}$$

where $U(\mathbf{r}, t)$ is the periodic crystal potential and $V(\mathbf{r}, t) = -e\Phi(\mathbf{r}, t)$. The microscopic dielectric ($\hat{\varepsilon}$-operator) is defined through

$$V = \hat{\varepsilon}^{-1} V_{ext}, \qquad V_{ext} = \hat{\varepsilon} V \tag{A3.2}$$

In particular we consider a monochromatic external potential with wave-vector $\mathbf{Q}$ and and frequency ω

$$V_{ext}(\mathbf{r}, t) = V_{ext}(\mathbf{Q}, \omega)e^{i(\mathbf{Qr} - \omega t)} + cc, \tag{A3.3}$$

where $\mathbf{Q} = \mathbf{q} + \mathbf{G}$ and $\mathbf{q}$ is within the first Brillouin-zone and $\mathbf{G}$ is a vector of the reciprocal lattice. Due to the periodicity of the crystal potential the induced charge distribution additionally includes contributions from other reciprocal lattice vectors even if $\mathbf{Q}$ is small (socalled local field contributions),

$$V_{ind}(\mathbf{r}, t) = V_{ind}(\mathbf{Q}, \omega)e^{i(\mathbf{Qr} - \omega t)} + \sum_{\mathbf{G}' \neq \mathbf{G}} V_{ind}(\mathbf{Q}', \omega)e^{i(\mathbf{Q}'\mathbf{r} - \omega t)} + cc. \tag{A3.4}$$

$\mathbf{Q}' = \mathbf{q} + \mathbf{G}'$. Reasoning along the same lines, the total potential in (A3.1) is coupled to the external potential by (A3.2) which becomes a matrix equation

$$\Phi_{ext}(\mathbf{q} + \mathbf{G}, \omega) = \sum_{\mathbf{G}'} \varepsilon_{\mathbf{GG}'}(\mathbf{q}, \omega)\, \Phi(\mathbf{q} + \mathbf{G}', \omega). \tag{A3.5}$$

For a crystal $\varepsilon_{\mathbf{GG}'}(\mathbf{q}, \omega)$ is the analogue of the Jellium $\varepsilon_\ell(\mathbf{q}, \omega)$.

Four steps are necessary to obtain the microscopic dielectric matrix [84,85]:
(a) First, the correction of the electron density operator is calculated to first order in the total field $V(\mathbf{r}, t)$, $\hat{\rho} = \hat{\rho}_0 + \hat{\rho}_1$, where $\hat{\rho}_0$ describes thermal equilibrium.
(b) The induced charge density is obtained from $n_1(\mathbf{r}, t) = Sp[\hat{\rho}_1(t)\delta(\mathbf{r} - \hat{\mathbf{r}})]$.
(c) Poisson equation $\Delta\Phi_{ind} = en_1(\mathbf{r}, t)/\epsilon_0$.
(d) $\Phi_{ind}[\Phi]$ is a (linear) functional of the total potential. When writing $\Phi = \Phi_{ind} + \Phi_{ext}$ in the form of (A3.5) the dielectric matrix can be read-off as

$$\varepsilon_{\mathbf{GG}'}(\mathbf{q}, \omega) = \epsilon_\infty \delta_{\mathbf{GG}'} - \frac{e^2}{\epsilon_0 \Omega \mid \mathbf{q} + \mathbf{G} \mid^2} \sum_{\alpha\alpha'} \frac{f(E_\alpha) - f(E_{\alpha'})}{E_\alpha - E_{\alpha'} - \hbar(\omega + i\delta)} (\alpha \mid e^{-i\mathbf{Q}'\mathbf{r}} \mid \alpha')\langle\alpha' \mid e^{i\mathbf{Qr}} \mid \alpha)$$

$$\tag{A3.6}$$

Ω is the crystal volume and ϵ_∞ accounts for core- states not explicitly contained in states numbered by α. For Bloch electrons $\alpha = (n, \mathbf{k})$, where n denotes the band index and $\mathbf{k}$ the wave-number. For a review see Sturm [86].

In an EELS-experiment the observed response $V(\mathbf{q} + \mathbf{G})$ has the same Fourier-components as the pertubation $V_{ext}(\mathbf{q}+\mathbf{G})$. It is convenient to define a macroscopic dielectric function

$$[\varepsilon_{macro}(\mathbf{q} + \mathbf{G}, \omega)]^{-1} = [\hat{\varepsilon}^{-1}]_{\mathbf{GG}}(\mathbf{q}, \omega). \tag{A3.7}$$

If local field effects are neglected

$$\varepsilon_{macro}(\mathbf{q} + \mathbf{G}, \omega) \approx \varepsilon_{\mathbf{GG}}(\mathbf{q}, \omega) \tag{A3.8}$$

(A3.6) leads to the Ehrenreich-Cohn result [83]. For free electrons the Lindhard–function (see [2,5]) is obtained

$$\varepsilon_L(\mathbf{q}, \omega) = 1 + \frac{3}{16x^3}\left(\frac{\hbar\omega_p}{E_F}\right)^2\left\{2x + [1 - (\frac{y - x^2}{2x})^2] \ln\left[\frac{y - x^2 - 2x}{y - x^2 + 2x}\right] - [1 - (\frac{y + x^2}{2x})^2] \ln\left[\frac{y + x^2 - 2x}{y + x^2 + 2x}\right]\right\},$$
$$\tag{A3.9}$$

where $x = q/k_F$ and $y = \hbar(\omega + i\delta)/E_F$. Explicit forms for the real and imaginary parts may be found, e.g. in [2,4].

According to translational symmetry of the interacting electron gas, momentum is conserved and $\varepsilon_\ell(0, \omega) = 1 - (\omega_p/\omega)^2$ is an exact result. Therefore, the plasma frequency as given by (1) is the exact bulk-plasmon frequency for Jellium at $q = 0$. (A3.9) is identical with the random phase approximation (RPA) which was worked out by Bohm and Pines (see [2]) to solve the equation of motion of the density operator.

It is well known that the effects of collisions in a degenerate electron gas cannot be taken into account merely by replacing ω by $\omega + i\gamma$ in the (collisionless) Lindhard function (A3.9). According to Mermin [87] the correct procedure is

$$\varepsilon_M(\mathbf{q}, \omega) = 1 + \frac{(1 + i\gamma/\omega)[\varepsilon_L(q, \omega + i\gamma) - 1]}{1 + (i\gamma/\omega)[\varepsilon_L(q, \omega + i\gamma) - 1]/[\varepsilon_L(q, \omega) - 1]}. \tag{A3.10}$$

Because of the complexity of the many body problem, knowledge of the exact dielectric function is still lacking. Approximate forms for the dielectric function are commonly written as

$$\varepsilon_L(\mathbf{q}, \omega) = 1 - \frac{v(\mathbf{q})\chi_0(\mathbf{q}, \omega)}{1 + v(\mathbf{q})G(\mathbf{q}, \omega)\chi_0(\mathbf{q}, \omega)}, \tag{A3.10}$$

where $v(\mathbf{q}) = e^2/\epsilon_0 q^2$ is the Fourier–transform of the Coulomb potential, $\chi_0(\mathbf{q}, \omega)$ is the Lindhard–susceptibility (of the noninteracting electron gas), $\varepsilon_\ell(\mathbf{q}, \omega) = 1 - v(\mathbf{q})\chi_0(\mathbf{q}, \omega)$, and $G(\mathbf{q}, \omega)$ is the socalled "local field function". The latter describes the short-range exchange and correlation effects which are responsible for the local depletion in the density around each electron. In this scheme the self-consistent potential in (A3.1) is given by

$$V = -e[\Phi_{ext} + (1 - G)\Phi_{ind}] \tag{A3.11}$$

which leads to a self-consistency equation

$$\Phi = \Phi_{ext} + v(\mathbf{q})\chi_0[\Phi_{ext} + (1 - G)\Phi_{ind}] \tag{A3.12}$$

from which (A3.10) is obtained.

In the RPA or standard SCFA, $G(\mathbf{q}, \omega) = 0$, yet the pair correlation function $g(\mathbf{r})$ becomes negative at small distances and the compressibility sum-rule is violated [2,4,16]. Reasonable approximations for $G(\mathbf{q}, \omega)$ can be found in [16,17]. For instance, in the Hubbard–approximation

$$G_H(\mathbf{q}, \omega) = \frac{1}{2}\frac{q^2}{q^2 + k_F^2}. \tag{A3.13}$$

In today's ab initio calculations exchange and correlation effects can be taken into account within the local density approximation which, in most metals, leads to satisfactory results, yet with enormous numerical efforts, see e.g. [20,21,88].

A4. Bulk Loss-Function

To relate the inelastic electron scattering probability to the dielectric function we start from the work done by a particle moving parallel to the z-axis with constant velocity $\mathbf{v} = (0, 0, v_0)$ and impact vector $\mathbf{r}_0 = (x_0, y_0, 0)$. According to the Maxwell-equations (22) the electron charge distribution

$$\rho_{ext}(\mathbf{r}, t) = -e\delta(\mathbf{r} - \mathbf{v}t - \mathbf{r}_0) \qquad (A4.1)$$

leads to longitudinal field components

$$D_\ell(\mathbf{q}, \omega) = \frac{2\pi i e}{q} \delta(\omega - q_\parallel v_0) \, \exp(-i\mathbf{q}_\perp \mathbf{r}_0). \qquad (A4.2)$$

For fast but nonrelativistic electrons, the reaction of the dielectric on the electron as well as contributions from $\mathbf{E}_t$ and $\mathbf{B}$ can be neglected. The work done per unit time by the electron is given by

$$
\begin{aligned}
\frac{dW}{dt} &= \int \mathbf{j}_{ext}(\mathbf{r}, t)\mathbf{E}_{ind}(\mathbf{r}, t)d^3\mathbf{r}, \\
&= -e\mathbf{v}\mathbf{E}_{ind}(\mathbf{v}t + \mathbf{r}_0, t), \\
&= -\int \frac{2i\pi e^2 \omega}{\epsilon_0 q^2} \left[\frac{1}{\epsilon_\ell(\mathbf{q}, \omega)} - 1\right] \delta(\omega - q_\parallel v_0) \frac{d^3 q d\omega}{(2\pi)^4}
\end{aligned} \qquad (A4.3)
$$

The real part of $\varepsilon_\ell(\mathbf{q}, \omega)$ is an even function with respect to frequency and, therefore, it drops-out from (A4.3). As a result we obtain

$$\frac{dW}{dt} = \int \int \hbar\omega P(\mathbf{q}, \omega) \frac{d^3 q \, d(\hbar\omega)}{(2\pi)^4} \qquad (A4.4)$$

with

$$P(\mathbf{q}, \omega) = \frac{2\pi e^2}{\epsilon_0 \hbar q^2} \, Im\left[-\frac{1}{\epsilon(\mathbf{q}, \omega)}\right] \delta(\hbar\omega - \hbar q_\parallel v_0). \qquad (A4.5)$$

The loss-function $P(\mathbf{q}, \omega)$ can be interpreted as the rate for excitation of "photons" with energy $\hbar\omega$ and momentum q.

ENLIGHTENMENT ON LUMINESCENT MATERIALS

C. R. Ronda

Philips GmbH Forschungslaboratorien Aachen
P.O. Box 1980
D-52021 Aachen, Federal Republic of Germany

ABSTRACT

Luminescent materials applied in present devices (like television sets, fluorescent lamps, or X-ray detectors) are high tech materials which perform at their physical limits. The mechanisms, leading to luminescence are understood to a high extent. In this contribution, the most important mechanisms are elucidated. Emphasis is also on application of luminescent materials.

I. INTRODUCTION

Very generally, luminescence can be defined by generation of light beyond thermal equilibrium. This implies that an external source of energy is necessary to induce luminescence. Many kinds of energy carriers are possible: electromagnetic radiation (visible light, UV light, X-rays or even heat when charge carriers are trapped in the material), electrons, etc.

Luminescent materials, also called phosphors, mostly are solid inorganic materials consisting of a host lattice, usually intentionally doped with impurities (see fig. 1). The absorption of energy takes place via either the host lattice or on impurites. In addition, the excitation energy can be transferred through the lattice. In most cases, the emission takes place on impurities.

Luminescent materials are applied widely. Many displays are based on luminescent materials, excited either by electrons (cathode-ray tubes, electroluminescence displays) or UV light (plasma display panels). In addition, in fluorescent lamps, phosphors are used to convert the UV light, generated by the Hg-discharge, into (mostly) visible light.

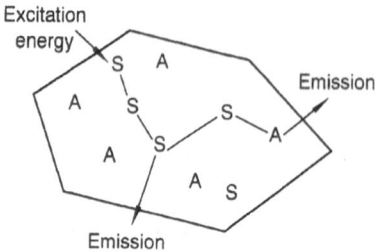

Fig. 1. Luminescent material, excited by UV light or electrons, containing activator ions (ions showing the desired emission) and sensitizing ions (on which UV excitation takes place).

Another well known application of luminescent materials are X-ray intensifiers. In these devices, phosphors are used to convert X-ray into light, adapted to photographic films.

In this contribution, the physics underlying luminescent materials will be treated. Emphasis is both on luminescent mechanisms and applications of luminescent materials. Finally, trends in research on luminescent materials will be elucidated.

II. LUMINESCENCE AND EXCITATION MECHANISMS

In this section, luminescence and luminescence excitation mechanisms will be discussed. We will start with mechanisms, leading to luminescence. To this end, we distinguish center luminescence, charge transfer luminescence and donor-acceptor pair luminescence. Emphasis will be on factors determining the shape of the emission spectra as well. Thereafter we will discuss excitation mechanisms. Here we distinguish host-lattice excitation and center excitation. In addition, energy transfer between ions will be treated. Finally, excitation of emission by application of an electric field will be elucidated.

II.A. Center luminescence

In case of center luminescence, the emission is generated on an optical center, as opposed to e.g. band emission.

One speaks of characteristic luminescence, when, in principle, the emission could also occur on an ion in vacuum, i.e. when the optical transition involved a transition between electronic states of one ion. Common to all these transitions is that the excited state can be described as a Frenkel exciton: a strongly localized electron-hole pair.

Characteristic luminescence can consist of relatively sharp emission bands, but also of broad bands. Broad emission bands are observed when the character of the bonding in ground and excited state differs considerably. This is the case for many $d \rightarrow d$ transitions (on transition metal ions), but also for $d \rightarrow f$ (on rare-earth ions) emission and for emission on s^2 ions, like Tl^+, Pb^{2+} or Sb^{3+}. Sharp emission bands are characteristic for optical transitions between electronic states with bonding character being (almost) the same for ground- and excited state, but also for optical transitions between electronic states that hardly participate in the chemical bonding (e.g. $f \rightarrow f$ transitions on rare-earth ions).

An example for a broad $d \rightarrow d$ emission band is the emission of Mn^{2+} in Zn_2SiO_4, see fig. 2.

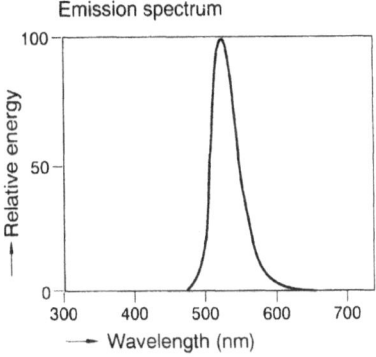

Fig. 2. Emission spectrum of Mn^{2+} in Zn_2SiO_4

Zn_2SiO_4:Mn is used as green primary in plasma display panels, the emission being located at the top of the colour triangle. The emission is generated by a d → d optical transition on the Mn^{2+}-ion with d^5 electronic configuration. This transition is spin forbidden, consequently the decay time of the emission is in the ms range (see also below). The optical transition leading to emission is $^4T_1$ → $^6A_1$. The emission generated reflects the properties of the ion in the crystal field of the host lattice. For many ions, the single ion luminescence consists of broad bands, due to coupling to lattice phonons. The energetic position of the emission bands depends strongly on the crystal field strength, i.e. chemical environment.

An example for d → d emission, consisting of a few relatively sharp bands is the emission of Mn^{4+} in $Mg_4GeO_{5.5}F$, fig. 3. In this case, the optical transition consists of a spin-flip transition within the $t_{2g})^3$ manifold (^2E →$^4A_2$ transition), i.e. hardly changing the character of the bonding.

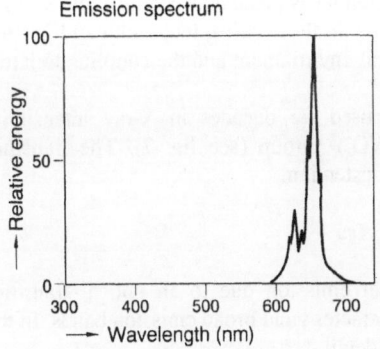

Fig. 3. Emission spectrum of Mn^{4+} in $Mg_4GeO_{5.5}F$

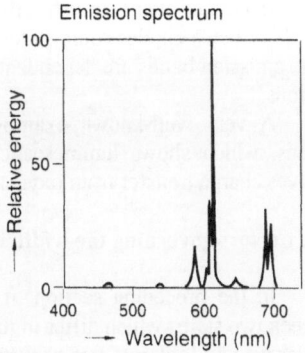

Fig. 4. Emission spectrum of $Y(V,P,B)O_4$:Eu^{3+}

341

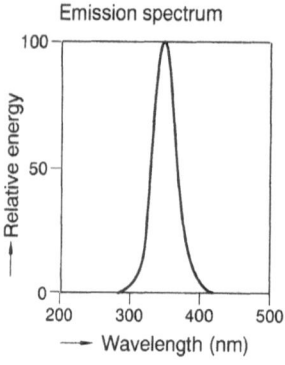

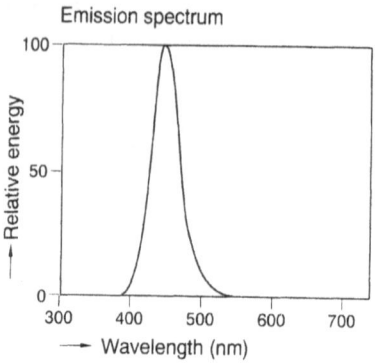

Fig. 5. Emission spectrum of $BaMgAl_{10}O_{17}$:Eu^{2+}

Fig. 6. Emission spectrum of $BaSi_2O_5$:Pb

Most rare earth ions show sharp emission bands, due to optical transitions in the f-manifold, e.g. Tb^{3+} ($4f^8$) and Eu^{3+} ($4f^6$), see fig 4. These transitions are spin- and parity forbidden and therefore slow as well (ms).

However, for a number of rare-earth ions, broad emission bands are known, due to d → f emission, e.g. Eu^{2+} ($4f^7$) or Ce^{3+} ($4f^1$), see fig. 5. These transitions are allowed and consequently very fast (µs or faster). Quite a few very important commercial phosphors are based on rare-earth ions, see also below.

An example of emission of an s^2 ion is given in fig. 6. It is the emission spectrum of $BaSi_2O_5$:Pb. The luminescence of s^2 ion is interesting, due to the interactions possible in the excited state. For Pb, spin-orbit coupling results in excited states $^3P_0$, $^3P_1$, and $^3P_2$. At higher temperature, emission is from the $^3P_1$ level and therefore rather fast. The phosphor emits in the UV and is used in suntanning lamps.

II.B. Charge transfer luminescence

In case of charge transfer, the optical transition takes place between different kinds of orbitals or between electronic states of different ions. In these cases, too, width and position of the emission bands are dependent on the chemical environment and the coupling to lattice phonons.

A very well-known example is $CaWO_4$, used for decades in X-ray intensifying screens, which shows luminescence from the $(WO_4)^{2-}$ group (see fig. 7). The transition involves charge transfer from oxygen ions to the tungsten ion.

II.C. Factors governing the width of emission spectra

In the preceding section, it was stated that emission due to an optical transition between two states which differ in their bonding character yield broad emission bands. In this chapter we will look into this phenomenon in more detail.

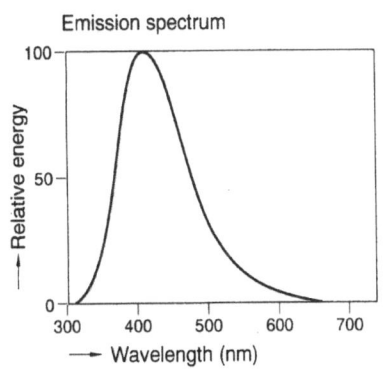

Emission spectrum

Relative energy

100

50

0

300   400   500   600   700

Wavelength (nm)

$\hbar\omega_e$ .... $S_e\hbar\omega_e$

E_a E_e

Δ

$\hbar\omega_g$ .... $S_g\hbar\omega_g$

Q_g Q_e

Fig. 7. Emission spectrum of CaWO₄   Fig. 8. Configuration coordinate diagram

The emission (and absorption) of light is such a fast process that during the optical transition the atoms virtually do not move. This leads to the classical Franck-Condon principle, which states that the most probable optical transition involves the same nuclear configuration in both the ground and excited state. In order to describe such optical transitions one generally uses the so-called configuration coordinate diagrams (CCD).

A CCD is a graphical representation of the electronic energy of the ground state and an excited state of the ion of interest. The wavefunction of the ion is written as $\varphi(r_i,Q)$, where r_i represents the coordinate of the electrons i and the ligand field is represented by one parameter only, the configuration coordinate Q. The value of this parameter oscillates about its average value Q_o, with lattice frequency ω. The equilibrium nuclear configuration, the average value of Q and the frequency of the lattice vibration are, in general, dependent on the electronic state. In fig. 8 a CCD is presented. In this figure, Q_g and Q_e are the equilibrium positions in ground and excited state, respectively. Near the minima of the parabola, the

vibrational states are harmonic oscillators with energies $(n_g + 1/2)\hbar\omega_g$, for the ground state and $(n_e + 1/2)\hbar\omega_e$, for the excited state. The most probable optical absorption corresponds to a vertical transition, without a change of Q (the Franck-Condon principle) and occurs at an energy $E_a = \Delta + S_e\hbar\omega_e$. However, the Franck-Condon principle cannot be used as a selection rule, except for the case that Q_g is equal to Q_e and the phonon frequencies in ground and excited state are the same. This means that in general the absorption spectrum consists of a band centered around E_a. In the same manner, the emission spectrum generally consists of a broad band centered around $E_e = \Delta - S_g\hbar\omega_g$. The transition Δ is the zero-phonon transition, corresponding to the optical transition between two completely relaxed states. The spectral position of this line is the same in absorption and emission.

The quantity $E_a - \Delta$ is the energy difference between the position of the maximum of the absorption band and the position of the zero-phonon line. The Huang-Rhys parameter S_e is defined as $(E_a - \Delta)/\hbar\omega_e$. and equals the average number of phonons involved in the optical transition. Therefore this quantity measures the electron-phonon interaction. In the same way, S_g equals $(\Delta - E_e)/\hbar\omega_g$.

From the CCD, we observe that the energy difference between the maximum of the absorption band and the emission band (Stokes' shift) is given by $S_e\hbar\omega_e + S_g\hbar\omega_g$. The Stokes shift is determined by the electron-phonon interaction.

343

Any change in bonding character (i.e. in charge distribution) is communicated to the host lattice by electron-phonon interaction. For this reason, lattice relaxation is expected to be large in ionic or polar lattices and less in more covalent lattices.

In order to get convenient expressions, in general further assumptions are made, resulting in equations of limited validity. It is frequently assumed that the electronic energies $E_g(Q)$ and $E_e(Q)$ have a similar dependence on Q, so that the phonon energies in ground- and excited state ($\hbar\omega_g$ and $\hbar\omega_e$) are the same. In such a case $S_e = S_g = S$, the Stokes' shift is $2S\hbar\omega$ and the emission and the absorption band are mirror images about the zero-phonon line.

In case of strong coupling ($S > n$) and on neglecting the dependence of the transition probability on Q (Condon approximation), the shape of the emission band is given by [1]:

$$I_{em}(h\nu) = (2\pi S(\hbar\omega)^2 \coth(\hbar\omega/2k_bT))^{-1/2} \cdot \exp(-(\Delta-h\nu-S\hbar\omega)^2/(2S(\hbar\omega)^2\coth(\hbar\omega/2k_bT))) \qquad (1)$$

where $I_{em}(h\nu)$ is the emission intensity at energy $h\nu$. The shape of the absorption band in this limit can be calculated analogously:

$$I_{ab}(h\nu) = (2\pi S(\hbar\omega)^2 \coth(\hbar\omega/2k_bT))^{-1/2} \cdot \exp(-(h\nu-\Delta-S\hbar\omega)^2/(2S(\hbar\omega)^2\coth(\hbar\omega/2k_bT))) \qquad (2)$$

In fig. 9, examples are given of emission spectra in the weak, medium and strong coupling case [1].

Many phosphors applied either show very sharp emission bands or are characterized by rather broad emission bands. In the first case, the emission does not involve lattice relaxation and generally spoken zero-phonon lines are observed, although coupling of f-states to vibrational modes is observed as well [2], beit that this coupling does virtually not result in lattice relaxation. In the second case, lattice relaxation is very important.

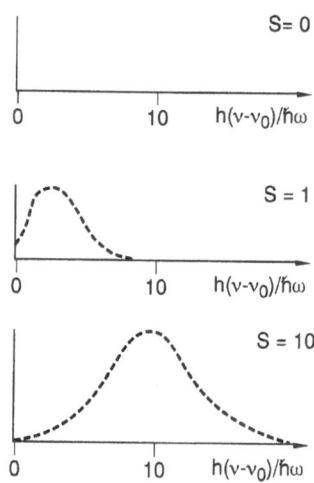

Fig. 9. Shape of an absorption band as a function of the Huang-Rhys factor S.

II.D. Donor-acceptor pair emission

In the preceding section, optical transitions on single centers have been discussed. In this section, we will treat optical transitions taking place on two localized centers will be discussed.

In semiconductors like ZnS, luminescence can occur which involves optical transitions on donors and acceptors. In this case, the electrical charges on the two localized states change as a consequence of the optical transitions. This so-called donor-acceptor pair luminescence mechanism is operating in the blue and green emitting phosphor used in colour television (ZnS:(Ag,Cl) and ZnS:(Cu,Au,Al), respectively).

We now first discuss the mechanism for the simple case in which the lattice relaxation can be neglected. Then we extend the treatment to systems including lattice relaxation.

The mechanism, leading to donor-acceptor pair luminescence is depicted in fig. 10.

The four figures correspond to the following 4 steps:

I. Absorption of light with the bandgap energy E_g : $E_1 = E_g$ (3)

II. Capture of an electron by the donor, yielding the donor ionization energy E_d, but costing Coulomb energy, due to the presence of an ionized negatively charged acceptor at distance R:
$E_2 = E_d - e^2 / \varepsilon_0 R$ (4)

III. Capture of a hole by the acceptor, yielding the acceptor ionization energy E_a:
$E_3 = E_a$ (5)

IV. The actual emission of light with photon energy E_L: $E_L = E_1 - E_2 - E_3 =$
$E_L = E_g - (E_a + E_d) + e^2 / \varepsilon_0 R$ (6)

in which ε_0 is the static dielectric constant in the host lattice, therefore this treatment yields the so-called zero-phonon lines only (see below).

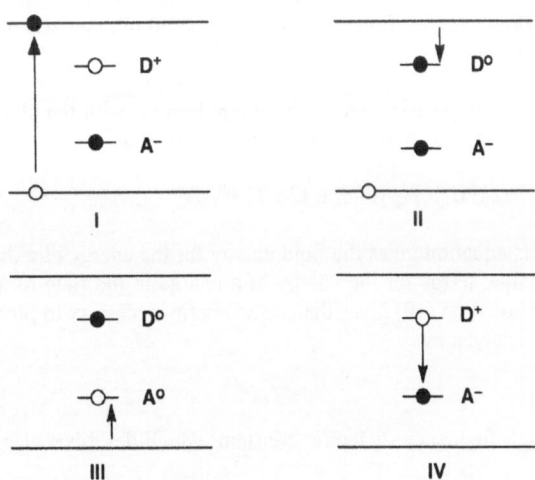

Fig. 10. Processes leading to donor-acceptor pair luminescence

Donor-acceptor pair emission has a number of fascinating properties. As the distance between donor and acceptor is determined by the crystal structure of the host, a number of discrete sharp luminescence peaks is observed in the emission spectrum; for a more general review, the reader is referred to [3].

Moreover, the emission colour is dependent on the excitation density. The transition probability increases with increasing wavefunction overlap, i.e. for donor-acceptor pairs with a shorter distance R. At high excitation density, the mean distance between neutralized donors and acceptors decreases, resulting in a blue shift of the emission spectrum. In addition, the decay time of the emission decreases on decreasing mean donor-acceptor distance and therefore also with increasing excitation density for the same reason.

A shortcoming of the treatment as given above is that equation (6) gives the position of the emission lines for zero-phonon line transitions only, i.e. only for those cases where no lattice polarization occurs as a consequence of the optical transitions. We will now derive an expression for the position of the emission bands for those cases where lattice polarization cannot be neglected, i.e. for those cases where, immediately after the optical transition, the lattice polarization is not in equilibrium with the new electric field. The treatment will be completely classically, no quantum effects are taken into account.

In vacuum, the total energy of a system of electric charges is given by:

$$E_{tot} = 1/(8\pi)\int \mathbf{F}^2 \, dV \qquad (7)$$

where F is the electric field, due to the electric charges.

In a dielectric, when the polarization is in equilibrium with the electric field, the total energy is given by:

$$E_{tot} = 1/(8\pi)\int \mathbf{D.F} \, dV \qquad (8)$$

where $\mathbf{D}$ is the dielectric displacement vector:

$$\mathbf{D} = \mathbf{F} + 4\pi\mathbf{P} \qquad (9)$$

and P is the polarization vector. Equation (9) is only valid in cases where the polarization is in equilibrium with the electric field.

In general, also when the polarization is not in equilibrium with the electric field, the total energy of a system of electric charges is given by the equation:

$$E_{tot} = 1/(8\pi)\int \mathbf{F}^2 \, dV + 1/(2\alpha_e)\int \mathbf{P}^2_e \, dV + 1/(2\alpha_i)\int \mathbf{P}^2_i \, dV \qquad (10)$$

The first term in this equation gives the field energy for the energy of a dipole (polarization) in the field of a charge, terms for the energy of a charge in the field of another charge etc. The second term in equation (10) gives the elastic energy necessary to produce the electronic polarization $\mathbf{P}_e$. α_e is given by:

$$\alpha_e = (\varepsilon_\infty - 1)/(4\pi) \qquad (11)$$

where ε_∞ is the high frequency dielectric constant, which describes electronic polarization only.

The third term gives the elastic energy, necessary to produce the lattice polarization P_l. α_l is given by:

$$\alpha_l = (\varepsilon_0 - \varepsilon_\infty) / (4\pi) \tag{12}$$

in which the static dielectric constant ε_0 describes both electronic and lattice polarization.

Next we derive the electric field F in the case that the lattice polarization is not in equilibrium with the electric field, for a system of electric charges e_i, placed at positions R_i. From the Maxwell equation $\nabla D = 4\pi\rho = \Sigma_i\, e_i\, \delta(R - R_i)$, D is equal to:

$$D = \Sigma_i\, e_i (R - R_i) / |R - R_i|^3 \tag{13}$$

Using the material equation $P_e = (\varepsilon_\infty - 1) / (4\pi)\, F$, equation (9) can also be written as:

$$D = F + 4\pi(P_e + P_l) = \varepsilon_\infty F + 4\pi P_l \tag{14}$$

This equation can also be written as: $F = 1 / \varepsilon_\infty * (D - 4\pi P_l)$ (15)

The integrals that appear in equation (10) are easily calculated to be:

$$\int_0^\infty (R - R_i).(R - R_j) / (\ |R - R_i|^3.\ |R - R_j|^3\ )\, dV = 4\pi / R_{ij} \tag{16}$$

where R_{ij} is the distance between the charges i and j. For i = j, integration starts at R_{ii} and results in $4\pi / R_{ii}$, where R_{ii} is the radius of the sphere in which the charge R_i is localized.

Now we have derived all equations, necessary to calculate the position of the emission (and absorption) bands in case of donor-acceptor pair luminescence with lattice relaxation.

From the equations derived above, we now calculate the total energy of a completely relaxed initial state ($D^0 + A^0$), a completely relaxed final state ($D^+ + A^-$) and the total energy of these states, immediately after the optical transition.

From (10) and the equations relating P_e and P_l to the electric field F (all polarizations are in equilibrium with the electric field) we readily obtain:

$$E_1 = E_{10} \tag{17}$$

where E_{10} is the sum of the (eigen)energies of the localized centers D^0 and A^0 in the completely relaxed initial state.

In exactly the same way, we obtain for the total energy of the completely relaxed excited state:

$$E_2 = E_{20} - e^2 / (\varepsilon_0 R) \tag{18}$$

where E_{20} is the sum of the (eigen)energies of the localized centers D^+ and A^- in the completely relaxed final state.

Next, we calculate the total energy of the unrelaxed final state ($D^+ + A^-$). The electric field, **F** immediately after the optical transition, is given by equation (15), **D** in this equation is given by equation (13), $\mathbf{P}_{13}$ is the lattice polarization in the initial state and is given by :

$$\mathbf{P}_{13} = 0 \tag{19}$$

We obtain for **F**:

$$\mathbf{F} = 1/\varepsilon_\infty [e \mathbf{A}_1 - e \mathbf{A}_2] \tag{20}$$

$\mathbf{A}_i$ is given by:

$$\mathbf{A}_i = (\mathbf{R} - \mathbf{R}_i)/(\mathbf{R} - \mathbf{R}_i)^3 \tag{21}$$

Insertion of P_{13} and F into (10) yields:

$$E_3 = E_{30} - (1/\varepsilon_\infty) e^2 / R \tag{22}$$

In the same way, we obtain for the unrelaxed initial state, immediately after the optical transition:

$$E_4 = E_{40} - (1/\varepsilon_\infty - 1/\varepsilon_o) e^2 / R \tag{23}$$

From the calculations we also find that $E_{40} - E_{10} = E_{30} - E_{20}$. This corresponds to equal relaxation energies in ground and excited states, as expected. Now all relevant energies have been determined, and we apply the equations derived to our donor-acceptor case (see fig. 11).

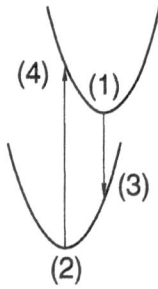

$$E_1 = E_{10}$$

$$E_2 = E_{20} - \frac{e^2}{R} \cdot \frac{1}{\varepsilon_0}$$

$$E_3 = E_{30} - \frac{e^2}{R} \left[\frac{1}{\varepsilon_\infty}\right]$$

$$E_4 = E_{40} - \frac{e^2}{R} \left[\frac{1}{\varepsilon_\infty} - \frac{1}{\varepsilon_0}\right]$$

Fig. 11. CCD for donor-acceptor pair luminescence in case of lattice relaxation

For the energies of absorption E_a, emission E_e (maxima of the bands) and the zero-phonon energy Δ we obtain :

$$E_e = \Delta° + e^2 / (\varepsilon_0 R) - E° - (1 / \varepsilon_0 - 1 / \varepsilon_\infty) e^2 / R \qquad (24)$$

$$E_a = \Delta° + e^2 / (\varepsilon_0 R) + E° + (1 / \varepsilon_0 - 1 / \varepsilon_\infty) e^2 / R \qquad (25)$$

$$\Delta = \Delta° + e^2 / (\varepsilon_0 R) \qquad (26)$$

in which $E°$ and $\Delta°$ are energy terms that do not depend on the distance R between the localized centers. Please note that the energy term $E° + (1 / \varepsilon_0 - 1 / \varepsilon_\infty) e^2 / R$ is always larger than zero as this energy term has been calculated using integral equations using the squared electric field. This term equals $S\hbar\omega$ in the CCD diagram! In this case S is dependent on the distance between donor and acceptor.

The absorption and emission spectra are determined completely by S(R) and both dielectric constants. The calculations, using dielectric constants are not very accurate for small distances. This means that our model will not be very accurate for the shortest distances R.

In fig. 12, pair spectra are given at T = 0 K with two absorption and emission bands. In fig. 12a it is assumed that $1 / \varepsilon_\infty > 2 / \varepsilon_0$. In this case, on decreasing R the bands become narrower, the emission band shifts to higher energy and the absorption band to lower energy. In fig. 12b it is assumed that $1 / \varepsilon_\infty < 2 / \varepsilon_0$. In this case on decreasing R the bands become narrower, and both the emission band and the absorption band shift to higher energy.

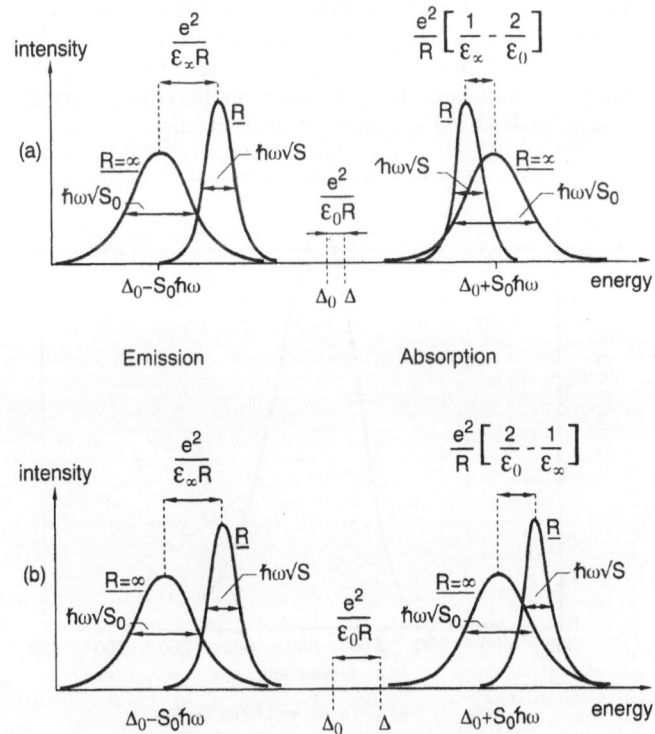

Figs. 12. Donor-acceptor pair absorption and emission spectra

349

Now, we apply the considerations given above to ZnS:Cu,Al. The emission spectrum of ZnS:Cu,Al (sphalerite) is given in fig. 13. We observe a broad, structureless band, extending at the high energy side to values of about 3.0 eV. The emission is nevertheless due to donor-acceptor pair luminescence, as a.o. deduced from the (small) blue shift of the emission band with increasing excitation density, see [4,5]. Please note that the spectral position of the zero-phonon line, due to emission on pairs with the shortest distances corresponds to about 3.0 eV, as deduced from the emission spectrum extending to 3.0 eV. Interestingly, in [6], for ZnS:Cu without compensating ion both broad emission and zero-phonon line emission are reported, the zero phonon line being located at 2.96 eV.

First we calculate the position of the zero-phonon emission, using the values for energy gap and donor and acceptor ionization energies. The donor and acceptor ionization energies are taken from [7]. Cu is acceptor, with ionization energy 0.95 eV and Al is donor, with ionization energy 0.25 eV. The bandgap of ZnS (sphalerite) is 3.68 eV. The shortest donor-acceptor distance in ZnS:Cu,Al equals 0.382 nm. The static dielectric (8.0) constant has been taken from [8]. Using (6), we obtain a value of 2.95 eV for the spectral position of the zero-phonon line. This is remarkably close to the value deduced above, in view of the fact that at short distances the dielectric constant is likely to be smaller than the bulk value and that other interactions are expected to be important as well (e.g. v.d. Waals). The Coulomb stabilization energy for the closest pair equals 0.47 eV. If the emission band consisted of superimposed zero-phonon lines, the emission band would have a maximum width of about this value. This is not the case, the total width of the emission band is about 0.92 eV (fig. 13). This again indicates lattice relaxation to be important, consequently, the treatment, developed above must be used in analyzing the observations. Due to the absence of structure in the emission spectrum, this treatment can be qualitatively only.

From the Stokes Shift of about 0.62 eV and the phonon energy of 44 meV [8], we calculate a FWHM of about 0.46 eV. The observed value is 0.36 eV. The difference observed suggests an overestimated lattice relaxation.

In our treatment we assumed that only the closest pairs contribute to the emission process in ZnS:Cu,Al. This is in agreement with the fact that the blue-shift of the emission band and the decrease of the luminescence decay time are only slightly dependent on the excitation density [5]. Nevertheless, this treatment is not likely to be very exact, as pairs at short distances cannot be described by mere Coulomb interaction.

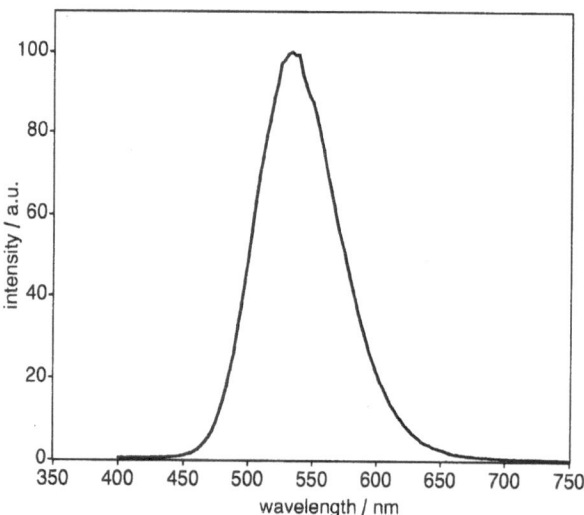

Fig. 13. Emission spectrum of ZnS:Cu,Al

II.E. Exchange induced luminescence

There is a number of possibilities to relax the selection rules for optical transitions. For parity forbidden transitions, coupling to phonons or excited states with odd symmetry enhances the transition probability. For spin forbidden transitions, spin-orbit coupling or in case of magnetic ions, exchange interaction between such ions can enhance the transition probability.

Consider a system with one kind of magnetic ions with spin S, in which exchange interaction occurs between ions in pairs. The Hamiltonian H for the exchange interaction in such pairs is written as:

$$H = -2J \, S_a.S_b \tag{27}$$

in which J is the exchange interaction strength, and S_a and S_b are the spins of ion **a** and **b**, respectively.

Due to the coupling, in the ground state the ion pair has spin states from $S_a + S_b \ldots.$ | $S_a - S_b$|. In a spin forbidden transition on ion **a** to serve as an example, the spin changes by one unit. Suppose the spin on ion **a** after the optical transition is $S_a - 1$.
In the excited state, the ion pair has spin states running from: $(S_a - 1) + S_b \ldots.|(S_a - 1)- S_b|$.
Now, due to the exchange interaction, there are spin states in ground- and excited state with the same value for the total spin, i.e. optical transitions between such states are spin allowed.

Now we apply these considerations to Zn_2SiO_4:Mn. Zn_2SiO_4:Mn is a green emitting phosphor. The Mn^{2+} emission is rather sharp and is located at the top of the colour triangle. By using this phosphor in displays, a very wide colour gamut can be obtained. However, the decay time of the emission ($^4T_1 \rightarrow {}^6A_1$ transition) is too long for television applications (> 5 msec). It has been known for a long time that on increasing the Mn^{2+} concentration, a shortening of the decay times occurs, accompanied by a red shift of the emission band [9] and even recently, this material has been studied in detail [10]. In [10], the existence of Mn^{2+}-pairs has been proposed as the origin of the relaxation of the spin selection rule. However, the sheer existence of pairs does not result in such a relaxation. In addition, for an increasing Mn^{2+} concentration a reduction in crystal field strength is expected, because Mn^{2+} is larger than Zn^{2+}. In that case, for this emission band, a blue shift is expected, contrary to the observation. We will show that these observations can be explained on assuming exchange interaction between Mn^{2+}-pairs.

In the ground state, both Mn^{2+} ions have spin $S_a = S_b = 5/2$, as a result the total spin of the ion pair has as possible values S = 0, 1, 2, 3, 4 and 5 (all singly degenerated with respect to the total spin). In the excited state, one of the Mn^{2+} ions still has spin 5/2, but the other ion has spin 3/2. The total spin of an ion pair with one excited Mn^{2+} ions therefore has as possible values S' = 1, 2, 3 and 4 (again all singly degenerate with respect to the total spin).
The operator P^B_{ex} for exchange-induced transitions on a pair of ions **a,b** is given by:

$$P^B_{ex} = {}_{ijkl} \Sigma \, \pi^B_{a(ik),b(jl)} \, S_{a(ik)}.S_{b(jl)} \tag{28}$$

in which B refers to the polarization of the light and in which the electron orbitals are specified by the indices ijkl.

The transition moment is given by [11]:

$$< \Gamma_y \, S_a' \, S_b \, S' \, M_S \, '| P^B_{ex} | \, S_a \, S_b \, S \, M_S > =$$

$1/(2S_b) \;_{i,k,\, j=1} \Sigma \; \pi^B_{a(ik),\, b(jj)} < \Gamma_y \, S_a \, '\|S_{a(ik)}\| \, S_a> * <S_b\| \, S_b\| \, S_b> \delta_{S\,S'} \, \delta_{M_S\,M_{S'}} (-1)^{S+1} *$

$* \, W \, (S_a \, S_b \, S' \; S_a' \, S_b' \, 1)$ (29)

in which the 6-j symbol W has the value 0, -1/15, $1/5\sqrt{(2/7)}$, $-1/5\sqrt{(3/7)}$, $1/(3\sqrt{7})$, 0 for S'= 0, 1, 2, 3, 4, 5, respectively. From equation (29), it follows that the selection rules are $\Delta S = 0$ and $\Delta M_S = 0$.

The total oscillator strength is given by:

$f = f_0 \;_{S'} \Sigma \; \alpha_{S'} |W_{S'}|^2$ (30)

and depends on the occupation probability $\alpha_{S'} \cong (2S' + 1) \exp(- E_{S'}/kT)$ of the ion pair.

The energy $E_{S'}$ depends, in turn, on the exchange interaction $- 2 \, J \, S_a . S_b$:

$E_{S'} = -J \, [S' \, (S' + 1) - S_a (S_a + 1) - S_b (S_b + 1)]$ (31)

In fig. 14, the relative oscillator strength f/f_0 is given as a function of the parameter J/T. We observe that the exchange interaction gives an only moderate modification of the relative oscillator strength.

In this evaluation, we did not take the reduced matrix elements into account explicitly. Therefore, the absolute changes in oscillator strength, which are strongly influenced by the strength of the exchange interaction, can be much larger.

From equation (31) it is deduced easily, that the energy of the $^4T_1 \rightarrow {}^6A_1$ emission band is changed by 5J with respect to the situation with no exchange interaction, due to the exchange interaction between the Mn^{2+} ions. This result is independent of the number of ions in the Mn^{2+} cluster and of the total spin state occupied. As the exchange interaction between Mn^{2+} ions is antiferromagnetically (d^5 configuration, J<0), this would result in a blue shift of the emission band on going from isolated ions to ion pairs.

In this treatment, we assumed implicitly that the strength of the exchange interaction is the same in both ground and excited state. This is not necessarily the case, as pointed out in [11]. On assuming a larger exchange interaction in the excited state (for example due to a more pronounced super exchange in the excited state, via oxygen states), even a red shift of the emission band on going from single ions to ion pairs can be expected [11]. In such a case, at least at higher temperatures, a fourfold splitting of the emission band can be expected. The energy difference between the peaks is determined by the differences in exchange interaction in ground and excited state. In table 1, the exchange induced shift of the emission bands with S = 1, 2, 3 and 4 is given as a function of the coupling constant for exchange interaction in ground (J) and excited state (J'), calculated using equation (31). In fig. 15, the shift of the emission bands with S = 1, 2, 3 and 4 is given.

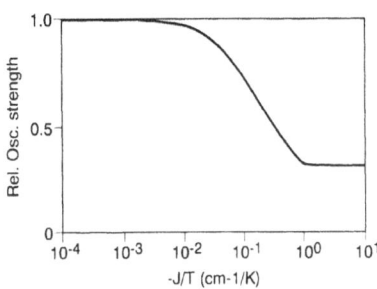

Fig. 14. Relative oscillator strength f/f_0 as a function of J/T

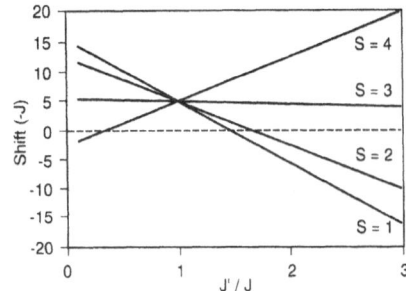

Fig. 15. Shift of the emission bands with S = 1, 2, 3 and 4

Table 1. Exchange induced shift for
spin states with S = 1,2,3, and 4 as
a function of the exchange
parameters J' and J

Spin state	Exchange induced shift
S = 1	$(21/2)J' - (31/2)J$
S = 2	$(13/2)J' - (23/2)J$
S = 3	$J'/2 - (11/2)J$
S = 4	$-(15/2)J' + (5/2)J$

We observe that a red shift will occur only for J'/J> 1.56. For not too small values of J, magnitude and in some cases even sign of the shift will be dependent on the temperature. For J' = J, the emission bands with S' = 1, 2, 3 and 4 all lie at the same energy. In this case, the shift is temperature independent and even independent of the number of exchange coupled Mn^{2+} ions (equation 31).

In [10], a red shift of the Mn^{2+} emission is observed on increasing the Mn^{2+} concentration. The maximum of the emission band shifts from 519 nm to 526 nm, this corresponds to a shift in energy of about 250 cm^{-1}. From the observed red shift, we conclude that the ratio of exchange coupling constants in excited and ground state J'/J is larger than 1.56. Similar effects were observed by Ferguson et al. [11], for Mn^{2+} pairs in perovskite fluorides.

Due to the strong dependence of the energy of the $^4T_1$ state on the metal-ligand distance, broad spectra are obtained. Therefore, the emission bands with S = 1, 2, 3 and 4 are not expected to be resolved in the spectra.

The decay time measured for single ions and ion pairs in Zn_2SiO_4:Mn is almost the same at 4.2K [9]. At 300K, they are different by about one order of magnitude (15 ms for single ions and 1.75 ms for pairs [10]). Almost the same values were reported in [9]. Such a large variation in the decay time of the transition on ion pairs cannot be explained with the occurrence of coupling of the spins of Mn^{2+} ions pairs only (see fig. 14). As possible origin we mention higher order exchange terms or spin-orbit coupling *of the total spin state of the ion pair* to an electronic quartet state, originating from Mn^{2+}. The theory, describing this latter effect has been developed in [12]. Both mechanisms are, in this form, not possible on single ions as they rely on the temperature dependent occupation of the total spin states of the *ion pair*.

The decay of the emission on isolated ions indeed is almost independent of the temperature [9,10], in accordance with the fact that this optical transition is spin forbidden.

II.F. Energy transfer

An excited ion can decay radiatively or non-radiatively. An until now not treated way of returning to the ground state is the transfer of the excitation energy by transfer of the excitation energy to another ion (either of the same kind or not), see also fig. 1. Such a transfer can be written as a chemical reaction:

$$S^* + A \rightarrow S + A^* \tag{32}$$

In this event, the excitation energy is transferred from S^* to A. Note that we used the classic notation, S is the sensitizer, A the activator and an asterisk indicates the excited state.

Energy transfer can occur when the energy differences between ground- and excited states of S and A are equal and when a suitable interaction between both systems exists. The interaction can be both electric or magnetic in nature. In addition, exchange interaction can play a role. In this latter case, wavefunction overlap is necessary. The interaction Hamiltonian is written as:

$$P_{SA} = 2\pi / \hbar \mid <S, A^* \mid H_{SA} \mid S^*, A > \mid^2 \int g_s(E) g_a(E) \, dE \tag{33}$$

The matrix element represents the interaction, H_{SA} is the interaction Hamiltonian between the initial state $\mid S^*, A>$ and the final state $\mid S, A^*>$. The integral represents the spectral overlap, calculated using the normalized optical line shape functions $g_s(E)$ and $g_a(E)$. From (33) we deduce that no energy transfer is possible without spectral overlap.
Evaluation of this integral equation yields more simple expressions, see table 2 and the approximate range of transfer for these interaction modes is given in table 3.

Table 2. Equation form for energy transfer as a function of the type of interaction. K_{dd}, K_{dq}, and K_{ex} are constants, the spectral overlap is given by A_1 - A_3, respectively, R is the distance between sensitizer and activator ion, τ is the decay time of the sensitizer ion and L is a constant describing wavefunction overlap.

interaction type	equation form
electric dipole-electric dipole	$P_{SA} = K_{dd} A_1 / (\tau R^6)$
electric dipole- electric quadrupole	$P_{SA} = K_{dq} A_2 / (\tau R^8)$
exchange	$P_{SA} = K_{ex} A_3 \exp(-2R/L)$

Table 3. Range of energy transfer for different interaction modes

Mode of transfer	Range (nm)
electric dipole-electric dipole	3.5
electric dipole- electric quadrupole	0.8
exchange	0.6

A more quantitative analysis of the equations shows that energy transfer from a broad-band emitter to a line absorber is only possible for nearest neighbours. Energy transfer from a line emitter to a broad-band emitter is possible over much longer distances (some 2 nm). Energy transfer between two broad bands is possible at even larger distances (up to 3.5 nm).

Now, mechanisms leading to luminescence will be discussed. To this end, we will distinguish between luminescence excited by absorption of energy on an optical center and luminescence excited by absorption of energy in the host lattice, followed by energy transfer to the luminescing ions. CRT and X-ray phosphors all operate according to the second mechanism, i.e. via host-lattice excitation. In addition, the theory used in describing the underlying physics is very similar. Finally, quite a few of them are used in both kinds of devices. On the other hand, almost all phosphors, applied in fluorescent lamps are excited by absorption on an isolated optical center. An obvious reason for this difference is the very small penetration depth of UV photons in case of band absorption (less than 1 μm), leading to surface losses.

II.G. Lattice excitation

After absorption of electrons or high energy photons (e.g. X-ray quanta), impinging on the phosphor material, in the lattice secondary charge carriers, i.e. electron-holes pairs are generated. The electron-hole pairs thermalize, finally resulting in electron-hole pairs with bandgap energy. After thermalization, the excitation is transferred to an activator (or sensitizer), ultimately leading to emission.
For each absorbed electron or high energy photon, a large number of electron-hole pairs will be generated. Each electron-hole pair can give rise to emission of one photon on the activator ion.
Robbins has treated these processes more quantitatively [8]. The energy efficiency of the overall process is given by:

$$\eta = (1 - r_b)h\nu_{em}\eta_t\eta_a\eta_{esc} / \beta E_g \qquad (34)$$

in which r_b is the fraction of backscattered electrons, $h\nu_{em}$ is the mean energy of the photons emitted, η_t is the efficiency with which the energy of the thermalized electron-hole pairs is transferred to the activator ions, η_a is the quantum efficiency of the activator ions, η_{esc} is the escape probability (the ratio between the number of photons generated internally and emitted externally), and finally βE_g equals the energy, needed to generate one thermalized electron-hole pair. In this formalism, β is a pure number, whereas E_g is the bandgap energy.
Backscattering is neglectable for X-rays, whereas for electrons, the backscatter coefficient r_b is in the order of 0.1 - 0.2.

In practice, phosphors which operate at physical limits are of interest only. This condition implies the transfer efficiency, the activator quantum efficiency and the escape probability to be unity. Equation (34) then simplifies to:

$$\eta_{max} = (1 - r_b)\, h\nu_{em} / \beta E_g \qquad (35)$$

This expression implies that the energy efficiency is determined by the mean energy of the emission of the activator ion and the product βE_g. The particles, used in exciting the luminescence lose their energy by impact ionization and the generation of optical phonons. The average energy needed to create an electron-hole pair can be written as:

$$\beta E_g = E_i + E_{op} + 2E_f \qquad (36)$$

in which E_i is the ionization threshold, E_{op} the average energy lost in generating optical phonons and E_f is the threshold energy for the generation of electron-hole pairs.

The ratio of the energy needed to generate optical phonons with frequency $h\nu_{op}$ and impact ionization is proportional to R, given by [8]:

$$R = (1/\varepsilon_\infty - 1/\varepsilon_0)\,(h\nu_{op})^{1.5} / (1.5\, E_g) \qquad (37)$$

in which ε_∞ and ε_0 are the optical and the static dielectric constants of the phosphor host lattice, respectively. The dependence of β on R is given in fig. 16.

The number β is found to vary between about 2.5 and 10 for a number of host lattices. We observe that in order to obtain host lattices with small β, resulting in highly efficient phosphors, the value of R should be small as well. This condition implies a low optical phonon frequency or a small difference between the optical and the static dielectric constant. In table 4, for a number of well-known phosphor materials, the relevant data are given. We observe a good agreement between the energy efficiencies observed and the maximum efficiencies predicted.

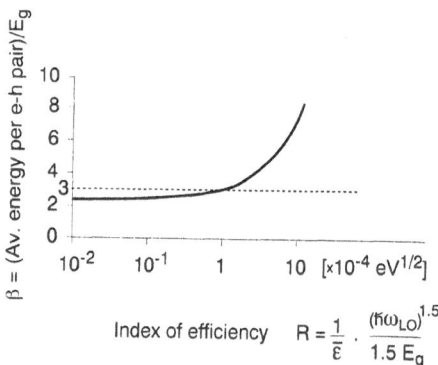

Fig. 16. Dependence of β on R

Table 4. Physical constants of a number of efficient host lattice excited phosphors. η_{the} is the maximum energy efficiency calculated, η_{exp} is the energy efficiency observed.

Phosphor	$h\nu_{op}$ (eV)	E_g(eV)	$h\nu_{em}$(eV)	β	η_{the}	η_{exp}
CsI:Tl	0.011	6.4	2.25	2.5	0.14	0.14
ZnS:Ag	0.044	3.8	2.75	2.9	0.25	0.20
ZnS:Cu	0.044	3.8	2.3	2.9	0.21	0.17
CaS:Ce	0.047	4.8	2.3	3.0	0.16	0.22
CaS:Mn	0.047	4.8	2.1	3.0	0.15	0.16
La$_2$O$_2$S:Eu	0.057	4.4	2.0	3.9	0.12	0.11
Y$_2$O$_3$:Eu	0.068	5.6	2.0	4.6	0.07	0.08
YVO$_4$:Eu	0.116	3.7	2.0	7.5	0.07	0.07

II.H. Phosphors excited by vacuum ultraviolet and soft X-ray excitation

In the preceding sections, mechanisms for exciting the luminescence by absorption on an optical center and by high energy particles (X-rays or electrons) were discussed. Now we discuss excitation by photons with intermediate energy: excitation by vacuum ultraviolet radiation or soft X-rays (with photon energy 10 - 100 eV). In the literature, quantum efficiency data of a number of well known phosphors is published, see a.o. [13,14]. It is observed that the quantum efficiency increases to values larger than one for photon energies above about 20 eV, as given in table 5. This effect is interpreted as being due to Auger interband transitions [14]. Surprisingly, we found that for those materials, for which the measurements have been extended into the low energy region, the quantum efficiency has a minimum at excitation energy about 2.5 times the value of the bandgap [15]. This is due to a very small penetration depth, leading to non-linear energy loss processes (in addition to the desired Auger processes). At higher energy, the quantum efficiency rises again. In table 5, we have also given the photon efficiency (number of photons generated per eV excitation energy). By comparison with the value βE_g, deduced from the treatment by Robbins [8], we see that, though host-lattice processes definitely play a role, the excitation mechanism is not identical, as the photon efficiency in case of VUV excitation is higher in all systems investigated. At present, this phenomenon is not understood.

Table 5. Luminescence characteristics for a number of phosphors under vacuum excitation. E_g: Band gap energy, Min.: excitation energy for which quantum efficiency has its lowest value. P.E.: number of photons generated per electron volt excitation energy, P.E. $_{Rob}$ is the number of photons generated per electron volt excitation energy in case of cathode-ray excitation.

Phosphor	E_g (eV)	Min. (eV)	Min. / E_g	P.E. (1/eV)	P.E. $_{Rob}$ (1/eV)
YVO$_4$:Eu	3.7	9.0	2.4	0.075	0.036
Zn$_2$SiO$_4$:Mn	5.5	14.0	2.5	0.07	0.027
ZnO:Zn	3.4	9.0	2.6	0.063	0.061
Y$_2$O$_3$:Eu	5.6	14.0	2.5	0.067	0.039
ZnS:Ag	3.8	9.0	2.4	0.10	0.091
Gd$_2$O$_2$S:Eu				0.099	0.049

II.I. Electroluminescence

One speaks of electroluminescence when the luminescence is excited by an electrical current, transported through the luminescent material. AC and DC electroluminescence are distinguished. In the first mechanism electrical energy is applied to the phosphor material by a breakthrough current, in the second mechanism, one relies on injection of a current via electron- and hole transporting layers. Usually, relatively high voltages are applied in the breakthrough cases (> 100 V) over relatively thin layers (1000 nm), whereas in the latter case at relatively low voltages (< 20 V), luminescence can be observed already. In the latter case, the device has a diode-characteristic (light emitting diodes).

In case of AC electroluminescence (ACEL), electrons (holes) are accelerated in a solid by the electric field applied. The charge carriers can easily lose energy by phonon emission (exciting lattice vibration, i.e. heat). This means that a high electric field is necessary, as only in this case the energy, taken up from the field can be higher than the losses due to phonon emission.

Only for a few materials, the ACEL process seems to be clear. In case of ZnS:Mn (which is used in commercially available devices), the luminescent centers are excited by impact excitation. The energy efficiency, obtainable with this material is estimated to be:

$$\eta = h\nu_{em} \, \sigma \, N \, / \, eF \qquad (38)$$

in which $h\nu_{em}$ is the photon energy of the emitted radiation, σ the cross-section for impact excitation, N is the optimum concentration of luminescent centers and F is the electric field applied. $1 / (\sigma N)$ is the mean distance that an electron travels through the luminescent material between two impact excitation events.

The cross-section is not known a priori. In case of ZnS:Mn we approximate it by using atomic dimensions, i.e. $\sigma = 10^{-16}$ cm^2. The other (typical) values are: $h\nu_{em} = 2$ eV, N = 10^{20} cm^{-3} and F = 10^6 V/cm. It follows that the energy efficiency equals about 2%, this is in very good agreement with experiment. In this treatment, however, we have used a number of simplifications. We did not account for the Stokes shift. Moreover, we neglected light trapping effects in the thin layers. All these phenomena further reduce the energy efficiency. The energy efficiency is not likely to be improved significantly. This is mainly due to the low value for the cross-section, because N cannot be chosen too large in view of concentration quenching.

The mean energy the charge carrier has taken up from the electric field between two impact excitation events equals $eF / \sigma N$, neglecting any losses due to phonon emission. The minimum pathway able to excite an activator ion L_{crit} equals $L_{crit} = h\nu_{exc} / eF$. We remark that L_{crit} is dependent on the electric field strength.

We now formulate an expression for the energy efficiency of the electroluminescence process, analogous to the one describing the energy efficiency under cathode-ray excitation: (on assuming the activator efficiency and the escape probability to be unity):

$$\eta = (h\nu_{em} / h\nu_{exc}) \, \sigma N \, L_{crit} \qquad (39)$$

This expression shows $\sigma N \, L_{crit}$ to be the transfer efficiency. In case of cathode-ray excitation, this figure can be unity, as it is for efficient cathode-ray phosphors. In case of ZnS:Mn its optimal value is calculated to be about 0.02 only. This is the main reason for the low energy efficiency of this material.

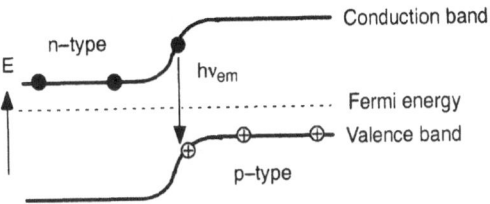

Fig. 17. DC electroluminescence generated by recombination of electrons and holes in a junction of a p- and n-type conductor of the same material

Despite large research activity, no substances have been found that are as efficient as ZnS:Mn. Other EL materials identified are ZnS:Tb (green, but with a low Tb-concentration (about 0.1%) due to solubility limits), ZnS:Sm (red), SrS:Ce (blue-green, with energy efficiency about 1%). Very recently, alkaline earth thiogallates, doped with Ce^{3+} were proposed by Planar (a company producing EL panels) [16], emitting bluish green.

In case of DC electroluminescence, the light is generated in the junction between an n- and p-type semiconducting material. The electrons are transported to the junction via the n-type semiconductor, the holes via the p-type semiconductor. In the very simple case of fig. 17, the photon energy of the light emitted equals the bandgap energy.

In case of doping, the energy of the photons emitted can be lower than the bandgap energy as well. DC electroluminescence can be efficient in so-called direct gap materials, but also in doped materials. Apart from a search for more efficient materials, the structure of the LED is optimized as well.

III. QUANTUM EFFICIENCY AND COLOUR OF THE EMISSION

In the preceding chapter, we discussed the energy efficiency of cathode-ray phosphors under the assumption that the quantum efficiency of the activator emission was unity. In this chapter we will look into more detail in the factors determining the activator quantum efficiency. These considerations also apply to activator ions in lamp phosphors. In addition, we will discuss factors determining the colour of the emission. Both factors are very important for practical applications of luminescence.

III.A. Factor influencing the activator efficiency

As argued in the preceding paragraph, a change in the chemical bonding as a consequence of the optical transition, results in lattice relaxation, causing a change in equilibrium distance between metal and ligand ion in ground- and excited state and therefore in

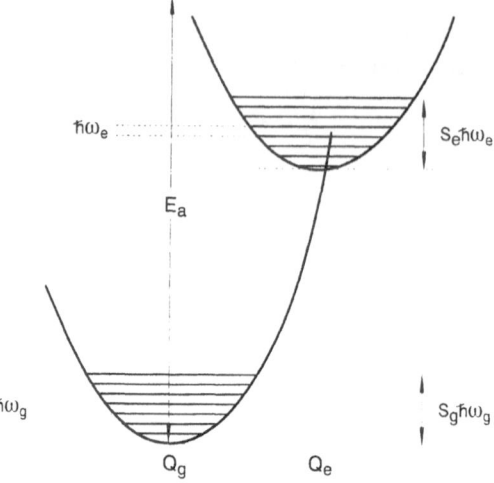

Fig. 18. CCD of a system with strong lattice relaxation

broadened emission bands. In case of strong lattice relaxation, the parabola describing the potential energy of the excited state can cross the corresponding parabola of the ground state near its minimum (fig. 18). In that case the system can return to the ground state (in many cases thermally activated) without light emission.

The metal-ligand distance can increase or decrease as a consequence of optical absorption. When the bonding increases in the excited state, the metal-ligand distance will decrease. A well known example is the Mn^{2+} ion. The ground state has the $t_2)^3e)^2$ configuration. The excited state has $t_2)^4e)^1$. The d-electrons occupy antibonding orbitals, the t_2 electrons of the π-type and the e electrons of the σ-type. As σ bonds are stronger than π bonds, the bonding character is stronger in the excited state, i.e. in the excited state the metal-ligand distance is smaller. In case of decreasing metal-ligand distance after optical absorption, incorporation of the luminescent ion at too large a lattice site (with metal-ligand larger than that of the luminescent ion) will minimize the lattice relaxation effect. The too small luminescent ion already induces a shorter metal-ligand distance in the ground state, reducing the lattice relaxation effect in the excited state. On the other hand, incorporation of such an ion at too small a lattice site will cause effective quenching of the emission: the too large metal ion induces an increase of the metal-ligand distance, counterbalanced by elastic forces in the lattice. In the excited state the metal-ligand distance can decrease considerably. Completely analogously, ions for which the metal-ligand distance increases in the excited state should be incorporated at small lattice sites (e.g. Eu^{3+}). For Eu^{2+}, to serve as an example, this effect is summarized in table 6 [17].

For transitions within the f-manifold, lattice relaxation is virtually absent. Indeed it is frequently observed that on excitation within the f-manifold, high quantum efficiencies are observed in cases where e.g. charge-transfer excitation already yields a low efficiency.

These considerations apply to all kinds of luminescence. However, there are some additional quenching mechanisms. In case of donor-acceptor pair luminescence, the donors or the acceptors can be ionized thermally, thus quenching the emission. Thermal quenching of the acceptors (Ag, Cu) is the reason for the fact that blue emitting ZnS:Ag,Al shows a more pronounced thermal quenching than green emitting ZnS:Cu,Al.

Table 6. Quenching temperature of
some Eu^{2+} phosphors as a function
of the size of the lattice site

Phosphor	Quenching temperature
$CaAl_{12}O_{19}$:Eu	420 K
$SrAl_{12}O_{19}$:Eu	460 K
$BaAl_{12}O_{19}$:Eu	670 K
$Ca_3MgSi_2O_8$:Eu	505 K
$Sr_3MgSi_2O_8$:Eu	520 K
$Ba_3MgSi_2O_8$:Eu	545 K

A third reason for luminescence quenching is (phonon assisted) energy transfer to so-called killer centers. In any crystal lattice defects are present, that can quench the emission, once excitation energy is transferred to these, usually unidentified, centers. This phenomenon generally sets an upper limit to the activator concentration. At too high activator concentrations, the excitation energy migrates towards these centers via activator states. In cases where a considerable lattice relaxation occurs, at low temperatures such a migration is not possible, due to too low a spectral overlap between absorption and emission band. At higher temperatures, the spectral overlap increases, resulting in phonon assisted energy transfer to killing centers, see e.g. [18].

Please also note that energy transfer is the principle underlying sensitization. In this case, the sensitizer emission is quenched in favour of the activator emission.

III.B. Factors determining the emission colour

Many luminescent ions show emission at different wavelengths in different host lattices. This phenomenon, once understood, opens the possibility to change, within certain limits, the emission colour. In this way the emission spectra (and excitation spectra) can be tuned towards own specifications.

In cases where at least one of the electronic states is involved in the chemical bonding, the coupling to the lattice has to be taken into account. In fig. 19, this situation is given for a d → f optical transition. d → f emission is observed for Eu^{2+} or Ce^{3+}. The energy difference between the d- and f-electrons is modified by the electronegativity of the chemical environment and the crystal field strength. In this way, e.g. for Eu^{2+}, emission can be obtained extending from the UV part to the red part of the optical spectrum, see table 7.

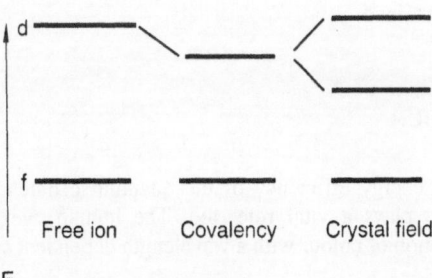

Fig. 19, showing the energy separation of the d- and f- electronic states as a function of the covalency and the crystal field strength.

Table 7, showing the spectral position
of the Eu^{2+} 5d →4f emission as a
function of the chemical environment.

Phosphor	Emission max. (nm)
(Sr,Eu)B$_4$O$_7$	365
(Sr,Eu)MgP$_2$O$_7$	392
(Ba,Eu)Al$_{12}$O$_{19}$	438
(Sr,Eu)MgSi$_2$O$_8$	462
(Ba,Eu)$_2$MgSi$_2$O$_7$	500
(Ba,Eu)Al$_2$O$_4$	505
(Ca,Eu)S	630

The spectral position of the emission lines due to transitions between f-electronic states does not vary very much on changing the host lattice. However, the relative emission intensity of the several optical transitions possible varies considerably on changing the host lattice.

As general remark, one can state that in cases where the rare-earth ion occupies a site with inversion symmetry, the selection rule states: $|\Delta J| = 0$, 1. In cases where $J = 0$, any transition to another state with $J = 0$ is forbidden as well. In such a case, ΔJ is necessarily $+1$. These transitions are all magnetic dipole transitions. In lattices without inversion symmetry there is also electric dipole emission. For these transitions, the selection rule is: $|\Delta J| \leq 6$. Here again, for initial- or final states with $J = 0$, other selection rules are operative. In such a case, for electric dipole transitions, $|\Delta J| = 2$, 4 or 6. We observe that the presence of an inversion center already opens up the possibility to tune the emission spectrum to a little extent. For Eu^{3+} with excited state $^5D_0$, the emission can be tuned from orange (590nm, in case of inversion symmetry, $^5D_0 \rightarrow {}^7F_1$ transition) to red (610 nm, without inversion symmetry, $^5D_0 \rightarrow {}^7F_2$ transition). More generally, these effects can be described by the Judd-Ofelt theory [19,20]. As a function of three parameters, all possible spectra can be calculated. However, a direct coupling to the chemical environment is lacking. Nevertheless, such calculations are useful. Apart from being able to calculate the relative intensities, these calculations can also be used to calculate subsequent optical transitions, i.e. quantum cutters. For Pr^{3+}, in principle a quantum efficiency of 198% can be obtained in the visible. The same kind of calculation has shown that for Tm^{3+}, no quantum cutter, yielding two visible photons can be obtained [15].

Finally, in case of donor-acceptor pair luminescence, both the donors and the acceptors and the magnitude of the bandgap strongly influence the spectral position of the emission colour to be obtained.

IV. EMISSION COLOUR

Apart from the energy efficiency or the quantum efficiency, in many devices the colour of the emission plays a vital role, too. The human eye contains three different receptors for the perception of colour with a wavelength dependent sensitivity: $p(\lambda)$, $a(\lambda)$ and $t(\lambda)$.

There is a number of ways to quantify the colour, using colour coordinates. In this contribution the C.I.E. (Commission International de L'Eclairage) is treated briefly. This system has been introduced in 1931. The system is based on three functions $\underline{x}$, $\underline{y}$ and $\underline{z}$ and given by:

$$\underline{x}(\lambda) = [1.359\ p(\lambda) + 0.101\ t(\lambda) - a(\lambda)] / 0.460 \qquad (40)$$

$$\underline{y}(\lambda) = p(\lambda) \qquad (41)$$

$$\underline{z}(\lambda) = t(\lambda) \qquad (42)$$

The values obtained are also known as the tristimulus values (see fig. 20).

Light with energy distribution $E(\lambda)$ is described by three integrals of the form:

$$X = \int_{380}^{780} \underline{x}(\lambda)E(\lambda)d\lambda \qquad (43)$$

and analogously for Y and Z

Colour perception is hardly influenced by the illumination level, therefore the ratio of the values X, Y and Z determine the impression of the colour perceived. These ratios are the colour coordinates in the C.I.E. system, and are defined as:

$$x = X / (X + Y + Z),\ y = Y / (X + Y + Z)\ \text{and}\ z = Z / (X + Y + Z) \qquad (44)$$

It follows that $x + y + z$ equals 1, therefore it is sufficient to characterize any colour by two coordinates x and y. Not all values of x and y are possible, all possible values lie in the so-called colour triangle (fig. 21). The border line (except for the baseline) is formed by the colour points of monochromatic radiation. In this figure, the colour points of daylight (D 65) and of the incandescent lamp (A) are given as well.

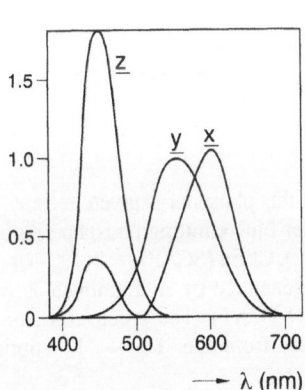

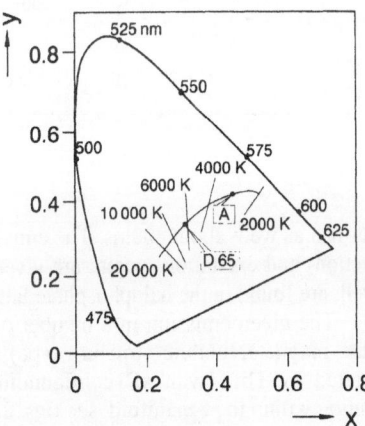

Fig. 20. Tristimulus values as a function of the wavelength

Fig. 21. Colour triangle in the CIE system

V. PHYSICAL ASPECTS OF SOME IMPORTANT LUMINESCENT MATERIALS

V.A. Phosphors used in fluorescent lamps

As mentioned already, almost all fluorescent lamp phosphors are excited by absorption on an optical center. In this section, physical aspects of fluorescent lamp phosphors will be elucidated, using a few examples. To this end, we will discuss a few phosphors, used in Philips colour 80 lamps. In such lamps three phosphors are used, emitting in the blue, the green and the red, respectively.

The blue emission is generated by $BaMgAl_{10}O_{17}$:Eu (BAM), a hexaaluminate. Eu is incorporated as Eu^{2+} in this lattice and is incorporated on Ba^{2+} sites. This ion has a $4f^7$ configuration. The absorption of the UV radiation, generated by the Hg-discharge, takes place on the Eu-ion ($4f^7 \rightarrow 4f^65d$ optical transition). The ground state of the Eu^{2+} ion is the $^8A_1$ state, allowed absorption takes place in a number of excited states. Therefore, the absorption strength of this ion is very large. After relaxation in the excited state, the emission taking place is the radiative return to the ground state. The decay time of the Eu^{2+} emission is

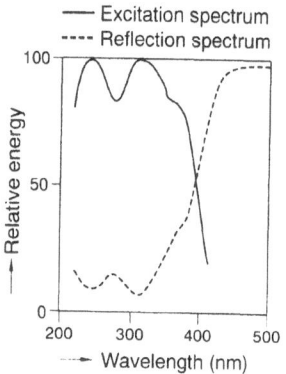

Fig. 22. Reflection- and excitation spectra of $BaMgAl_{10}O_{17}$:Eu

rather fast as well, about 800 ns. The emission spectrum of this phosphor is given in fig. 5, its reflection- and excitation spectra are given in fig. 22. Other blue emitting phosphors, in use as well, are found in the halophosphate lattices, e.g. $Sr_3(PO_4)_5$ Cl:Eu (SCAP).

The green emission in a number of lamp types is generated by an aluminate as well (of the magnetoplumbite structure type): $(Ce,Tb)MgAl_{11}O_{19}$(CAT). The green emission is generated by Tb^{3+} (with $4f^8$ configuration) and originates from the $^5D_4 \rightarrow {}^7F_5$ optical transition within the f-manifold, see figs. 23 and 24.

At low concentration, emission takes place from both the $^5D_3$ and the $^5D_4$ excited states to the ^7F states. At higher Tb-concentrations, cross-relaxation between Tb^{3+} ions results in a depopulation of the $^5D_3$ excited state on one ion by excitation of an electron within the ^7F multiplet on another ion (see fig. 24).

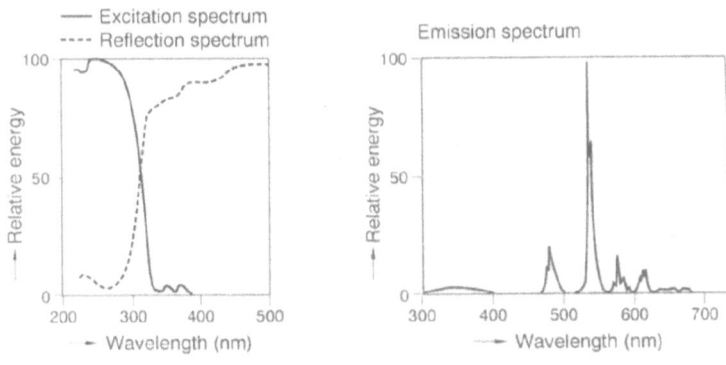

Fig. 23. Emission-, excitation- and reflection spectra of $(Ce,Tb)MgAl_{11}O_{19}$

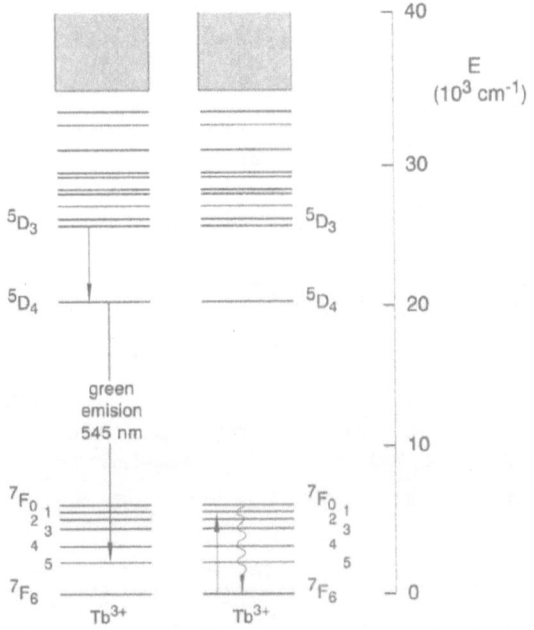

Fig. 24. Cross-relaxation between two Tb^{3+} ions

The emission is generated by a forbidden optical transition. Therefore the decay time of the emission is rather large (a few ms). The absorption of the UV radiation, generated by the Hg-discharge, by the Tb^{3+} in this compound is very low, due to the forbidden nature of the optical transition. Therefore, all green emitting phosphors, used in high quality fluorescent lamps, are codoped with Ce^{3+}. Ce^{3+} has a $4f^1$ configuration, the Hg-radiation is absorbed by the Ce-ion by a $4f \rightarrow 5d$ optical transition (an allowed transition). In CAT, part of the Ce^{3+} ions show emission in the UV, but most of the energy is transferred to the Tb^{3+} ions, which in turn show green emission. The energy tranfer involves transfer from a broad band emitter to a line emitter. Therefore, energy transfer takes place between neighbouring Ce^{3+} and Tb^{3+} ions only. Consequently, the Ce^{3+} and Tb^{3+} concentrations in this compound are rather high. Other green emitting phosphors, used in fluorescent lamps are $LaPO_4$:(Ce,Tb) (LAP) and $(Ce,Gd,Tb)MgB_5O_{10}$ (CBT). In CBT, the energy absorbed by the Ce-ions is transferred to the

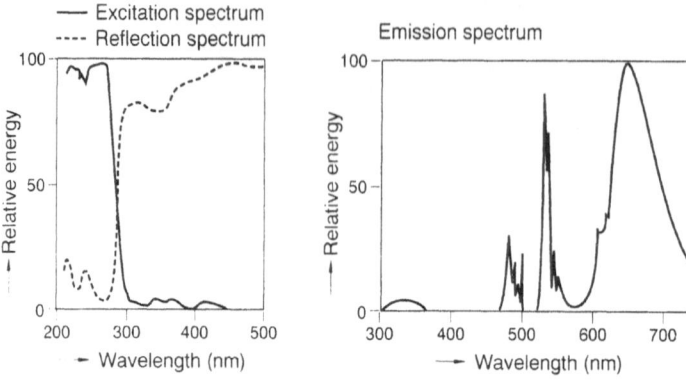

Fig. 25. Emission-, excitation- and reflection spectra of $(Ce,Gd,Tb)MgB_5O_{10}$:Mn

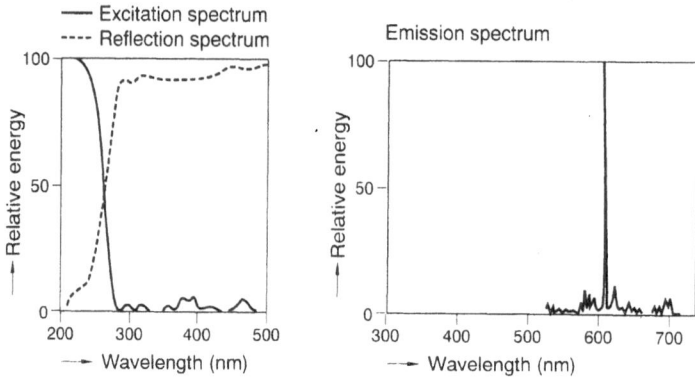

Fig. 26. Emission-, excitation- and reflection spectra of Y_2O_3:Eu

Gd-ions, which in turn transfer the energy to the Tb-ions which then show the green emission. On codoping CBT with Mn^{2+} (CBTM), in addition red emission can be obtained (see fig. 25). With such a phosphor colour rendering of fluorescent lamps can be improved beit at the cost of lamp efficiency.

The red emission is generated by Y_2O_3:Eu (YOX). In YOX, the Eu ion is present as Eu^{3+}. The emission spectrum is given in fig. 26. The emission is due to optical transitions within the $4f^6$ manifold. The main emission line at 611 nm originates from the $^5D_0 \rightarrow {}^7F_2$ transition. Due to the forbidden nature of the optical transitions, the decay time is large (a few ms). The excitation of the emission is due to a charge transfer transition: charge is transferred from the surrounding oxygen ions to the Eu^{3+} ion. This transition is allowed, nevertheless the absorption strength at 254 nm is moderate only, due to the fact that the absorption maximum is located at higher photon energy (fig. 26).

Though numerous studies have been performed in view of the high cost-price of this phosphor, no alternative for YOX has been found as yet, it is the only red emitting phosphor applied in lamps combining good colour rendering, high efficiency and maintenance.

V.B. Phosphors used in television sets

The physics, describing the excitation of emission generated by cathode-ray excitation, has been treated in section 2. In standard television applications, both the blue and the green emission is generated by donor-acceptor pair emission. As blue primary ZnS:Ag is used. Ag^+ (the donor) is compensated by either Cl^- or Al^{3+} (being the acceptor). ZnS:(Cu,Al) or ZnS:(Cu,Au,Al) is used as green primary. The red emission is generated by Y_2O_2S:Eu. In systems with a higher load (like in projection television (PTV)), these phosphors show a strong saturation: the output power is no longer proportional to the input power. This reduces the light output of such sets. To reduce the loss in light output at higher load, the electron beam is defocused. This in turn reduces the resolution.

A number of mechanisms leads to saturation: activator depletion, interaction between excited activators and excited state absorption, Auger interactions, but also thermal quenching. A very important reason for saturation is ground state depletion: no activator ions are present anymore than can be excited. Bril derived quite a long time ago that under stationary excitation saturation occurs when the generation rate of electron-hole pairs per unit volume I equals [21]:

$$I = N / (\tau\eta) \tag{45}$$

in which N is the activator ion density, τ is the decay time of the excited state and η is the transfer efficiency (of electron-hole pairs to the activator). Though under pulsed conditions, the underlying physics is more complicated, see e.g. [22], for our discussion it is sufficient to consider (45) only. The linearity of phosphors can be increased by increasing the activator ion density or reducing the activator decay time. In the zincsulfides, the activator concentrations are relatively low (in the order of 0.01%). In highly loaded devices, such as projection televisions, the excitation energy density per pulse is about 100 mJ / cm^2. On assuming a penetration depth of 6 μ m, the excited volume is about 6.10^{-4} cm^3. On assuming furthermore that per 10 eV excitation energy, one electron-hole pair is created, the number of electron-hole pairs generated in this volume is 1.6×10^{17}. The number of activator ions in this volume (for ZnS with activator concentration 0.01%) equals about 1.5×10^{15}. The decay time of the emission in ZnS for more distant pairs can be as large as about 50 ms. Insertion of these values in (45) clearly shows that ZnS must saturate under these conditions. One way out is increasing the activator concentration. The beneficial effect of an increasing activator ion density could be counterbalanced by concentration quenching. However, in rare-earth materials, rather high activator concentrations can be applied without running into concentration quenching (up to 10%). In fig. 27, both the linear efficiency (low excitation density) and the efficiency at excitation density 10 mJ / cm^2 is given for a number of green emitting phosphors, measured on settled powder screens [15].

We observe a much more linear behaviour for the highly doped phosphors. Nevertheless, large differences in linearity for the Tb-doped phosphors are found. Therefore, there are other effects influencing the saturation behaviour of phosphors. It has been shown that for Tb-doped phosphors, excited state absorption by the Tb^{3+} ions and interaction between excited Tb^{3+} ions (comparable to cross-relaxation) leads to saturation as well [23-25].

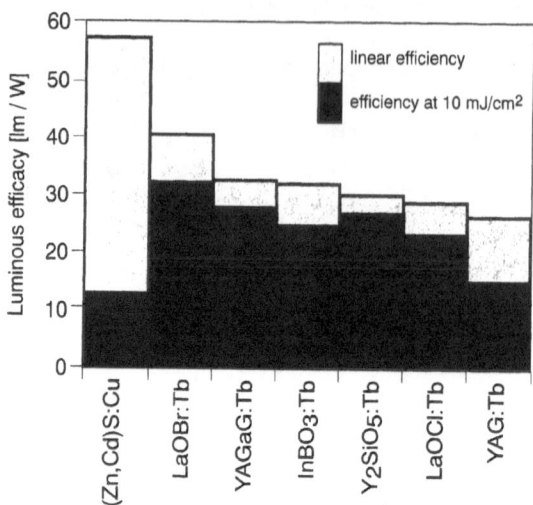

Fig. 27. Linear efficiency and efficiency at excitation density 10 mJ / cm² for a number of green emitting phosphors, measured on settled powder screens

From the discussion, given above, it follows that thermal quenching can be of importance as well, due to the very high load. The temperature of the phosphors can reach values up to 100 °C. Thermal quenching is the reason for Gd_2O_2S:Tb discarded as green primary in PTV. This phosphor had been used in earlier generation PTV sets.

Though as green primary, a number of phosphors is available, for blue there is no proven alternative for ZnS:Ag, despite large research activity. Although in this compound, the activator concentration can be increased to about 0.1%, the linearity is less than expected. At low concentration, saturation is due to ground state depletion. At higher energy other physical processes cause saturation: interaction between excited donor-acceptor pairs and neutral donors. The energy from the recombining D-A pair is transferred to a neutral donor, which then ionizes [26]. In this so-called Auger process, no luminescence is generated by the recombining D-A pairs. The probablity for this process increases with increasing donor and acceptor concentrations. Therefore, saturation of the luminescence output is an intrinsic property of ZnS:Ag.

Finally, there are no problems with the saturation of red emitting Y_2O_3:Eu, used in PTV sets. Any improvement in linearity would result in a reduced excitation current in the red PTV tube.

VI. DEPRECIATION OF LUMINESCENT MATERIALS

Almost all devices based in luminescent materials show a decrease in light output during their life. In this section, degradation processes occuring in or at the phosphors themselves (intrinsic degradation) will be elucidated.

In the literature, quite a number of processes, leading to phosphor depreciation have been identified [27-32]:
- a change in the oxidation state of the activator, possibly causing a change in UV absorption in case of lamp phosphors or a smaller value for the transfer efficiency from host lattice to

activator ions in case of host lattice excited phosphors. In addition, due to energy transfer from unchanged activator ions to the activator ions with changed oxidation state, a lower activator quantum efficiency may result
- a change in the oxidation state of the sensitizer, with similar consequences
- photolysis, due to the interaction of the host lattice with ionizing radiation, leading to:
    * the formation of colour centers in the lattice, resulting in structured additional absorption
    * a disordered host-lattice leading to unstructured additional absorption ('Urbach-tail')
    * a change in the chemical environment of the activator or sensitizer, leading to additional deexcitation channels, in this way reducing the light output
    * formation of metal clusters
    * in fluorescent lamps the presence of Hg and or HgO at the surface of phosphors, leading to additional absorption in the visible and the UV part of the optical spectrum. These phenomena have been dealt with by Tamatani et al. [29,30].

As an example we discuss the degradation of LaOBr:Tb under cathode-ray excitation, investigated in our laboratory a few years ago [32].

The absorption coefficient of LaOBr:Tb as a function of photon energy before and after charge deposition is given in fig. 28. In fig. 29, the absorption strength of aged LaOBr:Tb and LaO(Br,Cl):Tb is given as a function of the photon energy.

The absorption at about 5eV is due to the Tb^{3+} 4f → 5d absorption. The optical absorption at even higher energy is due to band absorption of the LaOBr:Tb matrix. We observe a structured absorption in the visible part of the spectrum, a.o. at the spectral position where Tb^{3+} shows its emission. This absorption consists of three well resolved peaks, with an intensity ratio independent of the degree of degradation of the material. In addition, we observe a broad structureless absorption just below the bandgap of this material. Finally, we observe that the Tb^{3+} absorption spectrum remains almost unchanged.

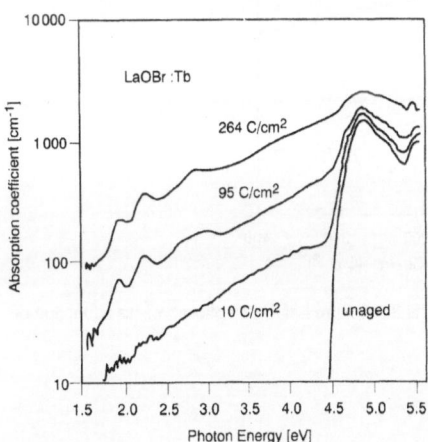
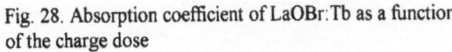
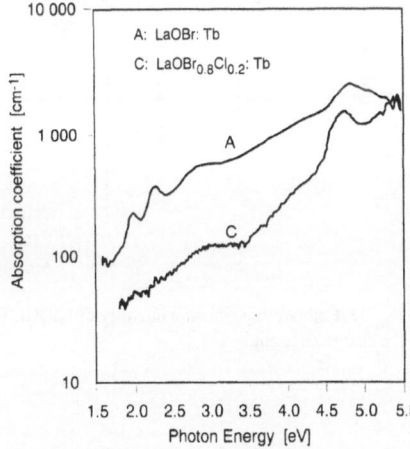

Fig. 28. Absorption coefficient of LaOBr:Tb as a function of the charge dose

Fig. 29. Absorption strength of aged LaOBr:Tb and LaO(Br,Cl):Tb as a function of the photon energy

369

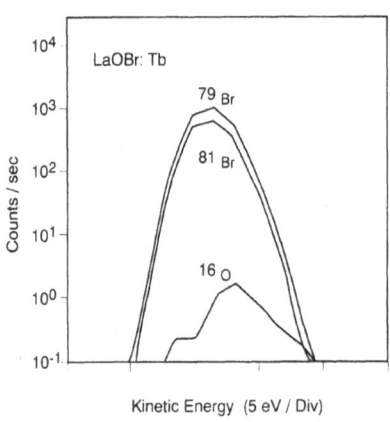

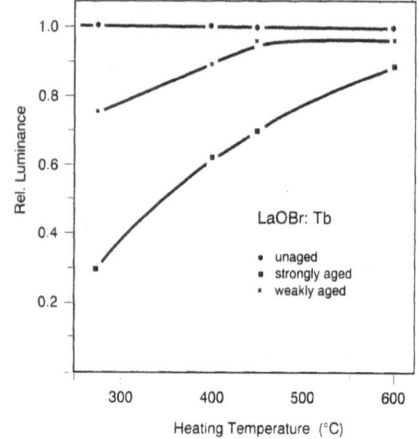

Fig. 30. Kinetic energy of positive ions, desorbing from LaOBr:Tb during electron bombardment

Fig. 31. Cathode-ray emission intensity as a function of annealing temperature

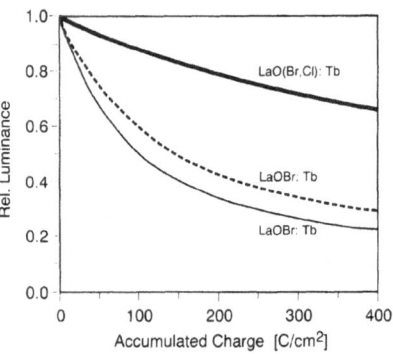

Fig. 32. Cathode-ray emission intensity of LaO(Br,Cl):Tb and two pure LaOBr:Tb phosphors as a function of the charge dose applied.

In mass-spectrometry experiments, performed during electron bombardment, we observed the release of Br^+ ions, with a kinetic energy of about 12 eV (and a much smaller amount of oxygen ions), i.e. an energy much higher than the band-gap energy (fig. 30).

Finally, on heating the material in an inert atmosphere, the energy efficiency of the material under cathode-ray efficiency is (almost) recovered (fig. 31).

The structured absorption in the visible part of the emission spectrum is assigned to F-center absorption. In LaOBr:Tb, the F-centers can originate from both Br- and O vacancies. Both sites have the same symmetry (C_{4v}). Based on the mass spectrometry experiments, the F-center absorption is assigned to Br vacancies. As the C_{4v} sites do not possess inversion symmetry, the parity selection rule is relaxed. The three absorption bands are assigned to the optical transition from the 1s electron in the F-center to the 2s excited state and the symmetry split 2p state of the F-center (the C_{4v} symmetry splits the 2p state into two states with A_1 and E symmetry, respectively). This interpretation is consistent with the fact that the absorption intensity ratios are independent of the degree of degradation of the luminescent material, as the optical transitions take place on an isolated center.

The release of Br^+ ions implies the localization of two photo-holes on one Br ion during irradiation. The mechanism, leading to localization of two photo-holes on one ion, has been described by Feibelmann and Knotek [33,34]. It relies on two important parameters: the width of the valence band (W) and the Coulomb interaction (U) between the two photo-holes. If U is larger than W, localization of two photo-holes on one Br ion can occur. This ion subsequently has a positive charge, at a site where the crystal lattice requires a negatively charged ion. This implies that the Br^- ion will have to leave its lattice site. (Note that the kinetic energy of 12 eV corresponds to about the Madelung energy.) In the bulk of the phosphor material, the Br^+ ions will move to interstitial sites. This explains why the original lumen output can be recovered on heating the material. At the surface of the phosphor material, Br^+ ions can leave the material.

The unstructured absorption, just below the bandgap is ascribed to the so-called Urbach tail absorption [35]. This absorption finds its physical origin in deviations from the periodic potential of the crystal host lattice. In this case, interstitial Br ions disturb the periodic potential of the host-lattice. After heating, the Br ions have diffused to regular lattice sites and the heated material consequently does not show Urbach tail absorption anymore.

On codoping LaOBr:Tb with a small amount of Cl (20%), a much more stable material has been obtained (fig. 32).

This effect is not completely understood. It is conceivable that doping LaOBr:Tb with Cl results in smaller diffusion coefficients for Cl and or Br ions, in analogy to the well known mixed alkali effect in glasses and hexa-aluminates. Another explanation is the following. From the desorption of Br^+ ions under electron bombardment, we deduce that U is larger than W in LaOBr. The width of the valence band has been measured to be about 4.3 eV. This means that U is larger than 4.3 eV, i.e. is at least approaching the value of the energy gap. In this region the dielectric constant is strongly dependent on the energy. Therefore small lattice variations can strongly influence the value of U (W is hardly influenced by codoping LaOBr:Tb with such a small amount of Cl). This could lead to the situation where W is larger than U, i.e. no ions are ejected from their lattice sites.

VII. OUTLOOK

After more than 50 years of research, luminescent materials have been developed which perform extremely well. Phosphors excited by UV light have a very high absorption of the incident UV (more than 90%, except for the red emitting phosphor) and a very high quantum efficiency (more than 90%). Phosphors applied in picture tubes have an energy efficiency close to the maximum, at least at low excitation density. Nevertheless, research on luminescent materials is still useful.

In the first place, the rare-earth based phosphors are rather expensive (150$/kg), this makes it worthwile to search for cheaper phosphor materials.

In addition, though the quantum efficiency of UV excited phosphors is almost 100%, the energy efficiency of such phosphors is much lower. Energy conservation does not prohibit quantum cutters, i.e. materials with generate more than one visible photon from one UV photon. In the literature, two mechanisms have been distinguished, leading to quantum cutting. In the first mechanism, the cascade involved takes place on the activator ion. A well known example is YF_3:Pr, with quantum efficiency 140% on excitation into the Pr^{3+} 5d level with 185 nm UV radiation [36,37]. On the other hand, no other ions or atoms are found, showing such a cascade. Unfortunately, the emission spectrum of Pr^{3+} is not suitable for use in high quality fluorescent lamps. In the second mechanism, the cascade takes place in the host-lattice. Examples are Y_2O_3:Eu or Zn_2SiO_4:Mn. In the substances found, quantum cutting takes place only on excitation using radiation with photon energy larger than about 20 eV. All discharges yielding photons of such a high energy are much less efficient than the Hg-discharge. Therefore, new materials have to be found, showing quantum cutting at lower excitation energy. Such materials would open the possibility to obtain energy efficient mercury free fluorescent lamps, but also plasma panel displays with higher efficiency.

CRT phosphors operate at physical limits at low excitation density. In higher loaded devices, saturation occurs, causing brightness and resolution losses. Research on materials for application in such devices (like projection television) is still going on.

EL phosphors are characterized by low energy efficiencies. This is due to the rather low values for the transfer integral (excitation of the activator ions by the charge carriers), which in turn is due to the low values for the cross-section for impact excitation and the impossibility to increase the activator concentration beyond the limit of concentration quenching. Any significant improvements involve application of other than (Coulomb) neutral traps or disentangling excitation and emission by exploiting energy transfer from sensitizer ions to activator ions (as in the case in some fluorescent lamp phosphors).

ACKNOWLEDGEMENTS

The author acknowledges the pleasant cooperation with Prof. Dr. C. Haas, State University of Groningen, the Netherlands, in the development of the theory used in describing donor-acceptor pair luminescence in polar lattices.

REFERENCES

1. J. Bourgoin and M. Lannoo, Point Defects in Semiconductors II, Springer Series in Solid-State Sciences 35, Springer Verlag Berlin, 1983
2. G. Blasse, Prog. Solid State Chem. 18, 79 (1988)
3. P. Dean 'Progress in Solid State Chemistry 8', edited by J.O. McCaldin and G. Somorjai, Pergamon Press, New York (1973)
4. M. Bredol, J. Merikhi and C. Ronda, Ber. Bunsenges. Phys. Chem. 96, 1770 (1992)
5. M. Bredol, J. Merikhi, I. Köhler, H. Bechtel and W. Czarnojan, J. Solid State Chem. 110, 250 (1994)
6. I. Broser, R. Broser and E. Birkicht, J. Lum. 40&41 331 (1988)
7. R.H. Bube, 'Photoconductivity of Solids', Wiley, New York (1960)
8. D.J. Robbins, J. Electrochem. Soc. 127, 2694 (1980)

9. A.L.N. Stevels and A.T. Vink, J. Lum. 8, 443 (1974)
10. C. Barthou, J. Benoit, P. Benalloul and A. Morell, J. Electrochem. Soc. 141, 524 (1994)
11. J. Ferguson, H.J. Guggenheim and Y. Tanabe, J. Phys. Soc. Japan 21, 692 (1966)
12. H.J.W.M. Hoekstra, P.R. Boudewijn, H. Groenier and C. Haas, Physica 121B, 62 (1983)
13. J.A. Berkowitz and J.A. Olsen, J. Lum. 50, 111 (1991)
14. E.L. Benitez, D.E. Husk, S.E. Schnatterly and C. Tarrio, J. Appl. Phys. 70, 3256 (1991)
15. C.R. Ronda, J. Alloys and Compounds, accepted for publication (1995)
16. W.A. Barrow, R.C. Coovert, E. Dickey, C.N. King, C. Laakso, S.S. Sun, R.T. Tuenge, R. Wentross and J. Kan, SID 93 Digest, 23, 761 (1993)
17. G. Blasse and A. Bril, Philips Technische Rundschau 31 (1970/1971)
18. C.R. Ronda, H.H. Siekman and C. Haas, Physica 144B, 331 (1987)
19. B.R. Judd, Phys. Rev. 127, 750 (1962)
20. G.S. Ofelt, J. Chem. Phys. 37, 511 (1962)
21. A. Bril, Physica 15, 361(1949)
22. D.B.M. Klaassen, T.G.M. van Rijn and A.T. Vink, J. Electrochem. Soc. 136, 2732 (1989)
23. K.J.B.M. Nieuwesteeg, R. Raue and W. Busselt, J. Appl. Phys. 68, 6044(1990)
24. K.J.B.M. Nieuwesteeg and R. Raue, J. Appl. Phys. 68, 6058 (1990)
25. R. Raue, K.J.B.M. Nieuwesteeg and W. Busselt, J. Lum. 48&49, 485 (1991)
26. R. Raue, M. Shiiki, H. Matsukiyo, H. Toyama and H. Yamamoto, J. Appl. Phys. 75, 481 (1994)
27. D.B.M. Klaassen, D.M. de Leeuw and T. Welker, J. Lum. 31/32 , 687 (1984)
28. D.B.M. Klaassen, D.M. de Leeuw and T. Welker, J. Lum. 37, 21 (1987)
29. T. Oguchi and M. Tamatani, J. Electrochem. Soc. 133, 841 (1986)
30. H. Ito, Y. Yuge, A. Taya, M. Tamatani and K. Terasima, 'Book of Invited Papers and Abstracts of the 6th International Symposium on the Science and Technology of Light Sources', 67 (1992)
31. H. Bechtel, C.R. Ronda and T. Welker, 'Book of Extended Abstracts of the 178th Meeting of the Electrochemical Society', 927 (1990)
32. C.R. Ronda, H. Bechtel, U. Kynast and T. Welker, J. of Appl. Phys. 75, 4641 (1994)
33. M.L. Knotek and P.J. Feibelman, Surf. Sci. 90, 78 (1979)
34. M.L. Knotek, Phys. Today 37, 24 (1984)
35. V. Sa-Yakanit and H.R. Glyde, Comm. Cond. Mater. Phys. 13, 35 (1987)
36. J.L Sommerdijk, A.Bril and A.W. de Jager, J. Lum. 8, 341 (1974)
37. W.W. Piper, J.A. de Luca and F.S. Ham, J. Lum. 8, 344 (1974)

TECHNIQUES OF ULTRAFAST SPECTROSCOPY

Eli N. Glezer

Gordon McKay Laboratory
Division of Applied Sciences
Harvard University
9 Oxford Street
Cambridge, MA 02138

ABSTRACT

This chapter begins with a general introduction to ultrafast spectroscopy, considers the limits of time and frequency resolution, and reviews the linear and nonlinear propagation of light pulses in a dispersive medium. Next, the basic elements of ultrashort laser pulse generation are described, including gain medium requirements, mode-locking mechanisms, compensation for group velocity dispersion, and pulse amplification. The last section deals with the measurement of ultrashort pulses, including joint time-frequency techniques, and also describes pulse-shaping in the frequency domain.

I. INTRODUCTION

I. A. Introduction to Ultrafast Spectroscopy

The development of lasers over the last thirty five years has greatly expanded the range of physical phenomena accessible to optical measurements. Even before the invention of the first laser, classical spectroscopy was a well-developed science and an important experimental tool. Classical spectroscopy involves measuring the transmission, reflection or emission of light by some medium, as a function of the frequency of the light. Laser spectroscopy has made possible measurements which could not be made with other light sources. The advantages of using a laser source are many: narrow linewidth, tunability, spatial coherence, temporal coherence, and high intensity. The tunability and narrow linewidth, (today as narrow as 1 KHz or even 1 Hz for some frequency-stabilized lasers), has greatly increased the achievable spectral resolution and has allowed, for example, such measurements as the Lamb shift, observed in the fine structure of the Hydrogen absorption spectrum [1]. But it is not the narrow linewidth alone that has made many of the new measurements possible: the high intensity of laser light, especially when the laser is operated in a pulsed mode, has made possible nonlinear spectroscopy, with such techniques as saturation spectroscopy, coherent Raman spectroscopy, photon-echo, multiphoton absorption, and transient-grating, just to name a few. For example, in the Hydrogen fine structure measurements, saturation spectroscopy was used as a clever tool to circumvent Doppler broadening by probing only atoms at rest, because only they would be simultaneously resonant with narrow-frequency light from counter-propagating beams.

Ultrafast spectroscopy is a relatively recent branch of laser spectroscopy. Lasers have produced pulsed outputs since the first laser (Ruby pumped by a Xenon flashlamp), and production of much shorter pulses by Q-switching was demonstrated shortly after that. Since then, there has been a progression to shorter and shorter pulses, from nanoseconds, to picoseconds, to femtoseconds (fs). The term 'ultrafast' is generally applied to phenomena that occur on a femtosecond- or picosecond- time scale. In calling anything 'ultra' there is a risk that soon something 'more ultra' will be discovered. However, in the case of ultrafast spectroscopy the term may be justified, at least for visible and infrared light. Today's shortest optical pulses – less that 10 fs in duration – consist of only a few cycles of the electromagnetic field and thus require a bandwidth that is a sizable fraction of the central frequency. Figure 1 shows the electric field of a 7-fs pulse of 800-nm wavelength light. Because the duration of a single cycle of visible light is about 2 fs, producing pulses much shorter than a few femtoseconds will require going beyond the visible spectrum into the ultraviolet or X-ray region.

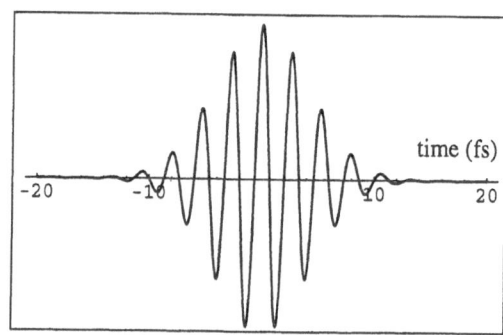

Figure 1. 7-fs pulse of 800-nm wavelength light.

Two questions fundamental to ultrafast optics are: how do we measure such short optical pulses when there are no electronic detectors with fast enough response times? and how do we use ultrashort pulses for studying ultrafast processes? The answer to the first question will be discussed in section III. A. The second question has been answered in a great variety of ways in clever experiments, but there is one general technique that includes a wide range of linear and nonlinear optical measurements: the pump-probe technique, illustrated in Fig. 2. In the simplest version, two ultrashort pulses are employed: a strong 'pump' pulse is used to excite the system of interest, and a weak 'probe' pulse that arrives at some later time is used to observe the resulting dynamics. By changing the time delay between the two pulses, the time evolution of the excited system can be mapped out. A more complicated pump may consist of multiple pulses, pulses of different frequencies, or specially tailored waveforms, and the probe may involve a variety of linear or nonlinear optical interactions, but the concept of separately inducing and monitoring dynamics still applies.

Ultrafast spectroscopy comes in two basic versions: 1. Spectroscopy in the time domain, and 2. Spectroscopy in frequency with time resolution. In the first case, the frequency information comes entirely from the temporal response of the system. An excellent example of this are impulsively driven lattice vibrations observed in time. Figure 3 shows oscillations in the reflectivity of an Antimony crystal as it responds to a 70-fs optical excitation, as

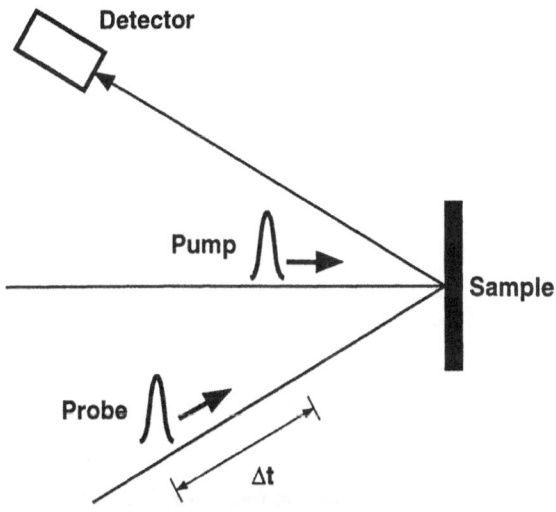

Figure 2. Pump-probe technique.

measured by varying the time-delay of a 70-fs probe [2]. By analyzing the frequency content of the response through a Fourier transform, shown in the inset, the authors identify which of the known phonon modes are driven by the excitation. In this case, only the A_{1g} symmetric breathing mode is excited, while other Raman-active modes are not. The authors are able to determine that the excitation is not through a Raman process, but instead is due to a near-instantaneous displacement of the potential surface seen by the atoms when electrons are transferred to higher energy states through direct absorption. As the lattice vibrates, the electronic bandstructure oscillates with it, causing the modulation in the reflectivity. In this example the frequency information in the THz range is obtained by simply Fourier transforming the temporal response.

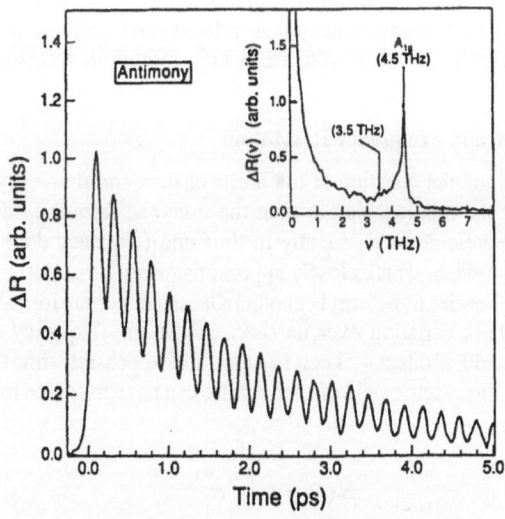

Figure 3. Example of time-domain spectroscopy (from [2]).

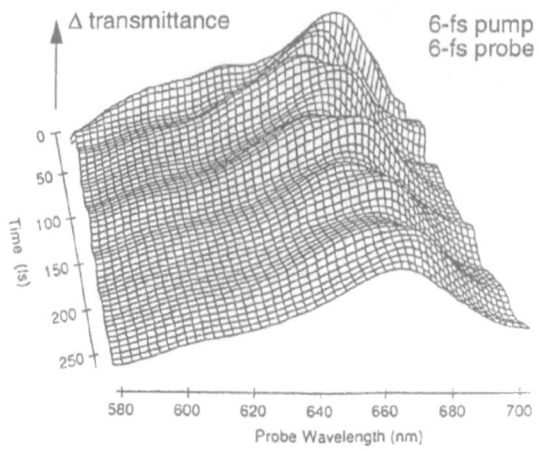

Figure 4. Example of spectroscopy with femtosecond time resolution (from [3]).

The second kind of ultrafast spectroscopy involves obtaining frequency information in the optical range by using a spectrometer to directly resolve the probe pulse into its spectral components. An impressive early example of this type of spectroscopy is the measurement of the vibration of an organic dye molecule in response to a 6-fs laser pulse excitation. Fig.4 shows the change in the transmittance of the dye solution as a function of both the pump-probe time-delay and the probe wavelength [3]. In fact, the entire wavelength range is covered by a single 6-fs probe pulse, which is spectrally dispersed after it is transmitted through the sample.

A 6-fs pulse in the visible range has a minimum bandwidth limit of over 50 nm. However, as we see in Fig. 4, it is certainly possible to do a measurement with higher frequency resolution than the minimum bandwidth of the probe pulse. This example may be a good backdrop for considering the question of fundamental limits of frequency and time resolution in ultrafast spectroscopy. What are these limits and how does the uncertainty principle enter into such measurements? A related, and perhaps more experimentally concrete question can be asked: if you use a high-resolution spectrometer, do you lose temporal resolution?

I. B. Limits of Time and Frequency Resolution

Before considering the question of the limits of time and frequency resolution, it may be useful to review the relationship between the time and frequency domains. Any pulse waveform can be represented equivalently in time and frequency domains. A very useful pulse shape is the Gaussian, which closely approximates many real laser pulses and also has the property that its Fourier transform is another Gaussian. A transform-limited pulse is one that has minimal phase variation over its spectrum (in the frequency domain) and has a minimal time-bandwidth product — i.e. it has the shortest possible time duration for a given spectral bandwidth. The electric field of such a pulse can be represented in the time domain as

$$E(t) \sim e^{-\frac{t^2}{2\sigma_t^2}} e^{i\omega_0 t} ,$$

(1)

and in the frequency domain as

$$E(\omega) \sim e^{-\frac{(\omega - \omega_0)^2}{2\sigma_\omega^2}}, \tag{2}$$

where σ_t characterizes the duration of the pulse, ω_0 is the center frequency, and σ_ω characterizes the width of the pulse spectrum. The pulse duration and spectral width are inversely related: $\sigma_t = 1/\sigma_\omega$. (Note that while the complex notation for the field is useful for representing amplitude and phase, only the real part of the complex quantity has physical significance.) The time and frequency domain representations are related by Fourier transforms:

$$E(\omega) = \frac{1}{\sqrt{2\pi}}\int_{-\infty}^{\infty} E(t)\, e^{-i\omega t}\, dt \tag{3}$$

$$E(t) = \frac{1}{\sqrt{2\pi}}\int_{-\infty}^{\infty} E(\omega)\, e^{i\omega t}\, d\omega. \tag{4}$$

These relationships hold generally and are not limited to Gaussian pulses. For illustration, two transform-limited Gaussian pulses are shown in both time and frequency in Fig. 5 and 6. Note that the shorter pulse has a broader spectral width.

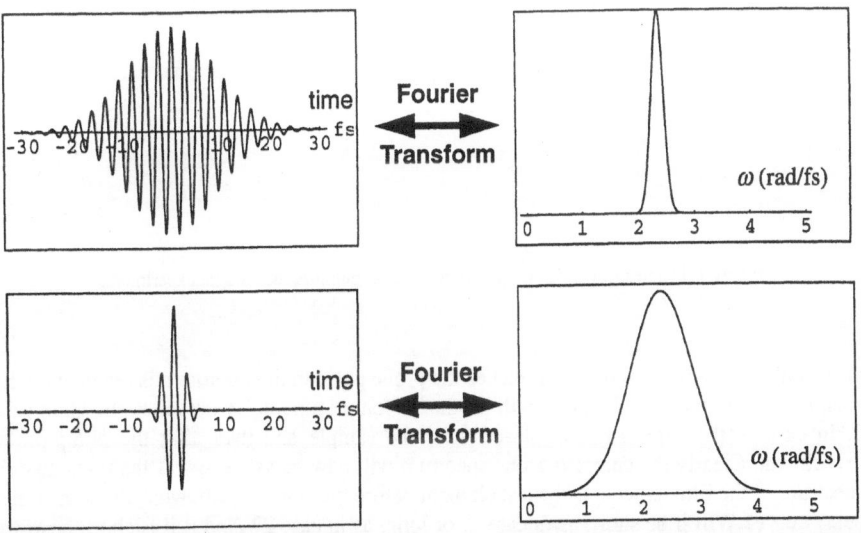

Figure 5. Time representation of two pulses.

Figure 6. Corresponding frequency representations.

Let's now conduct a thought experiment to provide a concrete example for the discussion of time and frequency resolution limits. Consider a pump-probe experiment performed on a material with an instantaneous response, where the medium is totally absorbing except when the presence of the pump pulse opens a window for the probe pulse. (Although bleaching and recovery of absorption in real materials cannot be instantaneous, it is possible to achieve a nearly instantaneous response for example by combining a nonlinear polarization rotation element between two crossed polarizers.) The detector in this experiment can be used to measure either the time-integrated probe intensity as a function of pump-probe time delay, or simply the spectrum of the transmitted probe. Case 1 of Fig. 7 shows the pump-probe setup and both the spectrum of the detected signal and the time delay curve. Now we add a spectral filter to the setup, case 2, which narrows the spectrum of the

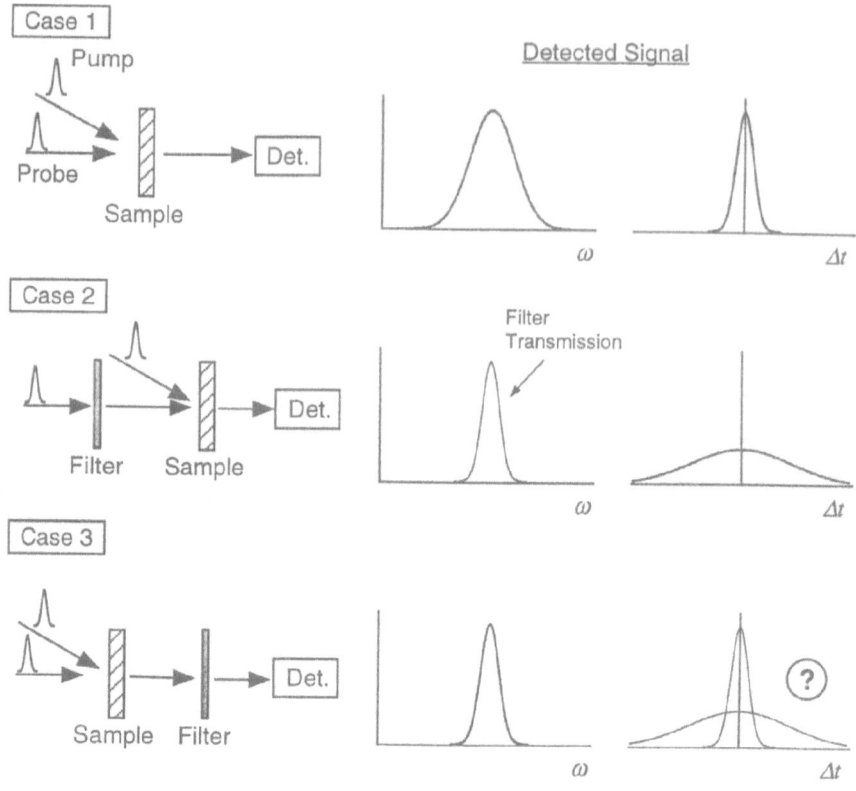

Figure 7. Limits of time and frequency resolution: pump-probe thought experiments.

probe pulse and correspondingly stretches the probe pulse in time before it is incident on the sample. The stretched probe pulse will produce a longer time-delay curve, as shown in case 2. In case 3, the same filter is placed after the sample and thus after the pump-probe experiment. Clearly the detected probe spectrum will now be as narrow as the transmission spectrum of the filter since it is the last element before the detector. But what about the time-delay curve? Will it be short, as in case 1, or long, as in case 2? And if it is short, have we somehow circumvented the uncertainty principle in getting the original time resolution for a narrower frequency band? It may be tempting to assume that the order of the sample and filter should not matter, and that the results of cases 2 and 3 should be the same. However, that assumption only holds for linear, time-invariant systems; the arrival of the pump pulse at the sample clearly disqualifies the time-invariance assumption.

To answer these questions let's consider in more detail what happens to the probe pulse in cases 2 and 3. Fig. 8 shows the temporal envelope of the probe pulse at several points in the experiment, as well as the pump-probe time-delay curve that would be measured. In case 2, the probe pulse is first stretched by the filter, but then only a short part of it is transmitted through the sample during the 'time window' opened by the pump pulse. (Note that this will actually make the spectral width of the probe broader than the filter transmission width). In case 3, the short pulse is transmitted through the sample and then stretched by the filter. However, the duration of the final probe pulse is not what is being measured. In fact, the duration of the transmitted pulse does not in any way affect the time-delay curve since the

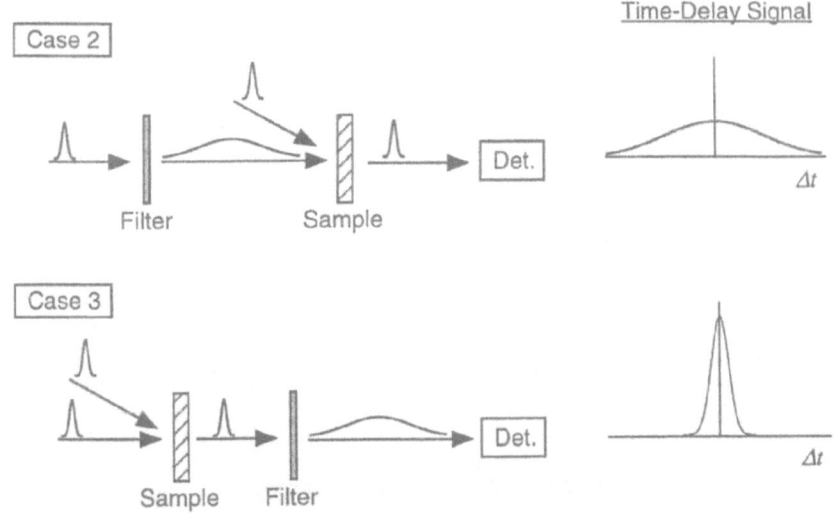

Figure 8. Pump-probe thought experiments: case 2 and 3 in more detail.

intensity of the pulse gets integrated by the detector anyway. Recall that the experimental time resolution is not due to a fast response time of the detector, but instead comes from scanning the pump-probe time-delay Δt. What matters is the duration of the probe pulse at the sample, and thus the time-delay curve in case 2 is long, but in case 3 is short. There is no loss in measured time resolution in case 3: $\Delta\omega_{filter}$ does not constrain the time resolution, no matter how narrow the filter bandwidth may be! We can replace the filter in case 3 with a spectrometer and an array of detectors, thus obtaining the same time resolution across the entire spectral range of the probe simultaneously. So what really limits the time-frequency resolution?

There is no fundamental limit to the time-frequency resolution in this measurement. Answering the posed question, we can say that using a high-resolution spectrometer does not destroy the temporal resolution in the pump-probe experiment. Instead, we need to focus our attention on the sample being studied, because it is the physical system itself that dictates the appropriate time and frequency resolution in the measurement.

Again let's consider a specific example: a fast, but not instantaneous, real saturable absorber as the sample in our thought experiment. The action of the saturable absorber is illustrated in Fig. 9. Before the arrival of the pump pulse the electrons are in their ground

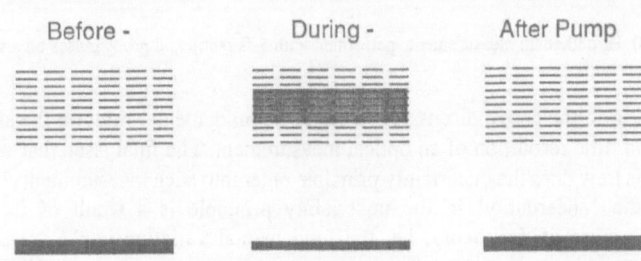

Figure 9. Very fast saturable absorber.

state, during the pulse they are promoted to higher lying states saturating the ability of the medium to absorb photons, and shortly after the pump pulse they return to the ground state. Fig. 10 shows two sets of hypothetical measurements performed with 5-fs pump and probe pulses. At 'negative time-delay' (–10 fs) the probe pulse arrives before the pump and thus sees the full absorption spectrum of the medium. At zero time-delay, the probe sees the full effect of the pump, which in the first case is a broad spectral hole in the absorption spectrum, whereas in the second case it is a set of spectrally narrow lines. At 10 fs after the pump pulse, the absorption has returned to its original state. It would appear that the second situation is an ideal case for employing a high resolution spectrometer in an ultrafast measurement. However, the second situation is impossible. A system cannot have sharp spectral features if it has such fast relaxation. This is why it is the system itself that dictates the appropriate time and frequency resolution in the measurement. A medium with a fast response must have a broad spectrum and thus a measurement would not profit from high frequency-resolution; similarly, sharp spectral features necessarily evolve slowly, and thus do not benefit from ultrafast time-resolution.

'Measurement': 5-fs pump/ 5-fs probe

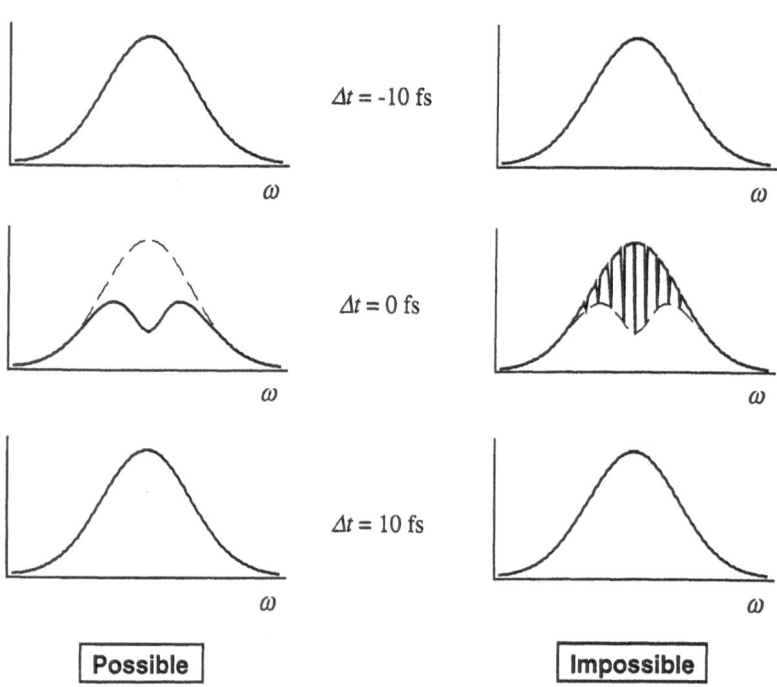

Figure 10. Hypothetical measurements performed with 5-fs pump and probe pulses on a very fast saturable absorber.

According to the above discussion, there is no fundamental limit on the simultaneous frequency and time resolution of an optical measurement. The final issue that was raised in Section I. A is how does the uncertainty principle enter into such measurements? In quantum mechanics, the 'uncertainty' in the uncertainty principle is a result of the inherently probabilistic nature of the theory, i.e. the fundamental variation in the measurement of physical observables represented by operators that do not commute. While there are mathematical parallels between quantum mechanics and time-frequency signal analysis, the

382

analogy cannot be taken too far. In a deterministic theory, such as the classical theory of electromagnetism, there really is no 'uncertainty principle', but rather just a time-bandwidth relation which says that a waveform and its Fourier transform cannot be made arbitrarily narrow simultaneously. But the waveform itself can be known exactly, and the resolution in a joint time-frequency representation (such as the one shown in Fig. 4) is unlimited. For a detailed and insightful review of joint time-frequency analysis and the connection to quantum mechanics see the review by Cohen [4].

I. C. Propagation of Light Pulses

The propagation of ultrashort pulses is governed by the same equations as the propagation of any light. However, some features of nonlinear and even linear propagation are more evident or more extreme for ultrashort pulses. In this section we will look at the effects of dispersion on a pulse, the meaning of group velocity in ultrashort pulse propagation, and the nonlinear effects of self-phase-modulation and self-focusing. We will also review the Wigner representation, which is useful in describing a pulse simultaneously in time and frequency.

1. Dispersion. Propagation of electromagnetic radiation is governed by the wave equation which is derived by combining Maxwell's equations. The wave equation for the electric field is

$$\nabla^2 \mathbf{E} - \frac{\mu \varepsilon}{c^2} \frac{\partial^2 \mathbf{E}}{\partial t^2} = 0, \tag{5}$$

and has the solutions

$$\mathbf{E} = \mathbf{E}_0 e^{i(\mathbf{k} \cdot \mathbf{x} - \omega t)}, \text{ where } k = \frac{\sqrt{\mu \varepsilon}}{c} \omega. \tag{6}$$

The electric field of the wave induces a polarization in the medium, proportional to the field, $P = \chi E$, where χ is the linear susceptibility and describes both the strength and phase of the polarization response. The electric field in the medium consists of the displacement field D, minus the response of the bound charges: $E = D - 4\pi P$. Writing D in terms of E, $D = E + 4\pi P = (1 + 4\pi \chi) E = \varepsilon E$, where ε is the dielectric constant. The dielectric constant is really not a constant, but a function of frequency, and characterizes the optical response of the medium. The complex index of refraction is defined as

$$\tilde{n} \equiv \sqrt{\mu \varepsilon}, \tag{7}$$

and if the magnetic permeability is $\mu = 1$,

$$\tilde{n} \equiv \sqrt{\varepsilon} . \tag{8}$$

The real and imaginary parts of k correspond to dispersion and absorption,

$$k = \frac{\tilde{n} \omega}{c} = \beta + i \frac{\alpha}{2}, \tag{9}$$

where α is the absorption coefficient, and is defined so that it gives the exponent of the decay of the intensity with propagation distance:

$$I = I_0 e^{-\alpha z}. \tag{10}$$

The dispersion of the medium is determined by β. If we consider propagation through a transparent medium of length L, then the field (in the frequency representation) will acquire the following frequency dependent phase:

$$E_{\text{out}}(\omega) = e^{i\phi(\omega)} E_{\text{in}}(\omega) = e^{i\beta(\omega)L} E_{\text{in}}(\omega).$$ (11)

It is useful to Taylor expand β near ω_0 to examine the effects of the coefficients on the pulse shape:

$$\beta(\omega) = \beta_0 + \beta'\Big|_{\omega_0}(\omega - \omega_0) + \frac{1}{2}\beta''\Big|_{\omega_0}(\omega - \omega_0)^2 + \dots.$$ (12)

The zero-order coefficient, β_0, is just a fixed phase shift that has no effect on the pulse shape. Notably, β' also has no effect on pulse shape, and is equivalent to propagation through vacuum. So β'' is the first term that alters the pulse shape; its effect on a transform-limited pulse is to introduce a 'chirp' — a spreading in time of the spectral components of the pulse. An example of the effect of chirp on a pulse is shown later in the top plot of Fig. 40.

2. Group Velocity. The group velocity determines the speed of propagation of the energy in a wave of a given frequency, and is thus an important physical quantity. It is defined as

$$V_g = \frac{\delta\omega}{\delta k} = [\beta']^{-1}.$$ (13)

We can put this into more useful form by differentiating $k = n\omega/c$

$$\frac{\delta k}{\delta\omega} = \frac{1}{c}\left[n + \frac{\delta n}{\delta\omega}\cdot\omega\right],$$ (14)

giving

$$V_g = \frac{c}{n + \dfrac{\delta n}{\delta\omega}\omega},$$ (15)

or in terms of wavelength,

$$V_g = \frac{c}{n - \dfrac{\delta n}{\delta\lambda}\lambda}.$$ (16)

The two dispersion regimes, $\delta n/\delta\omega > 0$ and $\delta n/\delta\omega < 0$ are called normal and anomalous dispersion, respectively. In the anomalous regime, V_g can become infinite. This occurs near a sharp resonance, and brings into question the physical meaning of an infinite group velocity. Of course energy does not propagate at an infinite velocity, so we must examine the assumptions behind the definition of V_g. In fact, the definition is only an approximation, and it breaks down when $k(\omega)$ varies too rapidly. The approximation requires that

$$\frac{\delta^2 k}{\delta\omega^2}\omega \ll \frac{\delta k}{\delta\omega}.$$ (17)

An explicit derivation of the group velocity of a waveform based on a plane wave decomposition can be found in [5].

3. Self-Phase-Modulation. Because of the high intensity of ultrashort pulses, nonlinear effects often play an important role in their propagation. Expanding the induced polarization in terms of field strength,

$$P = \chi^{(1)} E + \chi^{(2)} EE + \chi^{(3)} EEE + \dots , \tag{18}$$

the first term corresponds to ordinary linear propagation, and the rest describe the nonlinear response which introduces intensity dependent propagation and also couples together different frequencies. In general, the fields are vectors and the susceptibility coefficients are tensors. For example, the third order susceptibility, $\chi^{(3)}$, is a fourth order tensor. The range of nonlinear phenomena is very extensive, but here we will focus on just one particular third-order nonlinearity that leads to an intensity dependent modulation of the index of refraction.

Consider the following element of the third-order polarization:

$$
\begin{aligned}
P^{(3)}(\omega) &= 3\chi^{(3)} E(\omega) E^*(\omega) E(\omega) \\
&= 3\chi^{(3)} I(\omega) E(\omega)
\end{aligned}
\tag{19}
$$

where $I \equiv |E|^2$. Now combining this with the linear response, gives

$$
\begin{aligned}
P &= P^{(1)} + P^{(3)} \\
&= (\chi^{(1)} + 3\chi^{(3)} I) E \\
&= \chi_{\text{eff}} E
\end{aligned}
\tag{20}
$$

where χ_{eff} is an intensity-dependent 'effective susceptibility'. The index of refraction correspondingly becomes

$$n = \sqrt{\varepsilon} = \sqrt{1 + 4\pi\chi_{\text{eff}}}, \tag{21}$$

and can be expanded about its ordinary value if the nonlinear contribution is relatively small:

$$
\begin{aligned}
n &\approx \sqrt{1 + 4\pi\chi^{(1)}} + \frac{1}{2} \frac{4\pi 3\chi^{(3)} I}{\sqrt{1 + 4\pi\chi^{(1)}}} \\
&= n_0 + 6\pi \frac{\chi^{(3)}}{n_0} I \\
&= n_0 + n_2 I
\end{aligned}
\tag{22}
$$

This phenomenon is called the Kerr effect, and n_2 is referred to as the nonlinear index. The intensity dependence of the index of refraction produces a phase modulation throughout the duration of the pulse. We can calculate the phase delay, $-\phi$, due to propagation through a medium of length L, by relating it to the effective optical path, nL,

$$\frac{-\phi}{2\pi} = \frac{nL}{\lambda} \tag{23}$$

$$\phi = \frac{-2\pi}{\lambda} L (n_0 + n_2 I) . \tag{24}$$

The intensity-dependent phase modulation produces a time-dependent frequency shift

$$\Delta\omega = \frac{d\phi}{dt} = \frac{-2\pi}{\lambda} L n_2 \frac{dI}{dt} . \tag{25}$$

Figure 11 illustrates the frequency shift for a Gaussian pulse. For positive n_2, which is the case for most materials, the front part of the pulse is red-shifted (to lower frequencies), while the back part is blue-shifted. This results in a nearly linear frequency chirp across the central

385

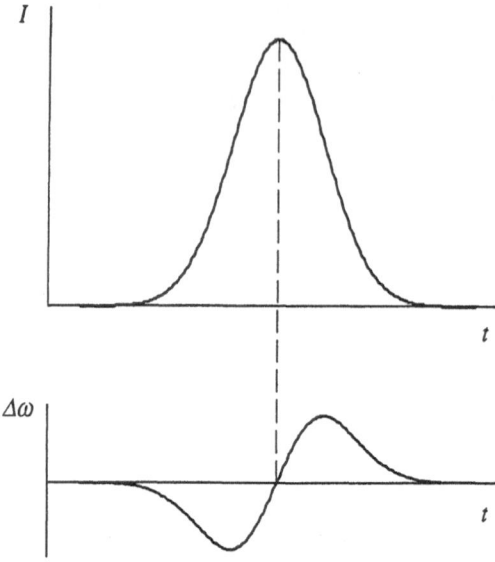

Figure 11. Self-phase-modulation.

part of the pulse. Self-phase-modulation plays an important role in mode-locking dynamics in ultrafast lasers, in propagation of solitons in optical fibers, and in white-light generation.

4. Self-Focusing. A closely related phenomenon to self-phase-modulation is self-focusing. Its source is the same third-order nonlinearity, i.e. the Kerr effect, except that it is the *spatial* variation in intensity, rather than temporal, that produces self-focusing. If the beam profile is given by $I(r)$, then the index of refraction is simply

$$n = n_0 + n_2 I(r) \quad . \tag{26}$$

A Gaussian beam propagating through this medium creates a lens for itself, as illustrated schematically in Fig. 12. Near its center, a Gaussian can be approximated by

$$I(r) = I_0 e^{-\frac{r^2}{a^2}} \approx I_0 \left(1 - \frac{r^2}{a^2} \right), \tag{27}$$

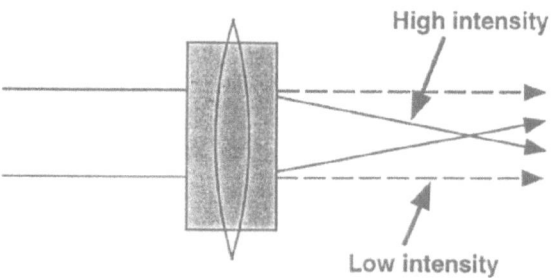

Figure 12. Self-focusing.

where a characterizes the beam width. This intensity distribution creates a quadratic variation in the index of refraction, resulting in a lens-like effect:

$$n(r) = n_0 + n_2 I_0 \left(1 - \frac{r^2}{a^2} \right). \tag{28}$$

Self-focusing is used in Kerr-lens mode locking, described in Section II. A.2.

5. Joint Time-Frequency Representation. While either time or frequency representation of a pulse offers a complete description, in some cases it can be insightful to use a joint time-frequency representation. Several such representations exist, (see for example [4]), but here we'll describe one that's referred to as the Wigner representation [6] or more specifically, the chronocyclic representation when it is applied to electromagnetic pulses (for a comprehensive review see [7]). The Wigner function is calculated from the field as represented in either frequency or time:

$$\begin{aligned}
W(t, \omega) &= \frac{1}{2\pi} \int_{-\infty}^{\infty} E\left(\omega + \frac{\omega'}{2} \right) E^*\left(\omega - \frac{\omega'}{2} \right) e^{-i\omega' t} d\omega' \\
&= \int_{-\infty}^{\infty} E\left(t + \frac{t'}{2} \right) E^*\left(t - \frac{t'}{2} \right) e^{-i\omega t'} dt'
\end{aligned} \tag{29}$$

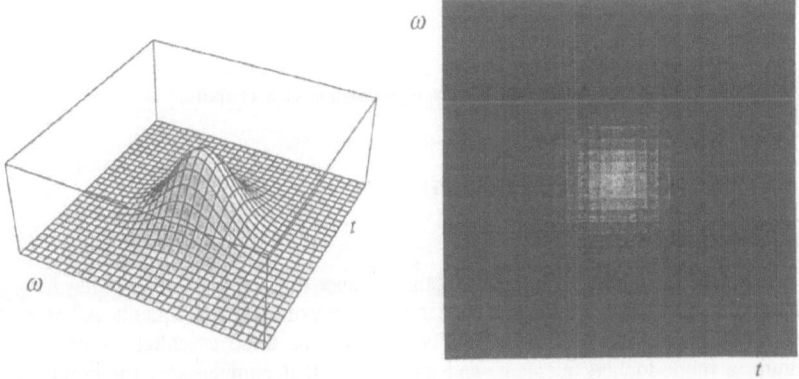

Figure 13. Wigner representation of a Gaussian pulse as a surface in 3-D and as a density plot.

The Wigner function attempts to represent the spectral and temporal distribution of the energy, or the 'chronocyclic intensity', as a two-dimensional real function (although it can take on negative values), while treating time and frequency on an equal footing. Some of its properties include: a time integral gives the intensity spectrum,

$$\int_{-\infty}^{\infty} W(t, \omega)\, dt = |E(\omega)|^2 = I(\omega), \tag{30}$$

integrating in frequency gives the intensity profile,

$$\frac{1}{2\pi} \int_{-\infty}^{\infty} W(t, \omega)\, d\omega = |E(t)|^2 = I(t), \tag{31}$$

and the area integral in both time and frequency gives the energy in the pulse,

$$\frac{1}{2\pi}\int_{-\infty}^{\infty}\int_{-\infty}^{\infty} W(t, \omega)\, dt d\omega = \text{Energy} \qquad (32)$$

In order to satisfy the limits imposed by Fourier relations, the Wigner function must be non-zero in a phase-space area larger than π. A transform-limited Gaussian pulse is shown in the Wigner representation in Fig. 13 as a 3-D plot and as a density plot. Figure 14 shows a chirped pulse, and demonstrates the immediate graphic appreciation of the pulse properties when seen in this representation. Finally, the Wigner representation provides a close connection to the joint time-frequency measurements discussed in Section III. A.

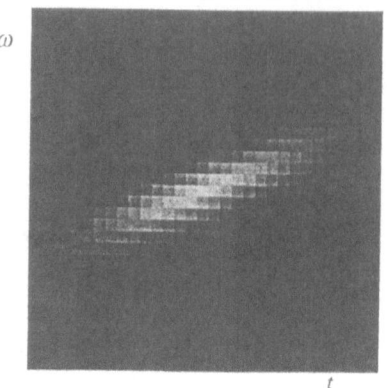

Figure 14. Wigner representation of a chirped pulse.

II. PRODUCING FEMTOSECOND PULSES

II. A. Femtosecond Lasers

Today there are many types of lasers that produce femtosecond pulses. They have some common principles of operation that appear in a variety of real embodiments. A femtosecond laser cavity, shown schematically in Fig. 15 contains three essential elements: a gain medium, a mode-locking element, and an element that compensates for group velocity dispersion (GVD) in the rest of the cavity. The gain medium has a broad gain bandwidth to allow a large number of modes (frequencies) to lase simultaneously. When properly locked in phase, these modes add up to a short pulse that circulates in the cavity. A fraction of this pulse is ejected out of the cavity through an output coupler, providing a train of pulses spaced by the round-trip time in the cavity.

Two examples of real femtosecond laser systems are shown in Fig. 16a & b. Fig. 16a

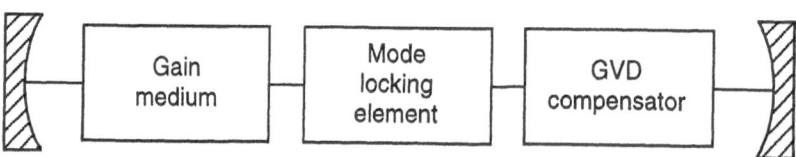

Figure 15. Basic elements of a femtosecond laser cavity.

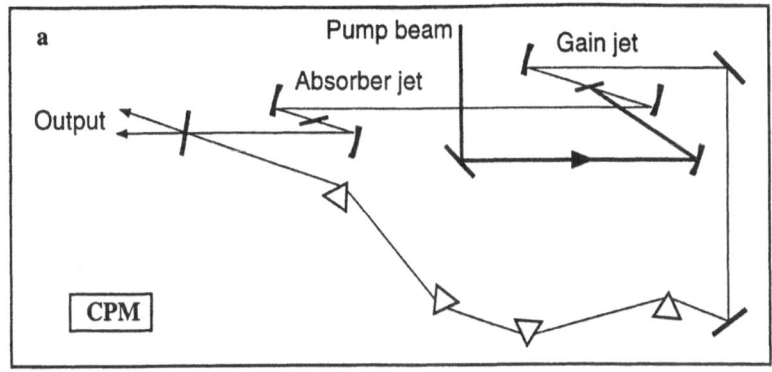

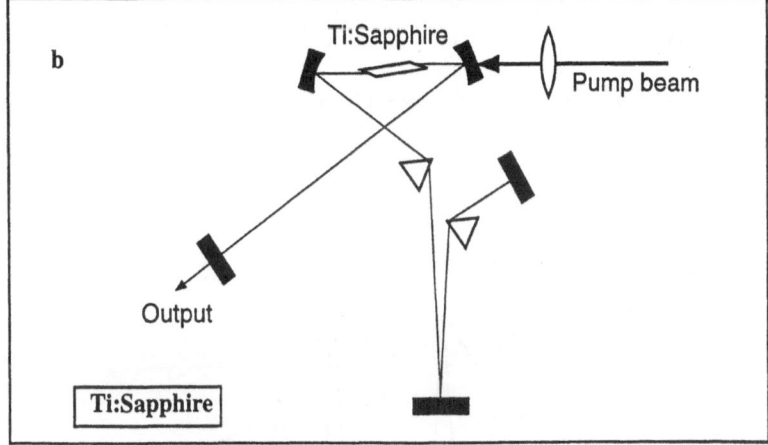

Figure 16. Two examples of a femtosecond laser: **a**. dye-based laser; **b**. solid-state laser.

is a dye-based laser system, which was the first type of laser to produce ultrashort pulses [8]. The gain medium is a jet of an organic dye in solution, most commonly Rhodamine 6G in ethylene glycol, pumped continuously by an Argon-Ion laser. The mode locking is provided by another jet of a different dye which acts as a saturable absorber. Finally, a sequence of prisms provides the GVD compensation. Fig. 16b is a solid-state based laser, where the gain medium in a doped crystal, in this case Titanium-doped Sapphire (Ti:Al$_2$O$_3$). This type of laser was first found to produce mode-locked pulses in 1991 [9], and since then there has been tremendous growth in solid-state ultrashort-pulse lasers based on a wide variety of materials, cavity types and geometries, and mode-locking methods. In the example shown in Fig. 16b, the Ti:Al$_2$O$_3$ crystal provides both the gain and mode locking in the laser. The prisms again provide the GVD compensation. In the following sections we will examine in more detail gain, mode locking, and GVD compensation in an ultrashort-pulse laser.

1. Gain Medium. The first requirement for the gain medium is that it have a broad gain spectrum. A narrow-band gain medium could not possibly produce ultrashort pulses, because a wide range of Fourier components is required to add up to a short pulse. A figure of merit has been suggested for comparing different gain media [10]:

$$M = \sigma \tau \Delta v, \tag{33}$$

where σ is the peak stimulated emission cross-section, τ is the lifetime of the inverted population and $\Delta\nu$ is the width of the fluorescence line. The larger the product $\sigma\tau$, the lower the pumping threshold. Note that the lifetime of the inverted population does not need to be short in order to produce ultrashort pulses. The larger the $\Delta\nu$, the shorter the pulse duration limit or alternatively, if the bandwidth is broader than necessary for the desired pulse duration, it can be used to provide tunability in frequency. Other desirable features in a gain medium include: convenient absorption band for pumping, photochemical stability, and of course mechanical stability.

As a contrast to ultrashort-pulse lasers, it may be useful to review gain in a CW (continuous wave) laser. The gain and loss spectra in an ideal CW laser are shown in Fig. 17. A number of cavity modes are shown within the gain spectrum, spaced by $\Delta\omega = 2\pi(c/2L)$, where $2L$ is the round-trip length of the cavity. The shown gain spectrum assumes that the lasing line is homogeneously broadened, so that spectral hole burning is not possible, and the line shape is fixed. A further assumption is that there is no spatial hole burning, so that the gain spectrum is constant throughout the gain medium. The main observation is that in steady state, the highest-gain mode cannot exceed the loss, and thus there will only be a single lasing mode! The other modes will see a gain smaller than the loss, and will not lase.

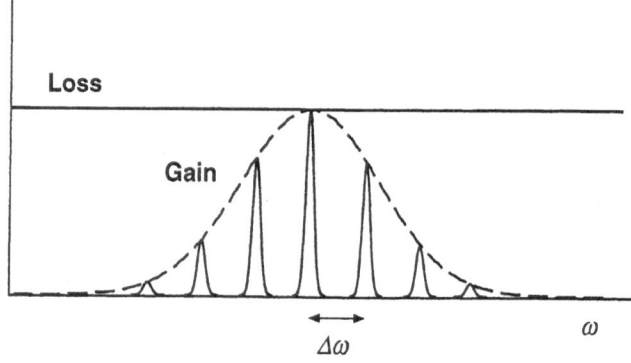

Figure 17. Gain and loss spectra in a CW laser.

Single frequency operation is desirable for narrow-linewidth CW laser, but is quite the opposite to what is required for producing ultrashort pulses. A femtosecond laser must have many lasing modes (typically on the order of 10^4-10^5), and these modes must all be locked in phase.

2. Mode Locking. The term 'mode locking' refers to establishing a desired fixed phase relationship between the different frequency modes in the laser cavity. The field can be written in the frequency domain as a sum of n cavity modes,

$$E(\omega) = \sum_n E_n(\omega_n) e^{i\phi_n}. \tag{34}$$

If the ϕ_n are random, then the laser output will just be a continuous fluctuating signal. If the ϕ_n are all zero on the other hand, then the modes will add up constructively to form a short pulse oscillating in the cavity. An additional effect of mode locking is to broaden the lasing spectrum to include a much larger number of modes.

Although 'mode locking' suggests a frequency domain perspective, the frequency description is not convenient for dealing with the propagation of ultrashort pulses inside the cavity because a very large number of modes is involved and they are nonlinearly coupled. For this reason a time-domain description is more useful and insightful.

The mode locking discussion in this section follows an excellent review article on the subject by Ippen [11]. The essential principle in mode locking is providing a mechanism for time dependent modulation of loss and/or gain. This modulation may be active, for example by including an externally driven modulator in the cavity, or it can be passive, by having an element in the cavity that lets the pulse itself vary the gain or loss in time so that the laser prefers to operate in a pulsed mode.

Figure 18 shows the gain and loss in an actively modulated cavity as a function of time. In the time window where the gain exceeds the loss a pulse will form as a result of the modulation. However, active mode locking is not used to produce ultrashort pulses because the speed of modulation is essentially limited by the response time of the modulator.

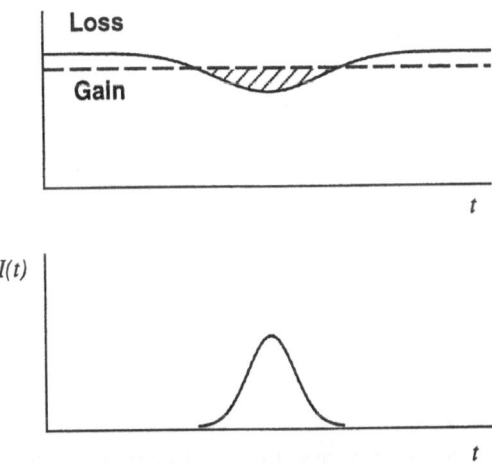

Figure 18. Active mode locking.

The basic approach in passive mode locking is to use a saturable absorber — an element whose ability to absorb is reduced at high intensity. The effect of the saturable absorber is to favor the propagation of a short burst of light over continuous lasing in the cavity. There are two classes of saturable absorbers, 'slow' and 'fast', characterized by their response time relative to the duration of the ultrashort pulse that is produced. In both cases, the response time is fast relative to the cavity round-trip time.

In the case of a slow saturable absorber, its action needs to be supplemented by the depletion of gain in the gain medium. Let's assume that a pulse somehow forms in the cavity, and examine the resulting gain and loss in the cavity. (Initial pulse formation will be discussed later). Figure 19 shows the pulse intensity envelope together with the gain and as a function of 'local time', $t - x/v_g$, so that the combined effect of two elements can be shown even though they are spatially and temporally separated. Despite the 'slow' recovery of both the absorption and the gain, the combined action leads to a fast modulation of the difference, as shown in the third graph.

The saturation of absorption results from exciting a large fraction of the valence electrons before they have a chance to relax to the ground state. Similarly, the depletion of the gain results from stimulating a large fraction of the inverted population to radiatively recombine, before the inversion can be reestablished by the pump source. Assuming the recovery is slow, the saturation of absorption and the depletion of the gain can be modeled as

$$l(t) = l_0 e^{-\sigma_a \int_{-\infty}^{t} |E(t')|^2 dt'}$$

(35)

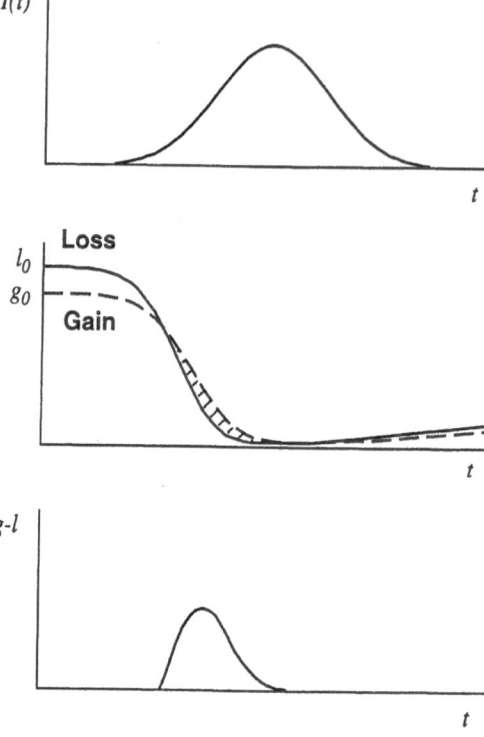

Figure 19. Passive mode locking with the combined action of 'slow' saturable absorption and gain depletion.

$$g(t) = g_0 e^{-\sigma_g \int_{-\infty}^{t} |E(t')|^2 dt'}$$ (36)

where l_0 and g_0 are the initial (or 'low intensity') loss and gain, and σ_a and σ_g characterize the saturation and depletion. The requirements for mode locking are that $l_0 > g_0$, so that there is no continuous lasing, and $\sigma_a > \sigma_g$, so that the absorption is saturated before the gain is depleted.

In the case of a fast saturable absorber, the gain medium does not need to play a role in the mode locking. This greatly expands the range of gain media that can be used in ultrafast lasers. Figure 20 shows a pulse and the resulting gain and loss in a laser with a fast saturable absorber and no depletion of gain.

In the limit of instantaneous response time of the absorber, the decrease in the loss simply follows the pulse intensity:

$$l(t) = l_0 - \gamma |E(t)|^2.$$ (37)

However, no real absorber has an instantaneous response. Absorption consists of electrons making transitions to excited states, and it takes time for them to relax to the ground state (at least a few picoseconds). Thus it would seem that the recovery timescale precludes the possibility of a nearly instantaneous saturable absorber, like the one in Fig 20. However, a clever new approach has been invented to get around the recovery time limit. It was first discovered experimentally [9], and even called 'magic mode locking' before the mode locking mechanism was understood.

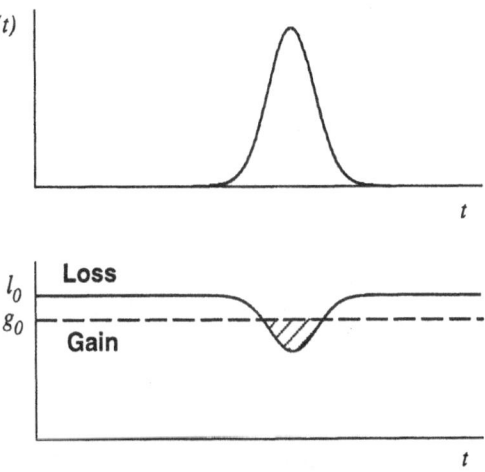

Figure 20. Fast saturable absorber.

The trick is to use nonlinear phase modulation and convert it into amplitude modulation. Phase modulation can be nearly instantaneous because it does not require absorption; by using a fast, nonresonant nonlinearity it is possible to modulate the real part of the susceptibility without affecting the imaginary part. For example, self-focusing and self-phase-modulation in a transparent material do not induce absorption — the imaginary part of the susceptibility remains zero. But how can we convert the induced changes in the index of refraction into amplitude modulation to produce an artificial 'instantaneous saturable absorber'? One approach is to combine self-focusing with a physical aperture in such a way as to select the self-focused propagation mode of a short intense pulse over the non-self-focused mode of continuous lasing. This type of mode locking is called Kerr-lens mode locking (KLM) and is the mechanism responsible for 'magic mode locking'.

Figure 21 shows two possible arrangements combining intensity-dependent focusing with an aperture [12]. The 'lens' is not a real lens, but the effect of self-focusing, and thus is only seen by the short, intense pulse; the low intensity continuous radiation is not affected. Of the two modes shown in each of the two cases, the mode for the pulse is more focused than the mode for continuous lasing. Both cases produce the smallest loss for the ultrashort-pulse mode and mimic the action of an almost instantaneous saturable absorber. Note that the nonlinear effect itself is not dissipative — amplitude modulation is achieved by coupling with other elements such as an aperture. An alternative configuration uses the pump laser beam as a gain aperture, avoiding the need for a loss aperture.

Self-phase-modulation is used in other mode-locking schemes with coupled cavities or cross-polarized modes in the same cavity, and is called additive pulse mode locking (APM) or coupled cavity mode locking (CCM). The most significant application of these mode-locking techniques has been in fiber-based lasers because self-focusing cannot be used in fibers, and also because the long fiber length, cavity stability, and good polarization control makes fiber lasers good candidates for APM. For more information on various types of mode locking see [11].

In our discussion of passive mode locking we have considered how a short pulse affects the loss and gain in the laser cavity. The discussed conditions are necessary for the ultrashort pulse to be a stable solution, but we have not considered whether the pulse will evolve from initial fluctuations. The ability of a laser to mode lock itself is called 'self-starting'. Whether

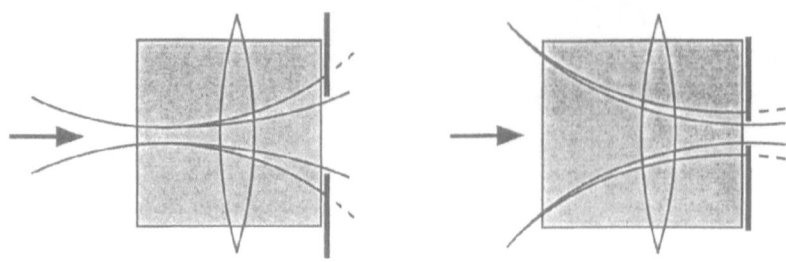

Figure 21. Examples of KLM with combinations of a self-focusing element and an aperture.

a laser is self-starting depends on whether an initial fluctuation will shorten within a cavity coherence time. If this does not happen, the fluctuation will be dispersed through random phase fluctuations, and the laser will not be self-starting. Pulse-shortening rate, a non-dimensional quantity, can be used to illustrate pulse duration limits and also to characterize the ability of a mode-locking mechanism to self-start. Figure 22 shows the pulse-shortening rate, $-\Delta\tau/\tau$, vs. $1/\tau$ for different mode-locking mechanisms (from [11]). Active mode locking is most effective for long pulses (small $1/\tau$), whereas a fast saturable absorber is most effective for short pulses (large $1/\tau$). So while a fast saturable absorber is ideal for sustaining ultrashort pulses, it usually requires an external perturbation to initiate the mode locking. In practice, this perturbation may be provided by the vibration resulting from a light tap on the laser platform, or a more sophisticated mechanism with feedback control that will automatically restart the mode locking if for some reason it is interrupted.

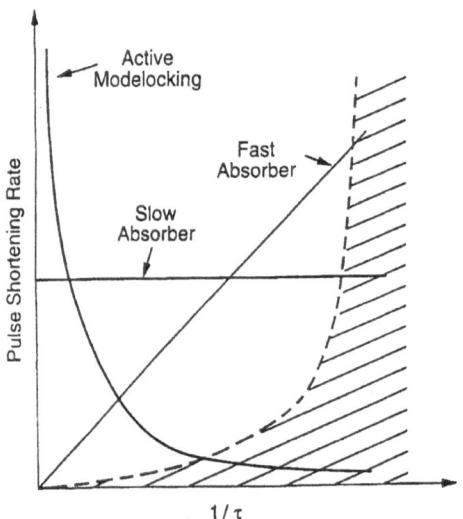

Figure 22. Pulse-shortening rate for different mode-locking mechanisms (from [11]).

3. GVD Compensation. In order to produce ultrashort pulses it is necessary to minimize the group velocity dispersion (GVD) in the laser cavity, so that the broad range of spectral components in the ultrashort pulse can all propagate at the same velocity (averged over a cavity round-trip). There are two effects that require compensation: the material dispersion in the cavity (a linear effect), and self-phase-modulation (a nonlinear effect).

Material dispersion in a solid state laser is usually dominated by the gain medium, but also includes contributions from all other elements in the cavity. Figure 23 shows the group velocity in fused silica glass as a function of frequency (expressed in units of photon energy). Other transparent materials produce very similar curves. GVD is just the derivative of the group velocity, $dV_g/d\omega$. The dispersion results from resonances that lie outside of the transparent range; the negative slope at visible frequencies is caused by electronic transitions further out in the ultraviolet, while the positive slope on the infrared side is due to optically active vibrational transitions in the far infrared. A balance between these effects occurs at the peak of the group velocity, where GVD is zero, at 1.3 μm (0.9 eV) in fused silica. For lasers operating in the visible or in the very near infrared, the GVD causes the 'blue' (higher frequency) part of the spectrum to travel more slowly than the 'red' (lower frequency) part.

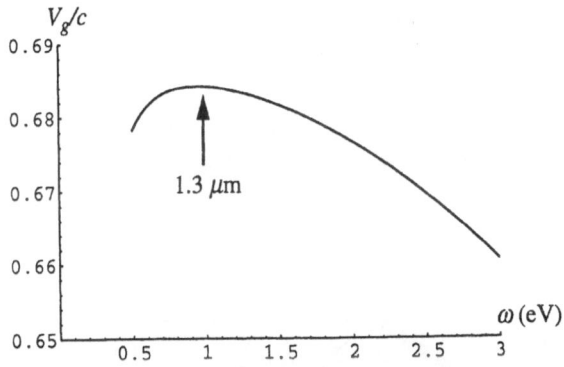

Figure 23. Group velocity in fused silica glass.

Self-phase-modulation introduces a frequency depended phase shift, which has an effect similar to group velocity dispersion (in addition to producing new spectral components). Because it is a nonlinear effect, it cannot just be generally represented as a simple curve in frequency, but depends on the duration, intensity, and spectrum of the pulse itself.

Two methods of compensating for GVD have been devised involving arrangements of gratings or prisms. Different colors take different paths as shown in Fig. 24. The basic idea is to set up a geometric arrangement to allow the blue part of the spectrum to 'catch up' to the red part. With grating pairs, this is easy to see intuitively, simply by comparing the optical path taken by different frequency components. With prisms the situation is much more tricky, as we'll see below.

To calculate the optical path through a pair of anti-parallel gratings, we can use simple geometry. Because the gratings are anti-parallel, and we use the first order diffracted light, all spectral components will be propagating in the same direction after the first pair of gratings. The second pair serves to recombine the spectral components back into a single beam, as well as doubling the temporal effect of the first pair. (Alternatively, a mirror can be used to send the spectral components back though the first pair of gratings.)

To compare path lengths for different frequencies, we can calculate the path length to a plane perpendicular to all the spectral components, and then (by symmetry) just double that to get the full path length. Figure 25 shows the path l, consisting of segments l_1 plus l_2, where

$$l_1 = \frac{L}{\cos \theta_r} \tag{38}$$

$$l_2 = l_1 \cos (\theta_i - \theta_r) \tag{39}$$

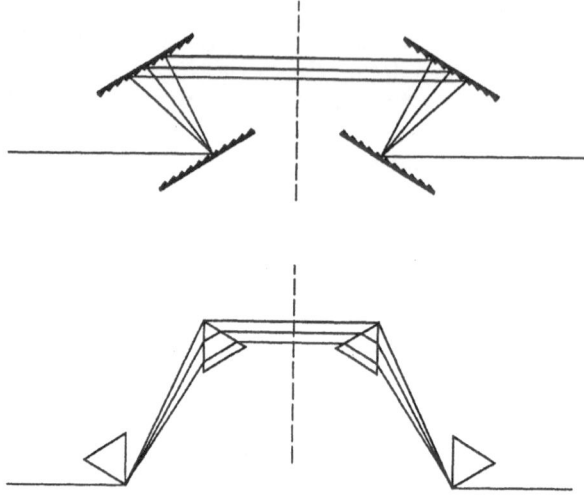

Figure 24. GVD compensation with gratings and prisms.

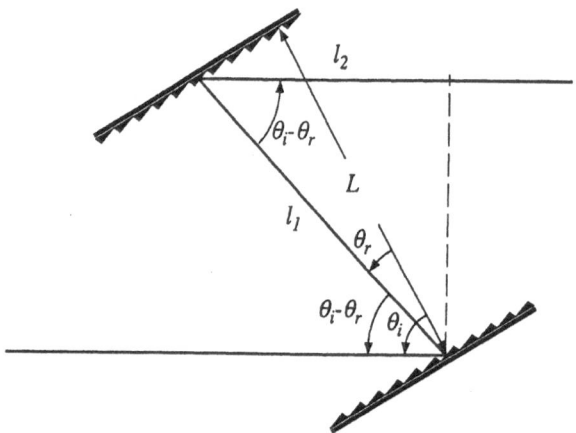

Figure 25. Schematic for calculating grating pair dispersion.

The reflected and incident angles θ_i and θ_r are related by

$$\sin \theta_r = -\sin \theta_i + \frac{m\lambda}{\Lambda},$$

(40)

where m is the order of the diffraction (in this case $m=1$), λ is the wavelength, and Λ is the groove spacing. For a proper choice of angles, l decreases almost linearly with ω, about a central frequency ω_0, giving a shorter path for the higher frequency components.

We now turn to prisms, to examine their effect on the dispersion, following the treatment of Fork, Martinez, and Gordon [13]. In Fig. 26 we assume that the prisms are used at the angle of minimum deviation for the central frequency. (At this angle the optical path

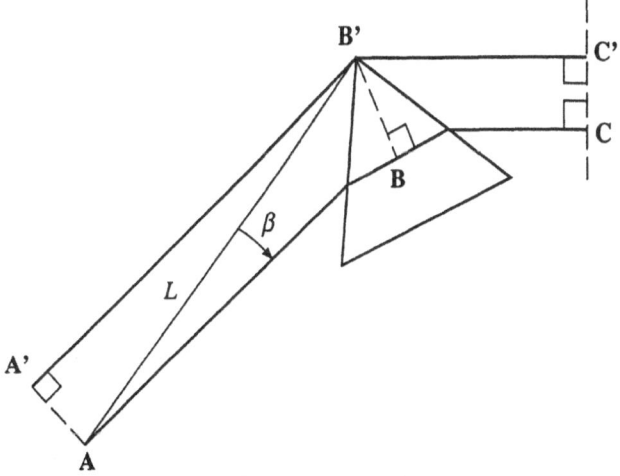

Figure 26. Schematic for calculating prism pair dispersion.

inside the prism is perpendicular to the angle bisector). Point A is the location of the apex of the first prism, where the beam is incident, and L is the distance to the apex of the second prism. To calculate the actual optical beam path ABC, we can use an equivalent fictitious path through the vertex of the second prism, A'B'C'. The two optical path lengths are equivalent, because they start and end at a common phase front. We now note that two *different* frequencies will have the same optical path length from the plane through BB' to the plane through CC'. So the frequency dependent part of the optical path can just be taken as the distance from A to B, or equivalently from A' to B'. This 'prism-pair' path is simply

$$l_{\text{p-p}} = L \cos [\beta(\omega)] \tag{41}$$

where β is the angle between the apex-to-apex connector and the actual beam path AB. The frequency dependence of the index of refraction affects the optical path length through β. The second pair of prisms simply doubles the effect of the first pair, while also recombining the spectral components into a single beam.

Without doing any calculations, we can see that because the index of refraction increases with frequency, a greater ω produces a smaller β, thus a larger $\cos[\beta]$ and hence a larger $l_{\text{p-p}}$. So the optical path *increases* with frequency! In other words, the derivative of the path length with frequency is positive. But this is the opposite of what we were interested in, and the opposite of the effect of gratings! So how is it possible for the higher frequencies to 'catch up' to the lower frequencies if the optical path length increases with frequency?

The short answer is that it is the second derivative of the phase that determines the dispersion, and not the first derivative of the optical path. The phase acquired by traveling through the four prisms is

$$\phi(\omega) = \frac{\omega}{\varepsilon} 2 l_{\text{p-p}} = \frac{\omega}{\varepsilon} 2 L \cos [\beta(\omega)] . \tag{42}$$

The condition for the desired dispersion (for the higher frequencies to 'catch up' to the lower frequencies) is

$$\frac{d^2 \phi}{d\omega^2} < 0 . \tag{43}$$

Note that it is possible to meet this condition despite having a path length that increases with frequency.

This short answer may seem unsatisfying, because the optical path length may be a more intuitive concept than phase. Let's examine in more detail the relationship between the effective optical path l_{eff}, phase ϕ, and group velocity V_g. For concreteness, we can consider the specific example of a pulse traveling through a block of transparent material of length l_0, with a frequency-dependent index of refraction n, so that $l_{eff} = n l_0$. The input field will acquire a frequency dependent phase,

$$E_{out}(\omega) = e^{i\phi(\omega)} E_{in}(\omega) . \tag{44}$$

The phase is related to the optical path by

$$\phi = k l_0 = \frac{2\pi}{\lambda} l_0 = \frac{2\pi}{\lambda_0} l_{eff} = \frac{\omega}{c} l_{eff}, \tag{45}$$

where λ is the wavelength in the material and λ_0 is the wavelength in vacuum. The first and second derivatives of the phase are:

$$\frac{d\phi}{d\omega} = \frac{1}{c} [l_{eff} + l'_{eff}\omega] , \tag{46}$$

$$\frac{d^2\phi}{d\omega^2} = \frac{1}{c} [2l'_{eff} + l''_{eff}\omega] . \tag{47}$$

Now we turn to the group velocity:

$$V_g = \frac{d\omega}{dk} = \left[\frac{dk}{d\omega}\right]^{-1} = \left[\frac{d\phi}{d\omega}\right]^{-1} l_0$$

$$= \frac{c l_0}{[l_{eff} + l'_{eff}\omega]} \tag{48}$$

The group velocity dispersion (GVD) is

$$\frac{dV_g}{d\omega} = \frac{-c l_0}{[l_{eff} + l'_{eff}\omega]^2} [2l'_{eff} + l''_{eff}\omega]$$

$$= \frac{-l_0}{\left(\frac{d\phi}{d\omega}\right)^2} \cdot \frac{d^2\phi}{d\omega^2} \tag{49}$$

The second derivative of the phase is often referred to as group delay dispersion (GDD). Thus we see that GVD and GDD are related by

$$GVD = \frac{-l_0}{\left(\frac{d\phi}{d\omega}\right)^2} \cdot GDD . \tag{50}$$

Table 1 shows the sign of the GVD, GDD, and first and second derivatives of the optical path for both prism and grating pairs. From eq. (49) it is apparent that the sign of the GVD depends on both the first and second derivatives of the optical path. In the case of gratings, the dominant term in the GVD is l'_{eff}, and thus the sign of the GVD is determined by the sign of l'_{eff}. In the case of prisms, on the other hand, at the appropriate angle, $l''_{eff}\omega$ can have a larger magnitude than $2l'_{eff}$, thus giving the counter-intuitive result of higher frequencies arriving earlier, despite traveling a longer optical path.

Table 1. GVD, GDD, and optical path derivatives.

	$\dfrac{dV_g}{d\omega}$	$\dfrac{d^2\phi}{d\omega^2}$	l'_{eff}	l''_{eff}
Gratings	+	-	-	small
Prisms	+	-	+	-

In practice, grating pairs are used when large dispersion is required, such as in stretching a pulse by a large factor before sending it into an amplifier, or compressing it afterwards, as will be discussed in the next section. Inside a laser cavity however, prisms are usually used because they produce much smaller loss and the small dispersion compensation they provide is sufficient.

An interesting new method for GVD compensation that does not use gratings or prisms was demonstrated by Stingl et al. [14]. Recognizing that reflection off dielectric mirrors can produce negative GDD (or positive GVD), they have designed specially coated multilayer mirrors that provide a nearly constant negative GDD across a very broad wavelength range (~200 nm). A Ti:Sapphire laser with mirror-controlled dispersion has produced pulses of only 8-fs duration. An advantage of this type of dispersion compensation is that unlike the prisms pairs, it is largely independent of the cavity alignment.

II. B. Amplification

Femtosecond pulses can be amplified outside of the laser cavity. Solid-state femtosecond lasers produce pulses that are typically a few nJ in energy, while dye-based femtosecond lasers produce pulses of even lower energy. The pulse repetition rate is determined by the cavity round-trip time, and is typically on the order of 100 MHz. Often, higher pulse energy at a lower repetition rate is desired. Before describing specific types of amplifiers, we review the concepts of gain narrowing and gain saturation, and how they apply to ultrashort pulses.

The gain of an amplifier is the ratio of the output intensity to the input intensity, and has a spectrum determined by the line shape of the amplifying medium. If the spontaneous emission line shape is given by $\alpha(\omega)$, then the gain over a distance z is

$$G(\omega) = \frac{I_{out}(\omega)}{I_{in}(\omega)} = e^{\alpha(\omega)z}. \tag{51}$$

If the line shape is Lorentzian (homogeneous broadening), then

$$\alpha(\omega) = \frac{N_2 - N_1}{1 + \dfrac{(\omega - \omega_0)^2}{\left(\dfrac{\Delta\omega}{2}\right)^2}}, \tag{52}$$

where $N_2 - N_1$ is the population inversion, ω_0 is the center frequency, and $\Delta\omega$ is the line width.

Amplification often reduces the spectral width of the input signal. This is called gain narrowing. To illustrate this effect, in Fig. 27 we plot the normalized gain spectrum for amplification by a factor of 10, 100, and 10,000. At high amplification, the gain narrowing can be significant. Because ultrashort pulses necessarily have a broad spectrum, the gain medium in the amplifier should have a gain bandwidth that is as broad as possible, otherwise

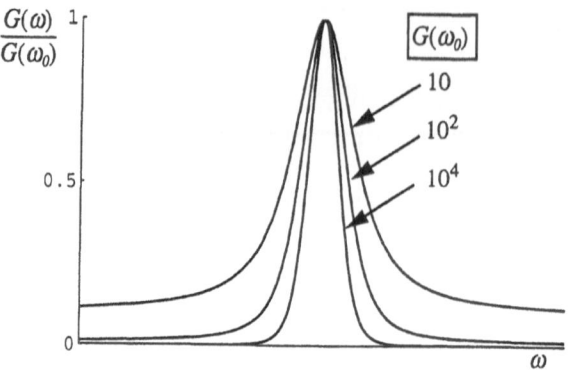

Figure 27. Gain narrowing in an amplifier.

the amplified pulses will have a longer duration than the input pulses.

Perhaps even more important than gain narrowing is the effect of gain saturation in an amplifier. For a homogeneously broadened medium, the following formula for saturation can be found in most laser textbooks:

$$\alpha = \alpha_0 \frac{1}{1 + I/I_{sat}} \tag{53}$$

where α_0 is the gain coefficient for a small signal, and I_{sat} is the saturation intensity. The gain coefficient decreases at higher intensity, producing smaller and smaller gain. Can this formula be used for ultrashort pulses? For example, is it possible to avoid saturation by stretching a pulse in time (thus reducing the intensity), then amplifying it, and then compressing it?

To answer this question we should examine what causes the gain saturation. The saturation occurs because stimulated emission in the amplifier reduces the population inversion and thus reduces the gain. With long pulses or continuous radiation the maximum intensity is limited by the pump rate, i.e. the rate at which the population inversion is being replenished. The underlying assumption here is that the system is in equilibrium: the rate of emission cannot exceed the pumping rate. This assumption holds as long as the pulse duration is much longer than the gain recovery time. (Note that in steady state, when the pump rate balances the decay rate, the gain recovery time equals the lifetime of the inverted population.)

With femtosecond and even picosecond pulses, the assumption of equilibrium is certainly false for almost all gain media. For example, in Ti:Sapphire the gain recovery time is three microseconds, so the pump rate is so slow that on the timescale of an ultrashort pulse it has no effect on the population inversion. In this opposite limit, where the pulse is much shorter than the population recovery time, the emission can certainly exceed the pump rate, and the only limit is that the total number of photons in the pulse cannot exceed the number of electrons in the inverted state. Thus, in amplifying ultrashort pulses, the proper measure of saturation is a saturation fluence, rather than a saturation intensity:

$$\alpha = \alpha_0 \frac{1}{1 + F/F_{sat}} \tag{54}$$

Thus we see that stretching an ultrashort pulse, even say by a factor of 1000, will not avoid saturation, because it is the fluence rather than the intensity that determines the saturation. (One can of course produce greater energy pulses by magnifying the beam and using an amplifier with a larger cross-section, thus reducing the fluence.) The saturation fluence for dyes in solution is several mJ/cm^2, whereas for doped solids, such as Ti:Sapphire, it is much greater, several J/cm^2 [15].

Yet there is a different reason why stretching the input pulse to an amplifier is useful: nonlinear effects, such as self-focusing, are not desirable in the amplifier, and can be avoided if a pulse is stretched in time. A pulse can be stretched reversibly by imposing a 'chirp' on it. The term 'chirp' refers to a linear spread in time of the frequencies in the pulse, and results from constant (non-zero) GDD. A negative chirp, where the higher frequency components are ahead of the lower frequency components, can be achieved with a grating pair, as discussed in the previous section. To recompress the pulse after the amplifier we would need to provide positive chirp. How can we impose a positive chirp on a pulse? One approach is to pass the pulse through a dispersive material. But if a large positive chirp is desired, the required material length may be too long.

An inventive approach to creating positive chirp uses a pair of gratings in a two-lens system. Figure 28 shows two lenses in a so called '4-F' or 'two-lens correlator' arrangement. This type of system thus finds great use in optical spatial information processing, for such operations as Fourier filtering, convolution, and correlation. Its main feature is that it performs a Fourier transform from the input plane to the second plane, and then again from the second plane to the output plane. For our purposes, we note that it is an inverting imaging system (from input to output planes), that all rays take equal length optical paths, and that the slope of the rays is 'inverted' from input to output.

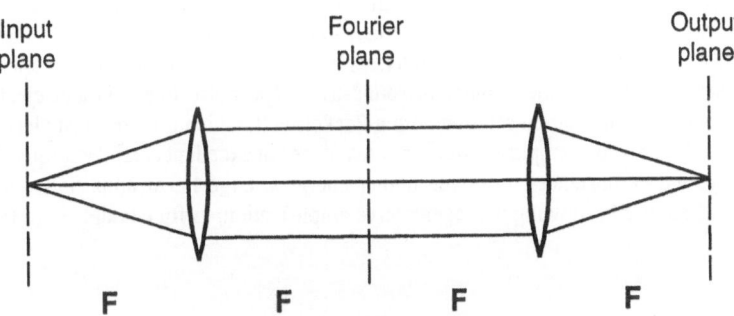

Figure 28. Two-lens correlator.

We now modify the two-lens system by adding two gratings and a mirror, as shown in Fig. 29. In this arrangement higher frequencies are made to take a longer path. The positions at which two spectral components ('red' and 'blue') have traveled equal paths are marked in the drawing as P_r and P_b. From those points to the mirror the 'blue' beam has a longer path than the 'red'. Note that if the second grating is placed at the output plane, then no dispersion is introduced. Thus by adjusting the position of the second grating one can control the amount of chirp. Positive chirp is obtained by placing the grating in front of the output plane, and negative chirp is obtained by placing the grating behind the output plane. This arrangement was originally designed to compensate for the dispersion in an optical fiber in the 1.3 – 1.6 μm wavelength range, where the material dispersion is negative [16].

In chirped pulse amplification (CPA) [17], such a method for introducing positive chirp is used to stretch an ultrashort pulse by more than 1000 times, the pulse is then amplified, and a grating pair is then used to recompress the pulse. Curved mirrors can be used instead of lenses in the stretcher to avoid spectral aberrations, and cylindrical rather than spherical optics can be used. A carefully designed stretcher-amplifier-compressor system that is dispersion compensated to fourth order (i.e. $d^2\phi/d\omega^2 = d^3\phi/d\omega^3 = d^4\phi/d\omega^4 = 0$) can stretch an

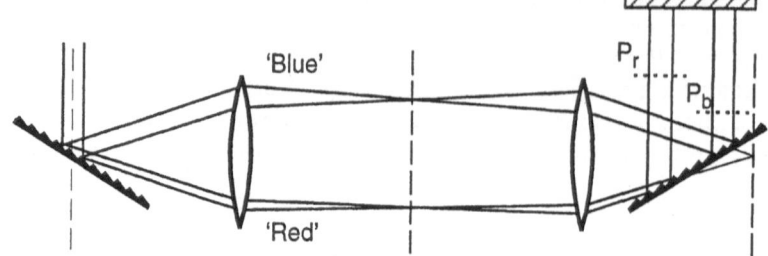

Figure 29. Pulse stretcher.

input pulse of 25 fs to 300 ps, amplify it, and recompress it back to about 27 fs [18].

Different types of amplifiers are used depending on the desired energy and pulse repetition rate. Figure 30 shows a representative selection of current state-of-the-art laser/amplifier systems producing sub-100-fs pulses. The 100 MHz and 1 GHz points are the direct (unamplified) output from the laser cavity of two Ti:Sapphire lasers, one optimized for maximum energy [19], the other optimized for compactness and high repetition rate [20]. The point at 400 KHz represents the output of CPA system with a regenerative amplifier pumped continuously by an Argon ion laser [21]. Similar amplifiers, pumped by Q-switched lasers, produce higher energy pulses in the low kHz range [22]. The regenerative amplifier is essentially a second laser cavity into which laser pulses are injected and later ejected after the gain is saturated after some number of round-trips. The pulse injection and ejection are accomplished by polarization rotation with a Pockels cell, which provides fast electro-optic switching. Regenerative amplifiers have the advantages of excellent beam mode quality, high efficiency, and compactness. For even higher energies, large linear amplifiers (single- or multi-pass) are used following the regenerative amplification, as for example in [23].

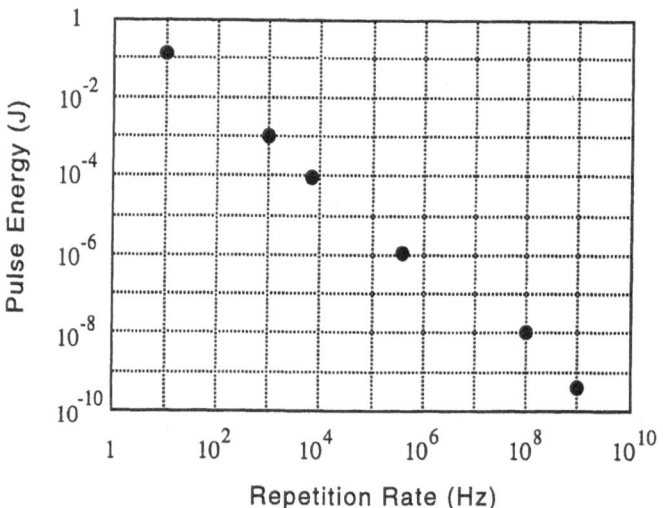

Figure 30. Lasers and amplifiers: state of the art for sub-100-fs pulses

III. MEASURING AND SHAPING FEMTOSECOND PULSES

We now return to a question posed in the introduction: how do we measure ultrashort optical pulses when there are no electronic detectors with fast enough response times? Since the development of ultrashort-pulse sources, a great deal has been learned about how such pulses can be characterized, so that recently it has even become possible to measure the exact electric field of an ultrashort pulse. In addition to detailed characterization, it is now possible to shape the pulses in frequency and time, producing, for example, controlled multiple-pulse sequences out of a single pulse, or imposing a specific phase variation on an originally unchirped pulse.

III. A. Pulse Characterization

1. Pulse Spectrum. The simplest characterization of a pulse is its spectrum. An example of a spectrum typical of the shortest pulses (about 10 fs) from Ti:Sapphire lasers is shown in Fig. 31. The width of the spectrum places a limit on the shortest possible pulse duration, as discussed in Section I.B. For a Gaussian pulse the limit is $\sigma_t = 1/\sigma_\omega$. At this point it may be useful to mention that pulse durations and spectral widths are often given as 'FWHM' — full width at half maximum of the intensity. For a Gaussian pulse

$$E(t) \sim e^{\frac{-t^2}{2\sigma_t^2}} \quad \text{and} \quad I(t) \sim e^{\frac{-t^2}{\sigma_t^2}}$$

$$\frac{I\left(\frac{\Delta t_{FWHM}}{2}\right)}{I(0)} \equiv \frac{1}{2} \tag{55}$$

$$\Delta t_{FWHM} = 2\sqrt{\ln 2}\,\sigma_t$$

Similarly, in frequency

$$\Delta\omega_{FWHM} = 2\sqrt{\ln 2}\,\sigma_\omega. \tag{56}$$

While it does set a limit on the shortest possible pulse duration, the pulse spectrum cannot be used to measure pulse duration because the spectrum contains no phase information.

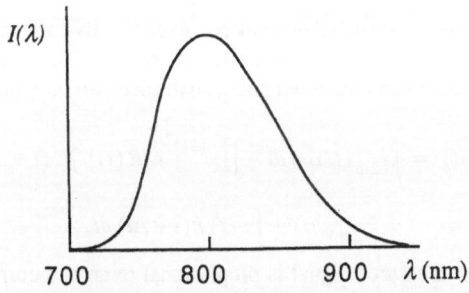

Figure 31. Typical spectrum of an ultrashort pulse from a Ti:Sapphire laser.

2. Temporal Characterization. To measure the pulse in time we can use the pulse itself. Figure 32 shows one simple technique that measures the temporal autocorrelation of a pulse. The pulse is split into two parts, one of which passes through an adjustable delay line. The two parts are then focused together onto a second-harmonic crystal, producing three beams at the second-harmonic frequency.

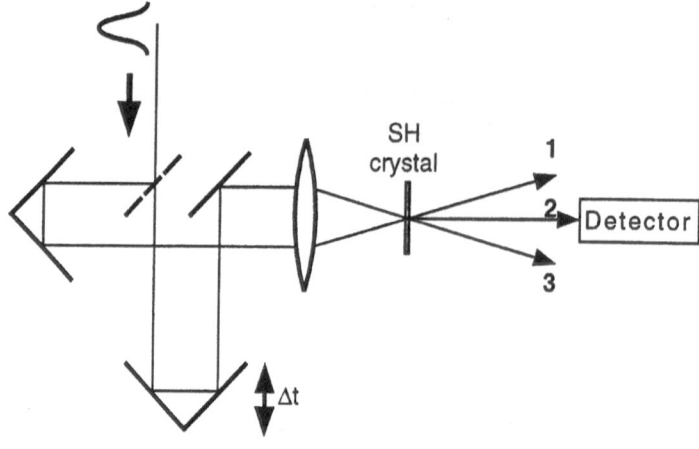

Figure 32. Non-collinear autocorrelation.

This arrangement is the so-called 'non-collinear' or 'background-free' autocorrelation. By measuring the energy in beam #2 as a function of the time-delay (Δt) we obtain an autocorrelation of the intensity profile of the pulse. To show how this occurs, we consider the field incident on the SH crystal:

$$E_{tot}(t, \Delta t) = \frac{1}{\sqrt{2}}E(t) + \frac{1}{\sqrt{2}}E(t + \Delta t).$$ (57)

The second harmonic field is proportional to the second-order susceptibility times the square of the incident field. (This is true only in the limit where only a small fraction of the incident energy is converted into the second-harmonic).

$$E_{2\omega} \sim \chi^{(2)} E^2_{tot}.$$ (58)

The second-harmonic intensity is proportional to the square of the second-harmonic field:

$$I_{2\omega}(t, \Delta t) \sim \left|\chi^{(2)}\right|^2 \left|E^2_{tot}\right|^2 = \left|\chi^{(2)}\right|^2 \left|E^2(t) + 2E(t)E(t+\Delta t) + E^2(t+\Delta t)\right|^2.$$ (59)

So far we have not made the spatial dependence explicit. By measuring the energy in beam #2, we effectively select only the second term in eq. (59). The detector response is much longer than the duration of an individual pulse, so the detector integrates the intensity in time:

$$F_{2\omega}(\Delta t) = \int I_{2\omega}(t, \Delta t)\, dt \sim \int \left|\chi^{(2)}\right|^2 4|E(t)|^2|E(t+\Delta t)|^2 dt$$ (60)

$$F_{2\omega}(\Delta t) \sim \int I(t)\, I(t+\Delta t)\, dt.$$ (61)

Thus we see that the measured signal is proportional to an autocorrelation of the intensity envelope of the pulse. Figure 33 shows the electric field of a 7-fs pulse, and the corresponding intensity autocorrelation.

An alternative pulse-duration measurement based on second-harmonic generation employs a colinear arrangement in which all 3 terms of eq. (59) are spatially coincident. As shown in Fig. 34, a Michelson interferometer is used to split the pulse into two, delay one part relative to the other, and then colinearly recombine the two. After the SH crystal, a filter

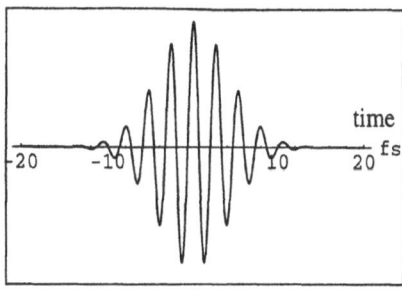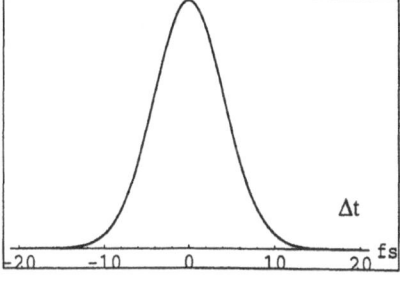

Figure 33. 7-fs pulse and the corresponding SH intensity autocorrelation

is used to block the beam at the fundamental frequency and transmit the SH beam. In this arrangement the 3 SH beams are colinear, and thus add coherently. The SH intensity is given by

$$I_{2\omega}(t, \Delta t) \sim \left|\chi^{(2)}\right|^2 \left|E^2(t) + 2E(t)E(t + \Delta t) + E^2(t + \Delta t)\right|^2$$

$$
\begin{aligned}
I_{2\omega}(t, \Delta t) \sim \left|\chi^{(2)}\right|^2 \{ &|E(t)|^4 + |E(t + \Delta t)|^4 + 4|E(t)|^2|E(t + \Delta t)|^2 \\
&+ E^2(t)\,[E^2(t + \Delta t)]^* + \text{c. c.} \\
&+ E^2(t)\,2E^*(t)\,E^*(t + \Delta t) + \text{c. c.} \\
&+ 2E(t)E(t + \Delta t)\,[E^2(t + \Delta t)]^* + \text{c. c.} \}
\end{aligned}
$$

(62)

where c.c. stands for complex conjugate. The detected integrated intensity is again a function of the time-delay, Δt,

$$F_{2\omega}(\Delta t) = \int I_{2\omega}(t, \Delta t)\, dt.$$

(63)

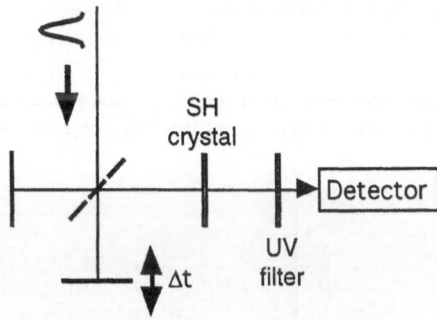

Figure 34. Interferometric autocorrelation.

Figure 35 again shows the electric field of a 7-fs pulse, and the corresponding 'interferometric autocorrelation', also referred to as a 'fringe-resolved autocorrelation'. The time-delay curve and even its envelope are not simple autocorrelations, but the width of the envelope is related to the width of the pulse. At $\Delta t = 0$ all the terms in equation (62) give non-zero contributions to the detected signal,

$$F_{2\omega}(\Delta t = 0) \sim 16 \int |E(t)|^4 dt,$$ (64)

while at $\Delta t = \pm\infty$ only the first two terms in equation (62) contribute, giving

$$F_{2\omega}(\Delta t = \pm\infty) \sim 2 \int |E(t)|^4 dt.$$ (65)

This gives a ratio of 8 between the peak at $\Delta t = 0$ and the background signal at long time-delay when the two pulses are well-separated in time; for this reason the interferometric autocorrelation is sometimes referred to as the '8-to-1 curve'.

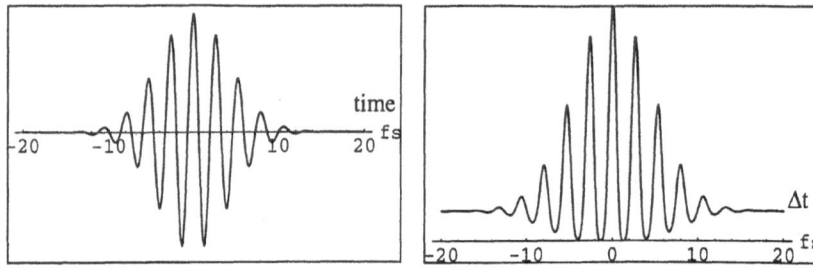

Figure 35. 7-fs pulse and the corresponding SH interferometric autocorrelation

One of the two autocorrelations described above is sometimes more appropriate than the other, depending on what properties of the pulse are of interest. The advantages of the interferometric autocorrelation include: the self-calibrating nature of the curve due to the presence of the fringes, the lack of temporal distortion due to the zero relative angle between the two beams (only significant for the very shortest pulses), and the 8-to-1 ratio, which serves as a check that the second-harmonic signal is not affected by other factors (such as saturation or self-focusing in the crystal). The main advantage of the intensity autocorrelation is that it is 'background-free', due to the non-colinear geometry. This is particularly useful in checking for small pre- or post-pulses, which can be more easily detected without a background which unavoidably has some noise.

Both autocorrelation techniques described above use a nonlinear crystal to produce a second-harmonic signal. We now ask the question: Is the nonlinear crystal necessary?

Figure 36 shows the same Michelson interferometer setup, except this time the SH crystal and the filter have been removed, so the detector now measures the integrated

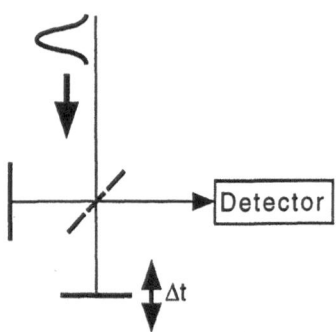

Figure 36. Interferometric time-delay setup without a nonlinear element.

intensity at the fundamental frequency, again as a function of time-delay (Δt). The intensity is proportional to the square of the sum of the fields,

$$I_\omega \sim |E(t) + E(t + \Delta t)|^2,$$ (66)

and the detected signal is thus:

$$F_\omega(\Delta t) = \int I_\omega(t, \Delta t)\, dt$$ (67)

$$F_\omega(\Delta t) \sim \int \{|E(t)|^2 + |E(t + \Delta t)|^2 + E(t)E^*(t + \Delta t) + E^*(t)E(t + \Delta t)\}\, dt.$$ (68)

The first two terms are just a constant 'background' independent of (Δt), while the last two are interference terms. Figure 37 shows a 7-fs pulse and the corresponding 'interference trace' as a function of time-delay. Figure 38 shows a transform-limited 14-fs pulse (of narrower spectral width), and its interference trace. The broader envelope of the time-delay trace corresponds to the longer pulse duration. From this example it is tempting to conclude that the linear interference trace can be used to measure pulse duration, and that the nonlinear crystal is unnecessary.

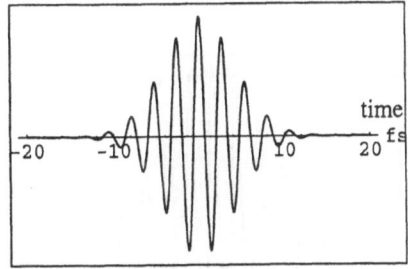

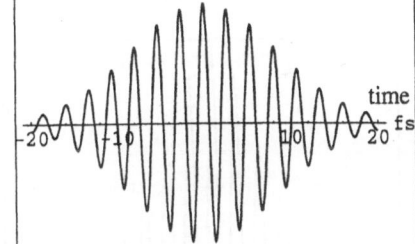

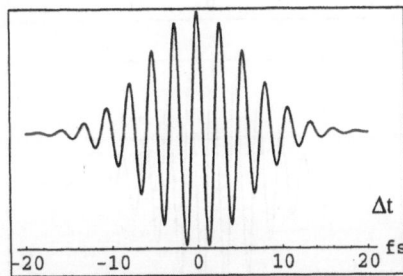

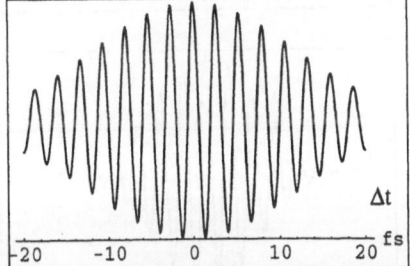

Figure 37. 7-fs transform-limited pulse and its linear interference trace.

Figure 38. 14-fs transform-limited pulse and its linear interference trace

But now we consider a different case. The original 7-fs pulse is passed through a dispersive medium, thereby stretching it in time while leaving its spectral width unaffected. The two pulses appear in Figs. 39 and 40, which also show the background-free SH, fringe-resolved SH, and the linear interference time-delay traces for both the 7-fs transform-limited pulse, and its 'stretched' or 'chirped' version   While both of the second-harmonic autocorrelations are broader for the longer pulse, the linear interference traces in the two cases are identical! This may seem counter-intuitive, because after all, why shouldn't a longer pulse produce a 'longer' interference trace?

A qualitative explanation for this is that the 'red' part of the pulse 'does not interfere'

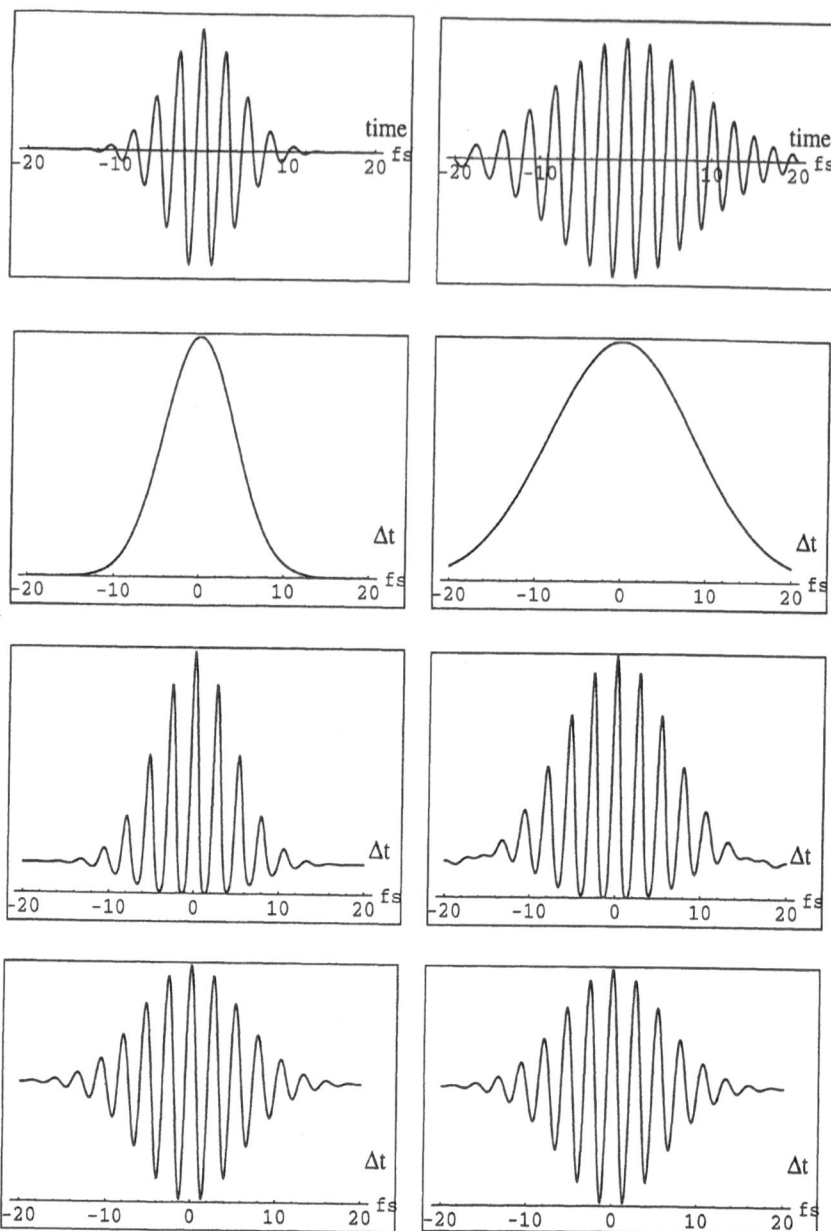

Figure 39. 7-fs pulse, with its SH intensity autocorrelation (AC), SH interferometric AC, and linear interference trace.

Figure 40. Same pulse after propagating through a dispersive medium, with its SH intensity AC, SH interferometric AC, and linear interference trace.

with the 'blue' part. If the front end of the pulse contains the lower frequency components, when it overlaps in time with the back end of its copy which contains the higher frequency components, there will be no significant constructive or destructive interference, i.e. the time-integrated intensity will not be significantly changed. While this explanation may give some insight, it gives only a rough justification since it mixes frequency and time descriptions of the pulse.

To prove that chirp does not affect the linear interference trace, we treat the pulse in its frequency representation. As the original pulse, $\tilde{E}(\omega)$, propagates through a dispersive medium, it acquires a frequency-dependent phase factor, producing $\tilde{E}(\omega)\,e^{i\phi(\omega)}$. We now consider the interference terms in eq. (68),

$$\int \{E^*(t)\,E(t+\Delta t) + \text{c. c.}\}\,dt.\tag{69}$$

Using the correlation theorem,

$$g_1(t) \otimes g_2(t) \equiv \int g_1(t+\Delta t)\,g_2^*(t)\,dt = \mathcal{F}^{-1}\{\tilde{g}_1(\omega)\,\tilde{g}_2^*(\omega)\},\tag{70}$$

(where the $\tilde{g}$ designates the Fourier transform of g), we rewrite the interference terms for the chirped pulse as

$$\int E_{\text{chirped}}(t+\Delta t)\,E^*_{\text{chirped}}(t)\,dt = \mathcal{F}^{-1}\{\tilde{E}(\omega)\,e^{i\phi(\omega)}\,\tilde{E}^*(\omega)\,e^{-i\phi(\omega)}\}$$

$$= \mathcal{F}^{-1}\{|\tilde{E}(\omega)|^2\}\tag{71}$$

$$= \int E(t+\Delta t)\,E^*(t)\,dt$$

Thus we have proved that the linear interference time-delay trace is completely determined by the intensity spectrum — and unaffected by the phase.

3. Joint Time-Frequency Characterization. We have seen that a nonlinear element can be used to obtain an autocorrelation of an ultrashort pulse, and thus provide a measure of its duration. When more detailed information about the electric field of a pulse as a function of time is desired, measurements that combine temporal and spectral resolution can be employed. The general approach is to use a pulse split into two in a time-delay arrangement, just as in the autocorrelations described above, but this time to measure the spectrum of the nonlinear signal rather than just its energy. This is illustrated schematically in Fig. 41.

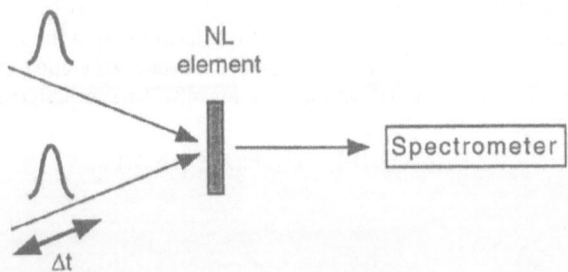

Figure 41. General approach to joint spectral-temporal pulse characterization.

One specific example of joint time-frequency analysis is called 'frequency-resolved optical gating' (FROG), [24] shown schematically in Fig. 42. The 'optical gating' refers to using one part of the pulse as a gate for the other. This is accomplished with a third-order nonlinear element, with the gate pulse (#2) used to rotate the polarization of the 'direct' pulse (#1). As can be seen in Fig. 42, the sequence of crossed polarizers will block pulse #1, unless its polarization is slightly rotated by the gate pulse. If the response of third-order element is essentially instantaneous, i.e. non-resonant, (a simple piece of glass is often used), then the

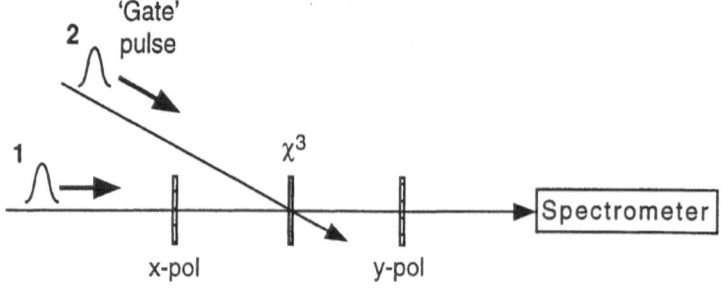

Figure 42. Frequency-resolved optical gating (FROG).

field in the transmitted beam will be proportional to the square of the magnitude of the electric field of the gate pulse. Note that although the transmitted signal is of the same frequency as the original pulses, it does involve a nonlinear process. Assuming the gate pulse is just a time-delayed version of the 'direct' pulse, the transmitted signal is

$$E_{\text{signal}}(t, \Delta t) \sim \chi^{(3)} E(t) \left| E(t + \Delta t) \right|^2. \tag{72}$$

The spectrometer performs a Fourier transform on the signal and the array of detectors in the spectrometer measures the time-integrated spectral intensity:

$$
\begin{aligned}
F_{\text{FROG}}(\omega, \Delta t) &\sim \left| \chi^{(3)} \right|^2 \left| E_{\text{signal}}(\omega, \Delta t) \right|^2 \\
&\sim \left| \chi^{(3)} \right|^2 \left| \int E_{\text{signal}}(t, \Delta t) \, e^{-i\omega t} dt \right|^2
\end{aligned}
\tag{73}
$$

One of the advantages of joint time-frequency characterization is that the signal (measured as a function of both frequency and time-delay) uniquely determines the electric field, $E(t)$. Although it is not possible to invert eqs. (72) and (73) to directly calculate the field from the FROG signal, the field can be obtained iteratively by successive approximation. This procedure is described in detail in [25]. Another advantage is that a plot of the FROG signal visually conveys information about the chirp of the pulse, and provides a close connection to the Wigner representation.

Joint time-frequency characterization is not restricted to using the specific third-order nonlinearity described above. For example, sum-frequency generation can be used in a 'spectrally-resolved autocorrelation', as shown in Fig. 43. The sum-frequency field is

$$E_{\text{signal}}^{2\omega}(t, \Delta t) \sim \chi^{(2)} E(t) E(t + \Delta t), \tag{74}$$

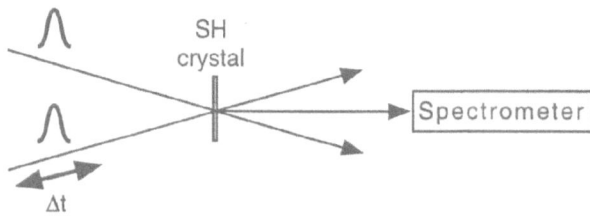

Figure 43. Spectrally-resolved autocorrelation.

and the measured spectrum is

$$F^{2\omega}(\omega, \Delta t) \sim \left|\chi^{(2)}\right|^2 \left|E_{\text{signal}}^{2\omega}(\omega, \Delta t)\right|^2$$

$$\sim \left|\chi^{(3)}\right|^2 \left|\int E_{\text{signal}}^{2\omega}(t, \Delta t) e^{-i\omega t} dt\right|^2 \tag{75}$$

Once again, the spectrally-resolved signal uniquely determines $E(t)$. An example of the spectrally resolved autocorrelation trace is shown in Fig. 43 together with the retrieved electric field of the 13-fs pulse (from [26]).

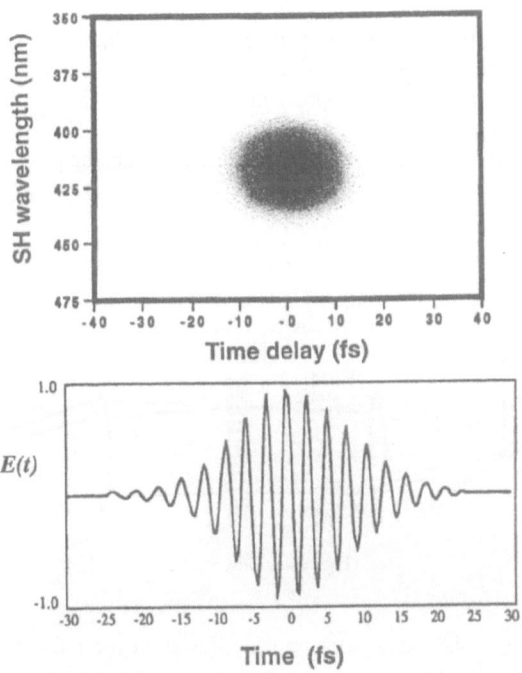

Figure 44. Example of a spectrally-resolved autocorrelation measurement and the electric field obtained from it (from [26])

III. B. Pulse Shaping

Now that we've answered the question of how ultrashort pulses can be measured, we are ready to examine how these pulses can be actively shaped or tailored. As a motivation for wanting to shape ultrashort pulses, consider two very different applications: coherent control and temporal information processing. 'Coherent control' refers to the optical manipulation of vibrational and electronic states of molecules and solids. Through precise control of the phase and amplitude of the electric field of an optical pulse or sequence of pulses, it becomes possible to achieve such goals as selectively breaking molecular bonds, controlling chemical reactions, and driving phase transitions in solids. For an introduction to coherent control see [27]. Temporal information processing takes advantage of the similarity between spatial and temporal domains to extend the techniques of spatial signal processing (such as correlation,

411

convolution, matched filtering) to processing temporal waveforms [28]. Furthermore, conversion of information between spatial and temporal domains makes possible such operations as encoding spatial information directly into a pulse sequence, retrieving time-encoded information into a spatial signal distribution, and identifying specific signatures in an optical data stream by correlating it with a spatial pattern [29]. Some very clever experimental implementations of these operations have recently been demonstrated [30], opening up an exciting new field of ultrafast spatio-temporal information processing.

How can we shape ultrashort optical pulses? Certainly no modulators can be driven fast enough to directly modulate the temporal envelope of the pulse. However, there is a very clever approach that does not require fast modulation at all — it involves shaping a pulse in its Fourier domain (i.e. in frequency). The trick is to somehow map the frequency domain onto a spatial coordinate, thus obtaining direct access to the frequency domain and making it possible to apply a spatial phase and amplitude mask to the pulse spectrum. Then, if the spectral components can be recombined into a single beam, the pulse will have been shaped by the Fourier transform of the applied mask. All this is accomplished by the optical arrangement shown in Fig. 45, first demonstrated by Weiner, Heritage and Kirschner [31]. (A somewhat different arrangement was used earlier for shaping picosecond pulses [32])

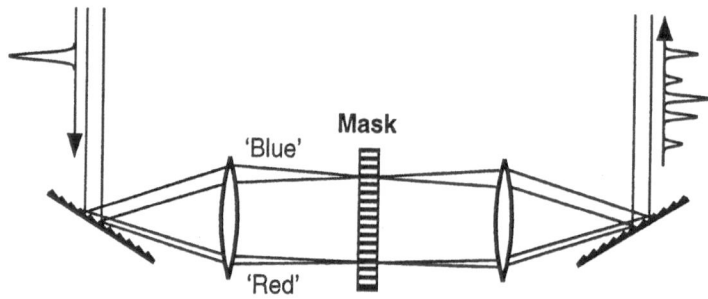

Figure 45. Pulse shaper.

Note that this is the same basic arrangement as in the pulse-stretcher discussed in Section II.B, in a zero-chirp configuration with the gratings positioned exactly in the input and output planes of the two-lens correlator. The role of the first grating is to disperse spectral frequencies into different directions (or 'spatial frequencies'), which are then mapped onto the transverse spatial coordinate by the Fourier transform action of the first lens. A phase and amplitude mask is applied in the Fourier plane. Finally, the shaped spectrum is recombined back into a single beam: the second lens transforms positions in the Fourier plane into spatial frequencies in the output plane, and the second grating brings the spectral components back into a single beam.

With such a pulse-shaping system almost any waveform can be produced that is within the bandwidth of the original pulse. If $E_{out}(t)$ is the desired waveform, then the required mask is

$$M(\omega) = \frac{E_{out}(\omega)}{E_{in}(\omega)} = \frac{\mathcal{F}\{E_{out}(t)\}}{\mathcal{F}\{E_{in}(t)\}}. \tag{76}$$

The constraint on the mask is $|M(\omega)| \leq 1$, since it is not an amplifier. Figure 46 shows an

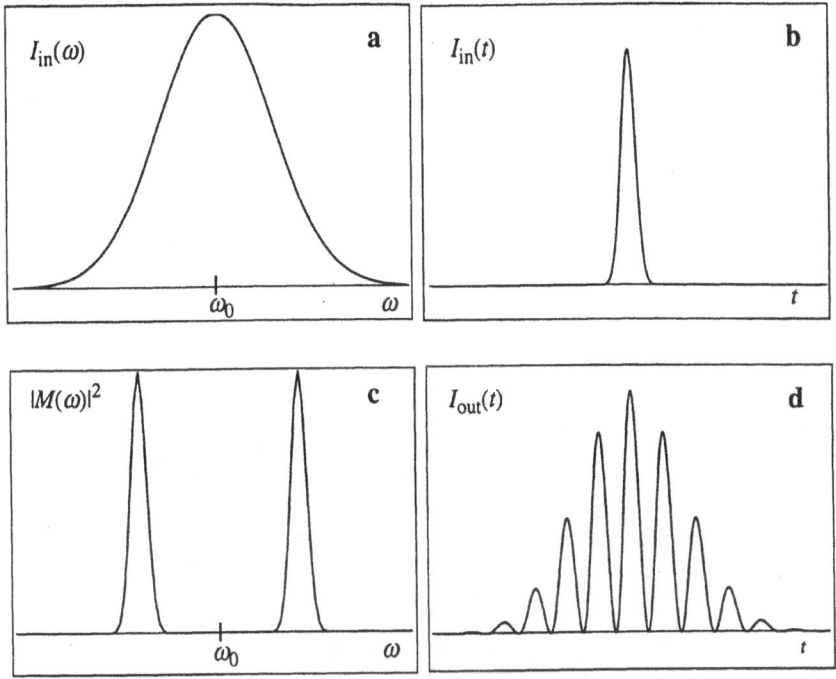

Figure 46. Example of pulse shaping. **a**. Spectrum of input pulse; **b**. Temporal intensity envelope of input pulse; **c**. Amplitude mask; **d**. Temporal intensity envelope of output pulse.

example of a Gaussian input pulse, shaped by a simple amplitude mask consisting of two Gaussian transmission peaks. The frequency filtering significantly lengthens the duration of pulse, and imposes a temporal modulation corresponding to the spectral spacing between the two transmission peaks. The result is a sequence of pulses inside a Gaussian envelope, with a total duration inversely proportional to the width of the transmission peaks.

The first masks were permanent phase and/or amplitude masks fabricated on glass substrates [31]. Subsequently, programmable masks have been designed which provide a great deal of flexibility in dynamically changing the waveform from pulse to pulse [33]. These masks are based on liquid crystal spatial light modulators (LC SLM's), in which the applied electric field controls the index of refraction of each pixel and thus provides control over phase delay across the spectrum. By combining a LC SLM with polarizers an amplitude mask can be assembled. In fact, both amplitude and phase control can be achieved in one mask by combining two SLM's and two crossed polarizers [34]. In this arrangement, two SLM's, one oriented in the x-plane, the other in the y-plane, are sandwiched between two polarizers at +45 and –45 degrees. Expressing the polarization of the electric field as a vector,

$$\mathbf{E} = \begin{bmatrix} E_x \\ E_y \end{bmatrix}, \tag{77}$$

we can use a matrix notation for the ±45° polarizers,

$$P_{45°} = \frac{1}{2}\begin{bmatrix} 1 & 1 \\ 1 & 1 \end{bmatrix} \qquad P_{-45°} = \frac{1}{2}\begin{bmatrix} 1 & -1 \\ -1 & 1 \end{bmatrix}, \tag{78}$$

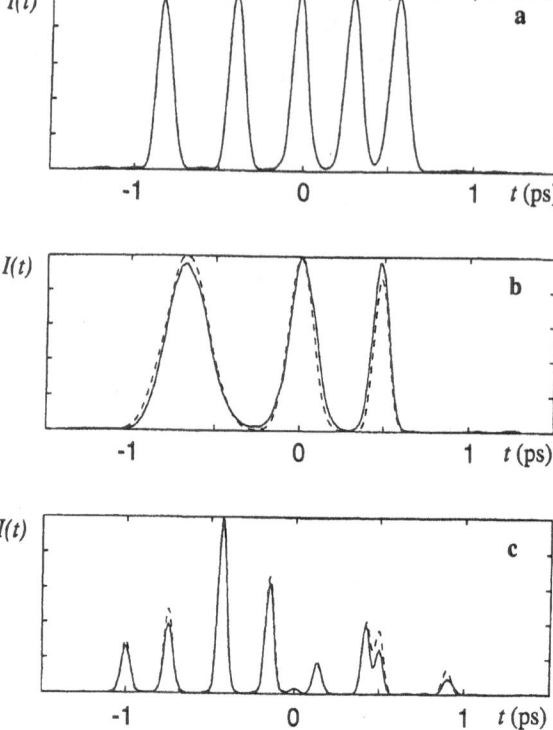

$I(t)$

-1   0   1 t (ps)

a

$I(t)$

-1   0   1 t (ps)

b

$I(t)$

-1   0   1 t (ps)

c

Figure 47. Examples of shaped pulses (from [34]).

and for x- and y- axis oriented masks,

$$M_x = \begin{bmatrix} e^{i\theta_x} & 0 \\ 0 & 1 \end{bmatrix} \qquad\qquad M_y = \begin{bmatrix} 1 & 0 \\ 0 & e^{i\theta_y} \end{bmatrix}. \tag{79}$$

Combining these four elements into a single mask matrix, we obtain

$$M = \begin{bmatrix} m & m \\ -m & -m \end{bmatrix}, \text{ where } m = \frac{1}{4}(e^{i\theta_x} - e^{i\theta_y}). \tag{80}$$

The output field is

$$\mathbf{E}_{out} = \begin{bmatrix} m & m \\ -m & -m \end{bmatrix}\begin{bmatrix} E_x \\ E_y \end{bmatrix} = 2m \begin{bmatrix} E_x + E_y \\ -(E_x + E_y) \end{bmatrix}, \tag{81}$$

polarized at $-45°$ as expected. By defining δ and σ as

$$\delta \equiv \frac{1}{2}(\theta_x - \theta_y); \qquad \sigma \equiv \frac{1}{2}(\theta_x + \theta_y), \tag{82}$$

we see that we have independent control over both the amplitude and phase of the output field:

$$m = \sin(\delta)\, e^{i\sigma}. \tag{83}$$

Each pixel of the mask, and thus each spectral component of the pulse, can be individually controlled in both amplitude and phase. The finite number of pixels places a limit on the spectral resolution of the mask.

Some examples of the kind of pulse shaping that is possible with programmable masks are shown in Fig. 32. The desired waveform is shown as a dashed line, and the measured waveform obtained with a 128-pixel phase and amplitude mask is shown as a solid line. The three waveforms shown are: a pulse sequence of decreasing separations, a pulse sequence with varying degrees of chirp, and an arbitrary pulse sequence.

The LC SLM is not the only available means for dynamic pulse shaping. An acousto-optic modulator (AOM) driven by microsecond radio-frequency electrical pulses can be used as a diffractive element in the Fourier plane of the pulse shaper [35]. This potentially offers the advantages of faster update times and no pixel gaps.

In addition to shaping pulses, the arrangement of Fig. 45 can be used for storage, recall, and processing of femtosecond waveforms. This is done by replacing the mask in the Fourier plane with a recordable hologram. The possibility of using holographic techniques for temporal pulse processing was first recognized and demonstrated by Weiner et al. [28]. These ideas have recently been extended to connect the spatial and temporal domains, by realizing that an ordinary hologram and one recorded with temporal signals inside a pulse-shaper are equivalent and interchangeable [29]. This is making possible such operations as encoding spatial information directly into a pulse sequence, retrieving time-encoded information as a spatial signal distribution, identifying specific signatures in an optical data stream, and high-speed all-optical data processing [30].

Acknowledgments

I thank Prof. Eric Mazur, Prof. Baldassare Di Bartolo, Dr. Shrenik Deliwala, and Aryeh Feder for insightful discussions during the preparation of these lectures, and Prof. Nicolaas Bloembergen, Paul Callan, Rich Finlay, Li Huang and Prof. Claus Klingshirn for their helpful comments on the manuscript.

I'm grateful to the Hertz Foundation for financial support throughout my graduate work.

REFERENCES

1. T. W. Hansch, I. S. Shahin, and A. L. Schawlow, Nature (Physical Science) **235**, 63 (1972).
2. T. K. Cheng et al., Appl. Phys. Lett. **57**, 1004 (1990).
3. H. L. Fragnito et al., Chem. Phys. Lett. **160**, 101 (1989).
4. L. Cohen, IEEE Proceedings **77**, 941 (1989).
5. J. D. Jackson, *Classical Electrodynamics*, John Wiley and Sons, New York (1975), Section 7.8.
6. E. Wigner, Phys. Rev. **40**, 749 (1932).
7. J. Paye, IEEE JQE **28**, 2262 (1992).
8. R. L. Fork, B. I. Greene, C.V. Shank, Appl. Phys. Lett. **38**, 671 (1981).
9. D. E. Spence, P. N. Kean, and W. Sibbett, Opt. Lett. **16**, 42 (1991).
10. F. Krausz et al., IEEE JQE **28**, 2097 (1992).
11. E. P. Ippen, Appl. Phys. B **58**, 159 (1994).
12. D. Huang et al., Opt. Lett. **17**, 511 (1992).
13. R. L. Fork, O. E. Martinez, and J. P. Gordon, Opt. Lett. **9**, 150 (1984).

14. A. Stingl et al., Opt. Lett. **19**, 204 (1994); Opt. Lett. **20**, 602 (1995).
15. P. Maine et al., IEEE JQE **24**, 398 (1988).
16. O. E. Martinez, IEEE JQE **23**, 59 (1987).
17. D. Strickland and G. Mourou, Opt. Comm. **56**, 219 (1985).
18. B. E. Lemoff and C. P. J. Barty, Opt. Lett. **18**, 1651 (1993).
19. B. E. Lemoff et al., Opt. Lett. **17**, 1367 (1992).
20. M. Ramaswamy-Paye et al., Opt. Lett. **19**, 1756 (1994).
21. T. B. Norris, Opt. Lett. **17**, 1009 (1992).
22. F. Salin et al., Opt. Lett. **16**, 1964 (1991).
23. C. P. J. Barty et al., Ultrafast Phenomena **7** (1994).
24. R. Trebino and D. J. Kane, J. Opt. Soc. Am. A **10**, 1101 (1993).
25. K. W. DeLong et al., Opt. Lett. **19**, 2152 (1994).
26. G. Taft et al., Opt. Lett. **20**, 743 (1995).
27. W. S. Warren, H. Rabitz, and M. Dahleh, Science **259**, 1581 (1993); S. Rice, Science **258**, 412 (1992).
28. A. M. Weiner et al., IEEE JQE **28**, 2251 (1992).
29. M. C. Nuss et al., Opt. Lett. **19**, 664 (1994).
30. M. C. Nuss and R. L. Morrison, Opt. Lett. **20**, 740 (1995).
31. A. M. Weiner et al., JOSA B **5**, 1563 (1988).
32. J. P. Heritage, A. M. Weiner, and R. N. Thurston, Opt. Lett. **10**, 609 (1985).
33. A. M. Weiner et al., Opt. Lett. **15**, 326, (1990).
34. M, M. Wefers and K. A. Nelson, Opt. Lett. **20**, 1047, (1995).
35. C. W. Hillegas et al., Opt. Lett. **19**, 737 (1994).

INTERACTION OF ULTRASHORT LASER PULSES WITH SOLIDS

Eric Mazur

Department of Physics
Harvard University
Cambridge, MA 02138

ABSTRACT

Beginning with some basic considerations in electromagnetic theory and solid state physics, I hope, in these four lectures, to develop some appreciation for the wonderfully rich electronic and optical properties of solids and to present an overview of current research in the area of the interaction of ultrashort laser pulses with solids.

The first two lectures are tutorials on the electronic and optical properties of solids and on energy transfer and relaxation in semiconductors. While some of the introductory material is treated in undergraduate courses, I will try to paint a broad picture and connect a number of facts that often remain disconnected. This introduction is followed by a survey of optical measurements of carrier and phonon dynamics in solids. I will conclude with an overview of recent experiments on electronic and structural changes induced by intense short laser pulses, including work done in my own research group.

I. ELECTRONIC AND OPTICAL PROPERTIES OF SOLIDS

In vacuum the frequency f and the wavelength λ of an electromagnetic wave are related by the speed c of light in vacuum,

$$f\lambda = c. \tag{1}$$

This yields a linear relation between the angular frequency $\omega \equiv 2\pi f$ and the wavevector $k \equiv 2\pi/\lambda$:

$$\omega = ck. \tag{2}$$

In a medium, the propagation of an electromagnetic wave is determined by the response of the material to electric and magnetic fields and is characterized by the

Spectroscopy and Dynamics of Collective Excitations in Solids
Edited by Di Bartolo, Plenum Press, New York, 1997

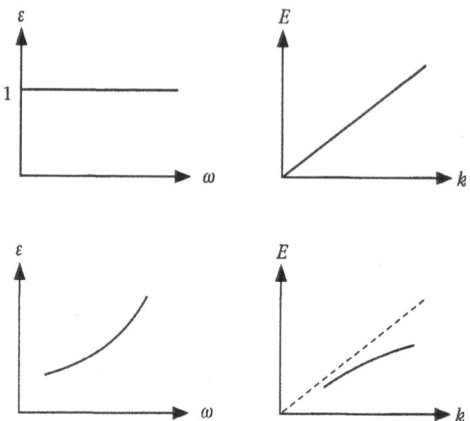

Figure 1. Dielectric function and energy-wavevector relationships in vacuum (top) and in a medium (bottom).

dielectric constant ε, the magnetic permeability μ, and the electric conductivity σ. Except for the magnetic permeability, which is nearly frequency independent at optical frequencies, the response of the medium depends on the frequency of the incident wave, and so dispersion occurs: waves of different frequencies propagate at different speeds. The frequency-dependent speed v of light in a medium is given by

$$v = \frac{c}{\mathrm{Re}\sqrt{\varepsilon(\omega)}} \equiv \frac{c}{n(\omega)}, \tag{3}$$

where $\varepsilon(\omega)$ is the frequency-dependent dielectric function and $n(\omega)$ the index of refraction of the medium. This frequency dependence results in a nonlinear relation between the angular frequency and the wavevector of the electromagnetic wave:

$$\omega = \frac{c}{\mathrm{Re}\sqrt{\varepsilon(\omega)}} k. \tag{4}$$

Figure 1 schematically shows the relationships between dielectric function and angular frequency, and between photon energy $E = \hbar\omega$ and wavevector, both in vacuum and in a medium. Throughout this part of my lectures I will use similar graphs to represent the optical properties of materials. The questions I will address are: why do different materials have different optical and electronic properties and what fundamental properties of solids are responsible for this behavior?

I. A. Propagation of Electromagnetic Waves through a Medium

The electromagnetic response of a material varies over the frequency spectrum because the charges present in the material respond at widely different frequencies. Roughly speaking, we can subdivide the charges into the following categories: ionic cores (the nuclei and core electrons at each lattice site), valence electrons, and free electrons. The ionic cores can form dipoles that tend to orient themselves along the direction of the applied external fields. This motion is usually limited to low

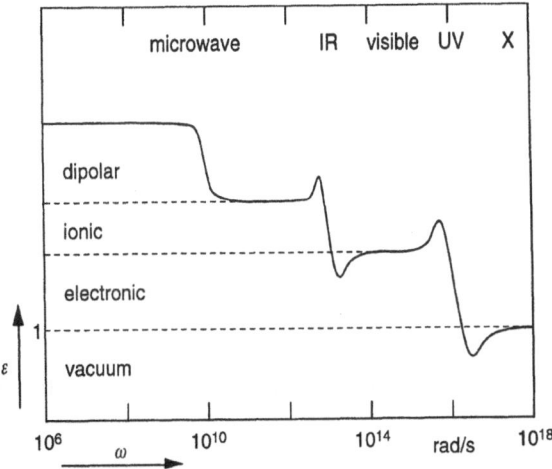

Figure 2. Schematic illustration of the various contributions to the dielectric constant across the electromagnetic spectrum.

frequencies and therefore only contributes to the polarization — and hence the propagation of the wave — at frequencies in the microwave region and below (see Fig. 2). At higher frequencies the dipoles can no longer follow the rapid oscillation of the applied field and their contribution to the dielectric constant vanishes. Lattice vibrations (displacement of the ionic cores) induced by the applied field contribute at frequencies up to the infrared region of the electromagnetic spectrum. In the visible and ultraviolet regions only the response of the free and bound valence electrons remain. Core electrons contribute at high frequency (10–1000 eV), but unless absorptions occur their contributions to the polarizability are generally small, and therefore the dielectric constant is close to 1 for x-rays. The dielectric function in Fig. 2 is very schematic — real materials generally show more structure and, depending on the type and number of charges present, some of the features shown may not be there.

Let us begin by analyzing the motion of a bound valence electron in response to an external driving field. If the field oscillates at frequency ω, the electron will oscillate at the same frequency with the phase and the amplitude of the oscillation determined by the binding and damping forces on the electron. The oscillation is described by the equation of motion of the electron [1]:

$$m\frac{d^2x}{dt^2} = -m\omega_0^2 x - m\gamma\frac{dx}{dt} - eE.$$

(5)

The first term on the right hand side of the equation represents a binding force with spring constant $k = m\omega_0^2$, where m and ω_0 are the mass and the resonant frequency of the bound electron, respectively. The second term is a velocity-dependent damping force and the third term is the driving force with E the applied field. Rearranging terms and assuming a sinusoidally varying applied field of amplitude E_o and frequency ω, we obtain an inhomogeneous second-order differential equation

$$m\frac{d^2x}{dt^2} + m\gamma\frac{dx}{dt} + m\omega_0^2 x = -eE_o e^{-i\omega t}.$$

(6)

419

The steady-state solution of this equation, representing the oscillating motion of the electron, must be of the form

$$x(t) = x_0 e^{-i\omega t}. \tag{7}$$

Substituting this into Eq. (6), we get for the amplitude of the motion

$$x_0 = -\frac{e}{m} \frac{1}{(\omega_0^2 - \omega^2) - i\gamma\omega} E_0. \tag{8}$$

As is to be expected, the amplitude of the electrons is maximal when the driving frequency is equal to the resonance frequency. The motion of the electron results in an oscillating dipole moment

$$p(t) = -ex(t) = \left(\frac{e^2}{m}\right) \frac{1}{(\omega_0^2 - \omega^2) - i\gamma\omega} E_0 e^{-i\omega t}. \tag{9}$$

In a sample with many bound electrons, the dipole moments of all the electrons contribute to a polarization

$$P(t) = \left(\frac{Ne^2}{m}\right) \sum_j \frac{f_j}{(\omega_j^2 - \omega^2) - i\gamma_j\omega} E(t), \tag{10}$$

where N is the total number of electrons, and f_j is the fraction of electrons having a resonant frequency ω_j and damping constant γ_j. In quantum mechanical terms, the factor Nf_j is the *oscillator strength*. This factor indicates how much each resonance contributes to the polarization.

The relation between the polarization P and the electric field E is usually written as $P(t) = \varepsilon_0 \chi_e E(t)$, with χ_e the dielectric susceptibility. The dielectric constant is given by $\varepsilon(\omega) = 1 + \chi_e$, so that

$$\varepsilon(\omega) = 1 + \frac{Ne^2}{\varepsilon_0 m} \sum_j \frac{f_j}{(\omega_j^2 - \omega^2) - i\gamma_j\omega} = \varepsilon'(\omega) + i\varepsilon''(\omega). \tag{11}$$

Figure 3 shows the frequency dependence of the dielectric constant for a single resonance and the resulting $E(k)$-behavior. At resonance, dissipation of energy is maximal since the amplitude of the electron motion is maximal. This dissipation of electromagnetic energy is what we call absorption and is reflected by the peak in the imaginary part of the dielectric constant at the resonance frequency. Note, also, that the real part crosses through zero near the resonance.

Let us next turn to the response of free electrons to an oscillating electromagnetic wave. Setting the binding force in Eq. (6) to zero we get

$$m\frac{d^2x}{dt^2} + m\gamma\frac{dx}{dt} = -eE_0 e^{-i\omega t}. \tag{12}$$

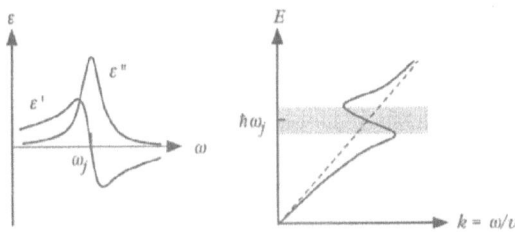

Figure 3. Frequency dependence of the dielectric constant near a resonance (left) and the resulting relation between energy and wavevector (right). The shaded region indicates the range of values for which absorption occurs.

Again, we obtain a solution of oscillating form,

$$x(t) = \frac{e}{m} \frac{1}{\omega^2 + i\gamma\omega} E_o e^{-i\omega t}, \tag{13}$$

but because there is no binding term, the motion does not exhibit any resonances. At low frequency, $\omega/\gamma \ll 1$, the applied electric field induces a time-varying current

$$J(t) \equiv \frac{dq}{dt} = -Ne\frac{dx}{dt} = \frac{Ne^2}{m}\frac{1}{\gamma - i\omega}E(t) \approx \frac{Ne^2}{m\gamma}E(t) \equiv \sigma E(t), \tag{14}$$

where N is the number of free electrons and σ the conductivity.

At high frequency the current can no longer keep up with the driving field and we may no longer ignore the imaginary part of the conductivity. Let us therefore again consider the oscillating dipole moment created by each electron:

$$p(t) = -ex(t) = -\left(\frac{e^2}{m}\right)\frac{1}{\omega^2 + i\gamma\omega}E_o e^{-i\omega t}. \tag{15}$$

For a sample with N free electrons, this gives a polarization

$$P(t) = -\left(\frac{Ne^2}{m}\right)\frac{1}{\omega^2 + i\gamma\omega}E(t) \equiv \varepsilon_0\chi_e E(t). \tag{16}$$

From this expression we obtain the free-electron contribution to the dielectric constant:

$$\varepsilon(\omega) = 1 - \left(\frac{Ne^2}{m\varepsilon_o}\right)\frac{1}{\omega^2 + i\gamma\omega} = \varepsilon'(\omega) + i\varepsilon''(\omega). \tag{17}$$

If the damping is negligible, $\gamma < <\omega$, the imaginary part of the free-electron contribution vanishes and the real part becomes

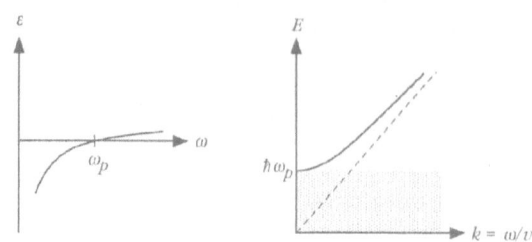

Figure 4. Dielectric function and $E(k)$ behavior for a plasma of free electrons with zero damping. The shaded region corresponds to a forbidden band of frequencies. Electromagnetic waves within this region are strongly attenuated.

$$\varepsilon'(\omega) = 1 - \frac{Ne^2}{m\varepsilon_0}\frac{1}{\omega^2} \equiv 1 - \frac{\omega_p^2}{\omega^2}, \tag{18}$$

where ω_p is called the plasma frequency. For frequencies below the plasma frequency, the dielectric constant is negative and so the index of refraction is purely imaginary resulting in strong attenuation of the electromagnetic wave (see Fig. 4). In the $E(k)$-plot this attenuation gives rise to a range of 'forbidden frequencies', or frequency gap below the plasma frequency. In this regime, however, the reflectivity is nearly one and incident electromagnetic waves do not penetrate into the plasma. Above the frequency gap the electromagnetic wave propagates through the medium and at high frequency the dielectric function approaches unity. The free electrons thus act like a high-pass filter: below the plasma frequency reflection occurs; above the plasma frequency the free electrons are transparent. For intrinsic semiconductors, the plasma frequency lies in the microwave or infrared part of the electromagnetic spectrum.

The effect of small, but nonzero damping is illustrated in Fig. 5. The damping results in a nonzero imaginary part at low frequency and a reduction of the frequency gap in the $E(k)$-plot. At high frequency the real part still approaches unity and the imaginary part vanishes.

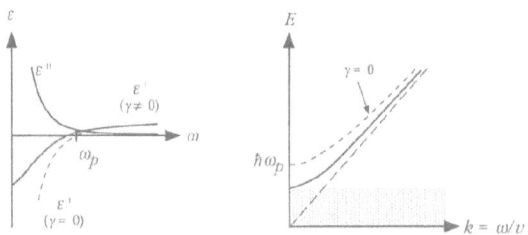

Figure 5. Dielectric function and $E(k)$ behavior for a plasma of free electrons with small nonzero damping. The effect of the damping is to reduce the forbidden band of frequencies.

Table 1. Bond types with typical bond energies E_{coh} and bond lengths a.

Type	Attractive Interaction	E_{coh} (eV)	a (nm)
covalent	overlap of charge distribution	3–10	0.2
ionic	electrostatic	6–9	0.2–0.3
metallic	delocalized overlap of charge distribution	1–5	0.3–0.4
van der Waals	induced dipole-dipole	0.01–0.1	0.4
Hydrogen	electrostatic	0.3–1	0.4

I. B. Chemical Bonds and Electronic Bands

The simple model given above describes the dielectric behavior for bound and free electrons. To what extent there are free or bound charges in a material strongly depends on the nature of the chemical bonds between the atoms in the material [2]. These bonds are due to an interplay between attractive and repulsive interactions between the atoms — the attraction makes it possible for atoms in a solid to have lower total energy than the same atoms when they are well separated, while the repulsion prevents the solid from collapsing onto itself. Table 1 provides an overview of the different types of bonds that can occur between atoms in a solid. The differences between these bonds are mainly due to differences in the mechanism responsible for attraction. The repulsive mechanisms are due to a combination of an electrostatic repulsion of the ionic cores, a repulsion due to the confinement of valence electrons, and a repulsion due to the Pauli exclusion principle.

The strongest bond is the covalent bond which is formed by an overlap of charge distribution between neighboring atoms. The lowering of the total energy comes about because the sharing of electrons makes it possible to fill bands. Figure 6a schematically illustrates covalent bonding between two atoms. If the two atoms get close enough for the atomic orbitals to overlap, then the wavefunctions of these orbitals can add with either the same or opposite phase. This results in a splitting of the energy level of the overlapping orbitals. The lower level corresponds to the wavefunctions adding in phase while the upper level corresponds to adding with opposite phase. If both of the original electronic levels in each atom were singly occupied, then in the ground state of the new system, the two electrons will occupy the lower level, which is lower in energy than the original level. Since this state is energetically more favorable than the state in which the two atoms are separated, the sharing of the two electrons results in a bond between the two atoms. If one or both of the electrons is excited to the upper level, then the new state no longer has

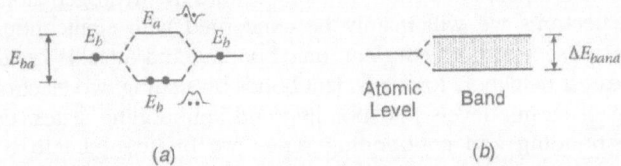

(a)   (b)

Figure 6. Energy level splitting in a covalent bond between two atoms (left) and energy-level broadening into a band in a covalently bonded solid (right).

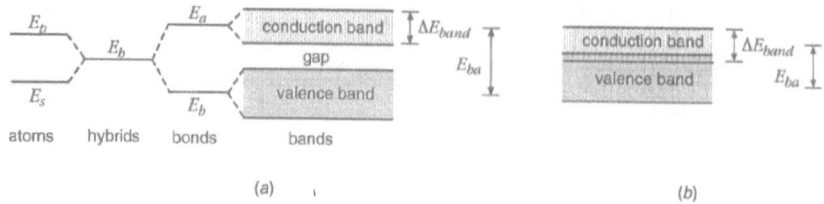

E_p

E_s

atoms

E_b

hybrids

E_a

E_b

bonds

conduction band

gap

valence band

bands

ΔE_{band}

E_{ba}

conduction band

valence band

ΔE_{band}

E_{ba}

(a)

(b)

Figure 7. Broadening of bonding and anti-bonding levels into valence and conduction bands. In insulators and semiconductors the bonding-antibonding splitting E_{ba} is larger than the width of the bands so the valence band is completely filled and the conduction band completely empty (left). In metals the two bands overlap resulting in a partially filled band (right).

a lower total energy than the state with two separated atoms. Thus, the lower level corresponds to a bonding state while the upper level corresponds to an antibonding state. In the case of a solid consisting of many atoms, the bonding and antibonding levels broaden into bands (see also Figs. 6b and 7a).

Covalently bonded solids are made up either of nonmetallic elements or a combination of a metallic and a nonmetallic element. The sharing of electrons between neighboring atoms results in strongly directed bonds, and as a consequence covalently bonded solids tend to be stiff and brittle. Moreover, due to the strong localization of the valence electrons covalent solids have zero or low conductivity.

Ionic bonds form when two neighboring atoms can fill their outer shell by exchanging one or more electrons. This results in a strong electrostatic attraction and a lower energy state. Ionic solids are always made up of a combination of metallic and nonmetallic elements, and the strong localization of the valence electrons results in a vanishingly small conductivity. Depending on the extent to which charge is transferred from one atom to another (as opposed to being shared), bonds can exhibit a broad spectrum of behavior between that of an ionic or covalent solid.

Metallic bonding arises from a more delocalized overlap of atomic orbitals in a solid. This type of bonding can only occur for a relatively large number of atoms and the delocalized nature of the bond results in materials that can be deformed and that are softer than covalent or ionic solids. The overlap of the orbitals of a particular atomic energy level leads to a broadening of this energy level into a band of energies. If the atomic energy levels are not fully occupied, then the total energy is lower for the atoms in a solid than for isolated atoms, as illustrated schematically in Fig. 7b. The partial occupation of the atomic levels results in a partially-filled energy band, and so the material exhibits metallic electronic characteristics [3].

Table 1 lists two more bonding mechanisms: the van der Waals bond and the hydrogen bond, both of which are extremely weak. The van der Waals bond is due to an induced dipole-dipole interaction. Hydrogen bonds are formed when covalently bonded hydrogen atoms in one molecule attract an atom in another molecule due to a dipole interaction (pointing toward the hydrogen atom).

In these lectures we will mainly be concerned with semiconductors which exhibit characteristics of both covalent ionic bonding and metallic bonding. On the one hand, nearest neighbors form covalent bonds by sharing two electrons, resulting in a splitting of atomic levels into bonding and antibonding states. On the other hand, these bonding and antibonding states are broadened into bonding and antibonding energy bands by more delocalized interactions with the other atoms in the solid. When the average bonding-antibonding splitting in the material is larger than the width of the energy bands, an energy gap will exist between the bonding and the antibonding bands (see Fig. 7a). If the atomic orbitals start out half-filled, then the ground state consists of a fully occupied bonding band and an empty antibonding band. In this case the material will not conduct electricity. However, when the average bonding-antibonding splitting is smaller than the width of the energy bands, the bonding and antibonding bands overlap (see Fig. 7b). The resulting partially-filled bands lead to metallic electronic characteristics even in the ground state. A semiconductor corresponds to the former case: an energy band gap separates the bonding (valence) and antibonding (conduction) bands. While semiconductors are insulators in the ground state, they start conducting when electrons are excited from the valence to the conduction band. Conductivity in a semiconductor is mediated both by the negatively charged electrons excited to the conduction band and by the positive holes left behind in the valence band by the excited electrons. The conducting electrons and holes are often referred to collectively as free charge carriers, or simply free carriers.

The existence of electronic energy levels and energy bands is inherently quantum mechanical. To determine the energy levels for a certain system one must know the potential energy function $U(\mathbf{r})$ for the electrons in the system and solve the Schrödinger equation

$$\nabla^2 \psi + \frac{2m}{\hbar^2}\left[E - U(\mathbf{r})\right]\psi = 0. \tag{19}$$

Here, ψ is the wavefunction of the electron, m its mass and E its energy. By substituting a certain potential energy function $U(\mathbf{r})$ and assuming a certain form for the wavefunction ψ, one can solve for the allowed energies E. Analytic solutions to the Schrödinger equation exist only for a few limited cases. Without providing any derivations, we will briefly review these cases because even though they do not correspond to any realistic situations, they do provide insight as to what gives rise to the electronic bandstructure of solids.

The simplest situation one can consider is that of a free particle. In that case the potential function is zero, $U(\mathbf{r}) = 0$, and the solution to the Schrödinger equation yields the standard parabolic relation between kinetic energy and momentum [4]:

$$E = \frac{\langle p \rangle^2}{2m}. \tag{20}$$

The energy of the free particle can take on any positive value, and hence the energy spectrum of the particle is a semi-infinite band (see Fig. 8a).

The energy of a particle confined to a one-dimensional infinite potential well,

$$\left.\begin{array}{ll} U(x) = 0 & (0 < x < a) \\ U(x) = \infty & (x \leq 0, x \geq a) \end{array}\right\} \tag{21}$$

can only take on discrete values [4],

$$E_n = \frac{n^2 \pi^2 \hbar^2}{2ma^2}, \tag{22}$$

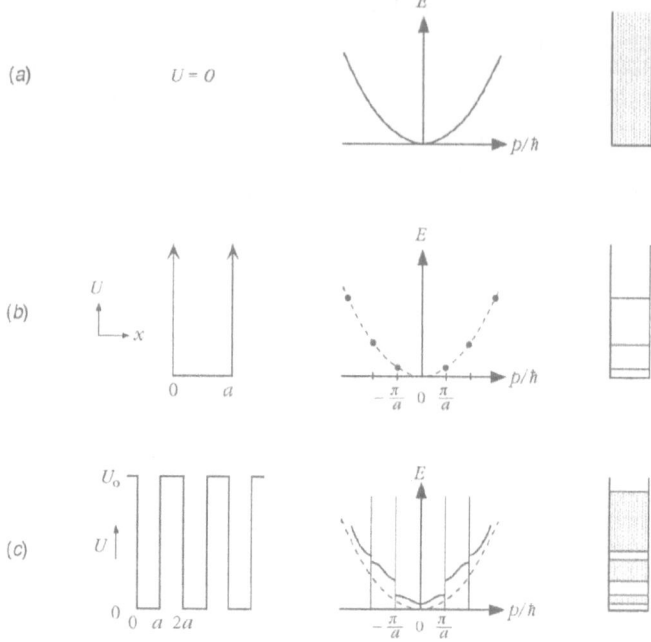

Figure 8. Solutions to the Schrödinger equation for three different potential energy functions: (*a*) free particle, (*b*) infinite square well, (*c*) periodic square well (Kronig-Penny model). The graphs on the right show the allowed values of energy.

where n is a positive integer. The energy spectrum exhibits a single level for each value of n (see Fig. 8*b*).

A solid can be idealized as a periodic potential of finite potential wells (see Fig. 8*c*). For this simplified situation the Schrödinger equation can still be solved analytically by substituting periodic Bloch functions [5]

$$\psi(\mathbf{k}, \mathbf{r}) = e^{i\mathbf{k}\cdot\mathbf{r}}u(\mathbf{k}, \mathbf{r}), \qquad (23)$$

where $\mathbf{k}$ is a constant vector and $u(\mathbf{r})$ is a function with the same periodicity as the lattice, *i.e.*, $u(\mathbf{k}, \mathbf{r} + \mathbf{R}) = u(\mathbf{k}, \mathbf{r})$ for any lattice vector $\mathbf{R}$. The resulting allowed values of energy are schematically illustrated in Fig. 8*c*. The periodic potential introduces a perturbation which segments and distorts the free-particle solution (Fig. 8*a*). Instead of the discrete values of energy found for a particle in a single, infinite well, we now find that the energy of a particle can take on values in a number of separate energy bands.

For a real solid the potential along a line of atoms is schematically illustrated in Fig. 9. Inside the solid the potential is periodic and has the same value at any two points separated by a lattice vector, except near the edge of the solid where the potential energy rises to zero. Near each nucleus the potential energy forms a deep well, closely approaching that of an isolated atom. In the region between the atoms, many atoms contribute to the potential energy and hence the total energy is lower than outside the solid.

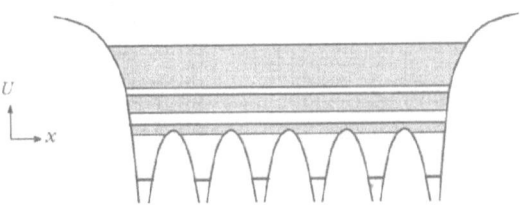

Figure 9. Electron potential energy along a line of atoms in a solid. The shaded regions indicate the allowed energy bands for the electrons.

The core electrons have energies that are deep in the potential energy wells near the atomic nuclei and the wavefunctions for these core electrons are nearly identical to those in an isolated atom. At higher energies, the wavefunctions extend throughout the solid and the allowed energy levels spread into bands. Whether or not a solid is conductive depends on the position of the Fermi level with respect to the bands. Since each level in an atom has two allowed states, N atoms contribute $2N$ states to each band. Therefore, if the number of electrons per unit cell is odd, the highest band is only half filled and the solid is a conductor. In the case of an even number of electrons per unit cell, the solid can still be a conductor if the bands overlap. If they don't the solid is an insulator or semiconductor. An example is shown in Fig. 10.

Figure 11 schematically illustrates the absorption coefficient of a typical semiconductor. The sharp rise in absorption in the visible region is the fundamental absorption edge: electrons are excited from the highest occupied band to a higher, unoccupied band. Below this absorption edge, ionic or partially ionic solids can absorb infrared radiation at specific resonances and the photon energy is converted to phonons (lattice vibrations). In addition, free carriers (electrons in the conduction band or holes in the valence band) can absorb radiation. Due to the very low density of free carriers in semiconductors, however, this contribution is small and limited to the microwave part of the electromagnetic spectrum. For metals, the free-carrier contribution is larger and shifts to higher frequency, while the fundamental absorption edge is absent. Conversely, for insulators, free-carrier absorption is absent and the fundamental absorption shifts toward higher frequency.

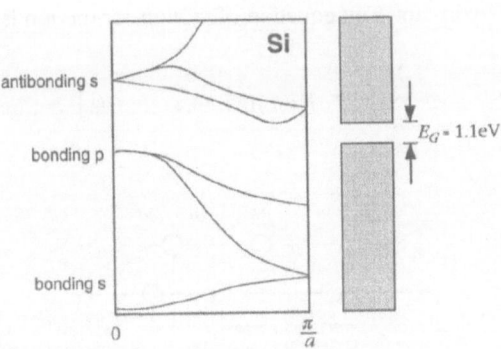

Figure 10. Bandstructure for silicon (left) and corresponding allowed energy bands (right). After Ref. [6].

427

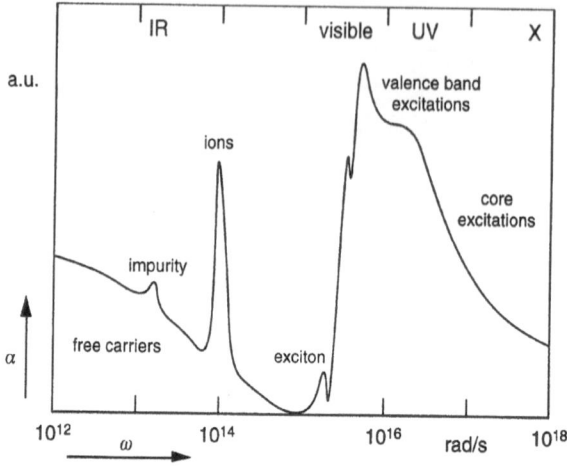

Figure 11. Absorption coefficient for a hypothetical semiconductor.

I. C. Phonons

In the previous two sections I assumed the ionic lattice to be fixed and immobile. In this section I consider collective motions of the ions. The ions can be displaced from their equilibrium positions and such disturbances can travel through the solid in the form of phonons which play an important role in the electronic and optical properties of solids because they can interact directly with electromagnetic waves.

Let us begin by considering a linear chain of identical atoms separated by a spacing a as illustrated in Fig. 12. The top of the drawing shows the atoms in their equilibrium position; at the bottom the atoms are displaced from their equilibrium position. Let us furthermore assume that only nearest neighbors exert forces on each other and that the interionic force obeys Hooke's law. The forces exerted on ion n by its two nearest neighbors are thus

$$\begin{aligned} F_{n-1,n} &= \gamma(u_{n-1} - u_n) \\ F_{n+1,n} &= \gamma(u_{n+1} - u_n) \end{aligned} \tag{24}$$

where γ is the force constant. The equation of motion for the ion is then

$$m \frac{d^2 u_n}{dt^2} = \gamma\left[u_{n-1} + u_{n+1} - 2u_n\right]. \tag{25}$$

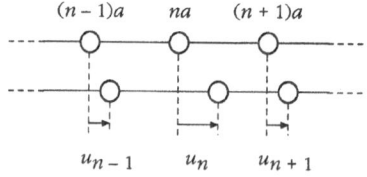

Figure 12. Vibrating linear chain of identical atoms spaced by a distance a.

428

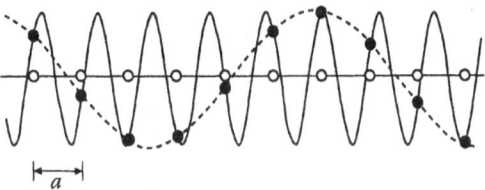

Figure 13. Oscillating chain of atoms showing instantaneous displacements. The solid curve conveys no information not given by the dashed one.

We look for solutions in the form of a traveling harmonic displacement wave (called a normal mode)

$$u_n(t) = Ae^{i(qna-\omega t)}, \tag{26}$$

where A is the amplitude of the displacement wave, q the wavevector, and ω the angular frequency. Substituting this into Eq. (25) we get

$$-m\omega^2 = \gamma[e^{-iqa} + e^{iqa} - 2] = -4\gamma \sin^2\left(\frac{qa}{2}\right), \tag{27}$$

so

$$\omega = \sqrt{\frac{4\gamma}{m}} \left| \sin\frac{qa}{2} \right|. \tag{28}$$

As Fig. 13 shows, we only need to consider displacement waves of wavelength larger than $2a$ — due to the discreteness of the chain, all waves of shorter wavelengths are equivalent to certain waves of longer wavelengths. This means we can restrict our analysis to small wavevectors:

$$\lambda \geq 2a \quad \Rightarrow \quad q \leq \frac{\pi}{a} \tag{29}$$

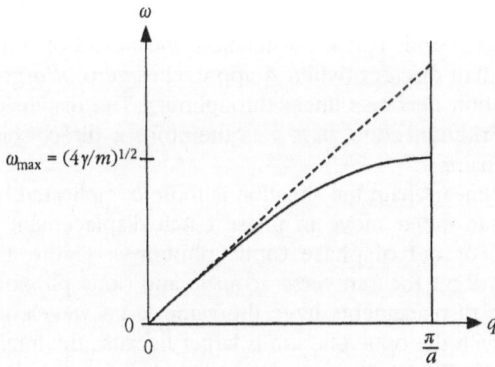

Figure 14. Dispersion of waves along a linear chain of atoms. The dashed line shows the result one would obtain for a continuous medium. The slope of the dashed line corresponds to the speed of sound waves in the medium.

small q   large q

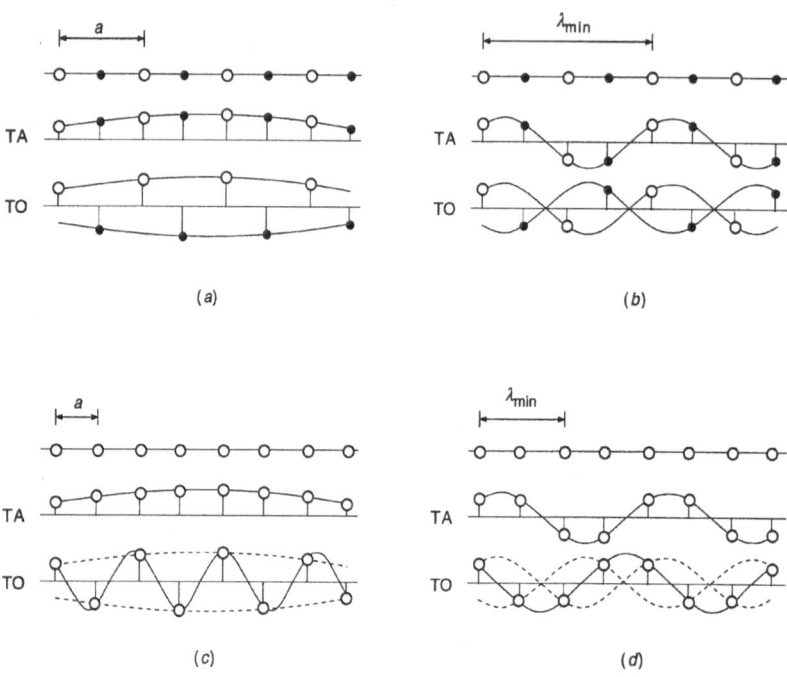

Figure 15. Waves on two-atom linear chains. Displacements are shown for (*a*) small and (*b*) large wavevector. The in- and out-of-phase waves correspond to acoustic and optic phonons, respectively. The bottom to graphs (*c* and *d*) show how the waves for a two-atom chain map onto waves of different wavevector for a one-atom chain.

Figure 14 shows the dependence of the displacement frequency on wavevector q (Eq. 28). For small wavevector, Eq. (28) becomes linear in the wavevector

$$\omega = \sqrt{\frac{4\gamma}{m}}\frac{qa}{2} = \sqrt{\frac{\gamma a}{m/a}}q = v_s q, \tag{30}$$

with v_s the speed of sound. This is the relation one would obtain if the chain were continuous rather than discreet (when a approaches zero, π/a goes to infinity and the dispersion relation becomes linear throughout). The dispersion of waves near the edge of the Brillouin zone at π/a is therefore a direct consequence of the granularity of the chain.

In a two-atom linear chain the situation is more complicated because the atoms of different kind can either move in phase (such displacement waves are called acoustic phonons) or out of phase (optic phonons). Figure 15*a* illustrates the displacements that occur for transverse acoustic and optic phonons of small wavevector. While both displacements have the same large wavelength, the potential energy associated with the optic phonon is larger because the interatomic bonds are much more distorted. The dispersion relation now has two branches (see Fig. 16*a*); for low wavevector the acoustic branch approaches zero, but because of the large distortion at low frequency, the corresponding energy for the optic branch is nonzero at zero wavevector.

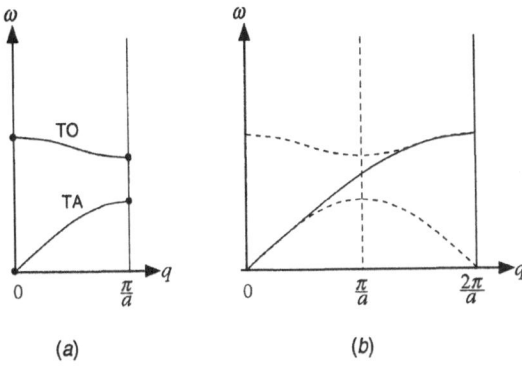

Figure 16. Dispersion relation for phonons (*a*) on a two-atom linear chain and (*b*) on a corresponding one-atom linear chain..

Figure 15*b* shows the displacements for the acoustic and optic phonons of the shortest possible wavelength (2*a*). The corresponding energies (see Fig. 16*a*) are slightly different. Figures 15*c* and 15*d* show how the cases illustrated in Figs. 15a and 15*b* relate to single-atom chain phonons: the optic branch vanishes as low wavevector optic phonons map onto large wavevector acoustic phonons. Note, in particular that the low-*q* TO phonon for the two-atom chain maps to a hiqh-*q* TA phonon on the one-atom chain (*cf*. Figs. 15*a* and *c*). Similarly, the TO and TA phonon modes at the edge of the Brillouin zone for the two-atom chain, are identical on the one-atom chain (*cf*. Figs. 15*b* and *d*), but are now in the middle of a Brillouin zone that is twice as wide (Fig. 16*b*).

I. D. Nonlinear Optical Interactions

In this section we consider the nonlinear optical properties of materials. In the presence of an electric field $E(t)$, atoms in a solid become polarized giving rise to a polarization of the solid. For small electric fields, the induced polarization is linear in the applied field:

$$P(t) = \chi^{(1)} E(t), \tag{31}$$

with $\chi^{(1)}$ the linear susceptibility. For large applied fields, however, the induced polarization becomes nonlinear in the applied field [7, 8]:

$$
\begin{aligned}
P(t) &= \chi^{(1)} E(t) + \chi^{(2)} E^2(t) + \chi^{(3)} E^3(t) + \ldots = \\
&= P^{(1)}(t) + P^{(2)}(t) + P^{(3)}(t) + \ldots
\end{aligned}
\tag{32}
$$

The first term represents the linear polarization, $P^{(n)}$ the *n*-th order polarization, and $\chi^{(n)}$ the *n*-th order nonlinear optical susceptibility. In general, the *n*-th order nonlinear susceptibility is not a scalar, but a tensor of rank $(n+1)$. For typical materials the electric field has to be of the order of atomic field strengths E_{at} before the second-order term becomes comparable to the linear term:

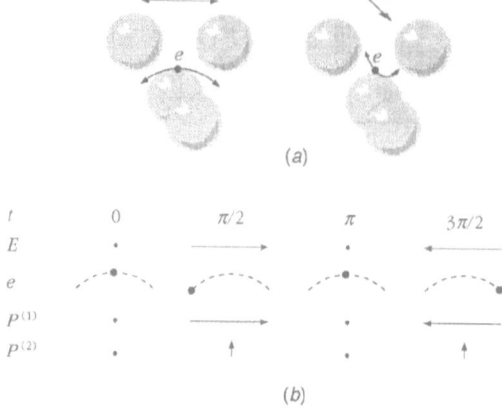

(a)

(b)

Figure 17. (a) An electron in the interstitial region between four atoms oscillates along a curved equipotential plane. (b) This oscillation causes a small second-order dipole in a direction perpendicular to the first-order dipole. The period of oscillation of the second-order dipole is one-half that of the first-order one.

$$\chi^{(n)} \approx \frac{\chi^{(1)}}{E_{at}^{n-1}}. \tag{33}$$

The nonlinear polarization can drive a new field. According to the wave equation that follows from the Maxwell equations, we have [1]

$$\nabla^2 E - \frac{n^2}{c^2}\frac{\partial^2 E}{\partial t^2} = \frac{4\pi}{c^2}\frac{\partial^2 P^{NL}}{\partial t^2}. \tag{34}$$

The second-order polarization, for instance, causes a driving term proportional to the square of the applied electric field, resulting in a new field at twice the applied frequency.

Consider an electron in the interstitial region between four atoms, as shown in Fig. 17a. Physically, the doubling of the frequency comes about because the charges move along a curved potential plane. For an electric field oscillating in the plane of the drawing, the electron moves along the dotted arc; in addition to being displaced in the horizontal direction, the electron also undergoes a small vertical displacement. This vertical displacement gives rise to a small second-order dipole moment perpendicular to the horizontal first-order dipole moment. As Fig. 17b shows, the period of oscillation of the second-order dipole moment is half that of the first-order dipole moment.

For systems with inversion symmetry, however, the second-order susceptibility vanishes. This can readily be seen by writing the second term of Eq. (32) in tensorial form:

$$\mathbf{P}^{(2)} = \chi^{(2)} : \mathbf{EE}. \tag{35}$$

Applying inversion, we get

$$-\mathbf{P}^{(2)} = \chi^{(2)} : (-\mathbf{E})(-\mathbf{E}), \tag{36}$$

and from Eqs. (35) and (36) we see that $\chi^{(2)} = -\chi^{(2)}$, which can only be satisfied if $\chi^{(2)}$ vanishes.

For systems that do not have inversion symmetry $\chi^{(2)}$ is not zero and an intense field will cause a second-order polarization. Let the applied field be of the form

$$E(t) = \tfrac{1}{2} E e^{i\omega t} + \text{c.c.} \tag{37}$$

The second-order polarization is then

$$P^{(2)} = \chi^{(2)} E^2(t) = \tfrac{1}{2} \chi^{(2)} E E^* + \tfrac{1}{4}\left\{\chi^{(2)} E^2 e^{-i2\omega t} + \text{c.c.}\right\}. \tag{38}$$

The second term on the right-hand side oscillates at frequency 2ω and can drive a new electromagnetic wave at double the incident frequency. This process is called second-harmonic generation.

If two oscillating fields of different frequency are present,

$$E(t) = \tfrac{1}{2} E_1 e^{i\omega_1 t} + \tfrac{1}{2} E_2 e^{i\omega_2 t} + \text{c.c.}, \tag{39}$$

the second-order polarization contains terms at frequencies $2\omega_1$, $2\omega_2$, at the sum-frequency $\omega_1+\omega_2$, at the difference frequency $\omega_1-\omega_2$, and at zero frequency. Because of dispersion the output beam at these new frequencies and the input beams at ω_1 and ω_2 travel at different velocities. To maximize output at any of the new frequencies it is therefore necessary to geometrically match the phases of the input and output beams [8].

Let us next briefly turn to the third-order polarization which 'mixes' four electric fields — three input fields generate one new output field. When three different input frequencies are present, 13 new frequencies can be generated. An example of a third-order effect is coherent anti-Stokes Raman spectroscopy (CARS). Two of the input frequencies are chosen such that their difference matches a resonant frequency in the system: $\omega_1 - \omega_2 = \omega_{res}$. The beating between the two input beams then coherently populates the upper level of the resonance. A third beam at frequency ω_3 then beats with the resonant oscillation in the system, generating a fourth beam at the anti-Stokes Raman frequency $\omega_a = \omega_1 - \omega_2 + \omega_3$. Note that this process is parametric, *i.e.*, the initial and final state is the same. The intensity of the coherent anti-Stokes beam is proportional to the population difference between the lower and upper levels of the resonance [8]. Hence, CARS can be used to measure population distributions.

II. ENERGY TRANSFER AND RELAXATION IN SEMICONDUCTORS

Next we turn our attention to the interaction between carriers, phonons, and photons in semiconductors and to optical excitation and relaxation processes. The types of questions I will address are: What are the absorption processes whereby the material gains energy from a laser pulse? How is the absorbed energy redistributed and how does the system progress to equilibrium? What happens for very large absorbed energy and what are the time scales for the various physical processes?

II. A. Interactions between Carriers and Photons

The interaction between electrons and photons is determined by the interaction Hamiltonian [9]

$$H_{int} = -\frac{e}{m_c}\mathbf{A} \bullet \mathbf{p}, \tag{40}$$

with $\mathbf{p}$ the momentum operator and $\mathbf{A}$ the single-photon vector potential operator:

$$\mathbf{A}(\mathbf{r},t) = \mathbf{A}(\mathbf{q},\omega)e^{i\mathbf{q}\bullet\mathbf{r}-\omega t}, \tag{41}$$

which is related to the electric field operator by $\mathbf{E} = -1/c(\partial\mathbf{A}/\partial t)$. The probability of an electron absorbing a photon is then given by the transition matrix element [10]

$$\langle k'n'|H_{int}|kn\rangle = -\frac{e}{mc}\int \psi_{n'}^*(\mathbf{k}',\mathbf{r})\mathbf{A}(\mathbf{q},\omega)e^{i(\mathbf{q}\bullet\mathbf{r}-\omega t)} \bullet \mathbf{p}\psi_n(\mathbf{k},\mathbf{r})d\mathbf{r} =$$

$$= -\frac{e}{mc}\mathbf{A}(\mathbf{q},\omega) \bullet \mathbf{P}_{n'n}(\mathbf{k})\, \delta_{\mathbf{k}',\mathbf{k}+\mathbf{q}}\, e^{-i\omega t} \tag{42}$$

where $\psi_n(\mathbf{k},\mathbf{r})$ is the wavefunction of the electron, and we have introduced the momentum matrix element

$$\mathbf{P}_{n'n}(\mathbf{k}) = \int \psi_{n'}^*(\mathbf{k}',\mathbf{r})\mathbf{p}\psi_n(\mathbf{k},\mathbf{r})\, d\mathbf{r}. \tag{43}$$

The Kronecker-delta in Eq. (42) represents momentum conservation. Note, however, that because the wavelength of light is much larger than the lattice constant, $\lambda \gg a$, the momentum of the photon is much smaller than that of the electron:

$$q = \frac{2\pi}{\lambda} \ll k \approx \frac{2\pi}{a} \tag{44}$$

The Kronecker-delta in Eq. (42) can therefore be replaced by $\delta_{\mathbf{k}'\mathbf{k}}$, and in an $E(k)$-plot the absorption of a photon by an electron results in a vertical transition (since the change in momentum is negligible on the scale of the reciprocal lattice vector k).

As a specific example let us consider the interaction of an electron near the top of the valence band in a direct-gap semiconductor (*i.e.*, a semiconductor for which the minimum in the conduction band and the maximum in the valence band occur at the same value of k). The photon leaves a hole behind in the valence band and creates a free electron in the conduction band, see Fig. 18. Neglecting the Coulomb interaction between the electron and the hole, the transition probability is given by [11]

$$W_{n'n} = \frac{2\pi}{\hbar}\left|\langle k'n'|H_{int}|kn\rangle\right|^2 \rho[\hbar\omega - (E_c - E_v)], \tag{45}$$

where n and n' denote the valence and conduction states, respectively, and ρ is the joint density of states, which is zero below the gap ($\hbar\omega < E_G = E_c - E_v$) and, for

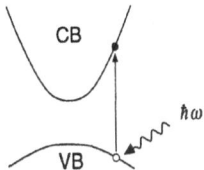

Figure 18. Vertical transition in a direct-gap semiconductor.

parabolic bands, proportional to $\sqrt{\hbar\omega - E_G}$ above the gap [11]. From Eq. (45) we thus see that the transition probability has the following dependence on frequency

$$W \propto A^2 \rho \propto \frac{1}{\omega^2}\sqrt{\hbar\omega - E_G}. \tag{46}$$

The absorption coefficient is given by the absorbed photon energy divided by the incident energy flux ϕ, so

$$\alpha = \frac{W\hbar\omega}{\phi} \propto \frac{\sqrt{\hbar\omega - E_G}}{\omega}. \tag{47}$$

Figure 19 shows the onset of the absorption for GaAs. The deviation from the theoretical dependence of Eq. (47) shown by the dashed curve in the figure, is the so-called Urbach tail [12].

In some semiconductors the minimum in the conduction band does not coincide with the maximum in the valence band. In such indirect-gap semiconductors, transitions near the band edge require absorption or emission of a phonon or an impurity scattering event and the phonon or the impurity provides the momentum change necessary to make the indirect transition (see Fig. 20). Because indirect transitions require a three-particle interaction, they are less likely to occur than direct transitions. Figure 21 shows the absorption near the band edge for Ge. At

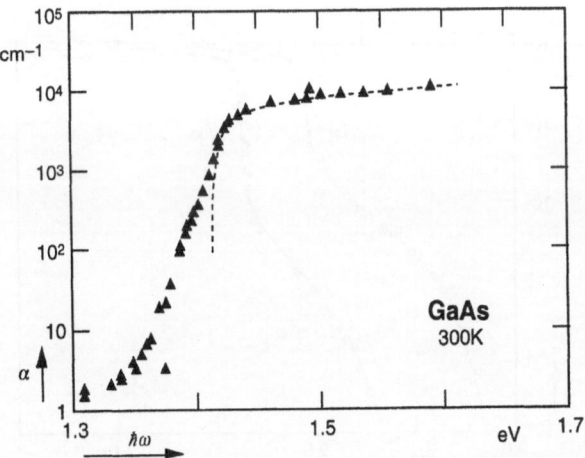

Figure 19. Onset of absorption near the bandedge for GaAs at 300 K. The dashed line represents the frequency dependence of Eq. (46). After Ref. [11].

435

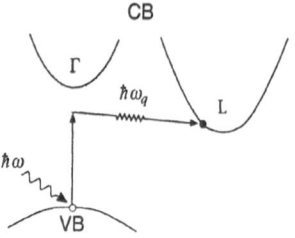

Figure 20. Optical transition in an indirect-gap semiconductor.

300 K absorption due to indirect transitions from the top of the valence band to the bottom of the L-valley at the edge of the Brillouin zone begins a little above 0.6 eV; at 0.8 eV direct transitions to the bottom of the Γ-valley (0.2 eV above the L-valley) take over.

II. B. Carrier Scattering

The excited carriers experience no 'collisions' with the periodically arranged ions — the solution to the Schrödinger equation already takes into account a periodic array of ions. They can, however, be scattered by phonons, which present a deviation from a strictly periodic arrangment, by impurities, and by other free carriers. Inelastic scattering by phonons usually dominates except at low temperature. Scattering by impurities and by other carriers are elastic processes, *i.e.*, the energy of the excited carriers remains the same.

To describe the scattering processes we introduce a (random) scattering potential $U_S(\mathbf{r},t)$. To find the scattering rate τ we need to evaluate the matrix element [10]

$$\left\langle \mathbf{k}'n' \left| U_s(\mathbf{r},t) \right| \mathbf{k}n \right\rangle = \int \psi_{n'}^{*}(\mathbf{k}',\mathbf{r}) U_s(\mathbf{r},t) \psi_n(\mathbf{k},\mathbf{r}) \, d\mathbf{r} \ . \tag{48}$$

Let us begin by considering scattering by a charged impurity. The carriers interact

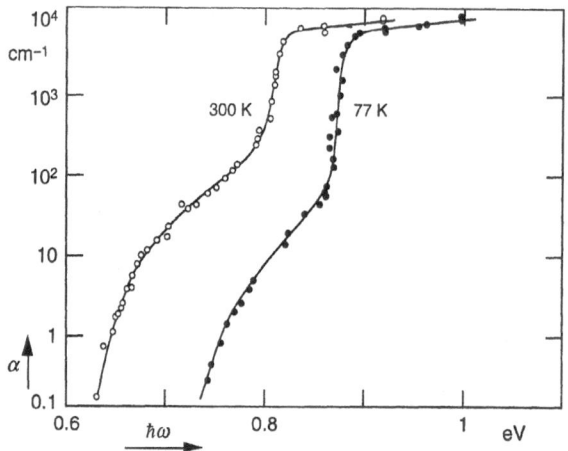

Figure 21. Onset of absorption near the bandedge for Ge at 77 and 300 K. After Ref. [11].

436

with charged impurities through the Coulomb potential

$$U_s(r) = \frac{q^2}{4\pi\varepsilon r}. \tag{49}$$

The scattering process is elastic, so the energy of the carrier does not change: $E(\mathbf{k}) = E(\mathbf{k}')$; the impurity merely changes the direction of the carrier ($\Delta\mathbf{k} \neq 0$). Solving the integral in Eq. (48) yields a scattering time $\tau \approx 10$ ps for a carrier energy $E = 0.5$ eV [10]. The rate scales with carrier energy as

$$\frac{1}{\tau} \propto E^{-\frac{3}{2}}. \tag{50}$$

In other words, carriers with low energy 'feel' the impurity more and are deflected over a large angle, whereas carriers with high energy undergo only slight deflection.

At high carrier density, the effect of impurity scattering is reduced by charges of opposite sign that are attracted by the impurity and screen the impurity potential. Solving the Poisson equation for a screened impurity, we find [10]

$$U_s(r) = \frac{q^2}{4\pi\varepsilon r} e^{-\frac{r}{L_D}}, \tag{51}$$

where L_D is the screening length for a nondegenerate electron gas:

$$L_D \equiv \sqrt{\frac{\varepsilon k T}{q^2 N}}, \tag{52}$$

with N the carrier density. As the carrier density goes up, the screening length becomes smaller and the scattering potential is reduced.

Scattering by phonons takes place through two different mechanisms: polar scattering and deformation potential scattering. Polar scattering occurs because phonons perturb the dipole moments between atoms. The effect is more pronounced for optical phonons (polar optical scattering) than for acoustic phonons (polar acoustic scattering or piezoelectric scattering) because optic phonons cause greater charge separation.

If a phonon causes a periodic displacement of the ions,

$$u(\mathbf{r},t) = A_q e^{i(\mathbf{q}\cdot\mathbf{r}-\omega t)}, \tag{53}$$

the dipole moments are perturbed by an amount $\delta p = q^* u$, with q^* the effective charge of the ionic cores. The field $E_{\delta p}$ of these perturbed dipoles causes a scattering potential $U_s(\mathbf{r},t) = e E_{\delta p}$ and substitution of this potential into the scattering matrix element (48) yields a scattering rate that is inversely proportional to the square of the wavevector q of the phonon [10]:

$$\frac{1}{\tau} \propto \frac{1}{q^2}. \tag{54}$$

This means that polar scattering is most pronounced for small phonon wavevectors causing the carriers to undergo only a small momentum change. Polar scattering therefore does not play a role in intervalley scattering.

In addition to perturbing the dipoles, phonons also perturb the bandgap. If the lattice constant a changes by a small amount, the bandgap will change by an amount

$$\delta E_G = D \frac{\delta a}{a}, \tag{55}$$

which is given by the deformation potential D. The change in lattice constant is related to the displacements of the ions by $\delta a \propto \partial u/\partial x$, and the scattering potential is thus given by

$$U_s(\mathbf{r},t) = D \frac{\partial u}{\partial x}. \tag{56}$$

Substituting this into the scattering matrix element (48) yields a scattering rate that is independent of phonon wavevector q. Deformation potential scattering is therefore responsible for intervalley scattering, which requires a large wavector phonon.

The final scattering process we need to consider is a binary collision between two carriers. Such a collision conserves total energy and momentum

$$\mathbf{k}_1 + \mathbf{k}_2 = \mathbf{k}_1' + \mathbf{k}_2'$$
$$E(\mathbf{k}_1) + E(\mathbf{k}_2) = E(\mathbf{k}_1') + E(\mathbf{k}_2') \tag{57}$$

but randomizes both the carrier momentum and energy. The scattering rate is

$$\frac{1}{\tau} \propto f(\mathbf{k}_2)\left[1 - f(\mathbf{k}_1')\right]\left[1 - f(\mathbf{k}_2')\right], \tag{58}$$

where $f(\mathbf{k}_2)$ is the probability that initial state $\mathbf{k}_2$ is occupied and $\left[1 - f(\mathbf{k}_1')\right]$ the probability that the final state $\mathbf{k}_1'$ is empty. These probability functions can be obtained from the Boltzmann equation [3].

From Eq. (58) we see that at $T = 0$ K near equilibrium a carrier at the Fermi energy, $E(\mathbf{k}_1) = E_F$, cannot scatter: the collision partner must have an energy below the Fermi energy, $E(\mathbf{k}_2) < E_F$, and because of conservation of energy one (or both) of the final states must therefore be below the Fermi energy (but these states are all occupied). For carriers above the Fermi energy we find a nonzero scattering rate $\tau^{-1} \propto \left(E(\mathbf{k}_1) - E_F\right)^2$ [3], and at temperatures above zero the rate scales with the square of the temperature $\tau^{-1} \propto (kT)^2$. At $T = 300$ K, for instance, carrier-carrier scattering rates are on the order of 0.1 ns. Away from equilibrium, carrier-carrier scattering rates increase rapidly as the carrier density increases. Monte Carlo calculations predict a $N^{-1/3}$ dependence on carrier density [10]. Experiments show scattering times of less than 20 fs at carrier densities on the order of 10^{18} cm^{-3} (see section III.A).

II. C. Optical Excitation Processes

Figure 22 illustrates four different excitation mechanisms that occur in semiconductors. In Fig. 22a, a single photon creates an electron-hole pair in a direct-

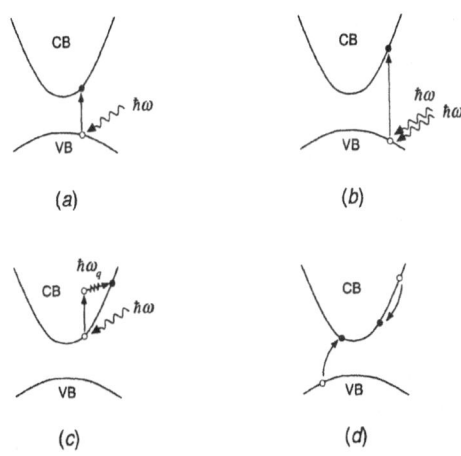

Figure 22. Excitation mechanisms in semiconductors.

gap semiconductor (in an indirect-gap semiconductor an additional phonon is required to conserve momentum). Since each photon creates a single free carrier pair, we can estimate the free-carrier density by dividing the absorbed energy by the volume into which this energy is deposited and the energy of a single photon. For a material of reflectivity R and absorption depth α^{-1}, and a laser pulse of incident fluence F, one-photon processes create a maximum carrier density

$$N \approx \frac{(1-R)F\alpha}{\hbar\omega}. \tag{59}$$

In practice the density of available states and band filling limit the carrier density that can be achieved with one-photon absorption.

Free carrier pairs can also be created through multiphoton processes. Multiphoton processes are much less likely to occur than single-photon processes, but their probability increases with increasing laser intensity. Table 2 lists the absorption depths for one-, two-, and three-photon absorption in GaAs. Figure 22b shows the creation of a carrier pair by two-photon absorption. This process populates different states, so it does not compete with single photon absorption and can help overcome band filling. On the other hand, it is less effective at creating carriers as two photons are required for each electron-hole pair.

Figure 22c shows the absorption of a photon by a carrier in the conduction band. Unless there is a resonance in the conduction band, free-electron absorption is

Table 2. Absorption depths for GaAs. Typical values for 1-mJ, 2.0-eV laser pulses of (*a*) 10-ps and (*b*) 100-fs duration. (*c*) Estimate from bandstructure.

Type	absorption coefficient	α^{-1} (μm) ps pulses[a]	α^{-1} (μm) fs pulses[b]
one-photon	$\alpha_1 \approx 4 \times 10^4$ cm^{-1}	0.2	0.2
two-photon	$\alpha_2 \approx 1 \times 10^{-1}$ cm/MW	100	0.5
three-photon	$\alpha_3 \approx 3 \times 10^{-1}$ cm^3/GW^2		0.1[c]

a three-particle process, requiring emission or absorption of a phonon or scattering off another electron. Free-carrier absorption doesn't generate new carriers, but it does increase the energy of the carriers. The interaction of free carriers with photons, given by the imaginary part of Eq. (17), has an absorption depth given by

$$\alpha = \frac{\omega_p^2}{Nc}\frac{\gamma}{\omega^2+\gamma^2},\tag{60}$$

where N is the free-carrier density. For a carrier density of 10^{21} cm^{-3} this yields an absorption depth $\alpha \approx 3 \times 10^2$ cm^{-1} which is relatively small compared to the absorption by carriers in the valence band.

If some of the free carriers have an excess energy above the conduction band minimum larger than the energy of the gap, impact ionization can occur (see Fig. 22d). The free carrier transfers some of its excess energy to an electron, creating a new electron hole pair. This process conserves the total energy of the free carriers, but increases their number.

Excitation of carriers in a semiconductor can thus be pictured as follows. Initially one-photon excitation generates electron-hole pairs. For intense laser pulses one-photon excitation will saturate due to band filling, but multiphoton excitation and free-carrier absorption will create more, and more energetic, carriers. These highly energetic carriers, in turn, can create additional carriers through impact ionization. Because of the interplay between these excitation mechanisms it is difficult to estimate the carrier density created by the laser pulse. The best one can do is to estimate an upper limit for the carrier density by replacing $\hbar\omega$ by E_G in Eq. (59). An additional difficulty is that this expression requires a value for the absorption depth which is known to vary with carrier density.

II. D. Relaxation Processes

A monochromatic laser pulse deposits free electrons in the conduction band at a number of well-defined energies. The carriers then quickly relax through a number of processes schematically illustrated in Fig. 23.

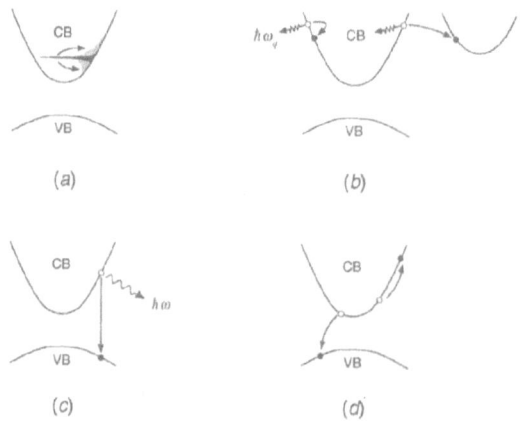

Figure 23. Relaxation processes in semiconductors.

Carrier-carrier scattering (Fig. 23a) causes thermalization of the nonequilibrium distribution created by the laser pulse takes place on a 10-fs time scale. This scattering process changes neither the total energy in the carriers nor the carrier density but by redistributing carriers over the band it helps overcome band filling during the excitation process.

Inelastic scattering of carriers by phonons — mostly emission of phonons by highly excited free carriers — reduces the energy of the carriers and heats the lattice (Fig. 23b). As we have seen in section II.B polar optical scattering is mainly responsible for small-wavevector intravalley scattering, while deformation potential scattering causes intervalley scattering. Both intravalley scattering as well as scattering from the central valley to side valleys takes place on a 100-fs time scale (see also section III.A). Scattering back from the side valleys to the central valley takes significantly longer because of the much higher density of states in the side valleys. It should be kept in mind that the energy of a typical LO phonon is generally small compared to the excess energy of the carriers. For instance, a carrier pair generated in GaAs (band gap: 1.4 eV) with a photon at 2.0 eV has 600 meV excess energy, whereas the LO phonon energy is 30 meV. Twenty phonons must therefore be emitted to rid the carrier pair of its excess energy, bringing the relaxation time due to LO phonon emission (or the lattice heating time) to approximately 2 ps.

Free electrons and holes can also recombine radiatively or nonradiatively. In a radiative recombination process, a hole and an electron of identical wavevector recombine emitting the energy difference in the form of a photon (Fig. 23c). This process, which lowers both the free carrier energy and the free carrier density, takes place on the 1-ns timescale. The nonradiative process shown in Fig. 23d, called Auger recombination, is the inverse of impact ionization: an electron and a hole recombine transferring momentum and energy to a third carrier. This three body process is proportional to N^3 and therefore important mainly at high carrier density. In fact, Auger recombination puts a cap on the carrier density, because its effect is to reduce the carrier density.

Another important mechanism that reduces the free carrier density in the region excited by the laser pulse is carrier diffusion out of the laser-excited region. If the laser spot size on the sample surface is much bigger than the absorption depth δ, then the carrier density varies only in the direction normal to the sample surface: $N(z) = N(0)e^{-z/\delta}$. In this case, the diffusion time is given by $\tau_{\text{diffusion}} = \delta^2/D_a$ [11]. The ambipolar diffusion constant D_a is related to the excited electron temperature T_e and the carrier mobility through the Einstein relation $D_a = k_B T_e \mu_a / e$ [11, 13], where k_B is the Boltzmann constant, μ_a a reduced carrier mobility for conduction electrons and valence holes, and e the magnitude of the electron charge. If the electron temperature varies in time, such as during and after femtosecond laser pulse excitation [14], the diffusion constant will also vary in time. This dependence of D_a on electronic temperature is further complicated by the implicit dependence of the carrier mobility μ_a on the electron temperature, which is not easily determined.

While an increased electron temperature resulting from femtosecond laser pulse excitation should speed up diffusion, carrier confinement slows it down [14]. This effect arises from the reduction in the bandgap in an excited semiconductor due to many-body interactions in the free carrier system [14, 15]. In laser pulse excitation, the spatial gradient of the light intensity in the material leads to a spatial gradient in the excited carrier density. The carrier-density gradient, in turn, causes a gradient in the bandgap, with a smaller bandgap coinciding with higher carrier density. Since, in the absence of diffusion, free carriers will tend towards the regions with smaller

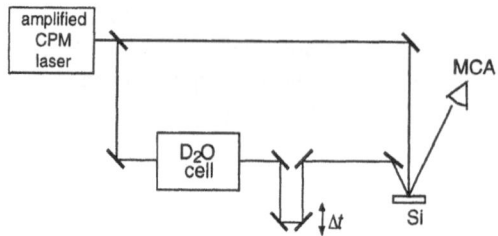

Figure 24. Schematic diagram of experimental setup to measure transient reflectivity. CPM laser = colliding pulse mode-locked laser; MCA: multi-channel analyzer. The D_2O cell serves to generate a broadband probe pulse. After Ref. [18].

bandgaps, this spatial gradient results in carrier confinement and therefore slows down diffusion [14]. Simulations and experimental measurements show that carrier-density relaxation occurs on a picosecond time scale [13, 14, 16, 17].

A number of other scattering mechanisms play a role in the distribution of the laser-deposited energy, such as scattering with plasmons (collective electronic excitations). A more detailed review of these processes can be found in Ref. [3].

III. OPTICAL MEASUREMENTS OF CARRIER AND PHONON DYNAMICS IN SOLIDS

In this section I will present a survey of relevant optical experiments on carrier and phonon dynamics in solids. Rather than being comprehensive, I will try to discuss the experiments that directly deal with the phenomena discussed in the previous section.

III. A. Carrier Dynamics

The first experiments to probe optical carrier generation in a semiconductor on the femtosecond scale were done by Shank and coworkers in 1983 [18]. Figure 24 shows a schematic diagram of the experiment. A 90-fs, 4-kJ/m^2 laser pulse of 2.0-eV photon energy strikes a Si wafer near normal incidence. The reflectivity of the central 15-μm of the 150-μm diameter excited region is probed with a broadband (1.2–2.8 eV) subpicosecond pulse. In the case of weak absorption, the reflectivity of the sample is determined by the index of refraction as follows:

$$R = \left(\frac{n-1}{n+1}\right)^2 \tag{61}$$

and from Eqs. (3) and (17) we see that for free carriers the index of refraction is given by

$$n = \sqrt{1 - \frac{\omega_p^2}{\omega^2}}, \tag{62}$$

with the plasma frequency given by Eq. (18): $\omega_p = \sqrt{Ne^2/(m\varepsilon_0)}$.

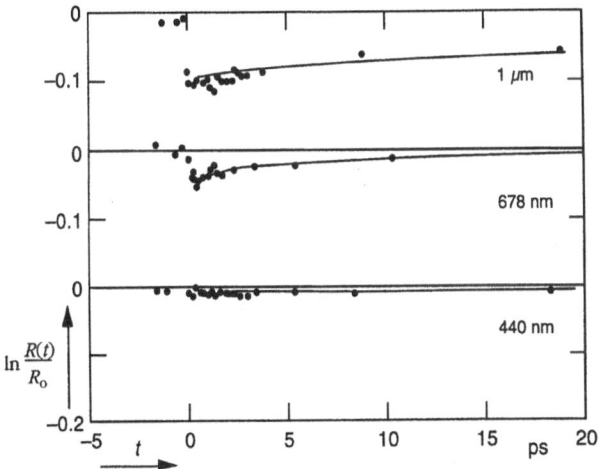

Figure 25. Changes in reflectivity at three different probe wavelengths for Si after femtosecond laser excitation. After Ref. [18].

Above a threshold incident energy of $E_{TH} \approx 1$ kJ/m^2 the sample was observed to undergo irreversible changes. Figure 25 shows the changes in reflectivity at three different wavelengths at an incident energy of 0.63 E_{TH}. The observed changes in reflectivity are attributed to changes in carrier density. As the carrier density increases, the plasma frequency goes up and, for frequencies above the plasma frequency, the index of refraction decreases. According to Eq. (61) this should in turn result in a decrease in reflectivity, which is indeed observed for $t < 1$ ps. Quantitative analysis of the reflectivity changes yields a carrier density $N \approx 5 \times 10^{21}$ cm^{-3}. The decay of the signal, caused by a decrease in carrier density in the probe region, is much slower than the measured Auger recombination rate for Si. This suggests that Auger recombination is screened at high carrier density [18].

By probing the transient absorption saturation of a 0.5-μm thin sample of GaAs,

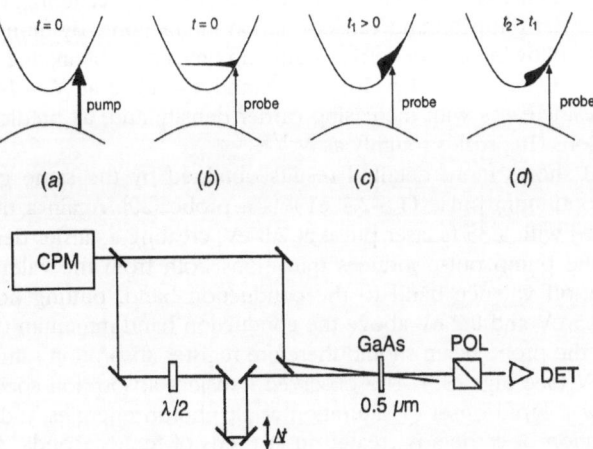

Figure 26. Schematic diagram of experimental setup to measure absorption saturation. CPM = colliding pulse mode-locked laser; $\lambda/2$ = half-wave plate; POL = polarizer; DET = detector. After Ref. [19].

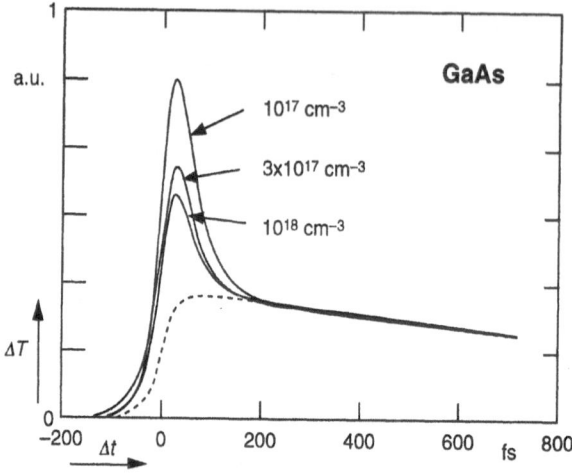

Figure 27. Change in transmittance for GaAs after femtosecond laser excitation. After Ref. [19].

Lin *et al.* were able to obtain a direct view of the dynamics of the excited carriers in the conduction band [19]. Figure 26 shows a schematic diagram of the experiment. An intense 35-fs pulse at 2 eV is used to saturate a transition from the valence band to the conduction band (Fig. 26*a*). A second pulse, equal to 20% of the pump pulse, is then used to probe the evolution of the population in the conduction band. Pump and probe have orthogonal polarizations and a detector is used to monitor the transmittance of the probe pulse through the sample. Right after excitation the population of the conduction band at the pump frequency is saturated and the probe passes through the sample registering an increased transmission on the detector (Fig. 26*b*). As the carriers thermalize the saturation diminishes and the transmission decreases (Fig. 26*c–d*). Figure 27 shows results obtained by Lin for three different carrier densities scaled so the exponential tails overlap. The dashed line shows a convolution of the laser pulse with a 1.5-ps exponential tail. The fast decay in transmittance ($\Delta t \leq 100$ fs) is due to carrier-carrier scattering, the slower 1.5-ps decay to carrier diffusion and carrier-phonon interactions. By fitting the transient saturation, the following carrier-carrier scattering times are obtained: $\tau \approx 30$ fs at $N = 10^{17}$ cm^{-3}; $\tau \approx 17$ fs at $N = 3 \times 10^{17}$ cm^{-3}; and $\tau \approx 13$ fs at $N = 10^{18}$ cm^{-3}. The scattering rate increases with increasing carrier density and, as predicted by Monte Carlo simulations [10], scales roughly as $N^{-1/3}$.

Figure 28 shows more detailed results obtained by the same group using a broadband continuum pulse (1.5-2.1 eV) as a probe [20]. Again a thin sample of GaAs is excited with a 35-fs laser pulse at 2.0 eV, creating a carrier density of about 10^{18} cm^{-3}. The pump pulse induces transitions both from the valence band and from the split-off valence band to the conduction band, putting nonequilibrium electrons at 0.5 eV and 0.2 eV above the conduction band minimum (see Fig. 28*a*). At zero delay the probe beam should therefore register absorption saturation around 2.0 and 1.7 eV (see Fig. 28*b*). The observed transient absorption spectra shown in Fig. 28*d* show a rapid onset of saturation at all photon energies, indicating that a broad distribution of carriers is created in just tens of femtoseconds. At short times ($\Delta t < 200$ fs), however, the spectra still reflect the initial population created by the pump pulse as evidenced by the two holes labeled '1' and '2,' corresponding to the transitions shown in Fig. 28*b*. As the electrons thermalize, we expect the holes in the

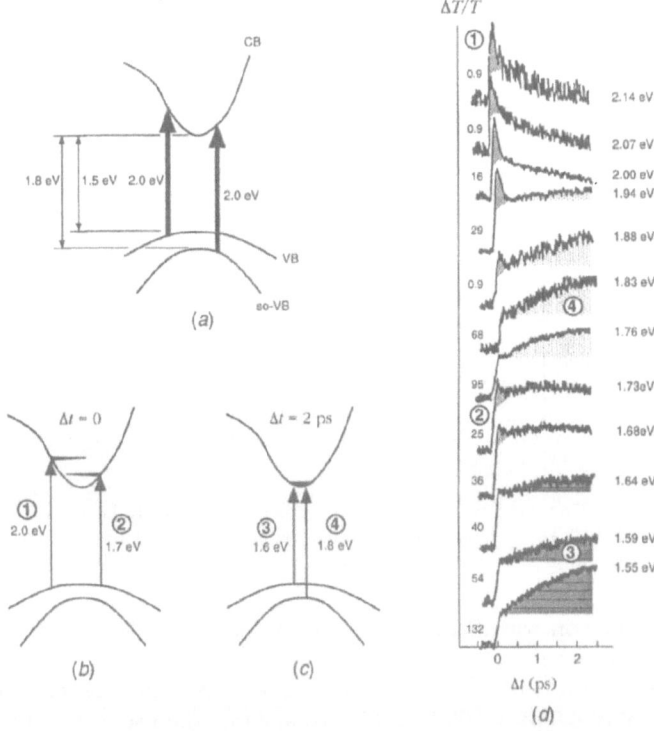

Figure 28. (*a*) Transitions induced by 2.0-eV photons in GaAs. (*b*) The populations created by the 2.0-eV photons create two spectral holes for photons at 1.7 and 2.0 eV (two additional holes for transitions from the split-off valence band lie outside the probe range). (*c*) As the electrons relax to the bottom of the conduction band, the holes shift to 1.6 and 1.8 eV. (*d*) Transient transmission of a broadband probe beam. The numbers 1–4 correspond to the transitions in (*b*) and (*c*). After Ref. [20].

absorption to disappear, which is indeed clearly visible in the spectra at 2.0 eV or higher. Finally, as the carriers accumulate at the bottom of the conduction band, the spectra show an increase in the transmission at 1.6 and 1.8 eV (see transitions labeled '3' and '4' in Figs. 28c and d).

The dynamics of the electrons observed in the spectra of Fig. 28 are due to a combination of carrier-carrier scattering and carrier relaxation (which, in turn, is due to carrier-phonon relaxation and diffusion). To separate these effects, Oudar and coworkers used a 0.5-ps laser pulse of 1.54 eV to excite electrons just 19 meV above the bandgap of GaAs [21]. Since the LO phonon energy is about 30 meV, relaxation through carrier-phonon scattering is not possible. The evolution of the transmittance of a probe pulse (1.49–1.57 eV) is shown in Fig. 29. The spectra show a hole in the transmittance at the pump frequency of 1.54 eV which disappears by about 4 ps. Modeling of the spectral hole yields a carrier-carrier relaxation time of 0.3±0.1 ps.

Carrier-phonon scattering in GaAs was studied using spontaneous Raman scattering [22]. A 0.6-ps pulse at 2.1 eV was used to create a free electron density of 2×10^{17} cm^{-3}. The free electrons relax back to the bottom of the conduction band via LO phonon emission. Using 2.1-eV photons, the excited electrons have an excess energy of about 0.6 eV, which is equivalent to the energy of 16 LO phonons. The

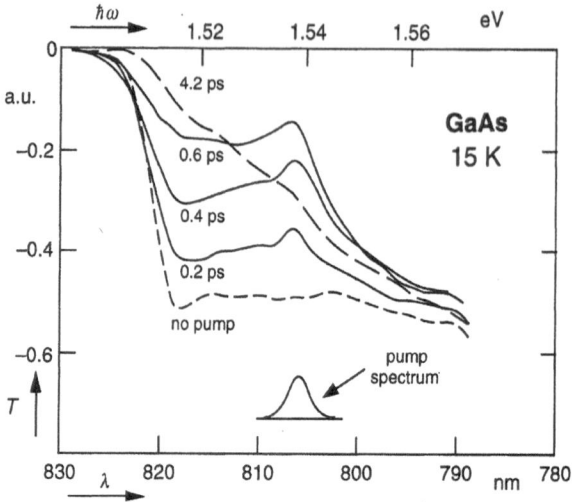

Figure 29. Transmission of a broad probe pulse for GaAs at 15 K after excitation of carriers just above the bottom of the conduction band (visible at 1.51 eV). Inset shows pump laser pulse spectrum. After Ref. [21].

phonon population was monitored by measuring the intensity of the anti-Stokes Raman signal (see Fig. 30).

Figure 31 shows that the anti-Stokes signal rises in about 2 ps and decays with a time constant of 4.0 ps at 300 K and 7.5 ps at 80 K. The rise in signal is caused by the increase in LO-phonon population that results from the LO-phonon emission of the excited carriers (see Fig. 30c). In the backscattering geometry used in this

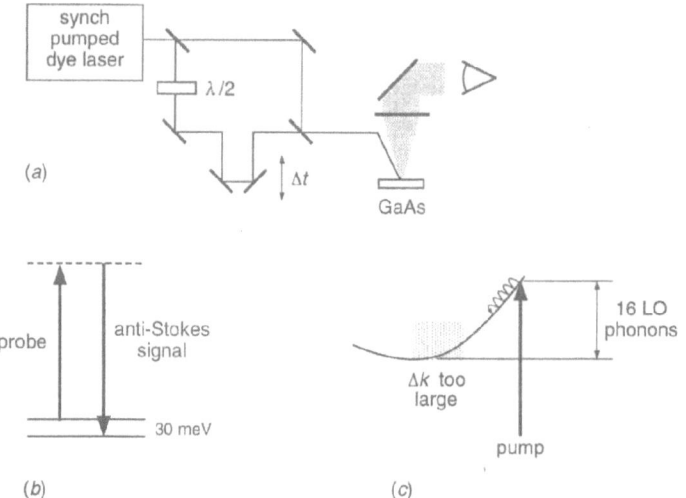

Figure 30. Measurement of the carrier-phonon scattering time after Ref. [22]. (*a*) Optical arrangement. (*b*) The probe pulse monitors the population in the LO-phonon mode. (*c*) The excited carriers relax down the side of the Γ-valley in the conductin band via a cascade of 16 LO phonons Only the first twelve of these phonons have a wavevector *q* of the right magnitude to be seen in the backscattering geometry shown in (*a*).

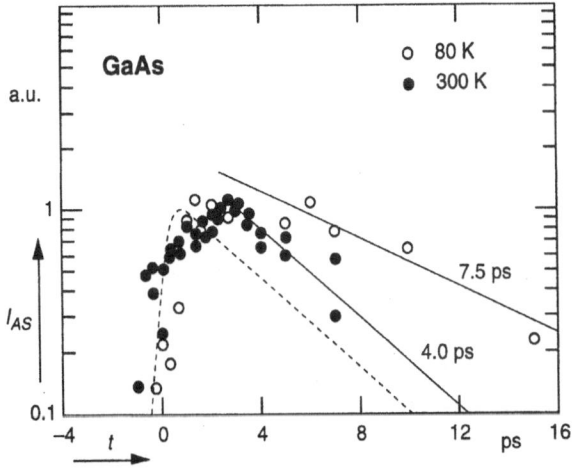

Figure 31. Evolution of spontaneous anti-Stokes Raman signal for GaAs at 80 and 300 K after subpicosecond laser excitation. After Ref. [22].

experiment, LO phonons of wavevector $q = 2k$, with k the photon wavevector, are probed; for 2.1-eV photons this corresponds to $q = 8.4 \times 10^5$ cm^{-1}. This means that only the first 12 of the 16 phonons needed for relaxation to the bottom of the conduction band are observed — the last four photons have a wavevector larger than $2k$. The 2-ps rise time therefore implies that the time for the emission of a single LO phonon takes (2 ps)/12 = 165 fs. The decay of the Raman signal reflects a decrease in the LO phonon population due to coupling of this mode to other (mainly acoustic) phonon modes. At low temperature this coupling is less effective.

Additional information on carrier-phonon dynamics was obtained using broadband absorption spectroscopy [23]. A 6-fs, 0.1-nJ broadband (1.85–2.15 eV) pump pulse was used to populate the entire bottom 600 meV of the conduction band. An indentical pulse was then used to probe the transmittance. Because of the spectral width of the pulse the transmittance is not affected by intravalley scattering — as long as the carriers remain in the Γ-valley, they cause absorption saturation. Scattering to the L and X valleys, however, reduces the population in the Γ-valley and therefore changes the transmittance (see Fig. 32). The measured transmittance is shown in Fig. 33. The data show a fast component with a 33-fs decay time due to fast scattering out of the central valley to the side valleys and a slower component due to carrier diffusion. At longer time scales (not shown) the signal increases again

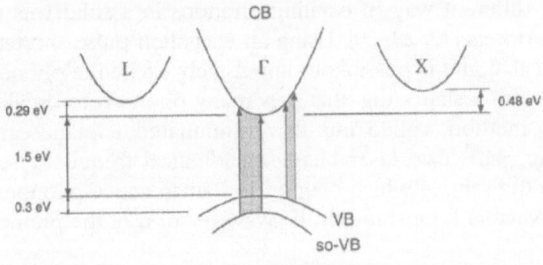

Figure 32. Region of Γ-valley of GaAs probed by broadband (1.85–2.15 eV) pulse. The relative position of the X and L valleys is also shown.

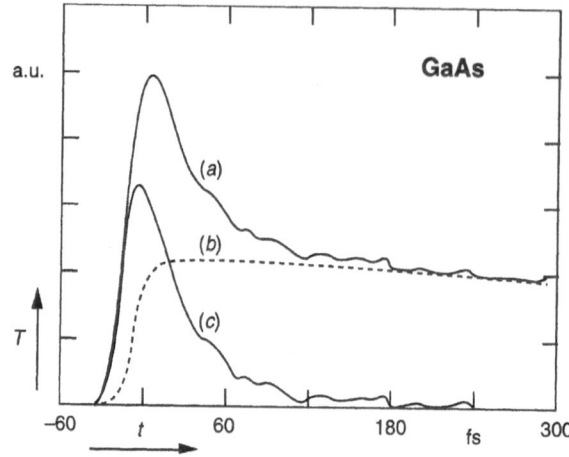

Figure 33. Transient transmittance of a broadband pulse for GaAs. (*a*) Measured signal. (*b*) Curve obtained by convoluting laser pulse envelope with a 1.5-ps exponential decay. (*c*) Transient signal obtained by subtracting (*b*) from (*a*).

due to the slower back-scattering from the side valleys to the central valleys. By cooling the sample to 35 K the bandgap increases by 0.14 eV and scattering to the X-valley is no longer possible. In this case the relaxation time of the fast component increases to about 80 fs indicating that the Γ to L intervalley scattering time is approximately 80 fs and the Γ to X scattering time about 55 fs (together these two scattering mechanisms yield the 33-fs decay time observed in Fig. 33).

III. B. Phonon Dynamics

Next we turn to the dynamics of phonons. A number of groups have studied phonon lifetimes in semiconductors using coherent anti-Stokes Raman spectroscopy [24, 25, 26]. Two picosecond laser pulses of about 2.2 eV are used to coherently drive phonons in the sample and a third laser pulse is used to monitor the decay of the coherent population created by the first two pulses. By delaying the third pulse with respect to the first two, the evolution of the coherent population can be observed. For GaAs at 77 K, the data show a 3-ps rise time (equal to the probe pulse duration) and a 7-ps decay time, indicating that the dephasing time of the coherent population is about 7-ps, close to the (6.3 ± 0.7)-ps phonon lifetime that follows from the spontaneous Raman line width [24].

An entirely different way of exciting phonons in a solid was demonstrated by Nelson and coworkers [27, 28, 29]. Using an excitation pulse shorter than the period of a phonon oscillation it is possible to impulsively excite the phonon oscillation: the laser pulse provides a sharp 'tug' that gets many phonons to oscillate in phase (see Fig. 34*a*). This method, called impulsive stimulated Raman scattering (ISRS), is illustrated in Fig. 34*b*. Two laser beams of identical frequency set up a standing wave pattern in the sample which couples to Raman active phonons. The difference in photon wavevectors **k** must match the wavevector **q** of the phonon:

$$\mathbf{k}_1 + \mathbf{k}_2 = \pm\mathbf{q}. \tag{63}$$

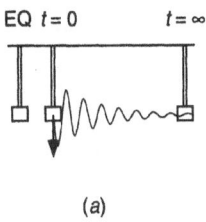

(a)

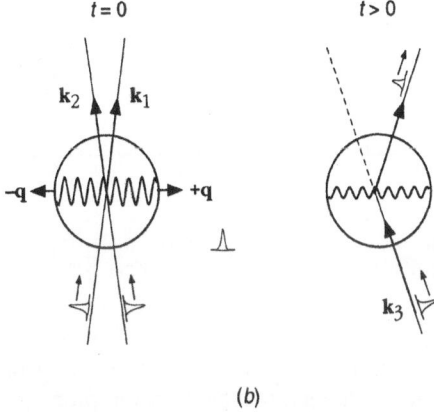

(b)

Figure 34. (*a*) Mass-on-spring model for impulsive stimulated Raman scattering (ISRS): at *t* = 0 the laser pulse provides a 'tug' which gets the atoms to oscillate coherently (*b*) Optical geometry for measuring ISRS signals. After Ref. [27].

As the phonons decay, the standing wave pattern decays. So, if a third beam strikes the decaying pattern, its scattering will reflect the amplitude of the phonon oscillation. The method is not very mode selective: since the dispersion of optic phonons is nearly flat (see Fig. 16*a*), all Raman active modes can satisfy Eq. (63). Figure 35 shows ISRS data for perylene at 18 K [27]. The beating in the signal occurs because of the beating between two Raman active modes of perylene at 80 and 104

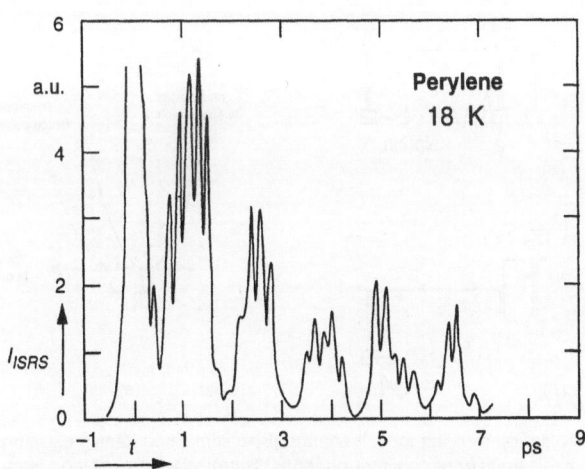

Figure 35. ISRS signal in Perylene at 18 K. After Ref. [27].

449

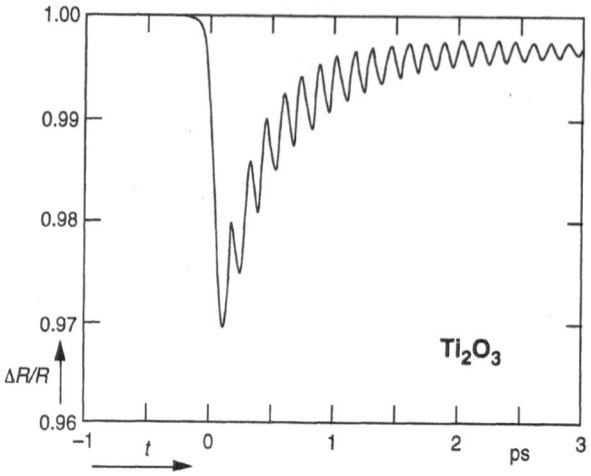

Figure 36. Transient reflectivity changes for Ti₂O₃ after excitation with a 0.1-pJ femtosecond laser pulse. After Ref. [32].

cm^{-1}. The signal decays with a time constant of about 9 ps.

In small bandgap materials a similar type of decaying phonon oscillation can also be observed in a much simpler, two-beam pump-probe geometry [30, 32]. A 60-fs, 10-pJ laser pulse at 1.98 eV is used to excite carriers in a sample and a 1-pJ pulse at the same frequency is used to measure changes in reflectivity. Figure 36 shows the large (3%) oscillations in reflectivity that occur in Ti₂O₃ following the pump pulse. In contrast to the ISRS data shown in Fig. 35, only a single-frequency oscillation at 7 THz is visible even though Ti₂O₃ has two additional Raman active modes around 9 THz. In Ti₂O₃ and other small-bandgap semiconductors and semimetals where this effect was observed, the excited mode always corresponds to a 'breathing' phonon mode (A_1 or A_{1g}). The observed oscillation is attributed to a displacive excitation of coherent phonons, as illustrated in Fig. 37b. Instead of coherently exciting phonons through an impulsive displacement of the ions (Fig.

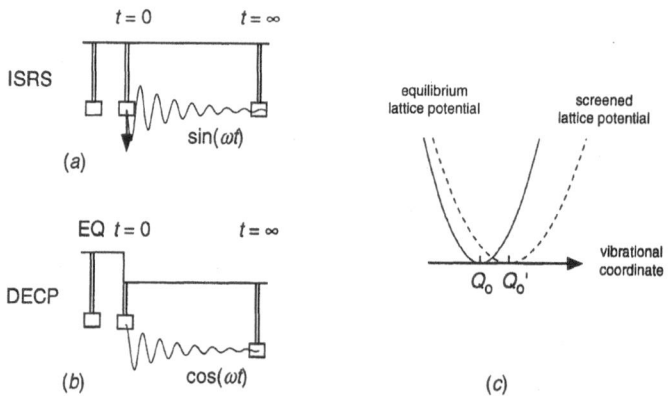

Figure 37. Mass-on-a-spring models for impulsive stimulated Raman excitation (a) and displacive excitation (b) of coherent phonons. The displacive excitation is caused by a sudden screening of the lattice potential by excited carriers. The screened potential has a different equilibrium position, displacing the ions in the lattice.

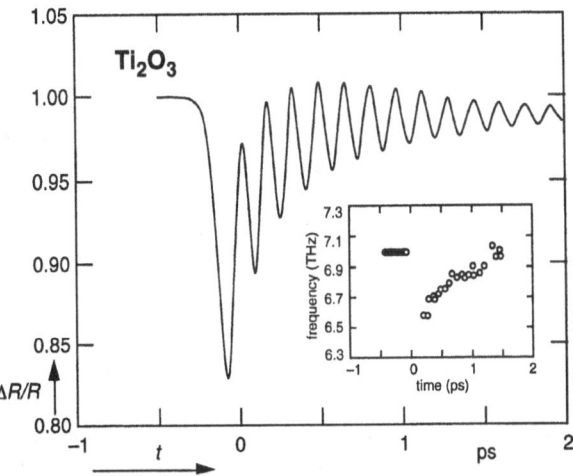

Figure 38. Transient reflectivity changes for Ti_2O_3 after excitation with a 0.1-nJ femtosecond laser pulse. Inset shows evolution of oscilllation frequency [31].

37*a*), this effect is caused by a sudden screening of the lattice potential by the photoexcited carriers (Fig. 37*c*). The screening results in a shift of the equilibrium coordinate of the ions, which in turn causes the ions to oscillate coherently (Fig. 37 *b*). Whereas impulsive stimulated Raman scattering has a sine-dependence on time, displacive excitation should have a cosine-dependence. This has indeed been verified experimentally [32].

As the pump intensity is increased, the reflectivity changes increase dramatically. Using an incident energy of 10 nJ with 2-eV, 70-fs pulses, reflectivity changes as large as 15% are observed (see Fig. 38) [31]. The inset shows the frequency of small segments of the oscillation: right after excitation the frequency of the phonon is almost 10% lower than in equilibrium, clearly showing a softening of the phonon mode (*cf.* the diminished curvature of the screened lattice potential in Fig. 37*c*).

IV. ELECTRONIC AND STRUCTURAL CHANGES INDUCED BY INTENSE SHORT LASER PULSES

As we continue to increase the pump intensity, irreversible changes occur: phonon modes soften to the point where the material loses its structure and undergoes a phase transition. In this final section I will present an overview of our recent work on laser-induced phase transitions in GaAs [33–35].

IV. A. Dielectric Properties during a Phase Transition

The field of laser-induced phase transitions in semiconductors dates back to the discovery in the 1970's that semiconductor crystals could be 'annealed' by irradiation with a short intense laser pulse [36]. Following the first experiments, two models were proposed to explain the structural change resulting from the laser excitation. One model, known as the thermal model, describes the structural change as a thermal melting process [37–39]. The thermal model assumes that the hot electrons rapidly equilibrate with the lattice by exciting lattice vibrations (through phonon

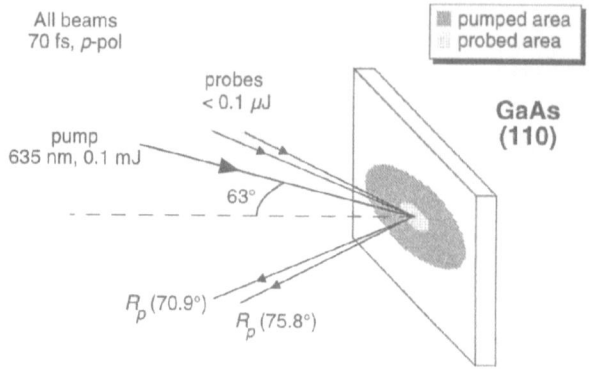

All beams
70 fs, p-pol

probes
< 0.1 μJ

pump
635 nm, 0.1 mJ

63°

pumped area
probed area

GaAs
(110)

R_p (70.9°)

R_p (75.8°)

Figure 39. Two-angle probe geometry for determining the dielectric constant of GaAs. After Ref. [34].

emission). With this assumption, the laser energy deposited in the material can be treated as though it is instantly converted to heat. If the incident laser pulse is strong enough, the irradiated part of the sample will heat up to the melting temperature and undergo a transition to the liquid phase as the latent heat of fusion is supplied.

The other model, known as the plasma model, attributes the structural change to destabilization of the covalent bonds resulting directly from the electronic excitation [40, 41]. The plasma model assumes a slow rate of phonon emission by the excited electronic system compared to the energy deposition time (*i.e.*, the laser pulse width). According to this model, the structural change is driven directly by the excited electronic system. If a high enough fraction of the valence electrons is excited from bonding states to antibonding states, the crystal becomes unstable, and a structural phase transition occurs.

With nanosecond and picosecond pulses the transition from solid to liquid is thermal: because of the fast coupling between carriers and phonons, the electrons and the lattice are in thermal equilibrium and the material simply 'melts.' A detailed account of these thermal transitions, including the heating of the solid, thermal diffusion, melting, motion of the interphase boundary and recrystallization can be found in Ref. [42]. Here, I will concentrate on the highly nonequilibrium dynamics in a semiconductor excited by intense femtosecond pulses.

Reflectivity and second-harmonic generation studies of femtosecond laser-excited semiconductors show evidence of rapid changes in the material (within a few hundred femtoseconds) following the excitation [43–48]. However, interpretation of reflectivity and second-harmonic generation results is difficult because these quantities do not directly yield the behavior of intrinsic material properties. In particular, the reflectivity at a specific wavelength, polarization, and incident angle depends on both the real and imaginary parts of the dielectric constant at that wavelength. Furthermore, the measured second-harmonic radiation depends on the dielectric constant at both the fundamental and second-harmonic wavelengths as well as on the second-order susceptibility.

In the absence of a direct determination of the dielectric constant, interpretation of reflectivity and second-harmonic data requires making assumptions about the functional form of the dielectric constant, for instance that changes in the dielectric constant induced by the excitation are dominated by the free carrier contribution to the optical susceptibility [48, 18, 49]. Under this assumption, the changes in the

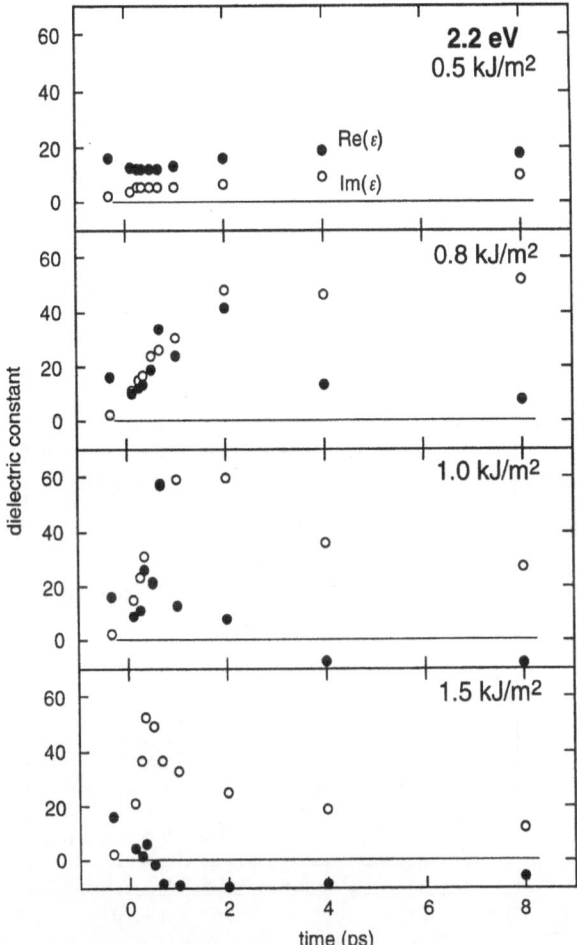

Figure 40. Dielectric constant at 2.2 eV vs. pump-probe time delay for four different pump fluences. ●: Re(ε), ○: Im (ε).

dielectric constant are described by the Drude-model formalism of section I.A (see Eq. 12). While this works well at lower excitation regimes, I will show that it leads to incorrect conclusions in the case of laser-induced disordering experiments.

We directly determined the time evolution of the real and imaginary parts of the dielectric constant using the two-angle probe technique shown in Fig. 39 [33, 34]. A 70-fs laser pulse at a photon energy of 1.9 eV is used to excite a 0.01-mm^2 area on the GaAs surface. The evolution of the reflectivity of the central 6% of the pumped area is measured at two different angles of incidence using 70-fs pulses at a photon energy of 2.2 eV. The probe beam fluence never exceeds 0.1 kJ/m^2 so as not to produce any detectable changes in the dielectric constant. To avoid cumulative damage effects, we translate the sample during data collection so that each data point is obtained at a new spot on the sample.

Each pair of reflectivity measurements is converted to a corresponding real and imaginary part of the dielectric constant by numerically inverting the Fresnel formula for reflectivity as a function of incident angle. Setting one of the probe beam angles

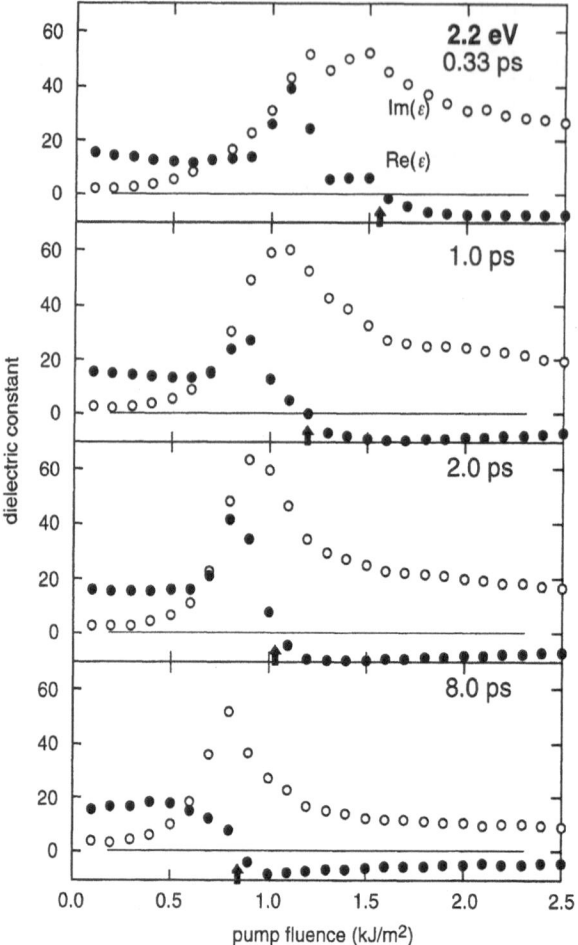

Figure 41. Dielectric constant at 2.2 eV vs. pump fluence for four different pump-probe time delays. ●: Re(ε), ○: Im (ε).

of incidence to the Brewster angle provides good sensitivity in distinguishing changes in Re(ε) from changes in Im(ε) because the p-polarized reflectivity at this angle is determined mainly by Im(ε) [50].

Figures 40 and 41 summarize the experimental values obtained for the dielectric function at 2.2 eV. In Fig. 40, Re(ε) (filled circles) and Im(ε) (open circles) are plotted vs. pump-probe time delay for four different excitation fluences; in Fig. 41, Re(ε) and Im(ε) are plotted vs. pump fluence at four different time delays. The change induced in the dielectric constant by the pump pulse excitation is completely different from that expected from the free carrier contribution to the optical susceptibility. In Fig. 40, at pump fluences near 1 kJ/m² for instance, Im(ε) starts at an initial value of about 2, rises to a peak near 60, and then drops to somewhere between 10 and 15, in strong contrast to the slight, featureless increase predicted by the Drude model. Re(ε), meanwhile, initially decreases slightly but then sharply increases before dropping through zero. Note that the zero-crossing of Re(ε) roughly coincides with the peak in Im(ε).

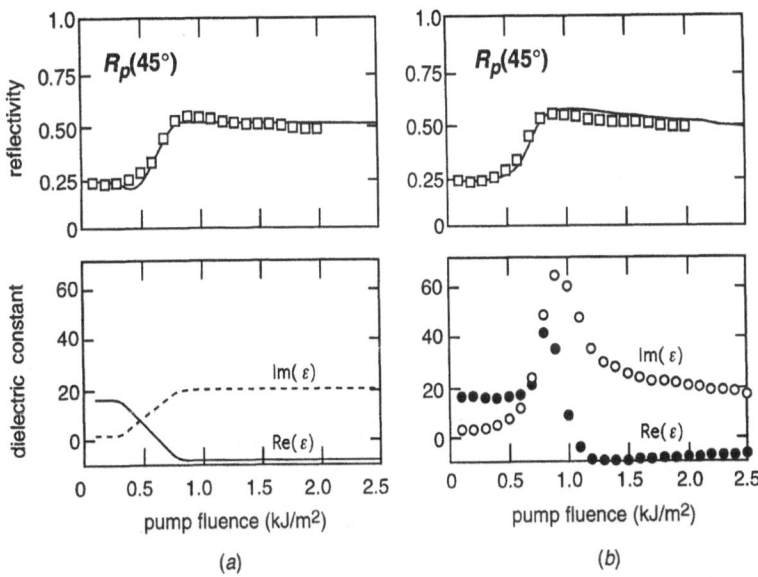

Figure 42. Comparison of (*a*) Drude-like dielectric function fitted to single-angle measurement of the reflectivity with (*b*) measured dielectric function. The top graphs show the *p*-polarized reflectivity at 45° incident angle. ——: reflectivity calculated from dielectric constant, □: measured reflectivity.

Figure 42 shows how misleading a single-angle reflectivity measurement can be. The top graph in Fig. 42*a* shows experimental values for the *p*-polarized reflectivity at a (single) incident angle of 45°, 2 ps after excitation with a 2.2-eV pump pulse. The bottom graph in Fig. 42*a* shows a Drude-like dielectric constant based on Eq. (17), adjusted to reproduce the measured 45°-reflectivity values as shown by the curve through the reflectivity data in the top graph in Fig. 42*a*. The excellent quality of the fit might lead one to conclude that the behavior of the dielectric constant is completely described by the Drude model. The measured dielectric constant, however, behaves very differently as can be seen in the bottom graph in Fig. 42*b*, which shows the experimentally determined dielectric constant at 2.2 eV plotted against pump fluence at a time delay of 2 ps (corresponding to the third graph in Fig. 41). If we use the experimentally determined dielectric constant values from this graph to calculate the corresponding *p*-polarized reflectivity at an incident angle of 45°, we obtain the curve through the data points in the top graph in Fig. 42*b* in excellent agreement with the experimental values. Note, however, how much the actual dielectric constant differs from that following from an assumed Drude-like behavior (bottom two graphs). Only the dielectric constant in Fig. 42*b* correctly accounts for the reflectivities at 45°, 70.9° and 75.8°; the Drude-like dielectric constant only correctly describes the 45° data. Notice, in particular, how the Drude assumption completely misses the strong aborption-like feature in the dielectric function which occurs at 1.0 kJ/m², precisely as the material undergoes a transition (see also Fig. 43).

The results shown in Figs. 40 and 41 indicate that a strong absorption peak comes into resonance with the probe frequency as a result of the excitation. The resonance behavior is most striking in Fig. 41 because the features are particularly clear when plotted versus pump fluence. This behavior must result from an interband absorption peak and not from a free carrier plasma resonance because the

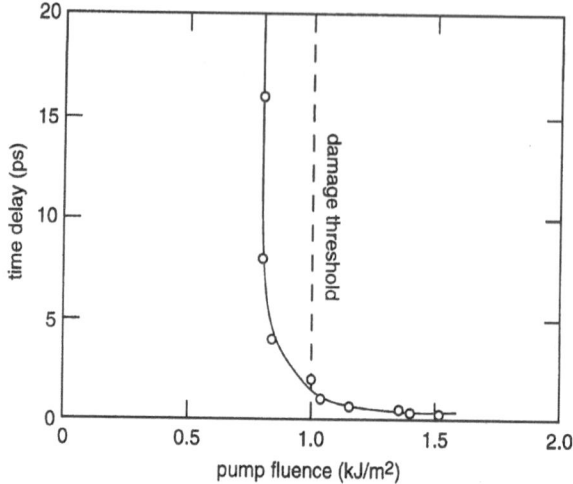

Figure 43. Pump-probe time delays at which Re(ε) = 0 for different pump fluences. The solid curve is drawn to guide the eye, and the dashed line corresponds to the damage threshold of 1.0 kJ/m². At fluences above this value, the induced changes in the material are irreversible while at fluences below the damage threshold, the induced changes are reversible.

zero-crossing in Re(ε) is accompanied by a peak in Im(ε) rather than by a steady increase. From the behavior of Re(ε), we can infer the time evolution of this interband absorption peak. Because Re(ε) is initially positive, the resonant frequency of the observed absorption peak evidently starts out higher than the probe frequency; it then sweeps down through the probe frequency as Re(ε) drops through zero.

The rate at which the resonant frequency of the absorption peak drops through the probe frequency depends on the strength of the excitation: the higher the pump fluence, the faster Re(ε) drops through zero. Figure 43 illustrates this dependence by showing the time delay at which Re(ε) crosses through zero plotted vs. pump fluence. For fluences around 2.0 kJ/m², the absorption peak comes into resonance with the probe frequency within a few hundred femtoseconds; at fluences just above 0.8 kJ/m², on the other hand, the absorption peak takes on the order of 10 picoseconds to come into resonance. For fluences below 0.8 kJ/m², Re(ε) never goes through zero, indicating that the excitation is not strong enough to bring the resonant frequency of the peak down to the probe frequency.

The dashed line in Fig. 43 at 1.0 kJ/m² indicates the threshold fluence for permanent damage to the sample. We determined this threshold by correlating pump pulse fluence with the size of damage spots on the sample measured through a microscope. Above the damage threshold the pump pulse induces irreversible changes in the sample while below the damage threshold the induced changes are reversible. Measurements taken several seconds after the excitation confirm that the dielectric constant eventually returns to its initial value for fluences below the damage threshold (see also Fig. 52). Note, however, that the absorption peak comes into resonance with the probe frequency even for pump fluences below this damage threshold.

Qualitatively, we can approximate the dielectric function of GaAs by a damped single harmonic oscillator with a resonant frequency equal to the average bonding-antibonding splitting [51], which in the ground state is about 4.75 eV [2, 52]. If we

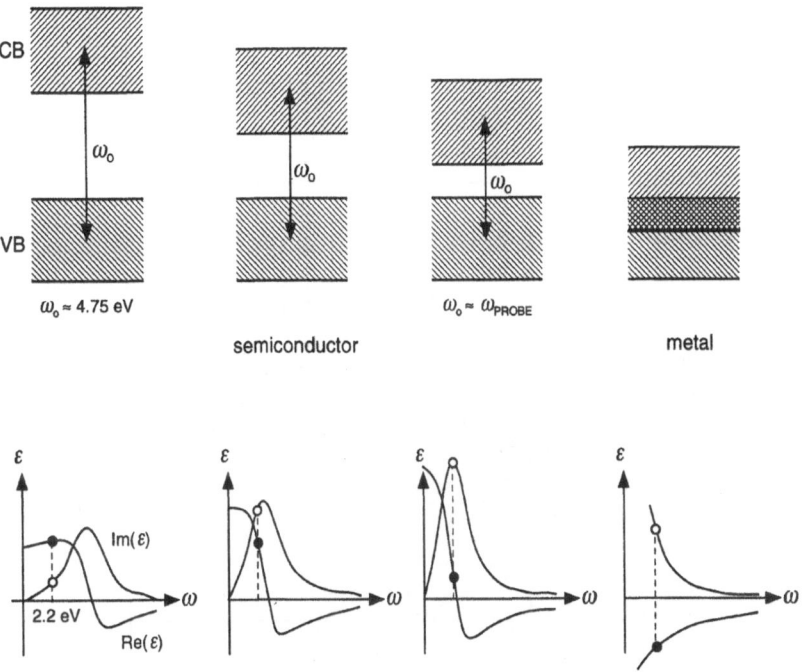

Figure 44. Schematic representation of the bandgap collapse. The pump pulse leads to a drop in the average bonding-antibonding splitting from its initial value of about 4.75 eV to below the probe photon energy of 2.2 eV. The drop in average bonding-antibonding splitting appears as a shift in the main absorption peak in the dielectric function, as illustrated at the bottom.

associate the absorption peak in the data with the harmonic oscillator peak corresponding to the bonding-antibonding splitting, then the observed sweeping of the aborption peak frequency through the probe laser frequency implies a drop in average bonding-antibonding splitting from 4.75 eV to below 2.2 eV. Note that this drop in the average splitting by more than a factor of two occurs even for fluences below the damage threshold, when the induced changes are reversible.

The laser-induced drop in the bonding-antibonding splitting is schematically illustrated in Figure 44. The average bonding-antibonding splitting $\Delta E_{b\text{-}a}$ starts out far above 2.2 eV, so the probe photon energy lies at the foot of the single-oscillator absorption peak where Im(ε) is small (step 1 in Fig. 44). After excitation, $\Delta E_{b\text{-}a}$ begins to decrease, leading to a downward shift in the resonant frequency of the single-oscillator absorption peak and therefore a rise in Im(ε) at 2.2 eV (step 2 in Fig. 44). As $\Delta E_{b\text{-}a}$ drops past 2.2 eV (step 3 in Fig. 44), Im(ε) goes through a peak. If $\Delta E_{b\text{-}a}$ drops far enough, the minimum in the conduction band will drop below the maximum in the valence band and the semiconductor will take on metallic properties (step 4 in Fig. 44).

This interpretation of the 2.2-eV data in terms of a drop in the average bonding-antibonding splitting allows us to predict qualitatively the behavior of the dielectric constant at different photon energies under similar excitation conditions. In particular, for a given excitation strength, the dielectric constant at a probe photon energy between 2.2 eV and 4.75 eV should exhibit resonance features at an earlier

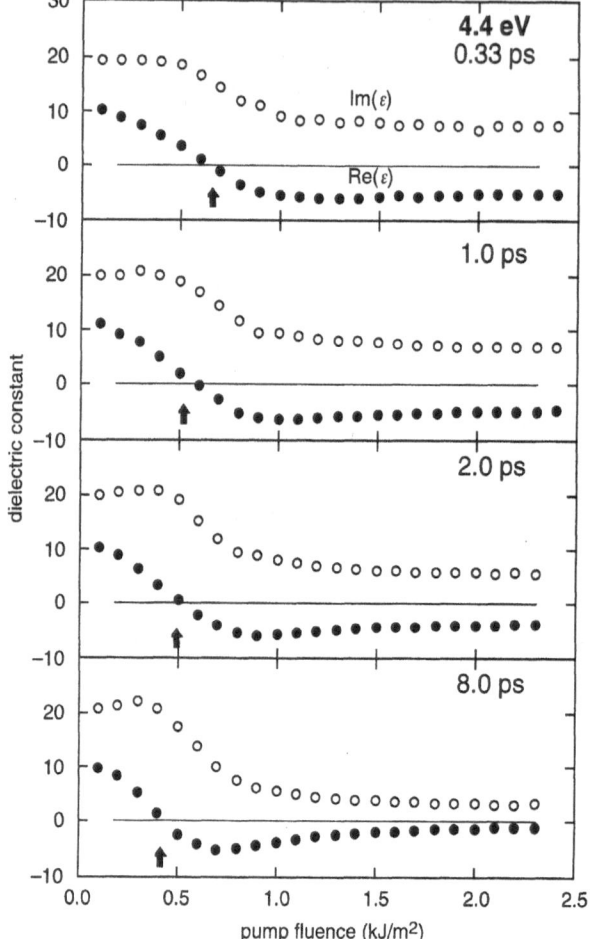

Figure 45. Dielectric constant at 4.4 eV vs. pump fluence for four different pump-probe time delays. ●: Re(ε), ○: Im (ε).

pump-probe time delay than the dielectric constant at 2.2 eV. Equivalently, for a fixed pump-probe time delay, the dielectric constant at a probe photon energy in the above range should exhibit resonance features at a lower pump fluence than the dielectric constant at 2.2 eV.

We verified this prediction by measuring the behavior of the dielectric constant using frequency-doubled probes at 4.4 eV, just below the initial value of the average bonding-antibonding splitting of GaAs. Figure 45 shows time-dependence of the 4.4-eV dielectric constant at the same four pump fluences shown in Fig. 41. Note that Im(ε) now starts near the peak and that the zero-crossing in Re(ε) occur at lower fluences than at 2.2 eV. This can also be seen in Fig. 46, which adds to Fig. 42 the corresponding points for the 4.4-eV data. At 4.4 eV, Re(ε) crosses zero for fluences as low as 0.5 kJ/m^2 well below the 0.8 kJ/m^2 required at 2.2 eV.

The behavior of the dielectric constant at 4.4 eV is indeed consistent with the proposed picture of a drop in the average bonding-antibonding splitting. Following the pump pulse excitation, $\Delta E_{b\text{-}a}$ drops from its initial value of about 4.75 eV first

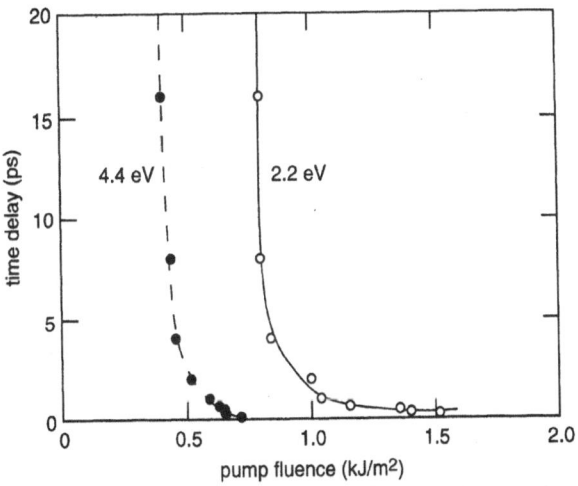

Figure 46. Pump-probe time delays at which Re(ε) = 0 for different pump fluences at both 2.2 eV and 4.4 eV. The curves are drawn to guide the eye. ●: 4.4 eV, ○: 2.2 eV.

past 4.4-eV and then continues down past 2.2-eV. A stronger excitation causes a faster drop through both probe frequencies. At pump fluences between 0.5 kJ/m^2 and 0.8 kJ/m^2, the excitation is strong enough to bring the resonant frequency of the absorption peak below 4.4 eV but not all the way down to 2.2 eV. Note that since 4.4 eV is close to the initial value of the average bonding-antibonding splitting, Im[ε(4.4 eV)] does not rise much above its initial value before coming down.

What underlying physical effects are responsible for this alteration of the band structure? To answer this question, we should examine two main sources of band structure modification: electronic screening and structural change. Through electron-hole pair generation, the pump pulse creates a large population of mobile charge carriers that can partially screen the ionic potential in the material. Since the average bonding-antibonding splitting increases with the strength of the ionic potential [2], electronic screening reduces the average bonding-antibonding splitting. A recent calculation shows that when 10% of the valence electrons are excited to the conduction band, the direct gap at the X-point in the band structure of GaAs will decrease by roughly 2 eV due to electronic screening and many-body bandgap renormalization while the direct gap at the Γ-point changes only slightly [53].

Because the strength of electronic screening increases with the free-carrier density, the effects due to screening should follow the carrier density instantaneously. Therefore, the effect of screening on the dielectric constant should be largest when the free carrier density is highest—*i.e.* immediately following the excitation. As Auger recombination and diffusion reduce the free-carrier density, the influence of electronic screening on the band structure should correspondingly decrease. However, as can be seen in Fig. 40, the changes in dielectric constant continue to grow for picoseconds following excitation, when the carrier density is already decreasing from its peak value. Clearly, electronic screening cannot, by itself, account for the observed behavior.

Since the band structure is determined by the crystal structure, changes in the atomic arrangement can also account for the observed collapse of the bandgap. Deformation of the diamond or zincblende structure in a semiconductor is known to lead to a collapse of the band gap and a semiconductor-metal transition [2, 54–56]. Just a 10% change in average bond length is enough to cause a semiconductor-metal transition [56]. Note that an ionic velocity as small as 25 m/s is already sufficient to achieve a 10% change in the GaAs bond length within 1 ps.

In semiconductors like GaAs the covalent bonds are stabilized by the valence electrons, so excitation of a large number of electrons from bonding valence states to antibonding conduction states can lead directly to lattice instability [40, 57, 58]. If the femtosecond laser pulse is intense enough to excite this critical density of electrons, the resulting instability causes the lattice to deform — a deformation which begins immediately following the excitation but continues long after the incidence of the pump pulse. The change in the dielectric constant accompanying the lattice deformation should therefore continue to increase in the picoseconds following the excitation, in agreement with the observed behavior of the dielectric constant.

IV. B. Nonlinear Optical Properties

To further investigate the possibility of structural changes following femtosecond laser excitation, we also studied the second-order susceptibility of GaAs [35]. Because of its sensitivity to crystal symmetry, second-harmonic generation has been used by a number of researchers to study laser-induced phase transitions in semiconductors [43, 45–48, 59–61]. The sensitivity of second-harmonic generation to the symmetry properties of a nonlinear crystal arises from the dependence of second-harmonic generation on the material's second-order optical susceptibility $\chi^{(2)}$, which reflects the symmetry group of the crystal [8]. A change in the material's symmetry properties, such as may occur in a phase transition, affects $\chi^{(2)}$ and results in a change in the detected second-harmonic signal. However, the detected second-harmonic signal depends on more material properties than just $\chi^{(2)}$. In particular, it depends also on the values of the linear optical susceptibility $\chi^{(1)}$ (or, equivalently, the linear dielectric constant ε) at both the fundamental frequency ω and the second-harmonic frequency 2ω of the probe beam used for second-harmonic generation [8]. One must therefore take into account the changes in $\varepsilon(\omega)$ and $\varepsilon(2\omega)$ to extract the behavior of $\chi^{(2)}$ from second-harmonic generation measurements. As I will show, the experimentally-determined changes in linear optical properties presented in the previous section have a significant effect on the detected second-harmonic signal.

As we have seen in section I.D, the second-order susceptibility $\chi^{(2)}$ determines the nonlinear polarization $\mathbf{P}^{(2)}(2\omega)$, induced by an electric field $\mathbf{E}(\omega)$, which in turn acts as a source for the second-harmonic field $\mathbf{E}(2\omega)$, see Eqs. (32–38). The dielectric constant influences $\mathbf{P}^{(2)}(2\omega)$ by affecting the orientation and magnitude of $\mathbf{E}(\omega)$ relative to the crystallographic axes. As an example, if $\mathbf{E}(\omega)$ is oriented in the y-z plane of a crystal and at an angle θ to the y-axis, then the nonlinear polarization produced through the xyz element of the $\chi^{(2)}$ tensor, is given by

$$P_x^{(2)}(2\omega) = \chi_{xyz}^{(2)} E(\omega) \cos\theta E(\omega) \sin\theta = \tfrac{1}{2} \chi_{xyz}^{(2)} E^2(\omega) \sin(2\theta). \qquad (64)$$

Changes in the dielectric constant affect the second-order polarization through the angle of refraction θ. If the beam at frequency ω strikes the crystal at some angle θ_i, then, through Snell's Law, $\varepsilon(\omega)$ determines the angle θ inside the material that appears in Eq. (64). Thus, if the dielectric constant changes, q changes. In addition,

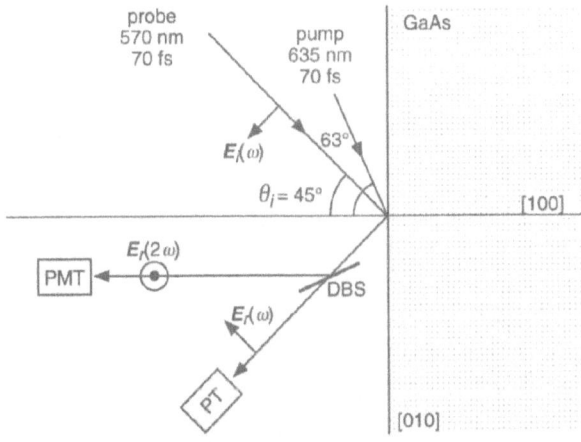

Figure 47. Probing geometry for second-harmonic generation measurements. A p-polarized probe beam is incident at 45° on a (100) GaAs sample in air. An ultraviolet mirror separates the reflected second-harmonic radiation, which is s-polarized, from the reflected fundamental radiation, which is still p-polarized. PMT: photomultiplier tube, PT: phototube, DBS: dichroic beamsplitter.

through the Fresnel formulas for reflection and refraction, $\varepsilon(\omega)$ also determines the magnitude of the field in the material for a given incident field. Note that the field magnitude $E(\omega)$ appearing in Eq. (64) is the field magnitude inside the material, so changes in reflectivity can have a significant effect on $\mathbf{P}^{(2)}(2\omega)$.

In addition to affecting the second-order polarization directly, the dielectric constant also affects the amount of second-harmonic radiation generated from a given $\mathbf{P}^{(2)}(2\omega)$. Firstly, the index of refraction, given by $n(\omega) = \mathrm{Re}\left[\sqrt{\varepsilon(\omega)}\right]$, determines the phase velocity for light at frequency ω. The efficiency of second-harmonic generation depends on $n(\omega) - n(2\omega)$, which determines the phase mismatch between the induced nonlinear polarization and the resulting second-harmonic field: the larger the phase mismatch, the less efficient is second-harmonic generation [8]. The phase mismatch is important especially in transmission because it determines the length scale over which second-harmonic radiation generated by the propagating fundamental (frequency ω) beam adds with the proper phase to the propagating second-harmonic beam generated earlier along the beam path. Another way in which $\varepsilon(\omega)$ and $\varepsilon(2\omega)$ affect second-harmonic generation is by determining the absorption depth at ω and at 2ω. The absorption depth can affect second-harmonic generation by limiting the interaction length over which this process takes place [8]. If the energy depleted from the fundamental beam by second-harmonic generation is negligible, as it is in our experiment, then the total amount of second-harmonic generation produced is proportional to the square of the interaction length. So if the absorption depth at one or both of the frequencies becomes smaller than the interaction length, production of second-harmonic radiation will decrease accordingly. Finally, by modifying the reflection of the second-harmonic radiation at the surface of a material, changes in $\varepsilon(2\omega)$ affect the amount of second-harmonic radiation that gets out of the material.

The second-harmonic measurements were carried out in a reflection geometry, as illustrated in Fig. 47 [35]. In this geometry, the reflected second-harmonic field amplitude is given by [7]

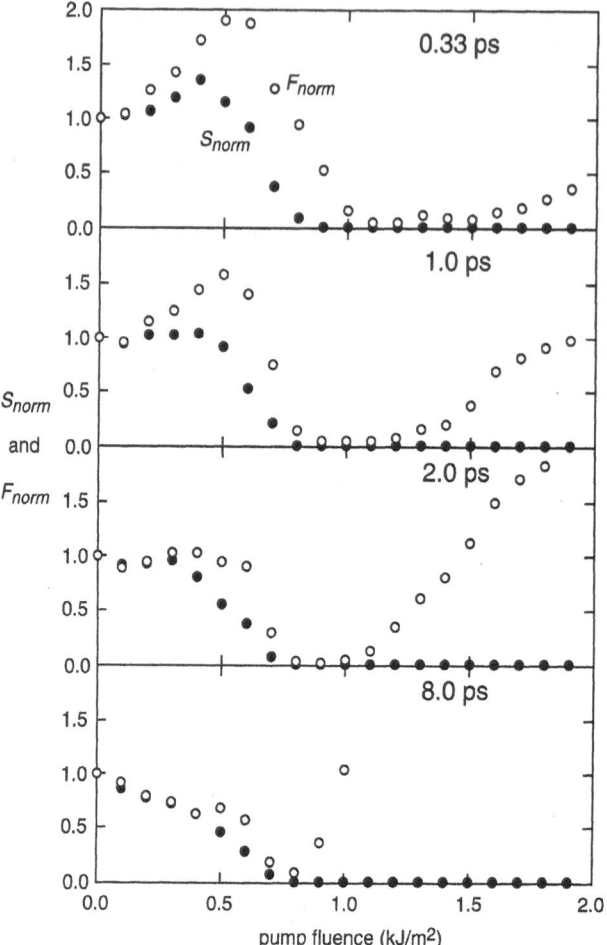

Figure 48. Normalized second-harmonic signal vs. pump fluence for four different pump-probe time delays. ●: measured second-harmonic signal, S_{norm} ; ○: second-harmonic signal calculated based solely on measured changes in dielectric constant, F_{norm} .

$$E_r(2\omega) = -4\pi P^{(2)}(2\omega)\left\{\frac{\sqrt{\varepsilon(\omega)-\sin^2\theta_i} - \sqrt{\varepsilon(2\omega)-\sin^2\theta_i}}{[\varepsilon(\omega)-\varepsilon(2\omega)]\left[\sqrt{\varepsilon(2\omega)-\sin^2\theta_i} - \cos\theta_i\right]}\right\}, \quad (65)$$

where

$$P^{(2)}(2\omega) = 2\chi^{(2)}E_i^2(\omega)\left\{\frac{4\sin\theta_i\cos^2\theta_i\sqrt{\varepsilon(\omega)-\sin^2\theta_i}}{\left[\varepsilon(\omega)\cos\theta_i + \sqrt{\varepsilon(\omega)-\sin^2\theta_i}\right]^2}\right\}. \quad (66)$$

The intensity of the reflected second-harmonic signal, and therefore the detected signal S, is proportional to the square of the field amplitude in Eq. (65). Note from

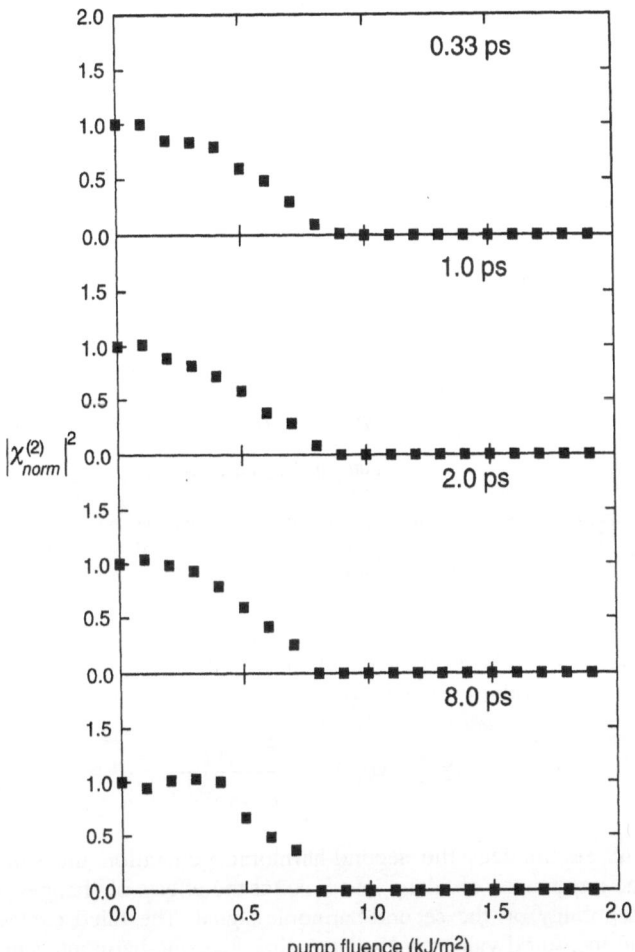

Figure 49. Square of the normalized second-order susceptibility vs. pump fluence for four different pump-probe time delays.

Eqs. (65) and (66) that we can separate the dependence of the detected signal on the dielectric constant from its dependence on the second-order susceptibility as follows:

$$\frac{S}{\left|E_i^2(\omega)\right|^2} = F\left[\theta_i, \varepsilon(\omega), \varepsilon(2\omega)\right] \times \left|\chi^{(2)}\right|^2,\tag{67}$$

where the function $F[\theta_i, \varepsilon(\omega), \varepsilon(2\omega)]$ depends only on the dielectric constant and incident angle and not on the second-order susceptibility. We can define the normalized second-harmonic signal $S_{norm}(\phi,t) \equiv S(\phi,t)/S(0,0)$, where $S(\phi,t)$ is the measured signal as a function of excitation fluence ϕ and pump-probe time delay t, and $S(0,0)$ is the second-harmonic signal detected in the absence of any excitation. Similarly, we define $F_{norm}(\phi,t) \equiv F(\phi,t)/F(0,0)$ and $\chi^{(2)}_{norm}(\phi,t) \equiv \chi^{(2)}(\phi,t)/\chi^{(2)}(0,0)$. Using the data in the previous section for $\varepsilon(\omega,\phi,t)$ and $\varepsilon(2\omega,\phi,t)$, with $\omega = 2.2$ eV and

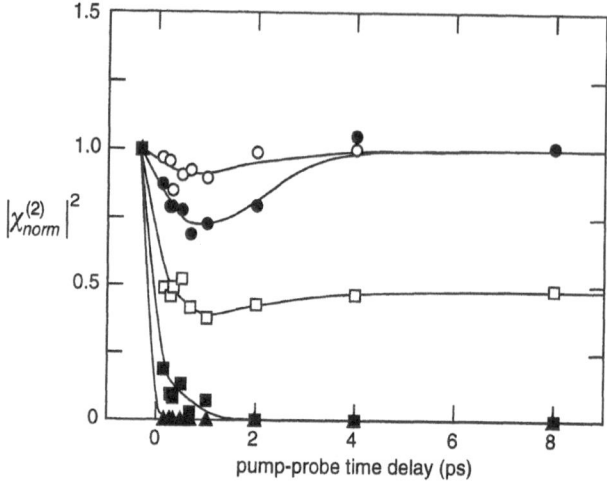

Figure 50. Square of the second-order susceptibility vs. pump-probe time delay for various pump fluences. The curves are drawn to guide the eye. ○: 0.2 kJ/m², ●: 0.4 kJ/m², □: 0.6 kJ/m², ■: 0.8 kJ/m², ▲: 1.5 kJ/m².

$2\omega = 4.4$ eV, we can calculate $F_{norm}(\phi,t)$. Then, with the measured values of $S_{norm}(\phi,t)$, we get from Eq. (67) values for $\left|\chi^{(2)}_{norm}(\phi,t)\right|^2$:

$$\left|\chi^{(2)}_{norm}(\phi,t)\right|^2 = \frac{S_{norm}(\phi,t)}{F_{norm}(\phi,t)}. \tag{68}$$

Figure 48 summarizes the second-harmonic generation measurements and highlights the importance of taking into account the effects of changes in the linear optical susceptibility on the second-harmonic signal. The filled circles in Fig. 48 represent the measured values of the normalized second-harmonic signal $S_{norm}(\phi,t)$ plotted versus pump fluence ϕ at the same four pump-probe delays as in Figs. 41 and 45. The open circles in the figure represent the values of $F_{norm}(\phi,t)$ obtained from the measured dielectric constant at 2.2 and 4.4 eV using Eqs. (65)–(67). Note, from Eq. (68), that $F_{norm}(\phi,t)$ is identical to $S_{norm}(\phi,t)$ if $\chi^{(2)}_{norm}(\phi,t)$ is held constant at its initial value of 1. In other words, $F_{norm}(\phi,t)$ (the open circles in Fig. 48) shows the changes that result in the second-harmonic signal *solely* from the behavior of $\chi^{(1)}$ following the excitation; it does not include any effects from changes in $\chi^{(2)}$. Clearly, the changes induced in dielectric constant have a significant impact on the measured second-harmonic signal. It is important to point out, however, that while the changes in $F_{norm}(\phi,t)$ play an important role in the behavior of $S_{norm}(\phi,t)$, $F_{norm}(\phi,t)$ never drops below the experimental noise and therefore cannot account for the vanishing of $S_{norm}(\phi,t)$. Thus, $\chi^{(2)}$ must go to zero for $S_{norm}(\phi,t)$ to reach zero.

Following Eq. (5), we can now obtain $\left|\chi^{(2)}_{norm}(\phi,t)\right|^2$ by dividing $S_{norm}(\phi,t)$ (the filled circles in Fig. 48) by $F_{norm}(\phi,t)$ (the open circles in Fig. 48). The resulting values of $\left|\chi^{(2)}_{norm}(\phi)\right|^2$ appear in Fig. 48, shown at the same four pump-probe time delays as in Fig. 49. Figure 50 illustrates the corresponding time dependence of $\left|\chi^{(2)}_{norm}(t)\right|^2$. The results exhibit a range of behaviors, depending on the excitation strength. At pump fluences of 0.8 kJ/m² and higher, $\chi^{(2)}$ goes to zero at a rate that increases with pump fluence: at 0.8 kJ/m² it takes about 2 ps to reach zero while at 1.5 kJ/m² it

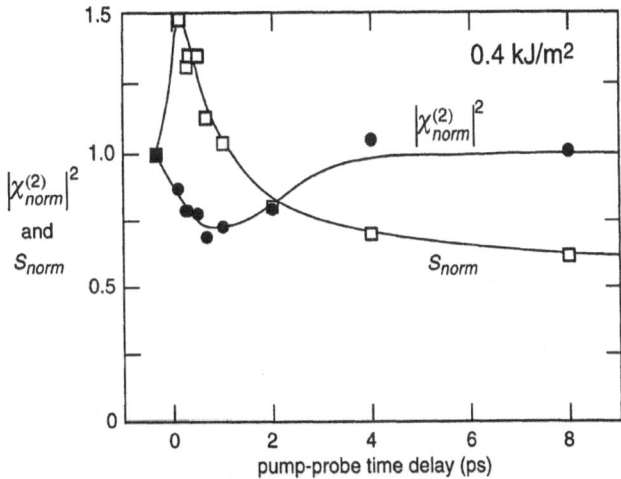

Figure 51. Normalized second-harmonic signal (□) and square of the second-order susceptibility (●) vs. pump-probe time delay at a pump fluence of 0.4 kJ/m². The curves are drawn to guide the eye.

reaches zero within a pump-probe time delay of 130 fs. In contrast, at pump fluences less than or equal to 0.6 kJ/m², $\chi^{(2)}$ undergoes a partial decrease, but it does not reach zero. For pump fluences below 0.5 kJ/m², $\chi^{(2)}$ recovers to its initial value on a time scale of a few picoseconds.

Figure 51 shows how important it is to take the changes in linear properties into account. At an excitation of 0.4 kJ/m² the second-harmonic signal first rises and then drops below its initial value within a few picoseconds. After correction for changes in the linear dielectric constant, however, we see that $\left|\chi^{(2)}_{norm}\right|^2$ first *decreases* and then *recovers* to its initial value within a few picoseconds.

We also measured both the second-harmonic signal and the linear reflectivity at

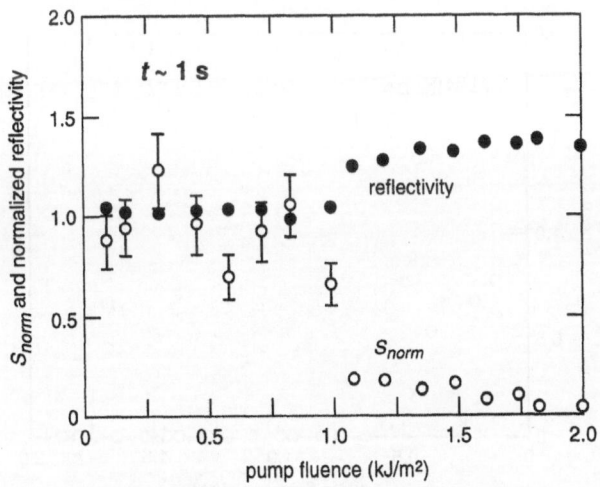

Figure 52. Normalized second-harmonic signal (○) and normalized 45° reflectivity (●) at a time delay of about a second, after the material has reached its final state.

465

a time delay of a few seconds (Fig. 52) as well as the second-harmonic signal at a pump-probe time delay of 100 ps (Fig. 53). Figure 52 shows a sharp demarcation at 1.0 kJ/m² in the final state of $\chi^{(2)}$. For excitation strengths below 1.0 kJ/m², both the second-harmonic signal and the linear reflectivity eventually return to their initial values. However, neither the second-harmonic signal nor the linear reflectivity ever returns to its initial value for pump fluences above 1.0 kJ/m². Thus, the changes induced in the material by the laser-pulse excitation are reversible if the pump fluence is below 1.0 kJ/m² but are irreversible if the pump fluence is above 1.0 kJ/m², consistent with inspection of the sample through a microscope. Figure 53 shows that for fluences greater than 0.6 kJ/m², once S_{norm} vanishes, it remains zero for at least 100 ps. Thus, the recovery of $\chi^{(2)}$ to its initial value at fluences between 0.6 and 1.0 kJ/m² occurs on a time scale which is orders of magnitude larger than the recovery times for fluences less than or equal to 0.5 kJ/m².

The data in Fig. 50 suggest three main regimes of behavior for $\chi^{(2)}$ following laser-pulse excitation. In the low-fluence regime, below 0.5 kJ/m², $\chi^{(2)}$ exhibits a partial drop but recovers to its initial value within a few picoseconds. At medium fluences, from roughly 0.8 to 1.0 kJ/m², $\chi^{(2)}$ drops to zero on a time scale between a few hundred femtoseconds and a few picoseconds and remains zero for over 100 ps but eventually also recovers to its initial value. In the high-fluence regime, above 1.0 kJ/m², $\chi^{(2)}$ drops to zero within a few hundred femtoseconds and never recovers to its initial value. While a clear boundary at 1.0 kJ/m² separates the medium- and high-fluence regimes, no clear boundary separates the low- and medium-fluence regimes. Rather, the behavior gradually changes from low-fluence behavior to medium-fluence behavior between 0.5 kJ/m² and 0.8 kJ/m².

At low-fluence, the behavior of $\chi^{(2)}$ is most likely dominated by electronic effects. The laser-pulse excitation of electrons from the valence to the conduction band can directly affect $\chi^{(2)}$ in a number of ways. Firstly, delocalized conduction electrons contribute little to $\chi^{(2)}$ in GaAs compared to localized valence electrons [62–64]. Thus, we expect the excitation of a high density of electrons from the valence band to the conduction band by the pump pulse (> 10^{21} cm⁻³) to cause a drop in $\chi^{(2)}$. Secondly, the excited free carriers can also reduce the interband contribution to

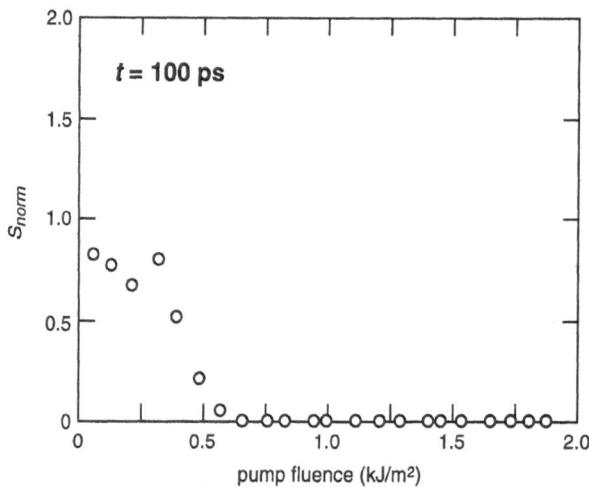

Figure 53. Normalized second-harmonic signal vs. pump fluence at a time delay of 100 ps.

$\chi^{(2)}$ by bleaching resonant transitions due to phase-space filling. In addition, electronic screening of the ionic potential by the free carriers will modify the electronic band structure [53], further affecting the interband contribution to $\chi^{(2)}$. Because the magnitude of these electronic effects should depend on the free-carrier density, we expect $\chi^{(2)}$ to return to its initial value as the excited free-carrier density relaxes through Auger recombination and diffusion following the excitation. The picosecond time scale for the observed recovery of $\chi^{(2)}$ in the low-fluence regime agrees with the time scale for Auger recombination and diffusion at free-carrier densities on the order of 10^{21} cm^{-3} [13, 14]. While the low-fluence recovery time is consistent with the expected behavior of the free carrier population, the source of the initial 1-ps delay between the excitation and the maximum change in $\chi^{(2)}$ in this fluence regime is unclear. One possible cause is that in this fluence regime the electronic excitation drives a slight structural distortion in the lattice, which is reversed when the carrier density decreases through recombination. If .it is a temporary distortion of the lattice that is responsible for the partial drop in $\chi^{(2)}$, the picosecond delay in the drop would result from the timescale of the atomic motion.

At medium and high fluence, the behavior of $\chi^{(2)}$ is not consistent with electronic effects. The recovery time for $\chi^{(2)}$ exceeds 100 ps which is much larger than typical electronic relaxation times of a few picoseconds. Moreover, the vanishing of $\chi^{(2)}$ at fluences greater than 0.6 kJ/m^2 cannot be accounted for by the roughly 10% valence band depopulation achieved by the pump pulse [64]. The behavior of $\chi^{(2)}$ is therefore most likely dominated by structural changes in the lattice, which could occur on the observed time scales [58]. Recovery times for reversible structural changes should be comparable to lattice relaxation times, which are much greater than 100 ps [65].

What kind of structural change is responsible for the vanishing of second-order susceptibility? While $\chi^{(2)} = 0$ in materials that have a center of inversion [8], the vanishing of $\chi^{(2)}$ does not necessarily mean that the material has taken on a true center of inversion within each unit cell. A loss of long-range order on the scale of the wavelength of light is sufficient to cause such a drop in $\chi^{(2)}$. Experiments show that the degree of amorphization induced by low-dosage ion implantation (1×10^{12} – 6×10^{15} cm^{-2} integrated flux of 80-keV Te$^+$ and S$^+$ ions) leads to a one-to-two order of magnitude drop in $\chi^{(2)}$ [61]. The extent of ionic motion required for GaAs to lose long-range order is much smaller than that required for GaAs to take on a local center of inversion in each unit cell. Given the time scales involved in the data, a loss of long-range order is the most likely explanation for the observed drop in $\chi^{(2)}$ to zero in the medium- and high-fluence regimes.

The second-harmonic data thus support the conclusion that the pump pulse induces an instability in the covalent bonding of GaAs that leads to structural change in the lattice. The instability, which results from the excitation of a critical density of electrons from bonding valence states to antibonding conduction states [40, 57, 58] occurs instantaneously with the generation of free carriers. Because the zincblende structure is no longer stable, the ions start to move away from their equilibrium positions and the material loses its long-range order causing $\chi^{(2)}$ to vanish. A stronger excitation causes a greater instability and, therefore, faster ionic motion and a faster drop in $\chi^{(2)}$ and a faster collapse of the bandgap.

In conclusion, our experiments on GaAs show three regimes of behavior: (1) at low fluences, below 0.5 kJ/m^2, the dielectric function shows a resonance that sweeps down through 4.4 eV; the second-order susceptibility exhibits a partial drop and a recovery to its initial value within a few picoseconds, (2) at medium fluences, between 0.8 and 1.0 kJ/m^2, the dielectric function shows resonance that sweeps down past 2.2 eV, while $\chi^{(2)}$ drops to zero but recovers to its initial value on a time scale greater than 100 ps, and (3) at high fluences, above 1.0 kJ/m², the changes occur more rapidly and the material undergoes permanent changes, in particular, $\chi^{(2)}$ vanishes and never recovers to its initial value.

The behavior of the dielectric function cannot be explained using a Drude model. Instead the data indicate a drop in the average bonding-antibonding splitting indicating that major changes occur in the electronic band structure of the sample. The second-harmonic data show a loss of long-range order, which suggests that structural changes occur following the laser-pulse excitation. These changes result form the destabilization of the covalent bonds and are very different from the thermal changes induced by picosecond or longer pulses.

Particularly interesting, at excitations between 0.8 and 1.0 kJ/m², the crystal undergoes sufficiently large structural changes to lose its long-range order but eventually reverts back to its original state. In this regime, also, the average bonding-antibonding splitting drops by more than a factor of two from 4.75 eV to beyond 2.2 eV, but it, too, recovers. The ability to induce such large transient changes in a semiconductor's structural, electronic, and optical properties may be relevant for the development of optoelectronic switching devices.

ACKNOWLEDGEMENTS

I thank Paul Callan and Li Huang for their help with the preparation of these lectures and the figures in this chapter and Prof. Claus Klingshirn and Eli Glezer for many valuable comments. The research described in section IV is sponsored by the Joint Services Electronics Program under contract number ONR N00014-89-J-1023.

REFERENCES

1. J.D. Jackson, *Classical Electrodynamics* (Wiley, New York, 1975).
2. W.A. Harrison, *Electronic Structure and the Properties of Solids: The Physics of the Chemical Bond*; (Dover, New York, 1989).
3. N.W. Ashcroft, N.D. Mermin, *Solid State Physics* (Saunders College, Philadelphia, 1976).
4. D.J. Griffiths, Introduction to Quantum Mechanics (Prentice Hall, New Jersey, 1995).
5. C. Kittel, *Introduction to Solid State Physics* (Wiley, New York, 1976).
6. R.F. Pierret, *Advanced Semiconductor Fundamentals*, Vol. VI of Modular Series on Solid State Devices (Addison Wesley, Reading, MA, 1990).
7. N. Bloembergen, and P.S. Pershan, *Phys. Rev.* **128**, 606 (1962).
8. R. Shen, *The Principles of Nonlinear Optics* (Wiley Interscience, New York, 1984).
9. E. Merzbacher, *Quantum Mechanics* (Wiley, New York , 1970).
10. M. Lundstrom, *Fundamentals of carrier transport*, Vol. X of Modular Series on Solid State Devices (Addison Wesley, Reading, MA, 1990).
11. K. Seeger, *Semiconductor Physics* (Springer Verlag, 1991).
12. C.F. Klingshirn, *Semiconductor Optics* (Springer Verlag, 1995).
13. E.J. Yoffa, *Phys. Rev. B* **21**, 2415 (1980).
14. H.M. van Driel, *Phys. Rev. B* **35**, 8166 (1987).
15. H. Kalt, M. Rinker, *Phys. Rev. B*, **45**, 1139 (1992).
16. A. Othonos, H.M. van Driel, J.F. Young, P.J. Kelly, *Phys. Rev. B*, **43**, 6682 (1991).
17. H.R. Choo, X.F. Hu, M.C. Downer, *Appl. Phys. Lett.*, **63**, 1507 (1993).

18. C.V. Shank, R. Yen, C. Hirlimann, *Phys. Rev. Lett.* **50**, 454 (1983).

19. W.-Z. Lin, J.G. Fujimoto, E.P. Ippen, R.A. Logan, *Appl. Phys. Letts.* **50**, 124 (1987).

20. W.-Z. Lin, R.W. Schoenlein, J.G. Fujimoto, *IEEE J. of Quan. Elec.* **24**, 267 (1988).

21. J.L. Oudar, D. Hulin, A. Migus, A. Antonetti, F. Alexandre, *Phys. Rev. Lett.* **55**, 2074 (1985).

22. J.A. Kash, J.C. Tsang, J.M. Hvam, *Phys. Rev. Lett.* **54**, 2151 (1985).

23. P.C. Becker, H.L. Fragnito, C.H.B. Cruz, J. Shah, R.L. Fork, J.E. Cunningham, J.E. Henry, C.V. Shank, *Appl. Phys. Letts.* **53**, 2089 (1988).

24. D. von der Linde, J. Kuhl, H. Klingenberg, *Phys. Rev. Lett.* **44**, 1505 (1980).

25. J. Kuhl, D. von der Linde, in *Picosecond Phenomena III*, Springer Ser. Chem. Phys. **23**, 201 (1982).

26. W.E. Bron, J. Kuhl, B.K. Rhee, *Phys. Rev. B* **34**, 6961 (1986).

27. S. de Silvestri, J.G. Fujimoto, E.P. Ippen, E.B. Gamble, Jr., L.R. Williams, K.A. Nelson., *Chem. Phys. Letts.* **116**, 146 (1985).

28. L. Dhar, J.A. Rogers, K. Nelson, *Chem Rev.* **94**, 157 (1994).

29. G.P. Wiederrecht, T.P. Dougherty, L. Dhar, K.A. Nelson, D.E. Leaird, A.M. Weiner, *Phys. Rev. B* **51**, 916 (1995).

30. T.K. Cheng, S.D. Brorson, A.S. Kazeroonian, J.S. Moodera, G. Dresselhaus, M.S. Dresselhaus, E.P. Ippen, *Appl. Phys. Letts.* **57**, 1004 (1990).

31. T.K. Cheng, L.H. Acioli, J. Vidal, H.J. Zeiger, G. Dresselhaus, M.S. Dresselhaus, E.P. Ippen, *Appl. Phys. Letts.* **62**, 1901 (1993).

32. H.J. Zeiger, J. Vidal, T.K. Cheng, E.P. Ippen, G. Dresselhaus, and M.S. Dresselhaus, *Phys. Rev. B* **45**, 768 (1992).

33. Y. Siegal, E.N. Glezer, E. Mazur, *Phys. Rev. B* **49**, 16403 (1994).

34. E.N. Glezer, Y. Siegal, L. Huang, E. Mazur, *Phys. Rev. B* **51**, 6959 (1995).

35. E.N. Glezer, Y. Siegal, L. Huang, E. Mazur, *Phys. Rev. B* **51**, 9589 (1995).

36. E.I. Shtyrkov, I.B. Khaibullin, M.M. Zaripov, M.F. Galyatudinov, R.M. Bayazitov, *Sov. Phys-Semicond.*, **9**, 1309 (1976).

37. R.F. Wood, C.W. White, R.T. Young, Eds. *Semiconductors and Semimetals* Vol. **23** (Academic Press, New York, 1984).

38. A.M. Malvezzi, H. Kurz, N. Bloembergen, in *Energy Beam-Solid Interactions and Transient Thermal Processing*, Eds. D. K. Biegelsen, G.A. Rozgonyi and C.V. Shank (Materials Research Society, Pittsburgh, 1985).

39. A.M. Malvezzi, in *Excited-State Spectroscopy in Solids*, Eds. U.M. Grassano, N. Terzi (North-Holland, Amsterdam, 1987).

40. J.A. Van Vechten, R. Tsu, and F.W. Saris, *Phys. Lett.* **74A**, 422 (1979).

41. J.A.. Van Vechten, in *Semiconductors Probed by Ultrafast Laser Spectroscopy*, Ed. R.R. Alfano (Academic Press, San Diego, 1984).

42. D.K. Biegelsen, G.A. Rozgonyi, C.V. Shank, Eds., *Energy Beam-Solid Interactions and Transient Thermal Processing*, Vol. **35** (Materials Research Society, Pittsburgh, 1985).

43. C. V. Shank, R. Yen, and C. Hirlimann, *Phys. Rev. Lett.* **51**, 900 (1983)..

44. M.C. Downer, R.L. Fork, and C.V. Shank, J. Opt. Soc. Am. B **2**, 595 (1985).

45. H.W.K. Tom, G.D. Aumiller, and C.H. Brito-Cruz, *Phys. Rev. Lett.* **60**, 1438 (1988).

46. S.V. Govorkov, I.L. Shumay, W. Rudolph, and T. Schroeder, *Opt. Lett.* **16**, 1013 (1991).

47. K. Sokolowski-Tinten, H. Schulz, J. Bialkowski, and D. von der Linde, Appl Phys A **53**, 227 (1991).

48. P. N. Saeta, J. Wang, Y. Siegal, N. Bloembergen, and E. Mazur, Phys. Rev. Lett. **67**, 1023 (1991).

49. H. Kurz, and N. Bloembergen, in *Energy Beam-Solid Interactions and Transient Thermal Processing*, Eds. D.K. Biegelsen, G.A. Rozgonyi and C.V. Shank (Materials Research Society, Pittsburgh, 1985).

50. D.L. Greenaway, and G. Harbeke, *Optical Properties and Band Structure of Semiconductors*; (Pergamon Press, Oxford, 1968).
51. M.L. Cohen, and J.R. Chelikowsky, *Electronic Structure and Optical Properties of Semiconductors*; (Springer-Verlag, Berlin, 1988).
52. D.E. Aspnes, G.P. Schwartz, G.J. Gualtieri, A.A. Studna, and B. Schwartz, *J. Electrochem. Soc.* **128**, 590 (1981).
53. D.H. Kim, H. Ehrenreich, and E. Runge, Solid State Commun. **89**, 119 (1994).
54. V.M. Glazov, S.N. Chizhevskaya, and N.N. Glagoleva, *Liquid Semiconductors*; (Plenum Press, New York, 1969).
55. W. Jank, and J. Hafner, *J. Non-Crystalline Solids* **114**, 16 (1989).
56. S. Froyen, and M. L. Cohen, Phys. Rev. B **28**, 3258 (1983).
57. R. Biswas, and V. Ambegoakar, *Phys. Rev. B* **26**, 1980 (1982).
58. P. Stampfli, and K.H. Bennemann, *Phys. Rev. B* **42**, 7163 (1990).
59. T. Schröder, W. Rudolph, S.V. Govorkov, and I.L. Shumai, Appl. Phys. A **51**, 1438 (1990).
60. J.-M. Liu, A. M. Malvezzi, and N. Bloembergen, in *Energy Beam-Solid Interactions and Transient Thermal Processing*, Eds. D. K. Biegelsen, G.A. Rozgonyi and C.V. Shank (Materials Research Society, Pittsburgh, 1985).
61. S.A. Akhmanov, V.I. Emel'yanov, N.I. Koroteev, and V.N. Seminogov, *Sov Phys Usp.* **28**, 1084 (1985).
62. N. Bloembergen, R.K. Chang, S.S. Jha, and C.H. Lee, *Phys. Rev.* **174**, 813 (1968).
63. S.A. Akhmanov, N.I. Koroteev, G.A. Paitan, I.L. Shumay, M.V. Galjautdinov, I.B. Khaibullin, and E.I. Shtyrkov, *Opt. Comm.* **47**, 202 (1983).
64. S.A. Akhmanov, M.F. Galyautdinov, N.I. Koroteev, G.A. Paityan, I.B. Khaibullin, E.I. Shtyrkov, and I.L. Shumai, *Bull. Acad. Sci. USSR, Phys. Ser.* **49**, 86 (1985).
65. F. Spaepen, in *Ultrafast Phenomena V*, Eds. G.R. Fleming and A.E. Siegman (Springer-Verlag, Berlin, 1986).

COLLECTIVE EXCITATIONS IN THE OPTICAL SPECTROSCOPY
OF MAGNETIC INSULATORS

Roger M. Macfarlane

IBM Research Division
Almaden Research Center
650 Harry Road
San Jose, California 95120-6099

ABSTRACT: The concept of collective excitations is essential to the understanding of the optical spectra of magnetic insulators. In particular magnons and Frenkel excitons play a major role in the absorption structure near the electronic origin of d-d absorption in transition metal ion compounds or of f-f absorption of rare-earth materials. Both pure exciton and exciton-magnon transitions are observed below the magnetic ordering temperature. These provide information on exciton dispersion and on exciton-magnon interactions. Two principal examples are used to illustrate these concepts: the antiferromagnets MnF_2 and Cr_2O_3.

I. INTRODUCTION

The optical spectra of transition metal magnetic insulators bear a close overall resemblance to those of their diluted paramagnetic counterparts, e.g., Cr_2O_3 and ruby (or Cr^{3+} in Al_2O_3). This is because the magnetic interactions between the transition metal ions are less than the crystal field interactions responsible for the structure of the d→d transitions. They typically result in magnetic ordering temperatures of $\approx 30-300K$. This is also the case for rare-earth magnetic insulators where the crystal field effects on the f-electron states are weaker than for the d-electron case and magnetic interactions are even weaker, resulting in magnetic ordering temperatures of a few degrees Kelvin. Despite the similarities in the optical spectra between diluted and stoichiometric materials, there are important consequences of magnetic ordering and the spectra must be interpreted in terms of the basic collective excitations: the magnons for excitations within the ground state spin system and Frenkel excitons in the case of optically excited states. These two elementary excitations are very similar in structure especially when the ground state is not a pure spin state.

The scope of these lectures is not to derive the Hamiltonian for the energy bands of magnetic insulators and solve it, but rather to outline some of the salient results that have been obtained and provide a bibliography in which the reader can find the necessary details. This field saw a period of rapid and active development from the late 1960s for which the review of Loudon [1] gives an excellent summary to the late 1980s covered by the review of Tanabe and Aoyagi [2].

The starting point for a discussion of the optical spectra of magnetic insulators is the energy level diagram for the isolated magnetic ion. For transition metal ions, this is provided by the now classic description of Tanabe et al. [3,4], to which the books by Griffith [5] and Sugano et al. [6] have added a great deal of depth and perspective. In this picture, the so-called "strong field" description, electrons are assigned to cubic t_{2g} and e_g orbitals. The cubic field part of the Hamiltonian is diagonal in this representation and the Coulomb interaction is added as a second term which can be almost as large. The diagonalization of these cubic crystal field and Coulomb interactions gives energy levels which were represented in the now celebrated "Tanabe-Sugano" diagrams [3]. These provide an almost essential framework for understanding the optical spectra of transition metal insulators.

For trivalent rare earth ions, optical spectra up to the UV region are largely determined by transitions within the $4f^n$ configurations. These f-electrons are well shielded from the crystal environment so that the free-ion basis states labeled by spin (S), orbital (L) and total (J) angular momentum are a natural starting point. Dieke and collaborators [7] assigned many of the spectra of rare earth ions in $LaCl_3$ to the quasi free-ion energy levels and produced and indispensable summary [8] which forms the starting point for a discussion of rare earth spectra in solids to this day. Modifications of this diagram to the case of rare earth ions in LaF_3 were made later by Carnall et al. [9].

The fine structure in the optical spectra of isolated transition-metal and rare-earth ions in crystals is determined by the splittings produced by the spin-orbit interaction and the crystal field appropriate to the exact site symmetry of the absorbing ions. These transitions are nominally parity forbidden, being $d^n - d^n$ for transition metal ions and $f^n \to f^n$ for rare earths, but become electric dipole allowed if the site symmetry is non-centrosymmetric, due to crystal field admixture of other configurations or charge-transfer states. Selection rules are based on the symmetry properties of the irreducible representations of the point group of the ion site which determine how the wavefunctions transform. For even electron systems, these are single group representations and for odd electron systems, double group representations.

In early work, the effect of the concentration of magnetic ions in a host crystal or even the result of incorporating them at a 100% level, was not always recognized. This is particularly true of rare earth magnetic insulators where such effects are smaller. For this reason, the details of some of the earlier work needs to be critically evaluated. At first, magnetic interactions between ions were approached via the pair interactions, a good example being the case of ruby [10,11] or for rare earths, Pr^{3+} in LaF_3 [12]. There are three major manifestations of the ion-pair coupling: one is to shift the spectral lines due to a crystal field perturbation since the dopant and host ions are not identical. This occurs whether the energy levels of the isolated ions are electronically degenerate or not. The second is the appearance of a splitting of the levels. If lines are sharp and the splittings (typically several cm^{-1} to hundreds of cm^{-1}) large enough, these can be easily resolved [10,11,13]. In other cases, the inhomogeneous broadening may obscure the splittings and techniques such as spectral holeburning [14] must be used to extract the interaction strengths. The third manifestation of pair interactions comes from emission studies where energy transfer [15] or upconversion [16] results from the pair interaction. Spectroscopy of pairs is often difficult and tedious because of the number and orientation of inequivalent pairs. The easiest cases are those on simple cubic lattices or those where the crystal structure makes one pair type dominate such as $CsCdBr_3$ [17]. The technology to move individual atoms on surfaces is now being mastered [18] and eventually one can hope to study the interaction between a single pair of ions using a combination of optics and scanning probe techniques. The study of ion-pair coupling is very basic and provides a microscopic understanding of spin-wave dispersion, exciton bandwidths and the collective excitations of magnetic insulators.

II. FORMALISM

A very brief outline of the basic formalism for determining the spectrum of collective excitations in magnetic insulators will now be given. Further details may be found in several excellent reviews [1,2,19−21] and in original works on the subject, for example [22−27]. The Hamiltonian for the stoichiometric magnetic insulator can be written as a sum of single ion ($\mathcal{H}_0$) and inter-ion (V) terms:

$$\mathcal{H} = \sum_{n=1}^{N} \sum_{i=1}^{P} \mathcal{H}_0(\vec{R}_{ni}) + \sum_{\vec{R}_{ni} \neq \vec{R}_{mj}} V(\vec{R}_{ni}, \vec{R}_{mj}) \tag{1}$$

where $\vec{R}_{ni}$ is the position of the ith of p ions in the nth of N unit cells. The single ion Hamiltonian includes the Coulomb interaction within an ion, the crystal field and also diagonal contributions from inter-ion exchange that are sometimes described as an effective field splitting of the Kramers' doublets of a magnetic species. The very interesting inter-ion term V is small compared to $\mathcal{H}_0$ and consists of off diagonal, so-called "transfer of excitation", terms which are a sum of multipole and exchange contributions. Diagonalization of $\mathcal{H}$ is facilitated by the formation of Bloch states labeled by a wavevector $\underline{k}$, a "sublattice-index" i and a "branch index" μ

$$|^\mu E_i(\underline{k}) > = \frac{1}{\sqrt{N}} \sum_{n=1}^{N} e^{i\underline{k} \cdot \underline{I}_{ni}} |^\mu E_{ni} >$$

$$= \frac{1}{\sqrt{N}} e^{i\underline{k} \cdot \rho_i} \sum_{n} e^{i\underline{k} \cdot \underline{I}_n} |^\mu E_{ni} > \tag{2}$$

Here ρ_i locates ions in a unit cell and $\underline{I}_n$ are primitive lattice translations. The sublattice is determined by the number of atoms in the unit cell of the magnetically ordered material. The simplest ferromagnet could have one sublattice and the simplest antiferromagnet or ferrimagnet could have two sublattices. The branch index is determined by the number of single ion states being considered. Each state contributes a branch to the excitation spectrum. The simplest case is that of a spin = ½ ground state where there is a single magnon branch. The Hamiltonian $\mathcal{H}$ is invariant under transformations of the symmetry groups of the wave vector $\underline{k}$ at all points in the Brillouin zone. This is the full magnetic space group rather than the simpler point group appropriate to the isolated ion case. Matrix elements of the Hamiltonian of Eq. (1) in the basis of the Bloch states of Eq. (2) can be written

$$^\mu H_{ij}(\underline{k}) = E_0 \delta_{ij} + < \mu \underline{k} i |V| \mu \underline{k} j >$$

$$= E_0 \delta_{ij} + \sum_{m,n} e^{i\underline{k} \cdot (\vec{R}_{ni} - \vec{R}_{mj})} < \mu R_{ni} |V| \mu R_{mj} > \tag{3}$$

The matrix elements $V_{ij}^{(\ell)}(\vec{r}_\ell)$ are the basic pair interactions between the ℓth neighbors separated by $\vec{r}_\ell = \vec{R}_{ni} - \vec{R}_{mj}$. As we have seen, a good measure of these can be obtained from isolated pairs in somewhat dilute systems, but they can also be considered as parameters to be determined by fitting to the details of optical spectra. At certain points in $\underline{k}$ space, there can be symmetry arguments that limit the number of pair interactions which contribute to the exciton band structure. In addition, the strong distance dependence of exchange interactions means that usually only a few near neighbor couplings need to be considered.

III. EXCITON BANDS AND OPTICAL SPECTRA: TWO EXAMPLES

Two of the most intensively investigated systems are manganese fluoride (MnF_2) [22-24] and chromium sesquioxide (Cr_2O_3) [25,26,28]. Their optical spectroscopy will be summarized here showing how collective excitations *viz* excitons, magnons and phonons, determine the optical response in important ways.

III.A. Manganese Fluoride

Manganese fluoride is a 2-sublattice antiferromagnet whose magnetic space group is $D_{4h}^{14}(D_{2h}^{12})$. The reduced symmetry in parenthesis arises from the antiparallel ordering of the spins at 67K along the tetragonal c-axis. The point group of the $\underline{k}$ vector at $\underline{k} = 0$ is D_{2h} and the Mn^{2+} site symmetry C_{2h}. These symmetries determine the optical selection rules at the various points in k-space. The overall spectrum of MnF_2 below the Néel temperature is shown in Fig. 1. In general scope, it strongly resembles the spectrum of an isolated Mn^{2+} ion because the ion-ion coupling is only tens of cm^{-1}. The individual bands are assigned to cubic crystal field terms with reference to the d^5 Tanabe-Sugano diagram [3,6]. Since the exact symmetry is lower, the details of the fine structure is more complicated than predicted by the cubic-field approximation. Each band consists of several zero-phonon lines together with multiphonon sidebands. The latter usually dominate the spectrum as is common for transition metals. Bands corresponding to transitions between different strong cubic field configurations ($t_{2g}^m e_g^n$) depend on the crystal field strength Dq and these exhibit higher exciton-phonon coupling and broader bands. An expansion of the energy scale of the absorption to the lowest $^4T_{1g}$ band in Fig. 2 shows the relative strength of the exciton and exciton-magnon bands on the low energy side with the broader phonon sideband to higher energies. The effect of collective excitations on the optical spectrum can be very clearly seen on this transition. This was historically the first example to be analyzed in detail [22-24] and it influenced much of the work which followed.

In the 2-sublattice MnF_2, there are 2 nearest neighbors (on the same sublattice) along the c-axis which couple ferromagnetically in the ground state, 8 second neighbors on opposite sublattices which couple antiferromagnetically and much weaker third neighbor interactions in the x,y direction. For an exciton branch μ, the interaction between same sublattice or translationally equivalent ions can be written [22]:

$$^\mu V_{11}(\underline{k}) = {}^\mu V_{22}(\underline{k}) = \sum_n {}^\mu V_{ii} \cos(\underline{k} \cdot \underline{r}_{n1}) \tag{4}$$

This gives rise to dispersion. Interaction between opposite sublattice ions is:

$$^\mu V_{12}(\underline{k}) = {}^\mu V_{21}(\underline{k}) = \sum_{ij} {}^\mu V_n \cos(\underline{k} \cdot \underline{r}_{n2}) \tag{5}$$

This would give rise to a Davydov splitting or splitting of the exciton branches at $k = 0$. For the lowest $^4T_{1g}$ excitons it is spin forbidden and any Davydov splitting is less than the 0.5 cm^{-1} linewidth of the excitons. Finally we have

$$^\mu E(\underline{k}) = {}^\mu E_0 + \underbrace{{}^\mu V_{11}(\underline{k})}_{\text{dispersion}} - \underbrace{{}^\mu V_{12}(\underline{k})}_{\text{Davydov splitting}} \tag{6}$$

In MnF_2 the magnon dispersion is given by [29]:

$$E_{mag}(\underline{k}) = 2SZ_2|J_2|(1 + 0.073/2S)[(1 + \varepsilon_k)^2 - \gamma_k^2]^{1/2} \tag{7}$$

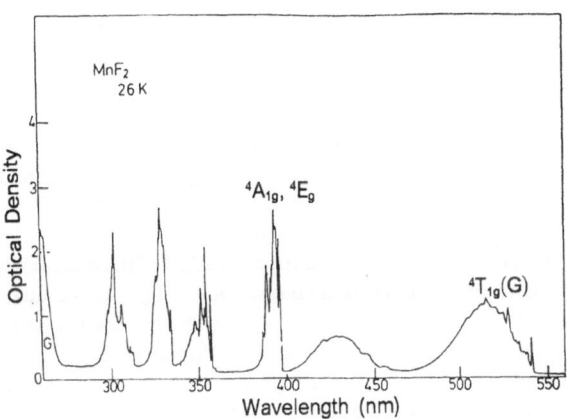

Fig. 1 σ polarized absorption spectrum of MnF$_2$ below the Néel temperature.

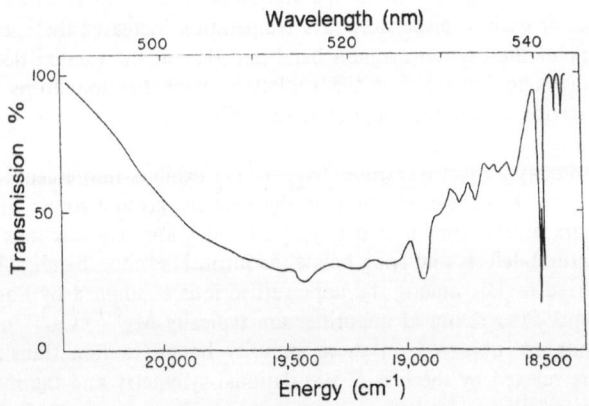

Fig. 2 Transmission of MnF$_2$ in the lowest $^4T_{1g}$ band showing at right two exciton lines, and the strong narrow exciton-magnon band together with the phonon sideband.

where

$$\gamma_k = \cos(ak_x/2)\cos(ak_y/2)\cos(ck_z/2)$$

$$\varepsilon_k = (g\beta H_A/2SZ_2J_2) - 2Z_1J_1/Z_2J_2 \sin^2(ck_z/2) \tag{8}$$

with $Z_1 = 2$, $Z_2 = 8$, $J_1 = 0.22\text{cm}^{-1}$, $J_2 = -1.22\text{cm}^{-1}$ and $g\beta H_A = 0.737\text{cm}^{-1}$.

The effect of magnons on the optical spectrum can be clearly seen in the origin .ls by 1.4 region of the $(t_{2g}^3 e_g^2)^6A_{1g} \rightarrow (t_{2g}^4 e_g)^4T_{1g}$ band around 540nm at 2K as shown in Fig. 3. Optical selection rules require $\Delta k \approx 0$ and σ-polarized ($E \perp c$) magnetic dipole transitions M1σ and M2σ occur to excitons of (Γ_1^+, Γ_2^+) symmetry while π-polarized transitions ($E \parallel c$) would occur to excitons of $(\Gamma_3^+ + \Gamma_4^+)$ symmetry. These latter are not observed here since they are probably broad and masked by the magnon and phonon structure. In principle, Davydov splittings could be seen between Γ_1^+ and Γ_2^+ but for the lowest $^4T_2$ excitons such interactions are spin-forbidden and small. The most interesting part of Fig. 3 is the appearance of exciton-magnon transitions [22,23] ('magnon sidebands') which are much stronger than the pure exciton transitions. This is because the two-center exciton-magnon transition corresponding to an exciton on one sublattice and a magnon on the other becomes electric dipole allowed since the ion-pair no longer has a center of symmetry and in addition is spin allowed [30]. Thus the exciton-magnon absorption dominates the zero-phonon line structure. The shape of the magnon sideband is determined by three main factors. The first is the combined exciton-magnon densities of states corresponding to (k,-k) combinations and which are determined by their dispersion. The second is the existence of selection rules based on the symmetry of the excitons and magnons at special points in the Brillouin zone [1,22]. For electric dipole sidebands, excitations at the X point contribute to the π polarized absorption and those at the Z and A points to the σ polarized absorption. The third factor is the exciton-magnon interaction [2,31] which modifies the shape and makes it difficult to detect the existence of small amounts of exciton dispersion. As temperature increases the energy difference between the peak of the exciton-magnon band and that of the exciton decreases as the magnon energy is renormalized and the sublattice magnetization drops [22]. This is additional evidence for the exciton-magnon nature of the bands.

The lowest energy optical transition $^6A_{1g} \rightarrow {}^4T_{1g}$ exhibits fluorescence since there is a large energy gap to the next level below, in this case the ground state. Fluorescence in magnetic insulators usually comes from traps [32] which are magnetic ions perturbed by chemical or structural defects and lying below the intrinsic exciton band. This is because energy is transferred rapidly among the unperturbed ions as implied by Eqs. (2) and (3). In the case of MnF$_2$ the chemical impurities are typically Mg^{2+}, Ca^{2+} or Zn^{2+} ions. Magnon sidebands are observed on these impurity bound exciton transitions but the selection rules are relaxed by the loss of translational symmetry and the modification of the point group of the perturbed ion so that mixed magnetic and electric dipole transitions are observed [32]. In pure crystals, intrinsic exciton emission is observed [33] with an intensity and lifetime which is very sensitive to the density of trapping centers.

The dynamical properties of the $^4T_{1g}$ excitons and magnons in MnF$_2$ have been probed in a number of experiments designed to measure the time dependence of the magnetization [34−36] or emitted light [37] following resonant excitation in the magnon sideband or exciton absorption. At 2K, exciton scattering between the sublattices is slow ($\sim 10^6$/s) and due to small spin admixtures on the two sublattices. It increases by several orders of magnitude up to 30K due to scattering by thermally excited magnons. The magnons on the other hand can cross sublattices rapidly at low temperatures since the

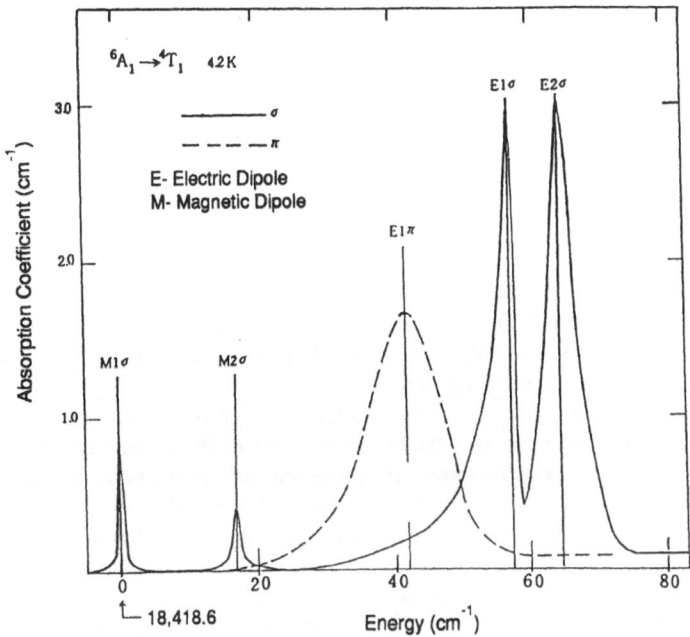

Fig. 3 Exciton (M1 and M2) and exciton-magnon (E1σ, E2σ and E1π) transitions in MnF₂ (after ref. 23).

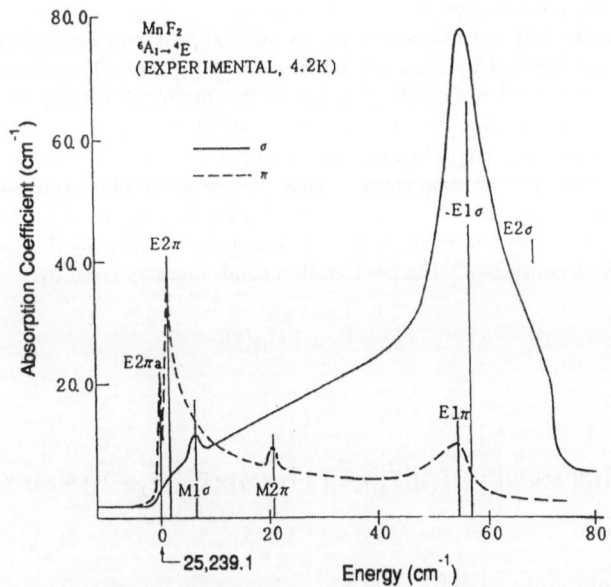

Fig. 4 Exciton (M) and exciton-magnon (E) transitions in MnF₂ showing the bound state E2πa resulting from negative exciton dispersion and E-M coupling (ref.24).

magnon eigenstates do not correspond so well to pure sublattice excitations. Exciton scattering in k-space is power dependent showing that it is controlled by exciton-exciton scattering [37].

Although the exciton dispersion is negligible for this transition, a large, negative dispersion (-74 cm^{-1}) has been observed for the $^6A_{1g} \rightarrow {}^4A_{1g}{}^4E_g$ excitons which results in the appearance of exciton-magnon bound states (Fig. 4) in the absorption around 25,240 cm^{-1} [24].

III.B. Chromium Sesquioxide

The second example is that of Cr_2O_3. This differs in important ways from the case of MnF_2 discussed above. It is a 4-sublattice system with a stronger ion-ion coupling giving a Néel temperature of 308K. The overall optical absorption spectrum (Fig. 5) is similar to that of the dilute analog ruby (Cr^{3+}:Al_2O_3) but with important differences in the region of the $t_{2g}^3 {}^4A_2 \rightarrow t_{2g}^3 {}^2E$ transitions corresponding to the well known R lines of ruby. These differences arise from the excitonic nature of the transitions. The magnetic space group of Cr_2O_3 is $D_{3d}^6(D_3^7)$ which is centrosymmetric in the paramagnetic state but non-centrosymmetric in the antiferromagnetic state. The site symmetry of the Cr ions is C_3 which lacks a center of inversion thereby allowing electric dipole transitions with $\Delta k \approx 0$. The four Cr^{3+} spins are collinear along the rhombohedral axis and are ordered $(+-+-)$ [38]. Thus the sublattice index i in eqn.(2) takes the values 1-4. The ^2E level has four components which are labelled by representations of the Cr site symmetry group as a_4, its conjugate level a_5 and the self conjugate a_6. In the absence of ion-ion coupling these levels are degenerate in pairs (a_4, a_5) and $2a_6$ corresponding to the two R-lines in ruby. Since there are 4 ions per unit cell there are a total of 16 exciton branches associated with transitions to ^2E. The strong diagonal exchange splits these into two groups with only the lower group being important since transitions to these are more allowed from the ground state by spin selection rules. They also give rise to sharper transitions because they are not broadened by spontaneous phonon emission. Exciton states at $k=0$ are labelled by representations of the D_3 group. Those arising from the (a_4, a_5) components ($\mu = 1$ in eqn.(2)) give rise to two doubly degenerate E excitons and those arising from the $2a_6$ components ($\mu = 2$ in eqn.(2)) give rise to four singly degenerate excitons ($2A_1$, $2A_2$) (Fig. 6). The simplest case turns out to be that for the $\mu = 1$ excitons (the two E branches). Here, to a good approximation, only the spin-allowed off diagonal matrix elements ($^1H_{13}(\underline{k})$) connecting the two fourth neighbor spin-up ions separated by the vector $\underline{r}_4$ and the equal ones connecting the spin-down ions $^1H_{24}(\underline{k})$ need to be considered. The $\mu = 1$ exciton bands are then given by

$$\varepsilon_{1,2} = E_1 \pm |^1H_{13}(\underline{k})| \qquad (9)$$

with

$$^1H_{13}(\underline{k}) = 2\,^1h_{13}^{(4)}(\underline{r}_4)[\cos \tfrac{1}{2}\underline{k}\cdot(-\vec{T}_1 + \vec{T}_2 + \vec{T}_3) + \cos \tfrac{1}{2}\underline{k}\cdot(\vec{T}_1 - \vec{T}_2 + \vec{T}_3) + \cos \tfrac{1}{2}\underline{k}\cdot(\vec{T}_1 + \vec{T}_2 - \vec{T}_3)]$$

(10)

where $E_1 = 13835$ cm^{-1} and $|^1h_{13}^{(4)}| = 15.3$ cm^{-1}.

The dispersion and Davydov splitting (i.e. the separation between the branches at $k=0$) are shown in Fig. 7 by the dotted curves. These E excitons are observed in σ polarized optical absorption (Fig. 8) where the 186 cm^{-1} Davydov splitting is clearly seen.

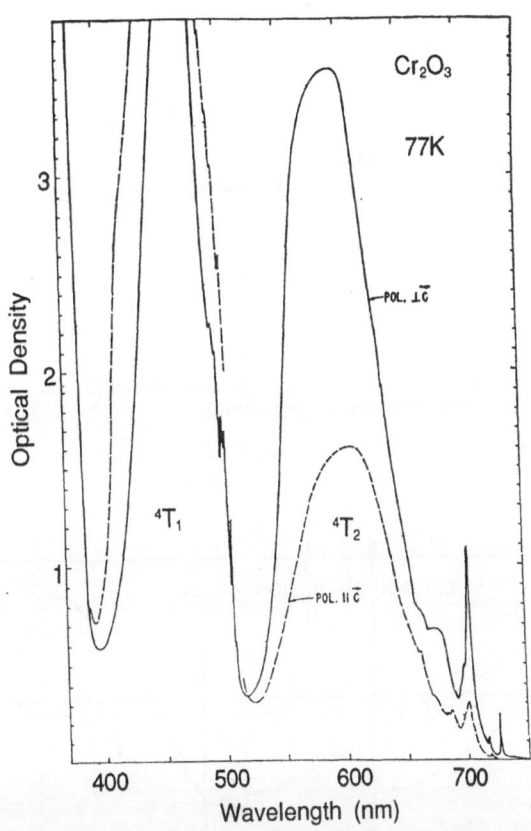

Fig. 5 Absorption spectrum of Cr_2O_3 (9.5μ thick) at 77K showing the quartet bands and the ^2E exciton and exciton-magnon bands at the low energy edge. (ref.39)

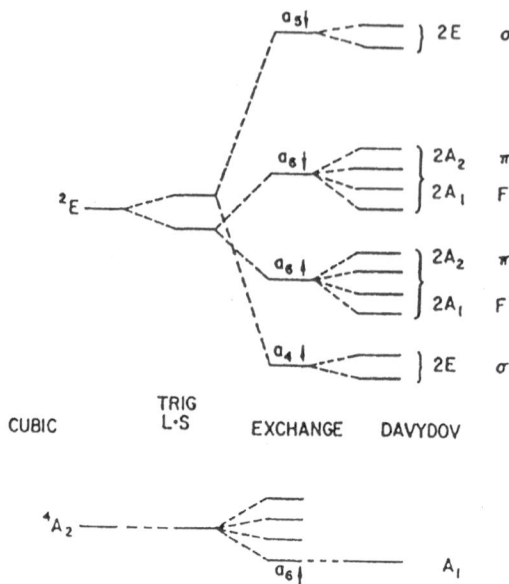

Fig. 6 Energy levels of isolated Cr^{3+} ions and the resulting excitonic levels due to ion-ion coupling in Cr_2O_3. F denotes forbidden transitions.

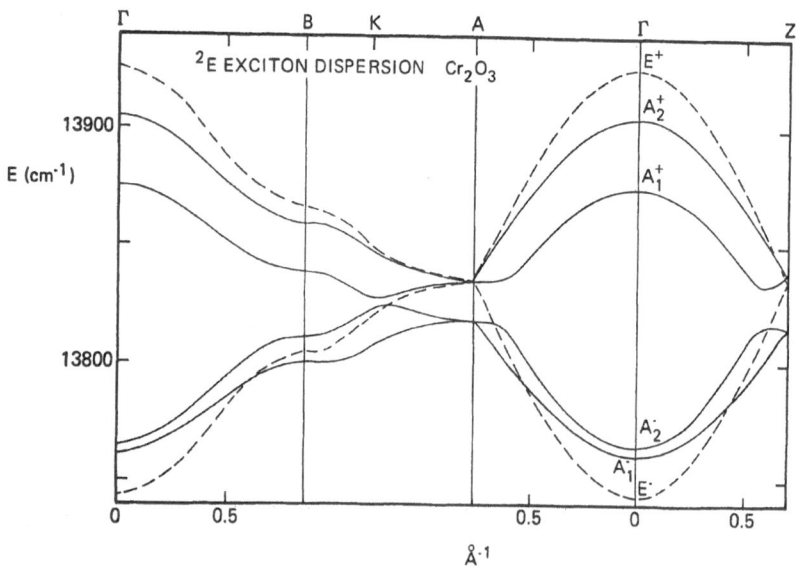

Fig. 7 Calculated exciton dispersion for the 8 lowest branches of ^2E excitons in Cr_2O_3. Dotted lines correspond to degenerate $\mu = 1$ branches and the solid lines to $\mu = 2$.

Similar considerations can be applied to the $\mu = 2$ branches, where retaining the spin-allowed terms $^2H_{13}(\underline{k})$ and $^2H_{14}(\underline{k})$ gives the dominant contributions to the exciton dispersion as:

$$\mathcal{E}_{1,2,3,4} = E_2 \pm {}^2|H_{13}(\underline{k})| \pm {}^2|H_{14}(\underline{k})| \qquad (11)$$

This form gives the exciton dispersion curves shown as solid lines in Fig. 7. Further details of the calculation can be found in Macfarlane and Allen [26]. Cr_2O_3 provides probably the best example of Davydov splittings in magnetic insulators. The exciton-magnon transitions are again a dominant feature of the optical absorption as shown in Fig. 8. The shape of the sidebands is rather well described by the combined exciton-magnon density of states [26]. In this situation the strong dispersion appears to mask the effects of exciton-magnon interaction.

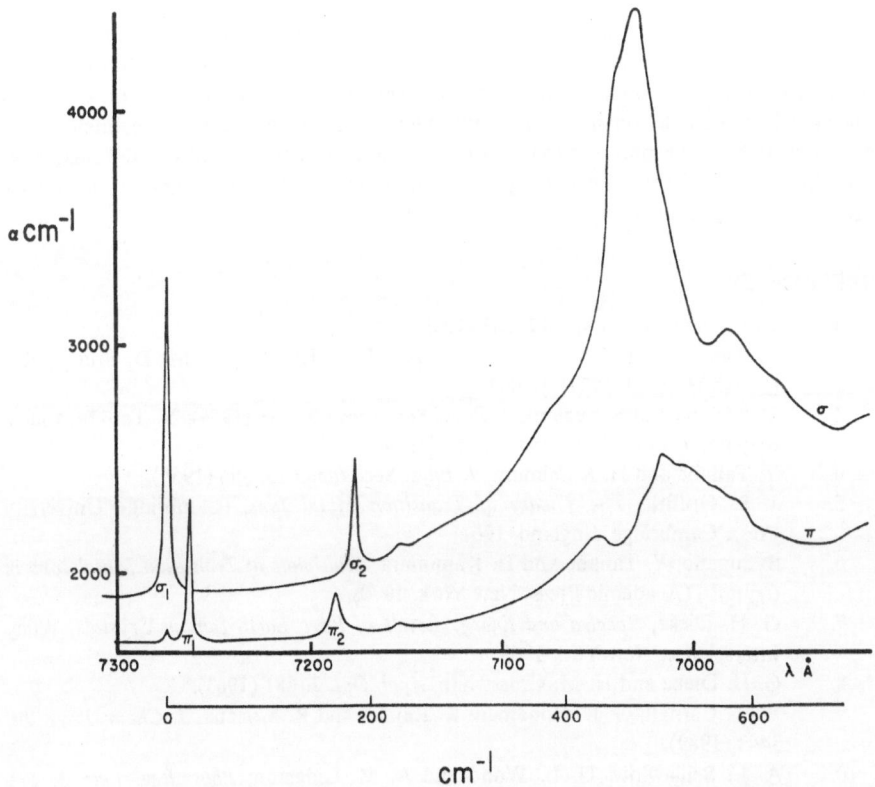

Fig. 8 $^4A_2 \rightarrow {}^2E$ exciton ($\sigma_1, \sigma_2, \pi_1, \pi_2$) and exciton-magnon ($\sigma, \pi$) absorption in Cr_2O_3 at 2K.

481

IV. OTHER CASES

In rare-earth magnetic insulators, the inter-ion coupling is smaller than for transition metals but can still lead to appreciable collective effects since the electron phonon coupling is generally weaker and the spectral lines sharper. Thus exciton dispersion is well documented in $GdCl_3$ and $Gd(OH)_3$ [40] and in $Tb(OH)_3$ [41,42]. A fuller discussion of these cases can be found in the review by Cone and Meltzer [21]. Ferromagnetic materials are not often insulators so rather few examples are known but they introduce some interesting properties where the effects of magnons on the optical spectra can be dramatic. This is because an exciton-magnon absorption cannot conserve spin by creating an exciton on an up-spin sublattice and a magnon on a down-spin sublattice. However a spin allowed transition can occur if a magnon is annihilated on the same sublattice as the exciton is created i.e. it corresponds to a hot sideband absorption. A particularly nice example of this strong, thermally induced exciton-magnon absorption is provided by the work of Day *et al.* [43] on K_2CrCl_4. Other examples of ferromagnetic systems are discussed by Tsuboi *et al.* [44,45].

V. CONCLUSION

The optical spectra of magnetic insulators often require a description in terms of collective excitations since the inter-ion coupling can have a magnitude which is large compared to the optical resolution and bandwidths, especially at low temperatures. Of particular interest in this class of materials are the Frenkel excitons and the exciton-magnon absorption in which a photon is absorbed, creating an exciton and a magnon of equal and opposite wave-vector. This is a qualitatively new transition not observed in the single ion or magnetically dilute case and it can have an intensity far exceeding that of the pure exciton. Two important examples - MnF_2 and Cr_2O_3 have been introduced. These show many of the important features encountered in a wide variety of other materials.

REFERENCES

1.   R. Loudon, *Adv. Phys.* **17**, 243 (1968).
2.   Y. Tanabe and K. Aoyagi, in *Excitons*, E. I. Rashba and M. D. Sturge, eds. (North Holland, 1987), p. 603.
3.   Y. Tanabe and S. Sugano, *J. Phys. Soc. Japan* **9**, 753 (1954); Y. Tanabe and S. Sugano, *J. Phys. Soc. Japan* **9**, 766 (1954).
4.   Y. Tanabe and H. Kamimura, *J. Phys. Soc. Japan* **13**, 394 (1958).
5.   J. S. Griffith, *The Theory of Transition Metal Ions*, (Cambridge University Press, Cambridge, England, 1961).
6.   S. Sugano, Y. Tanabe and H. Kamimura, *Multiplets of Transition Metal Ions in Crystals*, (Academic Press, New York, 1970).
7.   G. H. Dieke, *Spectra and Energy Levels of Rare Earth Ions in Crystals*, Wiley Interscience, New York 1968.
8.   G. H. Dieke and H. M. Crosswhite, *Appl. Opt.* **2**, 681 (1963).
9.   W. T. Carnall, G. L. Goodman, K. Rajnak and R. S. Rana, *J. Chem. Phys.* **90**, 3443 (1989).
10.   A. L. Schlawlow, D. L. Wood and A. M. Clogston, *Phys. Rev. Lett.* **3**, 271 (1959).
11.   L. F. Mollenauer and A. L. Schawlow, *Phys. Rev.* **168**, 309 (1968).
12.   J. C. Vial and R. Buisson, *J. de Phys.* **43**, L339 (1982).
13.   G. A. Prinz, *Phys. Rev.* **152**, 474 (1966); G. A. Prinz and E. Cohen, *Phys. Rev.* **165**, 335 (1968).

14. F. Ramaz, J. C. Vial and R. M. Macfarlane, *J. Lumin.* **53**, 244 (1992).
15. B. DiBartolo ed. *Energy Transfer Processes in Condensed Matter*, NATO ASI Series **114** (Plenum Press, New York, 1984).
16. F. Auzel, *C. R. Acad. Sci. (Paris)* **202**, 1016 (1966).
17. L. M. Henling and G. L. MacPherson, *Phys. Rev.* **B16**, 1889 (1977); R. B. Barthem, R. Buisson and R. L. Cone, *J. Chem. Phys.* **91**, 627 (1989).
18. D. Eigler and E. K. Schweizer, *Nature* **344**, 524 (1990).
19. D. D. Sell, *J. Appl. Phys.* **39**, 1030 (1968).
20. G. A. Gehring and K. A. Gehring, *Repts. Progr. Phys.* **38**, 1 (1975).
21. R. L. Cone and R. S. Meltzer, in *Spectroscopy of Solids Containing Rare Earth Ions*, A. A. Kaplyanskii and R. M. Macfarlane, eds. (North Holland, 1987), p. 481.
22. R. L. Greene, D. D. Sell, W. M. Yen, A. L. Schawlow and R. M. White, *Phys. Rev. Lett.* **15**, 656 (1965); D. D. Sell, R. L. Greene and R. M. White, *Phys. Rev.*, **158**, 489 (1967).
23. R. S. Meltzer, M. Lowe and D. S. McClure, *Phys. Rev.*, **180**, 561 (1969).
24. R. S. Meltzer, M. Y. Chen, D. S. McClure and M. Lowe-Pariseau, *Phys. Rev. Lett.*, **21**, 913 (1968);
25. J. W. Allen, R. M. Macfarlane and R. L. White, *Phys. Rev.*, **179**, 523 (1969).
26. R. M. Macfarlane and J. W. Allen, *Phys. Rev.*, **B4**, 3054 (1971).
27. J. B. Parkinson, *J. Phys.*, **C2**, 2012 (1969).
28. J. P. van der Ziel, *Phys. Rev. Lett.*, **18**, 237 (1967).
29. A. Okazaki, K. C. Turberfield and R. W. H. Stevenson, *Phys. Lett.*, **8**, 9 (1964).
30. Y. Tanabe, T. Moriya and S. Sugano, *Phys. Rev. Lett.*, **15**, 102 (1965).
31. Y. Tanabe, K. Gondaira and H. Murata, *J. Phys. Soc. Japan* **25**, 1562 (1968); S. Freeman and J. J. Hopfield, *Phys. Rev. Lett.* **21**, 910 (1968).
32. R. L. Greene, D. D. Sell, R. S. Feigelson, G. F. Imbusch and H. J. Guggenheim, *Phys. Rev.*, **171**, 600 (1968).
33. R. E. Dietz, A. Meixner, H. J. Guggenheim and A. Misetich, *Phys. Rev. Lett.*, **21**, 1067 (1968).
34. J. F. Holzrichter, R. M. Macfarlane and A. L. Schawlow, *Phys. Rev. Lett.* **26**, 652 (1971).
35. G. J. Jongerden, A. F. M. Arts, J. I. Dijkhuis and H. W. de Wijn, *Phys. Rev.*, **B40**, 9435 (1989).
36. M. L. J. Hollman, A. F. M. Arts, and H. W. de Wijn, *Phys. Rev.*, **B48**, 3290 (1993).
37. R. M. Macfarlane and A. C. Luntz, *Phys. Rev. Lett.* **31**, 832 (1973).
38. B. N. Brockhouse, *J. Chem. Phys.*, **21**, 961 (1956).
39. D. S. McClure, *J. Chem. Phys.*, **38**, 2289, (1963).
40. R. S. Meltzer and H. W. Moos, *Phys. Rev.* **B6**, 264 (1972).
41. R. L. Cone and R. S. Meltzer, *J. Chem. Phys.* **62**, 3573 (1975).
42. R. S. Meltzer, *Solid State Commun.* **20**, 553 (1976).
43. P. Day, A. K. Gregson and D. H. Leech, *Phys. Rev. Lett.* **30**, 19 (1972).
44. T. Tsuboi, M. Chiba and Y. Ajiro, *Phys. Rev.* **B32**, 354 (1985).
45. T. Tsuboi, M. Chiba, Y. Ajiro, K. Iio and R. Laiho, *J. Mag. & Mag. Matls.* **54-57**, 1395 (1986).

SPECTROSCOPY AND DEVELOPMENT OF
SOLID STATE LASERS AT NASA

Norman P. Barnes,[1] Clyde A. Morrison, [2] Elizabeth D. Filer, [3] Waldo J.
Rodriguez, [3] and Brian M. Walsh[4]

[1]NASA Langley Research Center
Hampton, VA 23681

[2]Army Research Laboratory
Adelphi, MD 20783

[3]Norfolk State University
Norfolk, VA 23504

[4]Boston College
Chestnut Hill, MA 02167

ABSTRACT

NASA Langley has developed expeditious methods of evaluating new laser materials and laser systems using quantum mechanical and energy transfer models. Veracity of these models has been tested against measured spectroscopic parameters as well as measured laser amplifier performance. In particular, these methods have been applied to the development of 2.0 μm laser such as the Ho:Tm laser materials.

I. INTRODUCTION

Expeditious methods of identifying and evaluating promising laser materials need to be found in order to avoid time-consuming and expensive empirical methods. This is especially true in the more complex laser systems such as the Ho:Tm laser systems where more design variables are available. For some common lanthanide series lasers, such as Nd:YAG, analysis is considerably simpler. In this case, Nd functions as both the major absorber as well as the active atom. Virtually all of the available pump bands relax quickly to the upper laser level with a quantum efficiency which is near unity. Once in the upper laser level, Nd lases to a virtually empty lower laser level. Thus, Nd acts like a nearly true four-level laser. Design variables are limited to the choice of the laser material (if other than YAG), the geometry of the laser material and the concentration of Nd, which is often limited to relatively low values. On the other hand, in the Ho:Tm laser system Tm usually functions as the absorber while Ho acts as the active atom. Rather than a rapid relaxation from the pump manifold to the upper laser manifold, the energy transfer process is highly complicated and involves both beneficial and deleterious processes. In addition, the lower laser level is thermally populated. However, there are more design variables, including: the concentration of both Ho and Tm in almost any concentration, the choice of the laser material, the geometry of the laser material, and the operating temperature. By adding more variables, empirical evaluation becomes very time consuming since there is a plethora of potential laser materials and the energy transfer process depends critically on the selection of the laser material.

Ho:Tm lasers were selected for nominal 2.0 μm operation primarily on efficiency considerations. 2.0 μm operation is desired since this is nominally in the eyesafe region of the spectrum. Ho:Tm lasers are compatible with AlGaAs laser diode pumping, providing for high efficiency and long term reliability. A two-for-one quantum efficiency is also possible, allowing a single pump photon to produce two laser photons. In addition, Ho has a useful stimulated emission cross section, leading to a reasonable gain in amplifiers as well as extraction fluences compatible with long life at reasonable laser induced damage thresholds.

Rather than constructing a single highly-complex model, four sequential models were developed. A quantum mechanical model utilizes spectroscopically-measured energy levels in a laser material to predict the level to level lifetimes or branching ratios. While this quantum mechanical model, developed at the Army Research Laboratory, could begin with only the lattice parameters and the energy centroids, a knowledge of the energy levels provides a closer link with measured parameters. Output from the quantum mechanical model is used to predict the energy transfer parameters, that is, the rate at which two atoms in a laser material can exchange a quantum of excitation. While energy transfer parameters can be measured spectroscopically, a full set of energy transfer parameters is difficult to measure accurately. Output of the energy transfer program is used in a rate equation model to determine into which manifolds of the Ho:Tm laser system the pump energy, or the quanta of excitation, appear as a function of time. Finally, the output of the rate equation model is used in an end-pumped laser amplifier model to predict both the gain and energy storage of a laser amplifier. Measured laser amplifier performance is then used to validate the models. Although some intermediate data is used to check the other models, it is the laser amplifier data which is the primary objective.

II. MODELS

A quantum mechanical model is used to predict all of the level to level lifetimes given the measured energy levels of a laser system. This model has been used with a great deal of success with several lanthanide series laser materials. It takes into account a central potential formed by the nucleus and inner shell electrons, the mutual repulsion of the electrons in the 4f subshell, the spin orbit interaction of the electrons and the Trees interactions. Crystal field effects are taken into account through the crystal field parameters, B_{ij}. Crystal field parameters are highly dependent on the crystal symmetry of the site of the particular lanthanide series atom. Output of the quantum mechanical model is a set of calculated energy levels to compare with the experimental energy levels, as well as the level to level lifetimes. Level to level lifetime calculations utilize both electric and magnetic dipole transitions.

Veracity of the quantum mechanical calculations can be assessed by comparing the computed and measured energy levels. An example of the agreement is given in Figure 1, where measured and calculated energy levels of the lowest two manifolds of Ho:YLF appear. Note the general agreement between the groupings of energy levels as well as the

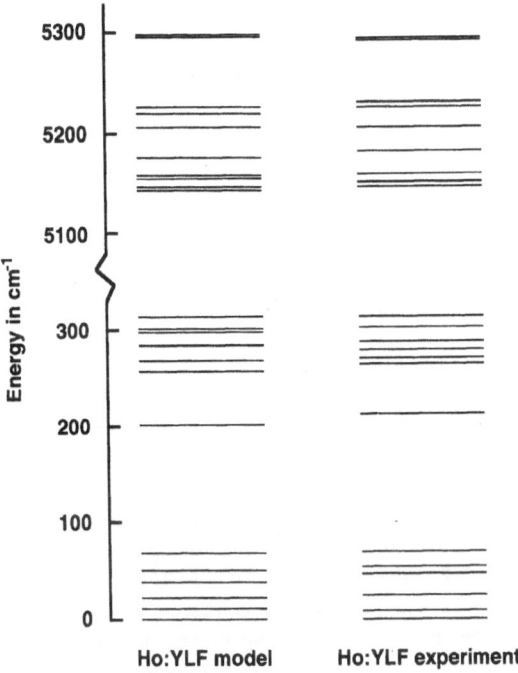

Figure 1. Calculated and experimental energy levels in Ho:YLF $^5I_7$ and $^5I_8$ manifolds.

position of the energy levels themselves. Typically, the agreement of measured and calculated energy levels is within 5 cm^{-1}.

An energy transfer model was developed to calculate the rate at which energy could be transferred between two atoms. If two atoms of the same type are involved and the same upper and lower manifolds are involved in the energy transfer, the process is often referred to as diffusion. For historical reasons, the atom which donates the quantum of excitation is referred to as the sensitizer while the atom which receives the quantum of excitation is referred to as the active atom. Since energy levels which are nearly in resonance enhance the energy transfer process, energy transfer calculations must sum over all energy levels in the upper and lower manifolds in the sensitizer as well as over all energy levels in the upper and lower manifolds of the active atom. Such a summation can involve 30,000 contributions. In addition, since active atoms and sensitizer atoms are distributed throughout the laser material, a summation over all distances must be performed.

While a classical approach to the energy transfer problem integrates over distances [1], the approach used here uses the spacing dictated by the crystal lattices to determine the strength of the energy transfer. Energy transfer process is usually a dipole-dipole interaction with the rate of the interaction depending on the distance between the interacting atoms to the inverse sixth power. Since the rate of the energy transfer process decreases rapidly with increasing distance, the exact choice of the nearest distance becomes critical. In the approach utilized here, all distances are set by the crystal lattice. Another aspect of the dipole-dipole interaction is the orientation of the interacting dipoles with respect to each another, shown in Figure 2. Rather than averaging over all possible orientations of the dipoles [1], the approach used here uses the dipole orientations set by the crystal lattices.

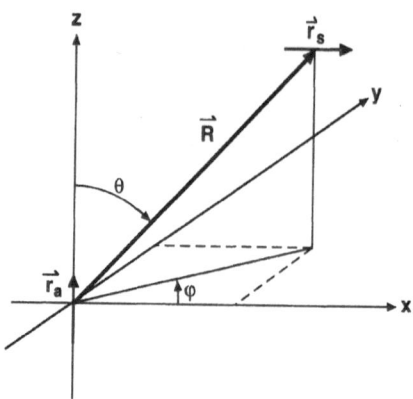

Figure 2. Dipole-dipole interaction between active atom and sensitizer.

Rate equations use the energy transfer parameters to predict the population densities of the various manifolds as a function of time and pumping rate. As the simplest example, consider the ground and first excited manifolds of Ho and Tm, as shown in Figure 3. Suppose one of the first excited manifolds, such as the Tm $^3F_4$ manifold, is pumped to a certain population density by a short pump pulse. It could spontaneously decay by emitting a photon, denoted by a lifetime τ_2, or could interact with the ground manifold of Ho, the $^5I_8$ manifold, and produce a Ho atom in the $^5I_7$ manifold while relaxing to the $^3H_6$ manifold, denoted by process P_{28}. A reverse process could also occur; that is, a Ho atom in the $^5I_7$ manifold could relax to the $^5I_8$ manifold while promoting a Tm atom from the $^3H_6$ manifold to the $^3F_4$ manifold denoted by process P_{71}. In this notation, the first subscript refers to the manifold which donates the quantum of excitation and the second refers to the manifold which receives the quantum of excitation. Rate equations for this simplified system also appear in Figure 3.

Solving this system of equations under the low excitation density approximation yields a closed form solution which allows a facile comparison with experiments. Figure 4 displays the agreement between the measured population densities as well as the fit to the above equation's solution. A rapid initial decay of the Tm $^3F_4$ manifold shows the energy sharing with the Ho $^5I_7$ manifold. After reaching quasi equilibrium, both manifolds tend to decay with a lifetime reflecting the relative population densities of both manifolds. Parameters obtained from the curve-fitting process agree well with the energy transfer parameters derived using the above model, as shown in Table I.

$$\frac{dN_2}{dt} = -\frac{N_2}{\tau_2} - P_{28} N_2 N_8 + P_{71} N_1 N_7$$

$$\frac{dN_7}{dt} = -\frac{N_7}{\tau_7} + P_{28} N_2 N_8 - P_{71} N_1 N_7$$

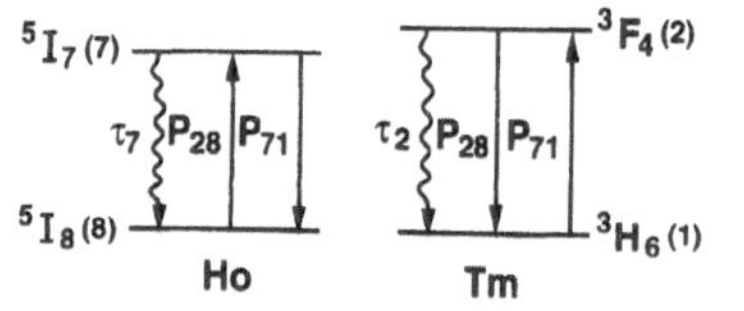

Table 1. Experimental and Predicted Values Energy Transfer Parameters.

Symbol	Process	Experiment	Predict
P_{28}	$Tm\,^3F_4 \rightarrow {}^3H_6; Ho\,^5I_8 \rightarrow {}^5I_7$	$0.698 \cdot 10^{-22}$	$0.829 \cdot 10^{-22}$
P_{71}	$Ho\,^5I_7 \rightarrow {}^5I_8; Tm\,^3H_6 \rightarrow {}^3F_4$	$0.106 \cdot 10^{-22}$	$0.104 \cdot 10^{-22}$

$^5I_7$ (7) $^3F_4$ (2)

τ_7 P_{28} P_{71} τ_2 P_{28} P_{71}

$^5I_8$ (8) $^3H_6$ (1)

Ho   Tm

Figure 3. Lowest two manifolds in Ho and Tm showing the basic energy transfer processes.

A more nearly complete description of the Ho:Tm laser system requires that the four lowest manifolds of both Ho and Tm be included in the energy transfer process. Even here, the principal of parsimony is applied. Only processes with relative large calculated energy transfer parameters are used. Processes used in the rate equation are shown if Figure 5. Of these, the commonly recognized processes are as follows: self quenching of the Tm pump manifold, $^3H_4$, produces two Tm atoms in the $^3F_4$ manifold, process P_{41}. This process is commonly referred to as the two-for-one process. Ho and Tm share quanta of excitation between the first excited manifolds, that is, Ho $^5I_7$ and Tm $^3F_4$, described above. Up conversion refers to a Ho atom in the $^5I_7$ manifold and a Tm atom in the $^3F_4$ manifold interacting to produce a Ho atom in the $^5I_5$ manifold, while the Tm atom relaxes to the $^3H_6$ manifold, process P_{27}. Up conversion is often associated with loss when the Ho atom in the $^5I_5$ manifold relaxes before the inverse process, process P_{51}, can occur.

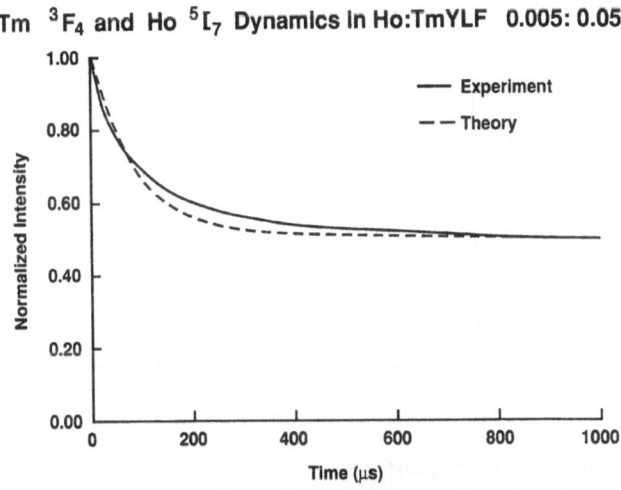

Tm $^3F_4$ and Ho $^5I_7$ Dynamics in Ho:TmYLF 0.005: 0.050

Figure 4. Experimental and predicted fluorescence from Tm $^3F_4$ manifold after initial excitation of this manifold.

Besides the energy transfer processes shown in Figure 5, all excited manifolds can decay, either radiatively or nonradiatively. Two of the faster decay rates are shown in Figure 5 as wavy lines. It requires 12 energy transfer processes to describe the manifold population dynamics as well as an equal number of decay rates. As such, a complete description of the dynamics of the Ho:Tm laser dynamics using measured parameters would be difficult since many of the parameters are difficult to measure experimentally. This is especially true of rates involving excited manifolds for both the upper and lower manifolds. Calculated energy transfer parameters are used exclusively rather than using a mixture of measured and calculated parameters.

An amplifier model was developed which takes into account the energy dynamics of the energy transfer processes described above and the saturation of the pump absorption as well. If a collinear pump and gain beam are used in an amplifier experiment, the pump energy will vary as a function of longitudinal position in the amplifier. In some situations, the absorbed energy varies approximately according to Beer's law absorption. However, in many practical Ho:Tm laser amplifier situations, the absorption saturates because of partial depletion of the Tm atoms in the ground level. Both effects have been taken into account in the laser amplifier model. Output of the amplifier model is the gain of the amplifier. Available stored energy is directly related to the gain.

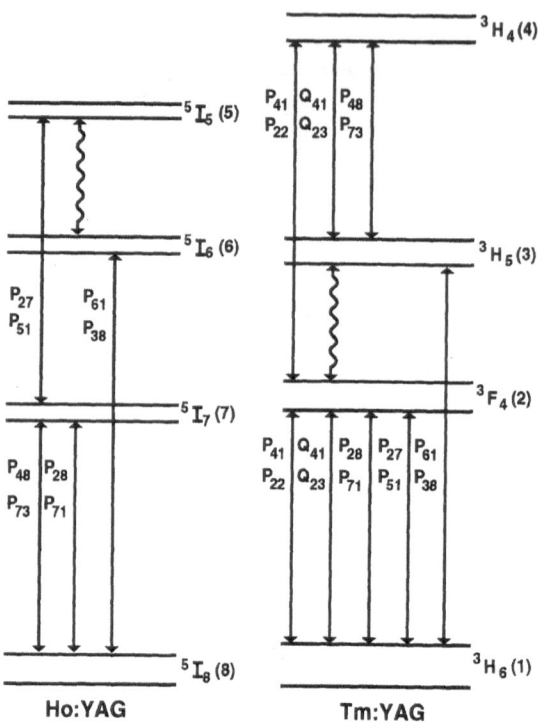

Figure 5. Energy transfer processes in lowest four manifolds of Ho and Tm including major nonradiative decay processes.

III. LASER AMPLIFIER EXPERIMENTS

An end-pumped laser amplifier was constructed to provide an experimental comparison for the models described above. Several Ho:Tm:YLF laser samples, with different Ho and Tm concentrations, were used in a series of experiments. A $Cr:BeAl_2O_4$ laser, tuned to the absorption peak of Tm in YLF, was used to provide a high-fluence pump source. Gain was measured by propagating a single-mode, continuous-wave Ho:Tm:YLF laser collinearly with the pump beam as shown in Figure 6. Dichroics and notch filters were used to separate any unabsorbed pump radiation from the amplified signal. A chopper, synchronized with the pump pulse, provided a no-signal level for the detector. Amplifier gain was measured as a function of time. Since the power of the probe Ho:Tm:YLF laser is very low, no saturation occurs so the peak gain measures the peak population density in the Ho $^5I_7$ manifold, the upper laser manifold. Peak gain is calculated by dividing the amplified signal, less the no-signal level, by the unamplified signal at the peak of the gain, less the no-signal level. Peak gain was then measured as a function of pump fluence for a variety of different Ho:Tm:YLF laser material samples.

A sample comparison of the predicted and measured gain shows good agreement. Experimentally measured gain is shown in Figure 7 as filled circles while the prediction of the models described above is shown as a solid line. Amplifier gain is negative at low absorbed energy densities, reflecting the quasi four-level nature of the Ho:Tm:YLF laser amplifier. At about 20 MJ/m³, the amplifier gain coefficient is 0.0; that is, the amplifier gain is unity, a condition of no amplifier loss or gain. This condition is referred to as

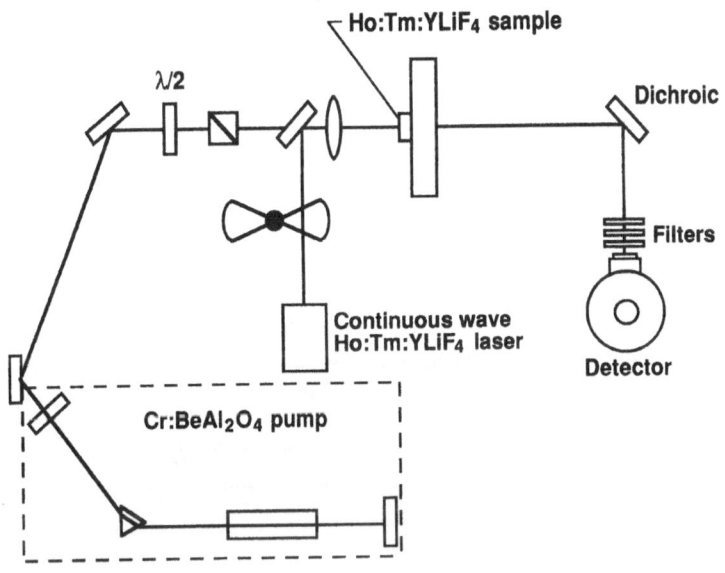

Figure 6. Experimental arrangement used to measure gain in end pumped Ho:Tm:YLF.

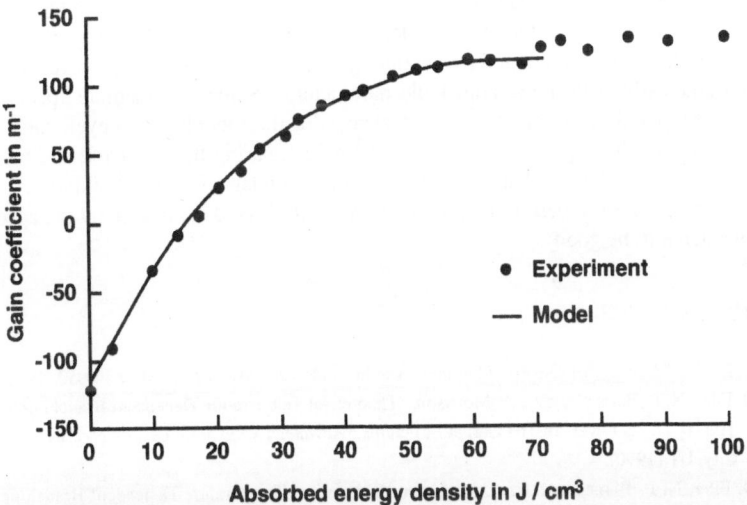

Figure 7. Measured and predicted amplification on Ho:Tm:YLF as a function of absorbed energy density.

optical transparency. Well above optical transparency, the gain tends to saturate, indicating that nearly the maximum number of Ho atoms possible are being excited to the upper laser manifold. Several other Ho:Tm:YLF samples were also evaluated using both 2.052 and 2.061 μm probes on both π and σ polarizations with similar results.

IV. HIGHER EFFICIENCY LASER MATERIALS

Quantum mechanical and energy transfer models can be used to find laser materials which show promise of higher efficiency when compared to commercially available laser materials. A survey of garnet and fluoride laser materials has been performed for both Ho and Tm lasers [2-4]. A result of the survey is that LuAG and LuLF were predicted to be superior laser materials when compared to their commercially available isomorphs, YAG and YLF. A reason for this is the stronger crystal field associated with LuAG and LuLF leads to a larger splitting of the lower laser manifold when compared to YAG and YLF. A larger splitting provides a smaller thermal occupation of the lower laser level and concomitant lower pump levels to achieve optical transparency. Substitution of Lu for Y also tends to change the dynamics of the laser system in a favorable direction. To test these predictions, Ho:Tm:LuAG, Tm:LuAG, and Ho:Tm:LuLF were obtained and experimentally evaluated.

Ho:Tm:LuLF has both a lower threshold and a higher slope efficiency when compared with Ho:Tm:YLF. Both laser materials were evaluated using laser diode pumping and under very nearly identical circumstances. When plotting the laser output energy versus the optical pump energy from the laser diodes, threshold and slope efficiency for a Ho:Tm:LuLF laser were 0.558 and 0.138, respectively. For a Ho:Tm:YLF laser, the same numbers were 0.582 and 0.120, respectively [5]. Ho:Tm:LuAG [6] and Tm:LuAG [7] have also been demonstrated as viable 2.0 μm laser materials.

V. SUMMARY

A series of four models has been used in sequence to predict the performance of 2.0 μm laser materials, a quantum mechanical model, an energy transfer model, a rate equation model, and a laser amplifier model. Through use of these models, performance of a laser amplifier can be predicted by beginning with the measured energy levels of a laser material and without utilizing adjustable parameters. Some intermediate spectroscopic data has been used to verify some of the aspects of the models. However, most of the comparison with the experiment was obtained by comparing the predicted and measured performance of various Ho:Tm:YLF laser amplifiers under a variety of situations. While space and time does not permit the presentation of all of the data, in general the agreement was considered to be good.

VI. REFERENCES

1. D. L. Dexter, "A theory of sensitized luminescence in solids," *J. Chem. Phys.*, 21: 836-850 (1953).
2. E. D. Filer, N. P. Barnes, and C. A. Morrison, "Theoretical Temperature Dependent Branching Ratios Of The $^5I_7$ To $^5I_8$ Levels In Ten Garnet Structures," *Advanced Solid State Laser Conference*, Salt Lake City, UT (1990).
3. E. D. Filer, N. P. Barnes, and C. A. Morrison, "Theoretical Temperature Dependent Branching Ratios And Laser Thresholds Of The $^3F_4$ To $^3H_6$ Level Of Tm^{3+} In Ten Garnets," *Proceedings Of Advanced Solid State Laser Conference*, 10: 189-200 (1991).
4. E. D. Filer, C. A. Morrison, and N. P. Barnes, "YLF Isomorphs For Ho And Tm Laser Applications," *Advanced Solid State Laser Conference*, Salt Lake City, UT (1994).
5. M. G. Jani, N. P. Barnes, K. E. Murray, G. E. Lockard, E. D. Filer, and G. J. Quarles, "Diode-Pumped Ho:Tm:LuLiF$_4$ Laser At Room Temperature," *CLEO Conference*, Baltimore, MD (1995).
6. N. P. Barnes, E. D. Filer, F. L. Naranjo, W. J. Rodriguez, and M. R. Kokta, "Spectroscopic and lasing properties of Ho:Tm:LuAG," *Opt. Lett.*, 18: 708-710, (1993).
7. N. P. Barnes, M. G. Jani, and R. L. Hutcheson, "Diode-pumped, room temperature Tm:LuAG Laser," *Appl. Opt.*, 34: 4290-4294 (1995).

NEW DEVELOPMENTS OF LASER CRYSTALS

A. A. Kaminskii

Institute of Crystallography
Russian Academy of Sciences
Moscow
RUSSIA

ABSTRACT

Several selected problems of modern physics of laser crystals were discussed, among which are: visible stimulated emission (SE) of Pr^{3+}-doped fluoride and oxide compounds for creation of highly-bright "white light"; self-frequency conversion phenomena; new potentialities of crystals doped with Nd^{3+} and Er^{3+} ions for UV and visible SE excitation; new mini-laser designs for opto-electronics; novel low-threshold room-temperature lamp-pumped 1.06 μm non-neodymium SE.

In the last few years efficient crystalline lasers have been developed which are capable of generating almost arbitrary wavelengths in the visible spectral range. This promotes the idea of using these lasers for the creation of a new generation of projection color TV and other display devices. We suggest that a good candidate for these purposes is the big family of visible Pr-lasers on the base of anisotropic $LiYF_4$, BaY_2F_8, and $YAlO_3$ type crystals. In particular, for obtaining high power (energy), their $^3P_0 \rightarrow \,^3H_4$ channel for blue SE is very attractive; the best channel for green SE is $^3P_0 \rightarrow \,^3H_5$, and for red laser action several intermanifold transitions $^3P_0 \rightarrow \,^3F_{2-4}$, $^3H_6$.

Ideas for using self-frequency conversion in laser crystals involve second-harmonic generation (SHG) and stimulated Raman scattering (SRS) phenomena. These suggestions have been implemented experimentally. In particular, the self-doubling of lasing and pumping frequencies in acentric crystals with second-order nonlinear susceptibility $\chi^{(2)}$ ($LiNbO_3$:MgO and $LaBGeO_5$ doped with Nd^{3+} and Er^{3+} ions); self-SRS of lasing and pumping frequencies in crystals with high value of $\chi^{(3)}$ ($KY(WO_4)_2$ and $KGd(WO_4)_2$ doped with Nd^{3+} and Er^{3+} ions); as well as multi wavelength cascading collinear and coaxial Strokes SRS ("laser rainbow") in crystals containing very active Raman vibrational optical modes in tetrahedral WO_4^{2-} and MoO_4^{2-} complexes.

New laser possibilites of Nd^{3+} and Er^{3+} ions, related to their high-lying $^{2S+1}L_S$ - manifolds and the principle regularities in the variational luminescence intensity charateristics of these manifolds, for different crystaline hosts ($LiYF_4$, BaY_2F_8, $YAlO_3$, $Y_3Al_5O_{12}$, etc.), have been examined and established. For this purpose we have performed the detailed calculations of the full set of squared reduced-marix elements $|<||U^{(t)}||>|^2$ for all $4f^3(Nd^{3+})$ and $4f^{11}(Er^{3+})$ intermanifold channels with energy up tp about 67000 cm^{-1} for Nd^{3+} and 97000 cm^{-1} for the Er^{3+} ions. The obtained data have helped us to excite new SE under the Xe-flashlamp pumping at 110 K for Nd^{3+}:BaY_2F_8, $\lambda_{SE} \approx 0.385$ μm ($^4D_{3/2} \rightarrow \,^4I_{11/2}$); for Er^{3+}:$LiLuF_4$, $\lambda_{SE} \approx 0.56$ μm ($^2H(2)_{9/2} \rightarrow \,^4I_{13/2}$); for Er^{3+} ions in $LiLuF_4$, BaY_2F_8, and $YAlO_3$, $\lambda_{SE} \approx 0.4$ μm ($^2P_{3/2} \rightarrow \,^4I_{13/2}$).

Spectroscopy and Dynamics of Collective Excitations in Solids
Edited by Di Bartolo, Plenum Press, New York, 1997

New types of miniature solid-state lasers with active elemens in the form of micron-thick single-crystal ribbons and sheets have been developed. It was reported that the first crystal used for this purpose was an orthorhombic mica-like $KY(MoO_4)_2$:Nd^{3+} compound, lasing at 300 K as a result of the $^4F_{3/2} \rightarrow {}^4I_{11/2}$ channel, when pumped with the Xe-flashlamp or with lasers. Active elements, in the form of a ribbon and a sheet, were made by the simple mechanical splitting of single-crystal boules.

Moreover, new results of low-threshold (≈ 0.5 J) pulsed one-micron SE at $^1D_2 \rightarrow {}^3F_4$ channel of Pr^{3+} ions in monoclinic $KY(WO_4)_2$ and $KGd(WO_4)_2$ crystals at room temperature and under Xe-flashlamp pumping were presented.

OPTICAL EXCITATION AND RELAXATION

OF SOLIDS WITH DEFECTS

G. Baldacchini

Centro Ricerche Energia di Frascati
C.P. 65, 00044 Frascati
Rome, Italy

Abstract. As color centers in alkali halides are the simplest of all the point defects in solids, they have been studied as a model-case since the beginning of this century. Color centers display an optical cycle that very often produces luminescence having a very high efficiency after excitation and relaxation. This optical characteristic has been utilized for the last 20 years to produce color center lasers, with a high spectral purity emission tunable from the visible to the near infrared. Among the laser active color centers, the F_A family is the most popular, with its very peculiar linear and nonlinear optical properties. Moreover, it is still supplying new experimental data that imply energy-transfer phenomena not contemplated by previous accepted models. An accurate description of the F_A centers in alkali halides is given together with a historical account of successes and failures, the latter due to chance human events, to limited technical choices, and to a lack of scientific curiosity.

I. INTRODUCTION

Under certain favorable physical conditions such as pressure and temperature, atoms and molecules tend to aggregate in an orderly way, eventually forming a macroscopic body known commonly as a crystal. This solid, which is the ultimate form of all matter at low temperatures, consists of the regular repetition all over its volume of a microscopic unit cell containing a fixed arrangement of the constituent atoms or molecules of the particular substance. Because of this intrinsic internal order, crystals display regular geometrical external forms well known since historical times. However, crystals are much more interesting for solid-state physics studies when they contain defects, which can be defined in general as a break down of the microscopic order. The simplest imperfections are those well localized in the lattice structure, such as impurities, vacancies etc., which are usually called point defects. Color centers in alkali halides are perhaps the best known point defects, mainly because they are the simplest. They were officially discovered in the middle of the last century and were the first to be studied systematically almost from the beginning of this century [1].

Spectroscopy and Dynamics of Collective Excitations in Solids
Edited by Di Bartolo, Plenum Press, New York, 1997

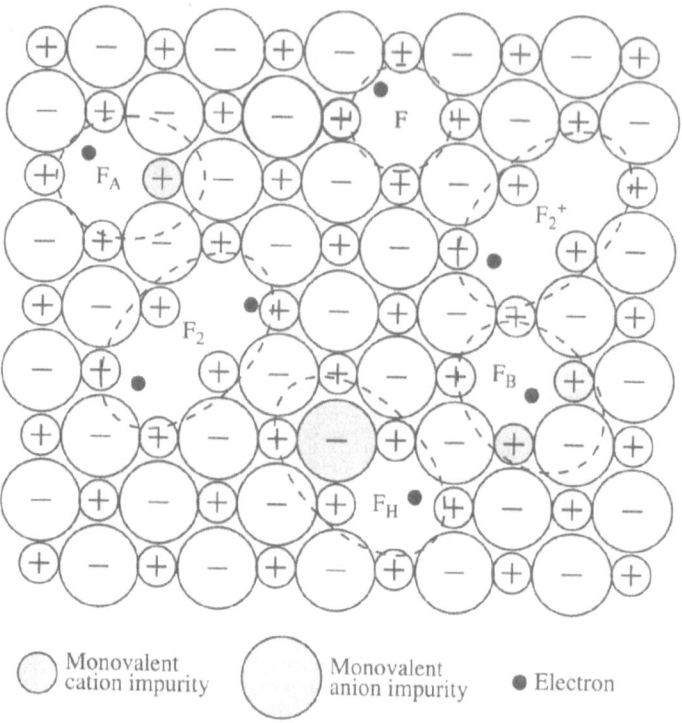

Figure 1. Structural models of the F center and of simple aggregated centers.

Color centers are formed by one or more vacancies with various degrees of ionization, which can also be aggregated to foreign ions. However, almost all of them have in common a basic building block called the F center, which is formed by an electron trapped in an anion vacancy (see Fig. 1). In all respects the F center is a prototype model for point defects in solids as the hydrogen atom is for atoms and molecules.

F centers display several properties, but the most notable are related to their optical transitions, which also originated their name, F≡ Farbe (in German) ≡ color. Indeed, when a crystal contains F centers, it displays a typical color that is a consequence of their absorption bands in the visible region of the electromagnetic spectrum. The whole optical cycle is represented in Fig. 2, where for the sake of simplicity, higher state transitions have been omitted. In normal conditions the F center is in the ground state (GS), 1, from where it can be excited by a photon of appropriate energy in the unrelaxed excited state (URES), 2. This transition changes the density of the negative electronic charge from a 1s-like to a 2p-like distribution, which is no longer in equilibrium with the ionic lattice surrounding the vacancy (Fig. 1). As a consequence, the first-neighbor ions move outward to reestablish the electrostatic equilibrium of the crystal. From the energy point of view, the F center relaxes very quickly from state 2 to the relaxed excited state (RES), 3. From this very interesting and complex state [2], the F center can reach the unrelaxed ground state (URGS), 4, with a radiative transition and, after a new relaxation process, the GS, thereby closing the optical cycle.

The energy diagram of Fig. 2 is common to most of the color centers, some of which are reported schematically in Fig. 1. The F center can be associated with a monovalent positive ion in a 100 position forming an F_A center. If the association is with two monovalent positive ions, the F_B center appears. Two F centers can be aggregated along the [110] direction with the formation of the so-called F_2 center. If such a center is ionized once, one gets the F_2^+ center. The F center can also be associated with a monovalent negative ion in a 110 position, with the formation of the F_H center. More complex color centers can be formed by continuing ionizations and associations.

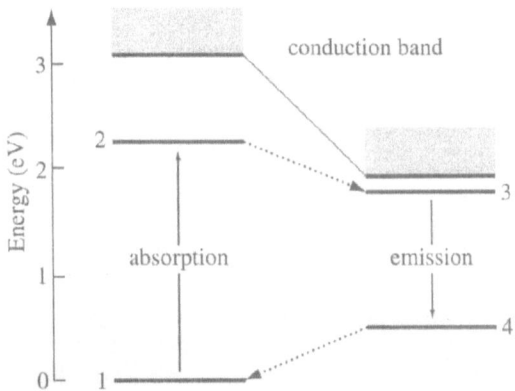

Figure 2. Main electronic states and optical cycle of the F center. The energy scale refers to the crystal host KCl.

Although the F-center-like model of Fig. 2 gives a fairly good general description of color centers, each center possesses properties, such as peak position and width of bands, relaxation rates, lifetimes, etc., that are peculiar to each kind of center. Figure 3 displays the structural model and the main optical properties of the color centers described. The absorption and emission bands can be quite different from each other, even inside the same family as in the case of the $F_A(I)$ and $F_A(II)$, or the $F_B(I)$ and $F_B(II)$ centers. Lifetimes can differ by one or two orders of magnitude, and the reorientational properties, implicit in most of the axial-symmetric centers, display different features [3-5].

The optical characteristics of some color centers have been used for the last two decades to produce efficient solid-state lasers, and a brief description of this subject is given in Section II. Section III is devoted to the very peculiar history behind the discovery of the F_A color center laser (CCL). Because of the central role played by the F_A center in CCL research and development, Section IV describes its main characteristics and some recent findings that have once again opened the debate on the understanding of this peculiar center. Conclusions and indications for future developments are given in Section V.

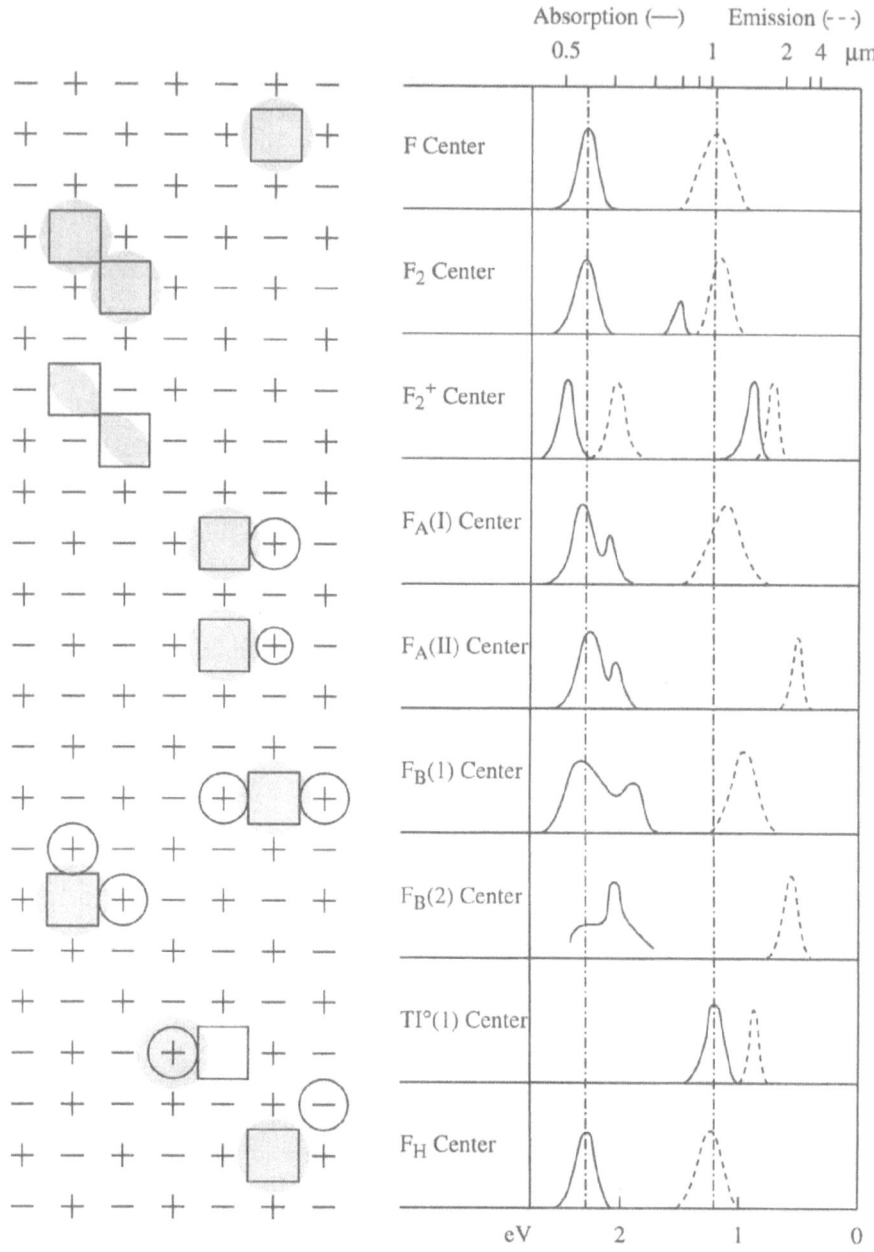

Figure 3. Structural models, absorption and emission behaviors of the F center and of simple aggregated centers at low temperature in KCl. The square is the halogen vacancy, the circle the foreign ion and the shaded area the approximate electron distribution.

II. COLOR CENTER LASERS

It is well known that the F center and also most of the color centers have a quantum efficiency approaching unity for the emission process at liquid nitrogen temperature (LN_2T) and temperatures below [4]. By raising the temperature, the thermal excitation from the RES to the conduction band becomes dominant, and the luminescence is eventually totally quenched at room temperature (RT) [6]. Moreover, during the optical cycle only two levels, the GS and the RES, are relatively populated because the relaxation transitions $2 \to 3$ and $4 \to 1$ occur in the ps range, while the RES lifetime is several orders of magnitude longer, reaching in a few cases some μs [6]. These properties are very favorable for laser action. Indeed, having defined the gain factor G (the ratio of the output to the input intensity of a beam of light passing through a colored crystal of thickness d) as

$$(1) \qquad G = \frac{I \text{ out}}{I \text{ in}} = \exp[(\alpha_g - \alpha_l)d] \ ,$$

where α_l is the loss coefficient, the amplification (gain) coefficient α_g, due to stimulated transitions, is given by [7]

$$(2) \qquad \alpha_g(\nu_0) = \frac{\lambda_0^2 \eta}{8\pi(1.07\Delta\nu)} \ \frac{N_3 - N_4}{n^2\tau} \ ,$$

where $\nu_0 = c/\lambda_0$ is the center and $\Delta\nu$ the width (FWHM) of the gaussian emission band, η is the quantum efficiency of the emission, n is the index of refraction of the host crystal, τ is the lifetime of the RES, and $N_{3,4}$ is the population of the levels $3, 4$ in the unit volume. In the case of color centers, as explained above, $N_4 \cong 0$, so the amplification coefficient is positive under any pumping intensity.

However, since this coefficient strongly depends on several parameters of the optical cycle, it can differ greatly from center to center. Table 1 reports the

Table 1. Characteristics of the luminescence and optical gain of a few color centers for a population inversion of 10^{16} cm^{-3} and at LN_2T.

Crystal	Center	λ_0 (μm)	τ/η (ns)	$\delta\nu$ (10^{13} Hz)	α_g cm^{-1}
KCl	F	1.0	600	6.3	0.04
KCl	F_2^+	1.7	80	1.69	3.5
KCl:Li$^+$	$F_A(II)$	2.7	200	1.45	4.5
KCl:Na$^+$	$F_B(II)$	2.5	200	1.57	3.3
KCl:Tl$^+$	Tl$^\circ(1)$	1.5	1600	2.00	0.12
NaCl:OH$^-$	$(F_2^+)_H$	1.6	150	3.29	0.77
KCl:Na$^+$:O$_2^-$	$(F_2^+)_{AH}$	1.8	150	2.8	1.5

value of the amplification coefficient for a few typical color centers by assuming the excitation number $N_3 = N_{RES} = 10^{16}$ center/cm^3, which is easily attainable with the available optical pumping sources. The first surprise in Table 1 is that the gain of the F centers is so small that it is practically impossible to observe any laser action at all. This is mainly due to a long lifetime and a broad emission band, which greatly decrease α_g in Eq. 2.

Anyway, without going into a detailed description of this field, which will be mentioned again in the next section, also from a historical perspective, it is possible today to entirely cover the spectral range from 0.5 to 4 μm with CCL emissions [8,9], and also have a modest but interesting laser emission at $\sim 5\,\mu$m [10]. By leaving aside the pulsed RT CCLs [8], the most common cw lasers working at LN$_2$T are reported in Fig. 4. The F_A and F_B centers have been described in the previous section; the Tl^0(1) center is similar to the F_A center (this is why it is often called F_A(Tl)), with the difference that the electron charge density sits more on the Tl than on the vacancy site (see Fig. 3). The F_2^+ centers are usually intrinsically unstable because they can capture an electron and can reorient under optical pumping. Hence, a great effort has been made to stabilize them through aggregation processes. The $(F_2^+)^*$ are F_2^+ centers perturbed by unknown impurities, the $(F_2^+)_A$ are associated with a positive ion, the $(F_2^+)_H$ with a negative ion, and the $(F_2^+)_{AH}$ with both negative and positive ions. However, most of these centers are neither stable enough nor easily produced, so only the F_A, Tl^0(1) and $(F_2^+)_H$ lasers have been widely used and commercialized up to now.

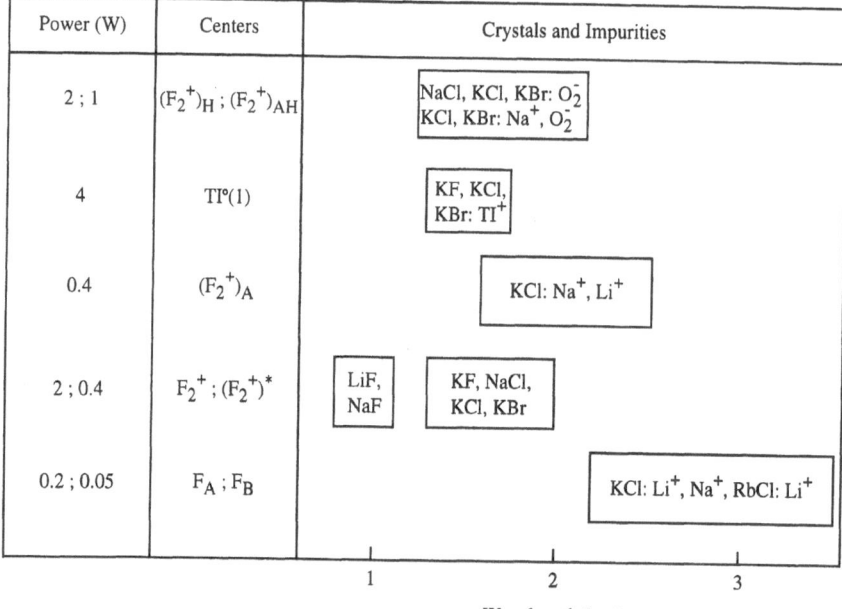

Figure 4. Maximum output power and tunability range in cw regime of the most common color center lasers working at LN$_2$T.

CCLs in general have some important characteristics, such as spectral pure emissions, wide continuous tunable bands, very short pulse regimes, and reasonably high powers, which have been put to work in several basic and applied fields; particularly in atomic, molecular, and solid-state high-resolution spectroscopy, in the detailed measurement of relaxation processes and in the study of signal propagation in optical fibers.

III. DISCOVERY OF THE F_A CENTER LASER

If we take all the lasing color centers reported in the preceding section, and consider their production, stability under pumping and when in storage, optical efficiency, and the availability of pumping laser sources, etc., the F_A center, which was the first to lase, was and still is the best. However, this assertion deserves further comment, especially in view of the human and scientific events involved in the discovery of the laser effect.

The first laser ever was discovered in July 1960 by T. Maiman, in ruby [11], contrary to the expectations of the time, i.e., laser action in a gas phase [12]. The He–Ne laser was discovered five months later [13]. From then on a frantic research expanded rapidly to include practically all the known luminescent materials in gas, liquid, and solid form, and the number of laser emissions increased exponentially in the first years after the original discovery [14].

As far as color centers in alkali halides are concerned, the first laser emission at 2.7 μm in pulsed regime was observed in Germany in 1965 with F_A(Li) centers in KCl [15]. This important achievement remained an isolated episode, and only a decade later was the work resumed in the USA with the implementation of a continuously tunable cw CCL in the range $2.5 < \lambda < 2.9\,\mu$m using the same active material [16]. Systematic research in this field then started in different laboratories around the world, and in a time span of $\sim$ 15 years, tens of different color center laser emissions were discovered, as briefly described in the preceding section.

At this point two questions arise spontaneously. How come the first laser effect in color centers was observed so early on, only five years after the discovery of the ruby laser, and in Germany? Why did it take so long, almost ten years, for the real start of CCL development, and why in the States? Satisfactory answers to these questions require going back in time for a brief historical survey.

Color center physics was started systematically in 1920 by R.W. Pohl in Gottingen, Germany; it remained mainly a German affair for more than a decade and was still dominated by the German school up to the beginning of World War II in 1940. At the time, the optical absorption properties of F and more complex color centers were well known [17,18]. There were also strong expectations in fluorescence after optical excitation [19], but the outbreak of the War caused a significant decrease in research in this field of solid-state physics, when not related to the War efforts. In Germany the situation was particularly bleak, especially after the War; only in 1948 was Pohl permitted to enter his laboratory again. By this time, color center physics was becoming an international field of research, with good schools in the USA, Russia, Japan and several European countries. Among other arguments, a lively debate started on the existence of the luminescence [20], which was, indeed, discovered soon after [21,22]. Moreover, the optical cycle of the F center, as shown in Fig. 2, was almost completely clarified by measuring the lifetime of the (relaxed) excited F center [23,6].

Meantime, the story of the F_A center was developing on a parallel line, although lagging a little behind. This peculiar defect, originally called A center, was discovered during World War II in Gottingen by St. Petroff, a Bulgarian guest of the Pohl group, who published his results only after the War, first in a Bulgarian report [24] and then in a German journal [25]. In due time the optical and photochemical properties of this center together with its [100] symmetry were established [26-28], and finally the lifetime of the RES of the F_A(Li) centers in KCl was also measured [29]. Thus, in 1964 the basic properties describing the optical cycle of the F_A center were reasonably well known. Soon after, in 1965, the first laser emission was announced by B. Fritz and E. Menke. This discovery took place at the University of Stuttgart, to the surprise of outsiders but not of the people working in the color center field, who knew very well that almost all the German school's knowledge of point defects in alkali halides was now concentrated in this city. Indeed, after the retirement of Pohl in 1954, Professor H. Pick, one of his former students, moved to a permanent position at the University of Stuttgart, bringing with him some people from the Pohl group and thus establishing a new school, which was the natural continuation of the no longer active Gottingen School. F. Luty, a diploma student of Pohl, was among those who moved to Stuttgart, where he made important contributions particularly in the field of F_A color centers [30]. So it is very clear that in Stuttgart, at the time of the invention of the first ruby laser and subsequent laser rush, there were all the necessary ingredients for successfully attempting a laser experiment, and no other similar place existed then anywhere else in the world. In the end the original idea sprung from discussions with W. Kaiser, who was a visiting professor at Stuttgart in 1963; it was then developed by B. Fritz, a coworker and intimate friend of Luty, and his diploma student E. Menke.

Fritz and Menke understood immediately that F_A centers were better candidates for laser emission than F centers, because they had a shorter lifetime and a narrower emission linewidth, (see Eq. 2 and Table 1). To obtain the laser emission, they followed the standard recipe of the time, similar to the formula used for the ruby laser. The sample of KCl containing F_A(Li) centers was shaped as a 3-cm-long rod with evaporated silver mirrors at both polished ends, cooled in a nitrogen glass cryostat, and illuminated by a xenon flash tube. Figure 5 reports the temporal shape of the laser pulse together with the pulse pumping light. However, the laser emission consisted of a series of irregularly spaced and shaped bands around 2.7 μm. After analyzing the laser emission in detail, Fritz and Menke also suggested the use of F centers as lasing defects because, although they had the previously described shortcomings, they possessed a longer lifetime and so, following their reasoning, it would be possible to have a much higher inversion population. In this expectation they were wrong, mainly due to a reason that will be clarified later. Unfortunately, they did not proceed further along this line of research, partly because laser technology was still in its infancy and partly because they were not really interested in this applied field since they probably had more basic research in mind. However, human events prevented further reconsideration of the laser in Stuttgart because Fritz died prematurely in 1969 and Menke left this research field after his degree.

At this stage apparently there were no people sufficiently interested in the subject, so this brilliant discovery was almost forgotten for several years. An attempt to measure the stimulated emission in the F centers in KCl was made in Rome by F. De Martini, who understood the laser field, U.M. Grassano, who had experience in the color center field, and a student, F. Simoni [31]. Grassano was a former student of Prof. G. Chiarotti, who became one of the founders of the Italian School on point defects after his work during 1955-1957 in the group of Prof. F. Seitz at Urbana, USA, and Chiarotti himself suggested that De Martini and Grassano investigate the F center potential for lasing effects. Moreover these people were also motivated to utilize the F center by the fact that its preparation was relatively simple, and its emission in an easily accessible region of the spectrum, ~ 1 μm, where the Nd:YAG laser also has an emission. They measured a slight gain for the F center, probably overestimated, but never tried to build a real laser, which would never have worked anyway, not only because the gain coefficient is very low, but also because of the effects of self-absorption. Indeed, the absorption from the RES is a consequence of the energy distance from the conduction band (optical gap), which for the F center is much smaller than the emission energy (see Fig. 2). This means that at high levels of excitation, as required for laser amplification, the stimulated emission will be absorbed by the electrons in the RES, with the final result of an increase in the loss coefficient α_l in Eq. 1 and thus a decrease in the gain factor G. On the contrary, such a phenomenon does not figure in the optical cycle of the F_A(Li)

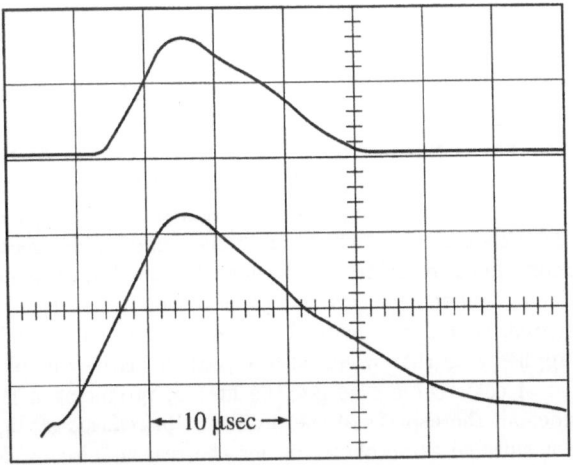

Figure 5. Oscilloscope traces of the 2.7 μm laser light (upper trace) and pumping light. The upper beam was triggered by the pumping light. Time scale = 5 μsec/div; crystal temperature: 70 K [15].

center because the optical gap is larger than the emission energy, as shown in Fig. 6 where this energy situation has been highlighted. This is known from the lack of thermal and electric field ionization [30], contrary to what happens to F centers, and from the measurement of a thermal activation energy of 0.28 eV from photoionization [32], which implies an optical gap larger than the emission energy of F_A(Li) centers of 0.46 eV. Finally, the absence of self-absorption in the F_A(Li) system was definitively proved by the functioning of the cw laser itself at the Bell Telephone Laboratories, Holmden, USA, in 1974 [16].

However, this story had started a few years before at the University of California, Berkeley, where Dr. L.F. Mollenauer, after his Ph.D. in the field of quantum optics at Stanford University under the guidance of Prof. A.L. Schawlow, was Assistant Professor in the group of Prof. C.D. Jeffries. The group was interested in producing samples containing strongly polarized nuclei [33] to be used as targets for high-energy elementary particles (electrons, photons, etc.) in

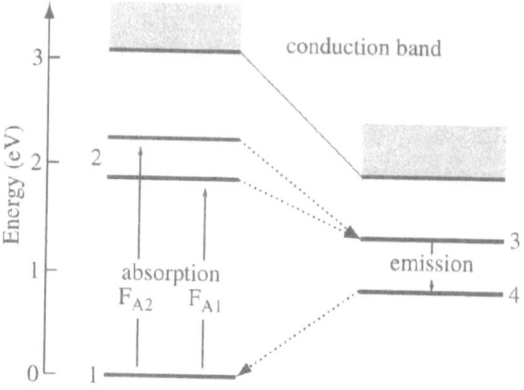

Figure 6. Main electronic states and optical cycle of the F_A(Li) center. The energy scale refers to the crystal host KCl.

the nearby Lawrence Radiation Laboratory (now Lawrence Berkeley Laboratory). Within this context, Jeffries suggested that Mollenauer polarize nuclei by optical pumping of F centers. The relative efforts taught him a lot about this new field and, in due time, he produced some very fine results [34,35]. At the end of 1971, the Faculty of Physics considered his appointment to a permanent position, i.e., tenure. After some discussion they decided negatively, and as a consequence he started to look for another post, which he finally found to his great satisfaction at the Bell Telephone Labs in Holmden, to where he moved in the summer of 1972. It was very clear to him that at this new place he could no longer deal exclusively with basic research, so before leaving Berkeley he decided to look for a topical field with an applied side that would also appeal to his new employers. From his fresh knowledge of color centers he got the idea of producing a laser device and began to consider all the aspects of the problem. He realized straightaway that F centers were not suitable as active media for the reasons already discussed, so he turned his attention to F_A centers that already showed a pulsed laser effect [15]. However, since these centers with their axial symmetry were completely new to him, he immediately contacted Luty, who had moved to the University of Utah in Salt Lake City in 1965, and who, at the time, was one of the most renowned experts in the physics of F_A centers [30]. Taking advantage of the availability of laser sources for pumping, the technology developed for optical cavities [36], and his own impressive technical ability, Mollenauer in collaboration with D.H. Olson produced the first cw CCL in 1974, later followed by more sophisticated models lasing with the same and other centers [37].

IV. DYNAMICS OF THE EXCITED F_A CENTER

It is evident from the previous sections that the F_A center has played an important role in the development of CCLs. At the same time it has also been of considerable interest for basic research on defects, since it is the simplest among the perturbed color centers. Indeed, the association of the F center with a substitutional alkali ion in a nearest-neighbor position (Figs. 1 and 3) produces an F center with reduced local symmetry, which is a consequence of strong static perturbation. However, although F_A centers have received particular attention in the past and their properties seemed well explained [30], there have been recent interesting results that do not fit the old models very well, and so a lively debate on this argument has been opened again. An interesting aspect of the history is that these new results could have been obtained long ago, but they were not because of a combination of limited cryogenic technology, polarized scientific curiosity, and most probably, casual mistakes.

Because of the axial symmetry of the F_A centers, their absorption band splits in two separate bands – F_{A2} double degenerate and F_{A1} single degenerate – due to transition moments perpendicular and parallel to the center axis, respectively (Figs. 3 and 6). Optically-induced reorientation can be used to align this dipole complex and to create dichroic absorption. The thermal behavior of the reorientation quantum efficiency η divides the F_A centers into two types: type-I centers with a thermally activated reorientation efficiency having a maximum value of 2/3; type-II centers with a temperature-independent value of $\eta = 1/2$. Figure 7 shows such properties for a few F_A centers [38], so F_A(Li) in KCl and RbCl were classified as F_A(II), while F_A(Li) in RbBr, F_A(Na) in RbCl and KCl, and F_A(K) in RbBr were classified as F_A(I) centers.

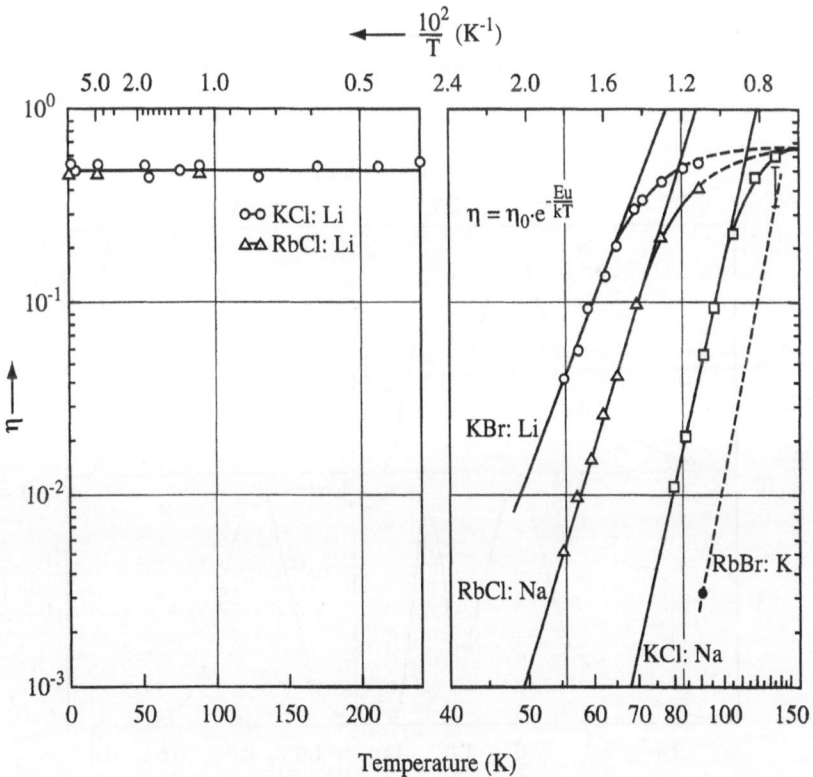

Figure 7. Reorientation quantum efficiency of various F_A centers induced by polarized optical pumping as a function of temperature [38].

The second marked difference between type-I and type-II centers is in the emission properties. Type-I centers give rise to an F-like broad emission band only slightly shifted to low energy with respect to the F emission band. Type-II centers show a much narrower emission band with a Stokes shift relative to the absorption band much larger than that of the F and $F_A(I)$ centers (see Fig. 3). The reason for the difference has been identified as the peculiar relaxed excited state present only in type-II centers. Instead of having the F-like relaxation of the excited electron into the normal vacancy-centered orbital, in the type-II center the relaxed configuration takes on the so-called saddle point shape, formed by two vacancies separated by an interstitial anion [30]. Thus, the ionic potential around the vacancy assumes the aspect of a double well, which can accomodate much better than a single well the two lobes of the 2p wavefunction, which also becomes more localized. As a consequence, the RES becomes deeper with respect to the conduction band and its lifetime shorter, which, as reported in the previous sections, are favorable conditions for laser action.

Because of the latter motivation and also for sake of scientific curiosity, several attempts have been made in the past to forecast the typology of F_A centers in various alkali halide crystals. Apart from a common and general consideration, which requires a foreign cation with a small diameter in order to have some form of movement, which is necessary for $F_A(II)$ centers, the various models have taken into account the dimensions of the ions [30], the effective charge parameter [39], and the residual covalent bonding [40]. These three phenomenological theories are not able to fully explain, perhaps with the exception of the third since not completely explored yet, the boundaries between

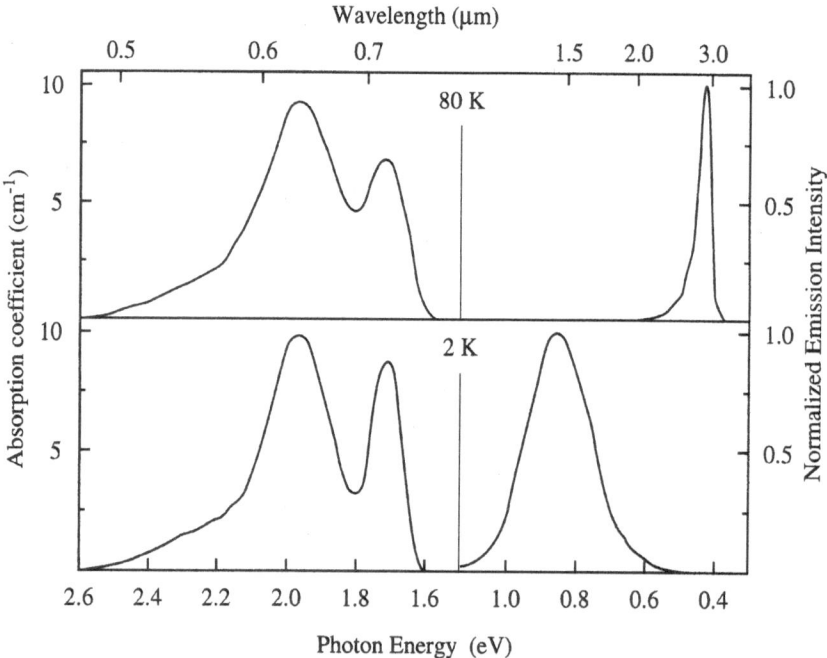

Figure 8. Absorption, left, and emission, right, at two temperatures of a RbCl:Li$^+$ crystal containing F_A centers. The luminescence has been excited by an He-Ne Laser.

type I and type II centers, but do give some general indications. In any case, none of them can explain in a direct way why both $F_A(I)$ and $F_A(II)$ centers can coexist in the same crystal host, as recently found in RbCl:Li$^+$ [41].

This particular crystal was under study [42] as part of a systematic campaign to determine the off-axis angle θ that the F_A dipole displays with respect to one of the crystal axes, and which in turn is a consequence o. the off-center ionic configuration typical of some substitutional ions [43]. The measurement of θ is based essentially on the photostimulated reorientational properties under selective pumping of the F_{A2} or the F_{A1} band. However, these two bands usually overlap, so low-temperature experiments are required for better spectral resolution during pumping. Figure 8 shows in the upper half the well-known absorption and emission spectra of F_A centers in RbCl:Li$^+$ at 80 K [44]; the overlapping of the two F_A absorption bands is evident. The crystal was then brought to liquid helium temperature (LHeT) and, as expected, we found much less overlapping between the F_{A2} and F_{A1} bands, as shown on the left of the lower half of Fig. 8. However, the reorientation quantum efficiency was lower with respect to that measured at 80 K, contrary to a previous measurement [38] reported in Fig. 7,

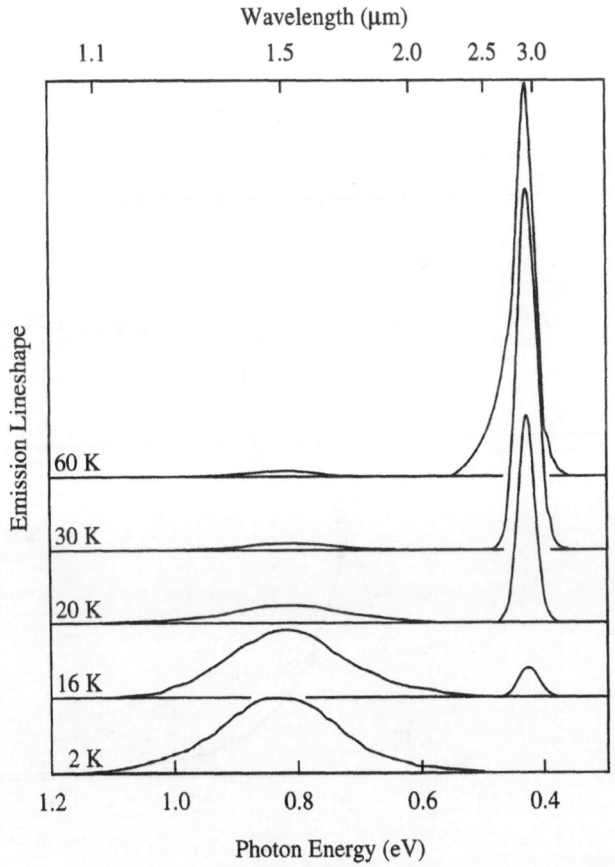

Figure 9. Evolution of the two emission bands of the F_A center in RbCl:Li$^+$ under an He-Ne laser excitation at different values of temperatures as indicated.

and to our great surprise we could no longer find any substantial luminescence at $\sim 2.9\,\mu\text{m}$, as measured at 80 K. The problem was not solved immediately because we were so sure that the luminescence should have been at $2.9\,\mu\text{m}$ that we lost considerable time in trying to find possible experimental errors of technical origin. Finally, we decided to extend our observation to a much broader range of wavelengths, and by using the same crystal and pumping excitation, we found new luminescence at $\sim 1.4\,\mu\text{m}$, as reported on the right of the lower half of Fig. 8.

At first sight we concluded that this new emission could have been originated by some sort of unknown impurity, and we started to study the dependence of the two emissions as a function of temperature in order to assemble more experimental material. Figure 9 shows the complete emission spectrum of the F_A center in RbCl:Li$^+$ at different temperatures from 60 to 2 K and under the excitation of the red line of an He-Ne laser of a few mW of power. It is immediately obvious that the two bands at 1.46 and $2.9\,\mu\text{m}$ are strongly related to each other. Indeed, while the emission at $1.46\,\mu\text{m}$ increases with decreasing temperature, the other at $2.9\,\mu\text{m}$ decreases until it almost disappears. The strong coupling between the two emission bands is demonstrated even further in Fig. 10 where the areas of their shape functions are reported as a function of temperature. These data show very clearly that there is an energy exchange between the two emissions, which are complementary to each other.

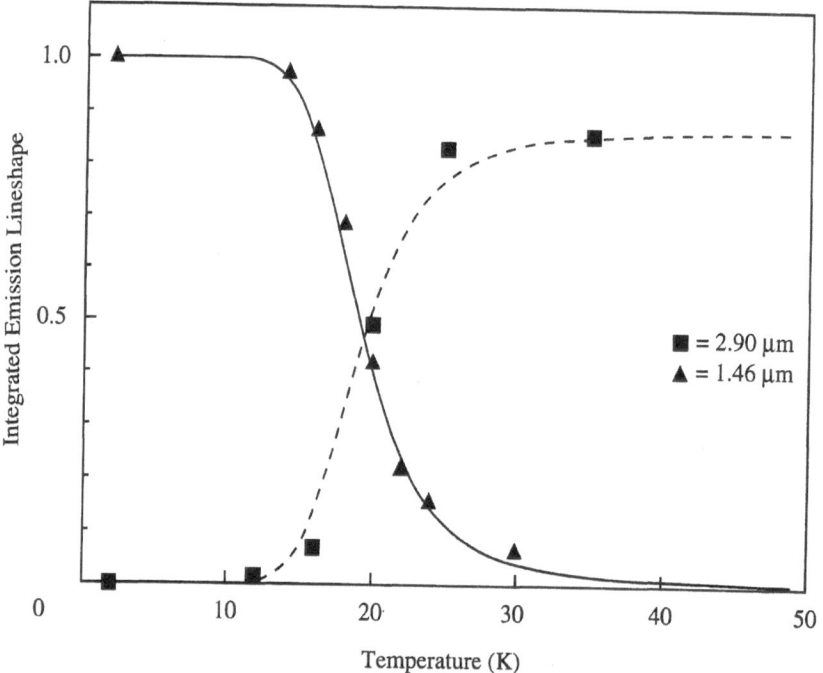

Figure 10. Temperature dependence of the area of the shape functions for the $2.90\,\mu\text{m}$ and for the $1.46\,\mu\text{m}$ emissions from F_A centers in RbCl:Li$^+$. The dashed and continuous curves have been calculated by Eqs. (5).

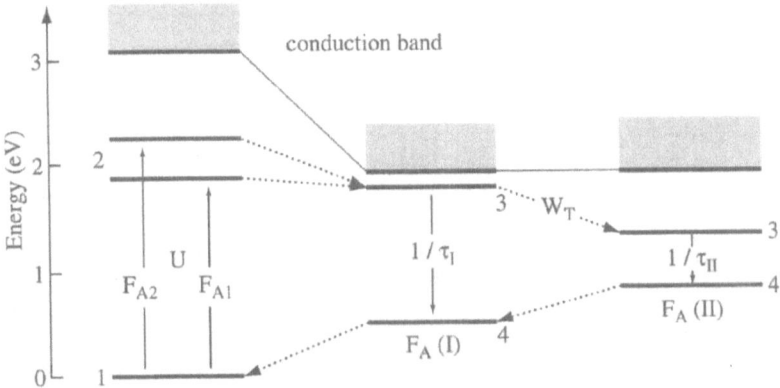

Figure 11. Optical cycle of the F_A center showing the relaxation to both the $F_A(I)$ and $F_A(II)$ RES, the latter thermally activated.

Such behavior can be interpreted with a model hinted at long ago [30], explicated schematically for F_A centers [45] and verified for F_B centers [45,46]. Referring to Fig. 11, the F_A center when excited either in the F_{A1} or F_{A2} absorption band relaxes immediately to the RES of type I (vacancy configuration) from where the normal $F_A(I)$ emission takes place. However, if the RES of type II (saddle point configuration) is energetically favorable, the F_A center can further relax towards it, from where the $F_A(II)$ emission takes place. Thus, in this model the end product of the excitation is decided solely by the energy configuration of the $F_A(I)$ and $F_A(II)$ RES. In practice, all the above mechanisms can be explained by introducing an activation energy E_A between type I and type II RES that produces a transition probability $W_T = \nu \exp(-E_A/KT)$. By denoting with n_I, n_{II}, and N the population of the *RES* of type I, type II, and of the ground state, respectively, the rate equations describing the optical cycle of Fig. 11 are

(3)
$$\begin{cases} \dfrac{dn_I}{dt} = NU - \dfrac{n_I}{\tau_I} - n_I W_T \\[2em] \dfrac{dn_{II}}{dt} = n_I W_T - \dfrac{n_{II}}{\tau_{II}} \end{cases}$$

with U the total pump rate out of the ground state, τ_I and τ_{II} the lifetimes of the two RESs, and ν the attempt frequency of the thermally activated transition $I \xrightarrow{T} II$. By bearing in mind that $N_0 = N + n_I + n_{II}$ is the total number of centers, and $U\tau_{II} \lesssim U\tau_I \ll 1$ under the present pumping conditions, the stationary solutions to a constant excitation are

$$n_I = \frac{N_0 U \tau_I}{1 + \tau_I W_T}$$

(4)

$$n_{II} = \frac{N_0 U \tau_I \tau_{II} W_T}{1 + \tau_I W_T}$$

which divided by τ_I and τ_{II} are directly proportional to the intensity of the luminescence of the $F_A(I)$ and $F_A(II)$ centers, respectively:

$$L(I) = C \frac{1}{1 + \tau_I \nu e^{-\frac{E_A}{KT}}}$$

(5)

$$L(II) = C \frac{\tau_I \nu e^{-\frac{E_A}{KT}}}{1 + \tau_I \nu e^{-\frac{E_A}{KT}}}$$

where C is a constant. Equations (5) have been used to fit the data of Fig. 10, yielding $E_A = 14.1\,\text{meV}$ and $\tau_I \nu = 5.3\,10^3$. The lifetime of the $F_A(I)$ center in RbCl:Li has not yet been measured, but if we take the value of the F center, $\tau_F = 0.75\,10^6\,\text{s}$ [47], which is expected to be similar to that of the $F_A(I)$ center, the attempt frequency is $\nu = 7\,10^9\,\text{s}^{-1}$. This value is much smaller than that of the typical phonon frequency in this host material, $\nu_{LO} = 5.4\,10^{12}\,\text{s}^{-1}$, indicating that other phonon modes or mechanisms are at work in the present thermally activated transition.

Type I and type II emissions in RbCl:Li$^+$, which are completely reversible with the temperature, should not be considered an anomalous effect but just the natural consequence of different ionic configurations around the perturbed F center. From the energy point of view, Fig. 11 is very explicit, and there is nothing against its validity for other analogous systems. As a matter of fact, similar behavior has been observed in F_B centers in KCl:Na$^+$ [45,46] and in F_A centers in mixed crystals KCl:KBr:Li$^+$ [48].

However, before accepting such a model even as a very general one, it was felt necessary to check its validity for other F_A centers. Hence, we searched for new candidate systems to submit to a close experimental investigation, and the first choice fell on the F_A centers in KF:Na$^+$. Indeed, such centers display at LN$_2$T an emission at $\sim 2.10\,\mu\text{m}$ that is characteristic of a type-II emission [49], while the reorientational properties are of type-I centers with anomalous deviations below 25 K [50].

Figure 12 shows the emission bands of a KF:Na$^+$ crystal containing F aggregated centers, measured at various temperatures and under excitation at 488 nm; namely, in one of the two F_A absorption bands [51]. The known $F_A(II)$ luminescence peaking at $\sim 2.10\,\mu\text{m}$, with a halfwidth of 0.08 eV, is observed at 150 K but vanishes cooling the crystal below 50 K. A new emission band at 0.92 μm appears below 100 K and increases with decreasing temperature. This luminescence has a halfwidth of 0.38 eV at 2 K, which is in good agreement with the values typically found for the emission of $F_A(I)$ centers in alkali halides [30]. Therefore, all the spectroscopic features make it reasonable to identify this new emission as type-I luminescence, which has never been observed before in KF. The two weak emissions at 0.600 and 0.720 μm are due to small numbers of F_B centers formed during the aggregation process, together with the more dominant F_A centers. A detailed study of F_B centers is reported elsewhere [52].

The temperature dependence of the emission intensities of the F_A(Na) defects in KF has been measured systematically in the range 2-200 K: the 0.92 μm luminescence dramatically drops with increasing temperatures and disappears at about 100 K, whereas the 2.10 μm band grows until $\sim$ 150 K, above which it is thermally quenched. However, the two bands, as also suggested by Fig. 12, do not show any anticorrelated behavior, at odds with the results in RbCl:Li$^+$, where the decrease of the F_A(I) emission was exactly compensated by the increase in the F_A(II) emission. In KF:Na$^+$ evidently, the dynamic process between the two RES is more complex and perhaps involves nonradiative transitions in the temperature range where no luminescence is observed.

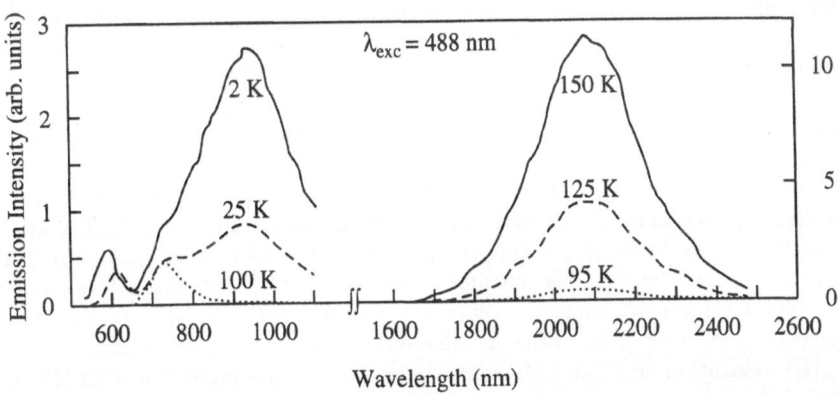

Figure 12. Emission spectra at various temperatures of aggregated color centers, namely F_A centers, in KF:Na$^+$ (0.25%) under laser excitation at $\lambda_{exc} = 488$ nm [51].

In order to clarify this last point, we started more detailed experimental studies, which have so far shown that interactions between F_A, F_B, and most probably F_C centers (an F center with three substitutional neighbor ions), substantially affect the optical cycle. We have found that the intensities of the two bands of Fig. 12, in a crystal containing mostly F_A centers, are indeed correlated as a function of temperature. On the contrary, when the crystal is heavily converted and possesses a large number of F_B centers, other emission bands also appear (noticeably one around 600 nm so intense at LHeT that it looks like a light), which have been interpreted as due to F_B centers [52].

The experimental studies with additively colored crystals of KF:Na$^+$ have produced a great wealth of results, most of which still await full understanding. However, the results on the F_A center support the idea, suggested by those on RbCl:Li$^+$, that the conversion between the two different excited states, which are strongly dependent on the ion configuration, i.e., electron-phonon interaction, is a common feature of F_A centers in alkali halides.

V. CONCLUDING REMARKS

Color centers in alkali halides were the first branch of solid-state physics to be studied systematically, and over the years, in parallel with the increasing knowledge of their basic properties, they have found several applications, the most notable being in the laser field. However, nowadays many experts consider the whole subject of color centers exhausted, especially in comparison with the new promising fields in the solid-state area. It is undeniable that the general interest in color centers has decreased in the last decade, as witnessed by the poor number of publications and attendances at conferences, but it is also true that, besides its important and unchanged didactical values, there is still a vast of interesting topics to be fully investigated. One example that has just been touched on in the previous section will now be discussed in the context of a much more general perspective.

The F_A center in alkali halides is the result of the association of an alkali ion impurity with an F center. With the condition that the impurity ion be smaller than the substituted ion of the host crystal, which is prescribed (how strictly?) by the finite available space, and excluding the Cs halides that crystallize in the simple cubic space lattice, there are 28 different specimens to be studied. Of these, 18 have been investigated with regard to their ion configuration and/or type of F_A center and/or dipole-axis angle, and for only two crystals have all three properties been determined. This situation is reported in Table 2 where the ion configuration refers to uncolored crystals, while the other two properties refer to crystals containing F_A centers. The question marks reflect the doubts arising from incomplete information. Due to the scattered results, it is possible to make only a few correlations. For instance, the on-center configuration is related to the type-I F_A center, while the off-center configuration is related to the type-II behavior and to a finite value of the off-axis angle. The correlation between $F_A(II)$ relaxation and Li^+ off-center behavior was proposed for $KCl:Li^+$ and $RbCl:Li^+$ some time ago [56], surfaced again more recently [43], and has been amply confirmed by the last data [41,51,58]. Moreover, besides the introduction of a new degree of freedom with the type I → type II transition, the data of Table 2 also agree with the theoretical expectation of a vibronic model, where the presence of an F center near an off-center Li^+ ion maintains an off-axis tilt of the Li^+ ion, although in a different symmetry position [67]. However, it is clear that Table 2, which does not contain the ten uninvestigated specimens, is rather empty. Hence, other experiments should be performed both to clarify doubtful existing data and to produce new information in order to validate the phenomenological trends just discussed and to clarify the model of the F_A center.

As far as the CCLs are concerned, it is known that they operate mostly at LN_2T, which scares most people away. It is expected that in the long run they will be substituted by stable and rugged solid-state lasers (SSL) operating at room temperature; however, they are still the only sources of laser emission in several regions of the NIR spectrum. Indeed, the cw vibronic SSLs do not emit efficiently enough beyond $\sim 2.3\,\mu m$ [68]. However, within these obvious limitations further developments of CCLs are definitively under way. Although new lasing color centers are not generally expected to be found in alkali halide crystals, but rather in materials such as diamond and various kinds of oxides, notable improvements are being made on more or less known systems. An F_2^+-like center in a crystal of $NaCl:OH^-$ has recently lased at $\sim 1.5\,\mu m$ in a quasi cw regime at room temperature [69]. Better optical pumping has been accomplished by using laser

Table 2. Alkali halide crystals with ions that have been investigated for their ion configuration, type of F_A center, and off-axis angle.

Crystal	Ion Configuration on/off center	F_A Center type I/II	Off-Axis Angle θ(Degree)
NaCl:Li$^+$	on[a]	I(?)[e]	
NaBr:Li$^+$	on[a]		
KF:Li$^+$		II[f]	14[r]
KF:Na$^+$		$(\text{I}^g \xrightarrow[T]{} \text{II}^f)^h$	1[s]
KCl:Li$^+$	off[b]	II[i]	5[t]
KCl:Na$^+$		I[i]	0[u]
KBr:Li$^+$	on[b,c]	I[i]	
KBr:Na$^+$		I[i]	
KI:Li$^+$		I[l]	
KI:Na$^+$		I(?)[m]	
RbF:Li$^+$		II[n]	
RbF:Na$^+$		II[n]	
RbCl:Li$^+$	off[d]	$(\text{I} \xrightarrow[T]{} \text{II}^{i,o})^p$	7[v]
RbCl:Na$^+$		I[i]	
RbBr:Li$^+$		I[i,o]	
RbBr:K$^+$		I[i]	
RbI:Li$^+$		I[o]	
RbI:Na$^+$		I(?)[q]	

[a]Ref. 53, [b]Ref. 54, [c]Ref. 55, [d]Ref. 56, [e]Ref. 57, [f]Ref. 49, [g]Ref. 50, [h]Ref. 51,58, [i]Ref. 30, [l]Ref. 59, [m]Ref. 60, [n]Ref. 61, [o]Ref. 44, [p]Ref. 41, [q]Ref. 62, [r]Ref. 63, [s]Ref. 64, [t]Ref. 65, [u]Ref. 66, [v]Ref. 42.

diodes with good results for laser emission around 1.5 μm [70], and ultrafast emission, pico- and subpico-second pulses with high energy in the range 1.4-2.4 μm, has been obtained by using new electro-optic devices [71]. New interest in the green pulsed laser emission of the F_3^+ center in LiF [72] also stimulated a careful investigation that indicates the possibility of a cw laser regime at LN$_2$T [73].

Anyway, leaving aside possible new breakthroughs in the field of CCLs, we should stress here the importance that CCLs have had up to now in molecular and atomic physics, in solid-state spectroscopy, and in optical pulse propagation in glass fibers [9,74]. Regarding the last topic, laser emission of the F_2^+ center in NaCl at $\sim 1.5 \mu$m produced the first optical soliton in a glass fiber [75], thus opening new possibilities for transmitting a huge amount of data via optical cable [76]. It is ironic that also in this case the CCL, which was pivotal in the initial discovery of the soliton and the subsequent studies on its propagation, has recently been substituted by the much simpler diode pumped erbium-doped fiber amplifier [77], which allowed the actual development of all-optical transmission lines [78].

At this point, after having discussed the present situation and the expected trends of the fundamental and applied physics of color centers in alkali halides, albeit limited to one aspect only and from a personal point of view, one cannot avoid making some observations that are also of historical and didactical interest.

The discovery of the first CCL is a classical example of interference between scientific development and the plain facts of life. The first laser emission was observed early in 1965 after careful planning, and in the right place [15]. However, the premature death of the main investigator prevented the continuation of the experiments when later on technical devices for laser development were more available. Ten years later, a tunable cw laser was implemented [16] because, among other reasons, more application-oriented results were required.

The story of the F_A center, although more intricate, is confined mainly to the basic research field. The $F_A(I) \rightarrow F_A(II)$ thermal-assisted conversion in RbCl: Li$^+$ (see Figs. 8,9 and 10) was not discovered earlier [38,44] for reasons that are not yet clear. Indeed, in Ref. 38 where the efficiency of reorientation is studied, measurements on RbCl:Li$^+$ are reported down to LHeT (Fig. 7), so a decrease in the efficiency should have been observed. In Ref. 44, apparently only the absorption band has been measured down to LHeT. In both cases the existence of the $F_A(I)$ center with its luminescence at $\sim 1.46 \, \mu$m (see Fig. 8) completely eluded the search. However, the dependence on the temperature of the $F_A(II)$ luminescence, which has a constant intensity down to ~ 25 K, can probably provide a technical explanation. Indeed, in order to reach low temperatures, most people used and still use a cryostat with a cold finger to which the sample is attached, usually at the bottom. However, it is well known by experts in cryogenics that, especially at very low temperatures, there can be several degrees of difference between the sample and the cryogenic liquid, usually liquid helium, used to cool the cold finger. In the previous experiments, a mistake of ~ 20 K could have been made. If this is not the case, some other technical problems or human errors could be blamed. However, it is also possible that scientific curiosity to look into the properties of the system at low temperature was not on the high-priority list, so the experiments were not carefully planned and performed. In the end all these unknown events delayed the discovery of a very important phenomenon for more than two decades [41].

In conclusion, beyond the interest in the subject of color centers in halkali halides, "which even after a century of activity continues to be very much alive as a result of the addition of modern experimental methods to old problems" [79], previous events teach everybody an important lesson. On the one hand, whether important results in physics are obtained or not depends not only on solid ideas and careful planning but also on human events. On the other hand, things can also go wrong because of scarce interest in pursuing scientific topics systematically. In this respect, the lack of due technological care in performing experiments might allow a high production of scientific results in a short time, but sometimes quantity is obtained to the detriment of quality.

Acknowledgments. I am indebted to Mladen Georgiev, Umberto M. Grassano, Fritz Luty, Linn F. Mollenauer, Hans J. Paus, and Herbert Welling for some of the information used in the historical section. I am grateful to Umberto M. Grassano and Augusto Scacco for discussions on various aspect of this work. Many thanks are also due to Rosa Maria Montereali for a critical reading of the manuscript and to Angelo Pace for general help during its preparation.

REFERENCES

[1] J. Teichmann and K. Szymborski, in *Out of the Crystal Maze*, L. Hoddeson, E. Braun, J. Teichmann, and S. Weart, eds, Oxford University Press, New York, 1992, Chap. 4.

[2] G. Baldacchini, in *Optical Properties of Excited States in Solids*, B. Di Bartolo, ed., Plenum Press, New York, 1992, p. 255.

[3] *Physics of Color Centers*, W.B. Fowler, ed., Academic Press, New York, 1968.

[4] G. Baldacchini, in *Advances in Nonradiative Processes in Solids*, B. Di Bartolo, ed., Plenum Press, New York, 1991, p. 219.

[5] G. Baldacchini, in *Nonlinear Spectroscopy of Solids: Advances and Applications*, B. Di Bartolo, ed., Plenum Press, New York, 1995, p. 395.

[6] R.K. Swank and F.C. Brown, *Phys. Rev.* **130**, 34 (1963).

[7] W.V. Smith and P.P. Sorokin, *The Laser*, McGraw Hill, New York, 1966, p. 135.

[8] T.T. Basiev, S.B. Mirov, and W.W. Osiko, *IEEE J. Quantum Electron* **24**, 1052 (1988).

[9] W. Gellermann, *J. Phys. Chem. Solids* **52**, 249 (1991).

[10] W. Gellerman and F. Luty, *Optics Commun.* **72**, 214 (1989).

[11] T.H. Maiman, *Nature* **187**, 493 (1960).

[12] A.L. Schawlow and C.H. Townes, *Phys. Rev.* **112**, 1940 (1958).

[13] A. Javan, W.R. Bennet, and D.R. Herriot, *Phys. Rev. Letters* **6**, 106 (1961).

[14] B.A. Lengyel, *Am. J. Phys.* **34**, 903 (1966).

[15] B. Fritz and E. Menke, *Sol. State Commun.* **3**, 61 (1965).

[16] L.F. Mollenauer and D.H. Olson, *Appl. Phys. Lett.* **24**, 386 (1974).

[17] R.W. Pohl, *Proc. Phys. Soc. London* **49**, extrapart 3 (1937).

[18] R. Hilsch, *Proc. Phys. Soc. London* **49**, extrapart 40 (1937).

[19] N.F. Mott and R.W. Gurney, *Electronic Processes in Ionic Crystals*, Dover Publ. Inc., New York, 1964, Chap. IV. This book was first published by Oxford University Press in 1940.

[20] F. Seitz, *Rev. Mod. Phys.* **26**, 7 (1954).

[21] Th. P.J. Botden, C.Z. van Doorn, and Y. Haven, *Philips Res. Repts.* **9**, 469 (1954).

[22] Ch. Becher and H. Pick, *Nachr. Akad. Wiss. Goettingen* **IIa**, 167 (1956).

[23] R.K. Swank and F.C. Brown, *Phys. Rev. Letters* **8**, 10 (1962).

[24] St. Petroff, Fond Nauchni Izdaniya, Sv. Kiril Slavyanob'lgarski, Universitet Varna, n. 52 (1946).

[25] St. Petroff, *Z. Physik* **127**, 443 (1950).

[26] H. Ohkura and T. Uchida, *J. Phys. Soc. Japan* **15**, 2114 (1960).

[27] F. Lüty, *Z. Physik* **165**, 17 (1961).

[28] K. Kojima, N. Nishimaki, and T. Kojima *J. Phys. Soc. Japan* **16**, 2033 (1961).

[29] G. Gramm, *Phys. Letters* **8**, 157 (1964).

[30] F. Lüty, in *Physics of Color Centers*, W.B. Fowler, ed. Academic Press, New York, 1968, Chap. 3; and references cited therein.

[31] F. De Martini, U.M. Grassano, and F. Simoni, *Opt. Commun.* **11**, 8 (1974).

[32] H. Ohkura and Y. Ohta, *Proceedings of the 3rd Photoconductivity Conference*, Stanford, 12-15 August 1969, E.M. Pell, ed., Pergamon Press, Oxford, 1971, p. 171.

[33] C.D. Jeffries, *Dynamic Nuclear Orientation*, Interscience Publisher Inc., New York, 1963.

[34] L.F. Mollenauer, S. Pan, and S. Yngvesson, *Phys. Rev. Letters* **23**, 683 (1969).

[35] .F. Mollenauer and G. Baldacchini, *Phys. Rev. Letters* **29**, 465 (1972).

[36] .W. Kogelnik, E.P. Ippen, A. Dienes, and C.V. Shank, *IEEE J. Quantum Electronics* **QE-8**, 373 (1972); and references cited therein.

[37] L.F. Mollenauer, in *Tunable Lasers*, L.F. Mollenauer and J. C. White, eds., Springer-Verlag, Berlin, 1987, Chap. 5; and references cited therein.

[38] B. Fritz, F. Lüty, and G. Rausch, *Phys. Stat. Sol.* **11**, 635 (1965).

[39] A.Y.S. Kung and J.M. Vail, *Phys. Stat. Sol.* B **100**, 621 (1980).

[40] L. Bosi and M. Nimis, *Phys. Stat. Sol.* B **156**, K 5 (1989).

[41] G. Baldacchini, E. Giovenale, F. De Matteis, A. Scacco, F. Somma, M. Casalboni, and U.M. Grassano, *Europhys. Lett.* **7**, 647 (1988).

[42] G. Baldacchini, E. Giovenale, F. De Matteis, A. Scacco, F. Somma, and U.M. Grassano, *Phys. Rev.* **B37**, 7014 (1988).

[43] A. Scacco, in *Defects in Insulating Materials*, O. Kanert and J.-M. Spaeth, eds., World Scientific, Singapore, 1993, p. 158; and references cited therein.

[44] H. Ohkura, *Prog. Theor. Phys. Suppl.* **46**, 11 (1970).

[45] Y. Yang, Ph.D. Thesis University of Utah, Department of Physics, (1984).

[46] Y. Yang and F. Lüty, *Proceedings International Conference on Defect in Insulating Crystals*, Salt Lake City, USA, 1984, pag. 496.

[47] L. Bosi, P. Podini, and G. Spinolo, *Phys. Rev.* **175**, 1133 (1968).

[48] K. Asami and M. Ishiguro, *Phys. Rev.* B **34**, 4199 (1986).

[49] L. Mollenauer, B.A. Hatch, D.H. Olson, and H.J. Guggenhein, *Phys. Rev.* B **12**, 731 (1975).

[50] W.C. Collins and I. Schneider, *J. Phys. Chem. Sol.* **37**, 917 (1976).

[51] G. Baldacchini, M. Cremona, R.M. Montereali, G. Giliberti, R. Scacco, U.M. Grassano, and A. Shpak, *J. Luminesc.* **58**, 278 (1994).

[52] G. Baldacchini, M. Cremona, R.M. Montereali, G. Giliberti, R. Scacco, U.M. Grassano, and A. Shpak, *J. Luminesc.* **60-61**, 548 (1994).

[53] R.J. Rollefson, *Phys. Rev.* **B5**, 3235 (1972).

[54] F. Bridges, *Crit. Rev. Solid State Sci.* **5**, 1 (1975).

[55] T. Watanabe, Y. Mori, and H. Ohkura, *J. Phys. Soc. Japan* **42**, 1787 (1977).

[56] K. Thormer and F. Lüty, *Phys. Stat. Sol.* (b) **90**, 277 (1978).

[57] C.R. Pollock and E. Georgiou, *Phys. Rev.* **39**, 10412 (1989).

[58] Ref. 52. and results not yet published.

[59] D.R. Foster and I. Schneider, *Phys. Rev.* B **33**, 8779 (1986).

[60] V. Topa, L. Matei, and T. Serban, *Phys. Stat. Sol.* **9**, 55 (1965).

[61] L.F. Mollenauer, private communication to J. Vail, *Phys. Stat. Sol.* B **100**, 621 (1980).

[62] P. Ketolainen, *Abstracts Int. Symposium on Color Centers in Alkali Halides*, Rome, Italy, 1968, p.160.

[63] F. De Matteis, M. Rossi, A. Scacco, F. Somma, G. Baldacchini, M. Cremona, and U.M. Grassano, *J. Phys. Chem. Solids* **51**, 1053 (1990).

[64] A. Scacco, C. Giliberti, U.M. Grassano, G. Baldacchini, M. Cremona, R.M. Montereali, and A. Shpak, *Abstracts EURODIM 94*, Lyon, France, 1994, p. 403, and *Radiation Effects and Defects in Solids*, **133-134**, (1995).

[65] G. Baldacchini, U.M. Grassano, A. Scacco, K. Somalah, and F. Somma, *Nuovo Cimento* **9D**, 1105 (1987).

[66] G. Baldacchini, E. Giovenale, F. De Matteis, A. Scacco, F. Somma and U.M. Grassano, *Nuovo Cimento* **10D**, 693 (1988).

[67] G. Baldacchini, U.M. Grassano, A. Scacco, F. Somma, M. Staikova, and M. Georgiev, *Nuovo Cimento* **13D**, 1399 (1991).

[68] A.A. Kaminskii, *Ann. Phys. Fr.* **16**, 639 (1991) and *Phys. Stat. Sol.* **148A**, 9 (1995).

[69] R.E. Matts, *Quantum Electron* **23**, 40 (1993).

[70] A. Konate and J.L. Doulan, *Abstracts EURODIM 94*, Lyon, France, 1994, p. 180, and *Radiation Effects and Defects in Solids*, **133-134**, (1995).

[71] W. Gellermann and K. Mollmann, *Abstracts EURODIM 94*, Lyon, France 1994, p. 30 and *Radiation Effects and Defects in Solids*, **133-134**, (1995).

[72] T. Tsuboi and H.E. Gu, *Appl. Opt.* **33**, 982 (1994).

[73] G. Baldacchini, M. Cremona, G. d'Auria, V. Kalinov, and R.M. Montereali, *Abstracts EURODIM 94*, Lyon, France 1994, p. 404 and *Radiation Effects and Defects in Solids*, **133-134**, (1995).

[74] G. Baldacchini, U.M. Grassano, and M. Tonelli, *Il Nuovo Saggiatore* **9**, n. 5/6, 63 (1994) (in italian).

[75] L.F. Mollenauer, R.H. Stolen, and J.P. Gordon, *Phys Rev. Lett.* **45**, 1095 (1980).

[76] E. Desurvive, *Phys. Today*, January 1994, p. 20.

[77] M. Nakazawa, Y. Kimura, and K. Suzuki, *Electron. Lett.* **25**, 199 (1989).

[78] L.F. Mollenauer and J.P. Gordon, *Nonlinear Spectroscopy of Solids: Advances and Applications*, B. Di Bartolo, ed., Plenum Press, New York, 1994, p. 451.

[79] F. Seitz, private communication, 1995.

NEWLY DEVELOPED SOLID STATE LASERS

Renata Reisfeld*

Department of Inorganic Chemistry
Hebrew University, 91904 Jerusalem, Israel

ABSTRACT

A new type of solid state lasers tunable in the visible range has been developed recently.

Incorporation of perylimide and pyrromethene dyes into glasses prepared by the sol gel method allows to design new types of visible, stable solid lasers. These can be prepared either in the form of slabs or rods or as waveguiding active media deposited on glass or polymer supports. The difference between the refractive indices of the film and its support determines whether waveguiding occurs in the film or a leaky waveguide laser is formed. The efficiency of the lasers obtained recently is up to 70 per cent when appropriate optical pumping is designed.

We discuss the chemical and physical aspects of the photostability of the dyes in glasses, phenomena responsible for non radiative relaxation of their excited state, and energy efficiencies of laser.

I. INTRODUCTION

Liquid dye lasers tunable in the visible range are known since the middle of sixties[1] They are based on organic colorants dissolved in various solvents. The light absorption from the pump source brings the dye molecule to its excited singlet state, the emission to the terminal vibrational state can then be tuned by using appropriate resonant carvities to emit laser emission in the spectral range of fluorescenct emission.

Since the introduction of organic dyes in gels ten years ago[2,3] much research work has been devoted to the application of these doped solids as gain media in laser cavities as they provide many advantages compared to liquid dye lasers[4-30]. Many types of gel matrices have been studied : from "totally" inorganic to mixed organic/inorganic, sometimes impregnated to fill up the porous volume (plastic host matrices have also been studied[31-36]. Doping procedures have been applied to both sol and gel. The most commonly used dyes were first rhodamines and now pyrromethenes. During all these years, the characteristics of laser emission have been constantly improved ranging from microjoule to millijoule energy

---

* Enrique Berman, Professor of Solar Energy

levels, the operating lifetimes increased from a few shots to several thousand shots. We shall give a short historical survey how these lasers have been developed. Describe the way they operate and give some insight of their future.

II. HISTORICAL DEVELOPMENT

Until recently, liquid dye lasers were the main systems used to achieve tunability in the visible, and the only commercial choice for tunable lasers between 400 and 660 nm. However, in the last few years an intensive effort was devoted to produce embedded organic dyes in various solid matrices, with the goal of achieving solid-state dye laser devices that may replace the liquid dye lasers; e.g. laser dyes were incorporated into polymers[37,38], silica-gels[39,15], xerogels[9], alumina gels[5,6], ormosils[19], and composite glasses[8,40,17,41]. A solid-state dye laser has advantages over a liquid dye laser by not being a volatile solvent, non-flammable, toxic, and by its compact size and mechanical stability. Still, for applications that require high powers, at either cw or pulsed high-repetition-rate operation, the problem of heat dissipation is a serious impediment for their utilization. In liquid dyes on the other hand, a jet or a flowing solution are handy practical ways of solving the heat problem. In both cases, photostability is a feature of prime importance in selecting a laser dye.

III. LASERS BASED ON SOL-GEL PROCESS

The principles of Sol Gel Process are as follows:

The sol-gel process consists of (i) preparing a homogeneous solution of easily purifiable precursor(s) generally in organic solvent miscible with water or the reagent used in the next step; (ii) converting the solution to the 'sol' form by treatment with a suitable catalyzer; (iii) allowing the sol to change into a 'gel' by polycondensation; (iv) shaping the gel (or viscous sol) to the finally desired forms or shape such as thin film, fibre or bulk and finally converting (sintering) the shaped gel to the desired ceramic material at temperatures generally much below (~500°C) much lower than those required in the conventional procedure of melting oxides together[42]:

In the sol-gel methods inorganic materials (glasses) are prepared from solutions containing metal compounds, such as sources of cations in the final oxides, water as hydrolysis agent and alcohols as solvents. In the solution metal compounds undergo hydrolysis and polycondensation, forming polymers or particles, and the solution becomes a sol. Further reaction connects the particles, solidifying the sol into a gel. The dried gels may form the final products. The maximum processing temperature may be lower than 100°C or 160°C and so one does not have to worry about the decomposition of the organic compounds imbedded in the gel which have been added as indispensable components. On the other hand, heating of pertinent dried gels to several hundreds degrees Celsius produces glasses and ceramics which have prescribed shapes, such as plate and cylinder, fiber and coating film. Thus in general, inorganic materials can be obtained without powder processing.

Dip-coating or spin-coating on glass, ceramic, metal, and plastic substrates by sol-gel method is very useful for modifying properties of the substrates with a large or small surface area and provides the substrates with new active functions. In the years around 1970, dip-coating was applied mainly to glasses, in order to modify their optical properties. In recent years, the dip-coating technique has again attracted much attention in developing various advanced materials.

The sol-gel method of fabricating this films offers potential advantages over traditional techniques, when a substrate is used for coating, the low processing temperature is particularly important in the application to electronic, optoelectronic, and photonic devices, because the substrate and other active elements on the substrate are not necessarily highly heat-resistant. Easy coating of large surfaces makes it possible to apply the sol-gel coating to display panel and windows as substrates. The small thickness may be advantageous for coating films for some optical and electronic devices. For other purposes, however, thicker films may be needed. also high optical; quality films can be provided[43]

The question whether an element is suitable for sol-gel processes or not cannot be answered very simply. However, there are some rules which can be drawn from the basics of the process. In order to form an inorganic network from a solution a network-forming step is required. this step mainly depends on the structure of the sol and can roughly be divided into two alternative mechanisms. The first is a mechanism which is based on the growth of molecules, leading to macromolecules which then grow together to an infinite network. this mechanism leads to the so-called polymerized gels and is very common in the acid-catalyses hydrolysis and condensation of tetraalkyl silicates. The other type is based on the aggregation of colloidal particles from a so-called colloidal sol and requires a fairly stable sol as intermediate. Otherwise the whole procedure would end up in a precipitation process with no sol phase to be identified. The network-forming step in these sols is the aggregation of particles to an infinite network. In the case of acid-catalyzed silica from alkoxides the polymerization process can be simple controlled by the limitation of water which leads to stable sols if enough unhydrolysed $\equiv$SiOR groups can be maintained to keep the average molecular weight small.

The influence of protons and water on the structure of SiO_2 sols has been intensively studied by Sakka[43] who showed by rheological analysis, that various structures of viscous sols can be obtained just by varying H^+ and H_2O concentration in the starting solution

The stabilization of sols is of high importance in sol-gel processing because it defines very strongly their processing properties which are of special importance for film formation. these are, for instance, the rheology, the maximum solid content and the particle size and distribution. Rheology is a complex parameter and depends on particle shape, temperature, solvent, concentration and particle interaction. the reduced viscosity η/c of non-interacting particles does not depend on concentration (Eq. 1).

$$\frac{\eta}{c} = \frac{k}{\rho} \tag{1}$$

k represents a constant and ρ the density of the particles. for an organic polymer solution, the intrinsic viscosity η_i is related to the average molecular weight M by Eq. (2),

$$i = k \cdot M\alpha \tag{2}$$

where k is a constant depending on temperature, solvent and chemistry of the polymer and α represents a parameter depending mainly on the polymer structure. Thus, a rough distinction can be made between "basic macromolecular forms", for example, $\alpha = 0$ for rigid spherical particles; $\alpha = 0.5$-1.0 for flexible chainlike molecules, and for rigid, rod-like molecules α becomes 1.0-2.0. The determination of α allows one to tailor processing properties, if the mechanisms of particle shaping can be controlled for example, the hydrolysis and condensation process of tetraethylorthosilicate, α can be selected for optimal fiber spinning.the process has been industrialyzed by Ashi Glass Co. for the production of high quality SiO_2 fibers. In this case hydrolysis under acid conditions leads to chainlike flexible polymers suitable for fiber drawing.

So for the SiO_2 system has received the highest scientific interest. This may be due to the fact that SiO_2 precursors in the form of alkylortho silicates have been readily available for almost 150 years, and, compared to almost all other common alkoxides that are relatively insensitive to moisture which means in this case, that the hydrolysis and condensation to gels take place pretty slowly (depending on concentration, type of alkoxides, and solvent, between hours and weeks). Therefore, it is possible to study the reaction kinetics and structure forming mechanisms of sols by condensation in details relatively conveniently compared to other systems. The reaction of tetraethyl orthosilicates as interesting precursors in sol-gel reactions are intensively discussed[44]. The films based on titanium dioxide have been also studied intensively as will be seen later in this chapter.

The sol structure also influences the maximum film thickness. For a given systems, the "cracking thickness" of coatings is defined by the system parameters only. The films can be transformed to dense layers by temperature treatment only if the thickness of them does not exceed certain limits defined by the intrinsic system parameters. In general, the film thicknesses to be obtained in a one-step coating process (spin or dip coating) does not exceed some tenths of micrometer. If the viscosity of a system is adjusted to obtain thicker films, cracks occur during drying. This problem can be circumvented by using composite materials such as ormocers vide infra

The ability to dope sol-gel derived silica hosts[2,3] with controlled amounts of a lasing species affords the possibility of developing a new generation of advanced tunable solid state lasers. Such lasers, as well as possessing the tunability common to all liquid lasers, can also take advantage of the superior physical properties of sol-gel derived silica. These properties include: low non-linear refractive index coefficient; low strain birefringence, coefficient of thermal expansion closed to zero; low temperature dependence of expansion coefficient and low impurity levels. However in the last few years intensive effort has been devoted to producing embedded organic dyes in various solid matrices, e.g. vide supra, with the goal being to achieve solid-state dye laser devices that may replace liquid dye lasers[45].

In 1989 we succeeded for the first time to prepare a photostable tunable laser by impregnating the orange perylene derivative (perylimide) dye "BASF-241" dissolved in MMA, into a silica-gel[8,40]. The method of Pope and Mackenzie which allows polymerization of MMA in the pores of the glass was here applied. The dye which is orders of magnitude more stable than the conventional laser dyes impregnated in the glass provided an efficient solid-state laser material. This laser was tunable in the range 568 - 583 nm.

Later we have followed the success of the previous work to further study the possibility of impregnating different dyes into the silica-gel - PMMA composite glass[17,18]. We started with the red perylimide dye (RPD hereafter). The fluorescence emission peak of this dye is centered at about 613 nm. This wavelength is important for medical photodynamic therapy (PDT) and diagnostics: Human blood and tissue absorption is small in the red, allowing the preferential absorption of light by a photo-active cancer therapeutic agent, such as hematoporphyrin derivative, which concentrates in tumors[46].The properties of RPD in this system were described in references 47-48.

Other research groups have also reported the introduction of some known laser dyes (such as rhodamine dyes) into different solid host materials; e.g. into polymers[37,38], silica-gels[15,39], xerogels[9], alumina gels[6], and ormosils (organic modified silicates)[21] and polymeric host developed recently in Los Alamos[49]. Most of the conventional laser dyes used exhibited severe photoinstability (bleaching) and their lasing output decreased rapidly with time of operation[6]. The perylimide dyes on the other hand exhibit an outstanding photostability in glass. Also the pyrromethene 1,3,5,7,8-pentamethyl-2-6-diethylpyrromethene-BF_2 complex (PM-567) has a quantum efficiency of 99.5% and a triplet extinction coefficient of one-fifth of that of Rhodamine 560. In solutions, these pyrromethene-based dye lasers have outperformed the rhodamine and coumarin dye in the same wavelength ranges under flashlamp and laser pumping.[36] The pyrromethene dyes exhibit reduced triplet-triplet (T-T) absorption over their fluorescence and lasing spectral region.

Hermes et al[36] incorporated this dye into a modified acrylic polymer. When pumped with 532 nm light of frequency doubled Nd:YAG this laser material was able to provide 30 mJ of energy with a loss of 34% after 20,000 shots at 3.33 Hz. Additional BF_2 complex laser dyes have been synthesized and their spectroscopic and laser characteristics studied in solution[50]. The 8-cyano-pyrromethene-BF_2 complexes showed the best performance with red emission and slope efficiencies of 48% in solution using the same pumping.

Recently pyrromethene 567 has been successfully incorporated into glasses prepared by hydrolysis and polycondensation of vinyltriethoxysilane using acetone as a solvent[29]. In a single shot of operation under excitation of frequency-doubled Q-switched Nd:YAG laser providing 532 nm 8 ns pulses of energy up to 10 mJ 30% efficiency was observed. The maximum energy output was 3 mJ/pulse. When the sample is pumped repeatedly the organic molecules are slowly degraded and the output energy lowered.

Laser properties of pyrromethene-doped ormosils were reported also by Dunn et al[26] using longitudinal pumping with a slope efficiencies compared with those of the lasers obtained in modified acrylic plastic[38] in which reported efficiencies of close to 60 per cent were obtained under sophisticated pumping arrangements. These results are very encouraging for the further development of solid state visible tunable lasers.

IV. PERYLIMIDE DOPED COMPOSITE GLASS LASERS

As described above lasers obtained by impregnation of the perylimide dyes into sol gel glasses where the dyes are enclosed in the pores of the glass seem to be so far the most photostable system. Their preparation procedure was described recently at the Conference on Sol Gel Optics (III) in San Diego[51] and comments of photostability of these dyes in glasses were made in reference 52.

While our main eforts went into the detailed elaboration of the preparation procedure and the understanding of photostability, we have used transverse pumping for excitation of the laser which is less efficient than the longitudinal pumping performed by our colleagues[36,50].

V. AN OUTLINE OF THE PREPARATION PROCEDURE FOR PERYLIMIDE DOPED GLASSES

The perylimide dyes were dissolved in a methylmethacrylate (MMA) monomer to form solutions of different concentrations in the range 10^{-6} - 10^{-3} mol/liter. Highly porous silica-gel bulk glasses (density about 0.7 g/cm³) were prepared by the sol-gel method, and dried by slow heating (100°C/day) from room temperature to 500°C. Then, the bulks were immersed in the dye-doped solution of the MMA monomer, which was simultaneously catalyzed by the addition of 2% benzoyl peroxide. The MMA-dye solution thus diffused into the silica-gel glass pores, and polymerized therein. After this process of dye impregnation, the bulks were re-immersed in an MMA-dye solution, which at this stage was catalyzed for full polymerization by 0.5% benzoyl peroxide, and kept in a sealed container at 40°C for about a week. The samples were then withdrawn, cleaned, and polished, to obtain parallel-piped slabs of approximate dimensions 10x10x3 mm³, with clear smooth surfaces.

The density of the composite glass was d = 1.447±0.005 g/cm³ and the refractive index n = 1.472±0.003.

VI. LASER EXPERIMENTS AND RESULTS

The light source used for pumping of the glass samples was: a frequency-doubled Nd:YAG laser (Lumonix HY600), emitting light pulses at wavelength 532 nm. Typical light pulses of the Nd:YAG laser were of ~8 ns full width at half maximum (FWHM) duration, 1-50 mJ/pulse energy, and 1-10 Hz repetition rate. The exciting beam was focused to a rectangular shape on the dye doped glass surface by a combination of spherical and cylindrical lenses. The excited surface region dimensions were 9x0.7 mm for excitation with the frequency-doubled Nd:YAG laser. Lasing was obtained in a cavity of either a combination of a ~100% reflecting metallic back mirror and a ~50% reflecting output coupler, or a combination of a grating of 1,200 grooves/mm at Littrow configuration as the back reflector, and the same output coupler.

Table 1 presents the laser parameters obtained experimentally in our laboratory for composite glass lasers[8,17,41,72] under excitation of Nd:YAG laser emitting at 532 nm (second harmonic) at a repetition rate of 10 Hz and the lifetime of 8 ns and energies ranging 1-50 mJ/pulse using transverse pumping. The results in the table correspond to our first work At present the energies are much higher (patent pending).

Table 1: The laser characteristics of perylimide dyes incorporated into glasses under excitation of second harmonic Nd:YAG laser 50 mJ/pulse with a transverse pumping.

Dye	RPD	BASF-241
Absorption coefficient	44,000 $M^{-1}cm^{-1}$	85,000 $M^{-1}cm^{-1}$
Lifetime radiative	5.74 ns	3.64 ns
Lifetime measured	5.3 ns	3.5 ns
Quantum yield	0.93	0.96
Tunability range	605-630 nm	575-590 nm
Threshold energy	90 μJ/pulse	60 μJ/pulse
Maximum output energy obtained	0.4 mJ/pulse	2.75 mJ/pulse
Energy flux	2 mJ/cm² per pulse	13.75 mJ/cm² per pulse
Peak power	75 KW per pulse	790 KW per pulse
Stability	500,000 pulses	long time, more than 1,000,000 pulses

VII. PHOTOSTABILITY MEASUREMENTS

The Dyes BASF 241 did not show appreciable photodegradation under excitation of one milion pulses, thus the photostability could be measured only with red perylimide dye RDP.

The lasing photostability of the red perylimide dye (RPD) in various solid matrices was measured under frequency-doubled Nd:YAG laser excitation. The RPD:composite glass laser intensity decayed to 50% of its initial value after approximately 20,000 pump pulses of 13 mJ/pulse. The output of RPD:ormosil glass and RPD:PMMA glass lasers decayed to 50% of their initial value after 1,200 and 1,000 pump pulses of the same energy, respectively. For Rhodamine-6G:silica-gel and Rhodamine-6G:ormosil glass lasers, the 50% decay occurred already after 1,000 and 300 pulses, respectively. The decay was non-exponential, indicating that the dye bleaching is not a single-photon process. The average laser output decay rates increased linearly with the pump energy.

Figures 1 and 2 show the comparison of the decay rate of the laser action of RPD 300 and Rh 6G in composite glass, ormocers and PMMA under the laser excitation of 13 mJ/pulse of the second harmonic output of Nd:YAG laser.

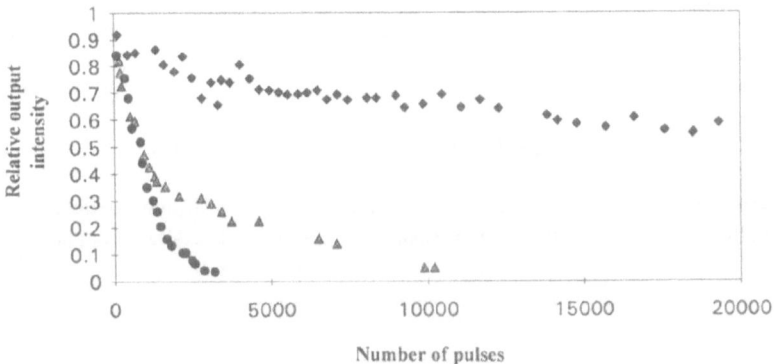

Fig. 1. Relative dye laser output of RPD 300 as function of number of pump pulses of 13 mJ each, for various matrices. Rhomboids-composite glass, triangles-ormosil, circles-PMMA[47].

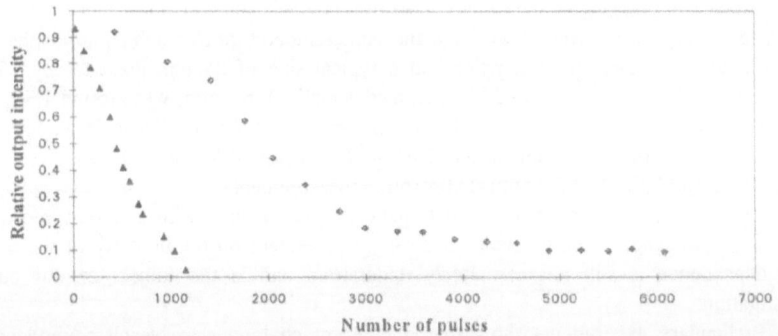

Fig. 2. Relative laser output of rhodamine-6G as function of the number of pump pulses of 13 mJ each, for various matrices. Rhomboids - silica-gel glass, triangles - ormosil glass[47].

Singlet-singlet excited state absorption of the RPD dye in the solid matrices was also measured between 550 and 730 nm. At ~600 nm the cross section was ~2x10⁻¹⁶ cm²/molecule. The excited-state absorption competes with the lasing, and is a main factor that limits the laser efficiency.

It should be noted, that the decay of the lasers output intensity as a function of the number of pulses is not exponential. This suggests that the dye bleaching is not a single-photon process.

To shed some more light on the dye bleaching mechanism we have measured the laser output decay for different pump light intensities[47]. Obviously the output decay becomes faster as the pump light intensity increases. The results show a linear dependence of the bleaching rate on the pump pulse energy, with a slope of 1.75×10^{-6} mJ^{-1}. The pump energy density is probably a more meaningful factor than the total pump energy, concerning the bleaching effect. The corresponding slope would be 1.25×10^{-7} cm^2/mJ.

The nearest sphere surrounding the perylimide dye molecules is a skeleton consisting of both PMMA and the composite glass. The silicate skeleton of the microphase appears only as a second neighbor. Thus, the effect of this skeleton on the dye photostability is strikingly large. In our opinion the incapsulation of the organic entity in the rigid glass phase allows a back transfer of the excited photoelectron which would be otherwise irreversible detached and cause photodecomposition.

VIII. NEW "ORGANIC DYE/XEROGEL MATRIX" COUPLES

Recently doped xerogels were obtained by hydrolysis-condensation of vinyltriethoxysilane or methyltriethoxysilane[28,53], under acid-catalyzed conditions with acetone as common solvent. The molar ratios of silicon alkoxide : water : acetone were respectively 1:3:3. HCl acidified water-acetone mixture was added to the alkoxide-acetone mixture. The solution was stirred vigorously for several hours at room temperature to ensure homogeneous mixing. After complete hydrolysis, a small amount (10^{-3} mol/l) of functionalized alkoxysilane was added to catalyze the condensation reaction. Dye solution was then added in an appropriate solvent. Pyrromethene 567, pyrromethene 580, pyrromethene 597, perylene orange, perylene red, neon red, rhodamine 6G, rhodamine B and xanthylium salt are readily soluble in a variety of solvents including acetone. The resulting clear sols were cast into cylindrical-shaped moulds which were then hermetically closed for gelation, in a drying oven, at 40°C. A slow gelation process was necessary to prevent possible cracking of the samples. After one week, the containers were opened and finally left to dry for a further 3 weeks at the same temperature to reach an atmospherically stable condition. such dried samples had a typical size of 25 mm diameter by 15 mm thickness. Bulk density, measured by Archimedes method in water, was around 1.3 g.cm^{-3}. The xerogels used for laser experiments were polished to obtain parallel surfaces with good optical quality (about 4 nm roughness). Xerogel thin layers (0.5 mm thick) were cut with a diamond saw and polished for optical absorption measurements.

The pyrromethene dyes in xerogels gave the best results with generally more than 50% slope efficiency. The efficiency is strongly dependent on the porosity of the matrix. This dependence is still not completely understood and is the subject of our current investigation.

Preliminary experiments show that it takes much longer to bleach a perylene red doped sample than a pyrromethene 597 one when submitted to a similar c.w. laser flux (531 nm).

A solution often proposed to increase the stability of the doping molecules is to covalently bond the dye to the solid matrix. Unfortunately this, to our knowledge, has always led to a dramatic decrease of fluorescence intensity preventing any laser action with such a sample. However, this method used with a relatively long spacer in order to retain fluorescent characteristics was applied to classical rhodamine B as this dye because it is relatively easier to graft it on the matrix than the perylene of pyrromethene dyes[53].

The active molecules (rhodamine B) were attached to the inorganic polymer backbone via a flexible spacer. This approach allows a high concentration of the incorporated active side groups without segregation. The dye containing monomer was prepared from rhodamine B isothiocyanate and (3-aminopropyl)triethoxysilane (APTES)[53]. The synthetic route was as follows: under inert atmosphere the amino groups (10% excess) of the silicon precursor react with the isothiocyanate groups of the dye in anhydrous ethanol. A small amount of triethylamine was used as catalyst. The reaction mixture was stirred overnight at room temperature. Ethanol was removed under reduced pressure. The crude product was washed with ether to eliminate excess APTES through three or four repeated decantations and dried in vacuo (yield: 50%).

IX. WAVEGUIDE LASERS

Numerous modern applications such as recording, communication printing, display etc. demand compact wave guiding lasers that can be tuned in the visible spectral range[54] Such a system was proposed by us about a decade ago suggesting theoretically an introduction of laser dyes into glass films[55,56]. Since then tunable lasers as waveguide amplifiers based on deposition of doped films on glasses have been reported by us [51,54,57-63].

X. RESONATORS FOR WAVEGUIDED LASERS

Distinctly from conventional lasers based on bulk optics, waveguide lasers offer the option of using *distributed feedback* (DFB) or *distributed reflectivity* (DBR). These configurations have several advantages as compared to resonators based on localized reflectors. The advantages were demonstrated most clearly in the case of semiconductor lasers[64-66]. The use of reflecting facets in that case results in high power concentration on the facets and limits the output power due to the danger of damaging the emitting surfaces. Moreover, the wide gain band of a semiconductor laser medium and the weak frequency selectivity of a Fabri-Perot resonator, causes the excitation (even for a slight excess over the threshold) of many modes whose spectral envelope amounts several nanometers. These problems are avoided in DFB structures which naturally provide high selectivity of wavelength. In addition, a periodic modulation of refractive index and/or width, provides the option of *distributed surface extraction* of radiation through the grating. These advantages of DFB layouts are even more apparent in the case of dye based medium. The bandwidth of gain here amounts tens of nanometers, so that wavelength control is a necessity.

The first demonstration of DFB took place more than two decades ago by Kogelnik and Shank[64]. Periodic modulation of the refractive index was achieved by several methods, including modulation of gain in the substrate medium by the use of optical interference pumping. Their results showed a drastic reduction of bandwidth, well below 1nm. In the case of interferometric pumping, tunability was achieved by the change of the intersecting angle of the pump beams. The utilization of DFB in semiconductor lasers is now a matter of routine, and the drastic reduction in linewidth achieved by this method, turns it into most widely used used in commercial lasers.

Turning to very recent reported work, Shamrakov and Reisfeld[60], and He et. al[67], in 1994, reported the operation of a leaky waveguide laser based on a sol-gel glass film and polymeric films respectively. Leaky-mode waveguided lasers are based on a guiding layer with refractive index lower than the supporting substrate. In a gain medium, such lossy modes can be steadily supported. Since the losses due to Fresnel reflection increase with the value of incident angle, the mode losses effectively discriminate the lowest order transverse mode in the case of thick film structure. The leaky wave mechanism here provided a natural way of outcoupling the radiation. The resonator cavity here had external mirrors.

Different types of planar resonators based on surface gratings can be designed. First or second order Bragg reflection can be used depending on whether the periodic structure should provide or not a means of outcoupling the radiation. The resonator structures can be further divided into linear or ring types. Ring resonator types have several advantages:

1. The broader area of light generation enlarges naturally the volume of the amplifying material. The enlarged area releases the need of high concentration of the pump radiation and reduces the danger of optical damage to the films.

2. The lithography resolution requirements are released in the case of ring type resonators since the period of the gratings is inversely proportional to $sin\theta$ where θ is the angle of incidence of light on each grating section.

3. A narrow ray of light inciding obliquely at a grating structure will be spatially spread out by the successive reflections, reducing the possibility of self-focusing, filamentation or other sources of light concentration.

The theory of distributed feedback lasers was originally postulated by Kogelnik and Shank[64], and further developed subsequently[68-70].

XI. TUNABILTY IN THIN-FILM LASERS

The large amplification bandwidths encountered in transparent media doped with organic dyes makes it an ideal source for the development of tunable sources. Although conventional ways of achieving tunability by means of external resonators are also viable.

A more attractive way of achieving tunability, which is proposed here for the first time, is to use the interaction of light propagating in the thin film with surface acoustic waves (SAW). Fortunately, *we were able to demonstrate the growth of sol-gel films on quartz substrates* a material which has well known piezo-electric properties, and found applications on diverse SAW devices. The basic principles involved in the interaction between guided light waves and SAWs are very well known. Basically, they were proved to function as TE-TM converters, wavelength filters, optical correlators and rf spectrum analyzers. In all cases light-SAW interaction provided a way of controlling the periodic perturbation on both frequency and amplitude. We proposed to use this principle in order to tune the film waveguided lasers.

The possibility of producing thin films of high optical quality by the sol-gel method has many attractive features. We have concentrate in the development of waveguided lasers based on organic-dye doping. Nevertheless, very significant preliminary results were obtained. Optical waveguides of high quality were demonstrated, guided modes and their propagation constants and losses were characterized. In a subsequent study, organic Rhodamine B dye was introduced into the sol-gel solution, the thin films were optically pumped by frequency doubled Nd: YAG laser light. An amplification factor of the order of 50dB/cm was measured followed by superradiant emission. These outcomes were published in references 57-59, 63. Furthermore, the high gain factor available, allowed the achievement of laser action in a resonator produced by simply cleaving two facets of the glass support. The graphs in Fig. 3 which are very recent measurements, are presented here for the first time and show a very typical lasing characteristic. Lasing was also indicated by a narrowing of the out coming pulse duration.

Here several sections of sol-gel thin films are deposited each of them doped with a different laser dye. Grating structures are inscribed on each part with different grating periods, selecting a given wavelegth from each dye medium by a distributed feedback

(DFB) mechanism. By steering the pump beam, each laser can be individually adressed, and practically the entire visible spectrum can be spanned. The applicability range of such a device would be very large indeed; it includes color displays, color printers, spectroscopy and sensing. The waveguided character of the generated emission makes it suitable for coupling into optical fibers, a property that by itself, brings with it a host of applications

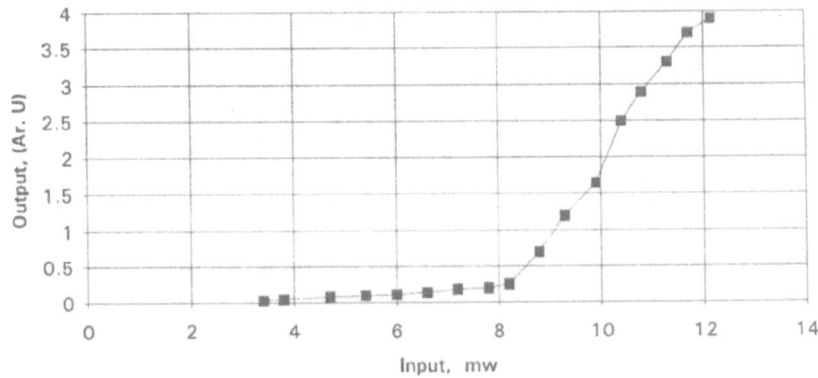

Fig. 3: Output of waveguide laser as a function of pump power showing lasing threshold. (Unpublished data obtained on March 1995 together with the group of Sh. Ruschin of Tel-Aviv University)

The high refractive index glass films at room temperature were prepared as follows[54,59,63]. The glass synthesis was performed by hydrolysis and subsequent copolymerization of titanium tetraethoxide Ti(OEt)$_4$ or titanium tetraisopropoxide Ti(OPr)$_4$ with the organically modified silanes ORMOSIL namely glycidyloxipropyltrimethoxysilane (GLYMO).

$$CH_2\text{--}CH\text{-}CH_2O(CH_2)_3\text{-}Si(OCH_3)_3$$

Glass films with refractive index up to 1.66 were thus obtained. In order to prepare films of lower refractive index of 1.48 and density of 1.68 gr/cm^3 we have used a procedure based on azeotropic distillation with benzene or toluene solution including the laser dye, tetraethoxysilane Si(OC$_2$H$_5$)$_4$ (TEOS), and triethoxyvinylsilane CH$_2$=CHSi(OC$_2$H$_5$)$_3$ (TEVS), and low molecular weight polymethylmethacrylate (PMMA). The block diagram of this procedure is given in Fig.4[58,60].

Table 2. presents examples of laser dyes introduced into the films (prepared in our laboratory) the lasing range, and the threshold of laser operation.

Fig. 5 presents laser emission as observed by narrowing of the fluorescencence at threshold energy, of the Lumogene LFR 300 provided by BASF. The spectroscopy of this dye in various solvents has been discussed in detail in reference 47.

Glass films from alumina, prepared by peptization of aluminum hydroxide doped by Rhodamine 6G, Rhodamine B and Oxazine 4 showed laser action with calculated conversion efficiency of 21%. The intensity decreased linearly with the shot number of exciting N$_2$ laser.

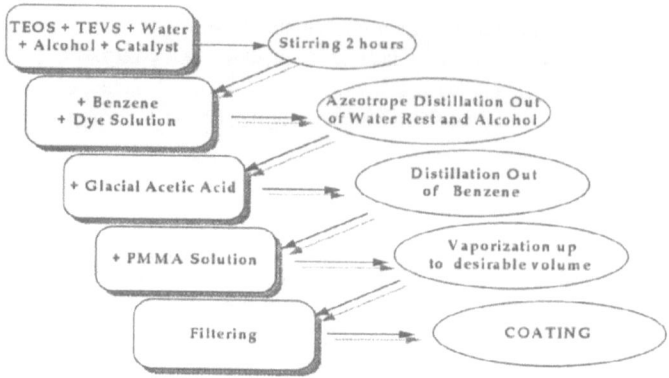

Fig. 4. Procedure for preparation of composite glass

Table 2. Spectral characteristics of the dyes used for glass films doping.

Dye	Abs. max, (nm)	Tuning range (nm)	Spontaneous spec.width (nm)	Laser Spec. width (nm)	Threshold (µJ/pulse)
Lumogen LFR 300	578	605-630	70	9	30
Lumogen LFO 240	525	568-583	30	3	50
R6G	546	560-610	44	11	50
DCM	472, 496	595-650	79	19	100
Rhodamine 610	560	585-635	45	10	40

Fluorescence intensity, photodegradation and kinetics of degradation of excited states of R6G in sol gel films was reported recently[72]. Photodegradation of the dye increased with the concentration as a result of interaction of the monomers with higher aggregates.

The coupling of leaky modes from lower refractive index of the doped film than the support has been calculated theoretically in reference 73, and a laser configuration based on active dielectric thin film, acting as an optical amplifier designed.

It has been predicted that the resonant beam width is dominated by the active medium width of the film, not by the resonator-mirror curvature. The calculations were based on the leaky-mode concept[74].

An investigation of oblique plane wave scattering in active dialectric films reveals the existence of anomalously large resonance that occur at discrete plane wave angles of incidence. This fact allows application of the active films with lower refractive index than the support as an efficient amplifier.

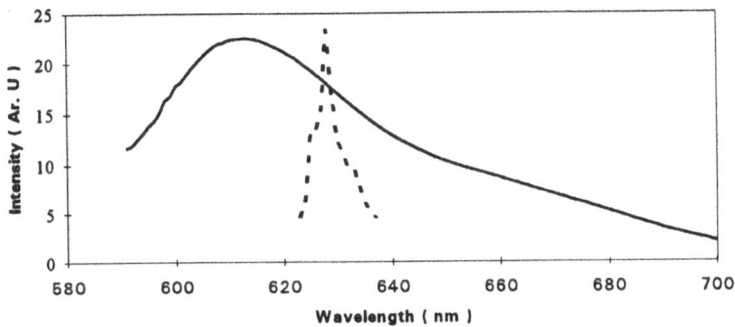

Figure 5: Fluorescence and laser emission of LFR 300 in a composite glass.

R = alkyl

R^x= O-aryl

Structural formula of LFR 300

XII. CONCLUSION AND PROSPECTIVES

The state of art of tunable lasers have to day reached such progress that a large workshop was held recently[71] at which a number of commecial companies have participated in addition to the scientific community indicating not only potential but also existing application.

Solid state dye laser is a rapidly expanding field. Much progress has bee made concerning the mechanical, thermal and optical properties of the matrix, and more suitable dyes have been identified, especially pyrromethene (597) for efficiency and perylene (red) for lifetime. Efficiency (several tens of percents) and tunability (several tens nanometers) have now obtained expected values for short operating times. The lifetime attained with a single area of the doped solid has been considerably increased (several tens of thousand pulses), yet this is certainly where future progress can be made. Grafting the dye to the host matrix could provide a solution. The results with the (photounstable) rhodamine B are preliminary indication that to the increased stability, but further study is needed. The recent results with rhodamine B show that the dye stability may be increased by covalent bonding to the silica network while retaining a strong fluorescent emission and therefor utilizable for tunable solid state de lasers. Attempts to graft more suitable dyes such as perylenes or pyrromethenes are in progress.

Acknowledgement

The work was supported by the US Army European Research Office and Night Vision Laboratory contract DAJA 45-90-C-055 and by the Israeli Ministry of Science. Dr.

Al Pinto of Night Vision is extremely helpful in providing important relevant information and discussing the current results.

We also thank D. Brusilovsky, R. Gvishi, H. Minti, D. Shamrakov, Y. Sorek and Mrs. I. Finkelstein for help in the experiments, and Prof. C.K. Jφrgensen, Z. Burshtein and S. Ruschin for helpful discussions.

REFERENCES

1. Dye Lasers, F.P. Schäfer editor Springer Verlag 1990.
2. Avnir D., Levy D. and Reisfeld R. "The nature of the silica cage as reflected by spectral changes and enhanced photostability of trapped Rhodamine 6G", *J. Phys. Chem.*, **88**, 5956-5959 (1984).
3. Avnir D., Kaufman V.R. and Reisfeld R. "Organic fluorescent dyes trapped in silica and silica-titania thin films by the sol-gel method, Photophysical, film and cage properties", *Journal of Non-Crystalline Solids*, **74**, 395-406 (1985)
4. Altshuler G.B., Bakhanov V.A., Dulneva E.G., Erofeev A.V., Mazurin O.V., Roskova V.P. and Tzekhomskaya T.S. "Laser based on dye-activated silica gel", *Op. spectrosc. (USSR)*, **62**, 1201-1203 (1987).
5. Kobayashi Y., Kurokawa T.,Imai Y. and Muto S, "Laser on silica-gel activated by dye", *Journal of Non-Crystalline Solids*, **105**, 198-200 (1988).
6. J. M. Mckiernan, S. A. Yamanaka, B. Dunn and J. I. Zink, "Spectroscopy and laser action of Rhodamine-6G doped allumino-silicate xerogels", *J. Phys. Chem.*, **94**, 5652-5654, (1990).
7. Pouxviel J.C., Dunn B. and Zink J.I., "Fluorescence study of aluminosilicate sols and gels doped with hydroxy trisulfonated pyrene", *J. Phys. Chem.*,**93**, 2134-2139 (1989).
8. Reisfeld R., Brusilovsky D., Eyal M., Miron E., Burstein Z. and Ivri J. "A new solid-state tunable laser in the visible", *Chemical Physics Letters*, **160**, 43-44 (1989).
9. Salin F., Lesaux G., Georges P., Brun A., Bagnall C. and Zarzycki J. "Efficient tunable solid-state laser near 630 nm using sulforhodamine 640-doped silica gel", *Optics Letters*, **14**, 785-787 (1989).
10. Altshuler G.B., Bakhanov V.A., Dulneva E.G., Mazurin O.V., Roskova G.P. "New laser media based on microporous glasses", *Sol-Gel Optics*, John D. Mackenzie, Donald R Ulrich, Eds., SPIE **1328**, 89-97 (1990).
11. Dunn B., Mackenzie J.D., Zink J.I. and Stafsudd O.M. "Solid-state tunable lasers based on dye-doped sol-gel materials", *Sol-Gel Optics*, John D. Mackenzie, Donald R. Ulrich, Eds., SPIE **1328**, 174-182 (1990).
12. Knobbe E.T., Dunn B., Fuqua P.D. and Nishida F. "Laser behavior and photostability characteristics of organic dye doped silicate gel materials", *Applied Optics*, **29**, 2729-2733 (1990).
13. McKiernan J.M., Yamanaka S.A., Dunn B. and Zink J.I. "Spectroscopy and laser action of Rhodamine 6G doped Aluminosilicate Xerogels", *J. Phys. Chem.*, **94**, 5652--5654 (1990).
14. Reisfeld R. "Theory and appication of spectroscopically active glasses prepared by the sol-gel method", *Sol-Gel Optics*, John D. Mackenzie, Donald R. Ulrich, Eds., SPIE **1328**, 29-39 (1990).
15. Whitehurst C., Shaw D.J. and King T.A. "Sol-gel glass solid state lasers doped with organic molecules", *Sol-Gel Optics*, John D. Mackenzie, Donald R. Ulrich, Eds., SPIE **1328**, 183-193 (1990).
16. Canva M., Georges P.,Brun A., Larrue d., Zarzycki Z. "Impregnated SiO_2 gels used as dye laser matrix hosts". Proceedings VI International Workshop on glasses and ceramics from gels, Seville, 6-11 Octobre 1991.

17. R. Gvishi and R. Reisfeld, "New solid glass laser for photodynamic therapy", J. Physique (Colloques C7), **1**, 199-202 (1991).

18. R. Reisfeld, "Solid state lasers tunable in the visible spectrum and nonlinear materials in glasses", J. Physique (Colloques C7), **1**, 415-417 (1991).

19. Altman J.C., Stone R.E., Nishida F. and Dunn B., "Dye activated ORMOSILS for lasers and optical amplifiers", *Sol-Gel Optics II*, John D. Mackenzie Ed. SPIE **1758**, 507-518 (1992).

20. Dunn B., Nishida F., Altman J.C. and Stone R.E. "Spectroscopy and laser behavior or Rhodamine-doped Ormosil", *Chemical Processing of Advanced Materials*, **84**, 941-951 (1992).

21. Hsin-Tah Lin, Bescher E., Mackenzie J.D. Hongxing Dai and Stafsudd O.M. "Preparation and Properties of laser dye-ORMOSIL composites", *Journal of Materials Science*, **27**, 5523-5528 (1992).

22. King T.A. "Lasers and ultrastructure processing", *Chemical Processing of Advanced Materials*, **90**, 997-1017 (1992).

23. Larrue D., Zarzycki J., Canva M., Georges P. and Brun A. "Semi-humid gels as matrices for laser media", *Sol-Gel Optics II*, John D. Mackenzie Ed., SPIE **1758**, 420-431 (1992).

24. Liu S. and Hench L.L. "Lasing characteristics of a porous gel silica matrix with 4PyPO-MePTS laser dye", *Chemical Processing of Advanced Materials*, **85**, 953-964 (1992).

25. Dai H., Lin H.T. and Stafsudd O.M. "Optical gain and laser action in rhodamine-doped solid-state PT-ORMOSIL composites", *Solid State Lasers IV*, G.J. Quarles and M.A. Woodall Eds., SPIE **1864**, , 50-56 (1993).

26. Dunn B., Nishida F, Toda R., Zink J.J., Allik T.H., Chandra S. and Hutchinson J.A. "Advances in dye-doped sol-gel lasers", Mat. Res. Soc. Symp. Proc. **329**, 267-277 (1994).

27. Lo D., Parris J.E. and Lawless J.L. "Laser and Fluorescence Properties of Dye-Doped Sol-Gel Silica from 400 nm to 800 nm", *Appl. Phys.* **B 56**, 385-390 (1993).

28. Canva M., Georges P., Perelgritz J.F., Brun A., Chaput F. and Boilot J.P. "Agile Solid state dye lagers", Supplement au Journal de Physique III, Colloque 4, volume 4, 369-372, (1994).

29. Canva M., Georges P., Perelgritz J.F., Brun A., Chaput F. and Boilot J.P. "Improved sol-gel Materials for efficient solid state dye lasers", Mat. Res. Soc. Symp. Proc. **329**, 279-284 (1994).

30. Rahn M.D. and King T.A. "Solid state dye doped sol-gel glass composite lasers", CLEO'94 proceedings, 389-390 (1994).

31. Peterson O.G. and Snavely B.B. "Stimulated emission from flashlamp-excited organic dyes in polymethyl methacrylate", *Appl. Phys. Lett.*, **12** n° 7, 238-239 (1968).

32. Itoh U., Takakusa M., Moriya T. and Saito S. "Optical grain of coumarin dye-doped thin film lasers", *Japan J. Appl. Phys.*, **16**, 1059-1060 (1977).

33. Gromov D.A., Dyumaev K.M., Manenkov A.A., Maslyukov A.P., Matyushin G.A., Nechitailo V.S. and Prokhorov A.M. "Efficient plastic-host dye Lasers", *J. Opt. Soc. Am.* **B 2**, 1028-1031 (1985).

34. Allik T.H., Chandra S., Hermes R.E., Hutchinson J.A., Soong M.L. and Boyer J.H. "Efficient and Robust Solid-State Dye Laser", OSA Proceedings on Advanced Solid-State Lasers, Pinto A.A. and Fan T.Y. Eds. (OSA Washington, DC, 1993) **15**, 271-273 (1993).

35. Allik T.H., Chandra S., Robinson T.R., Hutchinson J.A., Sathyamoorthi G. and Boyer J.H., "Laser performance and material properties of a high temperature plastic doped with pyrromethene-BF$_2$ dyes", *Mat. Res. Soc. Symp. Proc.*, **329**, 291-296 (1994).

36. Hermes, R.E., Allik T.H., Chandra S. and Hutchinson J.A. "High-efficiency pyrromethene doped solid-state dye lasers", *Appl. Phys. Lett.*, **63** (7), 877-879 (1993).

37. H.H.L. Wang and L. Gampel., "Simple, efficient plastic dye laser", *Optics Comm.*, **18** , 4 (1976).

38. Allik T.H., Chandra S., Robinson T.R., Hutchinson J.A.,Sathyamoorthi G. and Boyer J.H., "Laser performance and material properties of a high temperature plastic doped with phyrromethene-BF$_2$ dyes", *Mat. Res. Soc.Symp. Proc.*, **329**, 291-296 (1994).

39. G. B. Altshuler, V. A. Bakhanov, E. G. Dulneva, A. V. Erofeev, O. V. Mazurin, G. P. Roshova and T. S. Tsekhomskaya, "Laser on silica-gel activated by dye", *Opt. Spectroscopy*, **62**, 709 (1987).

40. R. Reisfeld, D. Brusilovsky, M. Eyal, E. Miron, Z. Burshtein and J.Ivri, "Perylene dye in a composite sol-gel glass:- a new solid-state tunable laser in the visible range", . Binational French-Israeli Workshop on Solid-State Lasers, George Boulon, Christian K. Jφrgensen, Renata Reisfeld Eds., SPIE **1182**, 230-239 (1989).

41. R. Gvishi, R. Reisfeld, Z. Burshtein, E. Miron, "New Stable Tunable Solid-State Dye Laser in the Red", 8th meeting on Optical Engineering in Israel, Tel-Aviv Israel: Optoelectronics and Applications in Industry and Medicine, Moshe Oron, Itzhak Shlodev, Itzhak Weissman, Eds., Proc. SPIE **1972**, 390-399 (1993).

42. R.C. Mehrotra, "Present status and future potential of the sol-gel process", *Structure and Bonding*, **77**, 1-36 (1992).

43. Sumio Sakka and Toshinobu Yoko, "Sol-gel-derived coating films and applications", *Structure and Bonding*, **77**, 89-118 (1992).

44. Helmut Schmidt, "Thin films, the chemical processing up to gelation", *Structure and Bonding*, **77**, 119-151 (1992).

45. R. Reisfeld and C.K. Jφrgensen, "Optical properties of colorants or luminescent species in sol-gel glasses", *Structure and Bonding, Springer-Verlag*, R. Reisfeld and C.K. Jφrgensen eds. 77, 207-256 (1992).

46. T. J. Dougherty, Yearly review, "Photosensitizers: therapy and detection of malignant tumors", *Photochem. and Photobiol.*, **45** (6) 879-889 (1987).

47. R. Reisfeld, R. Gvishi and Z. Burshtein, "Photostability and Loss Mechanism of Solid-State Red Perylimide Dye Lasers", *J. Sol-Gel Science and Technology*, **4**, 49-55 (1995).

48. R. Gvishi, R. Reisfeld, and Z.Burshtein, "Excited-state Absorption in Red Perylimide Dye in Solution", *Chem. Phys. Lett.*, **212** , 463-466, (1993).

49. R.E. Hermes, J.D. McGrew, C.E. Wiswall, S. Monroe and M. Kushina, A diode laser-pumped Nd:YAG-pumped polymeric host solid-state dye laser, *Appl. Phys. Comm.*, **11** (1), 1-6, (1992)..

50. Allik T.H., Hermes R.E., Sathyamoorti G. and Boyer J.H., "Spectroscopy and Laser Performance od New BF$_2$-complex Dyes in Solution", SPIE Proceedings, on Visible and UV Lasers, **2115** 240-248 (1994).

51. R. Reisfeld, "Film and Bulk Tunable Lasers in the Visible", *Sol-Gel Optics III*, John D. Mackenzie Ed., SPIE **2288**, 563-572,(1994).

52. C.K. Jφrgensen and R. Reisfeld, "Luminescence yields, minimal photodegradation of organic and Inorganic species in sol-gel glass and modalities of confinement", *Sol-Gel Optics III*, John D. Mackenzie Ed., SPIE **2288**, 208-215(1994)

53. Michael Canva, Arnaud Dubois, Patrick George and Alain Brun, "Perylene, pyrromethene and grafted rhodamine doped xerogels for tunable solid state laser", *Sol-Gel Optics III*, John D. Mackenzie Ed., SPIE **2288**, 298-309 (1994).

54. Y. Sorek, R. Reisfeld, I. Finkelstein, S. Rushin, Sol-gel glass wave guides prepared at low temperature, *Appl. Phys. Lett. 63*, 3256-3258, (1993).

55. R. Reisfeld, Energy transfer between dyes on glass surfaces and ions in glasses. *Chem. Phys. Lett.*, **95**, 95-96 (1983).

56. R. Reisfeld, "Increase of pumping efficiencies of glass lasers by radiative energy transfer", *Chem. Phys. Lett.*, **114**, 306-308 (1985).

57. R. Reisfeld, "Wave-Guided Sol-Gel Glass Lasers", *J. de Physique*, **4**, 281-284 (1994.).

58. R. Reisfeld, D. Shamrakov and Y.Sorek, "Spectroscopic Properties of Thin Glass Films Doped by Laser Dyes Prepared by Sol-Gel Method", *J. de Physique*, **4**, 487-490 (1994).

59. Y. Sorek, R. Reisfeld, I. Finkelstein and R. Ruschin, "Active glass waveguides prepared by sol-gel method", *Optical Materials* **4**, 99-101 (1994).

60 D. Shamrakov, R. Reisfeld, "Superradiant Laser Operation of Red Perylimide Dye Doped Silica-Polymethylmethacrylate Composite", *Chem. Phys. Letters*, **213**, 47-54 (1993).

61. D. Shamrakov and R. Reisfeld, "Super radiant film laser operation of perylimide dyes doped silica-polymethylmethacrylate composite", *J. Optical Materials,* **4**, 103-106 (1994).

62. Y Sorek, R. Reisfeld and R. Tenne, "The Microstructure of Titanium-Modified Silica Glass Waveguides Prepared by the Sol-Gel Method", *Chem. Phys. Letters*, **227**, 235-242 (1994).

63. Y. Sorek, R. Reisfeld, I. Finkelstein and R. Ruschin, "Light Amplification in a Dye-Doped Glass Planar Waveguide", *Appl. Phys. Letters,* **66**, 1169-1171 (1995).

64. C.V. Shank, J.E. Bjorkholm and H. Kogelnik, "Tunable distributed - fedback dye laser", *Appl. Phys. Lett.* **18**, 395 (1971).

65. H. Kogelnik and C.V. Shank, "Coupled mode theory of distributed feedback lasers", *J. Appl. Phys.*, **43**, 2327-2335 (1972).

66. A. Yariv, "Optical Electronics", 4th. Edition, Saunders publ. 1991.

67. G.S. He, C.F. Zhao, C-K Park and P.N. Prasad, "Dye film leaky waveguide laser", *Optics Commun.* **111**, 82 (1994).

68. W. Streifer, D.R. Scifers and R.D. Burnham, "Analysis of greating-coupled radiation inGaAs: GaAlAs and lasers waveguides", IEEE J. of Quant. Electron. *QE*-**12**, 422, (1976).

69. Y. Yamamoto, T. Kamiya and H. Yanai, "Improved coupled made analysis of corrugated waveguides and lasers", IEEE J. of Quant. Electron. *QE*-**14**, 245 (1978).

70. A. Hardy, D.F. Welch and W. Streifer, "Analysis of second-order gratings", IEEE J. of Quant. Electron. *QE*-**25**, 2096 (1989).

71. Proceedings of the Solid-State Dye Lasers Technology Workshop, 4 August 1994. Sponsored by Night Vision and electronic sensors directorate and Science Applications Int. Co. edited by T.H. Allik.

72. U. Narang, F. V. Bright and P. N. Prasad, "Characterization of rhodamine 6G-doped thin sol-gel films", *Appl .Spectroscopy*, **47**, 229-233 (1993).

73. M. J. Halmos, T. M. Fletcher and O. M. Stafsudd, "Active dielectric film laser", *Appl. Opt.*, **31**(21), 4132 (1992).

74. P. Yeh, *Optical Waves in Layered Media*, (Wiley, New York 1988), Chap.11.

ENERGY TRANSFER AND MIGRATION OF EXCITATION IN SOLIDS AND CONFINED STRUCTURES

F. Auzel

France Telecom, CNET Laboratoire de Bagneux,
B.P.107, F-92225, Bagneux Cedex, France

ABSTRACT

With the advances of Rare-Earth doped fibre amplifiers studies for telecommunication networks, a number of energy transfer processes barely noticeable in bulk materials are found to be considerably enhanced in the confined structure of optical fibres. Having recalled the basics of energy transfers, the following processes are discussed: radiative and non-radiative multiphonon transitions, APTE and cooperative up-conversion, photon avalanche. The relation of such processes with up-conversion lasers and optical amplification are finally discussed. Through this seminar, the current frontier in research in the Rare Earth (R.E.) doped fibre field is presented.

I. INTRODUCTION

Because of the energy confinement which arises in the optical fibre medium over long interaction length, many non-linear optical processes, otherwise negligible in bulk samples, may become prominent and simple linear processes such as absorption and spontaneous emission may become non-linear. In the first part of this seminar, we shall recall some basic properties of high aperture single mode fibre, then in the second part, some important multiphonon processes which had been over looked up to now in the fibre medium. In a third part, we recall the different types of up-conversion processes: APTE effect (Adition de Photons par Tranferts d'Energie), ESA (Excited State Absorption),cooperative effects, the photon avalanche effect. They are all connected to the necessary introduction of the Rare-Earth (RE) ions doping into the fibre since they provide their amplification properties. In a fourth part, we shall finally present some typical example of the roles of the above processes in various fibre up-conversion lasers and amplifiers.

Spectroscopy and Dynamics of Collective Excitations in Solids
Edited by Di Bartolo, Plenum Press, New York, 1997

II. THE OPTICAL FIBRE CONFINEMENT

II. A. Some Fibre Intrinsic Properties

1. The monomode condition In the following, we shall consider, for simplification, the ideal step index fibre. It is defined by a core medium of diameter a with index of refraction n_1, and by a cladding with index of refraction n_2. Defining a quantity V (the normalised frequency), by eq. (1) [1] as:

$$V = \frac{2\pi a}{\lambda}(n_1^2 - n_2^2)^{1/2} , \tag{1}$$

the monomode condition is:

$$V \leq 2.405 , \tag{2}$$

Equivalently this monomode condition for the core diameter is written:

$$2a < \frac{2.4\lambda}{\pi(NA)} , \tag{3}$$

where $(NA) = (n_1^2 - n_2^2)^{1/2} \sin\Theta_c$ is called the "Numerical Aperture" of the fibre and Θ_c is the

input critical half-angle.

As an example, a single mode fibre in the visible would have the following specifications:

$\lambda_c = 0.55\mu m$,and $(n_1 - n_2) = 8 \times 10^{-3}$, giving: 2a = 2.5 μm.

With such a small diameter, one understands why important pumping fluxes at Mw/cm^2 levels may be easily reached even with moderate pumping power of a few hundreds of mW. This is maintained over very long interaction lengths which can reach hundreds of meters. Comparing with usual optics, the focusing by a lens can provide such power density only over tens of micron; one can understand then the very particular advantage of fibre optics in general.

2. Transmission properties of fibres. The recent interest in Erbium (Er^{3+}) doped fibre amplifiers (EDFA) takes its roots in the recognised fact that the optical fibre transmission windows for silica or fluoride glasses (Fig.1) are in tune with the wavelengths of the two Er^{3+} transitions :

$${}^4I_{13/2} \rightarrow {}^4I_{15/2} \qquad \text{at 1.5 μm ,}$$

and

$${}^4I_{11/2} \rightarrow {}^4I_{13/2} \qquad \text{at 2.7 μm .}$$

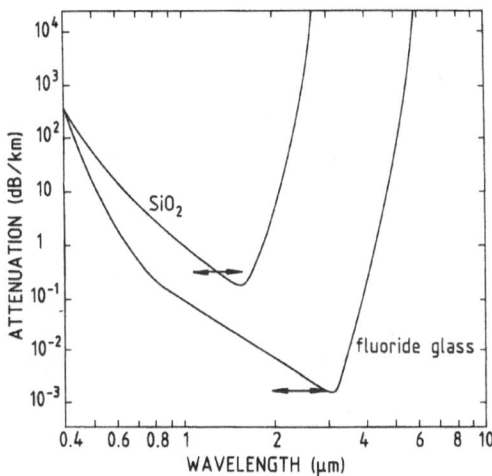

Figure 1. Theoretical absorption windows of undoped silica and fluoride fibres.

II. B. Spectroscopic properties of R.E. doped fibres : comparison with bulk samples

1. Absorption. Because high pumping fluxes are easily reached at levels of about 0.2 MW / cm^2, absorption saturation at a doping level of even 1000 ppm is readily obtained; it deforms the absorption spectra in comparison with bulk results; so studying a fibre, it is difficult to measure the real spectroscopic absorption coefficient and line shape of a R.E. transition as demonstrated on Fig., even at 50 mW input level [2].

2. Emission. When the fibre and pumping are long and large enough, emission spectra are deformed by amplification by stimulated emission (ASE) [2] (Fig.3). So, observed longitudinal emission spectra may be very different from bulk spectra and are parametrically dependent both on excitation level and fibre length. As such, the obtained spectrum may be different from the gain spectrum when the signal is larger than the spontaneous emission noise.

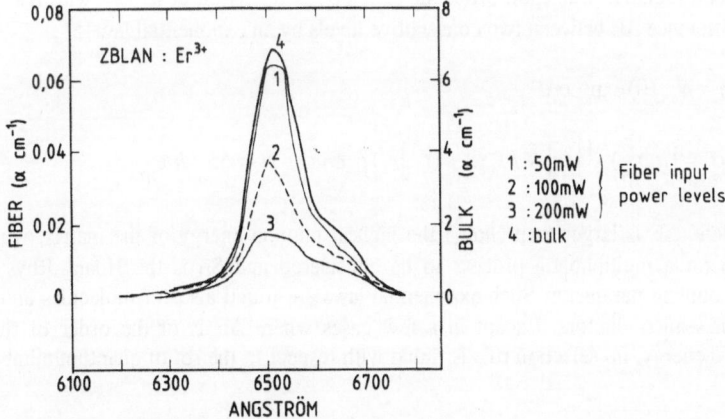

Figure 2. Comparison of bulk and fibre glass absorption spectra of Er^{3+} at different input.

III. RADIATIVE AND NON-RADIATIVE MULTIPHONON PROCESSES

Here, we shall deal with multiphonon non-radiative and radiative processes [3]. They can usually be considered in the first approximation as independent of the interactions between activator ions, that is, they are one-centre processes. Once an ion has been excited (the way in which it has been excited does not matter), it can lose energy non-radiatively in making a transition from the excited level to the one just below it. Experimentally it is found that the quantum efficiency of an excited state can be lower than expected from the one-phonon interaction for a given energy gap, even at low concentration (that is, without any possibility for energy transfer to sinks to take place).

A good description of this situation is given by the relation between quantum yields of levels and their difference in energy from the next lower level. Monochromatic excitation has shown that energy decay was effectively proceeding, by cascade [4].This comes from the fact

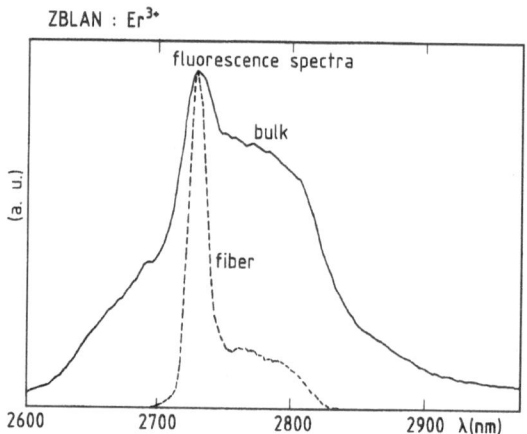

Figure 3. Emission spectrum at 2.7 μm of a bulk fluoride glass sample as compared with the ASE modified emission spectrum of the same glass fibre.

that the non-radiative transition probability (W_{NR}) is then well described with respect to the energy difference ΔE between two consecutive levels by an exponential law [5] :

$$W_{NR}(\Delta E) = W_{NR}(0)\exp-\alpha\Delta E \tag{4}$$

with [6] $\alpha = (\hbar\omega_m)^{-1}\left\{\ell n\left[\overline{N}/S_0(\overline{n}+1)^{-1}\right]-1\right\}$ and $\overline{N} = \Delta E/\hbar\omega_m$ (5)

Here, ΔE is larger than $(\hbar\omega_m)$ the highest phonon energy of the matrix, which is the condition for a multiphonon process to be considered and S_0 is the Huang-Rhys electron-phonon coupling parameter. Such exponential laws are found also for molecules and for deep centres in semiconductors. Except in a few cases where ΔE is of the order of the highest vibrational energy, no selection rule is found with respect to the set of quantum numbers of the levels.

Experimentally W_{NR} is usually obtained through one of the three following methods :
i) If W_{NR} is larger than W_R, the radiative transition probability, then a direct measurement of W_{NR} is obtained by a lifetime measurement for the level under consideration.

$$W_{NR} \cong 1/\tau \tag{6}$$

ii) When W_{NR} is smaller than W_R, then W_R is first estimated from an absorption measurement and W_{NR} is obtained by :

$$W_{NR} = \frac{1}{\tau} - W_R \tag{7}$$

iii) Assuming the validity of the rate-equation model, one can, by measuring intensity ratios solve the resulting system of equations for W_{NR}.

The form of eq (4) comes out from a statistical result and a lot of simplifications [3]. Two models can provide this simple result: i) the Nth order perturbation method [7] where electronic states are considered to be independent of the nucleus motion (the Huang-Rhys parameter, $S_0 \equiv 0$, case is called a M-process [8].); ii) the non adiabatic Hamiltonian method ($S_0 \neq 0$, called a Δ-process [8].). This last method recognises that in the Born-Oppenheimer approximation, the neglected term promotes the non-radiative transition. It can be viewed as the fact that a configuration diagram (Fig 4) is not completely valid and the neglected term allows a mixture of the two electronic states specially in the region where they are closer in energy. Two extreme cases can be considered : either there is a strong electron-phonon coupling ($S_0 > 5$) then in the Born-Oppenheimer approximation there can be a level crossing and non-radiative transitions can be viewed as a "short-circuit" between the two states, or for

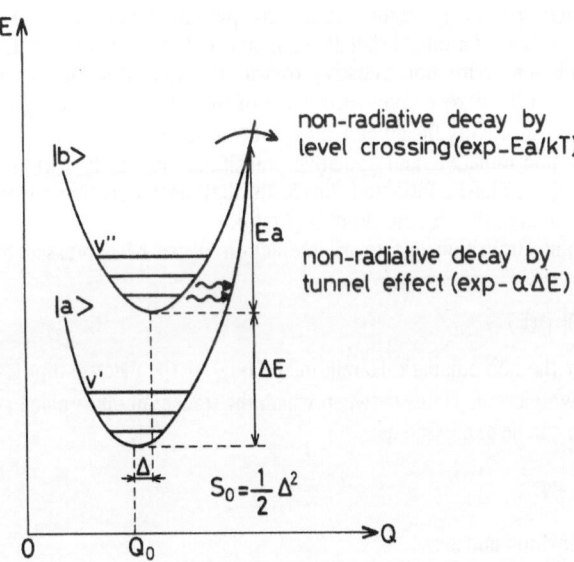

Figure 4. Configuration diagram for non-radiative and vibronic transitions in the Born-Oppenheimer energy scheme.

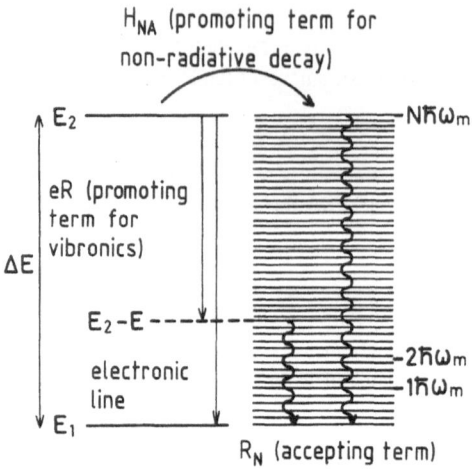

Figure 5. Both non-radiative and vibronic transitions are coupled to a quasi-continuum of phonons through two different promoting terms: respectively H_{NA} and $e\mathbf{R}$.

small coupling ($S_0 < 1$) there is a tunnelling of energy through the barrier arising between the two approaching states. This last case is the one for rare-earth ions and eq (4) is a good approximation for them. Transition metal ions have S_0 values somewhat in-between and it is safer to consider the full theory [3].

When states are not described by pure electronic wave functions but by Born-Oppenheimer states then mixed-nature transitions, partially vibrational and partially electronic (called vibronics) that is of a mixed radiative and non-radiative nature, are found in absorption, excitation or emissions pure non-radiative transitions, vibronics can be separated into two classes, according to the zero or non-zero value of the Huang-Rhys coupling parameter. This also gives rise to, respectively the so-called M and Δ processes.

For both non-radiative and radiative transition one deals with the coupling of an electronic state with a quasi-continuum. Fig 5, through an interaction which is either the non adiabatic Hamiltonian or the electric-dipole operator.

A probability transition due to an interaction H can be expressed by a Fermi Golden Rule as :

$$W_{21} = \frac{2\pi}{\hbar} \left| \langle 2|H| \rangle 1 \right|^2 \rho(E) \tag{8}$$

where H is either the non adiabatic Hamiltonian H_{NA} or the electric-dipole operator $e\mathbf{R}$; $|1\rangle$ and $|2\rangle$ are the two vibronic states between which the transition takes place ; ρ is the density of final states which can be expressed as:

$$\rho = R_N \delta(E_2 - E_1 - E) \tag{9}$$

for a radiative transition and as:

$$\rho = R_N \delta(E_2 - E_1) \tag{10}$$

R_N is usually given as a product of Franck-Condon factors corresponding to the so-called "accepting modes" which "receive" the energy given up by the excited electronic state [9].It can be shown in a simple way [3]that this R_N function can be described by statistics on the ways to distribute the electronic energy among the vibrational modes and can be expressed as:

$$R_N = e^{-S_0(2\bar{n}+1)} \frac{(\bar{n}+1)^{N/2}}{(\bar{n})^{N/2}} \sum_\ell \frac{S_0^2 \bar{n}(\bar{n}+1)^{\ell+N/2}}{\ell!(N+\ell)!} \tag{11}$$

which is the R_N of Huang and Rhys [10] and S_0 their electron-phonon coupling parameter. The temperature dependence is essentially contained in the phonon occupancy number, n

$$\bar{n} = (\exp(h\omega_m / kT) - 1)^{-1}, \text{then:}$$

$$R_N = e^{-S_0} \frac{S_0^N}{N!} (\bar{n}+1)^N e^{-2\bar{n}S_0} \left[1 + \frac{S_0^2 \bar{n}(\bar{n}+1)}{(N+1)} + \ldots \right] \tag{12}$$

For $S_0 \leq 1$ (the R.E. case), and not too high a temperature, we can retrain only the first term ; that is, for a "Pekarian function" at 0°K we associate a temperature dependence of the form

$$(\bar{n}+1)^N e^{-2\bar{n}S_0} \tag{13}$$

These factors give the variation with temperature proposed by Fong et al. [11]. In fact for small values of S_0, $\exp{-2S_0 \bar{n}}$ can be neglected ($\approx 4\%$ for LaF_3 at 400°K) and one is left with the usual dependence $(\bar{n}+1)^N$ as experimentally verified, for instance by Riseberg and Moos [12]. From this, the small-coupling approximation of eq (13) is equivalent even at temperature T to considering only phonon emission by $(n+1)^N$ or only absorption by $(\bar{n})^N$. This possibility arises from the rapid decrease of R_N with the number of phonons.

Since this simple reasoning underlying "Pekarian" functions can be applied to any phenomenon in which multiphonon jumps are involved, one can understand why when small coupling is considered, exponential gap laws are found whatever the process is. For strong coupling the Gaussian form of R_N gives the Arrhenius limit form for the non-radiative probability.

For small coupling, N!, coming from the R_N function, is developed by Stirling's Formula which finally the exponential term of eq (4). The precise knowledge of the phenomena comes only as a modification to the statistics on the filling of accepting modes so providing the gross features of the energetical behaviour. Then such forms apply as well for non-radiative transition as for radiative ones. For this last type of transition, the N=0 case is the so-called zero-phonon line the intensity of which decreases with coupling strength.

However, we have not yet considered the role of the matrix element in (8). It contains a so-called promoting term which differs according the considered Hamiltonian. For vibronics, it consists of a purely electronic term since it is caused by the electric dipole operator. This term can be simply factorized as in eq. (4) which then also gives the energetical shape of a vibronic transition. For a non-radiative transition, the promoting term is caused by promoting modes of vibrations taking their energy from the electronic energy difference ΔE; then eq.(4) takes the following practical form with reduced ΔE [13]:

$$W_{NR}(\Delta E) = W_{NR}(0) \exp{-\alpha(\Delta E - 2.4 \hbar\omega_p)} \tag{14}$$

where $\hbar\omega_p$ is the promoting mode energy.

When the concentration of active ions is increased, long before the appearance of new lines due to pairs or modifications in radiative transition probabilities, a migration of energy between the centres is found. We are going to study this now, assuming that the multiphonon decay and the radiative transition remain one-centre processes.

IV. APTE, ESA, COOPERATIVE, AND PHOTON AVALANCHE UP-CONVERSIONS

IV.A. Up-conversion processes

When active ions are situated at a sufficiently short distance for interactions between them to take place, two types of up-conversion processes may occur : i) summation of photon by energy transfers, [14], called the APTE effect (for Addition de Photon par Transfers d'Energie) [15];ii) cooperative effects either by sensitisation [16] or emission [17]. Both types are often mistaken one for the other because they present several similarities and may be simultaneously present in a given system for a given excitation.

Until 1966, all energy transfers between R.E. ions were considered to be such that the activator ion receiving the energy from a nearby sensitiser (S), was in its ground state before interaction. Then it was proposed [14] to consider cases where activators (A) were already in an excited state as shown on fig. 6.

This is evident afterwards because one has to realised that we can exchange only energy differences between ions and not absolute energy.

The reason for proposing such an up-going transfer was to point out that energy transfers then used [18] to improve the laser action of Er^{3+} by pumping Yb^{3+} in a glass matrix could also have the detrimental effect to increase reabsorption [14,19]. The simple proof of such an effect was to look for an up-converted green emission (from $^4S_{3/2}$ of Er^{3+}) while pumping Yb^{3+} ($^2F_{7/2} \rightarrow {}^2F_{5/2}$), which was effectively observed [14,20]. Of course the situation in fig. 7 could repeat itself several times at the activator site. This means that n-photon up-conversion by energy transfer is possible as demonstrated by the 3-photon up-conversion of 0.97 μm into blue light (0.475 μm) in the Yb^{3+} - Tm^{3+} couple [14]. Independently such IR $\rightarrow$ blue up-conversion was interpreted by Ovsyankin and Feofilov [16] as a 2-photon effect connected with a cooperative sensitisation of Tm^{3+} by two Yb^{3+} ions, because of saturation in an intermediate step reducing the intensity from a cubic to a quadratic law.

In order to clarify the terminology, a schematic comparison between the APTE effect and other 2-photon up-conversion processes namely: 2-step absorption, cooperative sensitisation, cooperative luminescence, second harmonic generation (SHG) and 2-photon absorption excitation is presented on Fig; 7, together with typical efficiencies.

Since we are dealing with non-linear processes, the usual efficiency, as defined in percentage, has no meaning because it depends linearly on excitation intensity. Values are then normalised for incident flux and given in (cm^2/W) units.

In confined structures, because some of the most efficient processes may start saturating due to the very high pump density, the lowest efficiency processes may dominate.

A simple inspection of the energy schemes involved shows that they differ at first sight by the resonances involved for in- and out-going photons : for the highest efficiency, photons have to interact with the medium a longer time which is practically obtained by the

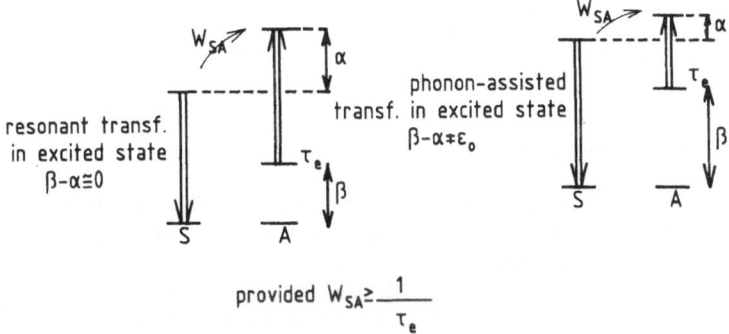

provided $W_{SA} \geq \dfrac{1}{\tau_e}$

Figure 6. Basic energy scheme for the elementary step of up-conversion by the APTE effect

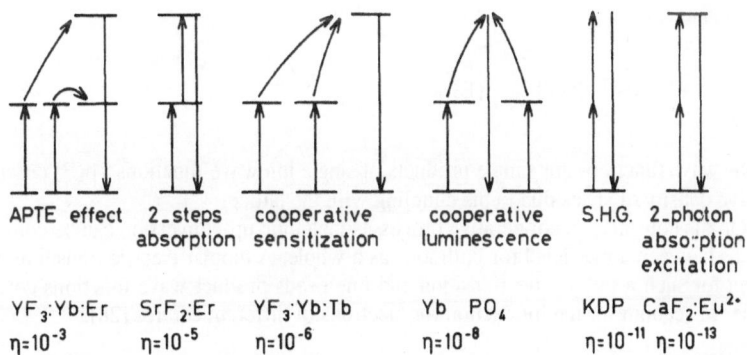

APTE effect	2-steps absorption	cooperative sensitization	cooperative luminescence	S.H.G.	2-photon absorption excitation
YF$_3$:Yb:Er	SrF$_2$:Er	YF$_3$:Yb:Tb	Yb PO$_4$	KDP	CaF$_2$:Eu^{2+}
$\eta=10^{-3}$	$\eta=10^{-5}$	$\eta=10^{-6}$	$\eta=10^{-8}$	$\eta=10^{-11}$	$\eta=10^{-13}$

Figure 7. Energy schemes for different 2-photon up-conversion processes with quantum efficiency in cm^2/W for specified materials.

existence of resonances. As shown, the APTE effect is the most efficient because it is the process which most approaches the full resonance case; consequently it is the more likely to be encountered.

Many times in literature up-conversion involving coupled ions are referred to as cooperative effects [21,22,23] without demonstration, when in fact as can be guessed from the relative positions of involved energy levels, APTE effects are involved [24,25]. The fact that the APTE effect and cooperative ones are often mistaken one for the other is linked to a number of common properties. For instance, for 2-photon up-conversion, both processes show a quadratic increase on excitation and on absorber concentrations ; both show an emission lifetime equal to half the absorber lifetime. However, as shown below, the difference is more basic, though sometimes difficult to establish experimentally except in special cases : when single ion resonances clearly do not exist or when diffusion between ions is prohibited by a too small concentration with still an interaction as in clusters of R.E. ions.

1. Up-conversion in a single ion level description (APTE) and in a pair-level one (cooperative effects). As seen in the introduction, up-conversion by energy transfer is just a generalisation of Dexter energy transfer [26] to the case of the activator being in a metastable state instead of being in its ground state ; this requires that the interaction between S and A (H_{SA}) be smaller than the vibronic interaction of S and A in order that both ions be described by single-ion levels coupled to the lattice. It is generally the case since for fully concentrated R crystals or for cluster, pairs level splitting is of the order of 0.5 cm^{-1} [27]; in a host with a smaller concentration this interaction can even be weaker, whereas one-phonon or multiphonon sidebands may modulate the level positions by several 100 cm^{-1}. Further, up-conversion requires that the transfer probability for the second step W_{SA} be faster than radiative and non-radiative decay from the metastable level that is $W_{SA} > \tau^{-1}$ with τ the measured lifetime. W_{SA} is obtained from

$$W_{SA} = \frac{2\pi}{\hbar} \left| < \Psi_S^e \Psi_A^0 | H_{SA} | \Psi_S^0 \Psi_A^e > \right|^2 \rho(E) ,$$ (15)

where the wave functions are simple products of single ion wave functions ; $\rho(E)$ describes, the dissipative density of states due to the coupling with the lattice.

On the contrary, all cooperative processes including up-conversion can be considered as transitions between a pair-level for both ions as a whole. A dipolar electric transition would be forbidden for such a two-centre transition and one needs product wave functions corrected to first order to account for the interaction for electrons of different centres [28].

$$\Psi_{pair} = \Psi^0(S)\Psi^0(A) - \sum_{s''\neq 0}\sum_{a''\neq 0} \frac{< s''a'' | H_{SA} | 00 >}{\delta_{s''} - 0 + \varepsilon_{a''} - 0} \Psi_{s''}(S)\Psi_{a''}(A) ,$$ (16)

for example for the ground state ; s", a" denotes intermediate states for S and A ; $\delta_{s''}$, $\varepsilon_{a''}$ denote their corresponding energies. Then any one-photon transition in the cooperative description involves already four terms in the matrix element which cannot be reduced to eq. (15)

APTE up-conversion does not correspond to the same order of approximation as cooperative processes; the later have to be considered practically only when the first type cannot take place.

Such is the case when real levels do not exist to allow energy transfers ; this is the case for Yb^{3+} - Tb^{3+} up-conversion [27] or when concentration is to small to allows efficient transfer by energy diffusion between sensitisers. Then cooperative up-conversion is likely to occur within clusters [29,30]. One may also look for crystal structures where the pair clustering is built in [31,32].

In order to illustrate the difference between APTE and cooperative up-conversion we shall discuss an example of a line-narrowing effect in n-photon summation as a mean to distinguish between both processes [24,33]. Irradiating Er^{3+} doped samples with IR photons at 1.5 μm leads to various visible emissions .

Room temperature IR F-centre laser excitation between 1.4 and 1.6 μm of Er^{3+}: YF_3 leads to emission bands from near IR to U.V. Such emissions may be ascribed to multiphoton excitation respectively of order 1 to 5, either by the APTE or by the cooperative type as

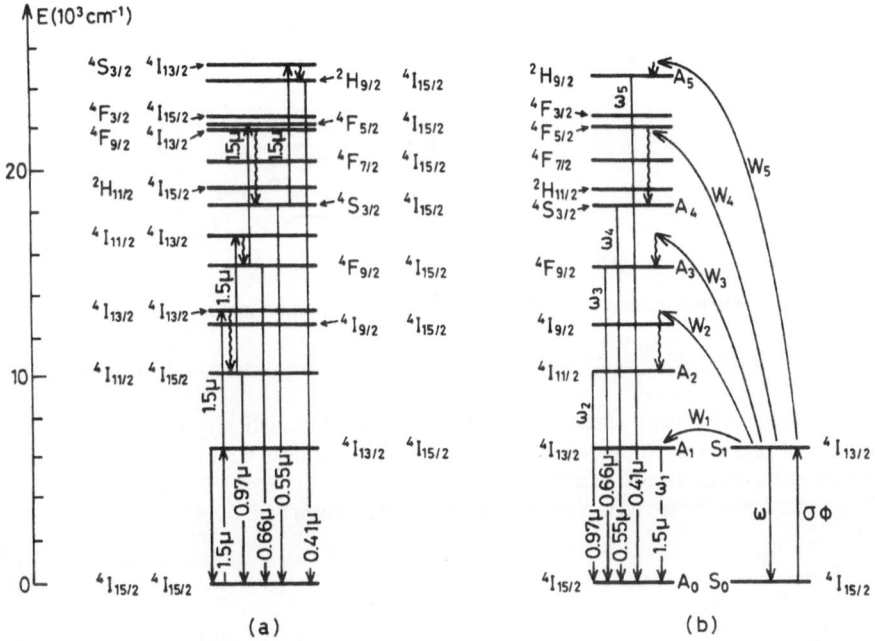

Figure 8. n-photon summation (n=1...5) in Er^{3+} systems by cooperative sequential pair absorptions (a) and by the APTE effect (b).

depicted respectively with energy levels of single ions (APTE) or with pair levels (Fig. 8 a and b).

Successive absorptions in fig. 8a involve a combination of several J states. The APTE effect, because of self matching by multiphonon processes, involves only J = 15/2 and J = 13/2 states (Fig. 8b).

Excitation spectra depicted in Fig. 9 show a striking behaviour : each spectrum presents the same spectral structure, but an increasing narrowing is clearly observed with process order. The structure reproduces the Stark structure of the $^4I_{15/2} \rightarrow {}^4I_{13/2}$ first excited terms as can be obtained by a diffuse reflectance spectra. This then is a direct proof of the validity of the APTE explanation, since a cooperative effect should show the convolution of all J states involved in the multiple absorption between pair-levels [33].

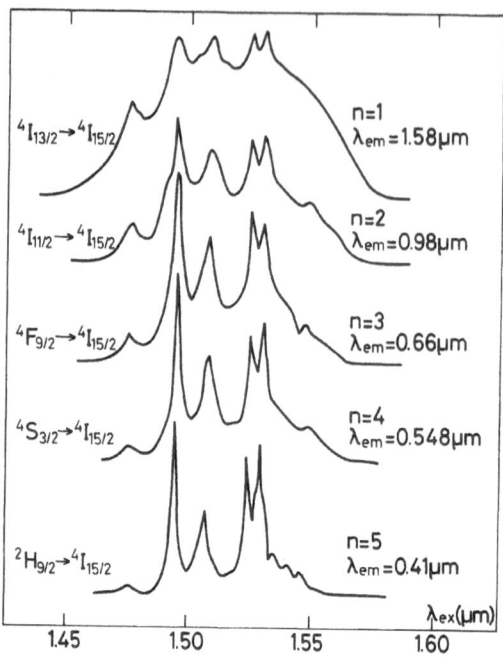

Figure 9. Excitation spectra for n-photon summation in $YF_3:Er^{3+}$ [33].

The spectral narrowing can be understood by a rate equation treatment where a higher excited population are neglected with respect to the lower ones in order to obtain a tractable development (weak excitation assumption).

The emitted power from an n-photon summation is then given by :

$$P_n(\lambda) = \frac{W_n \dots W_2}{(\omega_{(n-1)} \dots \omega_2)} P_1^n(\lambda) \tag{16}$$

with symbols of Fig. 8, and where $P_1(\lambda)$ is the line shape of $^4I_{15/2} \rightarrow {}^4I_{13/2}$ absorption [24].

2. ESA and the APTE effect. The probability for ESA in a two-step absorption (W_{13}) connecting a state E_1 to E_3 by the intermediate state E_2, is just given by the product of the probabilities for each step

$$W_{13} = W_{12} \cdot W_{23} \tag{17}$$

For ESA by an APTE effect we have also the product of two energy transfers probabilities. Calculating the rate for the same ESA transition with APTE, it follows

$$W_{13} = N_S^{*2} W_{SA1} W_{SA2} \tag{18}$$

where W_{SAi} are the energy transfer probabilities for each step i, and N_S^* the concentration of excited sensitisers which are given by

$$N_S^* = N_S W_{12}$$

Assuming all W_{ij} and W_{SA} to have same magnitude, as it is typical for R.E. ions, we have to compare

$$W_{13} = W_{12}^2 \quad \text{for a single ion ESA and} \tag{19}$$

$$W_{13} = W_{12}^2 N_3^2 W_{SA}^2 \text{ for ESA by APTE.} \tag{20}$$

Clearly the gain by APTE over one ion ESA can come from the product $N_S^2 W_{SA}^2$ which has to be > 1. This points to increase in sensitiser concentration (N_S) which is known to lead to "fast diffusion" [34] and to allow the use of rate equations in such multi-ions systems [35].

3. The photon avalanche effect. While looking for 2-steps absorptions (ESA) in Pr^{3+} doped $LaCl_3$ and $LaBr_3$ at low temperature (<40K) as a means to detect an IR photon by its energy summation with a more energetic photon performing ESA, it was found that the higher energy photon alone could, in the same time, give rise to up-conversion and reduce the transmission of the sample above a given intensity threshold [36] see Fig. 10. The effect was attributed to an increase of population on an excited state resulting from a cross-relaxation process. The starting process was initially not completely determined . In the Pr^{3+} case, the $^3H_{51} \to {}^3P_1$ absorption is initially very weak at low temperature because $^3H_5$ is about 2000 cm^{-1} above ground state (see Fig. 11) ; however above about 1 mW of excitation this transition is increased ; the cross-relaxation $({}^3H_6, {}^3H_4) \to ({}^3H_5, {}^3H_5)$ increasing the $^3H_5$ population which in turn reduces the transparency of the sample at the $({}^3P_1 - {}^3H_5)$ energy. Since the more the $({}^3P_1 - {}^3H_5)$ energy is absorbed the more the $^3H_5$ population is increased, the process was termed "photon avalanche"[36]. It is clearly a way to increase ESA in a sample.

Since then similar effects have been observed in Sm^{3+}, Nd^{3+}, Ni^{2+}, Tm^{3+} doped halide crystals [37,38,39,40].

There are three distinct aspects for the photon avalanche process non-linear behaviour:
i) transmission, ii) emission and iii) rise time on the pump power intensity with generally the existence of a critical pump threshold.

Particularly long rise times, from seconds to minutes [41], have been observed.

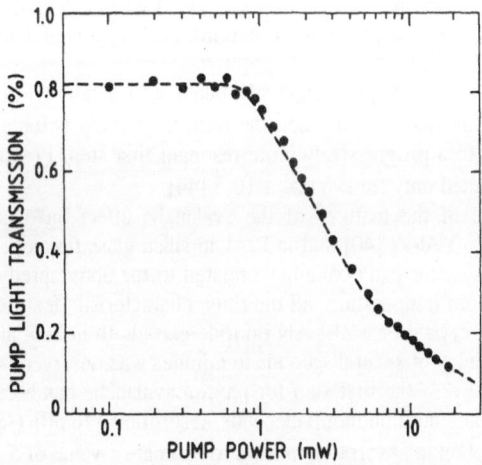

Figure 10. Decrease of transmission in a Pr^{3+}:$LaCl_3$ sample under $^3H_5$-$^3P_1$ pumping. After [36]

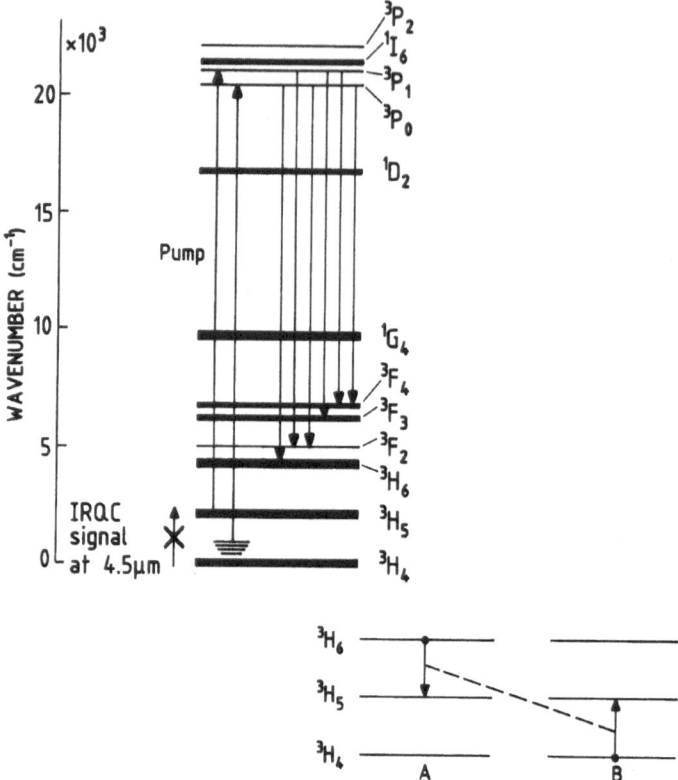

Figure 11. Pr^{3+} energy levels involved in the avalanche process,[36].

At this point it is worth discussing the notion of threshold for avalanche. Because of the complexity of the phenomena, it has been usually modelled by a simplified three-level system [42,43,44] with neglect of the first non-resonant absorption step (W_1); see Fig. 12. Taking into account only the second resonant absorption step (W_2) when calculating the population of the third level (p_3) versus W_2, (the pumping excitation), leads to a well defined non-linearity in p_3; curve (a) on Fig. 13.

When the first step (W_1) is explicitly taken into account [44], the "threshold" non linearity is progressively smooth out when the ratio $\beta = W_1/W_2$ is increased as shown on Fig. 13. This corresponds to a progressively more resonant first step. Practically a clear avalanche threshold can be expected only for β ratios $\leq 10^{-4}$ [44].

Recently some of the features of the avalanche effect have been observed at room temperature in Tm^{3+} : $YAlO_3$ [40] and in Pr^{3+} in silica glass fibres [45]. The lack of a clear threshold in this two systems can certainly be related to the above prediction. Up to now, only Er^{3+} has shown at room temperature, all the three characteristic features of avalanche as well when doping a $LiYF_4$ crystal or a ZBLAN fluoride glass both in bulk and in a fibre shape (see V.B.); even the long delay of several seconds to minutes was observed [46,47,48,49,50]

In the case of Er^{3+}, the first step for photon avalanche has been clearly identified and attributed to anti-Stokes multiphonon sidebands absorption [46,50] (see Fig. 14).Calculating the β ratio from mutiphonon absorption, allows to estimate a value of 5 10^{-6} [49,50] which, as observed, provides a clear threshold behaviour in the Erbium case [46].

Because the first absorption step, being of a multiphonon nature, is featureless, the excitation spectrum for avalanche provides directly with a single excitation beam the ESA spectrum of the resonant second absorption as shown on Fig.15 for the $^4I_{11/2}$-$^2G_{9/2}$ transition

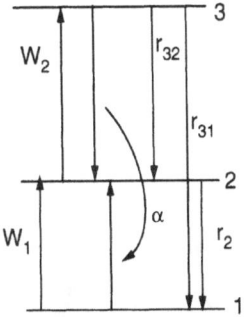

Figure 12. The avalanche simplified 3-level scheme; α is the cross-relaxation term; r_{ij} are the spontaneous emission terms.

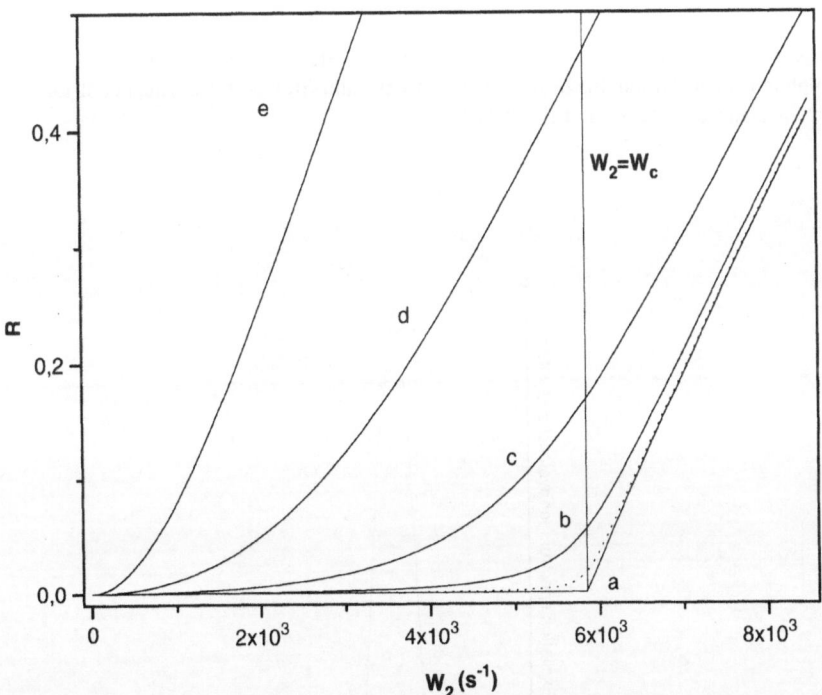

Figure 13. The third level normalised population $R = p_3(\beta, W_2 = W_c)/p_3(\beta = 1, W_2 = W_c)$ versus pumping term (W_2) with $\beta = W_1/W_2$ as parameter: (a) $\beta = 0$; (b) $\beta = 10^{-3}$; (c) $\beta = 10^{-2}$; (d) $\beta = 10^{-1}$; (e) $\beta = 1$. [44]

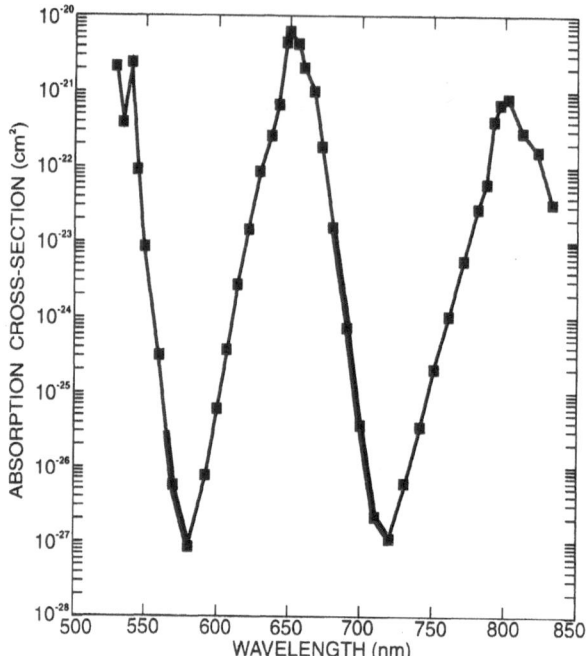

Figure 14. Absorption cross-section for a fluoride ZBLAN:Er glass taking into account the multiphonon contribution; the heavier lines show the anti-Stokes zones which contribute to W_1 for the avalanche processes in Erbium [48,51].

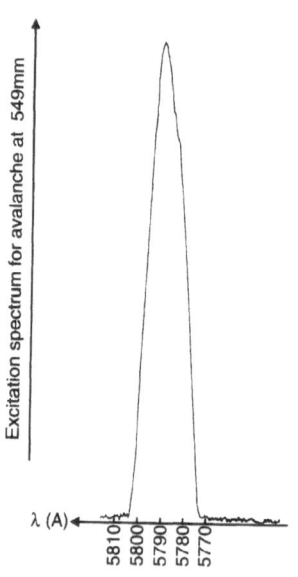

Figure 15. Excitation spectrum for the avalanche emission at 549nm in a ZBLAN:Er^{3+} glass showing the spectrum for the $^4I_{11/2}$-$^2G_{9/2}$ ESA transition,[48].

of Er^{3+}; this provides a new method with a single beam excitation to reach ESA spectra other wise difficult to obtain [48].

V. MULTIPHONON ABSORPTION AND UP-CONVERSION IN DOPED FIBRES

V.A Multiphonon absorption in doped fibres and background losses

As seen in chapter III, multiphonon vibronics absorption may behave exponentially versus energy in the case of weak electron-phonon coupling. Because the influence of the host lattice on the 4f electrons of the rare earth ions is weak, Huang-Rhys parameters (S_0) for such ions are of the order of 10^{-2}. So the effect of vibronic multiphonon absorption can be neglected in most cases. However, we had shown about twenty years ago[56] that this effect can be observed in bulk materials and cannot be always neglected because it is the root of multiphonon assisted energy transfers.

The observed α exponential parameter in eq. (4) for multiphonon Stokes excitations is found to be $9.4 \ 10^{-3}$cm in fluoride glasses (ZBLAN) at 300K [57]; this is much larger than the one for non-radiative decay which is $5.2 \ 10^{-3}$cm in the same glass[58].Even taking into account the role of promoting modes in the non-radiative case through eq.(14), cannot explain this difference. It has to be traced back to an effective phonon at a lower energy (≈ 330cm^{-1}) [57], instead of the maximum phonon energy of the glass host (580 cm^{-1}) generally considered as ruling multiphon processes in this glass.

In the very long sample that a simple length of doped fibre represents, such multiphonon absorption should not be neglected any more. Based on the exponential behaviour of the excitation spectra of Er^{3+}doped glass samples, we could determined the amount of "incompressible" absorption which do exist [59]between the electronic rare earth ions absorption and which had been solely attributed to some "compressible" defects linked to an unperfected chemical preparation. Fig. 16 shows the absorption loss of an Erbium doped fluoride fibre in comparison with observed background loss and the former theoretical limit which did not consider the multiphonon vibronic absorption.

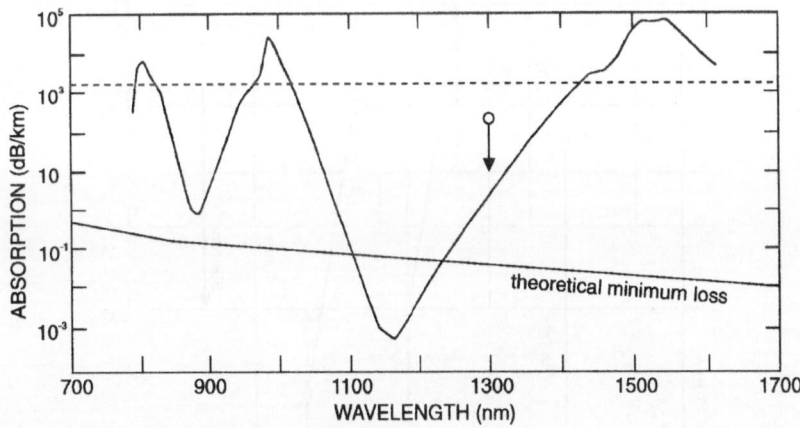

Figure 16. Absorption losses between 800 and 1600 nm for a 1000 ppm doped fluoride ZBLAN glass. The parts above the dotted line are the classically observed electronic absorption, those below it are the contributions from the multiphonon side-bands absorption [59]. (o) represents the measured background loss in the fibre [60]

In case the real background loss (wavelength independent) be weaker than the vibronic absorption, distributed amplifiers with kilometres length could be encompassed with pumping outside the rare earth electronic absorption. This would allow the use of pumping diodes at their most efficient wavelength that is also with a lower price.

V.B. Photon avalanche in doped fibres

Recently, the photon avalanche effect has been observed in a Pr doped silica fibre [45]and in an Er doped fluoride glass fibre [47,49] in both cases at room temperature. In the first case, only the transmission non-linearity is observed and not the up-conversion emission threshold. It was believed that the threshold was so low that it could not be observed .We think that this is explained by the too strong non-resonant to resonant absorption ratio as mentioned in IV.A.3. On the opposite in the second case, clear thresholds at 5mW and at 4mW of incident power at respectively 579nm [49]and 690 nm [47] are observed because in these last two cases, the first step is a weak anti-Stokes multiphonon absorption giving again a β ratio of about 10^{-6} much below the critical value of 10^{-4}. The involved energy scheme for both excitations is given on Fig.17; it shows both pumpings and the two types of involved cross-relaxations. Fig. 18 presents the typical threshold behaviour for the up-conversion emission. The long delay behaviour is displayed on Fig.19 showing near threshold the very long time of 3.5 s. which is largely in excess of any of the life-times of the metastable states of Erbium.

Also in the fluoride fibre case, yet not completely explained up-conversion spatial domains appear with periodic structures with periods ranging from few cm to mm and 100 µm [47,49], as shown on Fig.20.

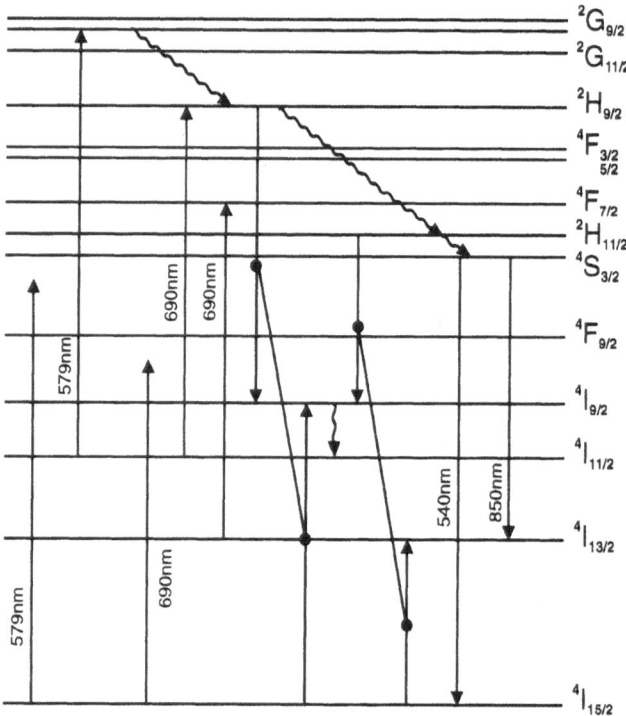

Figure 17. Er^{3+} energy scheme showing pumping transition at 579 an 690 nm; cross-relaxations between two ions are noted by vertical arrows connected by transverse lines; wavy lines denote non-radiative transitions.

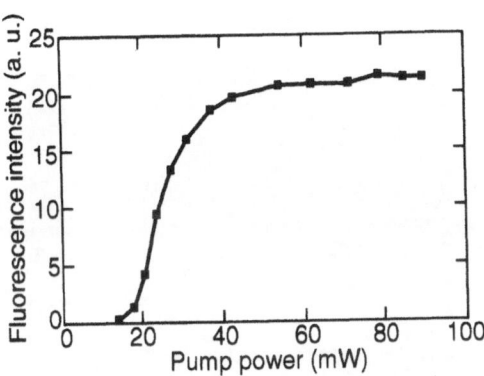

Figure 18. Up-conversion emission at 550 nm showing the existence of the avalanche threshold in a ZBLAN: Er^{3+} fibre observed from its extremity [49].

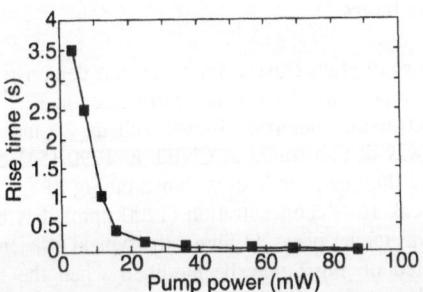

Figure 19. Time delay for the avalanche establishment versus incident pump power at 579 nm in a ZBLAN: Er^{3+} fibre, [49].

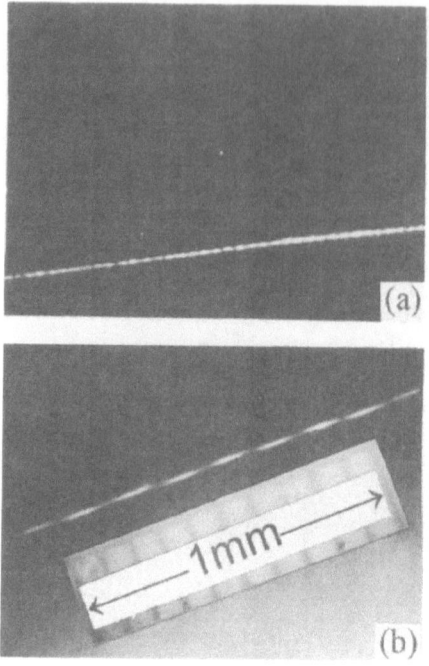

Figure 20. Spatial domains observed along the fibre above the photon avalanche threshold: (a) dot separation about 1mm; (b) microscope view of a 1mm avalanche dot, the scale is 100 μm per division.[47].

V.C. Up-conversion fibre lasers

1. AntiStokes lasers in glass fibres. Since the first demonstration in 1987 [51] of the feasibility of CW room-temperature 3-level rare-earth laser in Er^{3+} doped glass fibres, one could think that up-conversion pumped 3-level scheme could also be lased CW at room-temperature. This was demonstrated at CNET in 1990 [54] by Allain et al. The level scheme of the Ho^{3+} doped fluoride fibre laser is shown on Fig 21.

Because of the weak Ho^{3+} concentration (1200 ppm) it is believed that ESA within single ion levels system was taking place. In fibres, the typical efficiencies for the various two-photon processes presented on Fig.7 may be modified when the most efficient ones start saturating and ESA could overcome the APTE effect. However because clustering may some time takes place at much lower concentration, there is still some doubt about the very pumping process as in many of the following up-conversion pumping fibre lasers :

Er^{3+} doped glass fibre have also shown CW-room temperature 3-level laser emission at 0.54 μm when pumped at 0.801 μm [55]. Because the pumping wavelength is in the laser diode range there is some hope that, compact fibre laser could be obtained.

Besides these 2-photon up-conversion pumped lasers even a 3-photon one has recently been obtained in a Tm^{3+} doped fibre [56]. Pumping at 1.12 μm, emission at 480 nm is obtained CW at room-temperature with a differential efficiency of 18%; threshold being a rather low 30 mW.

Pr^{3+} doped fluoride fibre because of their low phonon energy have allowed anti-Stokes pumping lasers at blue, greens and red wavelengths in the same fibre laser medium [57], also in CW room temperature conditions.

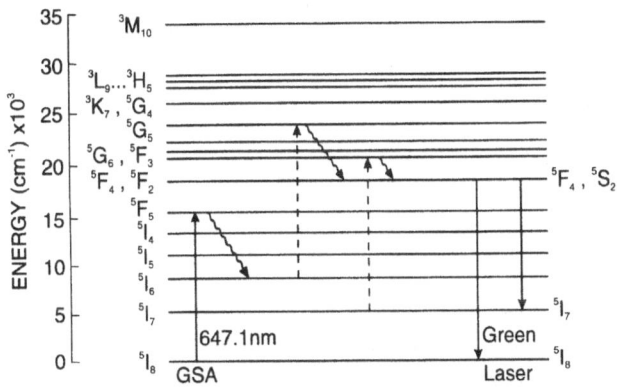

Figure 21. Energy scheme of the first CW, room temperature up-conversion laser [54].

2.Up-conversion laser with multiphonon absorption pumping. Besides pumping in the electronic rare-earth ions transitions, as shown above, up-conversion pumping was successful through multiphonon side-band pumping with energy mismatches as large as $1000 cm^{-1}$. It was the case for the Tm^{3+} fluoride fibre laser pumped at 1.06 μm in a 3-photon ESA process and lasing at 1.47 μm [62].It was a 4-level scheme laser so providing, at room-temperature, CW oscillation with a differential efficiency of 27%.

Finally, it is worth mentioning the up-conversion avalanche ASE (Amplification by Stimulated Emission) in an Er:ZBLAN glass fibre [61]. A CW ASE emission is obtained at 556 nm through an up-conversion avalanche process as described in V.B. under a 690 nm pumping which corresponds to an ESA pumping according to the energy scheme of Fig.17.

VI. CONCLUSION

Optically confined media doped with R.E. ions such as fibres have shown to provide long interactions lengths together with high pump densities. They so reveal various non-linear processes linked with the ion-ion, and ion-phonon interactions and otherwise neglected in bulk samples. Such approaches have already demonstrated the possibility of new up-conversion lasers. They are likely to present prospects for a whole new class of applications.

REFERENCES

1 see for instance:A.Cozanet, J. Fleuret, H. Maitre, and M. Rousseau, "Optique et Télécommunications", (Eyrolle, Paris 1981), p. 217.
2 F.Auzel, "Amplifiers and lasers with optical fibers", in "Defects in insulating materials", Ed.O.Kanert and J.-M.Spaeth, World Scientific,(Singapore, 1993), p.39.
3 F.Auzel, "Multiphonon processes, cross-relaxation and up-conversion in ion-activated solids, exemplified by minilaser materials", in "Radiationless Processes", Ed. B. Di Bartolo and V. Goldberg (Plenum Press, New York 1980.), p. 213.
4 F.Varsanyi, "Excitation of fluorescence with monochromatic light in rare-earth crystals", in "Quantum Electronics", Eds. P. Brivet and N. Bloembergen, (Dunod, Paris 1964.), p. 787.
5 M.J.Weber, Phys. Rev., 138, 54.(1973)
6 T.Miyakawa, and D.L. Dexter, Phys. Rev., B1, 70,(1971).
7 A.Kiel, "Multiphonon spontaneous emission in paramagnetic crystals", in "Quantum Electronics", Ed. P. Grivet and N. Bloembergen (Dunod, Paris 1964), p. 765.

8 T.Miyakawa, , "Luminescence of crystals. Molecules and Solutions", Ed. F. Williams, (Plenum Press 1973), p. 394.

9 F.K.Fong,, "Theory of Molecular Relaxation", (Wiley, New York 1975).

10 K.Huang,, and A. Rhys, , Proc. Roy. Soc., A204, 406,(1950).

11 F.K.Fong,, S.L. Naberhuis, and M.M. Miller, J. Chem. Phys., 56, 4020, (1972).

12 L.A.Riseberg, and H.W. Moos, Phys. Rev., 174, 429,(1968).

13 F.Auzel, "Advances in non-radiative processes in solid state laser materials", in "Advances in Non-Radiative Processes in Solids", Ed.B.Di Bartolo, (Plenum Press, New York 1991) p. 135.

14 F.Auzel, C.R. Acad. Sc. (Paris), 262, 1016 and 263, 819,(1966).

15 F.Auzel, Proc. IEEE, 61, 758,(1973).

16 V.V.Ovsyankin, and P.P. Feofilov, Sov. Phys. JEPT Lett., 4, 317,(1966)

17 E.Nakazawa, and S. Shionoya, Phys. Rev. Lett., 25, 1710,(1970).

18 E.Snitzer, and R. Woodcock, Appl. Phys. Lett., 6, 45,(1965).

19 F.Auzel, Ann. Télécom. (Paris), 24, 363,(1969).

20 F.Auzel, and O. Deutschbein, Z. Nat. (Germany), 24a, 1562,(1969)

21 J.P.Van der Ziel, L.G. Van Uitert, W.H. Grodkiewicz, and R.M. Mikulyak, J. Appl. Phys., 60, 4262.,(1986).

22 G.J.Kintz, R. Allen, and L. Esterowitz, Appl. Phys. Lett., 50, 1553,(1987).

23 E.Desurvire, J.R. Simpson, and P.C. Becker, Optics Lett., 12, 888,(1987).

24 F.Auzel, J. Lumin.., 31/32, 759,(1984).

25 F.Auzel, S. Hubert, and D. Meichenin, Appl. Phys. Lett., 54, 681,(1989).

26 D.L.Dexter, J. Chem. Phys., 21, 836,(1953).

27 V.V.Ovsyankin, , "Spectroscopy of collective states and cooperative transitions in disordered rare-earth activated solids", in "Spectroscopy of Solids Containing Rare-Earth Ions", Eds A.A. Kaplyanskii and R.M. Macfarlane, (North-Holland Amsterdam 1987.), p. 405.

28 M.Stavola, and D.L. Dexter, Phys. Rev., B20, 1867(1979).

29 J.C.Vial, R. Buisson, F. Madeore, and M. Poirier, J. de Phys., 40, 913.,(1979).

30 F.Auzel, D. Meichenin, F. Pellé, and P. Goldner, Optical Mat.,4, 35,(1994).

31 N.J.Cockroft, G.D. Jones, and R.W.G. Syme, J. Lumin., 43, 275,(1989).

32 F.Pellé, and P. Goldner, , Phys. Rev., B48, 9995,(1993).

33 F.Auzel, , Rare-Earth Spectroscopy, Ed. B. Trzebiatowska, J. Legendziewicz and W. Strek (World Scientific, Singapore 1985), p. 502.

34 F.Auzel, "Properties of highly populated excited states in solids : superfluorescence, hot luminescence, excited state absorption", in "Optical Properties of Excited States in Solids", Ed. B. Di Bartolo, (Plenum Press, New York 1992), p. 305.

35 F.Auzel, "Materials for Ionic Solid State Lasers", in "Spectroscopy of Solid State Laser-type Materials", Ed. B. Di Bartolo, (Plenum Press, New York 1987), p. 293.

36 A.W.Kueny, W.E. Case, and M.E. Koch, J. Opt. Soc. Am., B6, 639,(1989).

37 N.J. Krasutsky, J. Appl. Phys., 54, 1261,(1983).

38 W.Lenth, and R.M. Macfarlane, J. Lumin.,.45, 346,(1990).

39 U.Oetliker, M.J. Riley, P.S. May, and H.U. Güdel, J. Lumin., 53, 553,(1992).

40 H.Ni, and S.C. Rand, Opt. Lett., 17, 1222,(1992).

41 N.Pelletier-Allard, and R. Pelletier, Phys. Rev., B26, 4425,(1987)

42 M.F.Joubert, S.Guy and B.Jacquier Phys. Rev. B48, 10031, (1993).

43 A.W.Kueny, W.E.Case, and M.E.Koch, J. Opt. Soc. Am. B10, 1834, (1993).

44 P.Goldner and F.Pellé, submitted to Phys. Rev. (1994).

45 A.S.L.Gomes, G.S. Maciel, R.E. de Araujo, L.H.Acioli, and C. B. de Araujo, Opt. Commun. 103, 361, (1993).

46 F.Auzel, Y.H. Chen, and D. Meichenin, ICL'93, Storrs, CN, USA, 9-12 August 1993, and J. Lumin., 60/61, 692,(1994).

47 Y.H.Chen, and F. Auzel, Electron. Lett., 30, 323,(1994).

48 F.Auzel, and Y.H.Chen, J. Non-Crystal. Sol.,184, 57 , (1995).

49 Y.H.Chen, and F.Auzel,J. Phys.D: Appl. Phys.,28, 207, (1995).

50 F.Auzel,and Y.H.Chen, J.Lumin. to be published (1995).

51 R.J.Mears, L. Reekie, S.B. Poole, and D.N. Payne, Electron. Lett.,22, 159,(1986).

52 J.Y.Allain, M. Monerie, and H. Poignant, Electron. Lett., 25, 318,(1989).

53 T.J.Whitley, C.A. Millar, R. Wyatt, M.C. Brierley, and D. Szebesta, Electron. Lett., 27, 785,(1991).

54 S.G.Grubb, K.B. Bennett, R.S. Cannon, and W. F. Humer, Electron. Lett.,28, 124,(1992).

55 R.G.Smart, D.C. Hanna, A.C. Tropper, S.T. Davey, S.F. Carter, and D. Szebesta, Electron. Lett. 27, 1308,(1991).

56 F.Auzel, Phys. Rev., B13, 2809,(1976).

57 F.Auzel, and Y.H.Chen, Opt. Quant. Electon., 26, 559, (1994).

58 R.Reisfeld,in "Spectroscopy of Solid State Laser Type Materials", ed. B.Di Bartolo (Plenum New-York 1987), p.343.

59 F.Auzel, and Y.H. Chen, J. Lumin., 60/61, 101,(1994).

60 H.Ibrahim, D.Ronarc'h, M.Guibert, H.Poignant, L.Rivoallan, and J.F.Bayon, in "Proc. 8th Int. Symp.on Halide Glasses", (Perros Guirec, France,1992), p.463.

61 Y.H.Chen, and F.Auzel, Electron.Lett.,30, 1602, (1994).

62 T.Komukai, T. Yamamoto, T. Sugawa, and Y. Miyajima, Electron. Lett. 28, 830,(1992).

BROAD BAND CENTERS APPLIED FOR LASER MATERIALS: EXAMPLE OF TETRAHEDRALLY COORDINATED CENTERS

Georges Boulon

Laboratoire de Physico-Chimie des Matériaux Luminescents
Université Claude Bernard Lyon I
Unité de Recherche Associée au CNRS n° 442
43, boulevard du 11 Novembre 1918 - Bât. 205
69622 Villeurbanne cedex - France

ABSTRACT

This paper reports the main spectroscopic properties of what is now interpreted as Cr^{4+} and Mn^{5+} tetrahedrally active centers in crystals. Such systems are prospected for potential applications of tunable laser materials and saturable absorbers.

1. INTRODUCTION

Nowadays intensive research is being made on solid-state materials for application in laser cavities. Efforts are focused both on solid-state laser-type materials such as active oscillators or amplifiers for continued and pulsed operations and also on saturable absorbers such as passive mode-locking and Q-switching systems. In this approach, there is considerable interest n doped-single crystals which are characterized by absorption and emission broad bands in such a way that they can compete with dye-lasers or saturable dyes suitable for mode-locking. The essential reason for this tendency is governed by the higher stability of inorganic solid-state media rather than organic liquid ones. The most important consequence in the near future should be the appearance of compact all-solid-state sources. Among active centers, those belonging to transition metal ions are very well adapted to this objective. Optical properties are associated with 3d-configuration of activators which is strongly influenced by the crystalline environment of the host. In this family, electron-phonon interaction induces vibronic transitions between both electronic and vibrational states. The most important consequences is the large tunability of vibronic solid-state lasers in such a wide spectral range that they are also potential materials for the generation of extremely short bandwidth-limited mode-locked pulses in the picosecond and femtosecond time domains.

Until now, we must remember that transition metal ion-doped vibronic solid-state laser-type materials already exist and are commercialized in the $3d^n$ configuration with $Ti^{3+}(n=1)$, $Cr^{3+}(n=3)$ and $Co^{2+}(n=7)$. Inside the crystalline structure of the hosts, the activator ions are occupying octahedral or quasi-octahedral symmetry sites. We can mention Ti^{3+}-doped saphire ($Al_2O_3{:}Ti^{3+}$), alexandrite ($BeAl_2O_4{:}Cr^{3+}$), fluorides such as Cr^{3+}-doped LiCAF($LiCaAlF_6{:}Cr^{3+}$) and Cr^{3+}-doped LiSAF($LiSrAlF_6{:}Cr^{3+}$) and to a lesser extent $MgF_2{:}Co^{2+}$. Due to high local symmetry having an inversion center, all observed optical bands are interpreted under parity-forbidden intraconfigurational transitions. They are, in fact, only allowed from odd-vibrations coupling of the host and they are so-called forced-dipolar electric transitions characterized by oscillator strengths of roughly 10^{-7}-10^{-6} giving rise to stimulated emission cross-sections not higher than 10^{-19} cm^2, excepted Ti^{3+}-doped saphire laser. So laser gains are relatively weak.

This is the main reason why special attention is devoted to another kind of transition metal ion, more precisely Cr^{4+} and Mn^{5+} ($3d^2$ configuration) located in tetrahedral or quasi-tetrahedral crystalline environments having no inversion center. The consequence for optical spectra is the appearance of allowed electric dipole transitions. Thus the odd components of the local crystal field allow mixing between the wave-functions of the $3d^2$ electron configuration and those of higher energy with different parity. Such optical transitions are not forced by coupling with odd parity lattice phonons as in the case of ions in an octahedral environment and so, the intensities can be high.

Considering that the approach of octahedrally coordinated centers has been extensively related in the past, the main goal of this paper is to give only the approach with tetrahedrally coordinated centers applied to laser materials.

2. CR^{4+}-DOPED CRYSTALS : SPECTROSCOPIC DATA AND INTERPRETATION

The search for new Cr^{4+}-doped laser systems started only a few years ago after laser emission around 1.2 μm was discovered in chromium-doped forsterite (Mg_2SiO_4) [1]. That was the starting point of many discussions and controversies. At that time, it was thought that the emission spectra were due to Cr^{3+} ion in the octahedral environment of the Mg^{2+} substituted ion [2]. However, more complete studies have shown that chromium ions can be incorporated both as Cr^{3+} in the Mg^{2+} octahedrally coordinated centers by assuming compensation effects in the network and as Cr^{4+} in the Si^{4+}

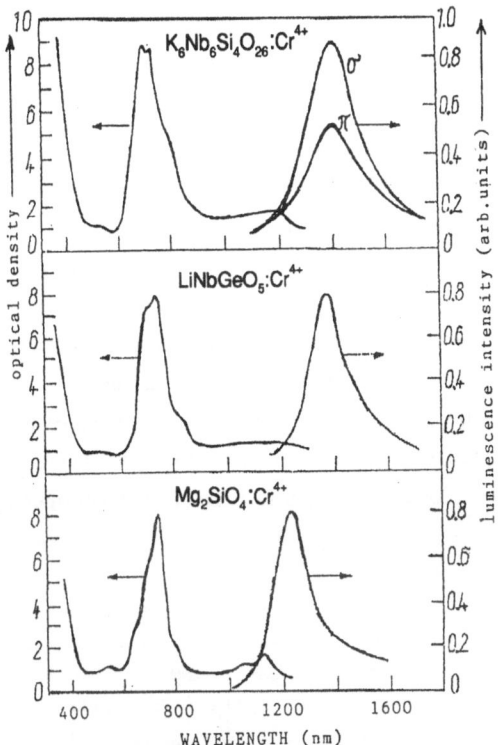

Fig. 1 - Shapes of the absorption and fluorescence spectra of Cr^{4+} ion-doped crystals at room temperature.

tetrahedrally coordinated centers of an Mg_2SiO_4 structure . The same interpretation was also made with chromium-doped $Y_3Al_2Al_3O_{12}$(YAG), Cr^{3+} substituting Al^{3+} octahedrally coordinated cation, and Cr^{4+} in the place of Al^{3+} tetrahedrally coordinated cation, with the additional introduction of Ca^{2+} for a charge compensation effect, two Al^{3+} neighbor cations being substituted by one Cr^{4+} and one Ca^{2+} cations.

Typical absorption and emission spectra are shown in Fig.1 at room temperature for three chromium-doped crystals : Mg_2SiO_4, $LiNbGeO_5$ and $K_6Nb_6Si_4O_{26}$ [3.4]. Vibronic broad bands are clearly seen with some shifts depending on the crystal field strength active on the center. Generally, laser tunability has been demonstrated at room temperature between 1130 and 1360 nm in Mg_2SiO_4:Cr^{4+} (stimulated emission cross-section $\sigma=1.9.10^{-19}$ cm^2 and life-time $\tau=2,7$ μs).and between 1360 and 1520 nm in YAG:Cr^{4+}-Ca^{2+} ($\sigma=3.3.10^{-19}$ cm^2 and $\tau=4,1$ μs). Although research of new Cr^{4+}-doped crystals is intense, these are the only two laser materials which are commercialized.

This spectroscopic information can be compared with those of other doped-crystals which have been reported in reference [5]. The similarities are very strong and we have to mention that laser application can take advantage of the tunability potential as can be seen in Figure 2 for different cristalline structures such as olivine, gehlenite, sillimanite and garnet.

The first argument in favor of the Cr^{4+} center is the strong absorption coefficient in the visible spectral range (few cm^{-1} to 15 cm^{-1} depending on the polarisation state of the pump beam) and also the high intensities of the emission spectra probably related to tetrahedral coordination of Cr^{4+} ions in sites having no inversion symmetry. In order to confirm the nature of chromium ions, two types of research have been attempted either by doping crystals with only one tetrahedrally coordinated center without any octahedrally coordinated center which perturb the interpretation of spectroscopic measurements, or by playing with the growth atmosphere and oxidizing annealing.

These two approaches have clearly pointed out the presence of Cr^{4+} ions. The spectra of $Ba_2MgGe_2O_7$, $Ca_2MgSi_2O_7$, $Bi_{12}SiO_{20}$, $CaGeO_3$ where only one 4+ oxidazion state exists look very similar to those of Cr^{4+}-doped crystals in which Cr^{3+}

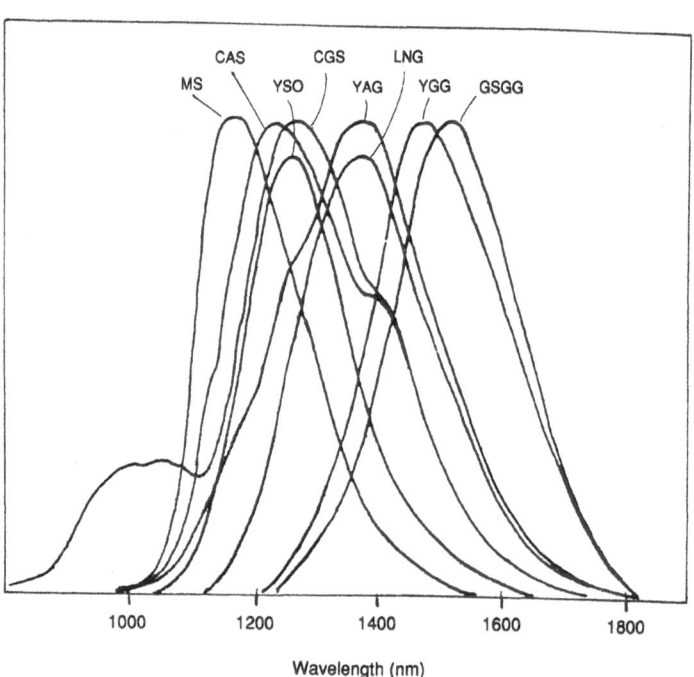

Wavelength (nm)

Fig. 2 - Evolution of the $^3T_2 \rightarrow {}^3A_2$ broad band emission at room temperature for different types of Cr^{4+}-doped crystals :
MS : Mg_2SiO_4 - CAS : $Ca_2Al_2SiO_7$ - YSO : Y_2SiO_5 - LNG : $LiNbGeO_5$
CGS : $Ca_2Ga_2SiO_7$
YAG : $Y_3Al_5O_{12}$ - YGG : $Y_3Ga_5O_{12}$ - GSGG : $Gd_3Sc_2Ga_3O_{12}$
The maximum is shifted to high wavelength when the distance Cr^{4+}-O^{2-} is increasing in agreement with Tanabe-Sugano diagram.

can be substituted. In addition two other experiments are very convincing : the evolution of the absorption spectra both of chromium-doped forsterite (Mg_2SiO_4) and chromium-doped glasses. The evolution of the absorption spectra with the growth atmosphere and oxidization annealing applied to forsterite shows bands around 860 and 1600 nm associated with Cr^{2+} ions at low oxigen pressures and at higher oxygen pressures, bands at around 500 nm and 700 nm are associated with Cr^{3+} ions where as Cr^{4+} ions are assigned by bands around 700 nm and 1000 nm as was expected [6-7].

The same type of conclusion has been obtained with alumino-silicate glasses [8]. Under reducing conditions, the absorption spectra show both $^4A_2 \rightarrow ^4T_2$ broad band near 650 nm and $^4A_2 \rightarrow ^4T_1$ broad band near 450 nm of Cr^{3+} ion. At room temperature, the $^4T_2 \rightarrow ^4A_2$ broad band is observed between 800 nm and 1400 nm with a maximum at 1100 nm and at 12 K, the usual $^2E \rightarrow ^4A_2$ line can be seen near 700 nm in excellent accord with Cr^{3+}-doped glass spectroscopy. Under weak oxidizing conditions to favor a 4+ valence state, absorption and emission drastically change. It is possible to recognize Cr^{4+}-broad bands between 600 nm and 1100 nm in absorption and between 1100 nm and 1700 nm in emission.

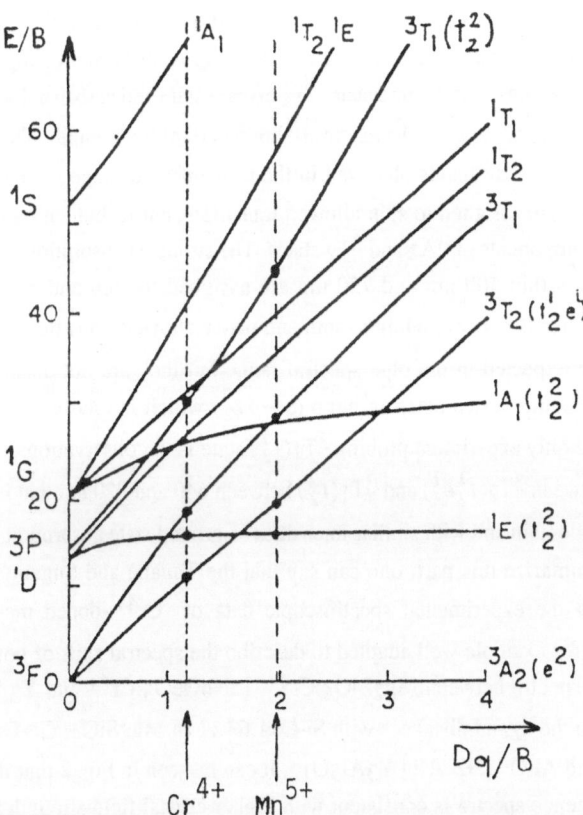

Fig. 3 - Tanabe-Sugano energy level diagram for $^3d_2$ ions in tetrahedral symmetry and average positions of the crystal field strength for Cr^{4+} and Mn^{5+} active centers.

If necessary, another argument has been presented by EPR relaxometry : the ground state is split into three components separated by about 2.1 and 0.4 cm^{-1} respectively confirming the electronic structure of the Cr^{4+} center [9-10].

As with transition metal ions, we need to interpret the absorption and emission spectra with the help of the Tanabe-Sugano diagram [11] which has been reported in Fig. 3 for $^3d_2$ ions in tetrahedral symmetry, also valuable for $^3d_8$ ions in octahedral symmetry. In this approximative schematic representation, the Cr^{4+}-centers correspond to rather low-crystal field intensity.

Within this model, the ground state is a spin triplet $^3A_2$ (e^2 electron configuration) and the first excited state is a triplet $^3T_2$ ($t_2^1 e^1$ electron configuration). Above this state we can see singlet ^1E (t_2^2 electron configuration) roughly independent of the crystal field strength and also another triplet state labelled $^3T_1$. The use of these states are obviously of the first importance to interpret the spectroscopic data. We wish to note at this point the great analogy with the evolution of the two first $^4T_2$ and ^2E energy levels of Cr^{3+} ion with a crossing point which leads to very wellknown spectral behavior of laser materials such as ruby Al$_2$O$_3$:Cr^{3+}, alexandrite BeAl$_2$O$_4$:Cr^{3+}, L AlF$_6$:Cr^{3+}, garnet Y$_3$Al$_5$O$_{12}$:Cr^{3+}, Gd$_3$Ga$_5$O$_{12}$:Cr^{3+} or Gd$_3$Sc$_2$Ga$_3$O$_{12}$:Cr^{3+}.

Now, by using configuration coordinate diagrams of Fig. 4, this is realistic way to understand tunable laser emission in Cr^{4+}-doped crystals at room temperature within the four-level scheme : $^3A_2$ ground state, $^3T_2$ excited state and associated vibronic levels, the Stokes shift depending on the electron-phonon coupling strength. These transitions correspond to the weak bands observed in the near infrared range around 800 nm and 1200 nm. They are assigned to spin allowed transitions but forbidden by rules between crystal field components of $^3A_2$ and $^3T_2$ states. The strongest absorption bands observed in the visible within 500 nm and 750 nm are assigned to spin and of $^3A_2$ and $^3T_1$ ($t_2^1 e^1$). Other bands corresponding to spin and parity allowed transitions to $^3T_1$ ($t_2^1 e^1$) state could be expected in the blue spectral range but they are not observed probably because they require a two-electron jump ($e^2 \rightarrow t_2^2$) which is totally excluded. To our knowledge the only experiment probing $^3T_1(t_2^2)$ state is the observation of excited state absorption between $^3T_2(t_2^1 e^1)$ and $^3T_1(t_2^2)$ between 640 and 820 nm that is to say in the same spectral domain and with similar intensities as ground state absorption [5-12].

To summarize this part, one can say that the Tanabe and Sugano model sheds some light on the experimental spectroscopic data of Cr^{4+}-doped materials. This description is for example well adapted to describe the spectral shift of both absorption and emission spectra between Mg$_2$SiO$_4$:Cr^{4+}, LiNbGeO$_5$:Cr^{4+} and Y$_3$Al$_5$O$_{12}$:Cr^{4+} with respect to the ligand distances with Si-O=1.63 $\mathring{A}$ in Mg$_2$SiO$_4$, Ge-O=1.739 $\mathring{A}$ in LiNbGeO$_5$, and Al-O=1.92 $\mathring{A}$ in Y$_3$Al$_5$O$_{12}$. It can be seen in Fig. 2 that the large shift of the fluorescence spectra is consistent with higher crystal field strength in Mg$_2$SiO$_4$ than in LiNbGeO$_5$, itself slightly higher than Y$_3$Al$_5$O$_{12}$ and Garnets.

We must add that the interest of Fig. 5 is to show some vibrational structures which need new complete experiments at low temperature to definitely assigne Zero-Phonon line of $^3A_2 \Leftrightarrow {}^3T_2$ vibronic broad band transition. The first observations seem to give the Zero-Phonon line positions at 9174 cm^{-1} (1094 nm) in Mg$_2$SiO$_4$:Cr^{4+}, 8709 cm^{-1} in Y$_2$SiO$_5$:Cr^{4+}, 7815 cm^{-1} (1280 nm) in YAG and 7752 cm^{-1} in Ba$_2$MgGe$_2$O$_7$.

We also want to say that this interpretation should be confirmed by both temperature life-time dependence and temperature fluorescence intensity dependence. We have not succeeded in describing them accurately but such experimental work illustrated in Fig. 6 has to be enhanced to be discussed. It is probable that the influence of ^1E level

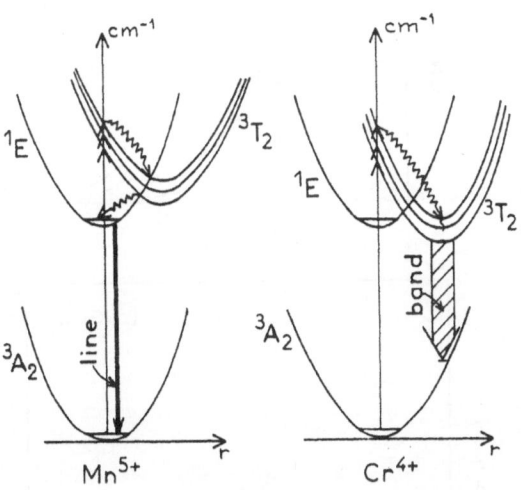

Fig. 4 - Configuration coordinate diagrams involving the ground state and the first excited states for Cr^{4+} and Mn^{5+}.

in the vicinity of $^3T_2$ level is strong, as it is shown in Fig. 3 and Fig. 7 by a simplified diagram, and the spectroscopic data would be described in terms of $^1E + {}^3T_2$ level mixing as we did for Cr^{3+}-doped garnets [13-14-15]. So, excited state dynamics of Cr^{4+}-centers should also be perturbed by such state mixing in the evaluation of quenching phenomena and thus of the radiative life-times and the stimulated emission cross-sections.

Present performances of Cr^{4+}-doped lasers based on Mg$_2$SiO^4 and Y$_3$Al$_5$O$_{12}$ have been given recently [16]. The tuning range for Mg$_2$SiO$_4$:Cr^{4+} is 1200-1300 nm with an output power of 0,7 W when the crystal is pumped with 7 W from a CW Nd^{3+}-doped YAG laser at 1064 nm. In the same pumping conditions, the tuning range of Y$_3$Al$_5$O$_{12}$:Cr^{4+} is 1410-1580 nm with a similar output power.

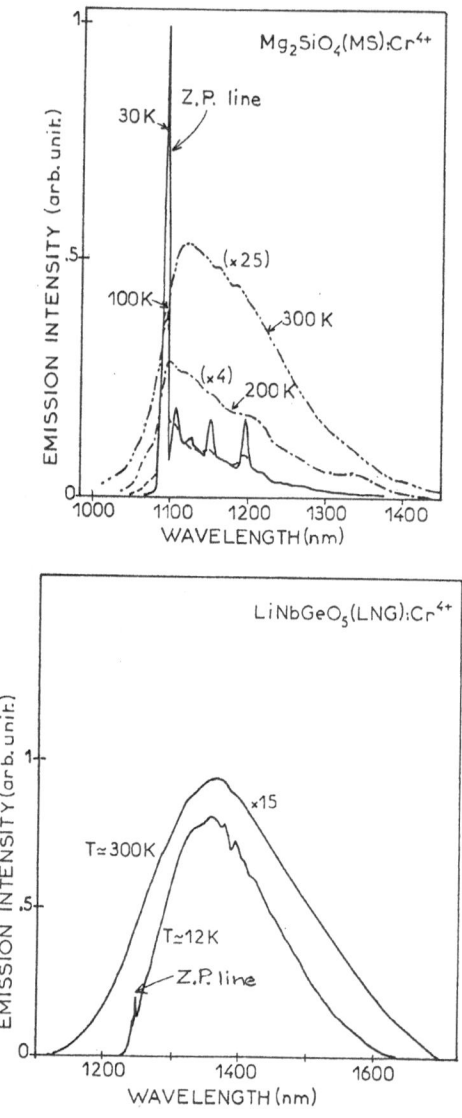

Fig. 5 - Low and high temperature emission spectra of Cr^{4+}-doped crystals with higher crystal field in Mg2SiO4 than in LiNbGeO5. Z.P. line means Zero-Phonon line of the $^3T_2 \rightarrow {}^3A_2$ transition

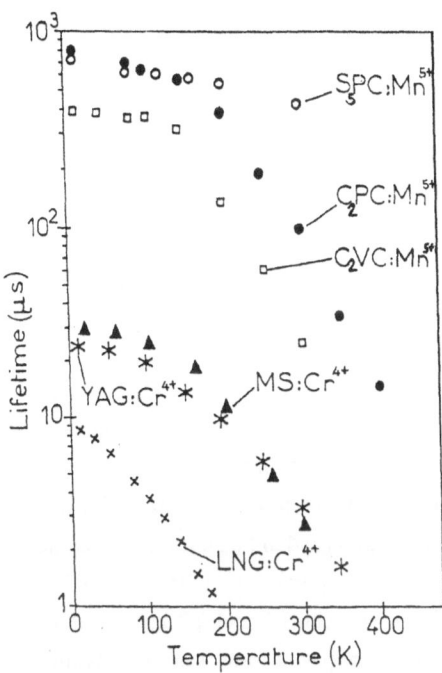

Fig. 6 - Temperature dependence of the infrared fluorescence decay times of Cr^{4+} and Mn^{5+} active centers in different crystals

YAG : $Y_3Al_5O_{12}$; MS : Mg_2SiO_4 ; LNG : $LiNbGeO_5$

S_5PC : $Sr_5(PO_4)_3Cl$; C_2PC : Ca_2PO_4Cl ; C_2VC : Ca_2VO_4Cl .

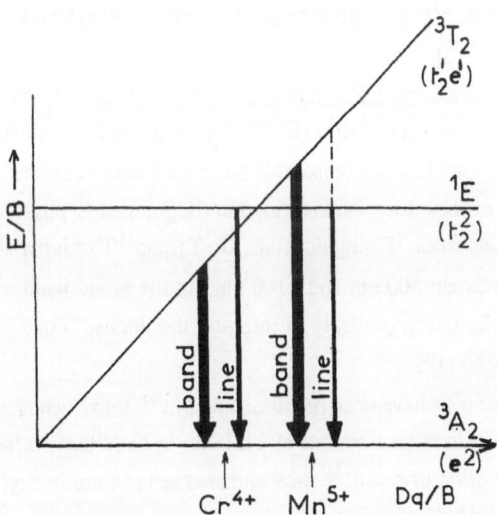

Fig. 7 - Simplified Tanabe-Sugano diagram for Cr^{4+} and Mn^{5+} ions to understand the emission processes

3. MN^{5+}-DOPED CRYSTALS : SPECTROSCOPIC DATA AND INTERPRETATION

The natural approach in the search of new laser materials is to find isoelectronic active centers allowing several possibilities of laser emission like spectral tunability, variation of the excited state life-times or higher output intensities. So it was logical tendency to extend Cr^{4+} ion spectroscopic properties to those of Mn^{5+} isoelectronic ion which can be substituted with 5+ cations in some hosts. First of all, we wanted to justify the reliability of the above interpretation to see any coherence in the spectroscopic properties and secondly, these new systems could offer possibilities for the laser application such as broad-band laser tunabilities in a more extended wavelength domain and also for their potentialities to be saturable absorbers.

The experimental data recorded for Mn^{5+} ion in Fig.8 for ground state absorption, in Fig. 9 for emission and in Fig. 6 for life-times are consistent with Tanabe-Sugano diagram by assuming higher crystal field strength. The average position of the usual Dq/B ratio is found between 1.9 and 2.2 instead of 1.2 to 1.6 for Cr^{4+} [5]. It means position on the other side with respect of the crossing point. So we can expect higher energy of the absorption spectra and consequently rather higher energy of the emission spectra with the occurence of the singlet ^1E metastable state, final state before emission and then longer lifetime from this level. All predictions were observed mainly in phosphates and vanadates where Mn^{5+} cation can be substituted in the place of P^{5+} or V^{5+} cations respectively. The positions of the highest ^1E energy component and its lifetime τ have been measured at 8700 cm^{-1} in Ca2PO4Cl (τ=100 μs at 300 K, 800 μs at 10 K), 8685 cm^{-1} in Ca2VO4Cl, 8572 cm^{-1} in Sr2VO4Cl, 8635 cm^{-1} in Sr5(VO4)3F (τ=450 μs at 300 K, 800 μs at 12 K) and 8498 cm^{-1} in Ba5(VO4)2 (τ=430 μs at 300 K, 1.1 ms at 11 K) [5-12].

It is worthwhile to know more about excited state dynamic in Mn^{5+} ion by recording excited state absorption (ESA). Fig. 8 gives ESA spectra for Sr5(VO4)3F promising laser crystal having comparable cross-section with ground state absorption one [17]. As it was expected by Tanabe-Sugano diagram, it is possible to point out spin allowed transitions from ^1E singulet state to $^1T_1$ and $^1T_2$ higher singulet states in the spectral range between 800 nm and 1100 nm but the broad band near 700 nm has not been ascribed. More, it is probable to interpret the intense broad band at 400 nm to a charge transfer band [12].

A few laser tests have been performed on Mn^{5+} active center at room temperature. We can mention gain measurements at 1150 nm in Ca2PO4Cl:Mn^{5+} with stimulated emission cross-section of σ=10^{-19} cm^2 and also at 1164 nm in Sr5(VO4)F:Mn^{5+} with the same value of σ [12].

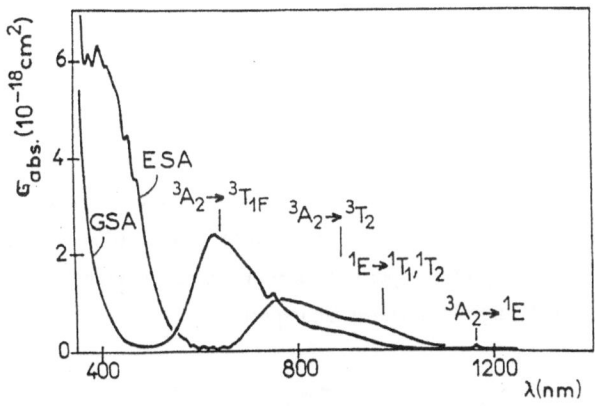

Fig. 8 - Ground state absorption (GSA) from $^3A_2$ level and excited state absorption (ESA) from ^1E level of Mn^{5+}-doped SVAP ($Sr_5(VO_4)_3F$ in σ polarization type.

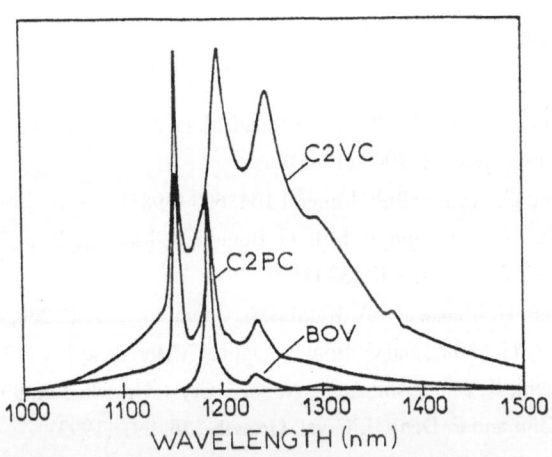

Fig. 9 - Emission spectra of Mn^{5+}-doped crystals at room temperature
C2VC : $Ca_2(VO_4)Cl$ C2PC : $Ca_2(PO_4)Cl$
BOV : $Ba_3(VO_4)_2$
Mixing between $^1E \rightarrow {}^3A_2$ line and $^3T_2 \rightarrow {}^3A_2$ broad band can be seen associated with vibrational structure.

CONCLUSION

What was puzzling a few years ago about new absorption and emission spectra in the near infrared is now well interpreted to the unusual 4+ valence state of chromium ion in tetrahedral coordination. This conclusion was confirmed by new data obtained with Mn^{5+} isoelectronic ion. Tanabe-Sugano diagram can be efficient to understand both ground state absorption excited state absorption and emission mechanisms. $Mg_2SiO_4:Cr^{4+}$ and $Y_3Al_5O_{12}:Cr^{4+}$ crystals are capable of producing CW output powers in excess of one watt when pumped with 7 W of 1064 nm power. Further investigations should improve the performances of these type of tunable solid state lasers and, then, propects for new host lattices are in progress.

ACKNOWLEDGMENTS

We are grateful to R. Moncorgé, H. Manaa and A. Kaminskii for helpful discussions on this field.

REFERENCES

[1] S.K. Gayen Petricevic, R.Alfano, K.Yamagishi, H. Anzai and Y. Yamaguchi, Appl. Phys. Lett. **52**, 1040 (1988)

[2] H. Rager and G. Weiser, Bull. Mineral **104**, 603 (1981)

[3] A. Kaminskii, A. Butaschin, B. Mill, G. Boulon, R. Moncorgé and J. Garcia-Solé, Phys. Stat. Sol. (a) **139** K 133 (1993)

[4] R. Moncorgé, H. Manaa and A. Kaminskii, Chem. Phys. Lett. **200**, 635 (1992)

[5] R. Moncorgé, H. Manaa and G. Boulon, Optical Materials **4**, 139 (1994)

[6] Y. Yamaguchi, K.Yamagishi ane Y. Nobe, J. Cryst. Growth **128**, 996 (1993)

[7] B. Hu, H. Zhu and P. Deng, J. Cryst. Growth **128**, 991 (1993)

[8] U. Hommerich, H. Eilers, W. M. Yen, J. S. Hayden and M. K. Aston, J. of Luminescence **60/61**, 119 (1994)

[9] J. Casas Gonzales, S.M. Jacobsen, K.R. Hoffman and W. M. Yen, Advanced Solid State Lasers Meeting, Hilton Head (1991), Proceed. series ed. G. Dubé and L. L. Chase **10**, 64 (1991)

[10] M.L. Meilman and M.G. Livoshitz, Advanced Solid State Lasers Meeting, Santa Fé (1992), OSA Proceed. series, eds. L. L. Chase and A. Pinto **13**, 17 (1992)

[11] S. Sugano, Y. Tanabe and H. Kamimura *in* Pure and Applied Physics **33** (1970), Academic Press, New York

[12] R. Moncorgé and H. Manaa, Ann. Chim. Fr., **20**, 241 (1995)

[13] M. Grinberg, A. Brenier, G. Boulon, C. Pédrini, C. Madej, J. of Luminescence **55**, 303 (1993)

[14] A. Brenier, G. Boulon, C. Pédrini and C. Madej, J. Appl. Phys. **71-12**, 6062 (1992)

[15] A. Brenier, A. Suchocki, C. Pédrini, G. Boulon and C. Madej, Phys. Rev. B, **46**, n°6, 3219 (1992)

[16] C. Pollock, D. Barker, J. Mass and S. Markgraf, IEEE J. of Selected Topics in Quantum Electronics **1**, n°1, 62 (1995)

[17] L. Merkle, Y. Guyot and B. Chai, J. Appl. Phys. **77**, 474 (1994)

COSMOLOGY: THE UNIVERSE IN EVOLUTION

G. V. Coyne, S.J.
Vatican Observatory
V - 00120 Vatican City State

ABSTRACT

There are three principal pieces of evidence which, when taken together, support the notion that the universe as a whole is evolving. These are: (a) the Hubble velocity-distance relationship; (b) the abundance of light elements in the universe; and (c) the three degree cosmic background radiation. The best explanation of these observations is that the universe began, if it had a beginning, in a hot dense state and that it has been expanding and cooling down ever since. In the course of that process, matter came to be out of energy, galaxies and stars formed, and you and I came to exist. In this process the emergence of organic material in general, and human beings in particular, seems to have required a very fine tuning of the evolutionary process itself. We do not have a definitive scientific explanation of that fine tuning and, therefore, we do not yet understand the linkage of the human being to cosmic evolution. But some interesting suggestions can be made.

INTRODUCTION

My intention in this paper is that it would serve as a kind of general background for discussion by presenting what we know of the origins and evolution of the physical universe as the matrix from which life has evolved. My intent, therefore, will be not to treat of the origins and evolution of life as such, although I will of necessity refer to life since its appearance in the universe is an important ingredient to understanding the nature of the universe itself. I will be speaking almost exclusively of Big Bang cosmologies since the evidence I will present points rather clearly in the direction of such evolutionary cosmologies as opposed to Steady State cosmologies. Note that I speak of Big Bang cosmologies in the plural because there are several versions, all of which have in common the following view of the origins and evolution of the universe: the universe began in a very dense and very high temperature state and with time it has been expanding and cooling down. Evolution, therefore, is an intrinsic and proper characteristic of the universe. Neither it nor any of its ingredients can be understood except in terms of evolution. Life has made a relatively late appearance considering the total age of the universe and there are three intriguing questions which it poses in terms of the evolution of the universe itself: (1) in terms of the evolution of the physical universe was it inevitable that life come to be; was it by chance; can it be understood; (2) is life unique to our planet; (3) is life at the level of intelligence and self-reflection an important factor in the future evolution of the universe.

DESCRIPTION OF A BIG BANG COSMOLOGY

Let us begin by describing the essential ingredients of a Big Bang cosmology and then present the observational evidence for it.[1] About fifteen billion (15 x 10^9) years ago[2] all of the matter and energy we know of was concentrated in a very small volume. It began to expand and cool rapidly. Let us concentrate for a moment on the first fractions of a second: at first there was only energy and then quarks, the building blocks of matter, came to be. It was only after one hundred thousandths (10^{-5}) of a second that protons and neutrons were formed. After 100 seconds the early-universe nucleosynthesis ended and the abundance of the elements in the universe, except for the nucleosynthesis which would take place in stars, was established. We will speak shortly of the abundance of the light elements in today's universe as evidence for the picture we are tracing. By this time the volume of the universe was about that of the solar system. In these early stages the number of atomic and sub-atomic particles was so high that the universe was a dense fog, opaque to energy transport. As the universe continued to expand and cool, more and more particles were able to combine to form atoms and molecules and the total number of scattering particles dropped rapidly. The fog began to lift and at an age of about three hundred thousand (3 x 10^5) years the universe became transparent. This is the epoch of the formation of the Cosmic Background Radiation (CBR) of which we will speak shortly. At this time the universe was about one thousandth (10^{-3}) its current volume. Although we are not certain as to whether clusters of stars first formed and then grouped into galaxies, the prevailing view is that galaxies began to form about one billion (10^9) years after the Big Bang and stars began to form in galaxies at about five billion (5 x 10^9) years. Although there are many theories, we are frankly quite ignorant about how galaxies formed. On the other hand we are quite knowledgeable as to the formation and evolution of stars. The Sun and solar system are relatively young, having been formed about 5 billion (5 x 10^9) years ago, but about five billion years after the first generation of stars were formed. The abundance of elements to form the terrestrial planets was provided by the nucleosynthesis which occurred in the first generations of stars. The Earth with the chemical abundances we know could not have been formed earlier in the life of the universe. The Sun will continue for another five billion years (5 x 10^9) in its steady output of energy derived from the conversion of hydrogen into helium. But then it will evolve into a red giant star and will engulf the Earth. The universe will either continue to expand and cool down or it will begin to collapse and heat up. Our current knowledge tells us that it is tantalizingly close to doing one or the other, but we do not know which.

OBSERVATIONAL EVIDENCE

Although there are a number of uncertainties about some of the details, the formation of galaxies, for instance, and about the exact time scales, the above description is very well established and represents a remarkable achievement since the beginning of this century due to wedding of theoretical and observational cosmology with high energy and elementary particle physics. I would like now to present the three principal observational facts which taken together give very high credence to a Big Bang cosmology such as that described above. These are: (1) the velocity-distance relationship for galaxies and clusters of galaxies, the so-called Hubble law; (2) the observed abundance of helium, lithium, beryllium, deuterium and other light elements; (3) the Cosmic Background Radiation.

Einstein assumed a static universe and invented the cosmological constant so that such a universe would be consistent with his Theory of General Relativity. Friedmann realized that such a universe was unstable and that the slightest perturbation would cause it to expand or contract. At about the time that these considerations were maturing V. M. Slipher, an observational astronomer at Lowell Observatory in Arizona, was collecting the first evidence that galaxies are actually moving apart. A few years later in 1929 Edwin Hubble, using all the measurements then available, showed that the amount of red-shift of the atomic lines in the spectrum of a galaxy was proportional to the galaxy's distance. If one interprets the red-shift as due to the recessional motion of the galaxy, then Hubble's law is exactly what one would expect to observe in an expanding universe. The name Big Bang applied to this phenomenon is not a very felicitous one. It was actually coined by Fred Hoyle, the principal proponent at that time of the competing Steady State cosmologies, in a clever, disparaging way, but it was so clever that it stuck. The problem is that Big Bang suggests an explosion which one must imagine as starting at a given place and sending matter out into space. This is not the case at all. In Einstein's General Relativity matter and space-time are intimately linked. In the Big Bang cosmologies it is the space-time framework that is expanding and this is revealed observationally by the increasing distance between galaxies. It is not that galaxies are moving away from one another; it is that the distance between them is increasing with time. And this effect will be seen by any observer wherever she/he is in the universe. To make the point, we might state that the Big Bang happened everywhere or nowhere; nowhere, if you consider that "when" it happened space-time did not exist; everywhere, if you wish to think of its results in today's expanding space-time structure.

It must be conceded that taken alone the Hubble law does not support Big Bang cosmologies against Steady State cosmologies. It is only necessary in a Steady State expanding universe to create matter as the universe expands so that its mean density remains the same. I see no difficulty with this, except that there are no observations which supporting a constant mean density. Some object to the continuous creation of matter but that is no more a problem for Steady State than it is for Big Bang cosmologies. If matter can be created once, it can be created many times (continuously). The mystery is still how it was created at all!

It is important at this point to note that the observational error associated with measuring a red-shift does not depend on distance, whereas the error in measuring distance increases enormously with increasing distance. For large distances we must make many assumptions about the uniformly mean brightness of certain kinds of objects, for instance, globular clusters, and we must apply assumed corrections for the aging of such objects. The distance limit for the so-called primary methods, where the assumptions I have just mentioned are rather reasonable, is about thirty million (3×10^7) light years, only two thousandths of the way back to the Big Bang. The distance errors cause an uncertainty in the slope of the Hubble law (the Hubble constant), which is roughly inversely proportional (a specific cosmological model with a given deceleration parameter must be assumed) to the age of the universe. Recent observations with Hubble Space Telescope (HST), for instance, indicate a change in the distance scale which would essentially decrease the currently estimated age of the universe by a factor of two. In that case we would have some globular clusters of stars, whose ages are known by methods completely independent of the Hubble law, with ages about twice that of the universe itself. A problem, indeed!

A second piece of evidence supporting a Big Bang cosmology is the observed abundance of the light elements in the universe. The observed abundances far exceed those that could be produced by stellar nucleosynthesis. It was actually by using nuclear physics data from the war effort that George Gamow in the 1940s began to consider the early universe as a thermonuclear reactor and thus offered the possibility to explain the formation of the light elements. The element production depends on the density and the range in density required to meet the observed abundances fits remarkably well with the otherwise known conditions in the early universe. All evidence indicates that the light elements were for the most part thus formed in the early universe and that the heavier elements were formed later and gradually in the thermonuclear furnaces of stars. This evidence appears to be incompatible with a Steady State universe.

The third piece of evidence is the observation of the Cosmic Background Radiation (CBR). Again Gamow predicted that this radiation should be observable as a result of the energy released when the early universe became transparent. As the universe expanded the temperature of this radiation would continuously decrease. And it would be thermal radiation, such as that given off by stars, but due to the expansion it would be at a very low temperature. The discovery of this radiation is somewhat serendipitous and interesting. In the 1940s Robert Dicke was developing the microwave radiometer, a device for detecting low levels of radiation, and the engineers in his group were receiving an unexplainable signal from all directions which they at first thought was some kind of instrumental noise. Recalling Gamow's prediction Dicke suggested that it might be the CBR and it was clearly identified to be such by Arno Penzias and Robert Wilson. Fossil radiation, if you will, had been detected from when the universe was only two hundred thousandths (2×10^{-5}) its current age. This remarkable discovery provided convincing evidence that we live in a Big Bang universe, one that is by its nature evolutionary. At the same time it provided some problems in that it was so remarkably uniform in all directions. How could the universe be so remarkably unperturbed at the time that it became transparent to radiation? And how could galaxies or stars ever come to be if there were no density anomalies whereby self gravity could begin its inevitable course of forming them?

Only in the past few years with the Cosmic Background Explorer (COBE) in space have small fluctuations (one part in about 100,000) in the CBR been discovered. With time these small fluctuations became the clumps in the universe that we call galaxies, stars and ourselves. On a small scale, of course, we see the universe as clumpy, and this is comforting since we are part of the clumps and we require other clumps to live on. But on a large scale out to the limits of the observed universe (about 10 billion light years) the average density of visible matter changes very little, about one tenth of a percent.

As a final piece of evidence that we live in an evolutionary universe, rather than a Steady State one, we may compare the more distant parts of the Universe with the less distant parts, realizing that since light travels at a finite velocity, we are seeing far back in time when we see distant objects. In principle this is a straight forward experiment; in practice it is quite difficult, since we must observe to large distances and, therefore, to very faint objects in order to detect evolutionary effects. But with modern large telescopes and with the HST we have succeeded. For instance, distant rich clusters of galaxies show many bluish galaxies in which star formation is still active; whereas nearby clusters show reddish galaxies in which star formation has ceased. It is suspected that collisions between galaxies, which would have been much more frequent in an earlier denser stage of the universe, may trigger enhanced star formation.

SOME PROBLEMS

The great success of Big Bang cosmologies during this century may conceal some of the principle problems still remaining. It would be useful to summarize these. We have alluded already to the determination of the distance scale and, therefore, to the time scale for the expanding universe. It is expected that with large ground-based telescopes and with the HST these problems will soon be resolved. We have also spoken of our ignorance about galaxy formation. Again with large telescopes we are getting increasingly closer to being able to actually observe the epoch of galaxy formation, some billions of years after the Big Bang, and so a great deal of direct observational detail on galaxy formation may be forthcoming. Two increasingly difficult problems are the quantity and nature of dark matter in the universe and whether the universe is open or closed. The two problems are related. From the observed rotation curves of spiral galaxies and from the gravitational binding of rich clusters of galaxies, it has become increasingly clear that about ninety percent of the gravitating matter in the universe does not radiate. What is the nature of this dark matter which is so predominant? We have already noted that the universe is tantalizingly close to being either open or closed. There are two approaches to trying to resolve which is the case. If we could measure the curvature of the Hubble law at large distances, then with certain reasonable theoretical assumptions, we could resolve whether the universe was open or closed. As we look at large distances, we are looking back in time and we can compare the observed expansion rate then to that in more recent times. We can, therefore, in principle, determine the rate of deceleration of the expanding universe and whether such a rate is sufficient to close the universe. The observational problem is precisely the one which we have discussed previously. Errors in the measurement of large distances will not allow us to determine accurately enough the curvature of the Hubble curve at large distances, or equivalently at long look-back times.

Another approach to determine the closure or not of the universe is to measure the mean density of the universe. If the mean density is greater than a critical value, estimated to be about five H atoms per cubic meter, then the universe is closed. This is a very sound approach, based on the simple working of gravity, but how are we to measure this mean density when it is estimated that ninety percent of the matter in the universe is not radiating? We are left with the uncertainty as to whether the universe will end at all and, if so, whether with a crunch or a whimper.

LIFE IN THE UNIVERSE, SOME CONSIDERATIONS

As I noted in the beginning, my intent has been to give a general background to eventual discussions by tracing the principal characteristics of a Big Bang cosmology. I would like now to locate the emergence of life in this panorama and propose a few considerations about it. Life is thought to have emerged about three billion (3×10^9) years ago in its first microscopic forms. This was about twelve billion years after the Big Bang and about seven billion years after the formation of the first stars. Why did it take so long for life to emerge? In order to provide the chemical abundances required for life it is estimated that three generations of stars were required. It is only through nucleosynthesis in stellar interiors that the heavier elements can be created and at the death of a star this material is regurgitated to form the matrix for a new generation of stars. The lifetime of a star depends upon its total mass and can vary from several millions of years for a very massive star to tens of billions of years for lower mass stars.[3] At any rate it took about ten billion years of stellar evolution to produce carbon, nitrogen, oxygen, etc. I repeat that the universe is by its nature evolutionary and it had to evolve to be big and old before it could contain us. I was tempted to say, "in order to contain us", but that would have introduced a finality which may not be justified scientifically.

The emergence of life in the universe poses, of course, a serious scientific problem to which I do not think an adequate answer has been given. Considering the fine tuning of the constants of nature and of physical laws that was required for life to emerge, we might ask how did it emerge at all. Life would have been impossible should anyone of several physical quantities had a different value.

One interesting case is the following one. One of the essential steps in stellar nucleosynthesis is the buildup of carbon 12 from helium. Two helium atoms form beryllium 8 which is unstable. However, before fissioning a certain amount of beryllium 8 captures another helium atom to form carbon 12 in an excited state, which then decays into a stable state of lowest energy by emitting a photon. But the capture of a helium atom by beryllium 8 is a resonant process, so that, if the energy level of excited carbon 12 were only slightly different, the rate of its formation would be greatly reduced. This, of course, would drastically effect not only the abundance of carbon 12 but also of oxygen, nitrogen and the other heavy elements essential to life, since these are produced in subsequent steps in stellar nucleosynthesis. To the best of my knowledge there is no theory which requires that the energy level of the excited state of carbon 12, spoken of above, must have the exact value it has. If it did not have that precise value, you and I would not be here.

This brings me back to the questions I asked at the beginning: (1) in terms of the evolution of the physical universe was it inevitable that life come to be; was it by chance; can it be understood; (2) is life unique to our planet; (3) is life at the level of intelligence and self-reflection an important factor in the future evolution of the universe. But these are consideration which go beyond the intent of this presentation and which, indeed, run the risk of carrying us off into non-scientific considerations.

REFERENCES

1. A more detailed description can be found in "The Evolution of the Universe" Peebles, J.E., Schramm, D.N., Turner, E.L., and Kron, R.C. in *Scientific American*, October 1994, Special Issue, p. 53.
2. Recent research has raised serious doubts about the age of the universe. It may vary from about 8 to 20 billion years. The problem is related to the distance scale of the universe and will be discussed later on.
3. For a classical treatment of this topic see: *Stellar Evolution and Nucleosynthesis* by Hubert Reeves (New York: Gordon and Breach, 1968).

UNIFICATION OF THE FUNDAMENTAL INTERACTIONS:

PROBLEMS AND PERSPECTIVES

G. Costa

Department of Physics, University of Padova
Istituto Nazionale di Fisica Nucleare, Padova
Via Marzolo, 8 – 35131 Padova (Italy)

ABSTRACT

The Standard Model for electroweak and strong interactions has been successfully tested, and it appears in good agreement with all available experimental
information. However, it cannot be considered as the final theory of elementary
particles; it still contains some elements of arbitrariness and some of its general
features remain unexplained. One of the most attractive ideas in going beyond
the Standard Model is the complete unification of all the fundamental interactions,
and many schemes of Grand Unified Theories have been proposed, some of which
contain a new theoretical ingredient: supersymmetry. While there are hints that
this may be the correct path, up to now there is no experimental evidence of signals of new physics. The present situation will be discussed and perspectives for
future investigations will be outlined.

INTRODUCTION

The main goal of Elementary Particle Physics is to arrive at a complete picture
of the microscopic world, in which all phenomena can be interpreted in terms of a
few fundamental interactions acting among elementary constituents.

In fact, the description of physical reality is not unique, since different ingredients are adopted according to the energy scale of the phenomena. The investigation
of the structure of matter shows a hierarchy of levels, each of which is, to some
extent, independent of the others. At a first level, ordinary matter is considered
as made of atoms composed of nuclei and electrons; at a smaller scale atomic nuclei reveal their structure, which is described in terms of nucleons (neutrons and
protons); at a much smaller scale, even nucleons appear to be composed of more
elementary constituents: the quarks.

At first sight, ordinary matter shows the presence of only three ingredients
at the third level of structure: electrons (e) and two kinds of quarks (u and d);
neutrinos (ν_e) appear in the β–decay of nuclei. In high energy reactions many
different states of matter are produced. One makes the following distinctions:

a) hadrons: they are bound states of quarks and they are separated into mesons
(bosons) and baryons (fermions);

by a function $\alpha(x)$ depending on the space–time co–ordinate x. Invariance under the local transformation

$$\psi(x) \to e^{i\alpha(x)}\psi(x) \ , \tag{3}$$

would require a peculiar modification in eq. (1):

$$\partial_\mu \to \partial_\mu + ie \, A_\mu(x) \ , \tag{4}$$

where e is the electron charge and A_μ is a four–vector which transforms according to

$$A_\mu(x) \to A_\mu(x) - \frac{1}{e}\partial_\mu \alpha(x) \ . \tag{5}$$

Then the expression of $\mathcal{L}_0(x)$ is replaced by

$$\begin{aligned} \mathcal{L}(x) &= \bar\psi(x)(i\gamma^\mu \partial_\mu - e\gamma^\mu A_\mu(x) - m)\psi(x) = \\ &= \mathcal{L}_0(x) - e\bar\psi(x)\gamma^\mu \psi(x) A_\mu(x) \end{aligned} \tag{6}$$

The modification consists in the addition of an interaction term, which couples the electron current

$$\mathcal{J}^\mu(x) = e\bar\psi(x)\gamma^\mu \psi(x) \tag{7}$$

to the four–vector field $A_\mu(x)$, which can be interpreted as the electromagnetic potential.

In order to complete the expression of the Lagrangian density (6) one has simply to add the free EM term $\mathcal{L}_0^{EM}$

$$\mathcal{L}^{EM}(x) = \mathcal{L}_0(x) + \mathcal{L}_0^{EM}(x) - \mathcal{J}^\mu(x)A_\mu(x) \tag{8}$$

where

$$\mathcal{L}_0^{EM}(x) = -\frac{1}{4}F_{\mu\nu}(x)F^{\mu\nu}(x) \tag{9}$$

where $F_{\mu\nu} = \partial_\mu A_\nu - \partial_\nu A_\mu$ is the EM field tensor. It is clear that invariance under eq. (5) prevents the presence of a mass term.

In conclusion, the case of QED teaches us that, while invariance under the global transformation (2) implies conservation of the electric charge, invariance under the local transformation (3) requires also the existence of a massless vector field.

Such conclusion can be extended to a more general case. Let us consider a quantum field theory (QFT) which is invariant under a gauge group G, i.e. the group transformations are imposed locally. If the group G is of order r, i.e. it has r generators, then invariance under G requires the introduction of r massless vector fields, transforming as the regular representation of G.

In order to build a QFT for the weak interactions we need, on phenomenological ground, three vector fields: there must be two vector bosons $W^\pm$ mediating the charged current interaction, and a neutral vector boson Z^0 coupled to the neutral current. The minimal symmetry group must have four generators, and the simplest choice is the group

$$G = SU(2) \otimes U(1) \tag{10}$$

If one relies on gauge symmetry, one encounters immediately the following difficulty: the vector fields have to be massless, and this cannot apply to the vector bosons $W^\pm$ and Z^0 which should be very heavy (even before the experimental determination of their masses, this fact was inferred from the short–range character of the weak interactions).

582

b) <u>leptons</u>: they have similar properties to those of electrons and neutrinos; they show no structure (up to present energies).

The analysis of the spectrum of hadrons revealed the existence of six different kinds of quarks (each of them is distinguished by a specific quantum number, called "flavour"). On the side of leptons, three charged states and three types of neutrinos were detected. Now we believe that there are three <u>families</u> of elementary particles, which share similar properties. These particles are listed in Table I, where also the values of the electric charge Q (in unit of the proton charge) and of the masses (in MeV=10^6 eV) are reported [1].

Table 1. Quarks and Leptons

	Q=2/3	Q=-1/3	Q=-1	Q=0
1st family	u (up) $m_u \sim 5$	d (down) $m_d \sim 10$	e $m_e = 0.511$	ν_e $m_\nu < 5 \times 10^{-6}$
2nd family	c (charm) $m_c \simeq 1500$	s (strange) $m_s \sim 200$	μ $m_\mu = 105.66$	ν_μ $m_\nu < 160 \times 10^{-3}$
3rd family	t (top) $m_t \simeq 175 \times 10^3$	b (bottom) $m_b \simeq 4.3 \times 10^3$	τ $m_\tau = 1.777 \times 10^3$	ν_τ $m_\nu < 29$

The mass spectrum looks rather random; in fact, the masses increase strongly in going from one to the next family, indicating a strong hierarchy. Just to have an idea of the order of magnitude, let us remember that the mass of the proton is $M_p = 938$ MeV, and note that the mass of the top–quark is like the mass of a nucleus of gold. The lighter particles are the neutrinos, for which only upper mass values are given; in principle, they could be massless, but this is still an open question.

Quarks and leptons interact among themselves, and one can distinguish four kinds of fundamental interactions:

1) <u>Electromagnetic</u>, acting among quarks and charged leptons;

2) <u>Weak</u>, acting among quarks, charged leptons and neutrinos;

3) <u>Strong</u>, acting only among quarks;

4) <u>Gravitational</u>, effective on all particles.

These interactions are specified by different properties, in particular the "range" and the "strength". They are described in terms of <u>fields</u>; the concept of field which is so useful at the classical level in the cases of the electromagnetic and gravitational forces, is successfully extended to quantum level.

Each field is specified by a given quantum, as indicated in Table 2. The quanta are all bosons; they have spin $S = 1$, except for the case of the graviton for which $S = 2$; they are massless, except for the bosons mediating the weak interactions, which are very heavy.

Table 2. Quanta of fundamental interactions

Interaction	Quanta	Spin	Mass
EM	photon (γ)	1	0
Weak	Weak bosons $(W^{\pm}, Z^0)$	1	large
Strong	gluons	1	0
Gravitational	graviton	2	0

On the other hand, quarks and leptons are fermions, all with spin $S = 1/2$. The concept of quantum field is extended also to these fermions, which in a way can be considered as the sources of the interactions. Consequently, they are interpreted as quanta of the corresponding fermion field.

Relativistic Quantum Field Theory (RQFT) provides a very powerful description; combining relativity with quantum requirements one obtains very important and general consequences.

First of all, for each particle there must exist an <u>antiparticle</u>, with exactly the same mass, and opposite sign of the electric charge and of the other relevant additive quantum numbers. Moreover, making use also of locality and causality, one can prove that all systems have to be invariant under the combined operation of PCT, where P stands for the space–inversion, C for the charge or particle–antiparticle conjugation and T for the time–reversal. The invariance does not hold, in general, for each of the separate operation. In fact, weak interactions show violation of parity and charge conjugation at a high degree. Even the invariance under CP appears to be slighly violated (together with time–reversal, since PCT must remain invariant) in some specific processes; this fact is of great interest also for cosmology.

The fundamental interactions exhibit different properties of invariance not only under the above discrete operations, but also under other kinds of symmetry, which in general imply continuous transformations. Symmetry is indeed another very important tool for investigating particle physics, as it will appear in the following sections.

ELECTROWEAK INTERACTIONS

We start from Quantum Electrodynamics (QED), which describes extremely well the electromagnetic interactions, and show how it can be extended in order to include the weak interactions.

The main ingredient is <u>gauge symmetry</u>, which implies invariance of the field Lagrangian under a set of <u>local</u> transformations. Global transformations, even if acceptable in principle, are not appropriate from the point of view of relativity, since they produce a rigid and simultaneous change in the whole field domain.

QED is the simplest gauge field theory, based on the Abelian group of phase transformations. If one considers the Lagrangian density for the free electron field $\psi(x)$:

$$\mathcal{L}_0(x) = \bar{\psi}(x)(i\gamma^{\mu}\partial_{\mu} - m)\psi(x) , \tag{1}$$

one immediately realizes that it is invariant under the global phase transformation

$$\psi(x) \rightarrow e^{i\alpha}\psi(x) , \tag{2}$$

where α is a real constant (the conjugate transformation is applied to the conjugate field $\bar{\psi}(x) = \psi^{+}(x)\gamma^0$). The invariance no longer holds if the constant α is replaced

Thus, on one side, gauge invariance does not seem to apply to the present case, since it implies massless vector fields; on the other side, it is a necessary ingredient of QFT, because it makes the theory "renormalizable".

The solution of the puzzle is based on the concept of <u>spontaneous symmetry breaking</u> [2] and on the Higgs mechanism [3], which have been used extensively in the study of the condensed matter systems. We limit ourselves here to the main results, referring to the existing literature on the subject [4,5,6] for a complete treatment.

The following situation may occur in a variety of systems: the Lagrangian of the system is invariant under a continuous symmetry group G while the ground state, i.e. the state of minimal energy, possesses a lower symmetry (described by a subgroup G' of G). In such case, one says that the symmetry has been broken spontaneously and, in the transition from a more to a less symmetrical configuration, a given number of <u>massless scalar bosons</u> appear (they are called Goldstone bosons); their number equals the number of the generators of G which are outside the subgroup G', i.e. it is equal to the difference of the $G - G'$ generators. However, if the system is interacting with a certain number of massless vector fields (which are the relativistic analogue of large–range forces), the would–be Goldstone bosons are absorbed by part of the vector fields, which become <u>massive</u> acquiring a longitudinal component at the expenses of the scalar bosons (a massless vector field has only two physical transversal degrees of freedom, while a massive vector field has three: two transversal and one longitudinal).

For the model based on the gauge group $SU(2) \otimes U(1)$, which is called the <u>Standard Model</u> of the electroweak interactions [7], the situation is represented in Table 3.

Table 3. Ingredients of the electroweak Standard Model

Gauge group	$SU(2)_I \otimes U(1)_Y$	$U(1)_Q$
Generators	I_1, I_2, I_3 (weak isospin) Y (weak hypercharge)	$Q = I_3 + \frac{1}{2}Y$
Massless vector fields	$A_\mu^{(i)}(i = 1, 2, 3); B_\mu$	$A_\mu = \cos\theta_W B_\mu + \sin\theta_W A_\mu^3$
Scalar fields	$(H^+, H^0), (\bar{H}^0, H^-)$	H
Massive vector fields		$W_\mu^\pm = \sqrt{\frac{1}{2}}(A_\mu^1 \mp iA_\mu^2)$ $Z_\mu = \sin\theta_W B_\mu - \cos\theta_W A_\mu^3$

The spontaneous breaking of the $SU(2) \otimes U(1)$ symmetry is assumed to occurs at a given energy scale v; its order of magnitude is fixed by the masses of the $W^\pm$ and Z^0 bosons. This phenomenon corresponds to a phase transition. In the more symmetrical phase, all the vector fields are massless, and there are two doublets of scalar field (H^+, H^0) and $(\bar{H}^0, H^-)$; in the other phase, three combination of the vector fields acquire mass absorbing three components of the scalar doublets.

At the end, there remain a massless vector field A_μ (i.e. the electromagnetic field) and a single massive scalar boson H, which is called Higgs boson. The residual gauge symmetry corresponds to the Abelian group $U(1)_Q$ of QED, where Q is the electric charge.

There are three parameters related to the sector of the vector fields of the theory: two coupling constants g and g', corresponding respectively to the groups $SU(2)_I$ and $U(1)_Y$, and the parameter v introduced above, which measures the energy scale of the phase transition. One usually replaces one of these parameters by the Weinberg angle θ_W, which appears in Table 3 in the definition of the physical neutral fields A_μ and Z_μ, and is defined by

$$\tan \theta_W = \frac{g'}{g} . \tag{12}$$

Two of these parameters can be related to low–energy quantities, such as the electric charge e and the Fermi coupling G_F of the weak interactions, as follows:

$$e = g \sin \theta_W$$
$$v = (\sqrt{2} \, G_F)^{-1/2} \tag{13}$$

From the value $G_F = 1.166 \times 10^{-5}$ GeV^{-2} (1 GeV$=10^3$ MeV), one obtains $v = 246$ GeV.

The third independent parameter is determined from the experimental analysis of the weak "neutral current" reactions [8]. Today it is known with great accuracy [9,10]:

$$\sin^2 \theta_W = 0.2320 \pm 0.0005 . \tag{14}$$

The Standard Model predicts the masses of the weak vector bosons $W^\pm$ and Z^0; to first order approximation one obtains the values:

$$M_W = \frac{1}{2} \, gv \simeq 80 \text{ GeV}$$
$$M_Z = \frac{M_W}{\cos \theta_W} \simeq 90 \text{ GeV} \tag{15}$$

However, higher order corrections have to be taken into account to make a comparison with the experimental determination. We recall that the $W^\pm$ and Z^0 were discovered in 1983 at CERN [11], thus providing a beautiful confirmation of the theoretical prediction. The present experimental values obtained at LEP are [9]:

$$M_W = 80.31 \pm 0.06 \text{ GeV}$$
$$M_Z = 91.1887 \pm 0.0022 \text{ GeV} . \tag{16}$$

Among other interesting results we would like to point out the following ones:
1) The higher order radiative corrections provide precision tests of the SM. In particular, one can obtain an indirect determination of the top–quark mass [8,12,13]:

$$m_t = 179 \pm 20 \text{ GeV} .$$

The first direct evidence of the top–quark was obtained in 1994 at Fermilab; the discovery was confirmed recently, and the experimetal result

$$m_t = 180 \pm 12 \text{ GeV}$$

was reported [14], in good agreement with the theoretical prediction.

2) From the experimental determination of the total and partial widths (into hadrons and charged leptons) of the Z^0 boson, one can deduce the number of light neutrinos (with $m_\nu < M_Z/2$). In fact, one can express the invisible width as follows:

$$\Gamma_{inv} = \Gamma_{tot} - \Gamma_{had} - \Gamma_{ch.lep} = N_\nu \Gamma_\nu . \tag{17}$$

Evaluating, within the Standard Model, the width Γ_ν into a neutrino–antineutrino pair, one gets the number of neutrino species [9,10]:

$$N_\nu = 2.987 \pm 0.016 \ .$$

Then there are three kinds of neutrinos, and therefore three families of fundamental particles. This result is in agreement with the astrophysical limit $N_\nu < 3.3$ (95% CL), obtained from nucleosynthesis [15].

3) Quarks and leptons are coupled to the $W^\pm$ bosons through charged currents. However, unlike the lepton case, quarks of different mass and flavour get mixed in the current. Specifically, the transition between u–quark and (d, s) quarks is given by

$$\bar{u}\gamma_5(1 - \gamma_\mu)(\cos\theta_c d + \sin\theta_c s) \ , \tag{18}$$

where θ_c is the Cabibbo angle [16]: $\sin\theta_c \simeq 0.22$.

In fact, in the case of only two families of quarks, the relation between the mass eigenstates and the "current" basis is performed by the unitary matrix

$$U = \begin{pmatrix} \cos\theta_c & -\sin\theta_c \\ \sin\theta_c & \cos\theta_c \end{pmatrix} \ , \tag{19}$$

which contains only a real parameter, i.e. the Cabibbo angle. In the case of three families, the corresponding unitary matrix contains complex elements which can be expressed in terms of 3 angles and 1 phase [17]. The appearance of a phase provides a mechanism for CP violation, which is a necessary ingredient for the understanding of the matter–antimatter asymmetry in the Universe [18,19].

In conclusion, the Standard Model describes with great accuracy, namely within 0.5%, all the electroweak phenomena. One important ingredient is still missing: the scalar Higgs boson. Its mass cannot be predicted by the model, because it is not fixed by the gauge couplings (g, g'), but it is related to an undetermined coupling. Its value can vary within the range

$$60 \text{ GeV} < M_H < 510 \text{ GeV} \ , \tag{20}$$

where the lower value is the experimental limit, and the upper one is obtained from indirect analyses [10]. Hopefully, the Higgs boson will be discovered at the planned accelerator facilities at CERN (LEP 2 and LHC).

STRONG INTERACTIONS

The development of the theory of strong interactions represents a very interesting chapter of Particle Physics. Several attempts, new ideas and experimental discoveries contributed to the present form of the theory which is called Quantum Chromodynamics and is a field theory based on the gauge group SU(3). A detailed analysis of the different steps which lead to the present formulation would be very stimulating, but it will take us to far away from the main path of the paper, which is to provide a general view of the recent developments. Therefore we shall limit ourselves to the principal aspects of the theory, referring to the literature on the subject [4,20,21] for more detailed information.

One of the key ingredient is the "colour" symmetry and the relative group SU(3). This symmetry was introduced in order to explain some peculiar properties of hadrons. The analysis of the spectrum of hadrons, specifically of the low-lying baryons, shows that baryons can be interpreted as bound states of three quarks, which are completely symmetrical, i.e. with respect to all the space, spin and flavour degrees of freedom. Since, on the other hand, the quarks are spin 1/2 fermions (as assumed in the quark model and indirectly confirmed by the experimental information), the hadronic states should be antisymmetric as required by the Fermi–Dirac statistics. The solution of this puzzle is based on the assumption that quarks possess an extra degree of freedom, described by the colour SU(3). In this way the antisymmetry of the baryon states is recovered by assuming that they are <u>totally antisymmetric</u> in the colour SU(3) degrees of freedom.

Moreover, in order to explain the spectrum of all hadrons (both baryons and mesons) one has to postulate, in addition, that physical states are only <u>colour singlets</u>.

The latter point requires that colour cannot appear explicitly, and then one has to look indirectly for its signature. By now different pieces of evidence have been established, but we shall quote here only a couple of well known examples.

The first is based on the comparison of the experimental value of the decay width of $\pi^0 \to \gamma + \gamma$, which is

$$\Gamma(\pi^0 \to \gamma + \gamma) = 7.7 \pm 0.6 \text{ eV} , \tag{21}$$

with the theoretical prediction

$$\Gamma(\pi^0 \to \gamma + \gamma) = \frac{1}{64\pi}\left(\frac{\alpha}{\pi}\right)^2 \frac{m_\pi^3}{f_\pi} \xi^2 = 7.6\xi^2 \text{ eV} , \tag{22}$$

where $\alpha = e^2/4\pi$ is the electromagnetic coupling constant, m_π is the pion mass and f_π the decay constant related to the charged pion decay $\pi^+ \to \mu^+\nu_\mu$. The above formula is obtained by assuming that the pion is composed by a quark–antiquark pair, and that only the <u>up</u> and <u>down</u> quarks contribute. One gets explicitly

$$\xi = N_c(Q_u^2 - Q_d^2) , \tag{23}$$

where the factor N_c is the number of colours of each quark, and $Q_u = 2/3$, $Q_d = -1/3$ are the electric charges of the u, d quarks. Then one obtains a good agreement between eqs. (21) and (22) provided $\xi = 1$, i.e. $N_c = 3$ as required by coloured $SU(3)$.

The second example is based on the analysis of the total cross section for electron–positron annihilation into hadrons. Experimentally, one can determine the ratio

$$R = \frac{\sigma(e^+e^- \to \text{ (hadrons)}}{\sigma(e^+e^- \to \mu^+\mu^-)} \tag{24}$$

as a function of energy. The theoretical prediction is based on the assumption that $\sigma(e^+e^- \to \text{hadrons})$ can be replaced by $\sigma(e^+e^- \to q\bar{q})$, where $q\bar{q}$ are all the possible quark–antiquark pairs which can contribute to the hadronic states at a given energy.

Then one gets

$$R = N_c \sum_q Q_q^2\left(1 + \frac{\alpha_s}{\pi}\right) , \tag{25}$$

where Q_q is the electric charge of the q–quark, and the sum is extended to all the quarks contributing to the reaction. The term proportional to $\alpha_s = g_s^2/4\pi$ (strong coupling constant) is a correction to the photon exchange contribution $e^+e^- \to \gamma \to q\bar{q}$. Good agreement is obtained between experimental data and theoretical predictions with $N_c = 3$.

It is important to point out the main features of QCD. The gauge group $SU(3)$ has 8 generators; correspondingly, there are 8 massless vector fields which are called <u>gluons</u> and which mediate the strong interactions among the quarks. The strength of the interaction between quarks and gluons is described by the coupling $\alpha_s = g_s^2/4\pi$.

Due to the non–Abelian character of the gauge group $SU(3)$, there are important differences between QCD and QED, which is based on the Abelian group $U(1)$. In QED, the vacuum polarization gives rise to a screening of the electric charge, so that the effective value of the coupling $\alpha = e^2/4\pi$ decreases in going away from the charged source. On the contrary, in QCD, due to the coupling of gluons among themselves (while photons do not interact directly), the effect is completely reversed: one gets anti–screening and the effective coupling decreases with increasing distance.

In general, the couplings α and α_s are not constants, but they depend on the momentum transfer q, which is related to the energy scale and to the distance. Specifically, α_s is a "running coupling" which decreases with increasing Q^2 ($Q^2 = -q^2 \geq 0$), i.e. going to very short distances less than the dimensions of hadrons, and increases with larger distances according to a $\log Q^2$ dependence.

The two different behaviours correspond to the limits:

$$\lim_{Q^2 \to \infty} \alpha_s(Q^2) = 0$$
$$\lim_{Q^2 \to 0} \alpha_s(Q^2) = \infty$$

(26)

In the first case, one reaches the asymptotic freedom, which means that quarks behave freely inside hadrons; in the second case, one gets colour confinement, i.e. quarks and gluons cannot be taken apart but they are confined inside hadrons.

The variation of α_s with Q has been tested experimentally, going from the value $\alpha_s \simeq 0.3$ for $|Q|$ about a few GeV to [10]

$$\alpha_s(M_Z) = 0.123 \pm 0.005$$

at $|Q| = M_Z$. This fact gives confidence to the feature of asymptotic freedom.

A special mechanism for explaining quark confinement has been recently proposed [22]. The mechanism is analogous to the condensation of electron Cooper–pairs in a superconductor: in this case the magnetic field is excluded from the superconductor (Meissner effect). In QCD a dual property becomes effective: the condensation of magnetic–like coloured particles pushes away from the vacuum the electric–like colour field; as a result, colour and consequently gluons and quarks are confined.

GRAND UNIFICATION

Combining the gauge theories of electroweak and of strong interactions, that we have outlines above, we obtained the so–called Standard Model (SM) which is a gauge field theory based on the group

$$G_S \equiv SU(3) \otimes SU(2) \otimes U(1) .$$

(27)

The request of a simpler and more unified picture lead to the question whether it is possible to reach a level of more complete unification going beyond the SM. In general, two different scenarios are possible:

i) matter exhibits an onion–like structure, and also quarks and leptons are composite;

ii) quarks and leptons are elementary, and one can extrapolate to higher and higher energies looking for Grand Unification.

So far there is no evidence for the first possibility, while there are some hints for the latter. In the following, we shall discuss the perspectives of the Grand Unification scenario.

The SM has been very successful from a phenomenological point of view, while it is not satisfactory from the theoretical side, since some of the following fundamental questions remain unexplained:

a) The electric charges of quarks and leptons could, in principle, assume any value: why the charge of the proton is perfectly identical to the charge of the positron ?

b) The SM contains too many free parameters; specifically, the masses of quarks and leptons are all independent: can one establish some symmetry relations ?

c) The unification of strong and electroweak interactions is incomplete: can one relate the values of the three couplings g, g' and g_s ?

d) The gravitational interactions have been completely left out: what is their rôle in particle physics ?

The idea of a grand unified theory (GUT) is to replace the gauge group G_S which is obtained as the direct product of three simple groups, by a simple group G containing G_S as a subgroup.

The simplest choice is based on the group $SU(5)$. The main features of the $SU(5)$–GUT are:

1) All the particles and antiparticles of a single family are classified in a representation of $SU(5)$ [23], specifically:

$$15 = \bar{5} + 10 . \tag{28}$$

In terms of the irreducible representations (m, n) of the subgroup $SU(3) \otimes SU(2)$, they are decomposed as follows:

$$\bar{5} = (\bar{3}, 1) \oplus (1, 2) \cdot$$
$$10 = (3, 2) \oplus (\bar{3}, 1) \oplus (1, 1) . \tag{29}$$

The classification of the members of the first family is shown in Table 4, where the (m, n) submultiplet is represented with m columns and n rows.

Table 4. Classification of the first family

$\bar{5}$		$(\bar{d}_1 \bar{d}_2 \bar{d}_3)$	$\begin{pmatrix} \nu_e \\ e^- \end{pmatrix}$
10	$\begin{pmatrix} u_1 u_2 u_3 \\ d_1 d_2 d_3 \end{pmatrix}$	$(\bar{u}_1 \bar{u}_2 \bar{u}_3)$	e^+

2) The electric charge $Q = I_3 + \frac{1}{2}Y$ is a generator of $SU(5)$, represented by a traceless matrix, and therefore one obtains:

$$3Q(\bar{d}) + Q(e^-) = 0 ,$$

i.e. in units of electron charge:

$$Q(d) = -Q(\bar{d}) = \frac{1}{3}Q(e^-) = -\frac{1}{3} \tag{30}$$

Since there are three colours, the down–quark is bound to have 1/3 of the electron charge. Moreover, since the up– and down–quarks are in the same SU(2)–doublet:

$$Q(u) = Q(d) + 1 = \frac{2}{3} . \tag{31}$$

Then:

$$Q(\text{Proton}) = Q(uud) = Q(e^+) . \tag{32}$$

3) Since the down–type quarks (d, s, b) are in the same irreducible representation $\bar{5}$ of the charged leptons (e, μ, τ), their masses can be related (at the GUT scale), i.e. in particular

$$m_b = m_\tau , \tag{33}$$

thus reducing the number of free parameters.

4) At the GUT scale M_U the three gauge couplings must coincide.

However, it is necessary to normalize consistently the generators of the groups, and it is then convenient to make use of the couplings

$$g_1 = \sqrt{\frac{5}{3}} g' \tag{34}$$

It is also convenient to work with the squared couplings:

$$\alpha_i = g_i^2/4\pi , \tag{35}$$

where

$$g_2 \equiv g , g_3 \equiv g_s . \tag{36}$$

How can one match the three couplings α_i ? It is necessary to take into account that the couplings are renormalized and they depend on the energy scale at which they are considered. Specifically, their scale variation is given by the solution of the so–called renormalization group equations (RGE) [24]. To first (one–loop) approximation one obtains the logarithmic dependence on the energy scale μ:

$$\frac{1}{\alpha_i(\mu)} = \frac{1}{\alpha_U} - \frac{b_i}{2\pi} \log \frac{\mu}{M_U} , \tag{37}$$

where b_i is a constant depending on the gauge group. The values of b_1, b_2, b_3 corresponding to the three subgroups $U(1)$, $SU(2)$ and $SU(3)$ of the SM group G_S are given in Table 5: F is the number of fermion multiplets and N_H the number of Higgs doublets. With three fermion families one has $F = 6$ and, in the minimal version of the SM, $N_H = 1$.

Table 5. Coefficients b_i of the SM

b_1	b_2	b_3
$\frac{2}{3}F + \frac{1}{10}N_H$	$-\frac{22}{3} + \frac{2}{3}F + \frac{1}{6}N_H$	$-11 + \frac{2}{3}F$

Starting from the electroweak scale $\mu = M_Z$, one can extrapolate the couplings α_i toward a grand unification scale M_U. As shown in Fig. 1, the three couplings get close around the value $M_U \approx 10^{14} \div 10^{15}$ GeV, but they do not meet at the same point.

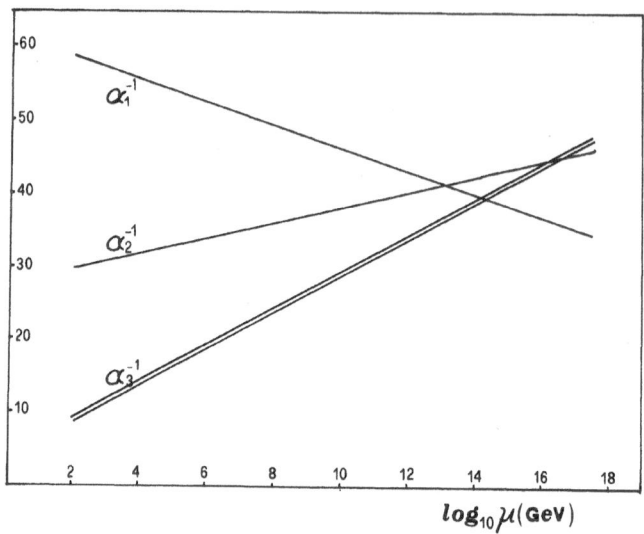

Fig. 1 – Running couplings in the SM

We would like to point out that this result is independent of the assumption that the GUT group is $SU(5)$, but depends only on the hypothesis that no other elementary particles, besides those present in the SM, appear between the scales M_Z and M_U. The conclusion is that the SM alone cannot lead to grand unification.

5) The partial unification of α_1 and α_2 allows to determine the quantity $\sin^2\theta_W$ which is a free parameter in the SM. In fact, one gets at the M_U scale:

$$\sin^2\theta_W = e^2/g^2 = \frac{g'^2}{g^2 + g'^2} = \frac{3\alpha_1}{3\alpha_1 + 5\alpha_2} \to \frac{3}{8}\ . \tag{38}$$

From the value at the unification point, one has to go down to the M_Z scale, making use of the running values $\alpha_1(\mu)$ and $\alpha_2(\mu)$. In this way, one obtains

$$\sin^2\theta_W(M_Z) = 0.210 \pm 0.003\ ,$$

which should be compared with the experimental value (14).

6) A general feature of GUT's is that quarks and leptons belong to the same multiplet, so that they can be transformed into each other. This property has a striking physical consequence: the proton, unlike what generally believed, becomes unstable.

This peculiar feature, which is quite general in GUT's, can be explained, in the present case of $SU(5)$, as follows. The gauge vector fields are in the "regular" representation of SU(5), i.e. in a 24–multiplet. Its decomposition with respect to $SU(3) \otimes SU(2)$ is:

$$24 = (8,1) \oplus (1,1) \oplus (1,3) \oplus (3,2) \oplus (\bar{3},2) \tag{41}$$

The first term (8,1) corresponds to the octet of coloured gluons, which mediate the strong interactions; the colour singlets (1,1)+(1,3) represent the electroweak vector bosons $W^\pm, Z^0$ and γ. The remaining terms individuate a new kind of particles, which are colour triplets and $SU(2)$ doublets.

The term (3,2) corresponds to the X- and Y–particles, which have electric charges Q=4/3 and 1/3, respectively; the last term $(\bar{3}, 2)$ indicates the antiparticles $\bar{X}, \bar{Y}$ with electric charges -4/3 and -1/3. These new particles are called <u>leptoquarks</u>, since they have the quantum numbers of lepton–quark pairs. It is this kind of particles which can mediate the proton decay, according to the scheme

$$P \equiv (uud) \to (e^+ \bar{d}d) \equiv e^+ \pi^0 \tag{42}$$

represented in Fig. 2.

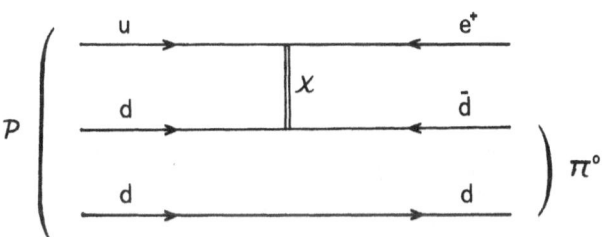

Fig. 2 - Graph contributing to proton decay

Other graphs with both X- and Y–exchange can contribute, and one has to consider all of them to make an estimate of proton lifetime. Here we are only interested in the order of magnitude. To this aim it is sufficient to note that the new interaction is proportional to the coupling

$$G_X \simeq \frac{\alpha_U}{M_X^2} . \tag{43}$$

With $\alpha_U \simeq 2.4 \times 10^{-2}$ and $M_X \simeq M_U \simeq 5 \times 10^{14}$ GeV, one obtains the value for the proton lifetime

$$\tau_P \simeq \frac{1}{m_P^5} \left(\frac{\alpha_U}{M_X^2} \right)^{-2} \simeq 3 \times 10^{30} \text{ years} , \tag{44}$$

to be compared with the experimental lower bound 6×10^{32} years [25]. Since the theoretical prediction is in disagreement with this bound, we can conclude that the minimal scheme of GUT based on $SU(5)$ is ruled out. However, this does not mean that the idea of GUT's has to be abandoned. Other non–minimal schemes, which are based on larger groups, such as, e.g. $SO(10)$, or which imply, in any case, the existence of other particles at intermediate energy scales, are still viable.

SUPERSYMMETRY

Some of the questions which are left open by grand unificatiom, find a possible solution with the introduction of a new kind of symmetry, namely supersymmetry.

The theoretical possibility of implementing this kind of symmetry in QFT was envisaged more than 30 years ago: it is the only non–trivial extension of space–time symmetry (invariance under the Poincaré group) which allows multiplets containing together particles with different spins and statistics [26,27].

The general advantages that one can obtain by requiring supersymmetry are the following:

a) unification of "matter" (spin $\frac{1}{2}$ particles) with "radiation", since vector bosons require spin $\frac{1}{2}$ partners;

b) "raison d'être" for elementary scalar particles, like the Higgs boson;

c) possibility of the unification of gravity with the other interactions;

d) improvement of the ultraviolet behaviour in QFT.

Supersymmetry requires that Nature is symmetric with respect to the exchange of half–integer spin particles (fermions) with integer–spin particles (bosons). Then each supermultiplet must contain the same number of fermionic and bosonic degrees of freedom.

As a consequence, each of the field theories considered previously can be replaced by the corresponding supersymmetric version, according to the following nomenclature:

$$SM \rightarrow MSSM$$

$$QCD \rightarrow susy\ QCD$$

$$GUT \rightarrow susy\ GUT\ ,$$

where MSSM stand for Minimal Supersymmetric Standard Model. In particular, the particle content of this theory is indicated in Table 6.

Table 6. Particle Content of the MSSM

SM particles	SUSY partners
quarks (S=1/2)	squarks (S=0)
leptons (S=1/2)	sleptons (S=0)
photon (S=1)	photino (S=1/2)
gluons (S=1)	gluinos (S=1/2)
gauge bosons	gauginos (S=1/2)
$W^\pm, Z^0$ (S=1)	
Higgs (S=0)	Higgsino (S=1/2)

We note, in particular, that supersymmetry requires the introduction of two distinct Higgs doublets, which gives mass, respectively, to the down–like and up–like quarks.

Up to now, no experimental evidence of the supersymmetric partners has been found. Therefore, it is assumed that SUSY is a good symmetry at very high energies, while it is broken at lower energies. As a consequence, susy–partners are expected to be heavy, with masses in the range $\gtrsim M_W$.

The negative searches give lower limits for the masses of squarks and gluinos [1]:

i) $m_{sq} > 218$ GeV (90% CL)   (for $m_{sq} \simeq m_{gl}$)

ii) $m_{gl} > 100$ GeV (90% CL)   (for any m_{sq})

It is interesting to point out that in the MSSM there are 5 physical Higgs bosons (3 neutral and two charged). Their masses are not all completely arbitrary; in the case of the lowest neutral boson, the mass is bound in the interval

$$60 \text{ GeV} < M_H < 130 \text{ GeV} ,$$

where the lower is an experimental bound, and the upper a theoretical one. Its discovery should be accessible at LEP 200.

Next we consider the case of susy–GUT's, and in particular the minimal version based on the group $SU(5)$. In this case there is an accurate prediction which appears to be in agreement with the experimental information.

We should point out that, in the case of MSSM, the rate of approach of the running coupling $\alpha_i(\mu)$ toward unification is slower than in the case of SM, due to the contribution of the susy–partners (of spin 0 and 1/2), which has opposite sign to that of the vector bosons. To first approximation, eq. (37) still holds, but the values of the coefficients b_i are modified according to Table 7.

Table 7. Coefficients b_i in the MSSM

b_1	b_2	b_3
$F + \frac{3}{10}N_H$	$-6 + F + \frac{1}{2}N_H$	$-9 + F$

In the present case, by assuming that supersymmetry is broken slightly above the elettroweak scale ($10^2 \div 10^3$ GeV), the couplings $\alpha_1(\mu), \alpha_2(\mu)$ and $\alpha_3(\mu)$ do meet at a point, as shown in Fig. 3, so there is good indication for Grand Unification.

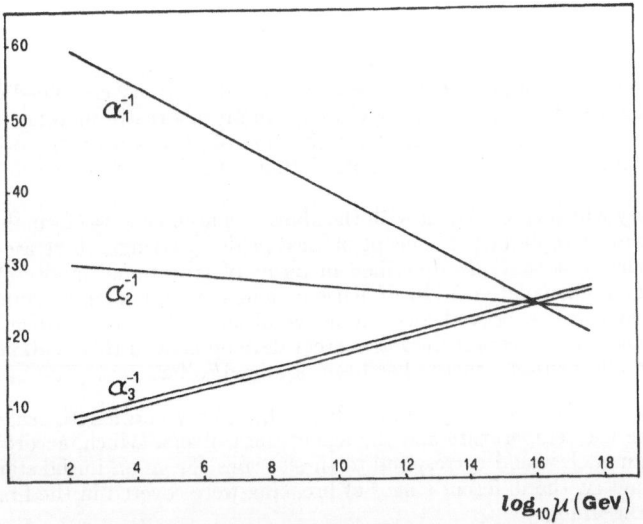

Fig. 3 – Running Couplings in MSSM

The scale of grand unification is around $M_U \simeq 10^{16}$ GeV. Scaling down the symmetric value $\sin^2 \theta_W = 3/8$ from M_U to M_Z, one obtains [10,12,28].

$$\sin^2 \theta_W(M_Z) = 0.2334 \pm 0.0035 , \tag{45}$$

which is in very good agreement with the experimental value (14).

Concerning the proton decay, going from the ordinary to the minimal supersymmetric version of the $SU(5)$ GUT, one obtains a longer lifetime. In fact, for the decay mode $P \to e^+ \pi^0$, one gets:

$$\tau(P \to e^+ \pi^0) \simeq 10^{35} \text{ years} , \tag{46}$$

which is beyond the present experimental sensitivity. Predictions for other modes, like

$$\tau(P \to K^+ \bar{\nu}) \simeq (2 \div 3) \times 10^{33} \text{ years} , \tag{47}$$

will be accessible to experimental test.

As the last point, we would like to mention the problem of gravitational interaction. In the realm of elementary particles such interaction is so weak, at least at ordinary energies, that it is completely negligible. In fact, the relevant coupling constant, expressed as a dimensionless quantity, has the value:

$$G_N m_p^2 \simeq 5.9 \times 10^{-39} , \tag{48}$$

which is very much smaller than the Fermi coupling G_F. One usually defines as "Planck mass" the quantity M_P defined by

$$G_N M_P^2 \simeq 1 \quad \to \quad M_P \simeq 1.22 \times 10^{19} \text{ GeV} . \tag{49}$$

It is at this scale that gravitation becomes important. This means that, in order to understand the rôle of gravitation in elementary particles, one has to go beyond the GUT scale ($10^{15} \div 10^{16}$ GeV), and in terms of distances one has to go down to the Planck scale:

$$\ell_P \simeq 1.6 \times 10^{-33} cm . \tag{50}$$

In such a situation, the requirements of general relativity combined with quantum mechanics, i.e. of Quantum Gravity, imply a drastic modification of the structure of space–time. Maybe the unification of gravity with the other fundamental interactions requires a new formulation of the laws of physics at very short distances.

A theory which is consistent with the above requirements was formulated about 10 years ago: it replaces the concept of local fields by strings; there are no point-like particles, but they are described in terms of vibrational modes of a string. The theory is formulated in 10 dimensions and it requires supersymmetry: the 4–dimensional low–energy theory should be obtained by compactification of the extra 6 dimensions. Even if there is a great development in this field, no realistic, experimentally testable, theory has been obtained so far.

As a conclusion, we compare in Table 8 the energy and length scales of unification with the temperature and the age of the Universe which, according to the Big–Bang model, would correspond to these scales. From an initial state of complete symmetry, the different stages of breaking were covered in the first instants after the Big–Bang.

Table 8. Scales of unification

Stage of unification	E(GeV)	ℓ(cm)	$T(^0K)$	t(sec.)
Electroweak	10^2	10^{-16}	10^{15}	10^{-8}
Susy–GUT	10^{16}	10^{-30}	10^{29}	10^{-37}
Inclusion of gravity	10^{19}	10^{-33}	10^{32}	10^{-43}

This indicates that the clue of fundamental questions of particle physics will be probably found in cosmology.

REFERENCES

1. Particle Data Group, L. Montanet et al., *Phys. Rev.* D50:1173 (1994).
2. J. Goldstone, *Il Nuovo Cimento* 19:154 (1961).
3. P.W. Higgs, *Phys. Rev. Letters* 13:508 (1964).
4. I.J.R. Aitchison and A.J.G. Hey, Gauge Theories in Particles Physics, Adam Hilger, Bristol and Philadelphia (1989).
5. H. Kleinert, Gauge Fields in Condensated Matter, *World Scientific Publishing Co.* (1989).
6. G. Costa, in Optical Properties of Excited States in Solids, edited by B. Di Bartolo, *Plenum Press*, New York (1992).
7. S.L. Glashow, *Nucl. Phys.* 22:579 (1961);
   S. Weinberg, *Phys. Rev. Letters* 19:1269 (1967);
   A. Salam, in Proceed. of the VIII Nobel Symposium, *ed. by N. Svartholm*; Almquist and Wiksells (1968).
8. U. Amaldi et al., *Phys. Rev.* D36:1385 (1987);
   G. Costa et al., *Nucl. Phys.* B297:244 (1988).
9. LEP Electroweak Working Group, preprint LEPEWWG/95–01 (March 1995).
10. Precision Tests of the Standard Electroweak Model, ed. by P. Langacker, *World Scientific Publ. Co.*, Singapore (1995).
11. G. Arnison et al., *Phys. Lett.* B122:103 (1983); *Phys. Lett.* B126:398 (1983);
    M. Banner et al., *Phys.Lett.* B122:476 (1983);
    P. Bagnaia et al., *Phys. Lett.* B129:130 (1983).
12. P. Langacker and M. Luo, *Phys. Rev.* D44:817 (1991).
13. J. Ellis, G.L. Fogli and E. Lisi, *Nucl. Phys.* B893:3 (1993).
14. F. Abe et al., *Phys. Rev. Lett.* 74:2626 (1995);
    S. Abachi et al., *Phys. Rev. Lett.* 74:2632 (1995).
15. K.A. Olive et al., *Phys. Lett.* B236:454 (1990);
    T.P. Walker et al., *Astrophys. Journ.* 376:51 (1991).
16. N. Cabibbo, *Phys. Rev. Lett.* 10:531 (1963).
17. M. Kobayashi and T. Maskawa, *Progr. Theor. Phys.* 49:552 (1973).
18. A.D. Sakharov, *JEPT Letters* 5:236 (1967).
19. S. Weinberg, Gravitation and Cosmology: Principles and Application of the General Theory of Relativity, *John Wiley*, New York (1972).
20. B.R. Martin and G. Shaw, Particle Physics, *John Wiley* (1992).
21. G. Costa, in Disorder Solids, edited by B. Di Bartolo, *Plenum Press*, New York (1989).
22. N. Seiberg and E. Witten, *Nucl. Phys.* B426:19 (1994); 431:484 (1994);
    G.P. Collins, Physics Today (March 1995).
23. H. Georgi and S.L. Glashow, *Phys. Rev. Letters* 32:438 (1974).
24. H. Georgi, H.R. Quinn and S. Weinberg, *Phys. Rev. Letters* 33:451 (1974).
25. R. Barloutaud, *Nucl. Phys. B* (Proc. Suppl.) 28A:437 (1992).
26. J. Wess and J. Bagger, Supersymmetry, Princeton Series in Physics, Princeton Univ. Press. (1983).
27. H.P. Nilles, *Phys. Rep.* 110C:1 (1984).
28. U. Amaldi, W. de Boer and H. Fürstenau, *Phys. Lett.* B260:447 (1991).
29. M.B. Green, J.H. Schwarz and E. Witten, Superstring Theory, Cambridge University Press, Cambridge (1987).

DETERMINATION OF TWO DIMENSIONAL ELECTRON GAS POPULATION ENHANCEMENT WITHIN ILLUMINATED SEMICONDUCTOR HETEROSTRUCTURES BY PERSISTENT PHOTOCONDUCTIVITY

E. A. Anagnostakis

Section of Solid State Physics
Department of Physics, University of Athens
GR 15771 Zografos
Athens
GREECE

The persistent photoconductivity (PP) effect exhibited by the upper, wide band gap side of a semiconductor heterojunction (HJ) is, for the first time, employed for the determination of the illumination-induced enhancement of the areal density of the two dimensional electron gas (2DEG) dwelling within the quantum well (QW) of the probed HJ.

The theory that processes the PP experimental measurements takes into account the photocurrent flowing within the HJ interface hosting the ionized donor depletion zone on the upper epilayer side and the QW formulated in the lower epitaxial layer.

Use is also made of the successive modifications induced by the absorbed photon dose to the populations of the persisting charge in the active part of the upper HJ epilayer, the ionized donors in its adjacent depletion zone, and the 2DEG in the QW of the lower HJ epitaxial layer.

The proposed model is applied to the characterization of typical molecular beam epitaxy $Al_x Ga_{1-x}As/GaAs$ heterodiodes mounted upon semi-insulating GaAs substrates. The PP measurements for the HJ $Al_x Ga_{1-x}As$ upper side are taken by means of the van der Pauw method at a temperature of 4 K, with a magnetic field of 0.5 T. The cumulative photon dose provided to the device evolves within six orders of magnitude.

The output of the model yields for the optoelectronic parameter of photoenhancement of the sheet concentration of the device 2DEG a monotonical increase in the range of 5×10^{11} to 10×10^{11} electrons/cm^2 versus the incoming total photon dose.

This work was done in collaboration with D.E. Theodorou. J. Knoester.

EXCITONS IN J-AGGREGATES: BEYOND THE HEITLER-LONDON APPROXIMATION

L. Bakalis

Institute of Theoretical Physics
University of Groningen
Nijenborgh 4, 9747 AG Groningen
THE NETHERLANDS

Currently, there is much interest in the optical properties of geometrically confined excitonic systems such as semiconductor quantum dots, wires and wells, and molecular aggregates and monolayers. Studying nanostructures provides insight into the way bulk crystal properties evolve from single atom (molecule) properties. In particular, the optical properties of molecular aggregates reveal interesting differences compared to those of the isolated constitutent molecules. Examples are the occurrence of superradiance and motional line narrowing.

The standard model to describe the optical response of molecular aggregates is the Frenkel exciton model for a system of N coupled two-level molecules, each with an optical transition. This model is usually treated within the Heitler-London approximation, in which only interactions that conserve the number of excitations are kept (hopping terms). In this approximation, the aggregate eigenstates fall into N+1 separate (exciton) bands that are distinguished by the total number of excitations on the chain. The complete Frenkel exciton Hamiltonian, however, also contains terms in which two neighboring molecules excite (or de-excite) simultaneously. Such interactions do not conserve the number of excited molecules and, in principle, mix all states in all the above odd (even) exciton bands, leading to a new excitonic band-structure and hence a different optical response. Even though this mixing is in practice counteracted by the large (optical) energy difference between the mixed states, important effects cannot *a priori* be excluded for J-aggregates of dye-molecules, in which the dipolar intermolecular interaction can be considerable.

In this contribution, we study the effect of the Heitler-London approximation in one-dimensional aggregates with nearest-neighbor interactions. For this system, we are able to solve the complete Frenkel exciton Hamiltonian, including the terms that do not conserve the number of molecular excitations, by taking advantage of methods developed earlier in the magnon literature. Consequences of the new level structure, transition dipoles, and selection rules on the (nonlinear) optical response will be discussed.

This work was done in collaboration with J. Knoester.

PHOTOCHEMISTRY OF VIBRATIONALLY EXCITED HCl IN RARE GAS MATRICES

K. Bammel

Fachbereich Physik
Institut für Experimentalphysik
Freie Universität Berlin
Arnimallee 14, D-14195 Berlin
GERMANY

Most of chemical reactions take place in a liquid or solid environment. Therefore, there exists a great interest in studying such processes, i.e. with matrix isolation spectroscopy. A mixture of HCl rare gas (1-.01%) is frozen out at temperatures between 20 and 70 K. The interaction between the HCl molecule and the rare gas atoms produces a change in the form of the intermolecular potential and creates a barrier. If the molecule dissociates after excitation one of the fragments should overcome the barrier. Thus the dissociation is hindered by the surrounded rare gas atoms and this leads to a decrease in the quantum efficiency in contrast to the gas phase (cage-effect). The aim is to measure the quantum efficiency as a function of excitation energy and to investigate the dynamics of the process. In order to achieve both a better mass selection with respect to HCl monomers, dimers and higher aggregates and an increasing Franck-Condon range in the first repulsive state ($A^1\pi$), the excitation is split, i.e. an IR excitation from $v''=0 \rightarrow 3.4$ in the ground state and a UV excitation to the repulsive state by means of an excimer or tunable dye laser. In order to detect the fragments, excimer or dye lasers are used.

This work was done in collaboration with P. Dietrich and N. Schwentner.

NONLINEAR WAKE IN THE RANDOM-PHASE APPROXIMATION

A. Bergara

Universidad del Pais Vasco
Facultad de Ciencias
Departamento Física de la Materia Condensada
Apartado 644, E-48080 Bilbao
SPAIN

We have formulated a diagrammatic analysis to study the many-body interactions between a moving ion passing through a solid with the electron gas embedded in a uniform distributed positive background. Using Feynman diagrams, a derivation of the first nonlinear correction (proportional to the second order on the ion charge) to the induced potential and electron density is presented. These magnitudes are evaluated numerically using the random-phase-approximation to describe the shielded interaction.

The first order induced electron density at the ion position can be approximated, in the high ion velocity regime by $\delta n/n_0 = \pi/v$, where v is the velocity of the incoming ion and n_0 is the conduction electron density. This induced density increases as the ion velocity decreases.

Obviously, only at $v > \pi$ the electronic density fluctuation verifies $\delta n < n_0$, where a first order analysis is useful and applicable. On the other hand, the first order induced electron density diminishes very rapidly as one separates very little from the ion.

It can be deduced that the nonlinear corrections are relatively smaller at higher densities, (at higher densities the kinetic energy of the electron gas increases and the ion potential becomes a smaller perturbation). On the other hand, as remarked before, nonlinear effects are less important as the ion velocity increases.

The nonlinear contribution to the induced potential makes the first minimum deeper yet, on the other hand, the position of the minimum is moved towards the position of the ion, having a direct consequence on the stopping of the particle.

This work was done in collaboration with J.M. Pitarke.

SCANNING TUNNELING OPTICAL SPECTROSCOPY ON NANOSTRUCTURES

K. Birkelund

Mikroelektronik Centret (MIC)
Bldg. 345 East
DK-2800 Lyngby
DENMARK

Combining optical excitation and scanning tunneling microscopy provides a technique called Scanning Tunneling Optical Spectroscopy (STOS). This technique offers the possibility of investigating the local electronic and optical properties of surfaces and subsurfaces by illuminating them with monochromatic light and measuring the photocurrent with the scanning tunneling microscope. It is expected that STOS can be used to spatially resolve the optical and electronic properties of nanostructures. Nanostructures as quantum well quantum dots consisting of CdS and HgS are today prepared by wet chemical methods and they exhibit size-quantum effects at room temperature. I am particularly interested in the investigation of these structures with STOS.

DYNAMICS OF LUMINESCENCE AND BLEACHING OF LOCALIZED STATES IN CdS$_{1-x}$Se$_x$

T. Breitkopf

Institut für Angewandte Physik
der Universität Karlsruhe
Kaiserstraße 12
Postfach 6980, D-76128 Karlsruhe
GERMANY

Investigations on the properties of localized carriers in the ternary CdS$_{1-x}$Se$_x$ mixed crystal are presented. We perform luminescence and pump-probe experiments under quasistationary excitation with picosecond pump pulses. We observe bleaching in the spectral region of localized states. In respect to the luminescence its maximum is shifted to the blue. The dynamics of the bleaching as well as the luminescence depend on localization depth; its decay becomes slower with deeper localization. The filling of localized states is observed as a blue shift of the luminescence band with rising excitation intensity and as a red shift with time, due to the decrease of carrier density by recombination. With further increasing pump intensity, a stimulated emission peak grows out of the low energy wing of the luminescence band. The appearance of the lasing process leads to a reduction of the bleaching in shallow localized states, while carriers in deeper states are unaffected.
This work was done in collaboration with A. Reznitsky, H. Kalt and C. Klingshirn.

INTERMOLECULAR INTERACTIONS IN SUPRAMOLECULAR SYSTEMS BASED ON CHLOROPHYLL-RELATED COMPOUNDS

A. Chernook

Institute of Physics 13303
TU Chemnitz-Zwickau
D-09107 Chemnitz
GERMANY

Various supramolecular systems have received considerable attention during the last two decades. Firstly, they have been used for testing various theoretical models of interchromophoric interactions. Secondly, the promising supramolecular materials for artificial photosynthesis, molecular electronics etc. have been created.

The chemical dimers of porphyrines and chlorins, heterogeneous oligomeric self-organized complexes, polymeric aggregates, and finally, solid layers of the complexes have been investigated in this work.

In all the systems the interactions between electronic transitions of the chromophores can be considered separately for the Soret and visible range.

Chemical dimers. Strong B-transitions in the Soret region of the porphyrine dimers (as well as strong visible Q(0,0) transitions in the chlorin dimer) are excitonically coupled. The coupling is well described in terms of the molecular exciton theory (Kasha, Gouterman). It is very sensitive to the conformational mobility of the subunits about the covalent bridge binding them. At 4.2 K, where the substantial inhomogeneous broadening occurs, an origin of the splitting hasn't been understood yet.

The interaction of weak visible transitions in the porphyrine dimers leads to the non-radiative energy transfer (ET). The ET manifests itself mainly in quenching of the luminescence of the donors in the heterodimers.

Oligomeric complexes. All electronic bands are shifted to the red due to formation of the coordination bonds between the dimers and the ligands. Component spectra in the Soret region are practically summed. The ET from dimers to ligands occurs within 10 ps. When redox potentials of the components favour electron transfer it occurs faster than the ET.

Polymeric self-organized aggregates. New components of the splitting (up to 1100 cm^{-1}) with respect to the oligomeric complexes appear in the Soret region. Visible absorption bands observe an additional long-wave shift (~50 cm^{-1}). Decay of the energy acceptor fluorescence in the aggregate consists of one component with the lifetime < 100 ps and another with lifetime of ~10 ns. The nature of the emitting centers in the aggregates is not understood yet.

Solid layers formed by the self-precipitation of some oligomers have almost the same properties as the polymeric aggregates. However, their absorption and fluorescence spectra lose the structure, and the long lifetimes of the ligands (acceptors of the electronic excitation energy) are reduced.

Acknowledgments. The work was partly supported by Volkswagenwerk Foundation. The investigations were performed in collaboration with C. von Borczyskowski, U. Rempel, E. Zenkevich, A. Shulga, G. Gurinovich, A. Starukhin, E. Sagun, A. Suisalu, K. Mauring, J. Kikas.

HIGH-RESOLUTION INFRARED SPECTROSCOPY OF GASEOUS SAMPLES

A. De Lorenzi

Dipartimento di Chimica Fisica
Università degli Studi di Venezia
Calle Larga S. Marta 2137
I-30123 Venezia
ITALY

A brief outline of high-resolution infrared spectroscopy of gases will be given. Attention will be focused on rovibrational bands of asymmetric rotors and on the algorithms employed in spectral interpretation and in data analysis. The description of the tunable diode laser spectrometer of the molecular spectroscopy group of the University of Venice will be followed by the presentation of some of the results that have been achieved in the study of molecules of atmospheric interest. In particular, the ν_6 band (C=CH deformation mode) of Vinyl-Chloride will be discussed for both natural and ^{37}Cl isotopically enriched species. Examples of other information that can be derived from high-resolution infrared spectra will be given employing data from the ν_5 band of CF_2HCl and the ν_4 band of CH_2=CHF.

COLLISIONAL EFFECTS ON THE LINESHAPE OF AMMONIA TRANSITIONS

M. De Rosa

Istituto di Fisica Atomica e Molecolare
Via del Giardino 7
I-56127 Pisa
ITALY

The knowledge of the lineshape parameters of molecular transitions and of their dependence upon temperature and pressure is of great interest, both in fundamental and applicative spectroscopic studies. It gives useful insights on the intermolecular forces, and it is also essential for modeling the composition of planetary atmospheres and for quantitative analysis of pollutants.

Pressure induced line broadening and shift has been measured as a function of temperature, in the range from 200 to 400 K, for a group of five transitions of the ν_2 band of ammonia near to 937 cm^{-1}. The results have been compared with semiclassical calculation relying on the impact approximation. A pretty good agreement has been found for the broadening, while some discrepancies are evident for the shift. However it is shown that an extensive and detailed investigation on the temperature dependence of the lineshape parameters is now possible and adequately reliable.

This work was done in collaboration with G. Baldacchini, A. Ciucci, F. D'Amato, F. Pelagalli, G. Buffa and O. Tarrini.

NUMERICAL AND ANALYTICAL CALCULATIONS OF THE SUPERHEATING FIELD FOR A SUPERCONDUCTING HALF-SPACE

J. Di Bartolo

Department of Physics
University of Virginia
Charlottesville, Virginia 22903
U.S.A.

Superconductivity is observed in some materials that are cooled below a critical temperature T_C. In the superconducting state, the sample expels magnetic fields (a phenomenon called the Meissner effect). However, if the applied magnetic field is greater than a certain critical field $H_C(T_C)$, superconductivity is destroyed and the sample becomes normal. In "semi-infinite" bulk samples (i.e. the sample is bounded on one side by a plane), superconductivity has been observed experimentally for magnetic fields larger than H_C. These metastable "superheated" states exist for $H_C < H < H_{sh}$. Our task was to numerically calculate H_{sh} as a function of κ, where κ is the ratio of the penetration depth to the correlation length of the sample.

In order to study the effect of magnetic fields on the superconductivity of a sample, we resorted to using an order parameter ψ which is related to the local density of superconducting carriers as $\rho = |\psi|^2$. According to Landau Theory, it is possible to expand the free energy of a superconductor in terms of ψ as long as T is near T_C. Taking functional derivatives of this free energy in order to minimize it results in the Ginzburg-Landau equations, two coupled non-linear differential equations which relate the magnetic vector potential to the order parameter in a superconductor. We used the relaxation method to numerically solve these equations in one dimension in Fortran77. For a given value of κ, we found the corresponding value of H_{sh} by looking for the highest value of H which allowed a superconducting solution.

Our results matched very well with Chapman's[1] analytical results for large κ, i.e. $H_{sh} \approx .71 + .36 \, \kappa^{-4/3}$. They also matched very well with Parr's[2] analytical results for small κ, i.e. $H_{sh} \approx .59 \, \kappa^{-1/2} + .39 \, \kappa^{1/2}$. Both these results match reasonably well with experimental data. We have recently completed another derivation of the small κ dependence of H_{sh} using the method of matched asymptotics.

This work was done in collaboration with A. Dolgert and A. Dorsey.

---

1 S.J. Chapman, "Superheating Field of Type II Superconductors", preprint
2 H. Parr, Zeitschrift für Physik B , **25** (1976) pp.359-361

MBE CLEAVED EDGE OVERGROWTH OF QUANTUM WIRES

H. Gislason

Mikroelektronik Centret, DTU
Bldg. 345 East
DK-2800 Lyngby
DENMARK

Cleaved edge overgrowth (CEO) is a molecular beam epitaxy (MBE) process which can produce laterally confined semiconductor nanostructures[1]. The basic idea of the CEO is to utilize the atomic layer precision of the MBE process to control the dimensions of a low-dimensional structure (dimensions < 10 nm!). First a multiple quantum well (MQW) structure is made by standard MBE growth on a (100) GaAs substrate. The (100) GaAs samples are then thinned to ~100-200 μm thickness and cleaved in ultra high vacuum in the MBE machine. The cleave exposes an exact (110) surface of GaAs and the final step is a MBE overgrowth on the cleaved edge. High quality MBE growth on (110) GaAs is not trivial. We have made growth tests on (110) GaAs substrates which indicates that an improved quality (110) GaAs/AlGaAs is achieved using molecular beams of As_2 instead of As_4 in the MBE growth. We are currently interested in the concept of T-shaped quantum wires[2-5]. Photoluminescence characterization at 4 K of our first CEO sample with T-shaped intersections between 70 Å thick (100) MQW's and a 70 Å (110) single quantum well (SQW) indicates formation of quantum wires ~14 meV below the (100) MQW luminescence and ~7 meV below the (110) SQW luminescence.

1 L.N. Pfeiffer, K.W. West, H.L. Stormer, J.P. Eisenstein, K.W. Baldwin, D. Gershoni, and J. Spector, Appl. Phys. Lett. **56** (17), 1697, 1990
2 A.R. Goñi, L.N. Pfeiffer, K.W. West, A. Pinczuk, H.U. Baranger, and H.L Stormer, Appl. Phys. Lett. **61** (16), 1956, 1992
3 W. Wegsheider, L.N. Pfeiffer, M.M. Dignam, A. Pinczuk, K.W. West, S.L. McCall, and R. Hull, Phys. Rev. Lett. **71** (24), 4071, 1993
4 R.D. Grober, T.D. Harris, J.K. Trautman, E. Betzig, W. Wegsheider, L.N. Pfeiffer, and K.W. West, Appl. Phys. Lett. **64** (11), 1421, 1994
5 T. Someya, H. Akiyama, and H. Sakaki, Phys. Rev. Lett. **74** (18), 3664, 1995

EXCITONS IN C_{60} STUDIED BY TEMPERATURE DEPENDENT OPTICAL SECOND-HARMONIC GENERATION

A.M. Janner

University of Groningen
Laboratory of Applied and Solid State Physics
Nijenborgh 4
9747 AG Groningen
THE NETHERLANDS

With a Second-Harmonic Generation (SHG) experiment the electric-dipole forbidden $^1T_{1g}$ excitonic state at $\hbar\omega=1.81$ eV of solid C_{60} can be probed[1]. We show that the SHG line shape depends strongly on the degree of rotational order. For decreasing temperature we observe a strong enhancement of the SH intensity, an overall blue shift, and below the rotational-ordering phase-transition temperature (260 K) the SH resonance splits into two peaks. The origin of the splitting will be discussed in terms of a possible Davydov splitting due to the four molecules per unit cell in the low temperature phase, and a mixing of the nearly-degenerate $^1T_{1g}$ and ^1G_g free molecule states because of the lower symmetry in the solid. The exciton band structure is calculated with a charge transfer mediated propagation mechanism as suggested by Lof et al.[2] and with one-electron (-hole) transfer integrals determined from band structure calculations. Comparison with our experimental SHG data leads to a reasonable agreement and shows that a mixing of $^1T_{1g}$ and ^1G_g states may explain the splitting at low temperature.

This work was done in collaboration with R. Eder, B. Koopmans, H. T. Jonkman and G.A. Sawatzky.

1 B. Koopmans, A.-M. Janner, H.T. Jonkman, G.A. Sawatzky, and F. van der Woude, Phys. Rev. Lett. **71**, 3569 (1993)
2 R.W. Lof, M.A. van Veenendaal, B. Koopmans, H.T. Jonkman, and G.A. Sawatzky, Phys. Rev. Lett. **68** 3924 (1992)

THE SIMULATION OF ENERGY-LEVEL SCHEMES OF SELECTED RARE EARTH OXYFLUORIDES

E. Kestilä

Department of Chemistry
University of Turku
FIN-20500 Turku
FINLAND

Optical absorption and luminescence spectra of stoichiometric rhombohedral rare-earth oxyfluorides, REOF, were measured in the temperature range 9-300 K. The energy-level schemes of REOF (RE = Nd^{3+}[1], Sm^{3+}[2], Eu^{3+}[3], and Dy^{3+}[1]) were simulated by using the phenomenological free-ion and crystal-field (c.f.) model. The model introduces one-, two-, and three-body interactions as well as the spin-orbit coupling and c.f. interactions using 14 free-ion and 6 c.f. parameters which are refined simultaneously.[†] The simulations reproduced the experimental energy-level schemes with satisfactory deviations between 8 and 16 cm^{-1}.

According to the simulations, the c.f. effect decreases from Nd^{3+} to Eu^{3+} due to the increasing nuclear charge of the RE ion as expected. For the heavier RE ions the c.f. effect seems to increase again which indicates the presence of additional interactions, e.g., the interaction between the 4f and upper orbitals (5d and 6s). In order to verify this observation more experimental data will be needed.

1 Kestilä, E. and Hölsä, J., Unpub. data (1995).

2 Kestilä, E., Porcher, P., Strek, W., and Hölsä, J., 8th Nat. Symp. Inorg. Anal. Chem., Turku, Finland, 1994, *Book of Abstracts*, p. 37.

3 Hölsä, J. and Kestilä, E., *J. Chem. Soc. Faraday Trans.*, **91** (1995) 1503.

† For Eu only the lowest $^{7}F_J$ term is considered in the calculation and only the c.f. parameters are resolved [3].

PLASMA MODES IN LAYERED METALLIC SYSTEMS

S. Kyrkos

Department of Physics
Boston College
Chestnut Hill, Massachusetts 02167
U.S.A.

Plasma modes are calculated for layered metallic jellium systems using the LDA ground state and TDLDA response. One model considered consists of few jellium layers on a bulk, with the density of electrons in each layer diminishing away from the bulk. We study the behavior of Multiple Modes, Surface Plasmons and the photoemision structure for the "expanded tail" model and compare the results with the calculations for a charged jellium surface. We consider also the multilayer metallic superlattice, in which enhancement of the strength of the multiple mode is expected.

This work was done in collaboration with K. Kempa.

RELAXATIONS AND LOSSES IN A PHOTOACOUSTIC RESONATOR

G. Lei

Department of Physics
Boston College
Chestnut Hill, Massachusetts 02167
U.S.A.

The photoacoustic resonant method has been applied to study the phenomenon of molecular relaxation in a cylindrical cavity. The profile of the acoustic resonance, excited by the vibrational mode v_3 of a gas sample SF_6 with a CO_2 laser, was measured as a function of pressure between 1 ~ 800 Torr. A general theoretical treatment of the effects of molecular relaxation, surface and volume losses, and nonideality of gas on the spectral characteristics of the resonance, was worked out in detail and was used to frame the discussion of the relaxation processes following the vibrational excitation. The analysis of the experimental data yields a value of vibrational relaxation time $(p\tau)_{V-T} = 0.21 \pm 0.01$ μs atm for the mode v_3, and a value $(p\tau)_{R-T} = 0.76 \pm 0.13$ ns atm for the rotational relaxation time of SF_6 at T=295 K.

This work was done in collaboration with B. Di Bartolo.

FORMATION OF LARGE BIPOLARONS

F. Luczak

Universitaire Instelling Antwerpen (UIA)
Departement Natuurkunde
Universiteitsplein 1
B-2610 Wilrijk
BELGIUM

An electron in a polar crystal lattice interacting with the longitudinal optical phonon modes is called a polaron. Since the effective electron-phonon interaction is attractive, it enhances the electron's effective mass. If a second electron in the phonon bath is taken into account, its interaction with the phonons might overcome the electron-electron repulsion. If so, the pair of electrons forms a stable bound state, called bipolaron. Due to the competition between the opposite forces the bipolaron exists only in a small region of electron-phonon coupling constant and electron-electron repulsion. It is the aim to contribute to the investigation of this stability region.

The ground state energy of the large bipolaron has been examined by use of a variational approach based on a canonical transformation, which was introduced in earlier studies of the bound polaron[1]. For the bipolaron the transformation reproduces both the oscillator strong coupling results and the weak coupling limit. The stability region has been compared with existing theories. The main advantage of the present method is the transparency of the variational results. To the best of our knowledge, the Ausak yields the lowest bounds ever predicted for two dimensions in the case of $\alpha > 4.5$. The theory has been analyzed with respect to materials with high ($\varepsilon_0 / \varepsilon_\infty$)-ratio.

This work was done in collaboration with F. Brosens and J.T. Devreese.

---

1 J.T. Devruse, R. Evrard, E. Kathenso, F. Brosens, Solid State Communications, 44, 10, 1435 (1982)

NEAR-INFRARED TO VISIBLE UPCONVERSION IN Cs_2NaYX_6: 10% Er^{3+} (X=Cl, Br)

S. Lüthi

Institut für Anorganische Chemie
Universität Bern
Freiestraße 3
CH-3012 Bern
SWITZERLAND

Cs_2NaYCl_6 and Cs_2NaYBr_6 doped with 10% Er^{3+} were synthesized and grown as single crystals using the Bridgeman technique.

Their f-f absorption spectra were measured and their upconversion luminescence behavior was studied under Ti:sapphire laser excitation in the near-infrared. Both $^4I_{15/2} \rightarrow$ $^4I_{11/2}$ and $^4I_{15/2} \rightarrow {}^4I_{9/2}$ excitations were used.

The highest phonon energies are about 260 cm^{-1} and 175 cm^{-1} in x = Cl, Br respectively. The upconversion and cross-relaxation behavior is essentially determined by two factors:

i) As a result of lower phonon energies multiphonon-relaxation processes are less competitive in chlorides and bromides than in fluorides.

ii) Despite the relatively large separation of the Er^{3+} ions within the lattice (7.6Å and 8.0Å in x = Cl, Br respectively) the high dopant concentration leads to several types of nonradiative energy-transfer processes.

The following trends are observed: a red shift of the f-f multiplet baricenters between chloride and bromide, a reduction of crystal-field splittings between chloride and bromide, and a strong suppression of multiphonon-relaxation ($^4F_{7/2}$ to $^4S_{3/2}$ via $^2H_{11/2}$) in bromide compared to chloride.

First, studies of the dynamics of the upconversion processes show both excited state absorption and energy-transfer upconversion to be active.

STRUCTURAL ASPECTS OF ENERGY TRANSFER IN CUBIC CRYSTALS

T. Luxbacher

Institut für Physikalische und Theoretische Chemie
Rechbauerstraße 12
8010 Graz
AUSTRIA

A recently developed discrete shell model[1,2] for cross-relaxation and energy-transfer processes in hexachloroelpasolites $Cs_2NaLn_xGd_{1-x}Cl_6$ (Ln = rare-earth ion, $x = 0.001,...,1$) is introduced. After deciding the energy-transfer mechanism, luminescent decay curves can be calculated for all concentrations and compared with experimental data. The radial and angular dependencies of the energy-transfer Hamiltonian have been considered with respect to the fcc lattice.

It is concluded that from luminescent decay measurements a distinction between different energy-transfer mechanisms is not possible. The question of an inhomogeneous dielectric even in crystals with cubic symmetry has been raised.

1 S. O. Vasquez, C. D. Flint, Chem. Phys. Letts. (1995), in press
2 T. Luxbacher, H. P. Fritzer, C. D. Flint, Chem. Phys. Letts. (1995), in press

300 V CATHODOLUMINESCENT EFFICIENCY AND DEGRADATION

Ronald O. Petersen

Motorola Inc.
Phoenix Corporate Research Laboratories
2100 East Elliot Road, Mail Drop EL508
Tempe, Arizona 85284
U.S.A.

With the possible commercialization of field emission display devices, low voltage phosphor performance has regained interest. Summarized in this presentation is 300 volt efficiency and electron degradation results for various RGB phosphor candidates.

As electrons strike the luminescent material, they are absorbed in the lattice, creating secondary electrons and holes by ionization. After absorbing the electron's energy, the lattice must transfer this energy to the luminescent center. Unfortunately, 300V electrons do not penetrate deeply the phosphor surface. This defective outer surface is the cause of energy loss due to color center formation and non-radiative transitions resulting in reduced luminous efficiency and increased degradation.

While the ZnS:Cu,Al phosphor host is the brightest green emitter at high voltage, it was found that etching the surface of the rare earth oxysulfides produced a very efficient green emission. The outer surface stoichiometry was analyzed by Auger analysis before and after etching. Before etching the surface was found to be more sulfate in nature and contained residual flux materials. After etching, the phosphor surface appeared to be stoichiometric. Similar effects were found by thermal annealing of LaOBr:Tb green. Before, the anneal efficiency was low. After the anneal, Auger analysis showed a stoichiometric surface and enhanced luminous efficiency.

Degradation of luminous efficiency under 300 V electron bombardment appears to be caused by several different mechanisms. The following summarizes several results:

Phosphor Material	Observation
Zn_2SiO_4:Mn	Loss of oxygen, amorphous surface formation
Y_2O_2S:Tb	Loss of sulfur
$Y_2Ga_3Al_2O_{12}$:Tb	Hydrolysis and color center formation
LaOBr:Tb	Hydrolysis and color center formation

Recent work on ZnS:Cu,Al has shown that the degradation in vacuum is dependent on the vacuum quality. Using an Auger system, composition was monitored as a function of coulombic aging. At 10^{-6} torr, decrease in brightness was correlated with decrease in sulfur and an increase in oxygen. It is postulated that the electron beam dissociates water in the vacuum into molecular species which subsequently react with the ZnS to form either ZnO or $ZnSO_4$, with a decrease in cathodoluminescent intensity. As the vacuum level is improved, the cathodoluminescent degradation rate is decreased. This rate depends on the type of gas present in the vacuum. Degradation is slower in oxygen partial pressure, and faster in the presence of water partial pressure.

Surface modification appears to decrease the degradation rate for several luminescent materials. Addition of a small amount of Hf to the surface of Zn_2SiO_4:Mn has been demonstrated to decrease degradation. Phosphor coating of a Y_2O_3:Eu phosphor also decreases the cathodoluminescent degradation. Finally, it appears that vacuum annealing of degraded surfaces results in some efficiency recovery.

In summary, 300 V efficiency of phosphor powder materials can be increased by reduction of surface defects. Likewise, degradation of said materials appears to be improved by altering the surface of the material.

APPLICATION OF SPECTRAL HOLE-BURNING IN COMBINATION WITH HOLOGRAPHIC RECORDING TECHNIQUES

B. Plagemann

Laboratory of Physical Chemistry
Universitätsstraße 22.
ETH-Zentrum
CH-8092 Zürich
SWITZERLAND

Several spectroscopic techniques make use of the nonlinear properties of solids in order to obtain high spectral resolution. Spectral hole-burning as one of them additionally offers the potential for applications in high density optical data storage. The ratio of inhomogeneous to homogeneous linewidth of the molecule 2,3-Dihydroporphine (Chlorin) doped into a polymer is in the order of 10^4 at liquid Helium temperatures. Excitation of molecules with a narrow band dye laser results in a tautomeric photochemical reaction and the formation of a spectral hole, representing digital information. In addition the spatial dimensions can be used in order to store two dimensional patterns or pictures. The use of holographic recording and readout techniques provides the possibility of background free defection. The formation of an absorption and refractive index grating - governed by the Kramers-Kronig relation - allows one to influence the diffraction properties of the hologram by controlling phase and frequency.

Experimental techniques and ideas about the control of phase and frequency are presented, which result in an increase in diffraction efficiency and the reduction of burning time and crosstalk between different holograms.

This work was done in collaboration with F. Graf and U. Wild.

LANTHANIDES PROBES SPECTROSCOPY AT ARARAQUARA - BRAZIL

S.J.L. Ribeiro

France Telecom CNET
Centre Paris B/ Laboratoire de Bagneux
196, Av. Henri-Ravera
BP 107, F-92225 Bagneux
FRANCE

At the Chemistry Institute of Araraquara our group has been working on the preparation and characterization of lanthanide ions containing luminescent materials. By spectroscopy (x-rays, UV-Vis, IR) and diffraction (x-rays and electrons) techniques we have been studying different classes of materials, ranging from amorphous (fluoride and oxide glasses, colloidal suspensions) to crystalline (fluoride and oxide ceramics and glass ceramics) systems.

In this talk we will present some of our results on Eu^{3+} containing SnO_2 based materials obtained by the so-called "sol-gel methodology". We will show how we have been utilizing Eu^{3+} spectroscopy and also vibrational spectroscopy to probe structural evolution occurring in the sol → gel → xerogel → crystals preparation route.[1]

---

[1] S.J.L. Ribeiro, C.V. Santilli, S.H. Pulcinelli, F.L. Fortes, and L.F.C.Oliveira, "Spectroscopic characterization of SnO_2 gels", J. of Sol-Gel Science and Technology **2**, 263-267 (1994)

FROG- A METHOD FOR ULTRASHORT PULSE MEASUREMENT

N. Schmitt

AG Schwentner
Fachbereich Physik
Institut für Experimentalphysik
Freie Universität Berlin
Arnimallee 14, D-14195 Berlin
GERMANY

In Frequency Resolved Optical Gating (FROG) the nonlinear optical Kerr effect is used to generate a signal similar to an autocorrelation function of the third order: the laser beam is divided in two parts. These two parts are overlapped again in a Kerr medium (sapphire crystal) with a determined delay in order to change the polarization plane of the beam in dependence of its intensity. The spectral distribution of the gained signal is measured for different delay times.

This setup delivers a fingerprint of the laser pulse which can be used for laser adjustment. Analytical values for shape and phase of the pulse can be obtained for further experimental calculations.

This work was done in collaboration with P. Dietrich and N. Schwentner.

ABSORPTION EDGE SINGULARITIES FOR A NONEQUILIBRIUM FERMI SEA

C. Tanguy

CNET Laboratoire de Bagneux
196 Avenue Henri-Ravera, BP 107
F-92225 Bagneux Cedex
FRANCE

Sharply edged carrier density profiles, high in the conduction band of semiconductors, can induce absorption edge singularities similar to those found in X-ray spectra of metals, also known as "Mahan excitons". These singularities, which have a power-law behavior, are the consequence of the interaction between the whole Fermi sea and the photocreated hole.

We address the case of a nonequilibrium Fermi sea $(\mu 1, \mu 2)$, stressing the similarities and differences with the usual situation for metals $(0, \mu)$. A new set of singularities is found at frequencies $\mu 2 + n(\mu 2 - \mu 1)$; the corresponding critical exponents are given exactly, using nonperturbative techniques[1].

This work was done in collaboration with M. Combescot.

---

1 C. Tanguy and M Combescot, submitted to Phys. Rev. B

THE FEMTOSECOND FIFTH-ORDER NONLINEAR RESPONSE OF NUCLEAR MOTION IN LIQUIDS

T. Steffen

Ultrafast Laser and Spectroscopy Laboratory
Dept. of Chemical Physics, University of Groningen
Nijenborgh 4
9747 AG Groningen
THE NETHERLANDS

The femtosecond fifth-order nonlinear response of CS_2 benzene and toluene is measured using nonresonant six-wave mixing[1]. This method provides information on the dephasing mechanisms (homogeneous/inhomogeneous) of the coherently excited nuclear motion that is not accessible in third-order experiments. From fs optical Kerr effect and transient phase-grating scattering experiments it is well known that rotational and translational motion of small molecules in liquids is inertial on a sub-100 fs time scale, i.e. the molecules cannot follow the impulsive excitation of the fs laser pulses immediately. These experiments provide information analogous to the free-induction decay in a resonant two-level system. Only higher-order nonlinear experiments alow to characterize the dephasing mechanisms of the nuclear motion. The interpretation of the results depends crucially on the coordinate dependence of the polarizability (non-Condon effects).

The experimental setup is described and first results are presented. The experiments were performed, using five ultrashort nonresonant laser pulses (duration 45 fs) of nearly equal intensity. The $\chi^{(5)}$-signal is emitted in the phase-matched direction and is well separated from third-order signals. The observed $\chi^{(5)}$ responses for all investigated liquids are also inertial: The signal vanishes, when the probe delay t or the delay T between the first and second pulse pair approaches zero, and the maximum is reached after a rise time of 110 to 150 fs. The shape of the signal as function of delay t does not depend of the delay T. In contrast to the observation of Tominaga and Yoshihara[1], the fifth-order signal of CS_2 as function of the probe delay t for fixed delay T is significantly different from the third-order grating signal. Both signals have a comparable rise time, but the fifth-order signal decays faster and is free of a diffusive tail at longer delays t. When the delay t is fixed and the other delay T is scanned, the signal reproduces the features of the optical heterodyned detected optical Kerr effect (OHD OKE): It decays nonexponentially for shorter than 2 ps and exponentially for a decay constant τ equal to the rotational diffusion time for longer delays.

A theoretical description of the probed material response is given. We discuss the role of higher-order electronic polarizabilities that in principle could mask the desired response. Results of model calculations for harmonic systems demonstrate that the response depends strongly on the coordinate dependence of the polarizability. In particular we show that there are no higher-order signals when the polarizability is only linear in the harmonic coordinate. When quadratic terms are also included in the calculations, an echo-type of signal at $t = T$ is predicted in the inhomogeneous limit.

This work was done in collaboration with K. Duppen.

1 K. Tominaga and K. Yoshihara, *Phys. Rev. Lett.*, **74** (1995) 3061

LASER PROCESSES AND OPTICAL NONLINEARITIES IN ZnSe-BASED HETEROSTRUCTURES

M. Umlauff

Institut für Angewandte Physik
der Universität Karlsruhe
Kaiserstraße 12
Postfach 6980
D-76128 Karlsruhe
GERMANY

Supported by the recent progress in epitaxial growth and doping of II-VI semiconductors, ZnSe-based heterostructures have become highly promising material systems for a commercial fabrication of laser diodes or other optoelectronic devices operating in the blue spectral range. Our interest in this field is concerning the identification of the microscopic origins of the optical gain and other nonlinear optical effects in these heterostructures with a focus on the role of excitons. Information about the underlying recombination process is drawn from a combination of different spectroscopic techniques. Gain spectra are determined from the variable stripe-length method and by pump and probe measurements, which simultaneously provide the nonlinear absorption changes in the spectral regime of the excitonic resonances. A lineshape analysis on the experimental data of the gain and of the luminescence is then performed according to theoretical models for the relevant processes. Furthermore, the redshifts of the stimulated emission at threshold and of the excitonic absorption lines with increasing lattice temperature are compared.

For epitaxial ZnSe/ZnSSe-layers without reduced dimensionality (so called double heterostructures) we find that at low lattice temperatures the optical gain sets in at excitation intensities of some $10KW/cm^2$ due to inelastic exciton-exciton scattering. Electron-hole plasma recombination takes place for higher pumping levels ($>500KW/cm^2$). In superlattices of the same material system, the gain arises at the spectral position of the so called M-band. The sharp resonance in the PLE-spectrum of the stimulated emission, at photon energies between the n=1 hh-absorption peak and the emission band could be a hint on the contribution of biexcitons in the recombination process.

In both types of samples, the spectral shifts of the gain maximum at threshold and of the excitation absorption lines as a function of temperature indicate exciton-electron scattering assisted recombination as the most probable laser mechanism in the intermediate temperature regime (100-200 K). In the room temperature regime the laser threshold increases to some MW/cm^2. Since for those excitation densities the excitonic features in the absorption spectra are already completely bleached, we conclude that the Mott transition occurs before lasing sets in.

This work was done in collaboration with J. Hoffmann, M. Kraushaar, H. Kalt and C. Klingshirn.

OPTICAL PROPERTIES OF Er^{3+} DOPED DISORDERED MELILITE LASER MATERIAL

B. Viana

Université Pierre et Marie Curie
École Nationale Supérieure de Chimie de Paris
11, rue Pierre et Marie Curie
F-75231 Paris Cédex 05
FRANCE

Several rare earth doped laser materials belonging to the melilite family have been grown at the Laboratory by the Czochralski process. Belonging to this family, the $Ca_2Al_2SiO_7$, $Ca_2Ga_2SiO_7$ and $SrLaGa_3O_7$ compounds present a strong disorder around the rare earth ions as cations with different charges and sizes surrounding the activator doping ions. This feature leads to an inhomogeneous broadening of the absorption and emission properties, favourable to the diode pumping system, as the temperature drift of the laser diode could be compensated by the broad absorption properties. Our interest is to compare the optical properties between materials belonging to the same family i.e. presenting the same space group and same mean distances between cations but with different phonon-ion coupling. In rare earth doped laser medium, phonon ion coupling can be considered as weak as there are no intensive phonon sidebands in the 4f-4f spectra. Nevertheless, one can consider a process in which multiple phonons are emitted when a large energy mismatch exists between the energy levels.

For instance, in $SrLaGa_3O_7$, highest phonon energy observed in the infrared spectrum is around 805 cm^{-1} while this value reaches 1025 cm^{-1} in $Ca_2Al_2SiO_7$ materials presenting higher frequency vibrations (v_{Si-O}). Emission vanishes when the non radiative probability increases. For the Er^{3+} doped melilite matrices, this effect on the $^4I_{11/2}$ emitting level is considerable as lifetimes drop from 1 ms in Er^{3+}:$SrLaGa_3O_7$ to 45 µs in Er^{3+}:$Ca_2Al_2SiO_7$.

In the presentation will be developed the effect of the medium to the infrared emission intensities either in the 2.7 µm range ($^4I_{11/2} \rightarrow {}^4I_{13/2}$ transition) or in the 1.55 µm range ($^4I_{13/2} \rightarrow {}^4I_{15/2}$ transition).

This work was done in collaboration with B. Teisseire, A.M. Lejus and D. Vivien.

RATE EQUATION MODELING OF Tm → Ho ENERGY TRANSFER IN LiYF$_4$

B. M. Walsh

NASA Langley Research Center
Mail Stop 474
Hampton, Virginia 23681-0001
U.S.A.

A detailed study of the spectroscopy and excitation dynamics of the trivalent lanthanides Tm^{3+} and Ho^{3+} in Yttrium Lithium Fluoride, LiYF$_4$ (YLF), has been done. YLF is a very versatile laser host that has been used to produce laser action at many different wavelengths when doped with trivalent lanthanides. Since the early 1970's YLF has been the subject of many studies, the main goal of which has been to produce long wavelength lasers in the eye safe 2 μm region. The temporal response of Tm and Ho in singly and co-doped YLF to pulsed laser excitation with a Ti:Al$_2$O$_3$ laser and a CoMgF$_2$ laser tuned to various wavelengths have also been studied. The energy transfer mechanisms of cross relaxation, upconversion, and resonant energy transfer between Tm^{3+} and Ho^{3+} ions have been modeled, and the model parameters extracted by a fitting procedure to the measured temporal response curves. Rate equation approaches to modeling are presented that result in the determination of rate constants for energy transfer processes.

The set of rate constants obtained for the transfer of energy from the Tm ^{3}F$_4$ to the Ho ^{5}I$_7$ show only a small variation with concentration of Tm and Ho in the different YLF:Tm,Ho samples studied. This is taken as a confirmation of the rate equation approach to extracting Tm → Ho and Ho → Tm transfer rates in YLF.

This work was done in collaboration with B. Di Bartolo.

SPECTROSCOPY OF EXCITED STATES IN CdSe QUANTUM DOTS

O. Wind

Institut für Angewandte Physik
der Universität Karlsruhe
Kaiserstraße 12
Postfach 6980
D-76128 Karlsruhe
GERMANY

The nonlinear optical properties of II-VI semiconductor nanocrystals in transparent matrices as a model for quasi zero-dimensional semiconductors are of increasing scientific interest in recent years. Using differential absorption spectroscopy their discrete energy level structure can be examined. For the case of CdSe quantum dots in glass we study samples containing nanocrystals with average radii below the excitonic Bohr radius. In this strong confinement regime the quantum-confined levels are well separated. Using a nanosecond pump-and-probe technique we investigate the differential absorption for low temperature in dependence on excitation energy. By size-selective excitation we find a complicated substructure showing the simultaneous bleaching of several hole states. From these findings we derive the size-dependent splitting of quantized levels which can be compared with recent theoretical calculations including a realistic valence band structure. The agreement between experiment and theory is good.

This work was done in collaboration with F. Gindele, U. Woggon and C. Klingshirn.

TWO-PHOTON SPECTROSCOPY OF ZnSe UNDER UNIAXIAL STRESS

J. Wrzesinski

Universität Dortmund
Lehrstuhl Exp. Physik II
D-44221 Dortmund
GERMANY

Resonances on the upper polariton branch of ZnSe are studied under uniaxial stress by two-photon excitation spectroscopy. The spectra are measured for uniaxial stress up to 1.2 kbar in configurations with stress directions [001], [111] and [11$\overline{2}$]. The polariton resonances are calculated by diagonalizing the Pikus-Bir-Hamiltonian and solving the macroscopic Maxwell equations. The stress-induced shifts and splittings of the resonances on the polariton dispersion curves are described by the hydrostatic, tetragonal and trigonal deformation potentials a, b, and d, respectively. The values for the deformation potentials resulting from a least square fit are a=-4.65 eV, b=3.72 eV and d=7.36 eV. Due to the much smaller linewidths (here: < 0.25 meV) of nonlinear resonances as compared to linear optical absorption or reflection lines, the deformation potentials are determined with high accuracy in this work.

NONLINEAR QUANTUM HYDRODYNAMIC MODEL FOR THE ELECTRON GAS

A. Bergara

Universidad del Pais Vasco
Facultad de Ciencias
Departamento Física de la Materia Condensada
Apartado 644
E-48080 Bilbao
SPAIN

It is well known that the hydrodynamic model of the electron gas is suitable when the electron density can be considered as the fundamental magnitude, small wavelength effects being less important. This is the case when the electron gas is excited by the passage of swift particles. In this paper we make use of the Quantum Hydrodynamical Model of the electron gas to derive, within this model, the nonlinear wake potential of the fast charged particles passing through matter, the Barkas effect, and the double-plasmon excitation probabilities. This is compared with the results obtained following the many-body schemes of the random-phase-approximation (RPA).

In conclusion, the obtained results coincide exactly with the RPA approximation for high velocities of the incoming ion, $v > 1.5$ (in atomic units) at usual metallic densities.

This work was done in collaboration with J.M. Pitarke and R. H. Ritchie.

OPTICAL PROPERTIES OF InGaAs/GaAs QUANTUM WELL HETEROSTRUCTURES: EXCITATION ENERGY DEPENDENCE

P. Borri

Laboratorio Europeo di Spettroscopie Non-Lineari
Università di Firenze
Largo E. Fermi, 2 (Arcetri)
I-50125 Firenze
ITALY

Strained-heterostructures such as InGaAs/GaAs quantum wells (QW) present several important physical properties like strain, interface segregation, and large heavy-light hole splittings, that are very different with respect to the most widely investigated GaAs/AlGaAs case and also very interesting for device-related applications.

We report a detailed study of the optical properties of InGaAs/GaAs quantum wells, performed by means of photoluminescence (PL) and photoluminescence excitation (PLE) spectroscopy, with particular attention for the excitation energy dependence.

The good quality of the samples is showed in the small linewidth and Stokes shift.

We find in the PL spectra a doublet structure consistent with a bound-exciton like recombination (BE) in addition to the free exciton peak (FE). This attribution is supported by the temperature and the excitation power dependence of these peaks. The BE structure has been systematically found in a very large set of samples: it increases with increasing well width up to $L_w \sim 100\text{-}150\text{Å}$ then it starts weakening for very wide wells; moreover for a fixed well width it increases as the Indium concentration increases. The comparison with C-doped samples allows us to conclude that the low energy structure is not associated with an exciton bound to impurities. The dependence of the BE intensity with the well width and the Indium concentration can suggest a localization due to disorder at the InGaAs/GaAs interface.

We have studied the temperature and excitation energy dependence of the PL doublet, finding an anomalous behaviour. The PLE spectra strongly depend on the detection energy, denoting that care has to be taken when interpreting the PLE profile as the probe of the absorption. In particular we find peaks and dips related to the PL intensity of only one of the two recombination bands. The PLE anomalies are quenched both when the temperature is increased or when an optical bias is used. The optical bias produces also a nonlinearity in the PL spectra which decays in the μs time scale.

This work was done in collaboration with Massimo Gurioli, Marcello Colocci, Faustino Martelli, Antonio Polimeni, Amalia Patanè and Mario Capizzi.

LIBS TECHNIQUE FOR ENVIRONMENTAL MEASUREMENTS

A. Ciucci

ENEA - CRAM "S.Teresa"
P.O. Box 316
I-19100 Lerici (SP)
ITALY

Time Resolved Laser Induced Breakdown Spectroscopy techniques have been developed in recent years as a powerful diagnostic system of atomic pollutant concentration in air.

The principle of this technique is very simple: focusing a laser beam of adequate energy and pulse duration on the surface of solid sample or inside a gas, a spark is produced. Monitoring the light emitted from the little plasma it is possible to identify atomic species present also as a trace ~ppm (5 ppb of Hg in air). LIBS has many advantages: the possibility to use a single wavelength'laser source for a multi-element analysis, fast data acquisition, quite good sensitivity over a wide range of pollutants (~1 ppm Ti, Fe, Cr, Cd, Cu, Pb, S, etc.). The only disadvantage is the impossibility of molecular study because the plasma destroys the original chemical composition.

LIBS is a valid alternative to traditional techniques for environmental measurements and can be applied both in laboratory and in hostile environments as urban and industrial dumpsites or heavy polluted zones.

This work was done in collaboration with V. Palleschi and S. Rastelli.

Yb^{3+}-DOPED NONLINEAR LiNbO$_3$: ELABORATION BY LASER HEATED PEDESTAL GROWTH AND OPTICAL PROPERTIES

G. Foulon

Université Claude Bernard-Lyon I
Laboratoire de Physico-Chimie des Matériaux Luminescent
URA 442 CNRS Bâtiment 205
43, boulevard du 11 Novembre 1918
F-69622 Villeurbanne Cédex
FRANCE

In order to reinforce the elaboration aspect of doped single crystals, the laser materials group of the laboratory has recently installed a Laser Heated Pedestal Grown system. This method presents several advantages in the way that, contrary to Czochralski, a few raw materials are required and no crucible is needed, avoiding pollution of the sample. The rapidity of this method allows us, moreover, to obtain a huge variety of host compositions and dopant rates. It is also important to note that the fiber shape of the samples leads to energy confinement favorable for nonlinear effects.

The main aim of this work is both growth and characterization of rare-earth doped nonlinear materials to obtain self-frequency doubling laser systems. Our first choice is Lithium Niobate (LiNbO$_3$) containing additives for reducing photorefractive effects and doped with Yb^{3+} activator ion. This one can be laser diode pumped around 980 nm into the $^2F_{5/2}$ excited state and can emit the $^2F_{5/2} \rightarrow {}^2F_{7/2}$ transition around 1060 nm. Overlapping of, on one hand, intrinsic nonlinear optical properties of the host according to its composition and, on the other hand, emission of the Yb^{3+} ion will be shown. For example, a coincidence between the wavelength of the phase matching in the green at 529 nm and those of the infra-red laser emission at 1058 nm especially for LiNbO$_3$: 1% Sc^{3+}: 1% Yb^{3+} has been found.

We will finally give our main research directions: creation of inversed ferroelectric domains and synthesis of other crystals containing Nb-O bonds giving high nonlinear susceptibility $\chi^{(2)}$.

This work was done in collaboration with A. Brenier, M. Ferriol, M.T. Cohen-Adad and G. Boulon.

INVESTIGATION OF STRIAE PATTERNS IN LASER CRYSTALS

J.C.A. Grossen

Department of Physics
College of William and Mary
Williamsburg, Virginia 23185
U.S.A.

This research involves studying the concentration variations of dopant ions in the striae patterns observed in crystals. These patterns may be due to the temperature variations during growth process. Materials to be investigated are 2 µm laser materials which are of great interest for atmospheric remote sensing measurements. Trivalent Thulium (Tm^{3+}) sensitized trivalent holmium (Ho^{3+}) doped crystals are used for 2 µm lasers. Tm is used to enhance Ho emission in the 2 µm wavelength region. Tm absorbs well at the wavelength ~780 nm which is suitable for diode laser pumping. The energy transfer process called "The Cross Relaxation Process" which takes place among Tm ions is described as follows: An excited Tm ion in the $^3H_4$ state decays to the $^3F_4$ state by exciting another Tm ion in the $^3H_6$ ground state to the $^3F_4$ state. Tm in the $^3F_4$ state then transfers its energy to the upper laser level $^5I_7$ of Ho enhancing the 2 µm emission from this level. The cross relaxation process has been found to be highly concentration dependent. However, even when high concentrations of Tm are examined, the decay of the $^3H_4$ manifold has always been found to have a nonexponential curve. The noneven concentration distribution in the striae patterns could contribute to this behavior of the Tm decay curve. Absorption, emission and lifetime measurements are being conducted on 5 to 11% at. Tm doped crystals using an excitation source with a small beam diameter (smaller than the width of striae).

This work was done in collaboration with G. Armagan and N.P. Barnes.

SPIN QUANTUM BEATS IN GaAs

S. Hallstein

Max-Planck-Institut für Festkörperforschung
Heisenbergstraße 1
D-70569 Stuttgart
GERMANY

The short laser pulses (FWHM 2ps) of a Ti:Sapphire laser are used to excite a GaAs sample with circularly polarized light perpendicular to an applied magnetic field (0-16Tesla). This excitation geometry (Voigt geometry) leads to a coherent excitation of the two energetically separated Zeeman states ($s_z = +\frac{1}{2}, s_z = -\frac{1}{2}$) in the GaAs conduction band. Due to the coherent superposition of the excited states a beating in the time-resolved luminescence signal with a frequency equal to the Larmor frequency can be observed if circularly polarized light is detected.

Since the oscillation frequency is directly related to the g-factor of the electrons via $g = \dfrac{\hbar}{B\mu_B}\omega_L$, it is possible to determine the electron g-factor with high precision. In accordance with theoretical predictions made by five level $k \cdot p$-theory[1] an anisotropy of the electron g-factor can be shown experimentally by aligning the magnetic field in the crystal directions [100] and [110]. The measured conduction band g-factor anisotropy turns out to be smaller than 50% of the calculated value.

Measurements with different excitation intensities show a strong increase of the electron g-factor with increasing carrier density. Until now it is not clear if this dependence has to be attributed to many particle effects (exchange interaction) or if it is due to the nonparabolicity of the conduction band. For a clear decision further experiments have to be performed.

This work was done in collaboration with W.W. Rühle.

1 P. Pfeffer and W. Zawadzki, Phys. Rev. B41, 1561 (1990)

REAL-TIME SPECTROSCOPY OF CAGE RELAXATION PROCESS APPLIED TO NO EMBEDDED IN RARE GAS CRYSTAL

C. Jeannin

Institut de Physique expérimentale
Faculté des Sciences, BSP
CH-1015 Lausanne-Dorigny
SWITZERLAND

By use of ultrashort pulses lasers, it is possible to resolve the dynamics of physico-chemical processes, such as photoinduced deformations in condensed media. Here, the case of NO embedded in a rare gas crystal (Argon) is presented. The setup and the preparation of the sample are described.

The pump and probe technique is used. The pump beam (195 nm) excites the first Rydberg state of the NO molecule. Therefore, the radius of the outer electron orbital becomes comparable to the NO-nearest rare gas atoms distance. As the outer shells of the host atoms are full, the interaction between the Rydberg electron and the rare gas atoms is repulsive, so that a "bubble" around the NO molecule is formed. This process, called cage relaxation, is nonradiative.

From classical spectroscopy, we know that the lifetime of the first Rydberg state is approximately 200 ns. Its relaxation can be observed by measurement of its fluorescence. The probe pulse (780 nm) depopulates the first Rydberg state to a higher level and fluorescence is observed for different delays between the pump and the probe pulses. From these measurements, we should be able to determine the duration of the cage relaxation process.

This work was done in collaboration with M. Chergui and M. Portella-Oberli.

APPLICATION OF THE SHELL MODEL FOR CROSS RELAXATION TO THE $^4G_{5/2}$ STATE OF Sm^{3+} IN $Cs_2NaSm_xGd_{1-x}Cl_6$

T. Luxbacher

Institut für Physikalische und Theoretische Chemie
Rechbauerstraße 12
8010 Graz
AUSTRIA

It has become conventional to treat energy-transfer processes in solids within the Inokuti-Hirayama approximation[1]. This model, which assumes a continuous distribution of acceptors surrounding the donor, is inapplicable to crystalline solids. The failure of the IH model is particularly severe at short donor-acceptor distances in high-symmetry crystals which dominate the energy-transfer processes. Attempts therefore to determine whether the energy transfer mechanism is due to dipole-dipole, dipole-quadrupole, quadrupole-quadrupole or exchange interactions by *fitting* experimental curves is likely to be misleading.

In this contribution the relaxation of the $^4G_{5/2}$ excited state of Sm^{3+} in $Cs_2NaSm_xGd_{1-x}Cl_6$ ($x = 0.001,...,1$) is considered. For $x = 0.001$ and $x = 1$ the decay curves are strictly exponential over the temperature range 10-300 K. At intermediate concentrations the decay is strictly non-exponential.

The results are interpreted in terms of a recently developed shell model[2,3] and analyses using a variety of assumptions concerning the energy-transfer process are compared.

A significant conclusion is that even when extensive, precise experimental data is available, unambiguous distinction between coupling mechanisms is not possible by *fitting* the experimental data to theoretical models.

1 M. Inokuti, F. Hirayama, J. Chem. Phys. **43** (1965) 1978
2 S. O. Vasquez, C. D. Flint, Chem. Phys. Letts. (1995), in press
3 T. Luxbacher, H.P. Fritzer, C. D. Flint, Chem. Phys. Letts. (1995), in press

A NEW QUANTUM EFFECT IN THE DYNAMICS OF A TWO-MODE FIELD COUPLED TO A TWO-LEVEL ATOM BY TWO-PHOTON PROCESSES

A. Napoli

Istituto di Fisica dell'Università degli Studi
Via Archirafi 36
I-90123 Palermo
ITALY

In the last years the successful development of new important experimental tools with which one can investigate fundamental features of the radiation-matter interaction[1], has rekindled the interest of theoreticians toward the construction of new exactly tractable models[2] generalizing the standard Jaynes-Cummings Hamiltonian[3]. Here we wish to study the coupling between a single effective two-level atom and two modes of the electromagnetic field of a lossless cavity. We assume that the atom can make only two-photon transitions between an upper level and a lower level, having identical parity, through a set of intermediate states. In addition we suppose that the two field modes have the same frequency. Under suitable carefully chosen conditions imposed on the atomic states and field polarization, we show that such a system may be described adopting the following rather simplified but exactly solvable effective Hamiltonian model:

$$H_{\text{eff}} = \hbar\omega_0 S_z + \hbar\omega \sum_{\mu=1}^{2} \alpha_{\mu}^{\dagger}\alpha_{\mu} + \sum_{\mu=1}^{2} (\lambda_{\mu}\alpha_{\mu}^{2}S_{+} + h.c.)$$

(1)

where we have neglected any dependence of the detuning on the field operators. In particular we investigate on the time evolution of the mean photon number in each field mode assuming the initially highly populated mode in a Fock state $|n\rangle$. Using a canonical approach by which the effective Hamiltonian may be reduced to a diagonal form, we demonstrate that the photon number of this mode decreases, in a characteristic time, up to $n/2$ determining the correspondent increase of the population in the other mode up to $n/2$. This photon "equipartition regime" lasts a period of time much longer than that required for the system to attain this condition. At the end of such an interval of time, characterized by the absence of macroscopic intermode energy exchanges, the dynamical behaviour of the system manifests a conspicuous sensitivity to the parity of the integer n. In fact this intensity-dependent "lethargy period" is followed by a complete transfer of the other $n/2$ photons to the initially empty mode if n is even, whereas we assist at an almost total repopulating of the initially empty mode when n is odd. From now on the dynamical behavior of the field reveals a peculiar periodicity, being dominated by net large n-parity dependent photon exchanges between the two modes. Thus the interaction mechanism leads to the possibility of distinguishing between two field states prepared at $t=0$ with a difference of even one photon, amplifying at a later evaluated time such a microscopic difference. We point out the quantum nature of this effect.

This work was done in collaboration with A. Messina.

1 S. Haroche, "Cavity Quantum Electrodynamics", Les Houches, 1990, J. Dalibard, J.M. Raimond and J. Zinn-Justin eds. (1992)
2 B.W. Shore and P.L. Knight, J. Mod. Optics, **40**, 1195 (1993); G. Benivegna and A. Messina, J. Mod. Optics, **41**, 907, 1994
3 E.T. Jaynes and F.W. Cummings, Proc. I.E.E.E., **51**, 89, (1963)

UV EMISSION IN NATURAL DIAMONDS

M.A.C. Neto

Departamento de Fisica
Universidade de Aveiro
Campus Universitário
3800 Aveiro
PORTUGAL

Diamond crystals are well known for many emitting centres, the most of them in the visible region of the EM spectrum. Some of these centres are coincident in natural and synthetic diamonds.

In this work we present some preliminary results of the luminescence of a new emitting centre in a natural diamond crystal. This centre when excited between 350 nm and 370 nm by a 150 watt Xenon arc-lamp, has an emission in the UV region with two ZPLs at 3.225 eV and 3.212 eV and a structural emission band with phonon coupling of 53 meV.

Because there is no change in the two ZPLs' relative intensities, when we change the excitation energy, we think that this centre has a similar behaviour to centres with two emitting levels, like the 417-418 nm and S1 (503-513 nm) centres.

This work was done in collaboration with E. Pereira and L. Pereira.

INFLUENCE OF THE CROSS-RELAXATION ON ESA-PROCESS IN Tm^{3+} DOPED FLUOROPHOSPHATE GLASS

G. Özen

Department of Physics
Faculty of Sciences and Letters
Istanbul Technical University
80626 Maslak-Istanbul
TURKEY

Spectral properties of UV and blue upconversion fluorescences in Tm^{3+} doped fluorophosphate glass with use of a tunable DCM dye laser for single wavelength pumping were studied at room temperature. Two emission bands centered at 363, 451 nm from the $^1D_2$ level and one emission band centered at 478 nm from the $^1G_4$ level were observed. Two photon absorption mechanisms responsible for these fluorescences was confirmed by quadratic dependence of fluorescence intensities on the incident pumping intensity. Excitation spectra and decay profiles of the same fluorescences also proved that the mechanism leading to these emissions was the excited state absorption (ESA) originating from the $^3F_4$ and the $^3H_4$ levels to the $^1D_2$ and the $^1G_4$ levels, respectively.

Influence of ($^3F_4,^3H_6 \rightarrow$ $^3H_4,^3H_4$) cross-relaxation process on the upconversion efficiency was evaluated from ratio of the 478 nm blue emission to 451 nm blue emission for different Tm^{3+} concentrations. Correspondingly lifetime of the luminescence originating from the $^3F_4$ was quenched as the dopant concentration was increased. Critical distance for this interaction was determined by fitting the decay profiles of the $^3F_4$ level to a theoretical model, and was found to be 16.9 Å for the sample having 0.2 mol.% TmF$_3$ concentration.

This work was done in collaboration with J.-P. Denis and F. Pelle.

A PHYSICAL CHARACTERIZATION OF THE GROUND STATE OF THE SPIN-BOSON INTERACTION

E. Paladino

Istituto di Fisica dell'Università degli Studi
Via Archirafi 36
I-90123 Palermo
ITALY

We present some characteristic properties of the ground state of a two-level system linearly coupled to a boson field[1]. Such properties are valid over the whole range of the parameters appearing in the Hamiltonian of the system. The knowledge of these exact results provides a physical basis for an accurate variational determination of the ground state[2]. A physical characterization for the ground state of the two-level unit interacting with one or two modes of the bosonic field under intermediate coupling is proposed. Our treatment is based on operator methods combining symmetry considerations with some general properties of the lowest energy state.

This work was done in collaboration with A. Messina.

1 A.J. Leggett, S. Chakravarty, A.T. Dorsey, M.P.A. Fisher, Anupan Garg e W. Zwerger, Rev. Mod. Phys., 59, 1, (1987)
2 G. Benivegna, A. Messina, E. Paladino, "Rigorous results for the ground state of a localized system coupled to a bosonic field", submitted for publication to Annals of Physics

NANOSCALE CARBON NANOTUBES STUDIED BY RAMAN SPECTROSCOPY

R. Provoost

Katholieke Universiteit Leuven
Laboratorium voor Vaste-Stoffysika en Magnetisme
Celestijnenlaan, 200D
B-3001 Leuven
BELGIUM

Due to the heterogeneous character of the deposit produced in the dc carbon arc discharge techniques, fullerene related structures are usually investigated by means of electron microscopy methods. We studied carbon nanotubes, generated from a dc arc discharge between carbon electrodes, by Raman spectroscopy to investigate the effect of the presence of Si in the arc plasma on the structural properties of the arc-derived nanotubes.

The first-order Raman spectra of the nanotubes we investigated show in addition to the E_{2g} zone center mode (G-mode), a weak shoulder at 1620 cm^{-1} and a band around 1350 cm^{-1} (D-mode). In contrast to the Raman results of Co-catalyzed nanotubes[1] for the Si-catalyzed nanotubes a downshift of the G-mode could be found. The phonon induced softening and the increase of the FWHM of the in-plane stretching mode might suggest that the D-band originates from size effects rather than from structural defects. In this assumption, the effect of the presence of Si in the core of the anode consists in reducing the microcrystallite planar size L_a. This could be derived from the change in the intensity ratio I_D/I_G, which is inversely proportional to L_a.

---

1 J.M. Holden, P. Zhou, X.-X. Bi, P.C. Ecklund, S. Bandow, R.A. Jishi, D. Das Chowdury, G. Dresselhaus, and M.S. Dresselhaus, Chem. Phys. Lett. 220, 186 (1994)
This work is supported by the Belgian Interuniversity Institute for Nuclear Sciences (IIKW) and by Concerted Action (G.O.A.) and Inter-University Attraction Poles (IUAP) Research Programs. Abstract submitted to the Annual Meeting of the Belgian Physical Society, May 1995.

OPTICAL STUDIES ON GaN SAMPLES

C. Ramos

Departamento de Fisica
Universidade de Aveiro
Campus Universitário
3800 Aveiro
PORTUGAL

Gallium nitride is a wide band gap semiconductor (~3.4 eV at room temperature) that makes it an interesting material for optoelectronic devices. In our GaN samples growth by MOVPE on 6H-SiC and sapphire, near band gap emission due to donor bound excitons and donor acceptor pair (DAP) recombination are usually observed with higher intensity ratio exciton to DAP on GaN growth on sapphire. This indicates that these samples are of better quality than GaN grown on 6H-SiC. GaN samples grown on 6H-SiC present, besides near band edge emission, luminescence bands peak at 2.95 eV, 2.64 eV, and 2.2 eV. These midgap emission bands are quite dependent on the homogeneity of the samples, and the 2.64 eV band is probably due to substrate emission. In this work a study of the midgap emission is made by temperature dependent steady state luminescence, time resolved spectroscopy and lifetime measurements.

This work was done in collaboration with T. Monteiro and E. Pereira.

ULTRA HIGH SPECTRAL DENSITY FLAT CONTINUUM LIGHT SOURCE FOR DYNAMIC SPECTROSCOPY

K. Ueda

Institute for Laser Science
University of Electro-Communications
Chofugaoka, Chofushi
Tokyo 182
JAPAN

We report the latest results on ultra-high spectral density and ultra-broad-band super continuum light source for ultra fast spectroscopy. The self phase modulation in optical channeling in rare gases generates super white light covering infrared (IR) to vacuum ultraviolet (VUV) with high spectral intensity 0.1 to 1 GW/nm. The most attractive feature of our experiments is the flat continuum over the whole light range. The high spectral intensity pulse is available for nonlinear spectroscopy in VUV region.

The excitation source is a Ti:sapphire laser system composed of oscillator, pulse stretcher optics, regenerative amplifier, multi-pass amplifier, and pulse compressor optics. The laser system generates 250 mJ output pulses of 125 fs with 10 Hz repetition rate.

Up to today, tight-focused optics has been used to generate the super continuum because we need high intensity at focal point. However, the super continuum generation does not happen only in the focal point. The spectral broadening in the self-phase-modulation is enhanced by the product of nonlinear refractive index, power density, and the interaction length. How can we realize a long-path interaction with high intensity? Nonlinear propagation of high peak power laser pulse introduces self-trapping phenomena even in rare gases with relatively small nonlinear refractive indices. Long-path interaction more than meters is possible.

As a result, we demonstrated the super-white light generation covering IR to VUV with flat continuum spectra. The shorter wavelength generation was limited by the energy loss due to the multi-photon absorption of the white light beams. The pulse duration of less than 400 fs was measured as a preliminary result by means of nonlinear parametric technique between the pumping laser and white light pulses.

Finally we would like to emphasize that our white light is more than 12th order of magnitude larger in its spectral intensity than the conventional SOR output. This is a real table-top ultra-high-power VUV source for ultra-fast dynamic spectroscopy.

This work was done in collaboration with H. Nishioka.

LUMINESCENCE OF ZnSe FILMS

G. Vilão

Departamento de Fisica
Universidade de Aveiro
Campus de Santiago

The effect of doping with Chlorine and Nitrogen in the emission of ZnSe films is studied. It is observed that the characteristic blue emission is accompanied by mid-band gap emission of excitonic origin even for low concentrations of dopant. This contrasts with the emission observed in the undoped samples, characteristic of donor-acceptor pair recombination.

Among the II-VI materials, ZnSe has a band gap of 2.82 eV that makes it a useful base for blue emission optoelectronic systems. Therefore it is necessary to grow both n- and p-type ZnSe with good quality, to make heterostructures that may be used as LASER or LED. Indeed both emission from excitons bound to the donor or the acceptor and from donor-acceptor pair (D-A) recombination has been found in this material in the blue spectral region[1]. However mid band gap emission is also present, and may reduce the blue emission as it represents an alternate decay path. This emission has not been studied in detail and thus the nature of the defects that cause it are not known[2].

This work was done in collaboration with E. Pereira.

1 H. Stanzl, K. Wolf, B. Hahn, W. Gebhardt, Journal of Crystal Growth 145, 918 (1994)
2 H. Stanzl, K. Wolf, S. Bauer, W. Kuhn, A. Naumov, W. Gebhardt, Journal of Electronic Materials 22, 501 (1993)

SUMMARY OF THE SCHOOL

C. Klingshirn

The concept of collective excitations is the basic means to handle the problem of a Hamiltonian involving 10^{23} particles. This concept proved to be extremely fruitful during the last decades and to be still very lively, since many new phenomena (e.g. those involving ultrashort time spectroscopy) can be explained only by using this concept, and since, on the other hand, optical and other techniques allow to investigate the collective excitations, their interactions and their dynamics.

Correspondingly the topic of the school was very widespread, but the various contributions were closely interconnected. In the following we shall stress this aspect and refer the reader to the collection of abstracts for a summary of the individual contributions.

The basic ideas of the collective excitations, their spectroscopy and their dynamics were outlined by Bloembergen, Di Bartolo, Mazur and Hvam.

The intrinsic collective excitations were covered by the following speakers:

-phonons:	Bloembergen, Di Bartolo, Riste, Sievers, Mazur
-excitons:	Bloembergen, Di Bartolo, Klingshirn, Zimmermann, Hvam, Prasad, Macfarlane, Mazur, Colocci
-plasmons:	v. Baltz
-magnons:	Riste, Gehring, Macfarlane

Localized centers were presented by Barnes, Macfarlane, Ronda, Baldacchini, Boulon, Reisfeld, Auzel and Kaminskii.

From the material side, the following, partly overlapping, groups of materials were covered:

-insulators:	Bloembergen, Riste, Barnes, Macfarlane, Ronda, Baldacchini, Auzel, Kaminskii
-semiconductors:	Bloembergen, Klingshirn, Zimmermann, Hvam, Riste, Mazur, Colocci
-metals:	Riste, v. Baltz
-composite materials:	Sievers, Reisfeld
-organic materials:	Bloembergen, Umeton, Prasad, Reisfeld
-magnetic materials:	Riste, Gehring
-ferroelectric materials:	Sievers, Gehring

Last but not least, the equipment and the spectroscopic techniques were presented, including:

-neutron scattering:	Riste
-Raman & Brillouin scattering:	Bloembergen
-linear and nonlinear laser spectroscopy:	Bloembergen, Di Bartolo, Klingshirn, Hvam, Umeton, Prasad, Sievers, Ronda, Baldacchini, Boulon, Reisfeld, Auzel, Kaminskii
-Ultrashort time-resolved spectroscopy:	Bloembergen, Zimmermann, Hvam, Prasad, Glezer, Mazur, Colocci
-spatial resolution:	Hvam

The topics were complemented by a series of short seminars and posters from the audience, by informal discussion meetings on various topics and by two contributions, which increased the scientific horizon of all the participants, namely on the unification of fundamental interactions by Costa and on Cosmology and Theology by Coyne.

The excursions contributed to the historical and cultural education, showing to all the participants examples of the many different cultures which were brought to Sicily by various people arriving there, and the results of different ways of interaction between them, namely of war (ruins) and of cooperation (highlights of human civilization). If this last aspect became clear to some participants, it was already worth attending the school.

THE PARTICIPANTS

KEY TO THE PHOTOGRAPH

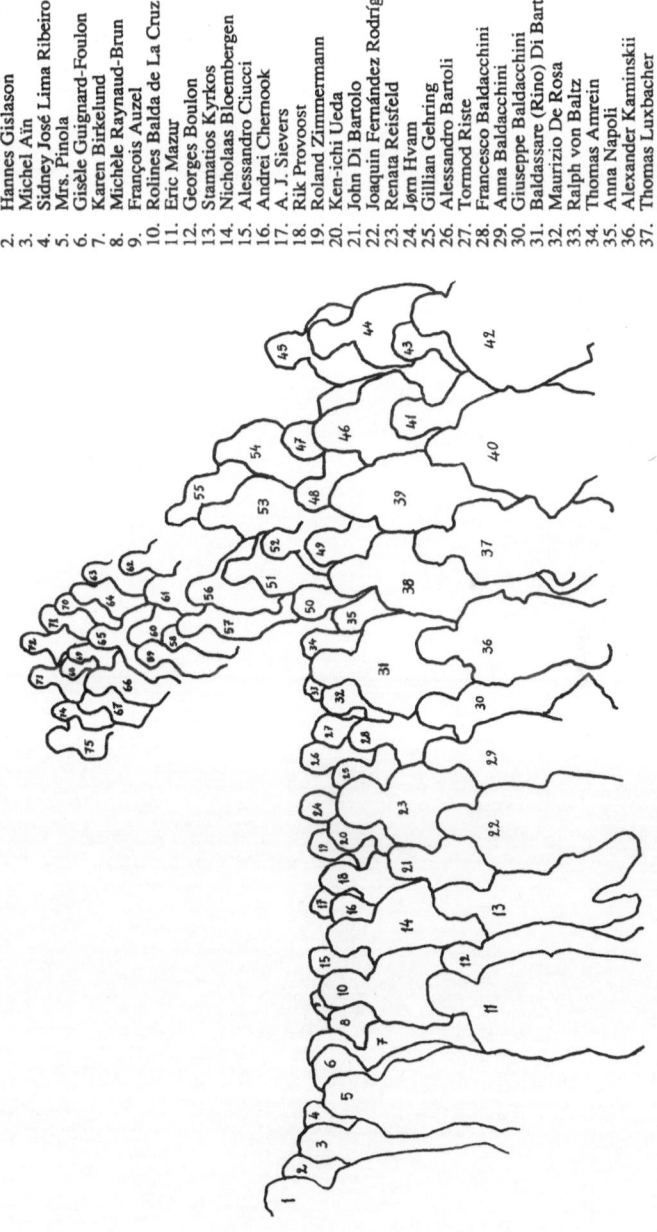

1. Emmanuel Anagnostakis
2. Hannes Gislason
3. Michel Aïn
4. Sidney José Lima Ribeiro
5. Mrs. Pinola
6. Gisèle Guignard-Foulon
7. Karen Birkelund
8. Michèle Raynaud-Brun
9. François Auzel
10. Rolines Balda de La Cruz
11. Eric Mazur
12. Georges Boulon
13. Stamatios Kyrkos
14. Nicholaas Bloembergen
15. Alessandro Ciucci
16. Andrei Chernook
17. A. J. Sievers
18. Rik Provoost
19. Roland Zimmermann
20. Ken-ichi Ueda
21. John Di Bartolo
22. Joaquin Fernández Rodríguez
23. Renata Reisfeld
24. Jørn Hvam
25. Gillian Gehring
26. Alessandro Bartoli
27. Tormod Riste
28. Francesco Baldacchini
29. Anna Baldacchini
30. Giuseppe Baldacchini
31. Baldassare (Rino) Di Bartolo
32. Maurizio De Rosa
33. Ralph von Baltz
34. Thomas Amrein
35. Anna Napoli
36. Alexander Kaminskii
37. Thomas Luxbacher
38. Elisabetta Paladino

39. Paola Borri
40. Stefan Lüthi
41. Giorgio Pozza
42. Markus Umlauff
43. Oliver Wind
44. Mrs. Klingshirn
45. Claus Klingshirn
46. Alessandra De Lorenzi
47. Andrea Cavalleri
48. Giulio Gambarota
49. Gabriele Ferrini
50. Ces Ronda
51. Cesare Umeton
52. Kailash Mishra
53. Gina Maria de Oliveira Vilão
54. Jörg Wrzesinski (?)
55. Carlos Augusto Xavier Ramos
56. Ronald Peterson
57. Eija Kestilä
58. Paras Prasad
59. Lisette Bakalis
60. Julie Grossen
61. Anna-Maria Janner
62. Angel García Adeva
63. Norman Barnes
64. Sascha Hallstein
65. Thomas Steffen
66. Roger Macfarlane
67. Miguel Angelo da Costa Neto
68. Franck Luczak
69. Katja Bammel
70. Thomas Breitkopf
71. Christian Tanguy
72. Bruno Viana
73. Bernd Plagemann
74. Catherine Jeannin
75. Angela Siraco

651

PARTICIPANTS

Michel Aïn
Laboratoire Léon Brillouin
CE-Saclay
F-91191 Gif-sur-Yvette Cedex
FRANCE
Tel   33 1 69085757
        33 1 69085993
Fax   33 1 69088261
Email   ain@bali.saclay.cea.fr

Low dimensional magnetism
Cuprates, Spin-Pierls distortion
Inelastic neutron scattering

Thomas Amrein
Philips GmbH
Forschunglaboratorien
Postfach 1980
D-52021 Aachen
GERMANY
Tel.   49 241 6003 408
Fax   49 241 6003 465
Email   amrein@pfa.philips.de

Luminescent materials (organic and
inorganic) for lamps and displays
New display technology

Emmanuel A. Anagnostakis
Postgraduate Faculty of Electronics and
Telecommunications of the Hellenic Army
22 Kalamakiou Ave.
GR-174 55 Alimos (Athens)
GREECE
Tel.   30 1 981 7810
Fax   30 1 984 3004

Optoelectronics
Semiconductor devices

François Auzel
Centre National d' Études
des Télécommunications
196, Ave. H. Ravera
F-92220 Bagneux
FRANCE
Tel.   33 1 4231 7202
Fax   33 1 4253 4930
Email   auzel@bagneux.cnet.fr

Rare-Earth spectroscopy for
telecommunications

Lisette Bakalis
Institute of Theoretical Physics
University of Groningen
Nijenborgh 4
9747 AG Groningen
THE NETHERLANDS
Tel.   00 31 50 634964
Fax   00 31 50 634947
Email bakalis@rugth11.th.rug.nl

(Non)-linear optical properties
and optical dynamics of the
molecular aggregates

Rolindes Balda de La Cruz
Departamento de Física Aplicada I
Escuela Tecnica Superior de
Ingenieros Industriales y de Telecomunicacion
Universidad del Pais Vasco
Alda. Urquijo, S/N
E-48013, Bilbao
SPAIN
Tel.   34 4 427 8055
Fax   34 4 441 4041

Optical spectroscopy of laser
materials
Photoacoustic spectroscopy

Giuseppe Baldacchini
ENEA, Centro di Frascati
Via E. Fermi, 27
I-00044 Frascati (Roma)
ITALY
Tel.   39 6 9400 5365
Fax   39 6 9400 5334

Optical spectroscopy of solid
state materials, color centers
and lasers
High resolution molecular
spectroscopy and trace gas detection

Katja Bammel
Fachbereich Physik
Institut für Experimentalphysik
Freie Universität Berlin
Arnimallee 14
D-14195 Berlin
GERMANY
Tel.   39 30 838 4294
Fax   39 30 838 3050
Email bammel@omega.physik.fu-berlin.de

Lasers, matrix isolation spectroscopy,
nonlinear optics

Norman Barnes
National Aeronautics and Space Administration
Langley Research Center, MS 474
Hampton, Virginia 23681-0001
U.S.A.
Tel.   804 864 1630
Fax   804 864 8809

Lasers, nonlinear optics,
spectroscopy

Alessandro Bartoli
Istituto Nazionale di Ottica
Largo E. Fermi, 6
I-50125 Firenze
ITALY
Tel.   39 55 23081
Fax   39 55 233 7755
Email bartoli@firefox.ino.it

Interaction of ultrashort laser
pulses with metals

Aitor Bergara Jauregui
Universidad del Pais Vasco
Facultad de Ciencias
Departamento Física de la Materia Condensada
Apartado 644
E-48080 Bilbao
SPAIN
Tel.   34 4 4647700
Fax   34 4 4648500
Email wmbvejaa@lg.ehu.es

Dynamical effects of the
of particles with matter
Many Body Effects
Low dimensional systems

Karen Birkelund
Mikroelektronik Centret (MIC)
Bldg. 345 East
DK-2800 Lyngby
DENMARK
Tel.   0045 45255783
Fax   0045 45887762
Email karen@mic.dtu.dk

Optical spectroscopy of
semiconductor nanostructures

Nicholaas Bloembergen
Division of Applied Sciences
Harvard University
Pierce Hall 231
Cambridge, Massachusetts 02138
U.S.A.
Tel.   617-495-3336
Fax   617-495-9837

Nonlinear optics
Quantum opto-electronics

Paola Borri
Laboratorio Europeo di
Spettroscopie Non-Lineari
Università di Firenze
Largo E. Fermi, 2 (Arcetri)
I-50125 Firenze
ITALY
Tel.   39 55 2307710
Fax   39 55 224072
Email borri@vaxfi.fi.infn.it

Optical properties of
InGaAs/GaAs quantum well
heterostructures

Georges Boulon
Université Claude Bernard-Lyon I
Laboratoire de Physico-Chimie
des Matériaux Luminescents
URA 442 CNRS Bâtiment 205
43, boulevard du 11 Novembre 1918
F-69622 Villeurbanne Cédex
FRANCE
Tel.   33 72 448271
Fax   33 72 431130

Laser materials, crystal growth,
optical spectroscopy, nonlinear optics

Thomas Breitkopf
Institut für Angewandte Physik
der Universität Karlsruhe
Kaiserstraße 12
Postfach 6980
D-76128 Karlsruhe
GERMANY
Tel.   49 0 721 608 3555
Fax   49 0 721 607 593
Email tbreitko@ap-pc513a.physik.uni-karlsruhe.de

Localization of carriers in II-VI
ternary compounds
Many particle effects of localized
carriers

Andrea Cavalleri
Dipartimento di Elettronica
Università di Pavia
Via Abbiategrasso 209
I-27100 Pavia
ITALY
Tel.   39 382 505595
Fax   39 382 422583
Email cavaller@ipusp4.unipv.it

Ultrafast properites of III-V
semiconductors and fullerites
deposited in thin films

Andrei Chernook
Institute of Physics 13303
 TU Chemnitz-Zwickau
D-09107 Chemnitz
GERMANY
Tel.   49 0 371 5313064
Fax   49 0 371 5313060
Email tschernook@physik.tu-chemnitz.de

Chemical physics, molecular
spectroscopy ultrafast phenomena,
electron energy transport,
nanostructures

Alessandro Ciucci
ENEA - CRAM "S.Teresa"
P.O. Box 316
I-19100 Lerici (SP)
ITALY
Tel.   39 0 187 536291
Fax   39 0 187 536263
Email alec@risc.ifam.pi.cnr.it

Atomic spectroscopy for
environmental diagnostics, molecular
spectroscopy by TDL, fitting
techniques and atmospheric CO_2
monitoring

Marcello Colocci
Dipartimento di Fisica
Università di Firenze
Largo E.Fermi, 2
I-50125 Firenze
ITALY

Ultrafast phenomena
Semiconductor Heterostructures

Giovanni Costa
Università degli Studi
Istituto di Fisica "Galileo Galilei"
Via F. Marzolo, 8
I-35100 Padova
ITALY

Elementary Particles
High Energy Physics

George V. Coyne, S.J.
Vatican Observatory Research Group
Steward Observatory
The University of Arizona
Tucson, Arizona 85721
U.S.A.

Astrophysics, Cosmology

Alessandra De Lorenzi
Dipartimento di Chimica Fisica
Università degli Studi di Venezia
Calle Larga S. Marta 2137
I-30123 Venezia
ITALY
Tel.   39 41 2578610
Fax   39 41 2578594
Email delorenz@unive.it

High-resolution spectroscopy of gases

Maurizio De Rosa
Istituto di Fisica Atomica e Molecolare
Via del Giardino 7
I-56127 Pisa
ITALY
Tel.   39 50 888126
Fax   39 50 888136
Email maurizio@risc.ifam.pi.cnr.it

High resolution molecular spectroscopy

Baldassare (Rino) Di Bartolo
Department of Physics
Boston College
Chestnut Hill, Massachusetts 02167
U.S.A.

Luminescence spectroscopy
Photoacoustics
Femtospectroscopy

Daniel Di Bartolo[1]
Department of Physics
Boston College
Chestnut Hill, Massachusetts 02167
U.S.A.

Psychology

John Di Bartolo
Department of Physics
University of Virginia
Charlottesville, Virginia 22903
U.S.A.
Tel.   804 924 6597
Fax   804 924 4576
Email sjd2u@virginia.edu

Theoretical superconductivity
Motion of normal/superconductive
boundaries

657

Joaquin Fernández Rodríguez
Departamento de Física Aplicada I
Escuela Tecnica Superior de
Ingenieros Industriales y de Telecomunicacíon
Universidad del Pais Vasco
Alda. Urquijo, S/N
E-48013, Bilbao
SPAIN
Tel.   34 4 427 8055
Fax   34 4 441 4041

Optical spectroscopy of laser
materials
Photoacoustic spectroscopy

Gabriele Ferrini
Politecnico di Milano
Piazza Leonardo Da Vinci, 32
I-20133 Milano
ITALY
Tel.   39 2 23996170
       39 2 2392562
Fax   39 2 2392543
Email ferrini@mvlasa.mi.infn.it

Surface physics, interaction of laser
pulses with metals and
semiconductors

Giulio Gambarota
Physics Department
Boston College
Chestnut Hill, MA 02167
U.S.A.
Tel   617 552 3575
Fax   617 552 8478

Solid State Physics

Angel García Adeva
Departamento de Física Aplicada I
Escuela Tecnica Superior de
Ingenieros Industriales Y de Telecomunicacion
Universidad del Pais Vasco
Alda. Urquijo, S/N
E-48013, Bilbao
SPAIN
Email wubgaada@bi.ehu.es

Low-temperature anomalous
properties of amorphous solids
Non-perturbative
dynamical theories of nonlinear
spectroscopy

Gillian Gehring
Department of Physics
Sheffield University
Sheffield, Yorkshire S3 7RH
U.K.
Tel.   44 114 282 4299
Fax   44 114 272 8079
Email g.gehring@sheffield.ac.uk

Thin film magnetism - spin waves,
anisotropy, domains
Quantum spin chain - density matrix
renormalization group GMR in
granular materials

Hannes Gislason
Mikroelektronik Centret, DTU
Bldg. 345 East
DK-2800 Lyngby
DENMARK
Tel   454 52 55784
Fax   454 45 87762
Email hannes@mic.dtu.dk

MBE cleaved edged overgrowth of
GaAs/AlGaAs quantun wires
photoluminescence
Spectroscopy of quantum wires

Eli Glezer
Division of Applied Sciences
Gordon McKay Laboratory
Harvard University
9 Oxford Street
Cambridge, Massachusetts 02138
U.S.A.
Tel.   617 495 9616
Email eli_glezer@lucifer.harvard.edu
        glezer@fas.harvard.edu

Spectroscopy of condensed matter,
ultrafast spectroscopy, light induced
phase transitions

Oleg Gogolin
Georgian Republic
380060 Tbilisi
Bulatschaurskaja 6. App 6
Tel.   009958832372361
or Kaiserstr. 12
D-76128 Karlsruhe
GERMANY

Quantum dots, Q.W. superlattice

Julie C. A. Grossen
Department of Physics
College of William and Mary
Williamsburg, Virginia 23185
U.S.A.
Tel.   804 221 3549
Email jgrossen@physics.wm.edu

Spectroscopy of solid state
laser materials

Gisèle Guignard - Foulon
Université Claude Bernard-Lyon I
Laboratoire de Physico-Chimie
des Matériaux Luminescents
URA 442 CNRS Bâtiment 205
43, boulevard du 11 Novembre 1918
F-69622 Villeurbanne Cédex
FRANCE
Tel.   33 72 44 80 00
Fax   33 72 43 11 30

Laser materials, crystal growth,
non linear optics spectroscopy

Sascha Hallstein
Max-Planck-Institut für Festkörperforschung
Heisenbergstraße 1
D-70569 Stuttgart
GERMANY
Tel.   49 711 689 1664
Email hallstei@quasix.mpi-stuttgart.mpg.de

Picosecond spectroscopy of III-V
semiconductors in magnetic fields

Jørn Hvam
Mikroelektronik Centret, DTU Bldg. 345 east
DK 2800 Lyngby
DENMARK
Tel.   45 4525 5758
Fax   45 4588 7762
Email hvam@mic.dtu.dk

Ultrafast nonlinear optical
spectroscopy of semiconductors,
electro-optical switching and
sampling, near-field microscopy

Anna-Maria Janner
University of Groningen
Laboratory of Applied
and Solid State Physics
Nijenborgh 4
9747 AG Groningen
THE NETHERLANDS
Tel.   0031504922
Fax   003150 634879
Email janner@vsf1.phys.rug.nl

Optical second-harmonic
generation from strongly correlated
systems: thin films of C_{60}
(exciton propagation) and NiO

Catherine Jeannin
Institut de Physique expérimentale
Faculté des Sciences, BSP
CH-1015 Lausanne-Dorigny
SWITZERLAND
Tel.   41 21 692 36 65
Fax   41 21 692 36 05
Email cjeannin@ulys.unil.ch

Real-time spectroscopy in
condensed media, ultrafast lasers,
nonlinear optics

Alexander Kaminskii
Joint Open Laboratory on Laser Crystals
and Precise Laser Systems
Institute of Crystallography
Russian Academy of Sciences
Leninsky pr.59
Moscow 117333
RUSSIA
Tel.   7 095 135 2210
Fax   7 095 135 1011

Physics of laser insulating crystals

Eija Kestilä
Department of Chemistry
University of Turku
FIN-20500 Turku
FINLAND
Tel.   358 21 6336736
Fax   358 21 6336700
Email eijakes@polaris.utu.fi

The structure and optical
properties of rare-earth oxide
compounds

Claus Klingshirn
Institut für Angewandte Physik
der Universität Karlsruhe
Kaiserstraße 12
D-76128 Karlsruhe
GERMANY
Tel.   49 721 6083410
Fax   49 721 607593
Email ck@ap-pc513a.physik.uni-karlsruhe.de

Semiconductor optics, linear,
nonlinear, time-resolved
semiconductor epitaxy

Stamatios Kyrkos[2]
Department of Physics
Boston College
Chestnut Hill, Massachusetts 02167
U.S.A.
Tel.   617 552 4588
Fax   617 552 8478
Email kyrkos@bcvms.bc.edu

Electronic structure of metals,
response functions: metal surfaces

Antonino La Francesca[3]
143 Pine Street
Medfield, Massachusetts 02052
U.S.A.

Gang Lei
Department of Physics
Boston College
Chestnut Hill, Massachusetts 02167
U.S.A.
Tel.   617 552 3596
Fax   617 552 8478
Email leig@bcvms.bc.edu

Solid-state spectroscopy
Photoacoustics
Thin-film electroluminescence

Frank Luczak
Uniuversitaire Instelling Antwerpen (UIA)
Departement Natuurkunde
Universiteitsplein 1
B-2610 Wilrijk
BELGIUM
Tel.   0032 3 820 24 77
Fax   0032 3 820 22 45
Email luczak@nat2.uia.ac.be

Electron-phonon interaction, polarons,
bipolarons, excitons and relations to
high TC superconductivity

Stefan Lüthi
Institut für Anorganische Chemie
Universität Bern
Freiestraße 3
CH-3012 Bern
SWITZERLAND
Tel.   031 631 42 54
Fax   031 631 39 93
Email luethi@iac.unibe.ch

Near-infrared to visible upconversion
in ternary rare-earth halides, cooperative
bistability, electronic energy level
structure and correlation crystal
field effects

Thomas Luxbacher
Institut für Physikalische und
Theoretische Chemie
Rechbauerstraße 12
8010 Graz
AUSTRIA
Tel.   43 316 873 8239
Fax   43 316 873 8225
Email luxbacher@ptc.tu-graz.ac.at

Energy transfer and vibronic
spectroscopy in rare-earth
hexachloroelpasolites

Roger M. Macfarlane
IBM Almaden Research Center K69/802
650 Harry Road
San Jose, California 95120
solids, U.S.A.
Tel.   408 927 2428
Fax   408 927 2100
Email macfarla@almaden.ibm.com

High resolution spectroscopy of solids,
coherence, spectral holeburning,
photon echoes, rare-earth
spectroscopy, photoionization in
photorefractive materials

Eric Mazur
Division of Applied Sciences
Harvard University
Pierce Hall 225
Cambridge, Massachusetts 02138
U.S.A.
Tel.   617 495 8729
Fax   617 495 9837
        617 495 1229
Email mazur@physics.harvard.edu
WWW http://mazur-www.harvard.edu

Interactions of ultrafast laser pulses
with semiconductors; initiation of
reactions at metal surface using
femtosecond lasers
Structure &dynamics of Langmuir
mono-layers

Kailash C. Mishra
Osram Sylvania Inc.
Central Research
100 Endicott Street
Danvers, Massachusetts 01923
U.S.A.
Tel.   508 750 1575
Email mishra@hq.sylvania.com

Electronic structures calculations of
atoms, molecules and solids.

Anna Napoli
Istituto di Fisica dell'Università degli Studi
Via Archirafi 36
I-90123 Palermo
ITALY
Tel.   091 623 4243
Fax   091 617 1617
Email messina@ipacuc.cuc.unipa.it

Mattei-radiation interactions

Miguel Angelo da Costa Neto
Departamento de Fisica
Universidade de Aveiro
Campus Universitário
3800 Aveiro
PORTUGAL
Tel.   034 26666 / 034 370200
Fax   034 26666 / 034 24965
Email mangelo@ideiafix.fis.ua.pt

Luminescence of natural and
synthetic diamonds, diamond films

Gina Maria de Oliveira Vilão
Departamento de Fisica
Universidade de Aveiro
Campus de Santiago
3800 Aveiro
PORTUGAL
or Carris
Oiã
3770 Oliveira do Bairro
PORTUGAL
Tel.   351 34 370200 (ext. 3135)
Fax   351 34 24965
Email gvilao@ideiafix.fis.ua.pt

Luminescence in films
semiconductor II-VI (ZnSe)

Gönül Özen
Department of Physics
Faculty of Sciences and Letters
Istanbul Technical University
80626 Maslak-Istanbul
TURKEY
Tel.   90 212 285 3206
Fax   90 212 285 6386
Email feozen@cc.itu.edu.tr

Spectroscopy of rare-earht ions in
solids

Elisabetta Paladino
Istituto di Fisica dell'Università degli Studi
Via Archirafi 36
I-90123 Palermo
ITALY
Tel.   091 623 4243
Fax   091 617 1617
Email messina@ipacuc.cuc.unipa.it

Spin-phonon interactions

Ronald O. Petersen
Motorola Inc.
Phoenix Corporate Research Laboratories
2100 East Elliot Road, Mail Drop EL508
Tempe, Arizona 85284
U.S.A.
Tel.   602 413 5930
Fax   602 413 5934
Email a338aa@email.sps.mot.com

Luminescent materials for FED
display, and other emissive display
technologies

Bernd Plagemann
Laboratory of Physical Chemistry
Universitätsstraße 22
ETH-Zentrum
CH-8092 Zürich
SWITZERLAND
Tel.   41 1 632 4384
Email bepl@phys.chem.ethz.ch

Nonlinear spectroscopy,
spectral hole-burning,
optical data storage, coherent
excitations

Giorgio Pozza
I.C.T.I.M.A., C.N.R.
Corso Stati Uniti 4
I-35127 Padova
ITALY
Tel.   049 8295939
Fax   049 8295649
Email ajo@ictr.pd.cnr.it

Luminescent materials for laser
and particle detection

Paras Prasad
Department of Chemistry
State University of New York
Buffalo, New York 14214
U.S.A.

Theoretical modeling and
experimental studies of nonlinear
optical processes in organic
materials and polymers

663

Rik Provoost
Katholieke Universiteit Leuven
Laboratorium voor Vaste-Stoffysika
en Magnetisme
Celestijnenlaan, 200D B-3001 Leuven
BELGIUM
Tel.   32 16 327120
Fax   32 16 327983
Email rik.provoost@fys.kuleuven.ac.be

Raman spectroscopy
High-temperaturesuperconductors

Carlos Augusto Xavier Ramos
Departamento de Fisica
Universidade de Aveiro
Campus Universitário
3800 Aveiro
PORTUGAL
or Rua Numo Gonçalves, 114
Gafanha da Nazaré
3830 Ilhavo
PORTUGAL
Tel.   351 34 370200 (ext.3135)
Fax   351 34 24965
Email cxavier@ideiafix.fis.ua.pt

Optical spectroscopy of III-V
semiconductor materials

Michèle Raynaud-Brun
Commissariat á l'Energie Atomique-Saclay
DRECAM/SRSIM - Bâtiment 462
F-91191 Gif-sur-Yvette
FRANCE
Tel.   33 1 69087010
Fax   33 1 69088446
Email rayn@santamaria.saclay.cea.fr

Surface and interface plasmons in
metallic systems (theory)

Renata Reisfeld
Department of Inorganic and
Analytical Chemistry
The Hebrew University of Jerusalem
Jerusalem
ISRAEL
Tel.   972 2 6585323
Fax   972 2 6585319
Email renata@vms.hoji.ac.il

Spectroscopy of inorganic ions and
organic molecules, energy transfer,
radiative and nonradiative transitions,
nonlinear materials, solid-state lasers,
electrochromic and photochromic
glasses

Sidney José Lima Ribeiro
France Telecom CNET
Centre Paris B/ Laboratoire de Bagneux
196, Av. Henri-Ravera
BP 107
F-92225 Bagneux
FRANCE
Permanent Address:
Instituto de Química-Unesp
PoBox 355 Araraquara-sp-Brazil
Fax   55 162 227932
Email sidneyr@arq000.uearq.ansp.br

Luminescent materials, lanthanide
spectroscopy, glass structure

Tormod Riste
Institutt for Energiteknikk
P.O. Box 40
N-2007 Kjeller
NORWAY
Tel.   4763 806076
Fax   4763 810920
Email gerd@ife.no

Condensed matter physics,
Statistical physics, neutron scattering

Cees Ronda
Philips GmbH
Forschungslaboratorien
Weisshausstrasse 2
Postfach 1980
D-52021 Aachen
GERMANY
Tel.   49 241 6003 397
Fax   49 241 6003 465
Email ronda@pfa.philips.de

Luminescent materials: theory,
preparation and applications.

Niko Schmitt
AG Schwentner
Fachbereich Physik
Institut für Experimentalphysik
Freie Universität Berlin
Arnimallee 14
D-14195 Berlin
GERMANY
Email schmittn@sirius.physik.fu-berlin.de

Nonlinear optics, femtosecond laser
systems and spectroscopy

A. J. Sievers
Atomic & Solid State Physics Lab.
Clark Hall, Cornell University
Ithaca, New York 14853-2501
U.S.A.
Tel.   607 255 6422
Fax   607 255 6428
Email sievers@msc.cornell.edu

Linear and nonlinear far IR and IR
spectroscopy

Angela Siraco[4]
10 Spruce Circle
Andover, Massachusetts 01810
U.S.A.
Tel.   508 475 2051
Email menke@bcvms.bc.edu

Italian Language

Thomas Steffen
Ultrafast Laser and
Spectroscopy Laboratory
Dept. of Chemical Physics
University of Groningen
Nijenborgh 4
9747 AG Groningen
THE NETHERLANDS
Tel.   31 50 634344
Fax   31 50 634441
Email steffen@chem.rug.nl

Ultrafast optical dynamics in
condensed phases, ultrafast
spectroscopy, nonlinear optics, liquid
dynamics

Christian Tanguy
CNET Laboratoire de Bagneux
196 Avenue Henri-Ravera, BP 107
F-92225 Bagneux Cedex
FRANCE
Tel.   33 1 42 31 78 81
Fax   33 1 42 53 49 30
Email tanguy@bagneux.cnet.fr

Optical properties of semiconductors
with a special emphasis on excitonic
effects, highly excited semiconductors,
ultrafast phenomena/nonlinear optics

Ken-ichi Ueda
Institute for Laser Science
University of Electro-Communications
Chofugaoka, Chofushi
Tokyo 182
JAPAN
Tel.   81 424 83 2161
Fax   81 424 85 8960
Email ueda@ils.uec.ac.jp

Laser diode pumped solid-state lasers,
laser stabilization for gravitational
wave detection, ultra-high peak power
solid-state & gas lasers, laser
development for initial confinement
fusion

Cesare Umeton
Dipartimento di Fisica
Università degli Studi della Calabria
87036 Rende(CS)
ITALY
Tel.   39 984 493260
Fax   39 984 839389
Email umeton@fis.unical.it

Nonlinear optics of nonlinear
crystals, light induced chaos in
liquid crystals, polymer dispersed
liquid crystals

Markus Umlauff
Institut für Angewandte Physik
der Universität Karlsruhe
Kaiserstraße 12
Postfach 6980
D-76128 Karlsruhe
GERMANY
Tel.   49 721 608 3555
Fax   49 721 607 593
Email umlauff@ap-pc513a.physik.uni-karlsruhe.de

Laser process and optical
nonlinearities in II-VI semiconductor
heterostructures, laser spectroscopy of
semiconductors

Bruno Viana
Université Pierre et Marie Curie
École Nationale Supérieure de Chimie de Paris
11, rue Pierre et Marie Curie
F-75231 Paris Cédex 05
FRANCE
Tel.   33 1 44276707
Fax   33 1 46347489
Email viana@ext.jussieu.fr

Crystal growth of laser materials,
optical spectroscopy, materials
science

Ralph von Baltz
Institut für Theorie der Kondensierte Materie
Fakultät für Physik der Universität
Postfach 6980
D-76128 Karlsruhe
GERMANY
Tel.   49 721 683362
Email baltz@tkm.physik.uni-karlsruhe.de

Optical properties, plasmons,
photogalvanic effect,
heterostructures

Brian Walsh
Department of Physics
Boston College
Chestnut Hill, Massachusetts 02167
U.S.A.
Currently at:
NASA Langley Research Center
Mail Stop 474
Hampton, Virginia 23681-0001
U.S.A.
Tel.   804 864 7112

Oliver Wind
Institut für Angewandte Physik
der Universität Karlsruhe
Kaiserstraße 12
Postfach 6980
D-76128 Karlsruhe
GERMANY
Tel.   49 721 6083555
Fax   49 721 607593
Email oliver.wind@phys.uni-karlsruhe.de

Jörg Wrzesinski
Universität Dortmund
Lehrstuhl Exp. Physik II
D-44221 Dortmund
GERMANY
Tel.   0049 231 5600846 / 0049 231 7553534
Email wrzesinski@fkp.physik.uni-dortmund.de

Roland Zimmermann
Max Planck Arbeitsgruppe Halbleitertheorie
Hausvogteiplatz 5-7
D-10117 Berlin
GERMANY
Tel.   49 30 20366204
Fax   49 30 2384763
Email zim@semic.ag-berlin.mpg.de

Spectroscopy of rare earth solid-state laser materials, solid state laser development

Spectroscopy of II-VI and I-VII semiconductor, quantum dots, gain processes.

ZnO, GaN; nonlinear properties, exitons multiphoton transitions

Theory of optical properties in semiconductor quantum structures

[1] Administrative Secretary of the Course
[2] Scientific Secretary of the Course
[3] Assistant to the Director
[4] Administrative Assistant

INDEX